Lecture Notes in Mathematics 2252

More information about this subseries at http://www.springer.com/series/4809

Catherine Donati-Martin • Antoine Lejay • Alain Rouault
Editors

Séminaire de Probabilités L

Springer

Editors
Catherine Donati-Martin
Laboratoire de Mathématiques de Versailles
Université de Versailles-St-Quentin
Versailles, France

Antoine Lejay
Institut Elie Cartan de Lorraine
Vandoeuvre-lès-Nancy, France

Alain Rouault
Laboratoire de Mathématiques de Versailles
Université de Versailles-St-Quentin
Versailles, France

ISSN 0075-8434 ISSN 1617-9692 (electronic)
Lecture Notes in Mathematics
ISSN 0720-8766 ISSN 2510-3660 (electronic)
Séminaire de Probabilités
ISBN 978-3-030-28534-0 ISBN 978-3-030-28535-7 (eBook)
https://doi.org/10.1007/978-3-030-28535-7

Mathematics Subject Classification (2010): Primary 60GXX; 28D20, 37A05, 60G05, 60G55, 60G60, 60H35, 60J05, 60J05, 60J10, 60J27, 60J65, 65C30, 91G60, 94A17

This Springer imprint is published by the registered company Springer Nature Switzerland AG.
The registered company address is: Gewerbestrasse 11, 6330 Cham, Switzerland

Preface

Jacques Azéma passed away on January 5, 2019. After P.A. Meyer (January 2003) and M. Yor (January 2014), he is one of the founders of the Séminaire de Strasbourg, then Séminaire de Probabilités, who is deceased. The following lines do not claim to be exhaustive, Jacques being very discrete on his personal background. J. Azéma was born in 1939. After studies at Supaéro, he undertook under the impulse of D. Dacunha-Castelle a thesis in probability in collaboration with Marie Duflo and Daniel Revuz, under the direction of J. Neveu. He defended his thesis in 1969. At that time, he was assistant at the Institut Henri Poincaré. During this year full of events, to which Jacques participated, this trio is extremely productive (eight research notes for CRAS and six articles between 1965 and 1969). He was then assistant professor and full professor at Paris 6 and Paris 5 until his retirement in 2007. He was an eminent member of the Laboratoire de Calcul des Probabilités which became LPMA and now LPSM.[1] In addition to this original scientific production (see the few themes described in the testimonies), Jacques Azéma was editor of the *Séminaire* from Volume XVI to Volume XXXVII. For the volume dedicated to Marc Yor (Volume XLVII), he gave us a large interview in 2015 in which he told about the spirit of the *Séminaire* and his participation as editor. We had the pleasure to find again his humor and his eloquence. During many years, he had organized the Journées de Probabilités at the CIRM, Marseille, where young and less young probabilists had the opportunity for an annual meeting.

The Journées de Probabilités were held in Tours in 2018. This Volume L was already being prepared with some contributions to this conference when we learned the death of Jacques Azéma. We dedicate this Volume L to his memory.

Versailles, France — Catherine Donati-Martin
Vandoeuvre-lès-Nancy, France — Antoine Lejay
Versailles, France — Alain Rouault

[1] Jacques Azéma is remembered on the page ⟨http://www.lpsm.paris/laboratoire/jacques-azema/⟩.

Contents

Part I
In memoriam Jacques Azéma

Jacques Azéma in Saint-Flour, 1994, by courtesy of Lucien Birgé

Chapter 1
Un témoignage

Michel Émery

Au tout début des années 1970, j'ai fréquenté comme étudiant le Labo de Probas, où enseignait Azéma. C'étaient les années post-68 ; tout à l'université fonctionnait n'importe comment, se réinventant constamment, entre grèves et contestation. Il est arrivé un jour la tête bandée, pour s'être, disait-on, frotté de trop près à une matraque de CRS. Même dans cet univers surréaliste, il était atypique, apparaissant comme une sorte d'improbable Gaston Lagaffe totalement décontracté, donnant en espadrilles des cours qui semblaient complètement improvisés : « Ah, ben oui, là on pourrait peut-être faire comme ça ... Euh... Attendez... Non, finalement ça marchera pas, mais vous inquiétez pas, on va y arriver quand même... » Et l'on se retrouvait à la fin du cours tout étonné du chemin parcouru en ayant cru seulement flâner de-ci de-là.

Ayant quitté Paris en 73, je l'ai peu fréquenté. Je ne le voyais plus guère que lors des « grands séminaires », ancêtres des Journées de Probabilité, qui rassemblaient deux fois par an à Strasbourg des probabilistes plus ou moins proches de la « théorie générale des processus ». Je suivais ses travaux, régulièrement exposés par Meyer au « petit séminaire » hebdomadaire, ainsi que ceux de Nicole et de Marc. Je le revoyais aussi à Saint-Flour, toujours aussi décontracté. Je me le rappelle demandant « Qui êtes-vous ? » à un autre participant, qui se trouvait être l'un des conférenciers, ayant commencé son cours depuis plusieurs jours.

Plus tard, lorsque je suis entré à la Rédaction du Séminaire, je n'ai que peu travaillé avec lui : les rôles avaient été depuis plusieurs années répartis entre Marc et lui, Marc s'occupant du volume et lui des Journées de Proba, alors organisées chaque année au CIRM, officiellement par Marc et lui ; mais c'est à lui qu'incombaient la préparation, l'organisation et les rapports. Il avait un sens, digne d'un publiciste, de la formule concise qui fait mouche. Un jour, Marc et moi lui

M. Émery (✉)
IRMA, Strasbourg, France
e-mail: emery@unistra.fr

C. Donati-Martin et al. (eds.), *Séminaire de Probabilités L*, Lecture Notes in Mathematics 2252, https://doi.org/10.1007/978-3-030-28535-7_1

soumettons notre laborieuse énième mouture de la dédicace à Meyer et Neveu du Séminaire XXX. Il nous jette un regard apitoyé, se saisit d'un crayon, et pulvérise en deux lignes percutantes notre paragraphe qui sentait la sueur.[1] Plus récemment, à la mort de Meyer, lorsque chacun y est allé de son évocation ou de son témoignage, le sien m'a bouleversé : dans un véritable petit poème en prose,[2] loin de la réthorique et du pathos, il avait touché, par-delà les mots, ce que je ressentais sans avoir su l'exprimer.

[1] Séminaire de Probabilités XXX, Lecture Notes in Math. 1626, (1996).

[2] In Memoriam Paul-André Meyer – Séminaire de Probabilités XXXIX, Lecture Notes in Math. 1874 (2006).

Chapter 2
Un dimanche de juin avec Jacques

Nathanaël Enriquez

J'ai partagé mon bureau à Jussieu avec Jacques pendant 7 ans. Cela va vous étonner mais c'est lors de notre dernière conversation que nous avons eu l'un de nos échanges mathématiques les plus intenses, dont nous vous exposerons la teneur avec Aurélie et Pierre dans ce qui va suivre. Au fond, il n'y a peut-être là rien de si étonnant : comme pour nombre de choses, c'est dans l'urgence, lorsque nous savons que le temps est compté que nous nous mettons à les faire...

La raison du faible nombre de nos échanges mathématiques lors de la période qui aurait été censée être propice à ces échanges est en fait double. La première en est que nous avons partagé en réalité notre bureau à quart-temps : Jacques n'y était présent que les matins, et de mon côté, je n'étais pas un lève-tôt. Le plus souvent, quand j'arrivais dans le bureau, je le trouvais déjà à sa table, en train de fumer (avant que ce ne soit plus possible à Chevaleret) et de maugréer au sujet d'un article de math qui n'était pas rédigé à son goût. Alors, il me hélait : « Dis-moi, toi qui es normalien, tu dois comprendre ça tout de suite ! Ça doit être "trivial" ! ». Je me penchais alors sur l'article, bredouillais quelques phrases et là, il me décochait un : « Bon... en fait tu es nul, c'est à Biane (Philippe Biane, notre co-bureau) qu'il faut que je demande, lui, il est vraiment fort. » Le coup d'envoi de notre « journée » était lancé. Il rangeait son article et c'est là que nous en arrivons à la seconde raison de notre problème à trouver un moment pour parler de math. Il ne nous restait plus qu'une petite heure avant l'heure du déjeuner. Nous parlions bien sûr des potins du landerneau probabiliste, mais pas que... nous parlions, finance appliquée, je veux dire par là... investissements boursiers ! que j'ai découverts avec lui. Il épluchait chaque matin, au café en bas de chez lui, *Le Monde* et *Les Échos*. Alors, tout y passait, les entreprises de la « nouvelle économie », les « biotechs », le Nasdaq, les avis des « gourous » de Wall Street... Avec quelques jeunes collègues de l'époque,

N. Enriquez (✉)
Laboratoire Mathématiques d'Orsay, Université Paris-Sud, Orsay, France
e-mail: nathanael.enriquez@u-psud.fr

C. Donati-Martin et al. (eds.), *Séminaire de Probabilités L*, Lecture Notes in Mathematics 2252, https://doi.org/10.1007/978-3-030-28535-7_2

dont je tairai le nom, nous formions une sorte de société secrète et Jacques était notre gourou... Lors du déjeuner avec les membres du laboratoire, nous reprenions des conversations anodines. J'arrivais tout juste à le retenir pour un déca sur la terrasse de Jussieu, puis il rentrait chez lui, épuisé par les mauvaises nuits qu'il passait invariablement. Je voyais alors à regret, sa silhouette inimitable s'éloigner.

Le 3 juin dernier, nous avons passé l'après-midi ensemble grâce à ma femme qui l'a reconnu de loin, rue de Rivoli, toujours sa silhouette... J'aimerais vous faire partager notre dernier échange. Nous avons d'abord évoqué quelques éditions mémorables de l'École d'été de Saint-Flour, qu'il n'aurait manquée pour rien au monde. Il y eut évidemment l'année Pitman/Tsirelson/Werner. Jacques encore fasciné par le cours de Wendelin, mais peut-être plus encore par ses qualités humaines, sa personnalité et ses multiples talents, me dit : « Pour moi, il restera toujours le Petit Prince de Saint-Flour ». Cette déclaration fut prononcée avec une affection et une émotion qui tranchait tellement avec le blagueur impénitent qu'il était, que j'en suis resté sans voix quelques secondes.

Ces souvenirs nous ont amené à un registre plus grave, celui de son état de santé. Après m'en avoir fait part, il m'a dit : « J'ai eu la chance dans ma vie de faire partie d'une aventure collective exaltante, celle de l'éclosion des probabilités modernes. J'y ai rencontré des gens extraordinaires d'une telle intelligence et d'une telle génerosité ! » Il m'a dit également à quel point il avait été chanceux de croiser la route de Meyer et Neveu. En l'écoutant, j'ai dû faire une petite moue qui n'était certainement pas désobligeante pour ces deux grands noms des probabilités, mais il me semblait que ces propos ne rendaient pas hommage à sa propre pensée mathématique. Toujours est-il que sans même que j'ai à parler, il m'évoque un souvenir, en compagnie de Kai-Lai Chung, le grand probabiliste chinois de Stanford. Alors qu'ils étaient en train de contempler l'océan, face au Pacifique, Jacques fait part à Chung de son admiration pour ses 2 aînés, et Chung de lui répondre : « Pourquoi vous placez-vous en dessous d'eux ? Vous savez... vous n'avez pas à rougir de vos travaux ! » Peut-être que Chung pensait, en lui parlant, au magnifique travail de Jacques sur le retournement du temps...

Alors que nous marchions vers chez lui dans la rue des Francs-Bourgeois, il m'interroge sur mes recherches du moment. C'est là que je lui parle de notre travail en cours avec Pierre Calka et notre étudiante Aurélie Chapron. Je lui parle donc de Processus de Poisson Ponctuel sur une surface. À ma grande surprise, il me laisse continuer, là où il m'aurait arrêté quelques années plus tôt d'un : « Vous les normaliens, ça ne vous suffit pas les probabilités, il faut que vous parliez de variétés, de fibré et de tout le b.... » Je commence par lui énoncer le résultat classique sur le nombre moyen de sommets d'une cellule de la mosaïque Poisson-Voronoi dans le plan (égal à 6) qu'il connaissait. Je m'enhardis en lui parlant d'un processus ponctuel de grande intensité sur une surface qui a donc tendance à oublier l'effet de la courbure de la surface sur laquelle il vit. J'en arrive à lui dire que la correction à la constante 6 dans la limite des grandes intensités, contient l'information sur la courbure scalaire de la surface. La formule d'Euler appliquée à la mosaïque de Poisson-Voronoi permet alors de retrouver le théorème de Gauss-Bonnet. Et lui de me poser la question : « Très joli, mais ensuite, tu en fais quoi ? Il est déjà démontré le théorème de Gauss-Bonnet. » Et moi de me défendre : “Si l’on devait compter le nombre de théorèmes d’analyse démontrés par les probabilités avec 100 ans de retard. . . et puis, on a une nouvelle interprétation de la courbure scalaire en toute dimension”. Sa réponse fut : « Tu as raison, j’aime l’idée que les probas puissent offrir un autre éclairage sur les notions classiques des maths. » S’adressant à ma femme : « Il est fort ton mari ! » On est rue de Sévigné, nos routes se séparent. . .

Chapter 3
Mosaïque de Poisson-Voronoi sur une surface

Pierre Calka, Aurélie Chapron, et Nathanaël Enriquez

L'objet de cette note est d'indiquer la manière dont on peut étendre au cas d'une surface S, le célèbre résultat, trouvé indépendamment par Meijering en 1953 [4] et Gilbert en 1962 [3], énonçant que le nombre moyen de sommets d'une cellule ou de façon équivalente l'espérance du nombre de sommets d'une cellule typique d'une mosaïque de Poisson-Voronoi dans le plan est égal à 6. On montrera alors comment le résultat trouvé aboutit à une preuve du théorème de Gauss-Bonnet. Cette note est l'objet d'un chapitre de la thèse du second auteur dont le contenu est déposé sur arXiv [1] et traite du cas plus général des variétés de dimension quelconque.

Le premier hic dans une telle entreprise est que le mot « cellule typique » n'a pas de sens dans le contexte d'un espace qui n'est pas homogène. Pour généraliser le résultat du plan, nous devons fixer un point x_0 de la surface S, lancer un processus ponctuel de Poisson homogène d'intensité λ, soit $\mathcal{P}_\lambda$ sur la surface, puis considérer la cellule de x_0 que nous noterons $C(x_0)$ dans la mosaïque de Voronoi associée à $\mathcal{P}_\lambda \cup \{x_0\}$. Ainsi,

$$C(x_0) := \{y \in S : \forall x \in \mathcal{P}_\lambda,\ d(y, x_0) \leq d(y, x)\},$$

où $d(\cdot, \cdot)$ désigne la distance géodésique sur S.

P. Calka (✉) · A. Chapron
Université de Rouen Normandie, Laboratoire de Mathématiques Raphaël Salem Avenue de l'Université, Saint-Étienne du Rouvray, France
e-mail: pierre.calka@univ-rouen.fr; aurelie.chapron@univ-rouen.fr

N. Enriquez
Laboratoire Mathématiques d'Orsay, Université Paris-Sud, Orsay, France
e-mail: nathanael.enriquez@u-psud.fr

C. Donati-Martin et al. (eds.), *Séminaire de Probabilités L*, Lecture Notes in Mathematics 2252, https://doi.org/10.1007/978-3-030-28535-7_3

L'ensemble des sommets de $C(x_0)$, que nous noterons $\mathcal{V}(x_0)$, est donc défini par

$$\mathcal{V}(x_0) := \{y \in S : \exists x_1, x_2 \in \mathcal{P}_\lambda,\ d(y, x_0) = d(y, x_1) = d(y, x_2)\}.$$

Nous allons commencer par présenter une preuve du théorème de Meijering et Gilbert qui pourra ensuite se généraliser au cas d'une surface.

Proposition 3.1 *Lorsque S est le plan euclidien, pour toute intensité λ,*

$$\mathbb{E}[\mathrm{card}\{\mathcal{V}(x_0)\}] = 6.$$

Remarque Ce théorème peut se prouver de façon abstraite et non calculatoire, en utilisant l'homogénéité de l'espace, et en réinterprétant la quantité $\mathbb{E}[\mathrm{card}\{\mathcal{V}(x_0)\}]$, comme la limite de la moyenne du nombre de sommets par cellule dans une grande boîte du plan. On utilise alors le fait que tous les sommets sont de degré 3, la formule d'Euler pour un graphe, puis le fait que l'aire d'une boîte ne grandit que polynomialement ce qui permet de négliger les effets de bord (ce qui ne serait pas le cas dans l'espace hyperbolique par exemple).

Preuve En vertu de l'homogénéité du plan et du processus de Poisson que l'on considère, la loi de $\mathrm{card}\{\mathcal{V}(x_0)\}$ ne dépend pas de x_0. Nous considérerons donc désormais $\mathrm{card}\{\mathcal{V}(0)\}$. De plus, l'image par une homothétie ne changeant pas le cardinal des sommets d'une cellule, la loi de $\mathrm{card}\{\mathcal{V}(0)\}$ne dépend pas non plus de l'intensité λ. Nous calculons donc $\mathbb{E}[\mathrm{card}\{\mathcal{V}(0)\}]$ dans le cas d'un processus de Poisson d'intensité 1.

Si, pour toute paire de points $\{x_1, x_2\}$ du plan, on note par $\mathcal{B}_{\mathrm{circ}}(0, x_1, x_2)$ la boule circonscrite à 0, x_1 et x_2, la définition de $\mathcal{V}(0)$ donne

$$\mathbb{E}[\mathrm{card}\{\mathcal{V}(0)\}] = \mathbb{E}\left[\sum_{\{x_1,x_2\}\in\mathcal{P}} \mathbf{1}_{\mathcal{B}_{\mathrm{circ}}(0,x_1,x_2)\cap\mathcal{P}=\emptyset}\right].$$

Mais cette dernière espérance peut se réécrire grâce à la formule de Mecke-Slivnyack, qui vient du fait qu'un processus de Poisson conditionné à contenir des points donnés n'est autre que la réunion du processus de Poisson et des dits points. On obtient alors

$$\mathbb{E}[\mathrm{card}\{\mathcal{V}(0)\}] = \frac{1}{2}\int_{\mathbb{R}^2\times\mathbb{R}^2} P(\mathcal{B}_{\mathrm{circ}}(0, x_1, x_2) \cap \mathcal{P} = \emptyset)dx_1dx_2.$$

(*stricto sensu*, on aurait dû écrire $P(\mathcal{B}_{\mathrm{circ}}(x_0, x_1, x_2) \cap (\mathcal{P} \cup \{x_1, x_2\})) = \emptyset)$ mais cette quantité est égale à $P(\mathcal{B}_{\mathrm{circ}}(x_0, x_1, x_2) \cap \mathcal{P} = \emptyset))$. Quant au facteur $\frac{1}{2}$, il provient du fait que les paires de points $\{x_1, x_2\}$ qui contribuent ne sont pas ordonnées et que l'intégrale les compte deux fois.)

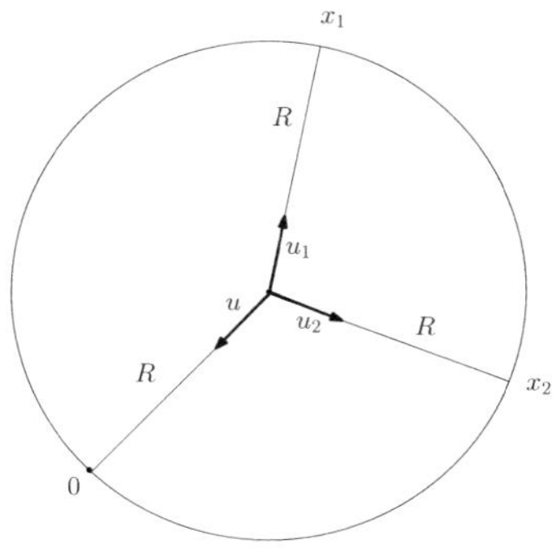

Fig. 3.1 Changement de variables dans le plan

D'après la définition des processus de Poisson, si l'on désigne $R_{\text{circ}}(0, x_1, x_2)$ le rayon de la boule circonscrite à 0, x_1 et x_2, on est donc amené à calculer

$$\mathbb{E}[\text{card}\{\mathcal{V}(0)\}] = \frac{1}{2} \int_{\mathbb{R}^2 \times \mathbb{R}^2} \exp(-\pi R_{\text{circ}}(0, x_1, x_2)^2) dx_1 dx_2.$$

Le calcul de cette intégrale nécessite un changement de variable. Les points x_1 et x_2 sont repérés grâce au centre du cercle circonscrit à $0x_1x_2$ qui, lui, est repéré en coordonnées polaires par $R_{\text{circ}}(0, x_1, x_2)$ et par un vecteur unité u. On a alors uniquement besoin de deux vecteurs unités u_1 et u_2 pour repérer x_1 et x_2 sur le cercle circonscrit à $0x_1x_2$. Les 4 coordonnées R, u, u_1, u_2 repèrent donc les deux points x_1 et x_2 (voir Fig. 3.1).

Calculons maintenant le jacobien. Pour écrire la matrice jacobienne, on décide que les 4 lignes correspondent respectivement aux dérivées partielles suivant R, u, u_1, u_2, et que les 4 colonnes correspondent, pour les deux premières, aux projections de la dérivée partielle de x_1 selon u_1 et $u_1^{\perp}$, et pour les deux dernières, aux projections de la dérivée partielle de x_2 selon u_2 et $u_2^{\perp}$, où pour un vecteur x, le vecteur $x^{\perp}$ désigne le vecteur obtenu à partir de x par une rotation de $\frac{\pi}{2}$.

Avec ces conventions,

$$\frac{dx_1 dx_2}{dR du du_1 du_2} = \left| \begin{vmatrix} \langle u - u_1, u_1 \rangle & \star & \langle u - u_2, u_2 \rangle & \star \\ R\langle u^{\perp}, u_1 \rangle & \star & R\langle u^{\perp}, u_2 \rangle & \star \\ 0 & R & 0 & 0 \\ 0 & 0 & 0 & R \end{vmatrix} \right|$$

$$= R^3 \left| \begin{vmatrix} \langle u - u_1, u_1 \rangle & \langle u - u_2, u_2 \rangle \\ \langle (u + u_1)^{\perp}, u_1 \rangle & \langle (u + u_2)^{\perp}, u_2 \rangle \end{vmatrix} \right|$$

$$= 2R^3 \text{ Aire } (uu_1u_2),$$

Fig. 3.2 Fonction Aire(uu_1u_2) (aire du triangle grisé)

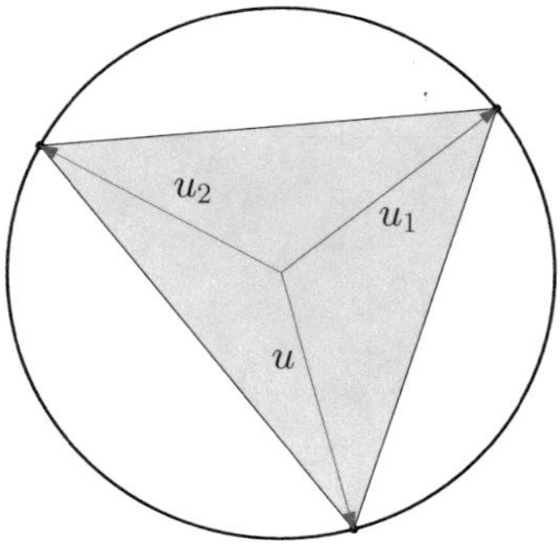

où Aire(uu_1u_2) désigne l'aire du triangle défini comme l'enveloppe convexe de u, u_1 et u_2 (voir Fig. 3.2). Ainsi,

$$\begin{aligned}
\mathbb{E}[\text{card}\{\mathcal{V}(0)\}] &= \frac{1}{2}\int_{\mathbb{R}_+\times[0,2\pi]^3} R^3\exp(-\pi R^2)2\,\text{Aire}(uu_1u_2)dRdudu_1du_2\\
&= \frac{1}{2}\int_{\mathbb{R}_+} R^3\exp(-\pi R^2)dR\int_{[0,2\pi]^3} 2\,\text{Aire}(uu_1u_2)dudu_1du_2\\
&= \frac{1}{2\pi^2}2\pi\int_0^{2\pi}\int_0^{2\pi-x} |\sin x+\sin y+\sin(2\pi-x-y)|dydx\\
&= 6.
\end{aligned}$$

(Le facteur $\frac{1}{2}$ disparaît dans la troisième égalité car les points sont ordonnés lorsque l'intégrale en u, u_1, u_2 est écrite comme une intégrale sur le domaine $0 < y < 2\pi - x$). ∎

Passons maintenant au cas d'une surface.

Théorème 3.1 *Soit S une surface et un point x_0 de S, lorsque l'intensité λ de $\mathcal{P}_\lambda$ tend vers l'infini,*

$$\mathbb{E}[\text{card}\{\mathcal{V}(x_0)\}] = 6 - \frac{K(x_0)}{3\pi}\frac{1}{\lambda} + o(\frac{1}{\lambda})$$

où $K(x_0)$ désigne la courbure de Gauss de la surface au point x_0.

Preuve On commence par remarquer que toutes les observations liées à l'homogénéité du processus ponctuel de Poisson, faites au début de la preuve précédente ne sont plus valides. En conséquence, nous considérons donc la cellule de x_0 et faisons tendre l'intensité λ vers l'infini.

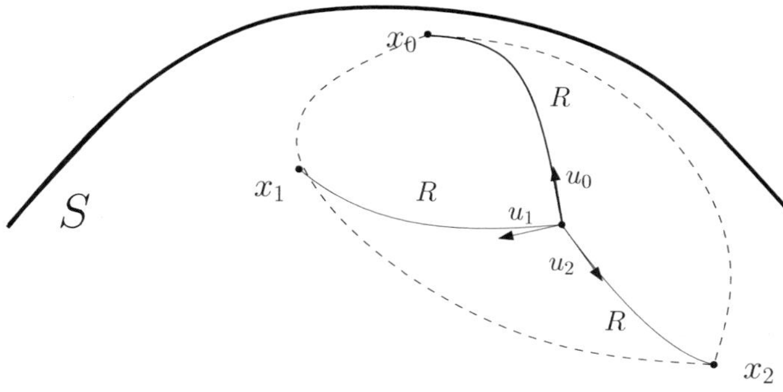

Fig. 3.3 Changement de variables sur S

Reprenant le calcul du dessus, nous pouvons écrire

$$\mathbb{E}[\mathrm{card}\{\mathcal{V}(x_0)\}] = \frac{\lambda^2}{2}\int_{S\times S} \exp(-\lambda\,\mathrm{Aire}_{\mathrm{circ}}(x_0, x_1, x_2))dx_1 dx_2$$

où $\mathrm{Aire}_{\mathrm{circ}}(x_0, x_1, x_2)$ désigne l'aire du disque circonscrit à $0, x_1, x_2$.

Nous sommes confrontés donc à nouveau à un calcul de jacobien, qui cette fois ne peut être effectué de façon exacte (voir le nouveau changement de variables en Fig. 3.3). Comme λ tend vers l'infini, la contribution prépondérante concerne les triplets dont le rayon circonscrit est petit. On va donc effectuer un développement limité du jacobien dans la limite des R tendant vers 0. Le premier ordre est le même que dans le plan. Pour connaître le terme suivant, nous réécrivons la matrice jacobienne. Le changement principal réside dans le fait que les termes en R de la matrice précédente, témoignant de l'éloignement linéaire des géodésiques dans le cas plan, sont remplacés par la fonction $J(R)$, qui satisfait $J(R) = R - \frac{K(x_0)}{6}R^3 + o(R^3)$.

$$\frac{dx_1 dx_2}{dR du du_1 du_2} = \left| \begin{vmatrix} \langle u - u_1, u_1\rangle & \star & \langle u - u_2, u_2\rangle & \star \\ J(R)\langle u^\perp, u_1\rangle & \star & J(R)\langle u^\perp, u_2\rangle & \star \\ 0 & J(R) & 0 & 0 \\ 0 & 0 & 0 & J(R) \end{vmatrix} \right|$$

$$= \mathrm{Aire}(u u_1 u_2) J(R)^3$$

$$= \mathrm{Aire}(u u_1 u_2) R^3 (1 - \frac{K(x_0)}{2} R^2 + o(R^2)).$$

Rappelons maintenant le résultat classique suivant, sur le développement de l'aire de petites boules en courbure $K(x_0)$, dû à Bertrand-Diquet-Puiseux en 1848, qui dit que

$$\mathrm{Aire}_{\mathrm{circ}}(x_0, x_1, x_2)) = \pi R^2(1 - \frac{K(x_0)}{12}R^2) + o(R^3).$$

On peut donc écrire

$$\begin{aligned}
&\mathbb{E}[\mathrm{card}\{\mathcal{V}(x_0)\}] \\
&= \frac{\lambda^2}{2}\int_{\mathbb{R}_+\times[0,2\pi]^3} R^3(1 - \frac{K(x_0)}{2}R^2 + o(R^2))\exp\left(-\lambda\pi R^2(1 - \frac{K(x_0)}{6}R^2) + o(R^3)\right) \\
&\qquad \mathrm{Aire}(uu_1u_2)dRdudu_1du_2 \\
&= 12\pi^2\lambda^2\int_{\mathbb{R}_+} R^3(1 - \frac{K(x_0)}{2}R^2 + o(R^2))\exp\left(-\lambda\pi R^2(1 - \frac{K(x_0)}{6}R^2) + o(R^3)\right)dR.
\end{aligned}$$

Il ne reste plus qu'à effectuer une méthode de Laplace dans l'intégrale pour obtenir le résultat annoncé. ■

Le théorème 3.1 se généralise à la dimension quelconque comme annoncé au début. Même si nous ne présenterons pas ce travail ici, le déterminant de taille n^2 se simplifie de manière assez spectaculaire, et après application de la méthode de Laplace dans l'intégrale, la courbure de Gauss se voit remplacée par la simple courbure scalaire. C'est ce que le troisième auteur a plaidé à Jacques Azéma lors de sa dernière conversation, pour lui dire que ce travail ne se limitait pas à une nouvelle preuve du théorème de Gauss-Bonnet mais proposait une vision probabiliste générale de la courbure scalaire...

Venons-en au dernier point de cette note. Comment déduire du Théorème 3.1, une preuve du théorème de Gauss-Bonnet ? Nous allons d'abord énoncer ce théorème qui, pour une surface compacte sans bord relie une quantité topologique, la caractéristique d'Euler de la surface, à une quantité géométrique, l'intégrale de la courbure de Gauss sur la surface :

Théorème 3.2 (Gauss-Bonnet [2, Section 4.5, Corollaire 2]) *Soit S une surface compacte sans bord, la caractéristique d'Euler $\chi(S)$ satisfait*

$$\chi(S) = \frac{1}{2\pi}\int_S K(x)d\sigma(x)$$

où $d\sigma$ désigne la mesure de surface sur S.

Preuve Considérons la mosaïque de Poisson-Voronoi associée à un processus de Poisson $\mathcal{P}_\lambda$ d'intensité homogène λ par rapport à σ. Notons F, E et V, le nombre total respectivement de faces, arêtes et sommets de la mosaïque. La relation d'Euler appliquée au graphe de Voronoi donne

$$\chi(S) = F - E + V.$$

Comme tous les sommets sont presque sûrement de degré 3, on peut faire partir de chaque sommet trois demi-arêtes et obtenir une partition de l'ensemble des arêtes. Ceci entraîne donc

$$E = \frac{3}{2}V.$$

De ces deux dernières équations, on déduit

$$\chi(S) = F - \frac{1}{2}V.$$

Cette dernière égalité étant vraie presque sûrement, en passant à l'espérance, on obtient

$$\chi(S) = \mathbb{E}[F] - \frac{1}{2}\mathbb{E}[V]. \tag{3.1}$$

Il nous reste donc à calculer les deux espérances du membre de droite. Remarquons d'abord que le nombre de faces F n'est rien d'autre que le nombre de points de $\mathcal{P}_\lambda$ dans S, ainsi

$$\mathbb{E}[F] = \mathbb{E}[\mathrm{card}(\mathcal{P}_\lambda \cap S)] = \lambda\sigma(S). \tag{3.2}$$

Maintenant, comme chaque sommet appartient à 3 cellules,

$$\mathbb{E}[V] = \frac{1}{3}\mathbb{E}\left[\sum_{x\in\mathcal{P}_\lambda} \mathrm{card}\{\mathcal{V}(x)\}\right].$$

En appliquant la formule de Mecke-Slivnyak, on obtient

$$\mathbb{E}[V] = \frac{\lambda}{3}\int_S \mathbb{E}[\mathrm{card}\{\mathcal{V}(x, \mathcal{P}_\lambda \cup \{x\})\}]d\sigma(x) \tag{3.3}$$

où $\mathcal{V}(x, \mathcal{P}_\lambda \cup \{x\})$ désigne les sommets de la cellule de x dans la mosaïque associée à $\mathcal{P}_\lambda \cup \{x\}$. Une version uniforme du Théorème 3.1, donne

$$\sup_{x\in S} \lambda \left(\mathbb{E}[\mathrm{card}\{\mathcal{V}(x, \mathcal{P}_\lambda \cup \{x\})\}] - 6 + \frac{3K(x)}{\pi\lambda}\right) \underset{\lambda\to\infty}{\to} 0. \tag{3.4}$$

En combinant (3.1), (3.2), (3.3) et (3.4), on obtient

$$\chi(S) = \lim_{\lambda \to \infty} \frac{1}{2\pi} \int_S K(x) d\sigma(x).$$

puis l'égalité du théorème, car l'intégrale ne dépend pas de λ. ■

Littérature

1. P. Calka, A. Chapron, N. Enriquez, *Mean Asymptotics for a Poisson-Voronoi Cell on a Riemannian Manifold* (2018). ArXiv:1807.09043
2. M. P. Do Carmo. *Differential Geometry of Curves and Surfaces* (Prentice-Hall, Upper Saddle River, 1976)
3. E. N. Gilbert. Random subdivisions of space into crystals. Ann. Math. Stat. **33**(3), 958–972 (1962)
4. J. L. Meijering. Interface area, edge length, and number of vertices in crystal aggregates with random nucleation. Philips Res. Rep. **8**, 270–290 (1953)

Chapter 4
Sur le retournement du temps

Sonia Fourati

Abstract This text is conceived to be an invitation to read or re-read one the major contribution of Jacques Azéma, who was my PhD advisor. This work, whose title is "théorie générale des processus et retournement du temps" had appeared in 1973. This tribute aims also at describing the influence that Jacques Azéma has had on my own work.

Après le DEA, je ne sais ni comment ni quand, je fus déclarée au « labo »—entendre le « Laboratoire de Probabilités de l'Université Paris 6-Jussieu »—officiellement, l'étudiante en thèse de Jacques Azéma. Je ne me souviens pas d'avoir choisi mon directeur, la réciproque était probablement plus proche des faits. Mais mon sort me convenait tout à fait. L'humour d'Azéma en première loge et pour quelques années, c'est un privilège qui s'apprécie!

En ce temps-là, c'est-à-dire à la fin de l'année 1981, Azéma travaillait, travaillait dur et même douloureusement, sur son papier « Sur les fermés aléatoires » [4]. Pour sujet de thèse, il m'a demandé de trouver une nouvelle preuve du « Théorème de Kesten » que voici :

« Pour tout temps t strictement positif et tout fermé aléatoire régénératif F, de mesure de Lebesgue nulle, la probabilité que t soit un élément de F est nulle ».

Cette nouvelle démonstration devait être une application des propriétés des noyaux de Lévy associés à un fermé aléatoire quelconque, qu'il avait obtenues et qu'on peut retrouver dans sa publication [4].

Je n'ai jamais trouvé cette nouvelle preuve. Mais mon lien avec les travaux d'Azéma ne se sont pas arrêtés là.

J'ai découvert son fameux travail « Théorie générale des processus et retournement du temps », paru dix ans plus tôt, en 1973 [3], en travaillant sur un nouveau sujet, que m'avait donné à traiter Erik Lenglart. Azéma était très attaché à ce texte et

---

S. Fourati (✉)
LMI-INSA de Rouen, Saint-Etienne-du-Rouvray, France

Sorbonne Université LPSM, Paris, France
e-mail: sonia.fourati@insa-rouen.fr

C. Donati-Martin et al. (eds.), *Séminaire de Probabilités L*, Lecture Notes in Mathematics 2252, https://doi.org/10.1007/978-3-030-28535-7_4

aujourd'hui, afin d'inviter à sa lecture, notamment le « jeune » probabiliste qui n'a pas connu la grande époque de la Théorie Générale des Processus, je vais tenter de résumer ici son propos.

Tout d'abord, Azéma reprend les objets, qu'il avait introduit dans [1] et [2], et qui permettait le retournement des processus et il le fait, à sa manière bien sûr. Il se donne un processus X qui prend ses valeurs dans un espace Lusinien E, X a une durée de vie finie mais aléatoire notée ζ, et on attribue à X, après le temps ζ, une valeur cimetière δ, qui est ajoutée à l'ensemble E. Les trajectoires sont supposées continues à droite et l'hypothèse supplémentaire de continuité à gauche sera envisagée aussi dans la suite.

Les opérateurs de translation $(\theta_t)_{t\geq 0}$ et de meurtre $(k_t)_{t>0}$ pour le processus X sont des fonctions définies sur l'espace des possibles Ω et qui vérifient l'identité, pour tous temps positifs s et t,

$$X_s(\theta_t(\omega)) = X_{s+t}(\omega), \quad X_s(k_t(\omega)) = X_s(\omega) \text{ si } s < t,\ X_s(k_t(\omega)) = \delta \text{ sinon}.$$

J.A. définit alors les opérateurs $\hat{k}_t$ et $\hat{\theta}_t$ pour tous temps t, de la manière suivante:

$$\hat{k}_t(\omega) := \theta_{(\zeta-t)^+}(\omega) \qquad \hat{\theta}_t(\omega) := k_{(\zeta-t)^+}\,.$$

Il établit que les opérateurs $(\hat{k}_t)_{t>0}$ constituent des opérateurs « de meurtre » pour le processus retourné $\hat{X}_t = X_{\zeta-t}$ si $t\in]0,\zeta] = \delta$ si $t > \zeta$ sont des opérateurs de translation pour $\hat{X}$. Cela signifie que pour tous temps s et t,

$$\hat{X}_s(\hat{k}_t(\omega)) = \hat{X}_s \text{ si } s \leqslant t \text{ et } \hat{X}_s(\hat{k}_t(\omega)) = \delta \text{ sinon}$$
$$\hat{X}_s(\hat{\theta}_t(\omega)) = \hat{X}_{s+t}(\omega)\,.$$

J.A. poursuit en introduisant la notion de temps de retour « co-prévisible » qui est la notion « retournée » des temps d'arrêt prévisibles.

Définition 4.1 On dit qu'une variable τ est un temps de retour co-prévisible si c'est une variable aléatoire à valeurs dans $[0,\zeta[\cup\{-\infty\}$ et qu'elle vérifie l'identité, pour tout réel positif t,

$$\tau(\theta_t) = \tau - t \text{ si } \tau \geqslant t \text{ et } \tau = -\infty \text{ sinon.}$$

Du fait de la dualité des opérateurs de translation et de meurtre citée plus haut, on ne s'étonnera pas que les temps d'arrêt prévisibles soient les temps aléatoires à valeurs dans $]0,\zeta]\cup\{+\infty\}$, qui vérifient l'identité

$$T(k_t) = T \text{ si } T \leqslant t \qquad \text{et} \qquad T = +\infty \text{ sinon}.$$

(On pourra aussi retrouver les temps d'arrêt prévisibles dans [6].)

Azéma poursuit son exposé en définissant la tribu co-prévisible, et rappelle celle de la tribu des « bien mesurables » (rebaptisée peu après « tribu optionnelle » par la communauté) :

Définition 4.2 La tribu co-prévisible $\mathscr{G}$ est la tribu sur $[0, +\infty[\times\Omega$ engendrée par les processus réels continus à droite, nuls hors de $[0, \zeta[$ vérifiant l'identité:

$$Z_s(\theta_t) = Z_{s+t} \quad \text{si} \quad 0 < s \leqslant s+t < \zeta \, .$$

La tribu des processus bien mesurables, est la tribu sur $[0, +\infty[\times\Omega$ engendrée par les processus réels càdlàg Z, nuls hors de $[0, \zeta[$, et tels que pour tous temps s et t,

$$Z_s(k_t) = Z_s \quad \text{si} \quad 0 \leqslant s < t < \zeta \, .$$

Ensuite vient l'énoncé de la projection sur ces tribus, cette projection est le pendant de l'espérance conditionnelle pour les variables aléatoires. Pour le premier énoncé qui suit, celui de la projection sur la tribu des bien mesurables, Azéma cite pour référence le travail séminal de Claude Dellacherie [6]. Le second énoncé est le sien.

Théorème 4.1 *Pour tout processus mesurable borné Z, il existe un processus « bien mesurable » (= « optionnel ») qui vérifie, pour tout temps d'arrêt, T ,*

$$\mathbb{E}(Z_T 1_{T\in[0,\zeta[}) = \mathbb{E}({}^o Z_T 1_{T\in[0,\zeta[}) \, ,$$

Il existe un processus co-prévisible $^{\hat{p}}Z$ tel que pour tout temps de retour τ,

$$\mathbb{E}(Z_\tau 1_{\tau\in[0,\zeta[}) = \mathbb{E}({}^{\hat{p}} Z_\tau 1_{\tau\in[0,\zeta[}) \, .$$

Si on pose que ^oZ et $^{\hat{p}}Z$ sont nuls hors de l'intervalle de vie $[0, \zeta[$, Ces deux processus sont uniquement déterminés (à une indistinguabilté[1] près) par ces propriétés.

J.A. note alors que la propriété de Markov forte de X, qu'il énonce ainsi :
pour toute variable aléatoire bornée, et pour tout temps d'arrêt T,

$$\mathbb{E}(z \circ \theta_T; T \in [0, \zeta[) = \mathbb{E}(P^{X_T}(z); T \in [0, \zeta[) \, ,$$

pour un noyau $(P^x)_{x\in E}$ sur (Ω, F).
implique la propriété suivante :
si Z est un processus co-prévisible alors la projection ^oZ est indistinguable d'un processus de la forme $f(X)$.

---

[1] Deux processus U et V sont dits *indistinguables* si l'ensemble $\{\omega; \exists t, U_t(\omega) \neq V_t(\omega)\}$ est de probabilité nulle.

Il suffit pour cela de remarquer qu'un processus co-prévisible Z se met sous la forme $Z = z \circ \theta$ pour une variable aléatoire z.

Ce dernier processus, $f(X)$, est à la fois mesurable par rapport à la tribu optionnelle et aussi par rapport à la tribu co-prévisible (à une indistinguabilité près). C'est alors que J.A. montre le théorème suivant :

Théorème 4.2 *Quand la propriété de Markov forte est vérifiée, les projections sur les tribus bien mesurable et co-prévisible commutent et, plus précisemment, pour tout processus mesurable borné Z , il existe une fonction mesurable f sur E, telle que les trois processsus ${}^{\hat{p}}({}^{o}Z)$ et ${}^{o}({}^{\hat{p}}Z)$ et $f(X)$ sont indistinguables.*

Le corollaire immédiat de cette commutation de projections est que si T est un temps d'arrêt OU si T est un temps de retour co-prévisible alors les tribus sur Ω qui décrivent l'une le passé $(\sigma(z \circ k_T, T \in [0, \zeta[))$ et l'autre le futur $(\sigma(z \circ \theta_T; T \in [0, \zeta[))$ sont indépendantes conditionellement à la tribu du « présent » $\sigma(X_T)$. Le processus continu à gauche retourné, $\hat{X}$, a donc une propriété markovienne. J.A. dit que $\hat{X}$ a la propriété de Markov « modérée ». En plus, en reprenant les arguments développés par Chung et Walsh [5], Azéma construit un noyau $(Q^x, x \in E)$ tel que, pour tout variable aléatoire z bornée, et tout temps de retour co-prévisible τ, l'identité suivante est vérifiée :

$$\mathbb{E}(z \circ k_\tau; \tau \in [0, \zeta[) = \mathbb{E}(Q^{X_\tau}(z); \tau \in [0, \zeta[) .$$

(On remarquera que si Z est un processus prévisible, il est bien mesurable et il se met sous la forme $Z_t = z \circ k_t$.)

Voici un exemple simple de processus de Markov modéré. Prenons d'abord le processus de Lévy, $X_t = 2t - N_t$ où N_t est un processus de Poisson standard, alors le processus $X_t - \inf_{s>t} X_s$ n'est pas fortement markovien mais $X_{t^-} - \inf_{s \geqslant t} X_s$ est bien modérément markovien (voir le commentaire page 29 de [7] pour plus de détails).

Comme pour la projection optionnelle, l'intérêt de l'existence de projection sur les tribus co-prévisibles est que l'on peut alors développer les techniques de la théorie du potentiel (résolvante, fonction harmonique...) avec « seulement » cette propriété de Markov modérée plutôt que forte. C'est ce point de vue de Jacques Azéma qui m'a énormément aidée, nettement plus tard, pour étudier l'existence ou non des points de croissance pour les trajectoires des processus de Lévy [7].

Mais revenons à la théorie de J.A. pour une dernière analyse.

La propriété de Markov se comprend comme l'indépendance du futur et du passé d'un processus X conditionnellement à son présent.

Elle devrait être, idéalement, parfaitement stable par retournement du temps. Comment donc « réparer » le fait qu'un retourné d'un processus de Markov fort n'est « que » modéré ?

Deux points de vue sont possibles pour cela.

(1) On peut renforcer la propriété de Markov forte afin que son retourné soit fortement Markovien aussi : c'est l'hypothèse CMF chez Azéma : Il suffit

de supposer que les trajectoires soient non seulement continues à droite mais aussi limitées à gauche, et que le couple (X^-, X) ait une propriété renforcée au sens suivant. On définit d'abord la tribu cooptionelle engendrée par les processus càglàd dont le régularisé à droite Z^+ est co-prévisible. La tribu des cooptionnels contient la tribu co-prévisible. (C'est le même schéma que pour la tribu optionnelle, qui est engendrée par les càdlàg dont le régularisé à gauche est prévisible). La propriété CMF de Markov consiste en ce que la projection d'un processus co-optionnel sur la tribu optionnelle est un processus de la forme $f(X^-, X)$ pour une fonction mesurable bivariée f. Azéma évoque bien que cette hypothèse est associée à celle de l'existence d'une topologie « cofine » sur l'espace E, selon les travaux de Walsh et Weil [11]. Il cite moins les travaux concernant la frontière de Martin, voir [8].

(2) La seconde option consiste à affaiblir la propriété de Markov forte en évitant de se tirer une balle dans le pied par le choix de la projection sur la tribu optionnelle (ou « bien mesurable ») :

En effet la propriété de Markov forte habituelle implique l'absence de « germes à droite » : je veux dire que les processus de la tribu « germe »

$$\cap_{\varepsilon>0}\sigma(t \mapsto f(X_{t+s}); s \in [0, \varepsilon[, f \text{ mesurable})$$

sont indistinguables de processus de la forme $f(X)$ (pour une fonction mesurable f).

Les germes à droite devenant des germes à gauche quand on retourne le temps, là se trouve le problème. La théorie de Lenglart, voir [9], donne des tribus alternatives à la tribu optionnelle, incluses dans celle-ci, et qui ne contiennent que le passé « exact » (ou « coupé net ») du processus X. Les germes à droite (qui représentent le « futur immédiat » de X) ne sont pas inclus dans cette tribu du passé de X.

Erik Lenglart montre que pour ces tribus, il existe aussi une projection et par voie de conséquence, on peut en déduire une propriété de Markov associée, dans le même esprit qu'Azéma, on pourra se reporter à [10] pour plus de détails.

Cette fois la propriété de Markov est parfaitement stable par retournement. Et toute l'artillerie de la théorie du potentiel pour étudier ces processus peut être étendue. Je le fis dans le cadre des processus de Markov liés aux fluctuations des processus de Lévy dans [7].

Epilogue

Azéma est parti le 5 Janvier. Je trie ses papiers, livres et autres notes mathématiques restés dans son bureau. Ses dossiers contiennent, entre autres, des cours manuscrits, du DEUG au DEA. Ces notes sont extrêmement soignées, d'une belle écriture et très peu de ratures.

Azéma, pourquoi cachais-tu tant cet aspect si consciencieux, cette peur de mal faire peut-être même, sous ton inimitable désinvolture et ton humour corrosif, extrêmement drôle parce qu'irréductiblement « uncorrect », si rare ?

Cet humour, qui risquerait de t'attirer encore plus d'ennuis aujourd'hui.

Azéma, j'ai mangé ce midi, à « l'Industrie », ton dernier bistrot. J'ai vu ta copine la barmaid, celle qui te câlinait tant. Elle est redevenue pas sympa et même pas belle... Le monde est décidemment moins intéressant sans toi,

Au revoir, euh, ...

Jacques.

Remerciements Je remercie les éditeurs du « Séminaire » de savoir préserver son aspect si particulier depuis toujours. En me permettant de faire revivre la muse, qui était aussi particulière que ce journal qu'elle contribua longuement à éditer, Catherine Rainer m'a aidée à surmonter l'angoisse de la feuille blanche et je l'en remercie aussi.

Littérature

1. J. Azéma, Quelques remarques sur les temps de retour. Trois applications.*Séminaire de Probabilités. VI*. Lecture Notes in Mathematics, vol. 258 (Springer, Berlin, 1972), pp. 35–50
2. J. Azéma, Quelques applications de la théorie générale des processus. Invent. Math. **18**, 293–336 (1972)
3. J. Azéma, Théorie générale des processus et retournement du temps. Ann. Sci. Ecole Norm. Sup. **6**(4), 459–519 (1973)
4. J. Azéma, Sur les fermés aléatoires. *Séminaire de Probabilités. XIX*. Lecture Notes in Mathematics, vol. 1123 (Springer, Berlin, 1985), pp. 397–495
5. K.L. Chung, J. Walsh, To reverse a Markov process. Acta Math. **123**, 225–251 (1970)
6. C. Dellacherie, Contributions à la téorie générale des processus: Capacités et processus stochastiques, in *Ergebnisse der Mathematik und ihrer Grenzgebiete*, Band 67 (Springer, Berlin, 1972)
7. S. Fourati, Points de croissance des processus de Lévy et théorie générale des processus. Probab.Theory Relat. Fields **110**, 13–49 (1998)
8. H. Kunita, T. Watanabe, Markov processes and Martin boundaries. Illinois J. Math. **9**, 485–526 (1965).
9. E. Lenglart, Tribus de Meyer et théorie des processus, in *Séminaire de Probabilités. XIV*. Lecture Notes in Mathematics, vol. 784 (Springer, Berlin, 1980), pp. 500–546
10. E. Lenglart, S. Fourati, Tribus homgènes et commutation des projections, in *Séminaire de Probabilités. XXI*. Lecture Notes in Mathematics, vol. 1247 (Springer, Berlin, 1987), pp. 276–288
11. J.B. Walsh, M. Weil, Représentation des temps terminaux et application aux fonctionnelles additives et aux systèmes de Lévy. Ann. Sci. Ec. Norm. Sup. **5**(4), 121–155 (1972)

Chapter 5
La martingale d'Azéma

Catherine Rainer

J'ai été très émue que les éditeurs de ce volume dédié à Jacques Azéma m'aient demandé d'y contribuer. Azéma était mon directeur de thèse dans la première moitié des années 1990. Ayant quitté à la fois le domaine et le milieu de la théorie générale des processus, je ne me sens pas vraiment habilitée à écrire un témoignage sur l'être humain que j'ai ainsi perdu de vue depuis longtemps ni d'écrire un article mathématique autour ou à la suite de ses recherches. Mais, en laissant divaguer mes souvenirs, il m'est venu rapidement l'envie de parler de la martingale d'Azéma, qui, avec le recul, me semble emblématique de cette époque et de la personnalité de son inventeur.

La martingale d'Azéma, telle qu'elle est définie dans l'article « Étude d'une martingale remarquable » d'Azéma et Yor [5], est la projection optionnelle d'un mouvement Brownien sur la filtration engendrée par son signe.
Plus précisément, considérons un mouvement Brownien standard en dimension 1, (B_t), tel que $B_0 = 0$. Pour tout $t \geqslant 0$, on pose

$$\operatorname{sign}(B_t) = \begin{cases} 1, & \text{si } B_t \geqslant 0, \\ -1 & \text{sinon.} \end{cases}$$

Soit $(\mathcal{F}_t)_{t \geqslant 0}$ la filtration rendue continue à droite et complétée, engendrée par le processus $(\operatorname{sign}(B_t))_{t \geqslant 0}$.

Définition 5.1 On appelle **martingale d'Azéma** le processus $(\mu_t)_{t \geqslant 0}$ tel que, pour tout $t \geqslant 0$, $\mu_t = E[B_t | \mathcal{F}_t]$.

C. Rainer (✉)
Université de Bretagne Occidentale, LBMA, Brest Cedex, France
e-mail: Catherine.Rainer@univ-brest.fr

C. Donati-Martin et al. (eds.), *Séminaire de Probabilités L*, Lecture Notes in Mathematics 2252, https://doi.org/10.1007/978-3-030-28535-7_5

Ce processus peut être calculé explicitement :

$$\mu_t = \sqrt{\frac{\pi}{2}}\,\mathrm{sign}(B_t)\sqrt{t - g_t}, \tag{5.1}$$

où $g_t = \sup\{s \leqslant t, B_s = 0\}$ est, pour tout $t \geqslant 0$, le dernier instant avant t où le mouvement brownien touche 0.
Du coup les trajectoires de (μ_t) ont l'allure suivante[1]:

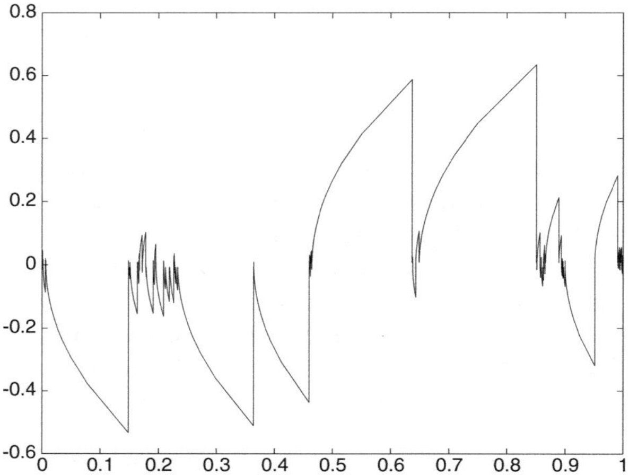

On peut définir aussi une « deuxième martingale d'Azéma » : $\nu_t := |\mu_t| - L_t^0$, où (L_t^0) est le temps local en zéro du mouvement brownien, qui représente la projection optionnelle du processus $(|B_t| - L_t^0)$ sur la filtration $(\mathcal{G}_t)$ engendrée par les zéros du mouvement brownien, i.e. vérifiant $\mathcal{G}_t = \sigma\{g_s, s \leqslant t\}$. Il est à noter que c'est cette deuxième martingale qui fut inventée en premier : elle apparait en 1984 à la 67ème page du gros article « Sur les fermés aléatoires » d'Azéma [4]. En effet, dans le chapitre 6 — qui s'appelle « Une sous-martingale remarquable » —, étant donné un fermé aléatoire H quelconque dans $\mathbb{R}_+$ et un temps local (L_t) dont il est le support, l'auteur définit une sous-martingale (Y_t) définie à partir du noyau de Lévy et des extrémités gauches de H, telle que $(Y_t - L_t)$ est une martingale locale s'annulant sur H. La martingale (ν_t) en est le cas particulier lorsque H est l'ensemble des zéros du mouvement brownien.

On montre facilement que les deux martingales (μ_t) et (ν_t) sont purement discontinues. Un premier intérêt en est donc de livrer des exemples de martingales purement discontinues qui n'appartiennent pas à la famille des processus de Poisson. De plus la première a la propriété remarquable de posséder les mêmes zéros et le même signe qu'un mouvement brownien.

[1]Merci beaucoup à Nathanael Enriquez, pour le prêt de ses graphiques tirés de [14] !

L'article d'Azéma-Yor [5] développe un grand nombre de résultats et de propriétés autour de la martingale (μ_t). Parmi celles-ci, c'est principalement la propriété de représentation chaotique qui a motivé l'intérêt accru qu'on lui a porté.

Définition 5.2 Soit $(X_t)_{t\geqslant 0}$ une martingale telle $\langle X, X\rangle_t = t$, et, pour $n \in \mathbb{N}^*$, $f : \Delta_n = \{0 < t_n < t_{n-1} < \ldots < t_1\} \to \mathbb{R}$ une fonction de carré intégrable. Alors, l'intégrale itérée

$$I_n(f) = \int_{\Delta_n} f(t_1, \ldots, t_n) dX_{t_1} \ldots dX_{t_n}$$

est bien définie. L'ensemble χ_n de ces variables $I_n(f)$ est appelé le **n-ième chaos généré par** (X_t), et on dit que (X_t) a la **propriété de représentation chaotique (PRC)** si l'espace engendré par l'union des chaos (où χ_0 est l'ensemble des variables constantes) est dense dans l'espace L^2 des variables de carré intégrable, i.e. si toute variable $X \in L^2$ peut s'écrire sous la forme

$$X = \sum_{n=0}^{+\infty} I_n(f_n),$$

avec $f_n \in L^2(\Delta_n)$.

La PRC est importante pour les probabilités quantiques, chaque martingale la possédant fournissant un exemple d'interprétation probabiliste de l'espace de Fock (voir par exemple Meyer [17]). Il est connu depuis Wiener [20] que le mouvement Brownien et les processus de Poisson compensés ont la propriété de représentation chaotique. La question (posée par Meyer dans [16]) est : y en a-t-il d'autres, et si oui, est-il possible de caractériser les martingales possédant la PRC ?

C'est dans le but d'apporter une réponse à cette question que, dans son article « On the Azéma martingales », Michel Émery introduit les « équations de structure ».

Définition 5.3 (Émery [11]) Soit (X_t) une martingale vérifiant $\langle X, X\rangle_t = t$. On dit que (X_t) vérifie une **équation de structure** s'il existe un processus prévisible (Φ_t) tel que

$$d[X, X]_t = dt + \Phi_t dX_t.$$

Après une première partie de résultats généraux sur les martingales vérifiant une telle équation, l'attention est portée sur une famille particulière de martingales, celles qu'Émery appelle « the Azéma martingales » : on y considère les équations de structure de la forme

$$d[X, X]_t = dt + (\alpha + \beta X_{t-}) dX_t, \tag{5.2}$$

avec α et β constantes. En voici les exemple les plus remarquables :

- Lorsque $\alpha = \beta = 0$, la solution de (5.2) est le mouvement brownien,
- pour $\beta = 0$ mais $\alpha \neq 0$, (X_t) est un processus de Poisson compensé.
- Une martingale très jolie est obtenue pour $\alpha = 0$ et $\beta = -2$. Elle vérifie $|X_t| = \sqrt{t}$ et change de signe en des temps totalement non prévisibles, juste comme il faut pour être une martingale.[2]

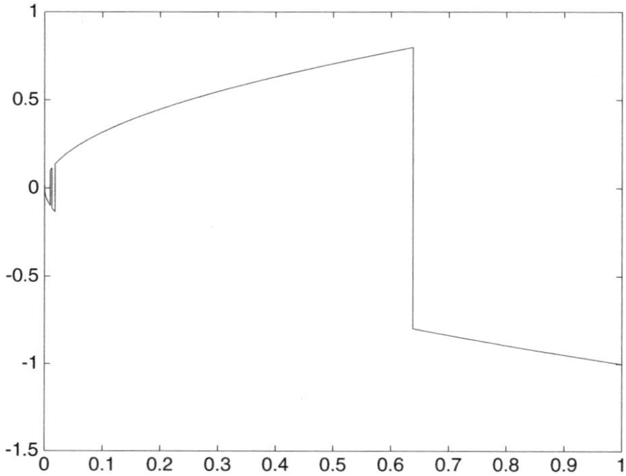

- L'autre cas intéressant est celui où $\alpha = 0$ et $\beta = -1$:

$$d[X, X]_t = dt - X_{t-}dX_t.$$

En effet la solution de cette équation n'est autre que la martingale d'Azéma (μ_t).

Le résultat central de l'article concernant ces martingales est

Théorème 5.1 *On considère l'équation de structure*

$$d[X, X]_t = dt + \beta X_{t-}dX_t, X_0 = x_0. \tag{5.3}$$

1. *Si $\beta \leqslant 0$, la solution de* (5.3) *est unique en loi, et c'est un processus de Markov fort.*
2. *Pour $-2 \leqslant \beta \leqslant 0$, la solution de* (5.3) *a la propriété de représentation chaotique.*

Corollaire 5.1 *La martingale d'Azéma (μ_t) a la propriété de représentation chaotique.*

[2]Même si c'est déjà un peu hors sujet, je ne peux m'empêcher d'évoquer le plaisir d'avoir retrouvé cette martingale de façon totalement inattendue en étudiant des jeux différentiels à information imparfaite [6].

A la recherche d'une caractérisation des martingales possédant la PRC, on peut voir facilement qu'une condition nécessaire en est la propriété de représentation prévisible : toute variable $Z \in L^2$ s'écrit sous la forme $Z = E[Z] + \int H_t dX_t$, avec (H_t) processus prévisible. Une question naturelle est alors de savoir si les deux propriétés sont équivalentes.
C'est ainsi que le sujet de thèse proposé par Azéma peu de temps après la publication de l'article d'Émery s'intitulait « Propriétés de représentations prévisible et chaotique ». Naïve et m'ennuyant un peu du calme train-train de mes études de mathématiques, c'est clairement le mot « chaotique » qui m'avait attirée (l'associant sûrement aussi à la théorie des catastrophes, très en vogue à cette époque). En réalité, sachant que la première martingale d'Azéma (μ_t) avait la PRC, ce sujet consistait à étudier si sa deuxième martingale (ν_t) l'avait aussi. La question s'est rapidement avérée beaucoup trop difficile, et la réponse restera sans doute pour toujours inconnue.

Alors que ma thèse se réorientait du coup vers d'autres extensions et variations de la théorie azémaéenne, Marc Yor a pris une place croissante dans nos discussions. J'ai eu alors le privilège d'être le témoin du couple Azéma-Yor. Leur symbiose était étonnante. En effet tout semblait les séparer : alors que Yor enrobait toute manifestation de sa virtuosité du message que, dans la vie, on n'était pas là pour rigoler mais pour travailler dur, Azéma cachait un esprit imperturbablement rigoureux derrière une nonchalance appuyée. Et pourtant je ne pense pas avoir vu d'autres personnes qu'Azéma à avoir réussi à faire aussi joyeusement rigoler Yor, et, inversement, Yor était peut-être la seule personne à rire aussi sincèrement des blagues volontiers mauvaises et politiquement incorrectes d'Azéma. Plus sérieusement je garde un beau souvenir de leur complicité mathématique. Par contre ce n'est qu'avec le recul, en me replongeant dans la vaste littérature, que j'entrevoie la richesse et générosité de cette époque, peut-être déjà un peu finissante, mais dont ils faisaient pleinement partie, où, autour de Paul-André Meyer, sur la grande place publique du Séminaire de Probabilité, des grands hommes dialoguaient par petits articles interposés.

...— pendant que je bricolais donc des formules de balayage pour fermés marqués, la liste des équations de structure s'allongeait et avec, celle des martingales à représentation chaotique (voir par exemple [3, 13], les thèses de David Kurtz et Anthony Phan [2]...). Émery a fini par construire un exemple contredisant la conjecture : la représentation prévisible n'entraine pas la représentation chaotique [12]. Alors que personne ne semble avoir trouvé de caractérisation des martingales possédant la PRC, la recherche de conditions suffisantes ou d'exemples ne s'est jamais arrêtée (voir par exemple récemment Di Tella-Engelbert [9]).

Ici se termine le vécu et fait place aux moteurs de recherche. Sur Mathscinet, parmi les 156 références pour les mots-clé « martingale » & « Azema » beaucoup se réfèrent à l'autre importante collaboration Azéma-Yor, le problème de Skorokhod. Mais on peut y découvrir aussi que, même actuellement encore, des articles portent — au moins en partie — sur la ou les martingales d'Azéma.

La majeure partie se situe dans le domaine des probabilités quantiques, où les martingales d'Azéma figurent parmi les modèles probabilistes de base.

Puis il y a la finance (le jour où, dans un colloque, un monsieur m'a demandé des références sur la martingale d'Azéma, pour l'appliquer à la finance, je n'ai pas donné suite, parce que je n'ai pas cru que c'était sérieux. J'en ai toujours un peu honte). En effet, un premier article de Dritschel et Protter [10] propose simplement de remplacer dans le modèle d'un marché financier le mouvement Brownien par des martingales d'Azéma. Ensuite le fait que le dernier zéro τ d'un processus puisse intervenir dans des modèles de ruine avec information partielle (voir Cetin & al. [8]), appelle la martingale (μ_t) comme outil indispensable. Parallèlement apparait aussi naturellement dans Gapeev & al. [15] la notion de « supermartingale d'Azéma » $P[\tau > t|\mathcal{F}_t]$, dont la référence la plus récente est de Aksamit et Jeanblanc [1] en 2017.

Finalement il y a toujours, régulièrement des articles qui s'intéressent à la martingale d'Azéma sans avoir besoin de justifier beaucoup le pourquoi [7, 14, 18, 19]... Alors que l'homme s'est éteint aujourd'hui, cela fait longtemps que sa martingale s'est envolée pour vivre sa propre vie mathématique, et, désormais devenue un classique de la culture probabiliste, elle est promise à durer.

Littérature

1. A. Aksamit, M. Jeanblanc, *Enlargement of Filtration with Finance in View. SpringerBriefs in Quantitative Finance* (Springer, Cham, 2017), p. x+150
2. S. Attal, A.C.R. Belton, The chaotic-representation property for a class of normal martingales. Probab. Theory Relat. Fields **139**(3–4), 543–562 (2007)
3. S. Attal, M. Émery, Martingales d'Azéma bidimensionnelles. (French) [Two-dimensional Azéma martingales] Hommage à P. A. Meyer et J. Neveu. Astérisque (236), 9–21 (1996)
4. J. Azéma, Sur les fermés aléatoires, in (French) [On random closed sets] *Séminaire de Probabilités, XIX*. Lecture Notes in Mathematics, 1983/84, vol. 1123 (Springer, Berlin, 1985), pp. 397–495
5. J. Azéma, M. Yor, Étude d'une martingale remarquable, in (French) [Study of a remarkable martingale] *Séminaire de Probabilités, XXIII*. Lecture Notes in Mathematics, vol. 1372 (Springer, Berlin, 1989), pp. 88–130.
6. P. Cardaliaguet, C. Rainer, On a continuous-time game with incomplete information. Math. Oper. Res. **34**(4), 769–794 (2009)
7. U. Cetin, Filtered Azéma martingales. Electron. Commun. Probab. **17**(62), 13 (2012)
8. U. Cetin, R. Jarrow, P. Protter, Y. Yildirim, Modeling credit risk with partial information. Ann. Appl. Probab. **14**(3), 1167–1178 (2004)
9. P. Di Tella, H.-J. Engelbert, The chaotic representation property of compensated-covariation stable families of martingales. Ann. Probab. **44**(6), 3965–4005 (2016)
10. M. Dritschel, P. Protter, Complete markets with discontinuous security price. Finance Stoch. **2**, 203–214 (1999)
11. M. Émery, On the Azéma martingales, in *Séminaire de Probabilités, XXIII*. Lecture Notes in Mathematics, vol. 1372 (Springer, Berlin, 1989), pp. 66–87
12. M. Émery, On the chaotic representation property for martingales, in *Probability Theory and Mathematical Statistics* (St. Petersburg, 1993), pp. 155–166, Gordon and Breach, Amsterdam, 1996
13. M. Émery, Chaotic representation property of certain Azéma martingales. Illinois J. Math. **50**(1–4), 395–411 (2006)

14. N. Enriquez, An invariance principle for Azéma martingales. Ann. Inst. H. Poincaré Probab. Statist. **43**(6), 717–727 (2007)
15. P.V. Gapeev, M. Jeanblanc, L. Li, M. Rutkowski, Constructing random times with given survival processes and applications to valuation of credit derivatives, in *Contemporary Quantitative Finance* (Springer, Berlin, 2010), pp. 255–280
16. P.-A. Meyer, Un cours sur les intégrales stochastiques, in (French) *Séminaire de Probabilités, X* (Seconde partie: Théorie des intégrales stochastiques, Univ. Strasbourg, Strasbourg, année universitaire 1974/1975). Lecture Notes in Mathemaics, vol. 511 (Springer, Berlin, 1976), , pp. 245–400
17. P.-A. Meyer, Éléments de probabilités quantiques. I–V (French) [Elements of quantum probabilities. I–V], in *Séminaire de Probabilités, XX, 1984/85*. Lecture Notes in Mathematics, vol. 1204 (Springer, Berlin, 1986), pp. 186–312
18. C. Rainer, Backward stochastic differential equations with Azéma's martingale. Stoch. Stoch. Rep. **73**(1–2), 65–98 (2002)
19. C. A. Tudor, J. Vives, Anticipating stratonovich integral with respect to the Azema's martingales. Stochastic Anal. Appl. **20**(3), 673–692 (2002)
20. N. Wiener, The homogeneous Chaos, Amer. J. Math. **60**, 897–936 (2002)

Part II
Regular Contributions

Chapter 6
Complementability and Maximality in Different Contexts: Ergodic Theory, Brownian and Poly-Adic Filtrations

Christophe Leuridan

Abstract The notions of complementability and maximality were introduced in 1974 by Ornstein and Weiss in the context of the automorphisms of a probability space, in 2008 by Brossard and Leuridan in the context of the Brownian filtrations, and in 2017 by Leuridan in the context of the poly-adic filtrations indexed by the non-positive integers. We present here some striking analogies and also some differences existing between these three contexts.

Keywords Automorphisms of Lebesgue spaces · Factors · Entropy · Filtrations indexed by the non-positive integers · Poly-adic filtrations · Brownian filtrations · Immersed filtrations · Complementability · Maximality · Exchange property

6.1 Introduction

In the present paper, we will work with three types of objects: automorphisms of Lebesgue spaces, Brownian filtrations and filtrations indexed by $\mathbf{Z}$ or $\mathbf{Z}_-$; the reason for choosing $\mathbf{Z}$ or $\mathbf{Z}_-$ and to rule out $\mathbf{Z}_+$ is that for discrete-time filtrations, the interesting phenomena occur near time $-\infty$.

6.1.1 General Context

Among the invertible measure-preserving maps, Bernoulli shifts form a remarkable class. Similarly, the *product-type* filtrations (generated modulo the null sets by sequences of independent random variables) are considered as a well-understood class. The Brownian filtrations (generated modulo the null sets by Brownian

C. Leuridan (✉)
Institut Fourier, Université Grenoble Alpes, Grenoble, France
e-mail: Christophe.Leuridan@univ-grenoble-alpes.fr

C. Donati-Martin et al. (eds.), *Séminaire de Probabilités L*, Lecture Notes in Mathematics 2252, https://doi.org/10.1007/978-3-030-28535-7_6

motions) form a natural and widely studied class of continuous-time filtrations, although less simple.

Measure-preserving maps considered here will be taken on diffuse Lebesgue spaces. Various equivalent definitions of Lebesgue spaces are available. A simple definition of a Lebesgue space is a probability space which is isomorphic modulo the null sets to the union of some sub-interval of [0, 1], endowed with the Lebesgue σ-field and the Lebesgue measure, and a countable set of atoms. Most of the time, the Lebsegue space considered is non-atomic, so the sub-interval is [0, 1] itself. The class of Lebesgue spaces includes the completion of every Polish space. See [11] to find the main properties of Lebesgue spaces or [8] to get equivalent definitions. Working on Lebesgue spaces provides non-trivial measurability results, existence of generators... We recall in Sect. 6.7 the definitions and the main properties of partitions, generators, entropy used in the present paper.

Similarly, the filtrations considered here will be defined on a standard Borel probability space $(\Omega, \mathcal{F}, \mathbf{P})$, i.e. $(\Omega, \mathcal{F})$ is the Borel space associated to some Polish space, to ensure the existence of regular conditional probabilities. Given two sub-σ-fields $\mathcal{A}$ and $\mathcal{B}$, the inclusion $\mathcal{A} \subset \mathcal{B} \mod \mathbf{P}$ means that for every $A \in \mathcal{A}$, there exists $B \in \mathcal{B}$ such that $\mathbf{P}(A \triangle B) = 0$. We say that $\mathcal{A}$ and $\mathcal{B}$ are equal modulo the null sets (or modulo $\mathbf{P}$) when $\mathcal{A} \subset \mathcal{B} \mod \mathbf{P}$ and $\mathcal{B} \subset \mathcal{A} \mod \mathbf{P}$. We do not systematically complete the σ-fields to avoid troubles when working with conditional probabilities.

6.1.2 Reminders on Filtrations Indexed by $\mathbf{Z}$ or $\mathbf{Z}_-$

We now recall some classical but less known definitions and facts on filtrations. Given a filtration $(\mathcal{F}_n)_n$ indexed by $\mathbf{Z}$ or $\mathbf{Z}_-$, one says that $(\mathcal{F}_n)_n$ *is product-type* if $(\mathcal{F}_n)_n$ can be generated modulo $\mathbf{P}$ by some sequence $(I_n)_n$ of (independent) random variables.

One says that $(\mathcal{F}_n)_n$ *has independent increments* if there exists a sequence $(I_n)_n$ of random variables such that for every n in $\mathbf{Z}$ or $\mathbf{Z}_-$,

$$\mathcal{F}_n = \mathcal{F}_{n-1} \vee \sigma(I_n) \mod \mathbf{P} \text{ and } I_n \text{ is independent of } \mathcal{F}_{n-1}.$$

Such a sequence $(I_n)_n$ is called a *sequence of innovations* and is necessarily a sequence of independent random variables.

One says that $(\mathcal{F}_n)_n$ is $(a_n)_n$*-adic* when it admits some sequence $(I_n)_n$ of innovations such that each I_n is uniformly distributed on some finite set with size a_n. One says that $(\mathcal{F}_n)_n$ is *poly-adic* when $(\mathcal{F}_n)_n$ is $(a_n)_n$-adic for some sequence $(a_n)_n$ of positive integers, called *adicity*.

One says that $(\mathcal{F}_n)_n$ is *Kolmogorovian* if the tail σ-field $\mathcal{F}_{-\infty} := \bigcap_n \mathcal{F}_n$ is trivial (i.e. contains only events with probability 0 or 1).

By the definition and by Kolmogorov's zero-one law, any filtration indexed by $\mathbf{Z}$ or $\mathbf{Z}_-$ must have independent increments and must be Kolmogorovian to be product-type. But Vershik showed in [22] that the converse is not true. A simple counter-example is given by Vershik's decimation process (example 2 in [22]). Actually Vershik worked with decreasing sequences of measurable partitions indexed by $\mathbf{Z}_+$ and this frame was translated into filtrations indexed by $\mathbf{Z}_-$ by Émery and Schachermayer [10].

6.1.3 *K*-Automorphisms

The Kolmogorov property for filtrations indexed by $\mathbf{Z}$ or $\mathbf{Z}_-$ has an analogue for dynamical systems, although the definition is less simple in this frame: one says that an automorphism T of a probability space $(Z, \mathcal{Z}, \pi)$ is a *K-automorphism* (or that T has completely positive entropy) if for every $A \in \mathcal{Z}$, one has $h(T, \{A, A^c\}) > 0$ whenever $0 < \pi(A) < 1$. This condition is nothing but the triviality of the σ-field

$$\Pi(T) := \{A \in \mathcal{Z} : h(T, \{A, A^c\}) = 0\},$$

called *Pinsker's factor*. Actually, the 'events' of Pinsker's factor can be seen as the 'asymptotic events'. Indeed, if γ is a countable generator of $(Z, \mathcal{Z}, \pi, T)$, then

$$\Pi(T) = \bigcap_{n\geq 0} \overline{\bigvee_{k\geq n} T^{-k}\gamma} = \bigcap_{n\geq 0} \overline{\bigvee_{k\geq n} T^{k}\gamma},$$

where the upper bar indicates completion with regard to π. To make the analogy clearer, set $\gamma = \{A_\lambda, \lambda \in \Lambda\}$. For each $x \in Z$, call $f(x) \in \Lambda$ the only index λ such that $x \in A_\lambda$. For every $k \in \mathbf{Z}$, the σ-field generated by $T^{-k}\gamma$ is the σ-field associated to $f \circ T^k$ viewed as a Λ-valued random variable on $(Z, \mathcal{Z}, \pi)$. Therefore, $\Pi(T)$ is the asymptotic σ-field generated by the sequence $(f \circ T^k)_{k\geq 0}$.

6.1.4 Content of the Paper

We have just viewed the analogy between the 'Kolmogorovianity' of a filtration indexed by $\mathbf{Z}_-$ and the K-property of an automorphism of a Lebesgue space.

The next section is devoted to a parallel presentation of analogous notions and results in the three following contexts: automorphisms of Lebesgue spaces, filtrations indexed by $\mathbf{Z}_-$ and Brownian filtrations. We investigate two notions—complementability and maximality—involving factors or poly-adic immersed filtrations or Brownian immersed filtrations according to the context. The results presented are essentially due to Ornstein and Weiss [17], Ornstein [16], and Thouvenot [21] for automorphisms of Lebesgue spaces; They come from [14] for filtrations indexed by $\mathbf{Z}_-$. They are due to Brossard, Émery and Leuridan [3–5] for Brownian filtrations.

Section 6.3 provides proofs of results on maximality which are not easy yo find in the literature. With some restrictions on the nature of the complement, complementability implies maximality. Section 6.4 is devoted to the proof of this implication. The converse was already known to be false for factors of automorphisms of Lebesgue spaces and for poly-adic immersed filtrations. In Sect. 6.5, we provide a counter-example in the context of Brownian filtrations. The construction relies on a counter-example for poly-adic immersed filtrations which is inspired by non-published notes of B. Tsirelson (About Yor's problem, Unpublished Notes. https://www.tau.ac.il/~tsirel/download/yor3.pdf).

In spite of the similitude of the notions regardless the context, some differences exist. In Sect. 6.6, we provide a non-complementable filtration (associated to a stationary process) yielding a complementable factor. This example is inspired by Vershik's decimation process (example 3 in [22]).

In Sect. 6.7, we recall the definitions and the main properties of partitions, generators, entropy used in the present paper.

6.2 Parallel Notions and Results

We now present the main results illustrating the analogies and also the differences between three contexts: factors of automorphisms of Lebesgue spaces, poly-adic immersed filtrations or Brownian immersed filtrations.

6.2.1 *Factors and Immersed Filtrations*

Given an invertible measure preserving map T of a Lebesgue space $(Z, \mathcal{Z}, \pi)$, we call *factor of* T, or more rigorously a factor of the dynamical system $(Z, \mathcal{Z}, \pi, T)$, any sub-σ-field $\mathcal{B}$ of $\mathcal{Z}$ such that $T^{-1}\mathcal{B} = \mathcal{B} = T\mathcal{B} \mod \pi$. Actually, the factor is the dynamical system $(Z, \mathcal{B}, \pi|_{\mathcal{B}}, T)$, which will be abbreviated in $(T, \mathcal{B})$ in the present paper. This definition of a factor is equivalent to the usual one.[1]

Given two filtrations $(\mathcal{U}_t)_{t\in\mathbf{T}}$ and $(\mathcal{Z}_t)_{t\in\mathbf{T}}$ on some probability space $(\Omega, \mathcal{A}, \mathbf{P})$, indexed by a common subset $\mathbf{T}$ of $\mathbf{R}$, one says that $(\mathcal{U}_t)_{t\in\mathbf{T}}$ is *immersed in* $(\mathcal{Z}_t)_{t\in\mathbf{T}}$ if every martingale in $(\mathcal{U}_t)_{t\in\mathbf{T}}$ is still a martingale in $(\mathcal{Z}_t)_{t\in\mathbf{T}}$. The notion of

[1]Actually, Rokhlin's theory ensures that if $\mathcal{B}$ is a factor of a Lebesgue space $(Z, \mathcal{Z}, \pi, T)$, then there exists a map f from Z to some Polish space E such that $\mathcal{B}$ is generated up to the negligible events by the map $\Phi : x \mapsto (f(T^k(x)))_{k\in\mathbf{Z}}$ from Z to the product space $E^{\mathbf{Z}}$. Call $\nu = \Phi(\pi) = \pi \circ \Phi^{-1}$ the image measure of μ by Φ. Then the completion $(E^{\mathbf{Z}}, \mathcal{B}(E^{\mathbf{Z}}), \nu)$ is a Lebesgue space, the shift operator $S : (y_k)_{k\in\mathbf{Z}} \mapsto (y_{k+1})_{k\in\mathbf{Z}}$ is an automorphism of $E^{\mathbf{Z}}$, and $S \circ \Phi = \Phi \circ T$.

Conversely, if $(Y, \mathcal{Y}, \nu, S)$ is a dynamical system and Φ a measurable map from Z to Y such that $\Phi(\pi) = \nu$ and $S \circ \Phi = \Phi \circ T$, then the σ-field $\Phi^{-1}(\mathcal{Y})$ is a factor of $(Z, \mathcal{Z}, \pi, T)$.

immersion is stronger than the inclusion. Actually, $(\mathcal{U}_t)_{t\in\mathbf{T}}$ is immersed in $(\mathcal{Z}_t)_{t\in\mathbf{T}}$ if and only if the two conditions below hold:

1. for every $t \in \mathbf{T}$, $\mathcal{U}_t \subset \mathcal{Z}_t$.
2. for every $s < t$ in $\mathbf{T}$, $\mathcal{U}_t$ and $\mathcal{Z}_s$ are independent conditionally on $\mathcal{U}_s$.

The additional condition means that the largest filtration does not give information in advance on the smallest one. We also make the useful following observation.

Lemma 6.1 *Assume that $(\mathcal{U}_t)_{t\in\mathbf{T}}$ is immersed in $(\mathcal{Z}_t)_{t\in\mathbf{T}}$. Then $(\mathcal{U}_t)_{t\in\mathbf{T}}$ is completely determined (up to null sets) by its final σ-field*

$$\mathcal{U}_\infty := \bigvee_{t\in\mathbf{T}} \mathcal{U}_t.$$

More precisely, $\mathcal{U}_t = \mathcal{U}_\infty \cap \mathcal{Z}_t \mod \mathbf{P}$ for every $t \in \mathbf{T}$. In particular, if $\mathcal{U}_\infty = \mathcal{Z}_\infty \mod \mathbf{P}$, then $\mathcal{U}_t = \mathcal{Z}_t \mod \mathbf{P}$ for every $t \in \mathbf{T}$.[2]

When one works with Brownian filtrations, the immersion has many equivalent translations. The next statements are very classical (close statements are proved in [1]) and they rely on the stochastic calculus and the predictable representation property of Brownian filtrations.

Proposition 6.2 *Let $(B_t)_{t\ge0}$ be a finite-dimensional Brownian motion adapted to some filtration $(\mathcal{Z}_t)_{t\ge0}$, and $(\mathcal{B}_t)_{t\ge0}$ its natural filtration. The following statements are equivalent.*

1. *$(B_t)_{t\ge0}$ is a martingale in $(\mathcal{Z}_t)_{t\ge0}$.*
2. *$(\mathcal{B}_t)_{t\ge0}$ is immersed in $(\mathcal{Z}_t)_{t\ge0}$.*
3. *For every $t \ge 0$, the process $B_{t+\cdot} - B_t$ is independent of $\mathcal{Z}_t$.*
4. *$(B_t)_{t\ge0}$ is a Markov process in $(\mathcal{Z}_t)_{t\ge0}$.*

Definition 6.3 When these statements hold, we say that $(B_t)_{t\ge0}$ is a Brownian motion in the filtration $(\mathcal{Z}_t)_{t\ge0}$.

Note the analogy between the following two results.

Theorem 6.4 (Ornstein [15]) *Every factor of a Bernoulli shift is equivalent to a Bernoulli shift.*

Theorem 6.5 (Vershik [22]) *If $(\mathcal{Z}_n)_{n\le0}$ is a product-type filtration whose final sigma-field $\mathcal{Z}_0$ is essentially separable, then every poly-adic filtration immersed in $(\mathcal{Z}_n)_{n\le0}$ is product-type.*

One-dimensional Brownian filtrations can be viewed as continuous time versions of dyadic product-type filtrations. In this analogy, the predictable representation

[2] The inclusion $\mathcal{U}_t = \mathcal{U}_\infty \cap \mathcal{Z}_t$ is immediate. To prove the converse, take $A \in \mathcal{U}_\infty \cap \mathcal{Z}_t$. Since $\mathcal{U}_\infty$ and $\mathcal{Z}_t$ are independent conditionally on $\mathcal{U}_t$, we get $\mathbf{P}[A|\mathcal{U}_t] = \mathbf{P}[A|\mathcal{Z}_t] = \mathbf{1}_A$ a.s., so $A \in \mathcal{U}_t \mod \mathbf{P}$.

property of the continuous-time filtration corresponds to the dyadicity of the discrete-time filtration. Yet, the situation is much more involved when one works with Brownian filtrations, and the following question remains open.

Open Question
If a filtration $(\mathcal{F}_t)_{t\geq 0}$ is immersed in some (possibly infinite-dimensional) Brownian filtration and has the predictable representation property with regard to some one-dimensional Brownian motion β (i.e. each martingale in $(\mathcal{F}_t)_{t\geq 0}$ can be obtained as a stochastic integral with regard to β), then is $(\mathcal{F}_t)_{t\geq 0}$ necessarily a Brownian filtration?

A partial answer was given by Émery (it follows from corollary 1 in [9]).

Theorem 6.6 (Émery [9]) *Let $d \in \mathbf{N} \cup \{+\infty\}$. Assume that the filtration $(\mathcal{F}_t)_{t\geq 0}$ is d-Brownian after 0, i.e. there exists a d-dimensional Brownian motion $(B_t)_{t\geq 0}$ in $(\mathcal{F}_t)_{t\geq 0}$ such that for every $t \geq \varepsilon > 0$, $\mathcal{F}_t$ is generated by $\mathcal{F}_\varepsilon$ and the increments $(B_s - B_\varepsilon)_{\varepsilon\leq s\leq t}$. If $(\mathcal{F}_t)_{t\geq 0}$ is immersed in some (possibly infinite-dimensional) Brownian filtration, then $(\mathcal{F}_t)_{t\geq 0}$ is a d-dimensional Brownian filtration.*

Here, the role of the stronger hypothesis that $(\mathcal{F}_t)_{t\geq 0}$ is Brownian after 0 is to guarantee that the difficulties arise only at time $0+$, so the situation gets closer to the filtrations indexed by $\mathbf{Z}$ or $\mathbf{Z}_-$, for which the difficulties arise only at time $-\infty$.

6.2.2 Complementability

By complementability, we will mean the existence of some independent complement, although we will have to specify the nature of the complement.

The following definition is abridged from [17].

Definition 6.7 Let $(Z, \mathcal{Z}, \pi, T)$ be a Lebesgue dynamical system and $\mathcal{B}$ be a factor of T. One says that $\mathcal{B}$ is complementable if $\mathcal{B}$ possesses an independent complement in $(Z, \mathcal{Z}, \pi, T)$, i.e. a factor $\mathcal{C}$ of T which is independent of $\mathcal{B}$ (with regard to π) such that $\mathcal{B} \vee \mathcal{C} = \mathcal{Z} \mod \pi$.

If $(Z, \mathcal{Z}, \pi, T)$ is the direct product of two dynamical systems $(Z_1, \mathcal{Z}_1, \pi_1, T_1)$ and $(Z_2, \mathcal{Z}_2, \pi_2, T_2)$, then $\mathcal{Z}_1 \otimes \{\emptyset, Z_2\}$ and $\{\emptyset, Z_1\} \otimes \mathcal{Z}_2$ are factors of $(Z, \mathcal{Z}, \pi, T)$ and each of them is a complement of the other one. Now, let us look at a counterexample.

Example 6.1 Let T be the Bernoulli shift on $Z = \{-1, 1\}^{\mathbf{Z}}$ endowed with product σ-field $\mathcal{Z}$ and the uniform law. The map $\Phi : Z \to Z$ defined by $\Phi((x_n)_{n\in\mathbf{Z}}) = (x_{n-1}x_n)_{n\in\mathbf{Z}}$ commutes with T, so $\Phi^{-1}(\mathcal{Z})$ is a factor of T. Call $p_0 : Z \to \{-1, 1\}$ the canonical projection defined by $p_0((x_n)_{n\in\mathbf{Z}}) = x_0$. Then the σ-field $p_0^{-1}(\mathcal{Z})$

is an independent complement of $\Phi^{-1}(\mathcal{Z})$, but this complement is not a factor. Actually, we will come back to this example after Definition 6.14 to show as an application of Theorem 6.21 that no factor can be an independent complement of $\Phi^{-1}(\mathcal{Z})$.

We now define the notion of complementability in the world of filtrations.

Definition 6.8 Consider two filtrations $(\mathcal{U}_t)_{t\in\mathbf{T}}$ and $(\mathcal{Z}_t)_{t\in\mathbf{T}}$ on some probability space $(\Omega, \mathcal{A}, \mathbf{P})$, indexed by a common subset $\mathbf{T}$ of $\mathbf{R}$. One says that $(\mathcal{U}_t)_{t\in\mathbf{T}}$ is complementable in $(\mathcal{Z}_t)_{t\in\mathbf{T}}$ if there exists a filtration $(\mathcal{V}_t)_{t\in\mathbf{T}}$ such that for every $t \in \mathbf{T}$, $\mathcal{U}_t$ and $\mathcal{V}_t$ are independent and $\mathcal{U}_t \vee \mathcal{V}_t = \mathcal{Z}_t \mod \mathbf{P}$.

Since independent enlargements of a filtration always produce filtrations in which the initial filtration is immersed, $(\mathcal{U}_t)_{t\in\mathbf{T}}$ needs to be immersed in $(\mathcal{Z}_t)_{t\in\mathbf{T}}$ to possess an independent complement.

We will use many times the next result, abridged from [14].

Proposition 6.9 *Keep the notations of the last definition. Let U be a random variable valued in some measurable space $(E, \mathcal{E})$, such that $\sigma(U) = \bigvee_{t\in\mathbf{T}} \mathcal{U}_t$, and $(\mathbf{P}_u)_{u\in E}$ a regular version of the conditional probability $\mathbf{P}$ given U. Assume that $(\mathcal{U}_t)_{t\in\mathbf{T}}$ is complementable in $(\mathcal{Z}_t)_{t\in\mathbf{T}}$ by a filtration $(\mathcal{V}_t)_{t\in\mathbf{T}}$. Then for $U(\mathbf{P})$-almost every $u \in E$, the filtered probability space $(\Omega, \mathcal{A}, \mathbf{P}_u, (\mathcal{Z}_t)_{t\le\mathbf{T}})$ is isomorphic to the filtered probability space $(\Omega, \mathcal{A}, \mathbf{P}, (\mathcal{V}_t)_{t\le\mathbf{T}})$.*

Let us give applications of the last result, that will be used in the present paper.

Corollary 6.10 (Particular Cases)

- *If a filtration $(\mathcal{U}_n)_{n\le 0}$ is complementable in $(\mathcal{Z}_n)_{n\le 0}$ by some product-type filtration, then for $U(\mathbf{P})$-almost every $u \in E$, $(\mathcal{Z}_n)_{n\le 0}$ is product-type under $\mathbf{P}_u$.*
- *If a filtration $(\mathcal{U}_t)_{t\ge 0}$ is complementable in $(\mathcal{Z}_t)_{t\ge 0}$ by some Brownian filtration, then for $U(\mathbf{P})$-almost every $u \in E$, $(\mathcal{Z}_t)_{t\ge 0}$ is a Brownian filtration under $\mathbf{P}_u$.*

Determining whether a 1-dimensional Brownian filtration immersed in a 2-dimensional Brownian filtration is complementable or not is often difficult. Except trivial cases, the only known cases are related to skew-product decomposition of the planar Brownian motion, see [5].

6.2.3 *Maximality*

The definition of the maximality requires a tool to measure the quantity of information. When one works with factors of an automorphism of a Lebesgue space, the quantity of information is the entropy. When one works with poly-adic filtrations, the quantity of information is the sequence of positive integers giving the adicity. When one works with Brownian filtrations, the quantity of information is the dimension of any generating Brownian motion. The statements below show hove these quantities vary when one considers a factor, a poly-adic immersed filtration, or a Brownian immersed filtration.

Remark 6.11 (Quantity of Information in Subsystems)

1. If $\mathcal{B}$ is a factor of $(Z, \mathcal{Z}, \pi, T)$, then $h(T, \mathcal{B}) \le h(T)$.
2. If a $(b_n)_{n\le 0}$-adic filtration is immersed in an $(r_n)_{n\le 0}$-adic filtration, then b_n divides r_n for every n.
3. If a m-dimensional Brownian filtration is immersed in a n-dimensional Brownian filtration, then $m \le n$.

The first statement is very classical. The second one is proved in [14]. The last one is classical and shows that the dimension of a Brownian filtration makes sense; a proof is given in the footnote.[3]

Let us give precise definitions, respectively abridged from [14, 16] and [3] or [4].

Definition 6.12 Let $(Z, \mathcal{Z}, \pi, T)$ be a Lebesgue dynamical system and $\mathcal{B}$ be a factor of T. One says that $\mathcal{B}$ is maximal if $(T, \mathcal{B})$ has a finite entropy and if for any factor $\mathcal{A}$, the conditions $\mathcal{A} \supset \mathcal{B}$ and $h(T, \mathcal{A}) = h(T, \mathcal{B})$ entail $\mathcal{A} = \mathcal{B}$ modulo null sets.

Definition 6.13 Let $(\mathcal{B}_n)_{n\le 0}$ be a $(b_n)_{n\le 0}$-adic filtration immersed in some filtration $(\mathcal{Z}_n)_{n\le 0}$. One says that $(\mathcal{B}_n)_{n\le 0}$ is maximal in $(\mathcal{Z}_n)_{n\le 0}$ if every $(b_n)_{n\le 0}$-adic filtration immersed in $(\mathcal{Z}_n)_{n\le 0}$ and containing $(\mathcal{B}_n)_{n\le 0}$ is equal to $(\mathcal{B}_n)_{n\le 0}$ modulo null events.

Definition 6.14 Let $(\mathcal{B}_t)_{t\ge 0}$ be a d-dimensional Brownian filtration immersed in some filtration $(\mathcal{Z}_t)_{t\ge 0}$. One says that $(\mathcal{B}_t)_{t\ge 0}$ is maximal in $(\mathcal{Z}_t)_{t\ge 0}$ if every d-dimensional Brownian filtration immersed in $(\mathcal{Z}_t)_{t\ge 0}$ and containing $(\mathcal{B}_t)_{t\le 0}$ is equal to $(\mathcal{B}_t)_{t\ge 0}$ modulo null events.

Let us come back to example of Sect. 6.2.2 in which T is the Bernoulli shift on $Z = \{-1, 1\}^{\mathbf{Z}}$ endowed with product σ-field $\mathcal{Z}$ and the uniform law. Since the map $\Phi : Z \to Z$ defined by $\Phi((x_n)_{n\in\mathbf{Z}}) = (x_{n-1}x_n)_{n\in\mathbf{Z}}$ commutes with T and preserves the uniform law on Z, the factor $(T, \Phi^{-1}(\mathcal{Z}))$ is a Bernoulli $(1/2, 1/2)$ shift like T itself. The factor $\Phi^{-1}(\mathcal{Z})$ is strictly contained in $\mathcal{Z}$ but has the same (finite) entropy as T, so it is not maximal. But every factor of T is a K-automorphism since T is. Hence, Theorem 6.21 will show that the factor $\Phi^{-1}(\mathcal{Z})$ is not complementable.

This example above can be abridged in the context of filtrations indexed by the relative integers: consider a sequence $(\xi_n)_{n\in\mathbf{Z}}$ of independent uniform random variables taking values in $\{-1, 1\}$. Then the sequence $(\eta_n)_{n\in\mathbf{Z}} := \Phi((\xi_n)_{n\in\mathbf{Z}})$ has the same law as $(\xi_n)_{n\in\mathbf{Z}}$. One checks that the inclusions $\mathcal{F}^{\eta}_n \subset \mathcal{F}^{\xi}_n$ are strict modulo

[3]Let Z be a n-dimensional Brownian motion and B be a m-dimensional Brownian motion in $\mathcal{F}^Z$. Then one can find an $\mathcal{F}^Z$-predictable process M taking values in the set of all $p \times n$ real matrices whose rows form an orthonormal family, such that $B = \int_0^{\cdot} M_s \mathrm{d}Z_s$. In particular, the m rows of each matrix M_s are independent and lie in a n-dimensional vector space, so $m \le n$.

P, although the tail σ-field $\mathcal{F}^{\xi}_{-\infty}$ are trivial and although $(\eta_n)_{n\in\mathbf{Z}}$ is an innovation sequence for $(\mathcal{F}^{\xi}_n)_{n\in\mathbf{Z}}$. Actually, one bit of information is lost when one transforms $(\xi_n)_{n\in\mathbf{Z}}$ into $(\eta_n)_{n\in\mathbf{Z}}$: for each $n_0 \in \mathbf{Z}$, the value ξ_{n_0} is independent of $(\eta_n)_{n\in\mathbf{Z}}$, and the knowledge of ξ_{n_0} and $(\eta_n)_{n\in\mathbf{Z}}$ is sufficient to recover $(\xi_n)_{n\in\mathbf{Z}}$. The paradox is that this loss of information is asymptotic at time $-\infty$ but invisible when one looks at $\mathcal{F}^{\xi}_{-\infty}$ and $\mathcal{F}^{\eta}_{-\infty}$.

The situation can be much more complex when one works with Brownian filtrations. For example, consider a linear Brownian motion W. Since W spends a null-time at 0, the stochastic integral

$$W' = \int_0^{\cdot} \operatorname{sgn}(W_s)\mathrm{d}W_s = |W| - L$$

(where L denotes the local time of W at 0) is still a linear Brownian motion. The natural filtration $\mathcal{F}^{W'}$ is immersed and strictly included in $\mathcal{F}^{W}$, therefore it is not maximal in $\mathcal{F}^{W}$. Actually, W' generates the same filtration as $|W|$ up to null events, so the Lévy transformation—which transforms the sample paths of W into the sample paths of W'—forgets the signs of all excursions of W, which are independent of $|W|$. Here, the loss of information occurs at each beginning of excursion of W, and not at time $0+$.

6.2.4 *Necessary or Sufficient Conditions for Maximality*

Given a finite-entropy factor, a poly-adic immersed filtration or a Brownian immersed filtration, one wishes to enlarge it to get a maximal one having the same entropy, adicity or dimension. This leads to the following constructions, abridged from [4, 14, 16]. In the next three propositions, the bars above the σ-fields indicate completions with regard to π or **P**.

Definition 6.15 Let $(Z, \mathcal{Z}, \pi)$ be a probability space, T be an invertible measure-preserving map on $(Z, \mathcal{Z}, \pi)$ and $\mathcal{B}$ be a factor with finite entropy. The conditional Pinsker factor associated to $\mathcal{B}$ is defined by

$$\mathcal{B}' := \{A \in \mathcal{Z} : h(T, \{A, A^c\}|\mathcal{B}) = 0\}.$$

where

$$h(T, \{A, A^c\}|\mathcal{B}) = \lim_{n\to+\infty} \frac{1}{n} H\Big(\bigvee_{k=0}^{n-1} \{T^{-k}A, T^{-k}A^c\}\Big|\mathcal{B}\Big).$$

Proposition 6.16 *The collection $\mathcal{B}'$ thus defined is the largest factor containing $\mathcal{B}$ and having the same entropy as $\mathcal{B}$. In particular, $\mathcal{B}'$ is maximal.*

Proposition 6.17 *Furthermore, assume $(Z, \mathcal{Z}, \pi, T)$ is a Lebesgue dynamical space, that T is aperiodic*[4] *and has finite entropy. Then for every generator γ of T,*

$$\mathcal{B}' = \overline{\bigcap_{n\geq 0} \Big(\mathcal{B} \vee \bigvee_{k\geq n} T^{-k}\gamma\Big)}.$$

Proposition 6.18 *Let $(\mathcal{B}_n)_{n\leq 0}$ be a $(b_n)_{n\leq 0}$-adic filtration immersed in some filtration $(\mathcal{Z}_n)_{n\leq 0}$. Then $(\mathcal{B}_n)_{n\leq 0}$ is immersed in the filtration $(\mathcal{B}'_n)_{n\leq 0}$ defined by*

$$\mathcal{B}'_n := \overline{\bigcap_{s\leq 0} (\mathcal{B}_n \vee \mathcal{Z}_s)}.$$

Moreover $(\mathcal{B}'_n)_{n\leq 0}$ is the largest $(b_n)_{n\leq 0}$-adic filtration containing $(\mathcal{B}_n)_{n\leq 0}$ and immersed in $(\mathcal{Z}_n)_{n\leq 0}$. In particular, $(\mathcal{B}'_n)_{n\leq 0}$ is maximal in $(\mathcal{Z}_n)_{n\leq 0}$.

Proposition 6.19 *Let $(\mathcal{B}_t)_{t\geq 0}$ be a d-dimensional Brownian filtration immersed in some Brownian filtration $(\mathcal{Z}_t)_{t\geq 0}$. Then $(\mathcal{B}_t)_{t\geq 0}$ is immersed in the filtration $(\mathcal{B}'_t)_{t\geq 0}$ defined by*

$$\mathcal{B}'_t := \overline{\bigcap_{s>0} (\mathcal{B}_t \vee \mathcal{Z}_s)}.$$

Moreover $(\mathcal{B}'_t)_{t\geq 0}$ is a d-dimensional Brownian filtration immersed in $(\mathcal{Z}_t)_{t\geq 0}$.

Warning
When $\mathcal{F}$ is a sub-σ-field and $(\mathcal{G}_n)_{n\geq 0}$ is a non-increasing sequence of sub-σ-fields of a probability space $(\Omega, \mathcal{T}, \mathbf{P})$, the trivial inclusion

$$\mathcal{F} \vee \Big(\bigcap_{n\geq 0} \mathcal{G}_n\Big) \subset \bigcap_{n\geq 0} (\mathcal{F} \vee \mathcal{G}_n)$$

may be strict modulo **P**. Equality modulo **P** holds when $\mathcal{F}$ and $\mathcal{G}_0$ are independent (see Corollary 6.38). Von Weizsäcker provides involved characterizations in [23] for equality modulo **P**. Therefore, the σ-fields $\mathcal{B}'$, $\mathcal{B}'_n$ and $\mathcal{B}'_t$ considered in Propositions 6.17, 6.18, and 6.19 can be strictly larger than the σ-fields $\overline{\mathcal{B} \vee \Pi(T)}$, $\overline{\mathcal{B}_n \vee \mathcal{Z}_{-\infty}}$ and $\overline{\mathcal{B}_t \vee \mathcal{Z}_{0+}} = \overline{\mathcal{B}_t}$ respectively.

[4] Aperiodicity of T means that $\pi\{z \in Z : \exists n \geq 1, T^n(z) = z\} = 0$. We make this assumption to ensure the existence of generator.

Note the analogy between the formulas in Propositions 6.17, 6.18, and 6.19. In these three contexts, we must have $\mathcal{B}' = \mathcal{B}$ up to null sets for $\mathcal{B}$ to be maximal. Moreover, applying the same procedure to $\mathcal{B}'$ leads to $\mathcal{B}'' = \mathcal{B}'$. Hence the condition $\mathcal{B}' = \mathcal{B}$ up to null sets is also sufficient for $\mathcal{B}$ to be maximal in the first two cases (factors of finite-entropy aperiodic Lebesgue automorphisms and poly-adic filtrations). But once again, the situation is more complex when one works with Brownian filtrations, since the filtration $\mathcal{B}'$ may be non-maximal. Here is a counter-example (the proof will be given in Sect. 6.3).

Example 6.2 Let X be a linear Brownian motion in some filtration $\mathcal{Z}$. Set

$$B = \int_0^{\cdot} \operatorname{sgn}(X_s)\mathrm{d}X_s,$$

and call $\mathcal{X}$ and $\mathcal{B}$ the natural filtrations of X and B. If $\mathcal{X}$ is maximal in $\mathcal{Z}$, then the filtration $\mathcal{B}'$ defined by Proposition 6.19 coincides with $\mathcal{B}$ up to null events. Therefore, the filtration $\mathcal{B}'$ (included in $\mathcal{X}$) cannot be maximal in $\mathcal{Z}$.

Actually, the maximality of Brownian filtrations is not an asymptotic property at $0+$, unlike the almost sure equality $\mathcal{B}' = \mathcal{B}$. To try to produce a maximal Brownian filtration containing a given Brownian filtration, one should perform the infinitesimal enlargement above at every time, but we do not see how to do that.

Yet, Proposition 6.25 in the next subsection shows that equality $\mathcal{B} = \mathcal{B}'$ ensures the maximality of $\mathcal{B}$ under the (strong) additional hypothesis that $\mathcal{B}$ is complementable after 0.

The next sufficient condition for the maximality of a poly-adic immersed filtration comes from [14].

Proposition 6.20 *Let $(\mathcal{B}_n)_{n\leq 0}$ be a $(b_n)_{n\leq 0}$-adic filtration immersed in $(\mathcal{Z}_n)_{n\leq 0}$. Let U be a random variable valued in some measurable space $(E, \mathcal{E})$, such that $\sigma(U) = \mathcal{B}_0$ and $(\mathbf{P}_u)_{u\in E}$ be a regular version of the conditional probability $\mathbf{P}$ given U. If for $U(\mathbf{P})$-almost every $u \in E$, the filtered probability space $(\Omega, \mathcal{A}, \mathbf{P}_u, (\mathcal{Z}_n)_{n\leq 0})$ is Kolmogorovian, then the filtration $(\mathcal{B}_n)_{n\leq 0}$ is maximal in $(\mathcal{Z}_n)_{n\leq 0}$.*

The assumption that $(\mathcal{Z}_n)_{n\leq 0}$ is Kolmogorovian under almost every conditional probability $\mathbf{P}_u$ echoes to the terminology of conditional K-automorphisms used by Thouvenot in [21] instead of the notion of maximality.

6.2.5 Complementability and Maximality

In the three contexts (factors of an automorphism of a Lebsgue space, poly-adic filtrations immersed in a filtration indexed by $\mathbf{Z}_-$, Brownian filtrations immersed in a Brownian filtration), we get very similar results.

The first one is stated by Ornstein in [16] as a direct consequence of a lemma stated in [17].

Proposition 6.21 *Assume that T with finite entropy. Let $\mathcal{B}$ be a factor of T. If $\mathcal{B}$ is complementable by some K-automorphism, then $\mathcal{B}$ is maximal.*

The second one comes from [14].

Proposition 6.22 *Let $(\mathcal{B}_n)_{n\le 0}$ be a $(b_n)_{n\le 0}$-adic filtration immersed in $(\mathcal{Z}_n)_{n\le 0}$. If $(\mathcal{B}_n)_{n\le 0}$ can be complemented by some Kolmogorovian filtration, then $(\mathcal{B}_n)_{n\le 0}$ is maximal in $(\mathcal{Z}_n)_{n\le 0}$.*

A particular case of the third one (in which the dimension of the Brownian filtrations are 1 and 2) can be found in [3] or [4].

Proposition 6.23 *Let $(\mathcal{B}_t)_{t\ge 0}$ be a Brownian filtration immersed in a Brownian filtration $(\mathcal{Z}_t)_{t\ge 0}$ with larger dimension. If $(\mathcal{B}_t)_{t\ge 0}$ can be complemented by some Brownian filtration, then $(\mathcal{B}_t)_{t\ge 0}$ is maximal in $(\mathcal{Z}_t)_{t\ge 0}$.*

The proofs of these three statements are rather simple, and present some similarities, although they are different. In the next section, we provide two different proofs of Proposition 6.21. The first one relies on Ornstein and Weiss' lemma (lemma 2 in [17]). The second one is a bit simpler but requires that T has finite entropy, and relies on Berg's lemma (lemma 2.3 in [2]). We also provide a proof of Proposition 6.23. Proposition 6.22 follows from Proposition 6.20 and Corollary 6.10.

The converses of the three implications above are false, but providing counter-examples is difficult. Ornstein gived in [16] an example of maximal but non-complementable factor in [16], but the proof is difficult to read. Two counter-examples of a maximal but non-complementable poly-adic filtration are given in [14]. In the present paper, we use a third example to construct a maximal but non-complementable Brownian filtration.

In the present paper, we will also use a small refinement of Proposition 6.23, using the notion of complementability after 0.

Definition 6.24 Let $(\mathcal{B}_t)_{t\ge 0}$ be a Brownian filtration immersed in a Brownian filtration $(\mathcal{Z}_t)_{t\ge 0}$ with larger dimension. One says that $(\mathcal{B}_t)_{t\ge 0}$ is complementable after 0 in $(\mathcal{Z}_t)_{t\ge 0}$ if there exists some Brownian filtration C immersed in $\mathcal{Z}$ and independent of $\mathcal{B}$ such that,

$$\forall t \ge 0, \mathcal{Z}_t = \bigcap_{s>0} (\mathcal{B}_t \vee C_t \vee \mathcal{Z}_s) \mod \pi.$$

Proposition 6.25 *Let $(\mathcal{B}_t)_{t\ge 0}$ be a d-dimensional Brownian filtration immersed in a Brownian filtration $(\mathcal{Z}_t)_{t\ge 0}$ with larger dimension. If $(\mathcal{B}_t)_{t\le 0}$ is complementable after 0, then the filtration provided by Proposition 6.19 is the largest d-dimensional Brownian filtration immersed in $(\mathcal{Z}_t)_{t\le 0}$ and containing $(\mathcal{B}_t)_{t\le 0}$. In particular, $(\mathcal{B}'_t)_{t\le 0}$ is maximal in $(\mathcal{Z}_t)_{t\le 0}$.*

6.3 Conditions for Maximality: Proofs

In this section, we provide proofs of the statements given in Sect. 6.2.4, except Proposition 6.18 which is proved in [14].

6.3.1 *Proof of Proposition 6.16*

By definition, $\mathcal{B}'$ is closed under taking complements. For every A and B in $\mathcal{Z}$, the partition $\{A \cup B, (A \cup B)^c\}$ is less fine that $\{A, A^c\} \vee \{B, B^c\}$, hence

$$h(T, \{A \cup B, (A \cup B)^c\}|\mathcal{B}) \leq h(T, \{A, A^c\} \vee \{B, B^c\}|\mathcal{B})$$
$$\leq h(T, \{A, A^c\}|\mathcal{B}) + h(T, \{B, B^c\}|\mathcal{B}).$$

One deduce that $\mathcal{B}'$ is closed under finite union.

But $h(T, \{A, A^c\}|\mathcal{B})$ depends continuously on A when $\mathcal{Z}$ is endowed with the pseudo-metric defined by $\delta(A, B) = \pi(A \triangle B)$ (see Proposition 6.72), so $\mathcal{B}'$ is a closed subset. Hence, $\mathcal{B}'$ is a complete σ-field.

The equalities $h(T, \{A, A^c\}|\mathcal{B}) = h(T, \{T^{-1}A, T^{-1}A^c\}|\mathcal{B}) = h(T, \{TA, TA^c\}|\mathcal{B})$ for every $A \in \mathcal{Z}$ show that $\mathcal{B}'$ is a factor.

Moreover, $\mathcal{B} \subset \mathcal{B}'$ since for every $B \in \mathcal{B}$, $h(T, \{B, B^c\}|\mathcal{B}) \leq H(\{B, B^c\}|\mathcal{B}) = 0$.

The sub-additivity of entropy shows that $h(T, \alpha|\mathcal{B}) = 0$ for every finite partition $\alpha \subset \mathcal{B}'$. Hence $h(T, \mathcal{B}') - h(T, \mathcal{B}) = h((T, \mathcal{B}')|\mathcal{B}) = 0$.

Last, let $\mathcal{A}$ be a factor containing $\mathcal{B}$ and having the same entropy as $\mathcal{B}$. Then for every $A \in \mathcal{A}$,

$$h(T, \{A, A^c\}|\mathcal{B}) \leq h((T, \mathcal{A})|\mathcal{B}) = h(T, \mathcal{A}) - h(T, \mathcal{B}) = 0,$$

so $\mathcal{A} \subset \mathcal{B}'$. The proof is complete.

6.3.2 *Proof of Proposition 6.17*

The proofs given here are inspired by the proofs of the similar results involving (non-conditional) Pinsker factor given in [7].

For every countable measurable partition α of $(Z, \mathcal{Z}, \pi)$ and for every integers $p \leq q$, we introduce the notations

$$\alpha_p^q := \bigvee_{k=p}^{q} T^{-k}\alpha, \quad \alpha_1^\infty = \bigvee_{k \geq 1} T^{-k}\alpha, \quad \mathcal{B}^\alpha := \bigcap_{n \geq 0} \Big(\mathcal{B} \vee \bigvee_{k \geq n} T^{-k}\alpha\Big).$$

Let us recall that the inclusion

$$\mathcal{B}^\alpha \supset \mathcal{B} \vee \bigcap_{n\geq 0}\Big(\bigvee_{k\geq n} T^{-k}\alpha\Big)$$

can be strict modulo **P**. We also note that the larger is n, the smaller is the partition

$$T^{-n}\alpha_1^\infty = \bigvee_{k\geq n+1} T^{-k}\alpha,$$

so $\mathcal{B}^\alpha$ is also the intersection of the non-increasing sequence $(\mathcal{B} \vee T^{-n}\alpha_1^\infty)_{n\geq 0}$.

We begin with the following lemma.

Lemma 6.26 *Let α and γ be countable measurable partitions of $(Z, \mathcal{Z}, \pi)$, with finite entropy. Then $H(\alpha|\alpha_1^\infty \vee \mathcal{B}^\gamma) = H(\alpha|\alpha_1^\infty \vee \mathcal{B})$.*

Proof The addition formula for conditional entropy yields for every $n \geq 1$,

$$H(\alpha \vee \cdots \vee T^{n-1}\alpha|\alpha_1^\infty \vee \mathcal{B}) = \sum_{k=0}^{n-1} H(T^k\alpha|T^k\alpha_1^\infty \vee \mathcal{B}) = nH(\alpha|\alpha_1^\infty \vee \mathcal{B}).$$

Replacing α with $\alpha \vee \gamma$ gives

$$U_n := H(\alpha_{-n+1}^0 \vee \gamma_{-n+1}^0|\alpha_1^\infty \vee \mathcal{B} \vee \gamma_1^\infty) = nH(\alpha \vee \gamma|\alpha_1^\infty \vee \mathcal{B} \vee \gamma_1^\infty).$$

Set

$$V_n := H(\alpha_{-n+1}^0 \vee \gamma_{-n+1}^0|\alpha_1^\infty \vee \mathcal{B}) \text{ and } W_n := H(\alpha_{-n+1}^0 \vee \gamma_{-n+1}^0|\mathcal{B}).$$

Since $U_n \leq V_n \leq W_n$ and

$$\lim_n W_n/n = h((T, \alpha \vee \gamma)|\mathcal{B}) = H(\alpha \vee \gamma|\alpha_1^\infty \vee \mathcal{B} \vee \gamma_1^\infty),$$

we get $\lim_n U_n/n = \lim_n V_n/n$. But

$$U_n = H(\alpha_{-n+1}^0|\alpha_1^\infty \vee \mathcal{B} \vee \gamma_1^\infty) + H(\alpha_{-n+1}^0 \vee \gamma_{-n+1}^0|T^n\alpha_1^\infty \vee \mathcal{B} \vee \gamma_1^\infty),$$
$$V_n = H(\alpha_{-n+1}^0|\alpha_1^\infty \vee \mathcal{B}) + H(\alpha_{-n+1}^0 \vee \gamma_{-n+1}^0|T^n\alpha_1^\infty \vee \mathcal{B}).$$

In these two expressions, each term in U_n is less or equal that the corresponding term in V_n. Since $H(\alpha_{-n+1}^0|\alpha_1^\infty \vee \mathcal{B}) = nH(\alpha|\alpha_1^\infty \vee \mathcal{B})$ we derive that

$$\lim_n n^{-1}H(\alpha_{-n+1}^0|\alpha_1^\infty \vee \mathcal{B} \vee \gamma_1^\infty) = H(\alpha|\alpha_1^\infty \vee \mathcal{B}).$$

Furthermore, we note that

$$\alpha_1^\infty \vee \mathcal{B}^\gamma \subset \bigcap_{n\geq 0} \left(\alpha_1^\infty \vee \mathcal{B} \vee T^{-n}\gamma_1^\infty\right),$$

so Proposition 6.60, Cesàro's lemma and addition formula for conditional entropy yield

$$\begin{aligned}
H(\alpha|\alpha_1^\infty \vee \mathcal{B}^\gamma) &\geq \lim_n H(\alpha|\alpha_1^\infty \vee \mathcal{B} \vee T^{-n}\gamma_1^\infty) \\
&= \lim_n H(T^n\alpha|T^n\alpha_1^\infty \vee \mathcal{B} \vee \gamma_1^\infty) \\
&= \lim_n n^{-1} \sum_{k=0}^{n-1} H(T^k\alpha|T^k\alpha_1^\infty \vee \mathcal{B} \vee \gamma_1^\infty) \\
&= \lim_n n^{-1} H(\alpha_{-n+1}^0|\alpha_1^\infty \vee \mathcal{B} \vee \gamma_1^\infty) \\
&= H(\alpha|\alpha_1^\infty \vee \mathcal{B}).
\end{aligned}$$

But the reverse inequality follows from the inclusion $\alpha_1^\infty \vee \mathcal{B} \subset \alpha_1^\infty \vee \mathcal{B}^\gamma$. Hence the equality holds. □

Lemma 6.26 yields one inclusion in the equality of Proposition 6.17, thanks to the next corollary.

Corollary 6.27 *For every countable measurable partition γ of $(Z, \mathcal{Z}, \pi)$, with finite entropy, $\overline{\mathcal{B}^\gamma} \subset \mathcal{B}'$.*

Proof Since $\mathcal{B}'$ is complete, one only needs to check that $\mathcal{B}^\gamma \subset \mathcal{B}'$. Let $A \in \mathcal{B}^\gamma$ and $\alpha = \{A, A^c\}$. Lemma 6.26 yields $h((T,\alpha)|\mathcal{B}) = H(\alpha|\alpha_1^\infty \vee \mathcal{B}) = H(\alpha|\alpha_1^\infty \vee \mathcal{B}^\gamma) = 0$, so $A \in \mathcal{B}'$. The inclusion follows. □

Lemma 6.26 will also help us to prove the next useful lemma.

Lemma 6.28 *Let α and γ be countable measurable partitions of $(Z, \mathcal{Z}, \pi)$, with finite entropy. Let $N \geq 0$ and η be a finite partition less fine that $\gamma_{-N}^N = \bigvee_{k=-N}^N T^{-k}\gamma$. Then*

$$H(\eta|(\mathcal{B}^\alpha)^\gamma) = H(\eta|\mathcal{B}^\gamma \vee \mathcal{B}^\alpha) = H(\eta|\mathcal{B}^\gamma).$$

Proof For every $n \geq 0$, $(\gamma_{-n}^n)_1^\infty = T^n\gamma_1^\infty$, so

$$H(\gamma_{-n}^n|T^n\gamma_1^\infty \vee \mathcal{B}^\alpha) = H(\gamma_{-n}^n|T^n\gamma_1^\infty \vee \mathcal{B})$$

by Lemma 6.26. When $n \geq N$, $\gamma_{-n}^{n} = \eta \vee \gamma_{-n}^{n}$, so

$$H(\gamma_{-n}^{n}|T^n\gamma_1^{\infty} \vee \mathcal{B}^{\alpha}) = H(\eta|T^n\gamma_1^{\infty} \vee \mathcal{B}^{\alpha}) + H(\gamma_{-n}^{n}|T^n\gamma_1^{\infty} \vee \mathcal{B}^{\alpha} \vee \eta),$$
$$H(\gamma_{-n}^{n}|T^n\gamma_1^{\infty} \vee \mathcal{B}) = H(\eta|T^n\gamma_1^{\infty} \vee \mathcal{B}) + H(\gamma_{-n}^{n}|T^n\gamma_1^{\infty} \vee \mathcal{B} \vee \eta).$$

Since $H(\beta|\mathcal{A} \vee \mathcal{B}^{\alpha}) \leq H(\beta|\mathcal{A} \vee \mathcal{B})$ for every countable measurable partition β and every σ-field $\mathcal{A} \subset \mathcal{Z}$, we get $H(\eta|T^n\gamma_1^{\infty} \vee \mathcal{B}^{\alpha}) = H(\eta|T^n\gamma_1^{\infty} \vee \mathcal{B})$. Letting n go to infinity yields $H(\eta|(\mathcal{B}^{\alpha})^{\gamma}) = H(\eta|\mathcal{B}^{\gamma})$. Since $\mathcal{B}^{\gamma} \subset \mathcal{B}^{\gamma} \vee \mathcal{B}^{\alpha} \subset (\mathcal{B}^{\alpha})^{\gamma}$, the result follows. □

Corollary 6.29 *Assume that T has finite entropy and admits a generator γ. Then for every countable measurable partition α , $\mathcal{B}^{\alpha} \subset \mathcal{B}^{\gamma} = (\mathcal{B}^{\gamma})^{\gamma} \mod \pi$.*

Proof The collection of all $A \in \mathcal{Z}$ such that

$$H(\{A, A^c\}|\mathcal{B}^{\gamma}) = H(\{A, A^c\}|\mathcal{B}^{\gamma} \vee \mathcal{B}^{\alpha}) = H(\{A, A^c\}|(\mathcal{B}^{\gamma})^{\gamma})$$

is a closed subset for the pseudo-metric defined by $\delta(A, B) = \pi(A \triangle B)$, and contains the algebra $\bigcup_{N \in \mathbf{N}} \sigma(\gamma_{-N}^{N})$ by Lemma 6.28 applied once to (α, γ) and once to (γ, γ). Therefore, these collection equals the whole σ-field $\mathcal{Z}$ itself. In particular, $H(\{A, A^c\}|\mathcal{B}^{\gamma}) = 0$ whenever $A \in \mathcal{B}^{\gamma} \vee \mathcal{B}^{\alpha}$ or $A \in (\mathcal{B}^{\gamma})^{\gamma}$. Hence $\mathcal{B}^{\gamma} \vee \mathcal{B}^{\alpha}$ and $(\mathcal{B}^{\gamma})^{\gamma}$ are contained in $\mathcal{B}^{\gamma}$ modulo the null sets. The result follows. □

We now achieve the proof of Proposition 6.17. Assume that T has finite entropy and that γ is a generator of T. We have to prove that $\mathcal{B}' \subset \overline{\mathcal{B}^{\gamma}}$. So let $A \in \mathcal{B}'$ and $\alpha = \{A, A^c\}$. For every $n \geq 0$, set

$$\mathcal{D}_n := \sigma\Big(\bigvee_{k \geq n} T^{-k}\alpha\Big).$$

Since, $\mathcal{D}_1 = \alpha_1^{\infty}$, the equality $H(\alpha|\alpha_1^{\infty} \vee \mathcal{B}) = h(T, \alpha|\mathcal{B}) = 0$ yields $\alpha \subset \mathcal{D}_1 \vee \mathcal{B} \mod \pi$, so $\mathcal{D}_0 = \mathcal{D}_1 \vee \mathcal{B} \mod \pi$. By applying T^{-n}, we get more generally $\mathcal{D}_n = \mathcal{D}_{n+1} \vee \mathcal{B} \mod \pi$. By induction, $\mathcal{D}_0 = \mathcal{D}_n \vee \mathcal{B} \mod \pi$ for every $n \geq 0$. Hence $\mathcal{D}_0 = \mathcal{B}^{\alpha} \subset \mathcal{B}^{\gamma} \mod \pi$, thanks to the last corollary, so $A \in \overline{\mathcal{B}^{\gamma}}$. We are done.

6.3.3 Proof of Proposition 6.19

Fix a d-dimensional Brownian motion B generating the filtration $\mathcal{B}$ modulo the null sets.

Let $t > \varepsilon > 0$. Since $\mathcal{B}$ is immersed in $\mathcal{Z}$, the Brownian motion $B^{(\varepsilon)} := B_{\varepsilon+\cdot} - B_{\varepsilon}$ is independent of $\mathcal{Z}_{\varepsilon}$, which is the terminal σ-field of the filtration

$(\mathcal{B}_\varepsilon \vee \mathcal{Z}_s)_{s\in[0,\varepsilon]}$. Therefore (see Corollary 6.38)

$$\begin{aligned}\mathcal{B}'_t &= \bigcap_{s\in]0,\varepsilon]} \left(\sigma((B^{(\varepsilon)}_r)_{r\in[0,t-\varepsilon]}) \vee \mathcal{B}_\varepsilon \vee \mathcal{Z}_s\right)\\ &= \sigma((B^{(\varepsilon)}_r)_{r\in[0,t-\varepsilon]}) \vee \bigcap_{s\in]0,\varepsilon]} \left(\mathcal{B}_\varepsilon \vee \mathcal{Z}_s\right)\\ &= \sigma((B^{(\varepsilon)}_r)_{r\in[0,t-\varepsilon]}) \vee \mathcal{B}'_\varepsilon \mod \mathbf{P}.\end{aligned}$$

Hence, the filtration $\mathcal{B}'$ has independent increments after ε, provided by the increments of B after ε, so $\mathcal{B}'$ is Brownian after 0.

Moreover, since $B^{(\varepsilon)}$ is independent of $\mathcal{Z}_\varepsilon$, the equality modulo $\mathbf{P}$ above shows that for every $t > \varepsilon > 0$, $\mathcal{B}'_t$ and $\mathcal{Z}_\varepsilon$ are independent conditionally on $\mathcal{B}'_\varepsilon$. This conditional independence still holds when $\varepsilon = 0$, since $\mathcal{Z}_0$ is trivial. Thus $\mathcal{B}'$ is immersed in $\mathcal{Z}$. By Proposition 6.6 (or corollary 1 in [9]), $\mathcal{B}'$ is a d-dimensional Brownian filtration.

For every $t > 0$, $\mathcal{B}_t \subset \mathcal{B}'_t \subset \mathcal{B}_t \vee \mathcal{Z}_t = \mathcal{Z}_t \mod \pi$. These inclusions modulo $\mathbf{P}$ still hold when $t = 0$ since $\mathcal{B}_0$ and $\mathcal{Z}_{0+} = \bigcap_{s>0} \mathcal{Z}_s$ are trivial. Since $\mathcal{B}$ is immersed in $\mathcal{Z}$, we deduce that $\mathcal{B}$ is immersed in $\mathcal{B}'$.

Last, let $t \geq 0$. For every $n \geq 1$,

$$\mathcal{B}''_t \subset \overline{\mathcal{B}'_t \vee \mathcal{Z}_{1/n}} \subset \overline{\overline{(\mathcal{B}_t \vee \mathcal{Z}_{1/n})} \vee \mathcal{Z}_{1/n}} = \overline{\mathcal{B}_t \vee \mathcal{Z}_{1/n}}.$$

If $A \in \mathcal{B}''_t$, then for each $n \geq 1$, one can find $B_n \in \mathcal{B}_t \vee \mathcal{Z}_{1/n}$ such that $\mathbf{P}(A \triangle B_n) = 0$; hence $A \in \mathcal{B}'_t$ since $B := \limsup_n B_n$ belongs to $\bigcap_{n\geq 1}(\mathcal{B}_t \vee \mathcal{Z}_{1/n})$ and $\mathbf{P}(A \triangle B) = 0$. The equality $\mathcal{B}''_t = \mathcal{B}'_t$ follows.

6.3.4 Proof of the Statements of Example 6.2

Assume that $\mathcal{X}$ is maximal in $\mathcal{Z}$. Since $\mathcal{B}$ is immersed $\mathcal{B}'$, we have only to check the inclusion $\mathcal{B}'_\infty \subset \mathcal{B}_\infty \mod \mathbf{P}$. The maximality of $\mathcal{X}$ in $\mathcal{Z}$ yields

$$\mathcal{B}'_\infty \subset \bigcap_{s>0} (\mathcal{X}_\infty \vee \mathcal{Z}_s) = \mathcal{X}_\infty \mod \mathbf{P},$$

so one only needs to check that $\mathbf{E}[h(X)|\mathcal{B}'_\infty] = \mathbf{E}[h(X)|\mathcal{B}_\infty]$ almost surely for every real bounded measurable functional h defined on the space $C(\mathbf{R}_+)$ of all continuous functions from $\mathbf{R}_+$ to $\mathbf{R}$. Since the topology of uniform convergence on compact subsets on the space $C(\mathbf{R}_+)$ is metrizable, it is sufficient to check the equality when h is continuous. In this case, the random variable $h(X)$ is the limit in $L^1(\mathbf{P})$ of the sequence $(h(X^{(n)}))_{n\geq 1}$, where T_n denotes the first zero of X after time $1/n$, and $X^{(n)}_t = X_{T_n+t}$ for every $t \geq 0$. Since B generates the same filtration as $|X|$

up to null sets, $\mathcal{B}_\infty \vee \mathcal{Z}_{T_n} = \sigma(|X^{(n)}|) \vee \mathcal{Z}_{T_n} \mod \mathbf{P}$. But $X^{(n)}$ is independent of $\mathcal{Z}_{T_n}$ since X is immersed in $\mathcal{Z}$, so

$$\mathbf{E}[h(X^{(n)})|\mathcal{B}_\infty \vee \mathcal{Z}_{T_n}] = \mathbf{E}[h(X^{(n)})|\sigma(|X^{(n)}|) \vee \mathcal{Z}_{T_n}] = \mathbf{E}[h(X^{(n)})|\sigma(|X^{(n)}|)] \text{ a.s..}$$

But $\sigma(|X^{(n)}|) \subset \mathcal{B}_\infty \subset \mathcal{B}'_\infty \subset \mathcal{B}_\infty \vee \mathcal{Z}_{1/n} \subset \mathcal{B}_\infty \vee \mathcal{Z}_{T_n}$, so

$$\mathbf{E}[h(X^{(n)})|\mathcal{B}'_\infty] = \mathbf{E}[h(X^{(n)})|\mathcal{B}_\infty].$$

The statements follow.

6.4 Complementability Implies Maximality: Proofs

We now prove Propositions 6.21, 6.23 and 6.25. The first two rely on very similar key lemmas.

6.4.1 Key Lemma for Factors of a Lebesgue Automorphism

Proposition 6.21 follows from the next lemma, which can derived from lemma in [17] or from lemma 2.3 in [2].

Lemma 6.30 *Let $\mathcal{A}, \mathcal{B}, C$ be three factors of T. Assume that:*

1. *$\mathcal{A} \supset \mathcal{B}$;*
2. *$h(T, \mathcal{A}) = h(T, \mathcal{B}) < +\infty$;*
3. *(T, C) is a K-automorphism and has finite entropy.*
4. *$\mathcal{B}$ and C are independent.*

Then $\mathcal{A}$ and C are independent.

First, we show how to deduce Proposition 6.21 from Lemma 6.30.

Proof (Proof of Proposition 6.21) Let C be an independent complement of $\mathcal{B}$ having the property K. Let $\mathcal{A}$ be a factor of T such that $\mathcal{A} \supset \mathcal{B}$ and $h(T, \mathcal{A}) = h(T, \mathcal{B})$. Then Lemma 6.30 yields that $\mathcal{A}$ and C are independent. But $\mathcal{Z} = \mathcal{B} \vee C \subset \mathcal{A} \vee C \subset \mathcal{Z}$, so $\mathcal{A} \vee C = \mathcal{B} \vee C$. Hence $\mathcal{A} = \mathcal{B}$ by the next lemma. □

We have just used the following general statement, which will also help us in the context of Brownian filtrations.

Lemma 6.31 *Let $\mathcal{A}, \mathcal{B}, C$ be three sub-σ-fields of any probability space $(Z, \mathcal{Z}, \pi)$ such that*

- $\mathcal{A} \supset \mathcal{B}$;
- $\mathcal{A}$ *and* C *are independent;*
- $\mathcal{A} \vee C = \mathcal{B} \vee C$.

Then $\mathcal{A} = \mathcal{B} \mod \pi$.

Proof Let $A \in \mathcal{A}$. Then $\sigma(A) \vee \mathcal{B} \subset \mathcal{A}$, so $\sigma(A) \vee \mathcal{B}$ is independent of C, and

$$\pi[A|\mathcal{B}] = \pi[A|\mathcal{B} \vee C] = \pi[A|\mathcal{A} \vee C] = \mathbf{1}_A \ \pi\text{-almost surely.}$$

Hence $A \in \mathcal{B} \mod \pi$. □

We now give two different proofs of Lemma 6.30. The second one relies on Pinsker's formula and is a bit simpler.

6.4.2 Proof of Lemma 6.30

The proof below is a reformulation of the proof given in [17].

Assume that the assumptions hold. Let α, β, γ be countable partitions generating $(T, \mathcal{A})$, $(T, \mathcal{B})$, (T, C), respectively. Since $(T, \mathcal{A})$, $(T, \mathcal{B})$, (T, C) have finite entropy, α, β, γ have also finite entropy. Given $n \geq 1$, set $\alpha_0^{n-1} = \alpha \vee \cdots \vee T^{-(n-1)}\alpha$, $\beta_0^{n-1} = \beta \vee \cdots \vee T^{-(n-1)}\beta$,

$$C_n = \sigma\Big(\bigvee_{q \in \mathbf{Z}} T^{-qn}\gamma\Big) \text{ and } \mathcal{D}_n = \sigma\Big(\bigvee_{k \geq n} T^{-k}\gamma\Big).$$

Then C_n is a factor of T^n, $\alpha_0^{n-1} \vee \gamma$ is a generator of $(T^n, \mathcal{A} \vee C_n)$ whereas $\beta_0^{n-1} \vee \gamma$ is a generator of $(T^n, \mathcal{B} \vee C_n)$.

Therefore, on the one hand,

$$\begin{aligned}
h(T^n, \mathcal{A} \vee C_n) &= H\Big(\alpha_0^{n-1} \vee \gamma \Big| \bigvee_{q \geq 1} T^{-qn}(\alpha_0^{n-1} \vee \gamma)\Big) \\
&= H\Big(\alpha_0^{n-1} \vee \gamma \Big| \bigvee_{k \geq n} T^{-k}\alpha \vee \bigvee_{q \geq 1} T^{-qn}\gamma\Big) \\
&= H\Big(\alpha_0^{n-1} \Big| \bigvee_{k \geq n} T^{-k}\alpha \vee \bigvee_{q \geq 1} T^{-qn}\gamma\Big) \\
&\quad + H\Big(\gamma \Big| \bigvee_{k \geq 0} T^{-k}\alpha \vee \bigvee_{q \geq 1} T^{-qn}\gamma\Big)
\end{aligned}$$

$$\leq H\Big(\alpha_0^{n-1}\Big|\bigvee_{k\geq n} T^{-k}\alpha\Big) + H(\gamma|\alpha)$$

$$= h(T^n, \mathcal{A}) + H(\gamma|\alpha) = nh(T, \mathcal{A}) + H(\gamma|\alpha).$$

On the other hand, by independence of $\mathcal{B}$ and C,

$$h(T^n, \mathcal{B} \vee C_n) = H\Big(\beta_0^{n-1}\Big|\bigvee_{k\geq n} T^{-k}\beta \vee \bigvee_{q\geq 1} T^{-qn}\gamma\Big)$$

$$+ H\Big(\gamma\Big|\bigvee_{k\geq 0} T^{-k}\beta \vee \bigvee_{q\geq 1} T^{-qn}\gamma\Big)$$

$$= H\Big(\beta_0^{n-1}\Big|\bigvee_{k\geq n} T^{-k}\beta\Big) + H\Big(\gamma|\bigvee_{q\geq 1} T^{-qn}\gamma\Big)$$

$$\geq h(T^n, \mathcal{B}) + H(\gamma|\mathcal{D}_n) = nh(T, \mathcal{B}) + H(\gamma|\mathcal{D}_n).$$

But $h(T^n, \mathcal{B} \vee C_n) \leq h(T^n, \mathcal{A} \vee C_n)$ since $\mathcal{B} \subset \mathcal{A}$. Putting things together and using the assumption $h(T, \mathcal{A}) = h(T, \mathcal{B}) < +\infty$ yields $H(\gamma|\mathcal{D}_n) \leq H(\gamma|\alpha)$. But $(\mathcal{D}_n)_{n\geq 1}$ is a decreasing sequence of σ-fields with trivial intersection since (T, C) has the property K, so $H(\gamma|\mathcal{D}_n) \to H(\gamma)$ as $n \to +\infty$. Hence, $H(\gamma) \leq H(\gamma|\alpha)$, so α and γ are independent. This conclusion is preserved if one replaces the generators α and γ by the supremum of $T^{-k}\alpha$ and $T^{-k}\gamma$ over all $k \in [\![-n, n]\!]$. Letting n go to infinity yields the independence of $\mathcal{A}$ and C.

6.4.3 Alternative Proof of Lemma 6.30

The inclusion $\mathcal{A} \supset \mathcal{B}$ and the independence of $\mathcal{B}$ and C yield

$$h(T, \mathcal{A}) + h(T, C) \geq h(T, \mathcal{A} \vee C) \geq h(T, \mathcal{B} \vee C) = h(T, \mathcal{B}) + h(T, C).$$

But $h(T, \mathcal{A}) = h(T, \mathcal{B})$, hence $h(T, \mathcal{A} \vee C) = h(T, \mathcal{A}) + h(T, C)$. Since (T, C) is a K-automorphism with finite entropy, Berg's lemma below shows that $\mathcal{A}$ and C independent.

Lemma 6.32 (Lemma 2.3 in [2]) *Let $\mathcal{A}$ and C be two factors of the dynamical system $(Z, \mathcal{Z}, \pi, T)$, such that $h(T, \mathcal{A} \vee C) = h(T, \mathcal{A}) + h(T, C) < +\infty$ and (T, C) is a K-automorphism. Then $\mathcal{A}$ and C independent.*

Proof Let α and γ be countable generating partitions of $(T, \mathcal{A})$ and (T, C), respectively. Set

$$\alpha_1^\infty = \bigvee_{k\geq 1} T^{-k}\alpha \text{ and } \gamma_1^\infty = \bigvee_{k\geq 1} T^{-k}\gamma$$

Then

$$h(T,\mathcal{A}) = H(\alpha|\alpha_1^\infty), \quad h(T,C) = H(\gamma|\gamma_1^\infty), \quad h(T,\mathcal{A}\vee C) = H(\alpha\vee\gamma|\alpha_1^\infty\vee\gamma_1^\infty).$$

But Pinsker's formula (Proposition 6.75 in Sect. 6.7 or theorem 6.3 in [18]) gives

$$H(\alpha\vee\gamma|\alpha_1^\infty\vee\gamma_1^\infty) = H(\alpha|\alpha_1^\infty) + H(\gamma|\mathcal{A}\vee\gamma_1^\infty).$$

So the assumption $h(T,\mathcal{A}\vee C) = h(T,\mathcal{A}) + h(T,C) < +\infty$ yields $H(\gamma|\gamma_1^\infty) = H(\gamma|\mathcal{A}\vee\gamma_1^\infty)$.

Thus for any partition $\delta\subset\mathcal{A}$ with finite entropy, $H(\gamma|\delta\vee\gamma_1^\infty) = H(\gamma|\gamma_1^\infty)$, so

$$H(\delta\vee\gamma|\gamma_1^\infty) = H(\delta|\gamma_1^\infty) + H(\gamma|\delta\vee\gamma_1^\infty) = H(\delta|\gamma_1^\infty) + H(\gamma|\gamma_1^\infty).$$

But we have also

$$H(\delta\vee\gamma|\gamma_1^\infty) = H(\gamma|\gamma_1^\infty) + H(\delta|\gamma\vee\gamma_1^\infty).$$

Hence $H(\delta|\gamma_1^\infty) = H(\delta|\gamma\vee\gamma_1^\infty)$.

Let $m\geq 0$ and n be integers. Applying the last equality to $\delta := \bigvee_{|k|\leq m} T^{n-k}\alpha$ yields

$$H\Big(\bigvee_{|k|\leq m} T^{n-k}\alpha\,\Big|\bigvee_{k\geq 1} T^{-k}\gamma\Big) = H\Big(\bigvee_{|k|\leq m} T^{n-k}\alpha\,\Big|\bigvee_{k\geq 0} T^{-k}\gamma\Big).$$

Since T preserves π, this is equivalent to

$$H\Big(\bigvee_{|k|\leq m} T^{-k}\alpha\,\Big|\bigvee_{k\geq n+1} T^{-k}\gamma\Big) = H\Big(\bigvee_{|k|\leq m} T^{-k}\alpha\,\Big|\bigvee_{k\geq n} T^{-k}\gamma\Big).$$

As a result, the entropy above does not depend on n. Letting n go to $-\infty$ and to $+\infty$, and using the fact that (T,C) is a K-automorphism, we get at the limit

$$H\Big(\bigvee_{|k|\leq m} T^{-k}\alpha\,\Big|C\Big) = H\Big(\bigvee_{|k|\leq m} T^{-k}\alpha\Big),$$

so the partition $\bigvee_{|k|\leq m} T^{-k}\alpha$ is independent of C. Letting m go to $+\infty$ yields the independence of $\mathcal{A}$ and C. □

6.4.4 *Proof in the Context of Brownian Filtrations*

The proof of Proposition 6.23 below may look suspiciously simple, but actually, it relies on non-trivial theorems of stochastic integration, namely the predictable representation property and the bracket characterization of multi-dimensional Brownian motions among local martingales. The immersion of a filtration into another one is a strong property, as shown for example by the characterizations for a Brownian filtration recalled in the introduction (Proposition 6.2). The key step is very similar to Lemma 6.30.

Lemma 6.33 *Let A, B, C be three Brownian motions in some filtration $\mathcal{Z}$. Assume that:*

1. *$\sigma(A) \supset \sigma(B) \mod \mathbf{P}$;*
2. *A and B have the same finite dimension;*
3. *B and C are independent.*

Then A and C are independent.

Proof Call p the dimension of A and B and q the (possibly infinite) dimension of C. Since B is a Brownian motion in $\mathcal{Z}$ and its own filtration, it is also a Brownian motion in the intermediate filtration $\mathcal{F}^A$. Hence, one can find an $\mathcal{F}^A$-predictable process H with values in the group of all orthogonal $p \times p$ matrices such that

$$B = \int_0^\cdot H_s \mathrm{d}A_s \text{ a.s..}$$

Since $H_s^\top H_s = I_p$ for every $s \geq 0$ (where $H_s^\top$ is the transpose of H_s), we have also

$$A = \int_0^\cdot H_s^\top \mathrm{d}B_s \text{ a.s..}$$

Looking at the components, we get for every $i \in [\![1, p]\!]$,

$$A^{(i)} = \sum_{j=1}^{p} \int_0^\cdot H_s(j, i) \mathrm{d}B_s^{(j)} \text{ a.s..}$$

For every $i \in [\![1, p]\!]$ and $k \in [\![1, q]\!]$, we get

$$\langle A^{(i)}, C^{(k)} \rangle = \sum_{j=1}^{p} \int_0^\cdot H_s(j, i) \mathrm{d}\langle B^{(j)}, C^{(k)} \rangle_s = 0 \text{ a.s.,}$$

since $\langle B^{(j)}, C^{(k)} \rangle = 0$ a.s., by independence of B et C. We derive that (A, C) is a $p + q$-dimensional Brownian motion in $\mathcal{Z}$, so A and C are independent. □

Deducing Proposition 6.23 from the last lemma involves almost the same arguments as deducing Proposition 6.21 from Lemma 6.30.

Proof Proof of Proposition 6.23. Let $\mathcal{Z}$ be a finite Brownian filtration, and $\mathcal{A}, \mathcal{B}, \mathcal{C}$ be three Brownian filtrations in $\mathcal{Z}$ such that $\mathcal{A}_t \supset \mathcal{B}_t$ for every $t \geq 0$, $\mathcal{A}$ and $\mathcal{B}$ have the same dimension, and $\mathcal{C}$ is an independent complement of $\mathcal{B}$ in $\mathcal{Z}$.

Let A, B, C be Brownian motions generating $\mathcal{A}, \mathcal{B}, \mathcal{C}$ modulo the null events. Then Lemma 6.33 applies, so A is independent of C. By Lemma 6.31, $\sigma(A) = \sigma(B) \mod \mathbf{P}$. But $\mathcal{F}^B$ is immersed in $\mathcal{F}^A$. Since the final σ-fields $\mathcal{F}^A_\infty = \sigma(A)$ and $\mathcal{F}^B_\infty = \sigma(B)$ coincide almost surely, we get $\mathcal{F}^A_t = \mathcal{F}^B_t \mod \mathbf{P}$ for every $t \geq 0$ by Lemma 6.1.

□

We now prove Proposition 6.25.

Proof Let $\mathcal{C}$ be a complement of $\mathcal{B}$ after 0, and $\mathcal{A}$ be a d-dimensional Brownian filtration immersed in $\mathcal{Z}$ and containing $\mathcal{B}$. Let A, B, C be Brownian motions in $\mathcal{Z}$ generating $\mathcal{A}$, $\mathcal{B}$, $\mathcal{C}$ respectively modulo the null events. Since $\mathcal{A}$ and $\mathcal{B}'$ are immersed in $\mathcal{Z}$, it is sufficient to prove the inclusion $\mathcal{A}_\infty \subset \mathcal{B}'_\infty$. Hence, given $s > 0$, we have to check that $\mathcal{A}_\infty \subset \mathcal{B}_\infty \vee \mathcal{Z}_s \mod \mathbf{P}$.

By Lemma 6.33, we know that A and C are independent Brownian motions in $\mathcal{F}^Z$. Thus $\mathcal{Z}_s$, $A_{s+\cdot} - A_s$ and $C_{s+\cdot} - C_s$ are independent. Let

$$\tilde{\mathcal{A}} := \mathcal{A}_\infty \vee \mathcal{Z}_s = \sigma(A_{s+\cdot} - A_s) \vee \mathcal{Z}_s \mod \mathbf{P},$$

$$\tilde{\mathcal{B}} := \mathcal{B}_\infty \vee \mathcal{Z}_s = \sigma(B_{s+\cdot} - B_s) \vee \mathcal{Z}_s \mod \mathbf{P},$$

$$\tilde{\mathcal{C}} := \sigma(C_{s+\cdot} - C_s).$$

Then $\tilde{\mathcal{A}} \supset \tilde{\mathcal{B}}$, $\tilde{\mathcal{B}}$ and $\tilde{\mathcal{C}}$ are independent, and $\tilde{\mathcal{A}} \vee \tilde{\mathcal{C}} = \tilde{\mathcal{B}} \vee \tilde{\mathcal{C}} \mod \mathbf{P}$, since

$$\mathcal{Z}_\infty \supset \tilde{\mathcal{A}} \vee \tilde{\mathcal{C}} \supset \tilde{\mathcal{B}} \vee \tilde{\mathcal{C}} = \mathcal{B}_\infty \vee \mathcal{C}_\infty \vee \mathcal{Z}_s = \mathcal{Z}_\infty \mod \mathbf{P}.$$

Hence Lemma 6.31 applies, so $\mathcal{A}_\infty \subset \tilde{\mathcal{A}} = \tilde{\mathcal{B}} = \mathcal{B}_\infty \vee \mathcal{Z}_s \mod \mathbf{P}$. □

6.5 A Maximal But Not-Complementable Brownian Filtration

To get such an example, our strategy is to construct a maximal but not-complementable filtration in a dyadic product-type filtration and to embed these filtrations in Brownian filtrations.

6.5.1 A Maximal But Not-Complementable Filtration in a Dyadic Product-Type Filtration

This subsection is devoted to the proof of the following lemma.

Lemma 6.34 *One can construct:*

- *a dyadic product-type filtration* $(\mathcal{Z}_n)_{n\le 0}$,
- *a poly-adic product-type filtration* $(\mathcal{U}_n)_{n\le 0}$ *immersed in* $(\mathcal{Z}_n)_{n\le 0}$,
- *a random variable* U *with values in some Polish space* $(E,\mathcal{E})$ *and generating* $\mathcal{U}_0$,

such that for $U(\mathbf{P})$*-almost every* $u \in E$, $(\mathcal{Z}_n)_{n\le 0}$ *is Kolmogorovian but not product-type under* $\mathbf{P}_u = \mathbf{P}[\cdot|U=u]$. *Therefore, the filtration* $(\mathcal{U}_n)_{n\le 0}$ *is maximal but non complementable in* $(\mathcal{Z}_n)_{n\le 0}$.

Proof We begin with a variant of an example given in [6], which was itself inspired from an unpublished note of B. Tsirelson (About Yor's problem, Unpublished Notes. https://www.tau.ac.il/~tsirel/download/yor3.pdf).

For every $n \le 0$, call K_n the finite field with $q_n = 2^{2^{|n|}}$ elements. Start with a sequence of independent random variables $(Z_n)_{n\le 0}$ such that for every $n \le 0$, $Z_{2n} = (X_n, Y_n)$ is uniform on $K_n^4 \times K_n^4$ and $Z_{2n-1} = B_n$ is uniform on K_n^4. By construction, the filtration $(\mathcal{F}_n^Z)_{n\le 0}$ is product-type and $(r_n)_{n\le 0}$-adic, with $r_{2n-1} = q_n^4$ and $r_{2n} = q_n^8$ for every $n \le 0$.

Since $|K_{n-1}| = 2^{2^{|n|+1}} = |K_n|^2$, one can fix a bijection between $K_{n-1}^4 \times K_{n-1}^4$ and the set $\mathcal{M}_4(K_n)$ of all 4×4 matrices with entries in K_n. Call A_n the uniform random variable on $\mathcal{M}_4(K_n)$ corresponding to Z_{2n-2} through this bijection, and set $U_{2n-1} = 0$ and $U_{2n} = Y_n - A_n X_n - B_n$.

For every $n \le 0$, (X_n, Y_n) is independent of $\mathcal{F}_{2n-1}^Z$ and uniform on $K_n^4 \times K_n^4$. Since the random map $(x,y) \mapsto (x, y - A_n x - B_n)$ from $K_n^4 \times K_n^4$ to itself is $\mathcal{F}_{2n-1}^Z$-measurable and bijective, (X_n, U_{2n}) is also independent of $\mathcal{F}_{2n-1}^Z$ and uniform on $K_n^4 \times K_n^4$ and is still an innovation at time $2n$ of the filtration $\mathcal{F}^Z$. Therefore, the filtration $(\mathcal{F}_n^U)_{n\le 0}$ is immersed in $(\mathcal{F}_n^Z)_{n\le 0}$, product-type and $(r_n/q_n^4)_{n\le 0}$-adic.

For each $n \le 0$, the set K_n^4 is a 4-dimensional vector space on K_n and also a $2^{|n|+2}$-dimensional vector space on the sub-field $K_0 = \{0,1\}$. Fix a basis $(e_n^i)_{1\le i\le 2^{|n|+2}}$ and for every vector $v \in K_n^4$, call $(v^i)_{1\le i\le 2^{|n|+2}}$ its components in this basis. Then each one of the uniform random variables B_n, X_n, Y_n and U_{2n} yields $2^{|n|+2}$ independent and uniform Bernoulli random variables, namely B_n^i, X_n^i, Y_n^i and U_{2n}^i with $i \in [\![1, 2^{|n|+2}]\!]$.

Define an increasing map $s : \mathbf{Z}_- \mapsto \mathbf{Z}_-$ by $s(0) = 0$, $s(2n-1) = s(2n) - 2^{|n|+3}$ and $s(2n-2) = s(2n-1) - 2^{|n|+2}$ for every $n \le 0$, and two filtrations $\mathcal{Z} = (\mathcal{Z}_n)_{n\le 0}$ and $\mathcal{U} = (\mathcal{U}_n)_{n\le 0}$ by the following equalities for every $n \le 0$ and $i \in [\![1, 2^{|n|+2}]\!]$:

$$\mathcal{Z}_{s(2n-2)+i} := \mathcal{F}_{2n-2}^Z \vee \sigma(B_n^1, \dots, B_n^i),$$

$$\mathcal{Z}_{s(2n-1)+i} := \mathcal{F}_{2n-1}^Z \vee \sigma(X_n^1, \dots, X_n^i),$$

$$\mathcal{Z}_{s(2n-1)+2^{|n|+2}+i} := \mathcal{F}^Z_{2n-1} \vee \sigma(X_n) \vee \sigma(Y_n^1, \dots, Y_n^i),$$
$$\mathcal{U}_{s(2n-2)+i} := \mathcal{F}^U_{2n-2},$$
$$\mathcal{U}_{s(2n-1)+i} := \mathcal{F}^U_{2n-1} = \mathcal{F}^U_{2n-2},$$
$$\mathcal{U}_{s(2n-1)+2^{|n|+2}+i} := \mathcal{F}^U_{2n-1} \vee \sigma(U^1_{2n}, \dots, U^i_{2n}).$$

Taking $i = 2^{|n|+2}$ in these formulas yields

$$\mathcal{Z}_{s(2n-1)} = \mathcal{F}^Z_{2n-1}, \quad \mathcal{Z}_{s(2n)} = \mathcal{F}^Z_{2n}, \quad \mathcal{U}_{s(2n-1)} = \mathcal{F}^U_{2n-1}, \quad \mathcal{U}_{s(2n)} = \mathcal{F}^U_{2n},$$

so the filtrations $\mathcal{Z}$ and $\mathcal{U}$ interpolate $\mathcal{F}^Z$ and $\mathcal{F}^U$.

Moreover, the filtrations $\mathcal{Z}$ and $\mathcal{U}$ are product-type, $\mathcal{Z}$ is dyadic, $\mathcal{U}$ is poly-adic and increases only at times $s(2n-1) + 2^{|n|+2} + i$ with $n \leq 0$ and $i \in [\![1, 2^{|n|+2}]\!]$. At such a time, Y^i_{2n} is an innovation of the filtration $\mathcal{Z}$ whereas U^i_{2n} is an innovation of the filtration $\mathcal{U}$, and also of the larger filtration $\mathcal{Z}$ since

- $U^i_{2n} = Y^i_n - (A_n X_n)^i - B^i_n$,
- Y^i_n is uniform on K_0 and independent of $\mathcal{Z}_{s(2n-1)+2^{|n|+2}+i-1}$,
- $(A_n X_n)^i + B^i_n$ is $\mathcal{Z}_{s(2n-1)+2^{|n|+2}}$-measurable hence $\mathcal{Z}_{s(2n-1)+2^{|n|+2}+i-1}$-measurable.

As a result, the filtration $\mathcal{U}$ is immersed in $\mathcal{Z}$.

The random variable $U = (U_n)_{n\leq 0}$ generates $\mathcal{U}_0 = \mathcal{F}^U_0$. Let us check that for $U(\mathbf{P})$-almost every $u \in E$, $(\mathcal{Z}_n)$ is Kolmogorovian but not product-type under $\mathbf{P}_u = \mathbf{P}[\cdot|U = u]$. By Corollary 6.10 and Proposition 6.20 of the present paper (propositions 3,4 and corollary 9 of [14]), the last two statements will follow.

First, we note that for every $n \leq 0$, $\mathcal{F}^Z_{2n} = \mathcal{F}^U_{2n} \vee \mathcal{F}^{X,Y}_n$, and that $\mathcal{F}^U_0$ and $\mathcal{F}^{X,Y}_0$ are independent. The independence follows from the equalities $U_{2n} = Y_n - A_n X_n - B_n$, the $\mathcal{F}^{X,Y}_0$-measurability of the random variables $(Y_n - A_n X_n)_{n\leq 0}$ since each A_n is a function of $Z_{2n-2} = (X_{n-1}, Y_{n-1})$: conditionally on $\mathcal{F}^{X,Y}_0$, the random variables $(U_{2n})_{n\leq 0}$ taking values in the additive groups $(K^4_n)_{n\leq 0}$ are independent and uniform since the random variables $(B_n)_{n\leq 0}$ are independent and uniform.

To show that $(\mathcal{Z}_n)$ is not product-type under $\mathbf{P}_u$, it suffices to show that the extracted filtration $(\mathcal{F}^Z_n)_{n\leq 0}$ is not product-type under $\mathbf{P}_u$. To do this, we check that the random variable Z_0 does not satisfy the I-cosiness criterion. Let $Z' = (X', Y')$ and $Z'' = (X'', Y'')$ be two copies of the process Z under $\mathbf{P}_u$, defined on some probability space $(\bar{\Omega}, \bar{\mathcal{A}}, \bar{\mathbf{P}}_u)$, such that both natural filtrations $\mathcal{F}^{Z'}$ and $\mathcal{F}^{Z''}$ are immersed in some filtration $\mathcal{G}$.

For every $n \leq 0$, define the copies A'_n, A''_n and B'_n, B''_n of the random variables A_n and B_n by the obvious way, and set $S_n = \{x \in K^4_n : A'_n x + B'_n = A''_n x + B''_n\}$. Then for $U(\mathbf{P})$-almost every $u \in E$, the equalities $Y'_n = A'_n X'_n + B'_n + u_{2n}$ and

$Y_n'' = A_n''X_n'' + B_n'' + u_{2n}$ hold $\bar{\mathbf{P}}_u$-almost surely. Therefore,

$$\mathbf{1}_{[Z'_{2n}=Z''_{2n}]} = \mathbf{1}_{[X'_n=X''_n\in S_n]} \leq \mathbf{1}_{[X'_n\in S_n]} \ \bar{\mathbf{P}}_u\text{-almost surely.}$$

But S_n is $\mathcal{G}_{2n-1}$-measurable whereas X_n' is uniform on K_n^4 conditionally on $\mathcal{G}_{2n-1}$ since $\mathcal{F}^{Z'}$ is immersed in $\mathcal{G}$. Moreover, $|S_n| \leq q_n^3$ when $A_n' \neq A_n''$. Hence

$$\bar{\mathbf{P}}_u[Z'_{2n} = Z''_{2n}|\mathcal{G}_{2n-1}] \leq \frac{|S_n|}{q_n^4} \leq \mathbf{1}_{[A'_n=A''_n]} + \frac{1}{q_n}\mathbf{1}_{[A'_n\neq A''_n]} \ \bar{\mathbf{P}}_u\text{-almost surely.}$$

Passing to the complements and taking the expectations yields

$$\bar{\mathbf{P}}_u[Z'_{2n} \neq Z''_{2n}] \geq \Big(1 - \frac{1}{q_n}\Big)\bar{\mathbf{P}}_u[A'_n \neq A''_n] = \Big(1 - \frac{1}{q_n}\Big)\bar{\mathbf{P}}_u[Z'_{2n-2} \neq Z''_{2n-2}].$$

By induction, one gets that for every $n \leq 0$

$$\bar{\mathbf{P}}_u[Z'_0 \neq Z''_0] \geq \prod_{k=n+1}^{0} \Big(1 - \frac{1}{q_k}\Big) \times \bar{\mathbf{P}}_u[Z'_{2n} \neq Z''_{2n}].$$

If, for some $N > -\infty$, the σ-fields $\mathcal{F}_N^{Z'}$ and $\mathcal{F}_N^{Z''}$ are independent, then one has $\bar{\mathbf{P}}_u[Z'_{2n} \neq Z''_{2n}] \to 1$ as $n \to -\infty$, so

$$\bar{\mathbf{P}}_u[Z'_0 \neq Z''_0] \geq \prod_{k\leq 0} \Big(1 - \frac{1}{q_k}\Big) > 0.$$

The proof is complete. □

6.5.2 *Embedding Dyadic Filtrations in Brownian Filtrations*

We start with the two filtrations provided by Lemma 6.34. By construction, the filtration $(\mathcal{Z}_n)_{n\leq 0}$ can be generated by some i.i.d. sequence $(\xi_n)_{n\leq 0}$ of uniform random variables with values in $\{-1, 1\}$.

The filtration $(\mathcal{U}_n)_{n\leq 0}$ is product-type and $(a_n)_{n\leq 0}$-adic for some sequence $(a_n)_{n\leq 0}$ taking values 1 and 2 only. Call $D \subset \mathbf{Z}_-$ the set of all $n \leq 0$ such that $a_n = 2$. The filtration $(\mathcal{U}_n)_{n\leq 0}$ can be generated by some sequence $(\alpha_n)_{n\leq 0}$ of independent random variables with α_n uniform on $\{-1, 1\}$ if $n \in D$, $\alpha_n = 0$ if $n \notin D$.

By immersion of $(\mathcal{U}_n)_{n\leq 0}$ in $(\mathcal{Z}_n)_{n\leq 0}$, each α_n is $\mathcal{Z}_n$-measurable and independent of $\mathcal{Z}_{n-1}$. So when $n \in D$, α_n can be written $\alpha_n = H_n\xi_n$, where H_n is some $\mathcal{Z}_{n-1}$-random variable taking values in $\{-1, 1\}$.

Fix an increasing sequence $(t_n)_{n\leq 0}$ of positive real numbers such that $t_0 = 1$ and $t_n \to 0$ as $n \to -\infty$ (e.g. $t_n = 2^n$ for every $n \leq 0$). By symmetry and independence of Brownian increments, one may construct a Brownian motion X such that for every $n \leq 0$, $\xi_n = \text{sign}(X_{t_n} - X_{t_{n-1}})$. Let Y be a Brownian motion, independent of X.

Since $\mathcal{Z}_{n-1} \subset \mathcal{F}^{X,Y}_{t_{n-1}}$ for every $n \leq 0$, one gets a predictable process $(A_t)_{0<t\leq 1}$ with values in $O_2(\mathbf{R})$ and two independent Brownian motions B and C in $\mathcal{F}^{X,Y}$ on the time-interval $[0, 1]$ by setting for every $t \in]t_{n-1}, t_n]$,

$$A_t = \begin{pmatrix} H_n & 0 \\ 0 & 1 \end{pmatrix} \text{ if } n \in D, \quad A_t = \begin{pmatrix} 0 & 1 \\ 1 & 0 \end{pmatrix} \text{ if } n \notin D,$$

and for every $t > 0$,

$$\begin{pmatrix} dB_t \\ dC_t \end{pmatrix} = A_t \begin{pmatrix} dX_t \\ dY_t \end{pmatrix}.$$

Theorem 6.35 *The filtration generated by the Brownian motion B thus defined is complementable after* 0*, maximal, but not complementable in* $\mathcal{F}^{X,Y}$.

Proof Complementability after 0

Let us check that C is a complement after 0 of B, or equivalently that

$$\forall s \in]0, 1], \ \mathcal{F}^{B,C}_1 \vee \mathcal{F}^{X,Y}_s = \mathcal{F}^{X,Y}_1.$$

Since $t_0 = 1$ and $t_m \to 0+$ as $m \to -\infty$, it is sufficient to check the equality when $s = t_m$ with $m \leq 0$. Since for every $n \geq m$, the process A coincides on each time-interval $]t_n, t_{n+1}]$ with an $\mathcal{F}^{X,Y}_{t_n}$-measurable random variable, the formula

$$\begin{pmatrix} dX_t \\ dY_t \end{pmatrix} = A_t^{-1} \begin{pmatrix} dB_t \\ dC_t \end{pmatrix}$$

enables us to recover (X, Y) from the knowledge of $((X_s, Y_s))_{0\leq s\leq t_m}$ and (B, C). The complementability after 0 follows.

Maximality

By Proposition 6.25, the maximality of B will follow from its complementability after 0 once we will have proved the equality

$$\mathcal{F}^B_1 = \bigcap_{s\in]0,1]} (\mathcal{F}^B_1 \vee \mathcal{F}^{X,Y}_s).$$

The intersection above, over all $s \in]0, 1]$, can be restricted to the instants $(t_m)_{m\leq 0}$.

It is now convenient to introduce the notations

$$\Delta X_n = (X_t - X_{t_{n-1}})_{t_{n-1}\leq t\leq t_n} \text{ and } \xi_n \Delta X_n = (\xi_n (X_t - X_{t_{n-1}}))_{t_{n-1}\leq t\leq t_n}.$$

Recall that $\xi_n = \text{sign}(X_{t_n} - X_{t_{n-1}})$. Therefore, $\sigma(\Delta X_n) = \sigma(\xi_n) \vee \sigma(\xi_n \Delta X_n)$, with $\sigma(\xi_n)$ and $\sigma(\xi_n \Delta X_n)$ independent by symmetry of Brownian increments. We define in the same way the random variables ΔY_n, ΔB_n, ΔC_n and $\alpha_n \Delta B_n$. Then

$$\Delta B_n = H_n \Delta X_n = \alpha_n \xi_n \Delta X_n \text{ and } \Delta C_n = \Delta Y_n \text{ if } n \in D,$$
$$\Delta B_n = \Delta Y_n \text{ and } \Delta C_n = \Delta X_n \text{ if } n \in D^c.$$

Moreover, when $n \in D$, $\alpha_n = \text{sign}(B_{t_n} - B_{t_{n-1}})$ is independent of $\alpha_n \Delta B_n = \xi_n \Delta X_n$.

Therefore, $\mathcal{F}_1^B = \mathcal{A} \vee \mathcal{B}$, with

$$\mathcal{A} = \sigma((\alpha_n)_{n \in D}), \quad \mathcal{B} = \sigma((\xi_n \Delta X_n)_{n \in D}) \vee \sigma((\Delta Y_n)_{n \in D^c}).$$

For $n \in \mathbf{Z}_-$, set $D_n = D \cap]-\infty, n]$, $D_n^c =]-\infty, n] \setminus D$, and

$$\mathcal{C}_n = \mathcal{F}_n^\xi, \quad \mathcal{D}_n = \sigma((\xi_k \Delta X_k)_{k \in D_n^c}) \vee \sigma((\Delta Y_k)_{k \in D_n}).$$

Then $\mathcal{F}_n^\xi \vee \mathcal{F}_{t_n}^C = \mathcal{C}_n \vee \mathcal{D}_n$.

The maximality of $\mathcal{F}^\alpha$ in $\mathcal{F}^\xi$ yields the equality

$$\mathcal{A} = \bigcap_{n \leq 0} (\mathcal{A} \vee \mathcal{C}_n) \mod \mathbf{P}.$$

By independence of B and C, the σ-fields $\mathcal{B}$ and $\mathcal{D}_0$ are independent, so Corollary 6.38 applies and the following exchange property holds

$$\mathcal{B} = \mathcal{B} \vee \mathcal{D}_{-\infty} = \bigcap_{n \leq 0} (\mathcal{B} \vee \mathcal{D}_n) \mod \mathbf{P}.$$

The three sequences $(\xi_n)_{n \leq 0}$, $(\xi_n \Delta X_n)_{n \leq 0}$ and $(\Delta Y_n)_{n \leq 0}$ are independent, so the σ-fields $\mathcal{A} \vee \mathcal{C}_0 = \mathcal{F}_0^\xi$ and $\mathcal{B} \vee \mathcal{D}_0 = \mathcal{F}_0^{\xi \Delta X, Y}$ are independent and Lemma 6.37 yields

$$\mathcal{F}_1^B = \mathcal{A} \vee \mathcal{B} = \bigcap_{n \leq 0} (\mathcal{A} \vee \mathcal{B} \vee \mathcal{C}_n \vee \mathcal{D}_n) = \bigcap_{n \leq 0} (\mathcal{F}_1^B \vee \mathcal{F}_n^\xi \vee \mathcal{F}_{t_n}^C)$$
$$= \bigcap_{n \leq 0} (\mathcal{F}_1^B \vee \mathcal{F}_{t_n}^{X,Y}) \mod \mathbf{P}.$$

This proves the maximality of B.

Non-complementability

Keep the notations introduced in the proof of the maximality and set $\xi := (\xi_n)_{n \leq 0}$, $\alpha := (\alpha_n)_{n \leq 0}$. Remind that ξ, $(\xi_n \Delta X_n)_{n \leq 0}$ and $(\Delta Y_n)_{n \leq 0}$ are independent families of independent random variables and that $\mathcal{F}_1^B$ is the σ-field generated by α, $(\xi_n \Delta X_n)_{n \in D}$ and $(\Delta Y_n)_{n \in D^c}$.

The filtration $(\mathcal{F}_{t_n}^{X,Y})_{n\leq 0}$ can be splitted into three independent parts, namely

$$\mathcal{F}_{t_n}^{X,Y} = \mathcal{F}_n^{\xi} \vee \sigma((\xi_k \Delta X_k)_{k\in D_n} \cup (\Delta Y_k)_{k\in D_n^c}) \vee \sigma((\Delta Y_k)_{k\in D_n} \cup (\xi_k \Delta X_k)_{k\in D_n^c}).$$

The second part is a function of B whereas the third part is independent of (ξ, B). By independent enlargement, we get that for $B(\mathbf{P})$-almost every $b \in C([0,1],\mathbf{R})$, the filtration $(\mathcal{F}_n^{\xi})_{n\leq 0}$ is immersed in $(\mathcal{F}_{t_n}^{X,Y})_{n\leq 0}$ under $\mathbf{P}[\cdot|B=b]$.

But α is some measurable function Φ of B and is also a function of ξ. Since ξ, $(\xi_n \Delta X_n)_{n\in D}$ and $(\Delta Y_n)_{n\in D^c}$ are independent, the law of ξ under $\mathbf{P}[\cdot|B=b]$ coincides with the law of ξ under $\mathbf{P}[\cdot|\alpha=\Phi(b)]$.

Since α generates the same σ-field as the random variable U of Lemma 6.34, we derive that for $B(\mathbf{P})$-almost every $b \in C([0,1],\mathbf{R})$, the filtration $(\mathcal{F}_n^{\xi})_{n\leq 0}$ is $(2/a_n)$-adic but not product-type under $\mathbf{P}[\cdot|B=b]$. But this filtration is immersed in $(\mathcal{F}_{t_n}^{X,Y})_{n\leq 0}$ under $\mathbf{P}[\cdot|B=b]$, hence by Vershik's theorem (Theorem 6.5 in the present paper), $(\mathcal{F}_{t_n}^{X,Y})_{n\leq 0}$ cannot be product-type so $\mathcal{F}^{X,Y}$ cannot be Brownian under $\mathbf{P}[\cdot|B=b]$. Thus, the Brownian filtration $\mathcal{F}^B$ is not complementable in $\mathcal{F}^{X,Y}$. □

Actually, the construction above also yields the remarkable counter-example below.

Proposition 6.36 *Keep the notation of Theorem 6.35 and set* $\beta_n = \mathrm{sign}(B_{t_n} - B_{t_{n-1}})$, $\eta_n = \mathrm{sign}(Y_{t_n} - Y_{t_{n-1}})$ *for every* $n \leq 0$, *so* $\beta_n = \alpha_n$ *if* $n \in D$ *whereas* $\beta_n = \eta_n$ *if* $n \notin D$. *Then the dyadic filtration* $\mathcal{F}^{\beta}$ *is maximal but non-complementable in the product-type quadriadic filtration* $\mathcal{F}^{\xi,\eta}$.

Proof For $n \in \mathbf{Z}_-$, set $D_n = D \cap]-\infty, n]$ and

$$\mathcal{A} = \sigma((\alpha_n)_{n\in D}), \quad \mathcal{B}' = \sigma((\eta_n)_{n\in D^c}), \quad C_n = \mathcal{F}_n^{\xi}, \quad \mathcal{D}'_n = \sigma((\eta_k)_{k\in D_n}).$$

The maximality of $\mathcal{F}^{\alpha}$ in $\mathcal{F}^{\xi}$ yields the equality

$$\mathcal{A} = \bigcap_{n\leq 0} (\mathcal{A} \vee C_n) \mod \mathbf{P}.$$

By Corollary 6.38, the independence of $\mathcal{B}'$ and $\mathcal{D}'_0$ yields

$$\mathcal{B}' = \mathcal{B}' \vee \mathcal{D}'_{-\infty} = \bigcap_{n\leq 0} (\mathcal{B}' \vee \mathcal{D}'_n) \mod \mathbf{P}.$$

But $\mathcal{A} \vee C_0 = \mathcal{F}_0^{\xi}$ and $\mathcal{B}' \vee \mathcal{D}'_0 = \mathcal{F}_0^{\eta}$ are independent. Hence, Lemma 6.37 yields

$$\mathcal{F}_0^{\beta} = \mathcal{A} \vee \mathcal{B}' = \bigcap_{n\leq 0} (\mathcal{A} \vee \mathcal{B}' \vee C_n \vee \mathcal{D}'_n) = \bigcap_{n\leq 0} (\mathcal{F}_0^{\beta} \vee \mathcal{F}_n^{\xi,\eta}) \mod \mathbf{P}.$$

Hence at time 0, the filtration $\mathcal{F}^\beta$ coincides modulo $\mathbf{P}$ with the larger filtration $(\mathcal{F}^\beta)'$ provided by Proposition 6.18 when the filtration $\mathcal{Z}$ is $\mathcal{F}^{\xi,\eta}$. Since $\mathcal{F}^\beta$ is immersed in $(\mathcal{F}^\beta)'$, these two filtrations coincide modulo $\mathbf{P}$ at every time. The maximality of $\mathcal{F}^\beta$ follows.

Conditionally on α, the filtration $\mathcal{F}^\xi$ is not product-type. By independent enlargements, conditionally on β, the filtration $\mathcal{F}^\xi$ is not product-type, so $\mathcal{F}^{\xi,\eta}$ is not product-type since $\mathcal{F}^\xi$ is immersed in $\mathcal{F}^{\xi,\eta}$. The non-complementability of $\mathcal{F}^\beta$ follows. □

Lemma 6.37 *Let $\mathcal{A}$, $\mathcal{B}$ be two sub-σ-fields and $(C_n)_{n\leq 0}$, $(\mathcal{D}_n)_{n\leq 0}$ be two filtrations of the probability space $(\Omega, \mathcal{T}, \mathbf{P})$. If*

$$\mathcal{A} = \bigcap_{n\leq 0}(\mathcal{A} \vee C_n) \mod \mathbf{P}, \quad \mathcal{B} = \bigcap_{n\leq 0}(\mathcal{B} \vee \mathcal{D}_n) \mod \mathbf{P},$$

and if $\mathcal{A} \vee C_0$ and $\mathcal{B} \vee \mathcal{D}_0$ are independent, then

$$\mathcal{A} \vee \mathcal{B} = \bigcap_{n\leq 0}(\mathcal{A} \vee \mathcal{B} \vee C_n \vee \mathcal{D}_n) \mod \mathbf{P}.$$

Proof Both sides of the equality to be proved are sub-σ-fields of $\mathcal{A} \vee \mathcal{B} \vee C_0 \vee \mathcal{D}_0$, so it is sufficient to prove that for every $Z \in L^1(\mathcal{A} \vee C_0 \vee \mathcal{B} \vee \mathcal{D}_0)$,

$$\mathbf{E}[Z|\mathcal{A} \vee \mathcal{B}] = \mathbf{E}\Big[Z\Big|\bigcap_{n\leq 0}(\mathcal{A} \vee \mathcal{B} \vee C_n \vee \mathcal{D}_n)\Big].$$

One may assume that $Z = XY$ with $X \in L^1(\mathcal{A} \vee C_0)$ and $Y \in L^1(\mathcal{B} \vee \mathcal{D}_0)$ since such random variables span a dense subspace in $L^1(\mathcal{A} \vee C_0 \vee \mathcal{B} \vee \mathcal{D}_0)$. Given Z as above, one has $\mathbf{E}[Z|\mathcal{A} \vee C_0 \vee \mathcal{B}] = X\mathbf{E}[Y|\mathcal{B}]$, so

$$\mathbf{E}[Z|\mathcal{A} \vee \mathcal{B}] = \mathbf{E}[X|\mathcal{A} \vee \mathcal{B}]\mathbf{E}[Y|\mathcal{B}] = \mathbf{E}[X|\mathcal{A}]\mathbf{E}[Y|\mathcal{B}],$$

since $\sigma(X) \vee \mathcal{A}$ is independent of $\mathcal{B}$. In the same way, one gets that for every $n \leq 0$,

$$\mathbf{E}[Z|\mathcal{A} \vee \mathcal{B} \vee C_n \vee \mathcal{D}_n] = \mathbf{E}[X|\mathcal{A} \vee C_n]\mathbf{E}[Y|\mathcal{B} \vee \mathcal{D}_n].$$

Thus, taking the limit as $n \to -\infty$ yields the result by the martingale convergence theorem and the assumption. □

The particular case where $\mathcal{B}$ and the σ-fields C_n are equal to $\{\emptyset, \Omega\}$ yields the following classical and useful result.

Corollary 6.38 *Let $\mathcal{A}$ be a sub-σ-field and $(\mathcal{D}_n)_{n\leq 0}$ be a filtration of the probability space $(\Omega, \mathcal{T}, \mathbf{P})$. If $\mathcal{A}$ and $\mathcal{D}_0$ are independent, then*

$$\mathcal{A} = \bigcap_{n\leq 0}(\mathcal{A} \vee \mathcal{D}_n) \mod \mathbf{P}.$$

6.6 A Complementable Factor Arising From a Non-complementable Filtration

In this section, we study an example deriving from a variant of Vershik's example 3 in [22], namely the uniform randomised decimation process.

6.6.1 *Definition of a Uniform Randomised Decimation Process*

We denote by $\{a, b\}^\infty$ the set of all infinite words on the alphabet $\{a, b\}$, namely the set of all maps from $\mathbf{N} = \{1, 2, \dots\}$ to $\{a, b\}$. We endow this set with the uniform probability measure μ: a random infinite word X is chosen according to μ if the successive letters $X(1), X(2), \dots$ form a sequence of independent and uniform random variables taking values in $\{a, b\}$.

We denote by $\mathcal{P}(\mathbf{N})$ the power set of $\mathbf{N}$, i.e. the set of all subsets of $\mathbf{N}$. Given $p \in]0, 1[$, we define the probability measure ν_p on $\mathcal{P}(\mathbf{N})$ as follows: the law of a random subset I of $\mathbf{N}$ is ν_p if $\mathbf{1}_I(1), \mathbf{1}_I(2), \dots$ form an i.i.d. sequence of Bernoulli random variables with parameter p. Equivalently, this means that $\mathbf{P}[F \subset I] = p^{|F|}$ for every finite subset F of $\mathbf{N}$. In this case, we note that almost surely, I is infinite with infinite complement. The law $\nu := \nu_{1/2}$ will be called the uniform law on $\mathcal{P}(\mathbf{N})$.

When A is an infinite subset of $\mathbf{N}$, we denote by $\psi_A(1) < \psi_A(2) < \dots$ its elements. This defines an increasing map ψ_A from $\mathbf{N}$ to $\mathbf{N}$ whose range is A. Conversely, for every increasing map f from $\mathbf{N}$ to $\mathbf{N}$, there is a unique infinite subset A of $\mathbf{N}$, namely the range of f, such that $f = \psi_A$. These remarks lead to the following statement.

Lemma 6.39 *Let I and J be independent random infinite subsets of $\mathbf{N}$ with respective laws ν_p and ν_q, and $R = \psi_I \circ \psi_J(\mathbf{N}) = \psi_I(J)$ be the range of $\psi_I \circ \psi_J$. Then $\psi_I \circ \psi_J = \psi_R$ and the law of R is ν_{pq}.*

Proof The equality $\psi_I \circ \psi_J = \psi_R$ follows from the remarks above. Let F be a finite subset of $\mathbf{N}$. By injectivity of ψ_I,

$$[F \subset R] = [F \subset I\,;\ \psi_I^{-1}(F) \subset J]$$

and $[F \subset I] = [|\psi_I^{-1}(F)| = |F|]$, therefore by independence of I and J,

$$\mathbf{P}[F \subset R \mid \sigma(I)] = \mathbf{1}_{[F \subset I]}\, \mathbf{P}[\psi_I^{-1}(F) \subset J \mid \sigma(I)] = \mathbf{1}_{[F \subset I]}\, q^{|\psi_I^{-1}(F)|} = \mathbf{1}_{[F \subset I]}\, q^{|F|}.$$

Thus $\mathbf{P}[F \subset R] = \mathbf{P}[F \subset I] q^{|F|} = (pq)^{|F|}$. □

Here is another property that we will use to define the uniform randomised decimation process on $\{a, b\}$, and also later, in the proof of Proposition 6.49.

Lemma 6.40 *Let X be a uniform random word on $\{a,b\}^\infty$. Let I be a random subset of* $\mathbf{N}$ *with law ν_p, independent of X. Then*

- *I, $X \circ \psi_I$, $X \circ \psi_{I^c}$ are independent*
- *$X \circ \psi_I$, $X \circ \psi_{I^c}$ are uniform random words on $\{a,b\}^\infty$.*

Proof Almost surely, I is infinite with infinite complement, so the random maps ψ_I and ψ_{I^c} are well-defined. The integers $\psi_I(1), \psi_{I^c}(1), \psi_I(2), \psi_{I^c}(2)\dots$ are distinct, so conditionally on I, $X(\psi_I(1))$, $X(\psi_{I^c}(1))$, $X(\psi_I(2))$, $X(\psi_{I^c}(2)), \dots$ are independent and uniform on $\{a,b\}$. The result follows. □

Definition 6.41 Call $\mathcal{P}'(\mathbf{N})$ the set of all infinite subsets of $\mathbf{N}$. A uniform randomised decimation process in the alphabet $\{a,b\}$ is a stationary Markow chain $(X_n, I_n)_{n\in\mathbf{Z}}$ with values in $\{a,b\}^\infty \times \mathcal{P}'(\mathbf{N})$ defined as follows: for every $n \in \mathbf{Z}$,

1. the law of (X_n, I_n) is $\mu \otimes \nu$;
2. I_n is independent of (X_{n-1}, I_{n-1}) and uniform on $\mathcal{P}(\mathbf{N})$;
3. $X_n = X_{n-1} \circ \psi_{I_n}$.

Such a process is well-defined and unique in law since the law $\mu \otimes \nu$ is invariant by the transition kernel given by conditions 2 and 3 above, thanks to Lemma 6.40. Moreover, $(I_n)_{n\in\mathbf{Z}}$ is a sequence of innovations for the filtration $\mathcal{F}^{X,I}$. Therefore, the filtration $\mathcal{F}^{X,I}$ has independent increments or is locally of product-type, according to Laurent's terminology [12].

This process is a kind of randomisation of Vershik's decimation process given in example 3 of [22]. Indeed, Vershik's decimation process is equivalent to the process that we would get by choosing the random sets I_n uniformly among the set of all even positive integers and the set of all odd positive integers. Although Vershik's decimation process generates a non-standard filtration, we will show that our randomised process generates a standard one.

Theorem 6.42 *The uniform randomised decimation process on the alphabet $\{a,b\}$ generates a product-type filtration.*

6.6.2 *Proof of Theorem 6.42*

We have seen that the filtration $\mathcal{F}^{X,I}$ admits $(I_n)_{n\in\mathbf{Z}}$ as a sequence of innovations. Each innovation has diffuse law. Therefore, to prove that the filtration $(\mathcal{F}_n^{X,I})_{n\le 0}$, or equivalently, the filtration $(\mathcal{F}_n^{X,I})_{n\in\mathbf{Z}}$ is product-type, it suffices to check Vershik's first level criterion (see reminders further and definition 2.6 and theorem 2.25 in [12]). Concretely, we have to check any random variable in $L^1(\mathcal{F}_0^{X,I}, \mathbf{R})$ can be approached in $L^1(\mathcal{F}_0^{X,I}, \mathbf{R})$ by measurable functions of finitely many innovations of $(\mathcal{F}^{X,I})_{n\le 0}$.

The innovations $(I_n)_{n\in\mathbf{Z}}$ are inadequate to do this, since the random variable X_0 is independent of the whole sequence $(I_n)_{n\in\mathbf{Z}}$, so functions of the $(I_n)_{n\in\mathbf{Z}}$ cannot approach non-trivial functions of X_0. Therefore, we will have to construct new innovations. The next lemma gives us a general procedure to get some.

Lemma 6.43 *Fix $n \in \mathbf{Z}$. Let Φ be some $\mathcal{F}^{X,I}_{n-1}$-measurable map from $\mathbf{N}$ to $\mathbf{N}$. If Φ is almost surely bijective, then the random variable $J_n = \Phi(I_n)$ is independent of $\mathcal{F}^{X,I}_{n-1}$ and uniform on $\mathcal{P}(\mathbf{N})$.*

Proof Let F be a finite subset of $\mathbf{N}$. Then, almost surely,

$$P[F \subset J_n | \mathcal{F}^{X,I}_{n-1}] = P[\Phi^{-1}(F) \subset I_n | \mathcal{F}^{X,I}_{n-1}] = (1/2)^{|\Phi^{-1}(F)|} = (1/2)^{|F|}.$$

The result follows. □

Actually, the proof of Theorem 6.42 is similar to the proof of the standardness of the erased-words filtration by Laurent [13] and uses the same tools, namely canonical coupling and cascaded permutations.

Definition 6.44 (Canonical Word and Canonical Coupling) The infinite canonical word C on the alphabet $\{a, b\}$ is $abab\cdots$, namely the map from $\mathbf{N}$ to $\{a, b\}$ which sends the odd integers on a and the even integers on b.

If x is an infinite word x on the alphabet $\{a, b\}$, namely a map from $\mathbf{N}$ to $\{a, b\}$, we set for every $i \in \mathbf{N}$,

$$\phi_x(i) = 2q - 1 \text{ if } x(i) \text{ is the } q\text{-th occurence of the letter } a \text{ in } x,$$

$$\phi_x(i) = 2q \text{ if } x(i) \text{ is the } q\text{-th occurence of the letter } b \text{ in } x.$$

Lemma 6.45 *By definition, the map ϕ_x thus defined from $\mathbf{N}$ to $\mathbf{N}$ is injective and satisfies the equality $x = C \circ \phi_x$. When each possible letter a or b appears infinitely many times in x, ϕ_x is a permutation of $\mathbf{N}$, (called canonical coupling by S. Laurent).*

Roughly speaking, if x is a typical word of $\{a, b\}^\infty$ endowed with the uniform law, the asymptotic proportions of a and b are $1/2$ are $1/2$, so ϕ_x is asymptotically close to the identity map.

Definition 6.46 (New Innovations and Cascaded Permutations) Let Ω' be the almost sure event on which

- each possible letter a or b appears infinitely many times in the infinite word X_0;
- each subset I_n is infinite.

On Ω', we define by recursion a sequence $(\Phi_n)_{n \geq 0}$ of random permutations of $\mathbf{N}$ and a sequence $(J_n)_{n \geq 1}$ of random infinite subsets of $\mathbf{N}$ by setting $\Phi_0 = \phi_{X_0}$ and, for every $n \geq 1$,

$$J_n = \Phi_{n-1}(I_n) \text{ and } \Phi_{n-1} \circ \psi_{I_n} = \psi_{J_n} \circ \Phi_n. \tag{6.1}$$

Let us check that the inductive construction above actually works Ω'.

On Ω', the map $\Phi_0 = \phi_{X_0}$ is bijective by Lemma 6.45.

Once we know that Φ_{n-1} is a random permutations of $\mathbf{N}$, the map $\Phi_{n-1} \circ \psi_{I_n}$ is a random injective map from $\mathbf{N}$ to $\mathbf{N}$ with range $\Phi_{n-1}(I_n) = J_n$. Therefore, J_n is infinite and the map Φ_n is well defined by Eq. (6.1): for every $k \in \mathbf{N}$, $\Phi_n(k)$ is the rank of the integer $\Phi_{n-1}(\psi_{I_n}(k))$ in the set J_n. Moreover, Φ_n is a permutation of $\mathbf{N}$.

Informally, the *cascaded permutations* $(\Phi_n)_{n \geq 0}$ are induced by $\Phi_0 = \phi_{X_0}$ and the successive extractions. More precisely, Eq. 6.1 is represented by a commutative diagramm which gives the correspondence between the positions of a same letter in different words.

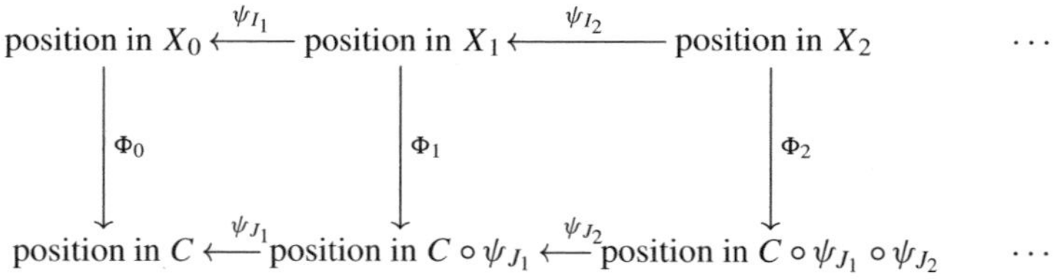

Here is a realisation of the first three steps. The boldface numbers form the subsets $I_1, J_1, I_2, J_2, \ldots$. Among the arrows representing ϕ_{X_0}, the plain arrows (from elements in I_1 to elements in J_1) provide the permutation ϕ_{X_0, I_1} by renumbering of the elements.

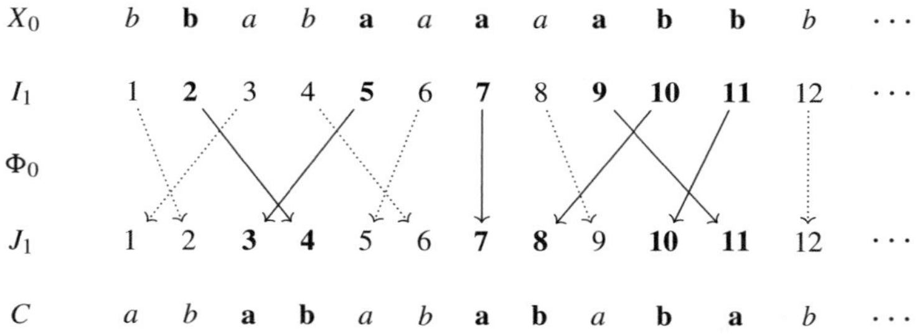

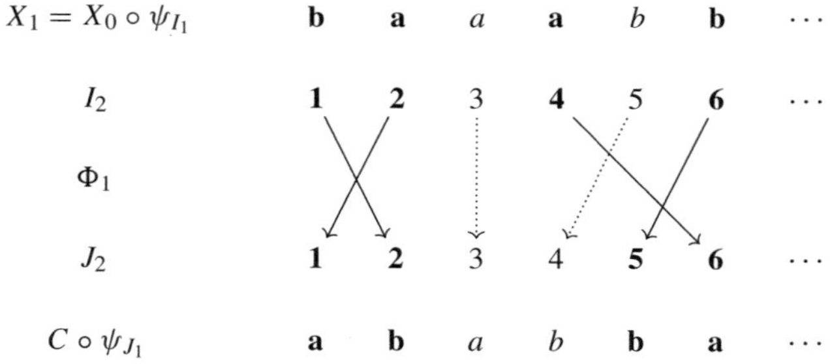

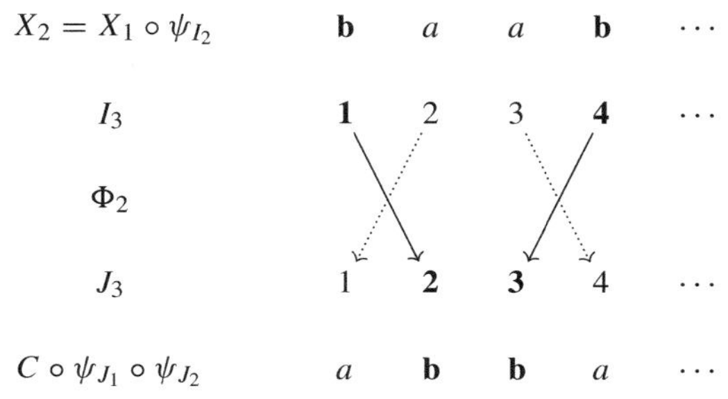

Lemma 6.47 *On the almost sure event Ω', the following properties hold for every $n \geq 1$,*

1. *J_n is independent of $\mathcal{F}_{n-1}^{X,I}$ and is uniform on $\mathcal{P}(\mathbf{N})$.*
2. *$\sigma(X_0, J_1, \dots, J_n) = \sigma(X_0, I_1, \dots, I_n)$.*
3. *the random map Φ_n is $\sigma(X_0, J_1, \dots, J_n)$-measurable.*
4. *$\phi_{X_0} \circ \psi_{I_1} \circ \cdots \circ \psi_{I_n} = \psi_{J_1} \circ \cdots \circ \psi_{J_n} \circ \Phi_n$.*
5. *$X_n = C \circ \psi_{J_1} \circ \cdots \circ \psi_{J_n} \circ \Phi_n$.*

Proof Since $\Phi_0 = \phi_{X_0}$, Properties 2, 3, 4, 5 above hold with n replaced by 0.

Let $n \geq 1$. Assume that Properties 2, 3, 4, 5 hold with n replaced by $n-1$.

Then by Lemma 6.43, Property 1 holds.

By definition and by the induction hypothesis, the random set $J_n = \Phi_{n-1}(I_n)$ is $\sigma(X_0, I_1, \dots, I_n)$-measurable. Conversely, since $I_n = \Phi_{n-1}^{-1}(J_n)$, the knowledge of Φ_{n-1} and J_n is sufficient to recover I_n, so Property 2 holds.

For every $k \in \mathbf{N}$, $\Phi_n(k)$ is the rank of the integer $\Phi_{n-1}(\psi_{I_n}(k))$ in the set $\Phi_{n-1}(I_n)$. Thus the random map Φ_n is a measurable for $\sigma(X_0, J_1, \dots, J_{n-1}, I_n) = \sigma(X_0, J_1, \dots, J_n)$. Therefore, Property 3 holds.

By induction hypothesis and by formula 6.1,

$$
\begin{aligned}
\phi_{X_0} \circ \psi_{I_1} \circ \cdots \circ \psi_{I_n} &= (\phi_{X_0} \circ \psi_{I_1} \circ \cdots \circ \psi_{I_{n-1}}) \circ \psi_{I_n} \\
&= (\psi_{J_1} \circ \cdots \circ \psi_{J_{n-1}} \circ \Phi_{n-1}) \circ \psi_{I_n} \\
&= (\psi_{J_1} \circ \cdots \circ \psi_{J_{n-1}}) \circ (\Phi_{n-1} \circ \psi_{I_n}) \\
&= (\psi_{J_1} \circ \cdots \circ \psi_{J_{n-1}}) \circ (\psi_{J_n} \circ \Phi_n) \\
&= \psi_{J_1} \circ \cdots \circ \psi_{J_n} \circ \Phi_n,
\end{aligned}
$$

so

$$
\begin{aligned}
X_n &= X_0 \circ \psi_{I_1} \circ \cdots \circ \psi_{I_n} \\
&= C \circ \phi_{X_0} \circ \psi_{I_1} \circ \cdots \circ \psi_{I_n} \\
&= C \circ \psi_{J_1} \circ \cdots \circ \psi_{J_n} \circ \Phi_n,
\end{aligned}
$$

which yields Properties 4 and 5.

Lemma 6.46 follows by recursion. □

The next result shows that the innovations $(J_n)_{n\geq 1}$ constructed above provide better and better approximations of X_n as $n \to +\infty$.

Lemma 6.48 *Fix $L \in \mathbf{N}$. Then $\mathbf{P}\big[X_n = C \circ \psi_{J_1} \circ \cdots \circ \psi_{J_n}$ on $[\![1, L]\!]\big] \to 1$ as $n \to +\infty$.*

Proof By equality 5, it suffices to check that, $\mathbf{P}(E_n) \to 1$ as $n \to +\infty$, where E_n is the event "Φ_n coincides on $[\![1, L]\!]$ with the identity map".

By Lemma 6.39, $\psi_{I_1} \circ \cdots \circ \psi_{I_n} = \psi_{A_n}$ and $\psi_{J_1} \circ \cdots \circ \psi_{J_n} = \psi_{B_n}$, where A_n and B_n are random subsets of $\mathbf{N}$ with law ν_{p_n}, where $p_n = 2^{-n}$.

Therefore, by Property 4 of Lemma 6.47, $\phi_{X_0} \circ \psi_{A_n} = \psi_{B_n} \circ \Phi_n$, so for each $k \in \mathbf{N}$, $\Phi_n(k)$ is the rank of the integer $\phi_{X_0}(\psi_{A_n}(k))$ in the set $\phi_{X_0}(A_n) = B_n$.

Thus, the event E_n holds if and only if the L first elements of the set $\phi_{X_0}(A_n)$ in increasing order are $\phi_{X_0}(\psi_{A_n}(1)), \ldots, \phi_{X_0}(\psi_{A_n}(L))$.

Set $\tau_{n,k} = \psi_{A_n}(k)$ for every $k \in \mathbf{N}$. Since the law of A_n is ν_{p_n}, the random variables $\tau_{n,1}, \tau_{n,2} - \tau_{n,1}, \tau_{n,3} - \tau_{n,2}, \ldots$ are independent and geometric with parameter p_n.

We have noted that

$$E_n = \big[\forall k \geq L+1,\ \phi_{X_0}(\tau_{n,1}) < \ldots < \phi_{X_0}(\tau_{n,L}) < \phi_{X_0}(\tau_{n,k})\big]$$

Roughly speaking, the probability of this event tends to 1 because ϕ_{X_0} is close to the identity map and the set A_n gets sparser and sparser as n increases to infinity. Let us formalize this argument.

Since X_0 is uniform on $\{a, b\}^\infty$, the random variables $(\eta_i)_{i\geq 1} = (\mathbf{1}_{[X_0(i)=b]})_{i\geq 1}$ form an i.i.d. sequence of Bernoulli random variables with parameter $1/2$. For every $t \in \mathbf{N}$, the random variable $S_t = \eta_1 + \cdots + \eta_t$ counts the number of b in the subword $X_0([\![1, t]\!])$, whereas $t - S_t$ counts the number of a in the subword $X_0([\![1, t]\!])$, so by definition of ϕ_{X_0},

$$\phi_{X_0}(t) = 2(t - S_t) - 1 \text{ if } X_0(t) = a,$$
$$\phi_{X_0}(t) = 2S_t \text{ if } X_0(t) = b.$$

Given $t_1 < t_2$ in $\mathbf{N}$, the inequality $\max(S_{t_1}, t_1 - S_{t_1}) < \min(S_{t_2}, t_2 - S_{t_2})$ implies $\phi_{X_0}(t_1) < \phi_{X_0}(t)$ for every integer $t \geq t_2$. Therefore,

$$E_n \supset \big[\forall k \in [\![1, L]\!],\ \max(S_{\tau_{n,k}}, \tau_{n,k} - S_{\tau_{n,k}}) < \min(S_{\tau_{n,k+1}}, \tau_{n,k+1} - S_{\tau_{n,k+1}})\big].$$

Thus it suffices to prove that for any fixed $k \in \mathbf{N}$,

$$p_{n,k} := \mathbf{P}\big[\max(S_{\tau_{n,k}}, \tau_{n,k} - S_{\tau_{n,k}}) \geq \min(S_{\tau_{n,k+1}}, \tau_{n,k+1} - S_{\tau_{n,k+1}})\big] \to 0 \text{ as } n \to +\infty.$$

Since X_0 is independent of $I_1, \ldots, I_n$, the sequence $(S_t)_{t\geq 1}$ is independent of the sequence $(\tau_{n,k})_{k\geq 1}$. Moreover, $(S_t)_{t\geq 1}$ has the same law as $(t - S_t)_{t\geq 1}$ and $S_{\tau_{n,k+1}} - S_{\tau_{n,k}}$ has the same law as $S_{\tau_{n,1}}$. Therefore, for every integer $x \geq 1$,

$$\begin{aligned}
p_{n,k} &\leq 2\mathbf{P}\big[S_{\tau_{n,k+1}} \leq \max(S_{\tau_{n,k}}, \tau_{n,k} - S_{\tau_{n,k}})\big] \\
&= 2\mathbf{P}\big[S_{\tau_{n,k+1}} - S_{\tau_{n,k}} \leq \max(0, \tau_{n,k} - 2S_{\tau_{n,k}})\big] \\
&\leq 2\mathbf{P}\big[S_{\tau_{n,k+1}} - S_{\tau_{n,k}} \leq x - 1\big] + 2\mathbf{P}\big[\tau_{n,k} - 2S_{\tau_{n,k}} \geq x\big] \\
&= 2\mathbf{P}\big[S_{\tau_{n,1}} \leq x - 1\big] + \mathbf{P}\big[|2S_{\tau_{n,k}} - \tau_{n,k}| \geq x\big].
\end{aligned}$$

On the one hand, the random variable $S_{\tau_{n,1}}$ is binomial with parameters $\tau_{n,1}$ and $1/2$ conditionally on $\tau_{n,1}$, so its generating function is given by

$$\begin{aligned}
\mathbf{E}[z^{S_{\tau_{n,1}}}] &= \mathbf{E}\big[\mathbf{E}[z^{S_{\tau_{n,1}}} |\sigma(\tau_{n,1})]\big] \\
&= \mathbf{E}\Big[\Big(\frac{1+z}{2}\Big)^{\tau_{n,1}}\Big] \\
&= \frac{p_n(1+z)/2}{1-(1-p_n)(1+z)/2} \\
&= \frac{p_n(1+z)}{1+p_n-(1-p_n)z} \\
&= \frac{p_n(1+z)}{1+p_n}\sum_{m=0}^{+\infty}\Big(\frac{1-p_n}{1+p_n}\Big)^m z^m
\end{aligned}$$

This yields the law of $S_{\tau_{n,1}}$, namely

$$P[S_{\tau_{n,1}} = 0] = \frac{p_n}{1+p_n},$$

$$P[S_{\tau_{n,1}} = s] = \frac{p_n}{1+p_n}\Big(\Big(\frac{1-p_n}{1+p_n}\Big)^{s-1} + \Big(\frac{1-p_n}{1+p_n}\Big)^{s}\Big) = 2p_n\frac{(1-p_n)^{s-1}}{(1+p_n)^{s+1}} \text{ if } s \geq 1.$$

Therefore, $\mathbf{P}[S_{\tau_{n,1}} = s] \leq 2p_n$ for every $s \geq 0$, so $\mathbf{P}\big[S_{\tau_{n,1}} \leq x - 1\big] \leq 2p_n x$.

On the other hand, $(2S_t - t)_{t\geq 0}$ is a simple symmetric random walk on $\mathbf{Z}$, independent of $\tau_{n,k}$ so

$$\mathbf{E}[2S_{\tau_{n,k}} - \tau_{n,k}|\sigma(\tau_{n,k})] = 0 \text{ and } \mathrm{Var}(2S_{\tau_{n,k}} - \tau_{n,k}|\sigma(\tau_{n,k})) = \tau_{n,k}.$$

Therefore,

$$\mathbf{E}[2S_{\tau_{n,k}} - \tau_{n,k}] = 0 \text{ and } \mathrm{Var}(2S_{\tau_{n,k}} - \tau_{n,k}) = \mathrm{Var}(0) + \mathbf{E}[\tau_{n,k}] = k/p_n,$$

so Bienaymé-Tchebicheff's inequality yields $\mathbf{P}\big[|2S_{\tau_{n,k}} - \tau_{n,k}| \geq x\big] \leq (k/p_n)x^{-2}$.

Hence, for every n and x in $\mathbf{N}$, $p_{n,k} \leq 4p_n x + (k/p_n)x^{-2}$. Choosing $x = \lceil p_n^{-2/3} \rceil$ yields $p_{n,k} \leq 4(p_n^{1/3} + p_n) + kp_n^{1/3}$. The result follows. □

To finish the proof of Theorem 6.42, we need to remind some standard facts about Vershik's first level criterion, namely definition 2.6, proposition 2.7 and proposition 2.17 of [12].

Let $\mathcal{F} = (\mathcal{F}_n)_{n \leq 0}$ be a filtration with independent increments (Laurent writes that $\mathcal{F}$ is locally of product-type). Given a separable metric space (E, d), one says that a random variable $R \in L^1(\mathcal{F}_0, E)$ satisfies Vershik's first level criterion (with respect to $\mathcal{F}$) if for every $\delta > 0$, one can find an integer $n_0 \leq 0$, some innovations $V_{n_0+1}, \ldots, V_0$ of $\mathcal{F}$ at times $n_0 + 1, \ldots, 0$ and some random variable $S \in L^1(\sigma(V_{n_0+1}, \ldots, V_0), E)$ such that $\mathbf{E}[d(R, S)] < \delta$.

The subset of random variables in $L^1(\mathcal{F}_0, E)$ satisfying Vershik's first level criterion (with respect to $\mathcal{F}$) is closed in $L^1(\mathcal{F}_0, E)$. If $R \in L^1(\mathcal{F}_0, E)$ satisfies Vershik's first level criterion, then any measurable real function of R also satisfies Vershik's first level criterion.

The first step of the proof is to check that for every $m \leq 0$, the random variable $(X_m(1), \ldots, X_m(L))$, taking values in $\{a, b\}^L$ endowed with the discrete metric, satisfies Vershik's first level criterion with respect to $(\mathcal{F}_n^{X,I})_{n \leq 0}$. Indeed, by stationarity, the construction of Lemma 6.46 can be started at any time n_0 instead of time 0. Starting this construction at time n_0 yields innovations $J_{n_0+1}^{n_0}, J_{n_0+2}^{n_0}, \ldots$ at times $n_0 + 1, n_0 + 2, \ldots$. Fix two integers $m \leq 0$ and $L \geq 1$. For every $n_0 \leq m$, the event

$$\big[X_m = C \circ \psi_{J_{n_0+1}^{n_0}} \circ \cdots \circ \psi_{J_m^{n_0}} \text{ on } [\![1, L]\!]\big]$$

has (by stationarity) the same probability as

$$\big[X_{m-n_0} = C \circ \psi_{J_1} \circ \cdots \circ \psi_{J_{m-n_0}} \text{ on } [\![1, L]\!]\big].$$

Lemma 6.48 ensures that this probability tends to 1 as $n \to +\infty$.

We derive successively that the following random variables also satisfies Vershik's first level criterion:

- X_m, valued in $\{a, b\}^\infty$ endowed with the metric given by

$$d(x, y) = 2^{-\inf\{i \geq 1 : x(i) \neq y(i)\}}.$$

- $(X_m, I_{m+1}, \ldots, I_0)$, valued in $\{a, b\}^\infty \times \mathcal{P}(\mathbf{N})^{|m|}$ endowed with the product of the metrics defined as above on each factor $\{a, b\}^\infty$ or $\mathcal{P}(\mathbf{N})$ identified with $\{0, 1\}^\infty$;
- any measurable real function of $(X_m, I_{m+1}, \ldots, I_0)$;
- any real random variable in $\mathcal{F}_0^{X,I}$.

The proof is complete.

6.6.3 A Non Complementable Filtration Yielding a Complementable Factor

We still work with the filtration generated by the uniform randomised decimation process $((X_n, I_n))_{n\in\mathbf{Z}}$ on the alphabet $\{a, b\}$. We call $\mathcal{P}''(\mathbf{N})$ the set of all infinite subsets of $\mathbf{N}$ with infinite complement, and we set $E = \{a, b\}^\infty \times \mathcal{P}''(\mathbf{N})$. Since $\nu(\mathcal{P}''(\mathbf{N})) = 1$, we may assume and we do assume that the Markov chain $((X_n, I_n))_{n\in\mathbf{Z}}$ takes values in E.

At each time n, we define the random variable $Y_n = \psi_{I_n^c}(X_{n-1})$ coding the portion of the infinite word X_{n-1} rejected at time n to get the word X_n. Of course, the knowledge of I_n, X_n and Y_n enables us to recover X_{n-1}: for every $i \in \mathbf{N}$, $X_{n-1}(i)$ equals $X_n(r)$ or $Y_n(r)$ according that i is the rth element of I_n or of I_n^c. We can say more.

Proposition 6.49 (Properties of the Sequences $(Y_n)_{n\in\mathbf{Z}}$ and $(I_n)_{n\in\mathbf{Z}}$)

1. *The random variables Y_n are independent and uniform on $\{a, b\}^\infty$.*
2. *The sequence $(Y_n)_{n\in\mathbf{Z}}$ is independent of the sequence $(I_n)_{n\in\mathbf{Z}}$.*
3. *Each X_n is almost surely a measurable function of $I_{n+1}, Y_{n+1}, I_{n+2}, Y_{n+2}, \ldots$.*

Note that Proposition 6.49 provides a constructive method to get a uniform randomised decimation process on $\{a, b\}$.

Proof The first two statements follow from a repeated application of Lemma 6.40. Since the formulas involving the processes I, X, Y are invariant by time-translations, one needs only to check the third statement when $n = 0$. For every $i \in \mathbf{N}$, call

$$N_i = \inf\{n \geq 1 : i \notin \psi_{I_1} \circ \cdots \circ \psi_{I_n}(\mathbf{N})\}.$$

the first time n at which the letter $X_0(i)$ is rejected when forming the word X_n. For every $n \geq 0$, $[N_i > n] = [i \in \psi_{I_1} \circ \cdots \circ \psi_{I_n}(\mathbf{N})]$; but by Lemma 6.39, the law of the range of $\psi_{I_1} \circ \cdots \circ \psi_{I_n}$ is $\nu_{2^{-n}}$, so $\mathbf{P}[N_i > n] = 2^{-n}$. Therefore, N_i is a measurable function of $(I_n)_{n\geq 1}$ and is almost surely finite. On the event $[N_i < +\infty]$, $X_0(i) = Y_{N_i}(R_i)$, where R_i is the rank of i in the set $\psi_{I_1} \circ \cdots \circ \psi_{I_{N_i}} \circ \psi_{I^c_{N_i+1}}(\mathbf{N})$. The proof is complete. □

We split each random variable I_n into two independent random variables, namely $U_n = \{I_n, I_n^c\}$ and $V_n = \mathbf{1}_{[1\in I_n]}$. The random variable U_n takes values in the set Π of all partitions of $\mathbf{N}$ into two infinite blocks. Given such a partition $u \in \Pi$, we denote by $u(1)$ the block containing 1 and by $u(0)$ its complement. Then $I_n = U_n(V_n)$ and each one of the random variables U_n, $U_n(0)$ and $U_n(1)$ carries the same information.

Call C the cylindrical σ-field on $E^{\mathbf{Z}}$ and π the law of $((X_n, I_n))_{n\in\mathbf{Z}}$. By stationarity, the shift operator T is an automorphism of $(E^{\mathbf{Z}}, C, \pi)$. The formulas defining U_n, V_n, Y_n from I_n and X_{n-1} are invariant by time-translations so the measurable maps Φ and Ψ yielding $(U_n)_{n\in\mathbf{Z}}$ and $((V_n, Y_n))_{n\in\mathbf{Z}}$ from $((X_n, I_n))_{n\in\mathbf{Z}}$ commute with T. Therefore, the σ-fields $\Phi^{-1}(C)$ and $\Psi^{-1}(C)$ are factors of T.

Theorem 6.50

1. *The factor* $\Phi^{-1}(C)$ *is complementable with complement* $\Psi^{-1}(C)$. *Therefore, if the Markov chain* $((Y_n, I_n))_{n\in\mathbf{Z}}$ *is defined on the canonical space* $(E^{\mathbf{Z}}, C, \pi)$, *then* $\mathcal{F}^U_\infty$ *is a complementable factor of* T *with complement* $\mathcal{F}^{V,Y}_\infty$.
2. *Yet, the filtration* $\mathcal{F}^U$ *is not complementable in the filtration* $\mathcal{F}^{X,I}$.

Proof By Proposition 6.49, $\mathcal{F}^U_\infty$ and $\mathcal{F}^{V,Y}_\infty$ are independent, and $\mathcal{F}^U_\infty \vee \mathcal{F}^{V,Y}_\infty = \mathcal{F}^{X,I}_\infty$ mod $\mathbf{P}$. Therefore, $\Phi^{-1}(C)$ and $\Psi^{-1}(C)$ are independent in $(E^{\mathbf{Z}}, C, \pi)$ and $\Phi^{-1}(C) \vee \Psi^{-1}(C) = C$ mod π: the factor $\Phi^{-1}(C)$ is complementable with complement $\Psi^{-1}(C)$.

Let $U = (U_n)_{n\leq 0}$. The random variable U takes values in $\Pi^{\mathbf{Z}_-}$. For every $u = (u_n)_{n\leq 0} \in \Pi^{\mathbf{Z}_-}$ and $n \leq 0$, call W^u_n the map from $\{0, 1\}^{|n|}$ to $\{a, b\}$ defined by

$$W^u_n(v_{n+1}, \ldots, v_0) = X_n \circ \psi_{u_{n+1}(v_{n+1})} \circ \cdots \circ \psi_{u_0(v_0)}(1).$$

By ordering the elements of $\{0, 1\}^{|n|}$ in the lexicographic order, one identifies W^u_n with an element of $\{a, b\}^{|n|}$.

Since $X_n = X_{n-1} \circ \psi_{I_n} = X_{n-1} \circ \psi_{u_n(V_n)}$ $\mathbf{P}_u$-almost surely, we have

$$W^u_n(v_{n+1}, \ldots, v_0) = W^u_{n-1}(V_n, v_{n+1}, \ldots, v_0) \ \mathbf{P}_u\text{-almost surely},$$

so W^u_n is the left half or the right half of W^u_{n-1} according V_n equals 0 or 1. Moreover, under $\mathbf{P}_u$, the random variable V_n is independent of $\mathcal{F}^{W^u,V}_{n-1}$ and uniform on $\{0, 1\}$.

Hence, under $\mathbf{P}_u$, the process $(W^u_n, V_n)_{n\leq 0}$ is a dyadic split-words process with innovations $(V_n)_{n\leq 0}$. The filtration of this process is known to be non-standard (see [20]). But one checks that $(V_n)_{n\leq 0}$ is also a sequence of innovations of the larger filtration $(\mathcal{F}^{X,I})_{n\leq 0}$ seen under $\mathbf{P}_u = \mathbf{P}[\cdot|U = u]$, so $(\mathcal{F}^{W^u,V}_n)_{n\leq 0}$ is immersed in $(\mathcal{F}^{X,I})_{n\leq 0}$ and $(\mathcal{F}^{X,I})_{n\leq 0}$ is also non-standard under $\mathbf{P}_u$.

If $(\mathcal{F}^U_n)_{n\leq 0}$ admitted an independent complement $(\mathcal{G}_n)_{n\leq 0}$ in $(\mathcal{F}^{X,I}_n)_{n\leq 0}$, this complement would be immersed in the product-type filtration $(\mathcal{F}^{X,I}_n)_{n\leq 0}$ thus standard. Therefore, for $U(\mathbf{P})$-almost every $u \in \Pi^{\mathbf{Z}_-}$, the filtration $(\mathcal{F}^{X,I}_n)_{n\leq 0}$ would be standard under the probability $\mathbf{P}_u$, by proposition 0.1 of [14]. This leads to a contradiction.

We are done. □

6.7 Annex: Reminders on Partitions and Entropy

We recall here classical definitions and results to make the paper self-contained. Most of them can be found in [19]. See also [18].

In the whole section, we fix a measure-preserving map T from a probability space $(Z, \mathcal{Z}, \pi)$ to itself, whereas α, β, γ denote measurable countable partitions of Z (here, 'measurable partition' means 'partition into measurable blocks'), and $\mathcal{F}, \mathcal{G}$ denote sub-σ-fields of $\mathcal{Z}$.

We will use the non-negative, continuous and strictly concave function φ : $[0, 1] \to \mathbf{R}$ defined by $\varphi(x) = -x \log_2(x)$, with the convention $\varphi(0) = 0$. The maximum of this function is $\varphi(1/e) = 1/(e \ln 2)$.

6.7.1 Partitions

Defining the entropy requires discretizations of the ambient probability space, that is why we introduce countable measurable partitions. Equivalently, we could use discrete random variables. We need a few basic definitions.

Definition 6.51 One says that β is finer than α (and note $\alpha \leq \beta$) when each block of α is the union of some collection of blocks of β, i.e. when $\sigma(\alpha) \subset \sigma(\beta)$.

Definition 6.52 The (non-empty) intersections $A \cap B$ with $A \in \alpha$ and $B \in \beta$ form a partition; this partition is the coarsest refinement of α and β and is denoted by $\alpha \vee \beta$.

Definition 6.53 More generally, if $(\alpha_k)_{k \in K}$ is a countable family of countable measurable partitions of Z, we denote by $\bigvee_{k \in K} \alpha_k$ the partition whose blocks are the (non-empty) intersections $\bigcap_{k \in K} A_k$ where $A_k \in \alpha_k$ for every $k \in K$; this partition is the coarsest refinement of the $(\alpha_k)_{k \in K}$; it is still measurable but it can be uncountable.

Definition 6.54 The partitions α and β are independent if and only if $\pi(A \cap B) = \pi(A)\pi(B)$ for every $A \in \alpha$ and $B \in \beta$.

Definition 6.55 We denote by $T^{-1}\alpha$ the partition defined by

$$T^{-1}\alpha = \{T^{-1}(A) : A \in \alpha\}.$$

If T is invertible (i.e. bimeasurable), we denote by $T\alpha$ the partition defined by

$$T\alpha = \{T(A) : A \in \alpha\}.$$

6.7.2 Fischer Information and Entropy of a Partition

Given $A \in \mathcal{Z}$, we view $-\log_2 \pi(A)$ as the quantity of information provided by the event A when A occurs, with the convention $-\log_2 0 = +\infty$.[5] With this definition,

[5]Taking logarithms in base 2 is an arbitrary convention which associates one unity of information to any uniform Bernoulli random variable.

the occurence of a rare event provide much information; moreover, the information provided by two independent events A and B occuring at the same time is the sum of the informations provided by each of them separately. The entropy of a countable measurable partition is the mean quantity of information provided by its blocks.

Definition 6.56 The Fischer information of the partition α is the random variable

$$I_\alpha := \sum_{A \in \alpha} (-\log_2 \pi(A)) \mathbf{1}_A.$$

The entropy of the partition α is the quantity

$$H(\alpha) = \mathbf{E}_\pi[I_\alpha] = \sum_{A \in \alpha} \varphi(\pi(A)).$$

Note that null blocks in α do not give any contribution to the entropy of a partition. Non-trivial partitions have positive entropy. Finite partitions have finite entropy. Infinite countable partition can have finite or infinite entropy.

The previous definition can be generalized as follows.

Definition 6.57 The conditional Fischer information of the partition α with regard to $\mathcal{F}$ is the random variable

$$I_{\alpha|\mathcal{F}} = \sum_{A \in \alpha} (-\log_2 \pi(A|\mathcal{F})) \mathbf{1}_A.$$

The conditional entropy of the partition α with regard to $\mathcal{F}$ is the quantity

$$H(\alpha|\mathcal{F}) = \mathbf{E}_\pi[I_{\alpha|\mathcal{F}}].$$

Remark 6.58 By conditional Beppo-Levi theorem,

$$\mathbf{E}[I_{\alpha|\mathcal{F}}|\mathcal{F}] = \sum_{A \in \alpha} \varphi(\pi(A|\mathcal{F})),$$

so

$$H(\alpha|\mathcal{F}) = \sum_{A \in \alpha} \mathbf{E}_\pi[\varphi(\pi(A|\mathcal{F}))].$$

Given any partition η into measurable blocks, we will use the following abbreviated notations: $H(\alpha|\eta) := H(\alpha|\sigma(\eta))$, $H(\alpha|\eta \vee \mathcal{F}) := H(\alpha|\sigma(\eta) \vee \mathcal{F})$.

Note that when $\mathcal{F}$ is the trivial σ-field $\{\emptyset, Z\}$, $I_{\alpha|\mathcal{F}}$ and $H(\alpha|\mathcal{F})$ are equal to I_α and $H(\alpha)$.

The following properties are very useful and are checked by direct computation, by using the positivity of Fischer information and the strict concavity of φ.

Proposition 6.59 (First Properties)

1. $I_{T^{-1}\alpha|T^{-1}\mathcal{F}} = I_{\alpha|\mathcal{F}} \circ T$ *so* $H(T^{-1}\alpha|T^{-1}\mathcal{F}) = H(\alpha|\mathcal{F})$.
2. $H(\alpha|\mathcal{F}) \geq 0$, *with equality if and only if* $\alpha \subset \mathcal{F} \mod \pi$.
3. $H(\alpha|\mathcal{F}) \leq H(\alpha)$. *When* $H(\alpha) < +\infty$, *equality holds if and only if* α *is independent of* $\mathcal{F}$.
4. *If* $\mathcal{F} \subset \mathcal{G}$, *then* $\mathbf{E}[I_{\alpha|\mathcal{F}}|\mathcal{F}] \geq \mathbf{E}[I_{\alpha|\mathcal{G}}|\mathcal{F}]$ *so* $H(\alpha|\mathcal{F}) \geq H(\alpha|\mathcal{G})$.
5. *If* $\alpha \leq \beta$, *then* $I_{\alpha|\mathcal{F}} \leq I_{\beta|\mathcal{F}}$ *so* $H(\alpha|\mathcal{F}) \leq H(\beta|\mathcal{F})$.
6. $H(\alpha \vee \beta|\mathcal{F}) = H(\alpha|\mathcal{F}) + H(\beta|\mathcal{F} \vee \alpha) \leq H(\alpha|\mathcal{F}) + H(\beta|\mathcal{F})$.

The last item above (addition formula above and sub-additivity of entropy) is used repeatedly in the present paper. We will also use the next result.

Proposition 6.60 (Monotone Sequence of σ-Fields) *Assume that* $H(\alpha) < +\infty$.

1. *If* $(\mathcal{F}_n)_{n\geq 0}$ *is a non-decreasing sequence of* σ*-fields, then*

$$H(\alpha|\mathcal{F}_n) \to H(\alpha|\mathcal{F}_\infty) \text{ where } \mathcal{F}_\infty = \bigvee_{n\geq 0} \mathcal{F}_n.$$

2. *If* $(\mathcal{D}_n)_{n\geq 0}$ *is a non-increasing sequence of* σ*-fields, then*

$$H(\alpha|\mathcal{D}_n) \to H(\alpha|\mathcal{D}_\infty) \text{ where } \mathcal{D}_\infty = \bigcap_{n\geq 0} \mathcal{D}_n.$$

Proof Given $A \in \alpha$, the martingale and backward martingale convergence theorems and the continuity of φ yield $\varphi(\pi(A|\mathcal{F}_n)) \to \varphi(\pi(A|\mathcal{F}_\infty))$ and $\varphi(\pi(A|\mathcal{D}_n)) \to \varphi(\pi(A|\mathcal{D}_\infty))$ as $n \to +\infty$. When α is finite, the convergences $H(\alpha|\mathcal{F}_n) \to H(\alpha|\mathcal{F}_\infty)$ and $H(\alpha|\mathcal{D}_n) \to H(\alpha|\mathcal{D}_\infty)$ follow by Remark 6.58.

The result can be extended to the general case by approximating α with finite mesurable partitions and using the equicontinuity of the maps $\alpha \mapsto \mathbf{E}[\alpha|\mathcal{F}]$, where $\mathcal{F}$ is any sub-σ-field of $\mathcal{Z}$. See Propositions 6.62 and 6.63 in the next subsection. □

6.7.3 Continuity Properties

Proposition 6.61 *The formula*

$$d(\alpha, \beta) = H(\alpha|\beta) + H(\beta|\alpha) = 2H(\alpha \vee \beta) - H(\alpha) - H(\beta)$$

defines a pseudo-metric on the set of all partitions of Z *with finite entropy. Moreover,* $d(\alpha, \beta) = 0$ *if and only if* $\sigma(\alpha) = \sigma(\beta)$ *modulo* π *(i.e. the non-null blocks of* α *and* β *are the same modulo* π*).*

Proof The triangle inequality follows from the inequality

$$H(\alpha|\gamma) \leq H(\alpha \vee \beta|\gamma) = H(\beta|\gamma) + H(\alpha|\beta \vee \gamma) \leq H(\beta|\gamma) + H(\alpha|\beta).$$

The other statements follow from Proposition 6.59. □

Proposition 6.62 *For the pseudo-metric d thus defined, the set of all finite measurable partitions of Z is dense in the set of all (measurable) partitions on Z with finite entropy.*

Proof Let $\alpha = \{A_n : n \geq 1\}$ be an infinite partition of Z with finite entropy. For every $n \geq 1$, set $\alpha_n = \{A_1, \cdots, A_n, (A_1 \cup \cdots \cup A_n)^c\}$. Since α is finer than α_n,

$$H(\alpha) \geq H(\alpha_n) \geq \sum_{k=1}^{n} \varphi(\pi(A_k)),$$

so $d(\alpha, \alpha_n) = H(\alpha) - H(\alpha_n) \to 0$ as $n \to +\infty$. □

Proposition 6.63 *Let $\mathcal{F}$ be a sub-σ-field of $\mathcal{Z}$. Then, for the pseudo-metric d, the map $\alpha \mapsto H(\alpha|\mathcal{F})$ is 1-Lipschitz.*

Proof Let α and β be two partitions of Z with finite entropy. Then

$$\begin{aligned} H(\beta|\mathcal{F}) - H(\alpha|\mathcal{F}) &\leq H(\alpha \vee \beta|\mathcal{F}) - H(\alpha|\mathcal{F}) = H(\beta|\mathcal{F} \vee \alpha) \\ &\leq H(\beta|\alpha) \leq d(\alpha, \beta). \end{aligned}$$

The result follows. □

Proposition 6.64 *Let $\alpha = \{A_1, \ldots, A_n\}$ and $\beta = \{B_1, \ldots, B_n\}$ be two finite measurable partitions of Z with the same finite number of blocks. For every A and B in $\mathcal{Z}$, set $\delta(A, B) = \pi(A \triangle B)$.Then*

$$d(\alpha, \beta) \leq \sum_{i=1}^{n} 2\varphi\big(\delta(A_i, B_i)/2)\big) + \sum_{i=1}^{n} \delta(A_i, B_i)/\ln 2.$$

In particular, the map $A \mapsto \{A, A^c\}$ is uniformly continuous for the pseudo-metrics δ and d.

Proof Fix $i \in [\![1, n]\!]$. Then the concavity of φ yields

$$\begin{aligned} \sum_{j \neq i} \pi(B_j)\varphi(\pi(A_i|B_j)) &\leq \pi(B_i^c)\varphi\Big(\sum_{j \neq i} \frac{\pi(B_j)}{\pi(B_i^c)}\pi(A_i|B_j)\Big) \\ &= \pi(B_i^c)\varphi\Big(\sum_{j \neq i} \frac{\pi(A_i \cap B_j)}{\pi(B_i^c)}\Big) \end{aligned}$$

$$
\begin{aligned}
&= \pi(B_i^c)\varphi\Big(\frac{\pi(A_i \cap B_i^c)}{\pi(B_i^c)}\Big) \\
&= \pi(A_i \cap B_i^c)[-\log_2 \pi(A_i \cap B_i^c) + \log_2 \pi(B_i^c)] \\
&\le \varphi(\pi(A_i \cap B_i^c)).
\end{aligned}
$$

But the concavity of φ also yields $\varphi(x) \le (1-x)/\ln 2$ for every $x \in [0, 1]$, so

$$
\pi(B_i)\varphi(\pi(A_i|B_i)) \le \pi(B_i)\pi(A_i^c|B_i)/\ln 2 = \pi(A_i^c \cap B_i)/\ln 2.
$$

Hence, by Remark 6.58,

$$
H(\alpha|\beta) = \sum_{i,j} \pi(B_j)\varphi(\pi(A_i|B_j)) \le \sum_i \varphi(\pi(A_i \cap B_i^c)) + \sum_i \pi(A_i^c \cap B_i)/\ln 2.
$$

A similar upper bound holds for $H(\beta|\alpha)$. Summing these two inequalities and using once again the concavity of φ yields the statement. □

6.7.4 *Entropy of a Measure-Preserving Map*

First, we define quantities $h(T, \alpha)$.

Proposition 6.65 (Definition and Formula for $h(T, \alpha)$)

(i) *The sequence $(H_n(T, \alpha))_{n\ge 0}$ defined by*

$$
H_n(T, \alpha) = H(\alpha \vee T^{-1}\alpha \vee \cdots \vee T^{-(n-1)}\alpha)
$$

is concave. Since $H_0(T, \alpha) = 0$, the sequence $(H_n(T, \alpha)/n)_{n\ge 1}$ is non-increasing so the limit $h(T, \alpha) = \lim_{n\to+\infty} H_n(T, \alpha)/n$ exists in $[0, +\infty]$.

(ii) *If $H(\alpha) < +\infty$, then $h(T, \alpha) = H(\alpha|\alpha_1^\infty)$, where $\alpha_1^\infty = \bigvee_{k\ge 1} T^{-k}\alpha$.*

Proof The first statement follows from the equality

$$
\begin{aligned}
H_{n+1}(T, \alpha) - H_n(T, \alpha) &= H(\alpha \vee T^{-1}\alpha \vee \cdots \vee T^{-n}\alpha) - H(T^{-1}\alpha \vee \cdots \vee T^{-n}\alpha) \\
&= H(\alpha|T^{-1}\alpha \vee \cdots \vee T^{-(n-1)}\alpha),
\end{aligned}
$$

and the fact that $H(\alpha|\mathcal{F})$ is non-increasing with regard to $\mathcal{F}$. Using Proposition 6.60 and Cesàro's lemma yields the second statement. □

Definition 6.66 The entropy of T is

$$
\begin{aligned}
h(T) &= \sup\{h(T, \alpha) : \alpha \text{ partition of } Z \text{ with finite entropy}\} \\
&= \sup\{h(T, \alpha) : \alpha \text{ finite measurable partition of } Z\}.
\end{aligned}
$$

These two quantities coincide thanks to Propositions 6.62 and 6.72.

Proposition 6.67 *For every $r \geq 1$, $h(T^r) = rh(T)$. If T is also invertible, one has also $h(T^{-1}) = h(T)$.*

Proof For every $n \geq 1$ and every partition α with finite entropy,

$$H_n(T^r, \alpha) \leq H_n(T^r, \alpha \vee \cdots \vee T^{-(r-1)}\alpha) = H_{rn}(T, \alpha).$$

Dividing by n and letting n go to infinity yields

$$h(T^r, \alpha) \leq h(T^r, \alpha \vee \cdots \vee T^{-(r-1)}\alpha) = rh(T, \alpha).$$

The inequalities $h(T^r) \leq rh(T)$ and $rh(T) \leq h(T^r)$ follow.

If T is invertible, the equalities $\alpha \vee \cdots \vee T^{-(n-1)}\alpha = T^{-(n-1)}(\alpha \vee \cdots \vee T^{n-1}\alpha)$ follow from Proposition 6.59 item 1 and yield $H_n(T, \alpha) = H_n(T^{-1}, \alpha)$, so $h(T^{-1}) = h(T)$. □

6.7.5 Generators

Countable generating partitions are useful to compute the entropy of invertible measure-preserving maps.

Definition 6.68 Assume that T is invertible. A countable measurable partition γ is generating (with regard to T) if the partitions $(T^k\gamma)_{k\in\mathbf{Z}}$ generate $\mathcal{Z}$ modulo the null sets.

Theorem 6.69 (Kolmogorov-Sinai Theorem) *If T is invertible and γ is a countable generator (with regard to T), then $h(T) = h(T, \gamma)$.*

In the next subsection, we will prove a conditional version of this classical theorem, namely Theorem 6.74.

Here is the basic example of generator.

Example 6.3 Let Λ be a countable set, $p_0 : (y_k)_{k\in\mathbf{Z}} \mapsto y_0$ the 0-coordinate projection from $\Lambda^{\mathbf{Z}}$ to Λ, $S : (y_k)_{k\in\mathbf{Z}} \mapsto (y_{k+1})_{k\in\mathbf{Z}}$ the shift operator on $\Lambda^{\mathbf{Z}}$, and μ any shift-invariant probability measure on $\Lambda^{\mathbf{Z}}$. Then the partition $\{p_0^{-1}\{\lambda\} : \lambda \in \Lambda\}$ is generating with regard to S.

The interesting fact is that many situations can be reduced to this particular case. The proof of next theorem is outlined in [11].

Theorem 6.70 (Rohlin's Countable Generator Theorem) *If $(Z, \mathcal{Z}, \pi)$ is a Lebesgue space, T is invertible and* aperiodic, *i.e. if $\pi\{z \in Z : \exists n \geq 1 : T^n(z) = z\} = 0$, then T admits a countable generating partition $\gamma = \{C_\lambda : \lambda \in \Lambda\}$. Moreover, the γ-name map Φ from Z to $\Lambda^{\mathbf{Z}}$, defined by $\Phi(z)_k = \lambda$ whenever $T^k(z) \in C_\lambda$, is invertible modulo the null sets, when $\Lambda^{\mathbf{Z}}$ is endowed with the* *probability measure* $\Phi(\pi)$. *The measure* $\Phi(\pi)$ *is shift-invariant, so* T *is isomorphic modulo the null sets to the shift operator on* $\Lambda^{\mathbf{Z}}$.

When T is invertible, ergodic and has finite entropy, Krieger's theorem ensures the existence of a finite generator with size at most $\lfloor 2^{h(T)} \rfloor + 1$. We do not use this refinement in the present paper.

Using the remark given in footnote in Sect. 6.2.1, one checks that if T is invertible and $(Z, \mathcal{Z}, \pi)$ is a Lebesgue space, then any factor of T admits a countable generating partition.

6.7.6 Conditional Entropy Given a Factor. Pinsker's Formula

Assume that T is invertible and that $\mathcal{B}$ is a factor of T. One may define the entropy of T given $\mathcal{B}$ as follows.

Proposition 6.71 (Definition and Formula for $h(T, \alpha|\mathcal{B})$**)**

(i) *The sequence* $(H_n(T, \alpha|\mathcal{B}))_{n \geq 0}$ *defined by*

$$H_n(T, \alpha|\mathcal{B}) = H(\alpha \vee T^{-1}\alpha \vee \cdots \vee T^{-(n-1)}\alpha|\mathcal{B})$$

is concave. Since $H_0(T, \alpha|\mathcal{B}) = 0$, *the sequence* $(H_n(T, \alpha|\mathcal{B})/n)_{n \geq 1}$ *is non-increasing so the limit* $h(T, \alpha|\mathcal{B}) = \lim_{n \to +\infty} H_n(T, \alpha|\mathcal{B})/n$ *exists in* $[0, +\infty]$.

(ii) *If* $H(\alpha) < +\infty$, *then* $h(T, \alpha|\mathcal{B}) = H(\alpha|\alpha_1^\infty \vee \mathcal{B})$, *where* $\alpha_1^\infty = \bigvee_{k \geq 1} T^{-k}\alpha$ *denotes the* σ*-field generated by the partitions* $(T^{-k}\alpha)_{k \geq 1}$.

Proof Since $T^{-1}\mathcal{B} = \mathcal{B}$, one has $H_n(T, \alpha|\mathcal{B}) = H(T^{-1}\alpha \vee \cdots \vee T^{-n}\alpha|\mathcal{B})$, so

$$\begin{aligned} H_{n+1}(T, \alpha|\mathcal{B}) - H_n(T, \alpha|\mathcal{B}) &= H(\alpha \vee T^{-1}\alpha \vee \cdots \vee T^{-n}\alpha|\mathcal{B}) \\ &\quad - H(T^{-1}\alpha \vee \cdots \vee T^{-n}\alpha|\mathcal{B}) \\ &= H(\alpha|\sigma(T^{-1}\alpha \vee \cdots \vee T^{-(n-1)}\alpha \vee \mathcal{B}). \end{aligned}$$

The statements follow, by Proposition 6.60 and Cesàro's lemma. □

Proposition 6.72 *Assume that* T *is invertible and that* $\mathcal{B}$ *is a factor of* T. *If* α *and* γ *are two partitions of* Z *with finite entropy, then* $h(T, \alpha|\mathcal{B}) - h(T, \gamma|\mathcal{B}) \leq H(\alpha|\gamma) \leq d(\alpha, \gamma)$. *Therefore, for the pseudo-metric* d, *the map* $\alpha \mapsto h(T, \alpha|\mathcal{B})$ *is* 1*-Lipschitz.*

Proof Set $\alpha_0^{n-1} = \alpha \vee T^{-1}\alpha \vee \cdots \vee T^{-(n-1)}\alpha$ and $\gamma_0^{n-1} = \gamma \vee T^{-1}\gamma \vee \cdots \vee T^{-(n-1)}\gamma$ for every $n \geq 1$. Then

$$\begin{aligned} H(\alpha_0^{n-1}|\mathcal{B}) - H(\gamma_0^{n-1}|\mathcal{B}) &\leq H(\alpha_0^{n-1} \vee \gamma_0^{n-1}|\mathcal{B}) - H(\gamma_0^{n-1}|\mathcal{B}) \\ &= H(\alpha_0^{n-1}|\mathcal{B} \vee \gamma_0^{n-1}) \end{aligned}$$

$$\leq \sum_{k=0}^{n-1} H(T^{-k}\alpha|\mathcal{B} \vee \gamma_0^{n-1})$$
$$\leq \sum_{k=0}^{n-1} H(T^{-k}\alpha|T^{-k}\gamma)$$
$$= nH(\alpha|\gamma).$$

Dividing by n and letting n go to infinity yields $h(T, \alpha|\mathcal{B}) - h(T, \gamma|\mathcal{B}) \leq H(\alpha|\gamma) \leq d(\alpha, \gamma)$. The result follows. □

Definition 6.73 The conditional entropy of T given $\mathcal{B}$ is the quantity

$$\begin{aligned} h(T|\mathcal{B}) &= \sup\{h(T, \alpha|\mathcal{B}) : \alpha \text{ partition of } Z \text{ with finite entropy}\} \\ &= \sup\{h(T, \alpha|\mathcal{B}) : \alpha \text{ finite measurable partition of } Z\}. \end{aligned}$$

These two quantities coincide thanks to Propositions 6.62 and 6.72.

Kolmogorov-Sinai theorem admits the following generalization.

Theorem 6.74 *If γ is a countable generator of T, then $h(T|\mathcal{B}) = h(T, \gamma|\mathcal{B})$.*

Proof For every integers $p \leq q$, set

$$\gamma_p^q = \bigvee_{k=p}^{q} T^{-k}\gamma.$$

Fix $r \geq 0$. Then for every integer $n \geq 1$,

$$\frac{1}{n} H_n(T, \gamma_{-r}^r|\mathcal{B}) = \frac{1}{n} H(T, \gamma_{-r-n+1}^r|\mathcal{B}) = \frac{n+2r}{n} \times \frac{1}{n+2r} H(T, \gamma_{-r-n+1}^r|\mathcal{B}).$$

Letting n go to infinity yields $h(T, \gamma_{-r}^r|\mathcal{B}) = h(T, \gamma|\mathcal{B})$. Thus, applying Proposition 6.72 any partition α of Z with finite entropy and yo γ_{-r}^r yields

$$h(T, \alpha|\mathcal{B}) - h(T, \gamma|\mathcal{B}) \leq H(\alpha|\gamma_{-r}^r).$$

But $H(\alpha|\gamma_{-r}^r) \to H(\alpha|\mathcal{Z}) = 0$ as $r \to +\infty$ since γ is a countable generator of T. Hence $h(T, \alpha|\mathcal{B}) \leq h(T, \gamma|\mathcal{B})$. The conclusion follows. □

Proposition 6.75 (Pinsker's Formula) *Assume that α and β have finite entropy. Let $\mathcal{A}$ and $\mathcal{B}$ be two factors generated by α and β. Set $\alpha_1^\infty = \bigvee_{k\geq 1} T^{-k}\alpha$ and $\beta_1^\infty = \bigvee_{k\geq 1} T^{-k}\beta$. Then*

$$h((T, \mathcal{A})|\mathcal{B}) = h(T, \mathcal{A} \vee \mathcal{B}) - h(T, \mathcal{B}),$$

or equivalently,

$$H(\alpha|\alpha_1^\infty \vee \mathcal{B}) = H(\alpha \vee \beta|\alpha_1^\infty \vee \beta_1^\infty) - H(\beta|\beta_1^\infty).$$

Proof For every integers $p \leq q$, define the partitions α_p^q and β_p^q like in the proof above. Then for every non-negative integer n,

$$\begin{aligned} H_{n+1}(T, \alpha \vee \beta) - H_{n+1}(T, \beta) &= H(\alpha_{-n}^0 \vee \beta_{-n}^0) - H(\beta_{-n}^0) \\ &= \sum_{k=0}^{n} H(T^k\alpha|\alpha_{-(k-1)}^0 \vee \beta_{-n}^0) \\ &= \sum_{k=0}^{n} H(\alpha|\alpha_1^k \vee \beta_{k-n}^k). \end{aligned}$$

By Proposition 6.60, $H(\alpha|\alpha_1^k \vee \beta_{-\ell}^k) \to H(\alpha|\alpha_1^\infty \vee \mathcal{B})$ as $k \to +\infty$ and $\ell \to +\infty$. Since the quantities $H(\alpha|\alpha_1^k \vee \beta_{k-n}^k)$ belong to the finite interval $[0, H(\alpha)]$, we get

$$h(T, \alpha \vee \beta) - h(T, \beta) = \lim_{n \to +\infty} \frac{1}{n+1}\big(H_{n+1}(T, \alpha \vee \beta) - H_{n+1}(T, \beta)\big) = H(\alpha|\alpha_1^\infty \vee \mathcal{B}).$$

Hence, the statement follows from Proposition 6.65, Theorem 6.69, Proposition 6.71 and Theorem 6.74. □

Acknowledgements I thank A. Coquio, J. Brossard, M. Émery, S. Laurent, J.P. Thouvenot for their useful remarks and for stimulating conversations.

References

1. S. Attal, K. Burdzy, M. Émery, Y. Hu, Sur quelques filtrations et transformations browniennes, in *Séminaire de Probabilités, IXXX*. Lecture Notes in Mathematics, vol. 1613 (Springer, Berlin, 1995), pp. 56–69
2. K. Berg, Convolution of invariant measures, maximal entropy. Math. Syst. Theory **3**(2), 146–150 (1969)
3. J. Brossard, C. Leuridan, Filtrations browniennes et compléments indépendants, in *Séminaire de Probabilités, XLI*. Lecture Notes in Mathematics, vol. 1934 (2008), pp. 265–278
4. J. Brossard, M. Émery, C. Leuridan, Maximal Brownian motions. Ann. de l'IHP **45**(3), 876–886 (2009)
5. J. Brossard, M. Émery, C. Leuridan, Skew-Product Decomposition of Planar Brownian Motion and Complementability, in *Séminaire de Probabilités, XLVI*. Lecture Notes in Mathematics, vol. 2123 (Springer, Berlin, 2014), pp. 377–394
6. G. Ceillier, C. Leuridan, Filtrations at the threshold of standardness. Probab. Theory Relat. Fields **158**(3–4), 785–808 (2014)
7. I. Cornfeld, S. Fomin, Y. Sinai, Ergodic theory, in *Grundlehren der mathematischen Wissenschaften*, vol. 245 (Springer, Berlin, 1982)

8. T. de la Rue, Espaces de Lebesgue, in *Séminaire de Probabilités XXVII*. Lecture Notes in Mathematics, vol. 1557 (Springer, Berlin, 1993), pp. 15–21
9. M. Émery, On certain almost Brownian filtrations. Ann. de l'IHP Probabilités et Stat. **41**, 285–305 (2005)
10. M. Émery, W. Schachermayer, On Vershik's standardness criterion and Tsirelson's notion of cosiness, in *Séminaire de Probabilités, XXXV*. Lecture Notes in Mathematics, vol. 1755 (Springer, Berlin, 2001), pp. 265–305
11. S. Kalikow, R. McCutcheon, An outline of Ergodic theory, in Cambridge Studies in Advanced Mathematics (Cambridge University, Cambridge, 2010)
12. S. Laurent, On standardness and I-cosiness, in *Séminaire de Probabilités, XLIII*. Lecture Notes in Mathematics, vol. 2006 (2011), pp. 127–186
13. S. Laurent, The filtration of erased-word processes, in *Séminaire de Probabilités, XLVIII*, Lecture Notes in Mathematics, vol. 2168 (2017), pp. 445–458
14. C. Leuridan, Poly-adic Filtrations, standardness, complementability and maximality. Ann. Probab. **45**(2), 1218–1246 (2017)
15. D. Ornstein, Factors of Bernoulli shifts are Beroulli shifts. Adv. Math. **5**, 349–364 (1971)
16. D. Ornstein, Factors of Bernoulli shifts. Isr. J. Math. **21**(2–3), 145–153 (1975)
17. D. Ornstein, B. Weiss, Finitely determined implies very weak Bernoulli. Isr. J. Math. **17**(1), 94–104 (1974)
18. W. Parry, *Entropy and Generators in Ergodic Theory* (W.A. Benjamin Inc., Benjamin, 1969)
19. K. Petersen, *Ergodic Theory* (Cambridge University, Cambridge, 1981)
20. M. Smorodinsky, Processes with no standard extension. Isr. J. Math. **107**, 327–331 (1998)
21. J.P. Thouvenot, Une classe de systèmes pour lesquels la conjecture de Pinsker est vraie. Isr. J. Math. **21**, 208–214 (1975)
22. A. Vershik, Theory of decreasing sequences of measurable partitions. Algebra i Analiz, **6**(4), 1–68 (1994). English Translation: St. Petersburg Math. J. **6**(4), 705–761 (1995)
23. H. von Weizsäcker, Exchanging the order of taking suprema and countable intersections of σ-algebras. Ann. Inst. H. Poincaré Sect. B **19**(1), 91–100 (1983)

Chapter 7
Uniform Entropy Scalings of Filtrations

Stéphane Laurent

Abstract We study Vershik and Gorbulsky's notion of entropy scalings for filtrations in the particular case when the scaling is not ϵ-dependent, and is then termed as *uniform scaling*. Among our main results, we prove that the scaled entropy of the filtration generated by the Vershik progressive predictions of a random variable is equal to the scaled entropy of this random variable. Standardness of a filtration is the case when the scaled entropy with a constant scaling is zero, thus our results generalize some known results about standardness. As a case-study we consider a family of next-jump time filtrations. We also provide some results about the entropy of poly-adic filtrations, rephrasing or generalizing some old results.

Keywords Standardness · Entropy

7.1 Introduction

This is the first paper about Vershik and Gorbulsky's theory of the entropy of filtrations written in the probabilistic language. It deals with the *scaled entropy* introduced in [21]. Our results focus on the case of *uniform entropy scalings*, and, because standardness is equivalent to zero entropy with a constant entropy scaling, they generalize the main properties about standardness.

In Sect. 7.2 we recall the definition of Vershik's standardness criterion. We use this criterion in Sect. 7.3 to give a new proof of the standardness criterion for the family of next-jump time filtrations studied in [12] (where I-cosiness was used to derive this criterion). In Sect. 7.4 we introduce the scaled entropy with uniform scalings. In Sect. 7.5 we pursue the work of Sect. 7.3 by studying uniform entropy scalings for the next-jump time filtrations; in fact, we just use our results to show that this problem comes down to the scaled entropy of a discrete measure which is studied in [14]. Section 7.6 deals with the exponential entropy for poly-adic

---

S. Laurent (✉)
Independent Researcher, Paris, France

C. Donati-Martin et al. (eds.), *Séminaire de Probabilités L*, Lecture Notes in Mathematics 2252, https://doi.org/10.1007/978-3-030-28535-7_7

filtrations. In this section, we mainly rephrase some old theorems by Vershik and, with a slight generalization, a theorem by Gorbulsky about the coincidence between the scaled entropy and the exponential entropy.

The relevance of the results provided by this paper is twofold. First, the general properties about standardness become particular cases of the general properties about the uniformly scaled entropy. Among these properties, Theorem 7.4.11 is one of the main results. It states that the scaled entropy of the filtration generated by the Vershik progressive predictions of a random variable equals the scaled entropy of this random variable, thereby considerably reducing the task of calculating the entropy of a filtration. The power of this theorem is well illustrated on the calculation of the scaled entropy for the next-jump time filtrations. Second, our results provide new knowledge about the known examples of non-standard filtrations. The results of Sect. 7.6 providing the scaled entropy of some non-standard split-word filtrations with the exponential scaling, are not new. When this scaled entropy has not the same value for two such filtrations, one can conclude that these two filtrations are not isomorphic. But thanks to our results of Sect. 7.4.3, we learn something more, namely that it is not possible to embed the filtration having the smallest entropy in, for example, an independent enlargement of the other one with a standard filtration.

7.2 Vershik's Standardness Criterion

In the probabilistic literature, *standardness* of a filtration $\mathcal{F} = (\mathcal{F}_n)_{n \leqslant 0}$ in discrete negative time is usually defined as the possibility to embed $\mathcal{F}$ in the filtration generated by a sequence of independent random variables (see [4, 9–11]). As long as the final σ-field $\mathcal{F}_0$ is essentially separable, standardness is known to be equivalent to Vershik's standardness criterion. In the present paper, we say that a filtration is *Vershikian* if it satisfies Vershik's standardness criterion, and we say that a filtration is *standard* if it is Vershikian and its final σ- field is essentially separable.

In this section we recall the statement of Vershik's standardness criterion and we state its main properties which are proved in [11]. In Sect. 7.4 we will see that these properties are particular cases of our results about the scaled entropy.

7.2.1 Vershik's Standardness Criterion

The *Kantorovich* distance plays a major role in the statement of Vershik's standardness criterion, as well as in the definition of the entropy. Given a separable metric space (E, ρ), the Kantorovich distance $\rho'(\mu, \nu)$ between two probability measures μ and ν on E is defined by

$$\rho'(\mu, \nu) = \inf_{\Lambda \in \mathcal{J}(\mu,\nu)} \iint \rho(x, y) \mathrm{d}\Lambda(x, y),$$

where $\mathcal{J}(\mu, \nu)$ is the set of joinings of μ and ν, that is, the set of probabilities on $E \times E$ whose first and second marginal measures are μ and ν respectively. In general, $\rho'(\mu, \nu)$ is possibly infinite, but ρ' defines a metric on the space E' of integrable probability measures on (E, ρ), when saying that a probability measure μ on (E, ρ) is integrable if the random variables $X \sim \mu$ satisfy $\mathbb{E}\big[\rho(X, x)\big] < \infty$ for some ($\iff$ for every) point $x \in E$, and such a random variable X is also said to be integrable. When E is bounded then every E-valued random variable is integrable. In general, the topology induced by ρ' on E' is finer than the topology of weak convergence. They coincide when (E, ρ) is compact, and (E', ρ') is itself compact in this case. We mainly use the fact that the metric space (E', ρ') is complete and separable whenever (E, ρ) is (see e.g. [2]).

In order to state Vershik's standardness criterion, one has to introduce the *Vershik progressive predictions* $\pi_n X$ of a random variable X (corresponding to the so-called *universal projectors*, or *tower of measures*, in [17] and [20]) and the iterated Kantorovich distance $\rho^{(n)}$ on the state space $E^{(n)}$ of $\pi_n X$. Let (E, ρ) be a Polish metric space. For a σ-field $\mathcal{B}$ on a given probability space, we denote by $L^1(\mathcal{B}; E)$ the space of integrable E-valued $\mathcal{B}$-measurable random variables. Let $\mathcal{F} = (\mathcal{F}_n)_{n \leqslant 0}$ be a filtration, and $X \in L^1(\mathcal{F}_0; E)$. The Vershik progressive predictions $\pi_n X$ of X with respect to $\mathcal{F}$ are recursively defined as follows: we put $\pi_0 X = X$, and $\pi_{n-1} X = \mathcal{L}(\pi_n X \mid \mathcal{F}_{n-1})$ (the conditional law of $\pi_n X$ given $\mathcal{F}_{n-1}$). Since X is integrable, for any $x \in E$ the conditional expectation $\mathbb{E}\big[\rho(X, x) \mid \mathcal{F}_{-1}\big]$ is finite almost surely. That shows that the conditional law $\mathcal{L}(X \mid \mathcal{F}_{-1})$ almost surely takes its values in E', the space of integrable probability measures on E. Moreover $\rho'\big(\mathcal{L}(X \mid \mathcal{F}_{-1}), \delta_x\big) = \mathbb{E}\big[\rho(X, x) \mid \mathcal{F}_{-1}\big]$ is an integrable real-valued random variable, and thus $\mathcal{L}(X \mid \mathcal{F}_{-1}) = \pi_{-1} X$ is an integrable E'-valued random variable. Thus, by a recursive reasoning, the n-th progressive prediction $\pi_n X$ is an integrable random variable taking its values in the Polish space $E^{(n)}$ recursively defined by $E^{(0)} = E$ and $E^{(n-1)} = (E^{(n)})'$, denoting as before by E' the space of integrable probability measures on any separable metric space E. Note that $(\pi_n X)_{n \leqslant 0}$ is a Markov process. We will denote by $\mathcal{F}^X$ the filtration it generates. The state space $E^{(n)}$ of $\pi_n X$ is Polish when endowed with the distance $\rho^{(n)}$ obtained by iterating $|n|$ times the construction of the Kantorovich distance starting with ρ: we recursively define $\rho^{(n)}$ by putting $\rho^{(0)} = \rho$ and by defining $\rho^{(n-1)} = (\rho^{(n)})'$ as the Kantorovich distance issued from $\rho^{(n)}$.

Finally, in order to state Vershik's standardness criterion, one introduces the *dispersion* disp X of (the law of) an integrable random variable X in a Polish metric space (E, ρ). It is defined as the expectation of $\rho(X', X'')$ where X' and X'' are two independent copies of X, that is, two independent random variables defined on the same probability space and having the same law as X. Now, Vershik's standardness criterion is defined as follows. Let $\mathcal{F}$ be a filtration, let E be a Polish metric space and $X \in L^1(\mathcal{F}_0; E)$. We say that the random variable X satisfies the *Vershik property*, or, for short, that X is *Vershikian* (with respect to $\mathcal{F}$) if $\operatorname{disp} \pi_n X \longrightarrow 0$ as n goes to $-\infty$. Then we extend this definition to σ-fields $\mathcal{E}_0 \subset \mathcal{F}_0$ and to the whole filtration as follows: we say that a σ-field $\mathcal{E}_0 \subset \mathcal{F}_0$ is *Vershikian* if each random variable

$X \in L^1(\mathcal{E}_0; [0,1])$ is Vershikian, and we say that the filtration $\mathcal{F}$ is *Vershikian*, or that $\mathcal{F}$ satisfies *Vershik's standardness criterion*, if the final σ-field $\mathcal{F}_0$ is Vershikian.

It is important to note that when $\mathcal{F}$ is *immersed*[1] in a bigger filtration $\mathcal{G}$ and $X \in L^1(\mathcal{F}_0; E) \subset L^1(\mathcal{G}_0; E)$, the Vershik progressive prediction $\pi_n X$ is the same considering either $\mathcal{F}$ or $\mathcal{G}$ as the underlying filtration. We refer to [4] or [10] for details about the immersion property. Consequently, in such a situation, there is no ambiguity in considering the Vershik property without specifying the underlying filtration. The filtration $\mathcal{F}^X$ generated by the Markov process $(\pi_n X)_{n\leqslant 0}$ is immersed in $\mathcal{F}$ (that means that this process is Markovian with respect to $\mathcal{F}$).

Later, we will use the two following lemmas about the Vershik progressive predictions. The proof of the first one is straightforward from the definitions and we leave it to the reader.

Lemma 7.2.1 *For any Polish space (E,ρ) and random variables $X, Y \in L^1(\mathcal{F}_0; E)$, the stochastic process $\big(\rho^{(n)}(\pi_n X, \pi_n Y)\big)_{n\leqslant 0}$ is a submartingale. In particular the expectation $\mathbb{E}\left[\rho^{(n)}(\pi_n X, \pi_n Y)\right]$ is increasing with n.*

Lemma 7.2.2 *Let $\mathcal{F}$ be a filtration. Let (E,ρ) and $(\tilde{E},\tilde{\rho})$ be two Polish metric spaces, and $f\colon E \to \tilde{E}$ a measurable function. If X is a $\mathcal{F}_0$-measurable random variable taking its values in E, then $\pi_n\big(f(X)\big) = f^n(\pi_n X)$ for some measurable function $f^n\colon E^{(n)} \to \tilde{E}^{(n)}$, and the two following properties hold:*

- *f^n is K-Lipschitz if f is K-Lipschitz;*
- *f^n is an isometry if f is an isometry.*

Proof The function f^n is inductively defined by $f^0 = f$ and $f^{n-1}(\mu) = f^n \star \mu$ (image measure). It is elementary to check the three claims asserted in the lemma.

□

7.2.2 Properties to be Generalized Later

Throughout this article, we denote by $V(X)$ the Vershik property for an integrable random variable X taking its values in a Polish space, when an underlying ambient filtration $\mathcal{F}$ is understood. We also denote by $V(\mathcal{E}_0)$ the Vershik property for a σ-field $\mathcal{E}_0 \subset \mathcal{F}_0$. We will see in Sect. 7.4 that $V(X)$ can be equivalently stated as $h_c(X) = 0$ where h_c is the scaled entropy of X with a constant scaling c. Then our results in Sect. 7.4 about the uniformly scaled entropy generalize the following propositions and theorem which are provided in [11].

Proposition 7.2.3 *Let $\mathcal{F}$ be a filtration, $n_0 \leqslant 0$ be an integer, and denote by $\mathcal{F}^{n_0]} = (\mathcal{F}_{n_0+n})_{n\leqslant 0}$ the filtration $\mathcal{F}$ truncated at n_0. Then $\mathcal{F}^{n_0]}$ is Vershikian if and only if $\mathcal{F}$ is Vershikian.*

---

[1] A filtration $\mathcal{F}$ is said to be immersed in a filtration $\mathcal{G}$ when every $\mathcal{F}$-martingale is a $\mathcal{G}$-martingale.

Proposition 7.2.4

(a) If $(\mathcal{B}_k)_{k \geqslant 1}$ *is an increasing sequence of sub-*σ*-fields of* $\mathcal{F}_0$ *then*

$$[\forall k \geqslant 1,\ V(\mathcal{B}_k)] \implies V\Big(\bigvee_{k \geqslant 1} \mathcal{B}_k\Big).$$

(b) For any Polish metric space (E, ρ) *and* $X \in L^1(\mathcal{F}_0; E)$*,*

$$V(X) \iff V\big(\sigma(X)\big).$$

Theorem 7.2.5 *For any* $X \in L^1(\mathcal{F}_0; E)$*, the filtration* $\mathcal{F}^X$ *generated by the Markov process* $(\pi_n X)_{n \leqslant 0}$ *is standard if and only if the random variable* X *satisfies the Vershik property.*

Proposition 7.2.3 is a consequence of Corollary 7.4.18. Proposition 7.2.4 is a consequence of Propositions 7.4.15 and 7.4.16. Theorem 7.2.5 is a consequence of Theorem 7.4.11.

We will use the two propositions and the theorem above in Sect. 7.3, before proving their generalization, and this is why we state them here. We will also provide generalizations of two other results: Theorem 7.4.19 generalizes the fact that a parametric extension of a Vershikian filtration is still a Vershikian filtration, and Theorem 7.4.22 generalizes the fact that the independent product of two Vershikian filtrations is a Vershikian filtration.

7.2.3 Vershik's Standardness Criterion in Practice

Vershik's standardness criterion may appear puzzling and complicated at first glance: calculating the progressive predictions $\pi_n X$ and the iterated Kantorovich distance $\rho^{(n)}$ on the strange state space of $\pi_n X$ does not appear easily practicable.

First note that $V(X)$ does not depend on the choice of the metric on the Polish space E in which X takes its values: this stems from the second claim of Proposition 7.2.4. Also note the importance of Theorem 7.2.5: property $V(X)$ is equivalent to standardness of the filtration $\mathcal{F}^X$ generated by the Markov process $(\pi_n X)_{n \leqslant 0}$. Thus, if we intend to show that standardness of $\mathcal{F}$ holds true, our task is reduced to only show $V(X)$ if we find X such that $\mathcal{F}^X = \mathcal{F}$.

Observe that any filtration $\mathcal{F}$ having an essentially separable final σ-field $\mathcal{F}_0$ can always be generated by a real-valued Markov process $(X_n)_{n \leqslant 0}$: just take for X_n any real-valued random variable generating the σ-field $\mathcal{F}_n$ for every $n \leqslant 0$. Vershik's standardness criterion can be rephrased to a more practical criterion by considering a generating Markov process $(X_n)_{n \leqslant 0}$, as we explain below and summarize in Lemma 7.2.6; but the practicality of the rephrased criterion depends on the choice of the generating Markov process. Firstly, thanks to this lemma, the strange state spaces of Vershik's progressive predictions $\pi_n X$ can be avoided when X is one of the random variable X_k of the Markov process $(X_n)_{n\leqslant 0}$. Let us explain this claim for $X = X_0$, which is enough to understand. We assume that X_n is distributed on a Polish space A_n for every $n \leqslant 0$. This guarantees the existence of a family of conditional laws $\mathcal{L}(X_{n+1} \mid X_n = x_n)$, $x_n \in A_n$ for every $n < 0$. Denote by ρ_0 the metric on A_0, and assume that X_0 is integrable. Then we recursively define a pseudometric ρ_n on A_n by setting

$$\rho_n(x_n, x_n') = (\rho_{n+1})'\big(\mathcal{L}(X_{n+1} \mid X_n = x_n), \mathcal{L}(X_{n+1} \mid X_n = x_n')\big)$$

where $(\rho_{n+1})'$ is the Kantorovich pseudometric derived from ρ_{n+1}. The ρ_n are more friendly than the $\rho^{(n)}$ appearing in Vershik's standardness criterion, and Lemma 7.2.6 states that there are some maps $\psi_n \colon A_n \to A_0^{(n)}$ such that $\pi_n X_0 = \psi_n(X_n)$ and

$$\rho^{(n)}\big(\psi_n(x_n), \psi_n(x_n')\big) = \rho_n(x_n, x_n')$$

for every $x_n, x_n' \in A_n$. Thus, in order for the Vershik property $V(X_0)$ to hold true, it suffices that $\rho_n(X_n', X_n'') \to 0$ in L^1 where X_n' and X_n'' are two independent copies of X_n. Moreover, Lemma 7.2.6 states that $\mathcal{F}^{X_0} = \mathcal{F}$ under the identifiability condition

$$\begin{aligned}&\forall n \leqslant 0,\ \forall x_n, x_n' \in A_n,\\ &\qquad [\,x_n \neq x_n'\,] \implies \big[\,\mathcal{L}(X_{n+1} \mid X_n = x_n) \neq \mathcal{L}(X_{n+1} \mid X_n = x_n')\,\big]\end{aligned} \tag{$\star$}$$

and then, by Theorem 7.2.5, standardness of $\mathcal{F}$ is equivalent to $V(X_0)$ under this condition.

Lemma 7.2.6 *Let $\mathcal{F}$ be the filtration generated by a Markov process $(X_n)_{n\leqslant 0}$, with X_n taking its values in a Polish space A_n, and assume that X_0 is integrable. Consider the pseudometrics ρ_n introduced above and the iterated Kantorovich metrics $\rho^{(n)}$ appearing in Vershik's standardness criterion.*

(1) There are some maps $\psi_n \colon A_n \to A_0^{(n)}$ such that $\pi_n X_0 = \psi_n(X_n)$ and

$$\rho^{(n)}\big(\psi_n(x_n), \psi_n(x_n')\big) = \rho_n(x_n, x_n')$$

for every $x_n, x_n' \in A_n$ and every $n \leqslant 0$.

(2) The Vershik property $V(X_0)$ is equivalent to $\mathbb{E}\big[\rho_n(X_n', X_n'')\big] \to 0$ where X_n' and X_n'' are two independent copies of X_n.

(3) Under the identifiability condition ($\star$), the ρ_n are metrics and the ψ_n are some isometries. Consequently $\mathcal{F}$ is generated by the process $(\pi_n X_0)_{n\leqslant 0}$.

(4) Under the identifiability condition ($\star$), the Vershik property $V(X_0)$ is equivalent to standardness of $\mathcal{F}$.

Proof Obviously $\pi_0 X_0$ is a $\sigma(X_0)$-measurable random variable, and $\pi_n X_0 = \mathcal{L}(\pi_{n+1} X_0 \mid \mathcal{F}_n)$ for $n < 0$ is a $\sigma(X_n)$-measurable random variable by the Markov property. Therefore, for each $n \leqslant 0$, the Doob-Dynkin lemma provides a measurable function ψ_n for which $\pi_n X_0 = \psi_n(X_n)$, and ψ_0 is nothing but the identity map. The equality in 1), relating $\rho^{(n)}$ and ρ_n, is obviously true for $n = 0$. Assuming $\rho^{(n+1)}\big(\psi_{n+1}(x_{n+1}), \psi_{n+1}(x'_{n+1})\big) = \rho_{n+1}(x_{n+1}, x'_{n+1})$, then the Kantorovich distance $\rho_n(x_n, x'_n)$ is given by

$$\rho_n(x_n, x'_n) = \inf_{\Lambda_{x_n, x'_n}} \int \rho^{(n+1)}\big(\psi_{n+1}(x_{n+1}), \psi_{n+1}(x'_{n+1})\big) \mathrm{d}\Lambda_{x_n, x'_n}(x_{n+1}, x'_{n+1}),$$

where the infimum is taken over all joinings Λ_{x_n, x'_n} of $\mathcal{L}(X_{n+1} \mid X_n = x_n)$ and $\mathcal{L}(X_{n+1} \mid X_n = x'_n)$, and then $\rho_n(x_n, x'_n)$ is also given by

$$\rho_n(x_n, x'_n) = \inf_{\Theta_{x_n, x'_n}} \int \rho^{(n+1)}(y_{n+1}, y'_{n+1}) \mathrm{d}\Theta_{x_n, x'_n}(y_{n+1}, y'_{n+1}),$$

where the infimum is taken over all joinings Θ_{x_n, x'_n} of $\mathcal{L}(\pi_{n+1} X_0 \mid X_n = x_n) = \psi_n(x_n)$ and $\mathcal{L}(\pi_{n+1} X_0 \mid X_n = x'_n) = \psi_n(x'_n)$, thereby showing $\rho^{(n)}\big(\psi_n(x_n), \psi_n(x'_n)\big) = \rho_n(x_n, x'_n)$. That shows 1), and 2) obviously follows.

The claim about the ρ_n in 3) is recursively shown too. It suffices to show that every ψ_n is injective. Assuming that ψ_{n+1} is injective and assuming $\mathcal{L}(X_{n+1} \mid X_n = x_n) \neq \mathcal{L}(X_{n+1} \mid X_n = x'_n)$, then, obviously,

$$\mathcal{L}\big(\psi_{n+1}(X_{n+1}) \mid X_n = x_n\big) \neq \mathcal{L}\big(\psi_{n+1}(X_{n+1}) \mid X_n = x'_n\big),$$

that is, $\psi_n(x_n) \neq \psi(x'_n)$, thereby showing 3). Finally, claim 4) stems from Theorem 7.2.5. □

Obviously we can similarly state Lemma 7.2.6 for X_k instead of X_0, for any $k \leqslant 0$. When the identifiability condition $(\star)$ does not hold, one can apply proposition 6.2 of [11], which claims that, in order to prove standardness of $\mathcal{F}$, it is sufficient to check that $V(X_k)$ holds true for every $k \leqslant 0$.

7.3 The Next-Jump Time Filtrations

In Sect. 7.5 we will study the scaled entropy of the next-jump time filtrations which are introduced in this section. Standardness of these filtrations has been characterized in [12] with the help of the I-cosiness criterion. In this section we provide a new proof of this characterization with the help of Vershik's standardness criterion (Sect. 7.2.1). More precisely, we will be in the context of Lemma 7.2.6 and the identifiability condition $(\star)$ will be fulfilled, and thus our main task will be to

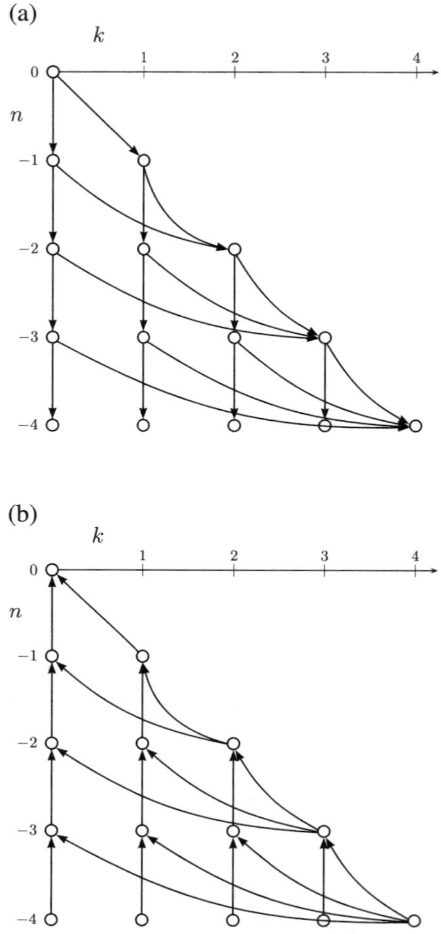

Fig. 7.1 Next-jump time process as a random walk. (**a**) Random walk from $n = 0$ to $n = -\infty$. (**b**) Random walk from $n = -\infty$ to $n = 0$

derive the metrics ρ_n of this lemma. This will be achieved in Sect. 7.3.2, after we introduce the next-jump filtrations in Sect. 7.3.1 as the filtrations generated by some random walks on the vertices of a Bratteli graph (shown on Fig. 7.1).

7.3.1 Next-Jump Time Process as a Random Walk on a Bratteli Graph

Our presentation of the next-jump time filtrations differs from the one given in [12]. Here we define these filtrations as those generated by a Markov process on the vertices of a Bratteli graph.

Let B be the $(-\mathbb{N})$-graded Bratteli graph shown on Fig. 7.1. At each level n, there are $|n|+1$ vertices labeled by $k \in \{0, \dots, |n|\}$, and the vertex labeled by k is connected to the two vertices at level $n-1$ labeled by k and $|n|+1$. A path on B is a sequence $(\gamma_n)_{n \leqslant 0}$ consisting of edges γ_n such that γ_n connects a vertex at level n to a vertex at level $n-1$ for every $n \leqslant 0$. The set of paths is denoted by Γ_B. When a path is taken at random in Γ_B we denote by V_n the label of the selected vertex at level n (thus $V_0 = 0$) and we are interested in the filtration $\mathcal{F}$ generated by the process $(V_n)_{n \leqslant 0}$. Since this causes no possible confusion, we identify a vertex to its label. We study the case when the process $(V_0, V_{-1}, \dots)$ is the Markov chain whose transition distributions are defined from a given $[0,1]$-valued sequence $(p_n)_{n \leqslant 0}$ satisfying $p_0 = 1$, by

$$\mathcal{L}(V_n \mid V_{n+1} = k) = (1 - p_n)\delta_k + p_n \delta_{|n|},$$

that is to say, given V_{n+1}, the vertex V_n is one of the two vertices connected to V_{n+1} and equals the extreme vertex $|n|$ with probability p_n.

In other words, if we consider that the set of paths Γ_B is $\{0,1\}^{-\mathbb{N}}$ by labelling the edges connecting a vertex v_n at level n to the vertex v_{n-1} at level $n-1$ by 0 if v_{n-1} and v_n have the same label and by 1 if v_{n-1} is labeled by $|n|+1$, then we are interested in the case when the paths are taken at random according to the independent product measure $\bigotimes_{n \leqslant -1}(1-p_n, p_n)$ by denoting by $(1-p, p)$ the Bernoulli probability measure with probability of success p.

The time-directed process $(V_n)_{n \leqslant 0}$ is Markovian too. The next-jump time process $(Z_n)_{n \leqslant 0}$ defined in [12] is obtained from V_n by putting $Z_0 = 0$ and $Z_n = -V_{n+1}$ for $n \leqslant -1$. Hence the filtration $\mathcal{F}$ generated by the Markov process $(V_n)_{n \leqslant 0} = (-Z_{n-1})_{n \leqslant 0}$ shares the same standardness status as the one studied in [12] because standardness is an asymptotic property (Proposition 7.2.3).

It is easy to see that $\Pr(V_n = |n|) = p_n$. We will say that the p_n are the *jumping probabilities* because one also has $p_n = \Pr(V_{n+1} \neq V_n)$ for every $n < 0$. It is shown in [12] that

$$\Pr(V_n = |k|) = (1-p_n)\cdots(1-p_{k-1})p_k \quad \text{if } 0 \leqslant |k| < |n|,$$

and the transitions kernels $P_n(v, \cdot)$ from $n-1$ to n are given by

$$P_n(v, \cdot) := \mathcal{L}(V_n \mid V_{n-1} = v) = \begin{cases} \delta_v & \text{if } 0 \leqslant v < |n|+1 \\ \mathcal{L}(V_n) & \text{if } v = |n|+1 \end{cases}. \tag{7.1}$$

Obviously the identifiability condition $(\star)$ defined in Sect. 7.2.3 cannot hold for $(V_n)_{n \leqslant 0}$ because $V_0 = 0$ is degenerate. But we will see in Lemma 7.3.3 that this condition holds for the process truncated at -1 when $p_{-1} \in]0,1[$ and $p_n < 1$ for every $n \leqslant -2$.

An important particular case is the one when $p_n = (|n|+1)^{-1}$. In this case, V_n has the uniform distribution on $\{0, \dots, |n|\}$ for every $n \leqslant 0$ and the filtration $\mathcal{F}$

generated by $(V_n)_{n\leqslant 0}$ is Kolmogorovian and not standard in this case. This results from the standardness criterion provided by Theorem 7.3.7, which was proved in [12] with the help of the I-cosiness criterion, and which is proved in the present paper with the help of Vershik's criterion.

When $p_n < 1$, the law of V_{n+1} is the law of V_n conditioned on $\{0, \ldots, |n| - 1\}$. Thus, when $p_n < 1$ for every $n \leqslant -1$, the law of V_n can be represented as the truncation of a measure μ on $\mathbb{N}$. For example, μ is the counting measure in the uniform case $p_n = (|n| + 1)^{-1}$. When the p_n are given, this measure is given by

$$\mu(-n) = \frac{p_n}{\prod_{k=n}^{-1}(1 - p_k)} = \frac{\Pr(V_n = n)}{\Pr(V_n = 0)}$$

for every $n \leqslant 0$, and

$$\mu\big(\{0, \ldots, -n\}\big) = \frac{1}{\prod_{k=n}^{-1}(1 - p_k)} = \frac{1}{\Pr(V_n = 0)},$$

thus μ is normalizable if and only if $\sum p_n < \infty$. In this case the law of V_n goes to the normalized version of μ and $\mathcal{F}$ is not Kolmogorovian, and in the other case V_n goes to ∞ and $\mathcal{F}$ is Kolmogorovian. This is due to the following proposition about the tail σ-field $\mathcal{F}_{-\infty}$, which is a rewriting of proposition 3.1 in [12], to which we refer for a detailed proof.

Proposition 7.3.1 *The sequence $(V_n)_{n\leqslant 0}$ goes to a random variable $V_{-\infty}$ when n goes to $-\infty$, and the tail σ-field $\mathcal{F}_{-\infty}$ is generated by $V_{-\infty}$. There are three possible situations:*

(1) if $\sum p_n = \infty$ then $V_{-\infty} = +\infty$ almost surely, therefore $\mathcal{F}$ is Kolmogorovian;
(2) if $\sum p_k < \infty$ then

a. either $V_{-\infty}$ is not degenerate, therefore $\mathcal{F}$ is not Kolmogorovian,
b. or we are in the following case

$$p_{n_0} = 1 \text{ and } p_n = 0 \text{ for every } n < n_0 \text{ for some } n_0 \leqslant 0 \qquad (*)$$

and then $V_{-\infty} = |n_0|$ almost surely, therefore $\mathcal{F}$ is Kolmogorovian and even standard.

Thus $\mathcal{F}$ is Kolmogorovian if and only if $\sum p_n = \infty$ or in case $(*)$. Standardness of $\mathcal{F}$ in case $(*)$ elementarily holds true because $\mathcal{F}_m = \{\varnothing, \Omega\}$ for every $m \leqslant n_0$ in this case.

7.3.2 Standardness of $\mathcal{F}$ Using Vershik's Criterion

Throughout this section, we denote by $(V_n)_{n\leqslant 0}$ the next-jump time process with jumping probabilities $(p_n)_{n\leqslant 0}$ and we denote by $\mathcal{F}$ the filtration it generates.

Discarding the elementary case (∗), it is shown in [12] with the help of the I-cosiness criterion that $\mathcal{F}$ is standard (Vershikian) if and only if $\sum p_n^2 = \infty$. In this section we derive again this result by using Vershik's standardness criterion. More precisely we will use the version of Vershik's standardness criterion given by Lemma 7.2.6. We firstly treat a particular case in lemma below.

Lemma 7.3.2 *If $p_n = 1$ for infinitely many n, then $\mathcal{F}$ is standard.*

Proof For every integer $k \leqslant 0$, define the random vector $X_k = (V_k, \dots, V_0)$ and denote by $\mathcal{B}_k = \sigma(V_k, \dots, V_0)$ the σ-field it generates. By the Markov property, the n-th progressive prediction $\pi_n X_k$ of X_k is measurable with respect to $\sigma(V_n)$ for every $n \leqslant k$, and $V_n = |n|$ almost surely when $p_n = 1$, therefore $\pi_n X_k$ is a degenerate random variable too, and $\operatorname{disp}(\pi_n X_k) = 0$. Consequently, $\mathcal{F}$ is Vershikian by Proposition 7.2.4(a). □

We also know by Proposition 7.3.1 that $\mathcal{F}$ is standard in the case when $p_n = 0$ for every $n < 0$. Then the following lemma will allow us to restrict our standardness study to the case when the identifiability condition (⋆) of Sect. 7.2.3 holds.

Lemma 7.3.3

(1) Let $(X_n)_{n \leqslant 0} = (V_{n-1})_{n \leqslant 0}$. The identifiability condition (⋆) holds when

$$p_{-1} \in]0, 1[\qquad \text{and} \qquad p_n < 1 \quad \text{for all } n < 0. \tag{7.2}$$

In this case, $\mathcal{F}$ is generated by the process $(\pi_n V_{-1})_{n \leqslant 0}$, and even more precisely, $\sigma(\pi_n V_{-1}) = \sigma(V_n)$ for every $n < 0$.

(2) If $p_{n_0} = 1$ for some $n_0 < 0$, then the process $\big(V_{n_0+n} - |n_0|\big)_{n \leqslant 0}$ is the next-jump time process with jumping probabilities $(p_{n_0+n})_{n \leqslant 0}$.

(3) If $p_{-1} = 0$, then the process $(W_{n-1})_{n \leqslant 0}$ defined by

$$W_n = \begin{cases} 0 & \text{if } V_{n-1} = 0 \\ V_{n-1} - 1 & \text{if } V_{n-1} > 0 \end{cases} \qquad \text{for } n \leqslant -1.$$

has the same distribution as $(V_{n-1})_{n \leqslant 0}$ where $(V_n)_{n \leqslant 0}$ is the next-jump time process with jumping probabilities $(p'_n)_{n \leqslant 0}$ given by $p'_n = p_{n-1}$ for every $n < 0$.

Proof For $v \neq v'$ in the state space of V_{n-1}, the conditional distributions $\mathcal{L}(V_n \mid V_{n-1} = v)$ and $\mathcal{L}(V_n \mid V_{n-1} = v')$ have different supports under (7.2), hence the first point follows. The equality $\sigma(\pi_n V_{-1}) = \sigma(V_n)$ under condition (⋆) is provided by Lemma 7.2.6. Checking the second and third points do not pose any difficulty. □

Thus, since standardness is an asymptotic property at $n = -\infty$ (Proposition 7.2.3), we will focus on the case when (7.2) holds, and this will allow us to use Lemma 7.2.6. In Lemma 7.3.4 we summarize the way we are going. Hereafter we denote by $\mathbb{V}_n = \{0, \ldots, |n|\}$ the state space of V_n and consider on $\mathbb{V}_n$ the n-th iterated Kantorovich metric ρ_n starting with the discrete 0–1 metric ρ_{-1} on $A_{-1} = \{0, 1\}$. That is,

$$\rho_n(v_n, v_n') = \inf_{\Lambda_{v_n, v_n'}} \int \rho_{n+1} \mathrm{d}\Lambda_{v_n, v_n'}$$

for every $n \leqslant -2$, where $\Lambda_{v_n, v_n'}$ is a joining of the conditional laws $\mathcal{L}(V_{n+1} \mid V_n = v_n) = P_{n+1}(v_n, \cdot)$ and $\mathcal{L}(V_{n+1} \mid V_n = v_n') = P_{n+1}(v_n', \cdot)$. Hereafter we denote by d_n the dispersion of V_n under ρ_n, defined by $d_n = \mathbb{E}[\rho_n(V_n', V_n'')]$ for two independent copies V_n' and V_n'' of V_n.

Lemma 7.3.4 *Under the identifiability condition* (7.2), *the filtration $\mathcal{F}$ is standard if and only if the Vershik property $V(X)$ holds for $X = V_{-1}$. Moreover, this property is equivalent to $d_n \to 0$.*

Proof Consequence of Lemma 7.2.6 and Lemma 7.3.3. □

In lemma below we provide a list of relations about the kernels P_n of the next-jump time Markov chain and the iterated Kantorovich distances ρ_n. We denote by $P_n(v, f)$ the expectation of a function f under the probability measure $P_n(v, \cdot)$. Recall that $P_{n+1}(|n|, \cdot)$ which occurs several times in the lemma is equal to the law of V_{n+1}. We use $P_{n+1}(|n|, \cdot)$ and not $\mathcal{L}(V_{n+1})$ in the lemma to emphasize that the derivation of the ρ_n only depends on the kernels P_n by nature.

Lemma 7.3.5 *Let $x \geqslant 0$ and $x' \geqslant 0$ be integer numbers.*

(1) If $n \leqslant -1$ and $x, x' \leqslant |n| - 1$, then $\rho_n(x, x') = \rho_{n+1}(x, x')$.
(2) If $n \leqslant -2$ and $x' \leqslant |n| - 1$, then $\rho_n(|n|, x') = P_{n+1}\big(|n|, \rho_{n+1}(\cdot, x')\big)$.
(3) If $n \leqslant -3$ and $x' \leqslant |n| - 2$, then $\rho_n(|n|, x') = \rho_{n+1}(|n + 1|, x')$.
(4) If $n \leqslant -1$, then $\rho_{n-1}(|n - 1|, |n|) = (1 - p_n)P_{n+1}\big(|n|, \rho_n(|n|, \cdot)\big)$.
(5) If $n \leqslant -2$, then $P_n\big(|n - 1|, \rho_{n-1}(|n - 1|, \cdot)\big) = (1 - p_n^2)P_{n+1}\big(|n|, \rho_n(|n|, \cdot)\big)$.
(6) For every $n \leqslant -1$, $P_n\big(|n - 1|, \rho_{n-1}(|n - 1|, \cdot)\big) = 2p_{-1}(1 - p_{-1}) \prod_{m=n}^{-2}(1 - p_m^2)$.

Proof (1) and (2) are easily get from the expression of $\mathcal{L}(V_{n+1} \mid V_n = v)$ given in Sect. 7.3.1. One obtains (3) as a consequence of (1) and (2) by using the relation

$$\Pr(V_n = k \mid V_{n-1} = |n - 1|) = (1 - p_n) \Pr(V_{n+1} = k \mid V_n = |n|) \tag{7.3}$$

valid for $0 \leqslant k < |n|$ and $n \leqslant -2$. One gets (4) by using (2) and (7.3). Finally, (5) is derived from (3), (4) and (7.3), and one obtains (6) by calculating the right member of (5) for $n = -2$ and then by applying (5) recursively. □

Lemma 7.3.6 *The dispersion of V_n under ρ_n is given by $d_n = 2p_{-1}(1 - p_{-1}) \prod_{m=n}^{-2}(1 - p_m^2)$ for every $n \leqslant -1$.*

Proof Because of $\mathcal{L}(V_{n+1}) = \mathcal{L}(V_{n+1} \mid V_n = |n|)$ we get

$$d_{n+1} = \mathbb{E}\big[\rho_n(|n|, V_{n+1}) \mid V_n = |n|\big]$$

for every $n \leqslant -2$ by equality (2) of Lemma 7.3.5, and then the assertion of the lemma is nothing but equality (6) of Lemma 7.3.5. □

Theorem 7.3.7 *The filtration $\mathfrak{F}$ is standard if and only if $\sum p_n^2 = \infty$ or in case $(*)$.*

Proof Case $(*)$ is treated in Proposition 7.3.1. Under the identifiability condition (7.2), we know that $\mathfrak{F}$ is standard if and only if $\prod_{n=-\infty}^{-2}(1 - p_n^2) = 0$ by Lemma 7.3.4 and by Lemma 7.3.6. If the identifiability condition (7.2) does not hold, there are two cases to be treated: either $p_n = 1$ for infinitely many n or $p_n = 1$ for finitely many values of n. In the first case, use Lemma 7.3.2. In the second case, the problem can be reduced to the case of the identifiability condition by using assertion (2) of Lemma 7.3.3 and the fact that standardness is an asymptotic property (Proposition 7.2.3). □

7.3.3 Iterated Kantorovich Distances

Denote by $\mathbb{V}_n = \{0, \dots, |n|\}$ the set of vertices at level n. The pseudometric spaces $(\mathbb{V}_n, \rho_n)$ are easily derived from relations (1), (3), (4) and (6) given in Lemma 7.3.5. Note that (1) means that the canonical embedding $(\mathbb{V}_n, \rho_n) \to (\mathbb{V}_{n-1}, \rho_{n-1})$ is an isometry, and this is a very particular situation (we mean this is not a general fact about the intrinsic pseudometrics on Bratteli graphs). The pseudometrics ρ_n are shown on Table 7.1.

This table is easily filled by successively and iteratively using the following equalities for $n \leqslant -2$:

$$\begin{cases} \rho_n(0, x) & = \begin{cases} 1 & \text{if } x = 1 \\ p_{-1} & \text{otherwise} \end{cases} \\ \rho_n(x, x') & = \rho_{n+1}(x, x') \qquad \text{for } x, x' < |n| \\ \rho_n(|n|, x) & = \begin{cases} \rho_{n+1}(|n+1|, x) & \text{if } x < |n+1| \\ (1 - p_{n+1})d_{n+2} & \text{if } x = |n+1| \end{cases} \end{cases}$$

Table 7.1 Intrinsic metrics $\rho_n(k, k')$ for $n = -1, -2, -3, -4, -5$

	k'					
k	0	1	2	3	4	5
0	0	1	p_{-1}	p_{-1}	p_{-1}	p_{-1}
1	1	0	$1 - p_{-1}$	$1 - p_{-1}$	$1 - p_{-1}$	$1 - p_{-1}$
2	p_{-1}	$1 - p_{-1}$	0	$(1 - p_{-2})d_{-1}$	$(1 - p_{-2})d_{-1}$	$(1 - p_{-2})d_{-1}$
3	p_{-1}	$1 - p_{-1}$	$(1 - p_{-2})d_{-1}$	0	$(1 - p_{-3})d_{-2}$	$(1 - p_{-3})d_{-2}$
4	p_{-1}	$1 - p_{-1}$	$(1 - p_{-2})d_{-1}$	$(1 - p_{-3})d_{-2}$	0	$(1 - p_{-4})d_{-3}$
5	p_{-1}	$1 - p_{-1}$	$(1 - p_{-2})d_{-1}$	$(1 - p_{-3})d_{-2}$	$(1 - p_{-4})d_{-3}$	0

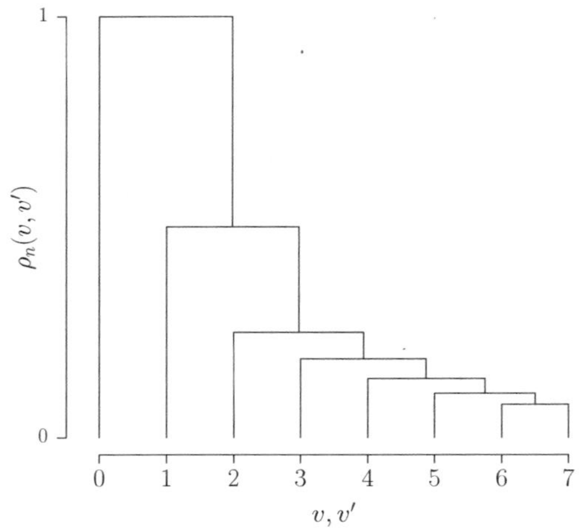

Fig. 7.2 The space $(\mathbb{V}_n, \rho_n)$

where the expression of d_n is given in Lemma 7.3.6 for every $n \leqslant -1$ and we set in addition $d_0 = 1$. It follows that the distance $\rho_n(v_n, v'_n)$ between two vertices v_n and v'_n at some level $n \leqslant -2$ is explicitely given when $v_n < v'_n$ by

$$\rho_n(v_n, v'_n) = \begin{cases} 1 & \text{if } v_n = 0 \text{ and } v'_n = 1 \\ p_{-1} & \text{if } v_n = 0 \text{ and } v'_n > 1 \\ (1 - p_{-v_n})d_{-v_n+1} & \text{if } v_n > 0 \end{cases}.$$

The ρ_n are metrics under the identifiability condition (7.2). The space $(\mathbb{V}_n, \rho_n)$ is an ultrametric space represented by the dendrogram shown in Fig. 7.2 (numerically, this figure shows the case $p_n \equiv \frac{1}{2}$ for $n < 0$).

7.4 The Uniformly Scaled Entropy

In this section we introduce the scaled entropy of filtrations by following Vershik and Gorbulsky [21], except that we use the probabilistic language and we restrict our attention to entropy scalings which are not ϵ-dependent (this is why we term them as *uniform* scalings). Theorem 7.4.11, our most significant result, generalizes Theorem 7.2.5.

A *uniform entropy scaling*, or, for short, a *uniform scaling*, or an *entropy scaling* or a *scaling*, is a bounded below function $c\colon (-\mathbb{N}) \to (0, \infty)$. Vershik and Gorbulsky more generally consider ϵ-dependent scalings $c \mapsto c(\epsilon, n)$.

The definition of the Vershik property $V(X)$ stated (in Sect. 7.2) by $\operatorname{disp} \pi_n X \to 0$ can be equivalently stated by:

$$\forall \epsilon > 0, \exists n \leqslant 0, \exists \mu \in E^{(n)}, \quad \mathbb{E}\big[\rho^{(n)}(\pi_n X, \mu)\big] < \epsilon. \tag{7.4}$$

In other words, the Vershik progressive prediction $\pi_n X$ can be approximated by a single value with probability as high as desired when $n \to -\infty$. When this property fails, it is natural to wonder about an optimal asymptotic approximation of $\pi_n X$. Roughly speaking, the scaled entropy compares the growth of the minimal entropy approximation of $\pi_n X$ with the given scaling. Its evident interest is its ability to distinguish locally isomorphic non-standard filtrations.

The definition of the scaled entropy of a filtration relies on a choice of a measure of entropy $H(\mu)$ for discrete probability distributions μ. Common choices include the well-known Shannon entropy

$$H(\mu) = -\sum \mu_i \log \mu_i,$$

and the max-entropy (or Rényi entropy of order 0)

$$H(\mu) = \log \#\{\mu_i \mid \mu_i \neq 0\}.$$

We will state the definition of the scaled entropy for a measure of entropy as defined below. The three conditions of this definition are fulfilled for the Shannon entropy and for the Rényi entropy of any order.

Definition 7.4.1 An application H associating a quantity $H(\theta) \in [0, +\infty]$ to a discrete probability measure θ is said to be a *measure of entropy* if it satisfies the following conditions.

1. *H is decisive*: $H(\theta) = 0$ if θ is concentrated on one point.
2. *H is relative to the maximum probability*: $H(\theta) \geqslant C(1 - \theta_{\max})$ for a certain constant $C > 0$, where $\theta_{\max}$ is the highest probability mass of θ.
3. *H is increasing*: $H(\theta') \leqslant H(\theta)$ when θ' is the image of θ under some map.

We simply say that $H(\theta)$ is the entropy of θ, and we denote by $H(X)$ the entropy of the law of a discrete random variable X.

7.4.1 *Definitions*

The definition of the scaled entropy of a filtration $\mathcal{F}$ has something similar to the definition of Vershik's standardness criterion: one begins by defining the scaled entropy for a $\mathcal{F}_0$-measurable random variable, then for a σ-field $\mathcal{B} \subset \mathcal{F}_0$, and finally for the filtration $\mathcal{F}$. It mainly involves the ϵ-entropy, defined below, of the Vershik progressive predictions $\pi_n X$ (introduced in Sect. 7.2). We say that a random variable is *simple* when it takes only finitely many values.

Definition 7.4.2 Let Y be an integrable random variable taking its values in a Polish metric space (E, ρ). The ϵ-entropy of Y is

$$H^{\epsilon}(Y) = \inf\left\{H(S) \mid \mathbb{E}[\rho(Y, S)] < \epsilon\right\}$$

where the infimum is taken over all simple $\sigma(Y)$-measurable random variables S taking values in E.

We define two scaled entropies in this definition, namely the *lower scaled entropy* and the *upper scaled entropy*. The scaled entropy defined by Vershik and Gorbulsky, in the case when the scaling does not depend on ϵ, is the upper scaled entropy. We think that the lower scaled entropy deserves to be defined in addition to the upper scaled entropy because each property we will give about the scaled entropy holds for both of them, and is proved in the same way.

Definition 7.4.3 Let $\mathcal{F}$ be a filtration. In (1) and (2) below we consider an integrable $\mathcal{F}_0$-measurable random variable X taking its values in a Polish metric space (E, ρ). In (2), (3) and (4) we consider an entropy scaling c.

(1) The ϵ-entropy of X (with respect to $\mathcal{F}$) at time n is $H_n^{\epsilon}(X; \mathcal{F}) = H^{\epsilon}(\pi_n X)$, shorter denoted by $H_n^{\epsilon}(X)$ when $\mathcal{F}$ is understood, where the n-th Vershik prediction $\pi_n X$ is considered as a random variable taking its values in the Polish space $E^{(n)}$ metrized by the n-th iterated Kantorovich metric $\rho^{(n)}$ (Sect. 7.2.1).

(2) The limits in $[0, \infty]$

$$h_c^-(X; \mathcal{F}) = \lim_{\epsilon \to 0} \liminf_{n \to -\infty} \frac{H_n^{\epsilon}(X)}{c(n)} \quad \text{and} \quad h_c^+(X; \mathcal{F}) = \lim_{\epsilon \to 0} \limsup_{n \to -\infty} \frac{H_n^{\epsilon}(X)}{c(n)}$$

are respectively called the *lower c-scaled entropy* of X and the *upper c-scaled entropy* of X, or, for short, the (lower/upper) *scaled entropy* of X when the scaling c is understood. Note that these limits exist because $H_n^{\epsilon}(X)$ increases as ϵ decreases. We shorter denote $h_c^-(X; \mathcal{F})$ by $h_c^-(X)$ and $h_c^+(X; \mathcal{F})$ by $h_c^+(X)$ when $\mathcal{F}$ is understood without ambiguity.

(3) For a σ- field $\mathcal{B} \subset \mathcal{F}_0$, the lower and upper c-scaled entropies of $\mathcal{B}$ with respect to $\mathcal{F}$ are defined as

$$h_c^-(\mathcal{B}; \mathcal{F}) = \sup_X h_c^-(X; \mathcal{F}) \quad \text{and} \quad h_c^+(\mathcal{B}; \mathcal{F}) = \sup_X h_c^+(X; \mathcal{F})$$

where the supremum is taken over all $\mathcal{B}$-measurable random variables X taking their values in the interval $[0, 1]$ equipped with its usual metric. We shorter denote $h_c^-(\mathcal{B}; \mathcal{F})$ by $h_c^-(\mathcal{B})$ and $h_c^+(\mathcal{B}; \mathcal{F})$ by $h_c^+(\mathcal{B})$ when $\mathcal{F}$ is understood without ambiguity.

(4) The lower and upper c-scaled entropies of the filtration $\mathcal{F}$ are defined as

$$h_c^-(\mathcal{F}) = h_c^-(\mathcal{F}_0; \mathcal{F}) \quad \text{and} \quad h_c^+(\mathcal{F}) = h_c^+(\mathcal{F}_0; \mathcal{F}).$$

In the sequel, we will use the notation h_c as a substitute for either h_c^- or h_c^+. That is, when we write a statement about h_c, that means this statement holds for both h_c^- and h_c^+.

As a first remark, note that, obviously, the scaled entropy $h_c(\mathcal{F})$ only depends on $\mathcal{F}$ up to isomorphism. Also note that $h_c(X) = h_{c'}(X)$ when the two scalings c and c' are equivalent at $-\infty$.

When the Vershik property holds for the random variable X, it is clear that $h_c^-(X) = h_c^+(X) = 0$ for the scaling $c(n) \equiv 1$ and consequently for any scaling c, in view of the statement (7.4) of the Vershik property and because of the decisiveness property of H (Definition 7.4.1). Therefore $h_c^-(\mathcal{F}) = h_c^+(\mathcal{F}) = 0$ for any scaling c when $\mathcal{F}$ is a Vershikian filtration. The converse is true, by virtue of the proposition below.

Proposition 7.4.4 *Let X be a $\mathcal{F}_0$-measurable random variable taking its values in a bounded Polish metric space (E, ρ) and such that $h_c^-(X) = 0$ for the scaling $c(n) \equiv 1$. Then X satisfies the Vershik standardness property (hence $h_c^+(X) = 0$ as well).*

Proof By the assumption $h_c^-(X) = 0$, one has $\liminf_{n\to-\infty} H^\epsilon(\pi_n X) = 0$ for every $\epsilon > 0$. Let $\delta > 0$. Take $n \leqslant 0$ such that $H^\delta(\pi_n X) < \delta$. There exists a simple $E^{(n)}$-valued random variable S such that $\mathbb{E}\big[\rho_n(\pi_n X, S)\big] < \delta$ and $H(S) < \delta$. Let p be the highest probability mass of S and $s \in E^{(n)}$ such that $p = \Pr(S = s)$. Then $p > 1 - \delta/C$, where C is the constant of condition 2 in Definition 7.4.1. Therefore $\mathbb{E}\big[\rho_n(S, s)\big] < \operatorname{diam}(E^{(n)})\delta/C \leqslant \operatorname{diam}(E)\delta/C$. Finally,

$$\mathbb{E}\big[\rho_n(\pi_n X, s)\big] \leqslant \mathbb{E}\big[\rho_n(\pi_n X, S)\big] + \mathbb{E}\big[\rho_n(S, s)\big] \leqslant \left(1 + \frac{\operatorname{diam}(E)}{C}\right)\delta.$$

Thus we get the statement (7.4) of the Vershik property for X. □

While the previous proposition is stated for a bounded metric space E, it is also true for an integrable random variable when E is unbounded. This stems from the fact mentioned in the following remark.

Remark 7.4.5 The notations $H_n^\epsilon(X)$ and $h_c(X)$ do not show the dependence on the metric ρ on the state space of X. But this is not important in view of Proposition 7.4.16 we will see later, which shows that $h_c(X) = h_c\big(\sigma(X)\big)$. Thus, we can replace ρ with $\rho \wedge 1$ in the definition of the scaled entropy $h_c(X)$, without altering its value. This allows to define $h_c(X)$ when X is non-integrable.

Remark 7.4.6 Using the previous remark, it is not difficult to see that we do not alter the value of $h_c(X)$ if we replace the definition of the ϵ-entropy $H^\epsilon(Y)$ with

$$\inf\Big\{H(S) \mid \mathbb{P}\big(\rho(Y, S) > \epsilon\big) < \epsilon\Big\},$$

and this ϵ-entropy allows to define $h_c(X)$ when X is non-integrable.

Remark 7.4.7 Using, as allowed by Remark 7.4.5, the bounded distance $\rho \wedge 1$ instead of ρ, it is easy to prove that we do not alter the value of $h_c(X)$ if we modify the definition of the ϵ-entropy $H^\epsilon(Y)$ by taking the infimum over all discrete random variables S instead of all simple random variables.

Remark 7.4.8 As already mentioned in the definition, the ϵ-entropy $H_n^\epsilon(X)$ is relative to the underlying filtration $\mathcal{F}$. It is important to note that it actually only depends on the filtration $\mathcal{F}^X$ generated by the Markov process $(\pi_n X)_{n\leqslant 0}$ of the Vershik progressive predictions of X. Indeed, it is easy to see that the value of $H_n^\epsilon(X)$ is the same whether we consider $\mathcal{F}$ as the underlying filtration or any filtration $\mathcal{E}$ *immersed* in $\mathcal{F}$ such that X is measurable with respect to the final σ-field $\mathcal{E}_0$ of $\mathcal{E}$, and the filtration $\mathcal{F}^X$ is the smallest such filtration (see [4]).

From the previous remark, it is easy to see that $h_c(\mathcal{F}) \leqslant h_c(\mathcal{G})$ when the filtration $\mathcal{F}$ is immersed in the filtration $\mathcal{G}$.

The above definition of $H^\epsilon(\pi_n X)$ is appropriate for deriving the general properties we will give. But for the calculation of $h_c(X)$ on a case-study, especially when we seek a lower bound of $h_c(X)$, it is generally better to use the alternative definition of $H^\epsilon(\pi_n X)$ given in the following lemma for the case when H is the Shannon entropy, and in the next lemma for the case when H is the max-entropy.

Lemma 7.4.9 *Assume H is the Shannon entropy. In the definition of $h_c(X;\mathcal{F})$, one can replace the ϵ-entropy $H^\epsilon(\pi_n X)$ with* $\inf -\sum \mu_n(P_j)\log\mu_n(P_j)$ *where μ_n is the law of $\pi_n X$ and the infimum runs over all finite partitions $\{P_j\}$ of the state space of $\pi_n X$ having form $\{A_i, C\}$ where $\mu_n(C) < \epsilon$ and each A_i is contained in an ϵ-ball of $(E^{(n)}, \rho^{(n)})$.*

Proof Denote by $H_0^\epsilon(\pi_n X)$ this value. We compare it to the ϵ-entropy $H^\epsilon(\pi_n X)$ as given in Remark 7.4.6. Let $\{P_j\} = \{A_i, C\}$ be a partition such as the ones described in the lemma. Define the function f by $f(x) = \operatorname{argmin}_{\Gamma_i}\rho^{(n)}(x,\Gamma_i)$. Then $\rho^{(n)}\big(x, f(x)\big) < \epsilon$ on $\cup A_i$, and that shows the inequality $H_0^\epsilon(\pi_n X) \geqslant H^\epsilon(\pi_n X)$. Conversely, take a function f taking only finitely many values and such that the μ_n-measure of the set $F := \{x \in E \mid \rho^{(n)}\big(x, f(x)\big) < \epsilon\}$ is greater than $1-\epsilon$. Let $\{\Gamma_i\}$ be an enumeration of $f(F)$. Set $A_i = f^{-1}(\Gamma_i) \cap F$. Then $\{A_i\}$ is a partition of F and $A_i \subset B(\Gamma_i, \epsilon)$ (ϵ-ball centered at Γ_i). Moreover, since the law of $f(\pi_n X)$ equals the weighted average $\mu_n(F) f\big(\mu_n(\cdot \mid F)\big) + \mu_n(F^c) f\big(\mu_n(\cdot \mid F^c)\big)$, the concavity of the Shannon entropy yields

$$
\begin{aligned}
H\big(f(\pi_n X)\big) &\geqslant \mu_n(F) H\Big(f\big(\mu_n(\cdot \mid F)\big)\Big) \\
&= -\sum \mu_n(A_i)\log\mu_n(A_i) + \mu_n(F)\log\mu_n(F) \\
&= -\sum \mu_n(P_j)\log\mu_n(P_j) - h(\mu_n(F))
\end{aligned}
$$

where $\{P_j\} = \{A_i, F^c\}$ and h is the binary entropy function defined by $h(\epsilon) = h(1-\epsilon) = \epsilon\log\frac{1}{\epsilon} + (1-\epsilon)\log\frac{1}{1-\epsilon}$. The function $\epsilon \mapsto h(\epsilon)$ is increasing for

$\epsilon < 1/2$. Hence, for ϵ small enough, one has $h(\mu_n(F)) \leqslant h(\epsilon)$ and the above inequality shows that $H^\epsilon(\pi_n X) \geqslant H_0^\epsilon(\pi_n X) - h(\epsilon)$. Finally,

$$H_0^\epsilon(\pi_n X) - h(\epsilon) \leqslant H^\epsilon(\pi_n X) \leqslant H_0^\epsilon(\pi_n X)$$

and the lemma follows because of $h(0^+) = 0$. □

Lemma 7.4.10 *Assume H is the max-entropy. Let X a random variable taking its values in a Polish space (E, ρ). Then the ϵ-entropy $H^\epsilon(X)$ as given in Remark 7.4.6 equals the minimal log-number of ϵ-balls in E such that X falls in the union of these balls with probability higher than $1-\epsilon$:*

$$H^\epsilon(X) = \min\left\{\log \#\{x_i\} \mid x_i \in E, \Pr\big(X \in \cup B(x_i, \epsilon)\big) > 1-\epsilon\right\}.$$

Proof Denote by $H_0^\epsilon(X)$ this value. For given balls $B(x_i, \epsilon)$, define the function f by $f(x) = \operatorname{argmin}_{x_i} \rho(x, x_i)$. Then $\rho\big(X, f(X)\big) < \epsilon$ if and only if $X \in \cup B(x_i, \epsilon)$, and that shows that $H_0^\epsilon(X) \geqslant H^\epsilon(X)$. Conversely, take a function $g\colon E \to E$ taking only finitely many values and such that $\Pr\Big(\rho\big(X, g(X)\big) > \epsilon\Big) < \epsilon$. Let $\{x_i\}$ be an enumeration of the image of g and define $f(x) = \operatorname{argmin}_{x_i} \rho(x, x_i)$. Then $\rho\big(X, f(X)\big) \leqslant \rho\big(X, g(X)\big)$, hence $\Pr\Big(\rho\big(X, f(X)\big) < \epsilon\Big) \geqslant \Pr\Big(\rho\big(X, g(X)\big) < \epsilon\Big)$. That shows the inequality $H_0^\epsilon(X) \leqslant H^\epsilon(X)$. □

7.4.2 Main Properties and Main Theorem

This section is devoted to prove the main theorem of this paper: Theorem 7.4.11, stated below, which generalizes Theorem 7.2.5. It will be applied in Sect. 7.5 to the study of the scaled entropy of the next-jump time filtrations. In this theorem and all other results of this section, a measure of entropy H as defined in Definition 7.4.1 is understood. We also provide two important properties of the scaled property: the heredity to the generated σ-field, that is to say, the equality $h_c(X) = h_c\big(\sigma(X)\big)$ (Proposition 7.4.16), and the left-continuity of $\mathcal{B} \mapsto h_c(\mathcal{B})$ (Proposition 7.4.15). These two properties generalize the two claims of Proposition 7.2.4.

Theorem 7.4.11 *Let $\mathcal{F}$ be a filtration, $X \in L^1(\mathcal{F}_0; E)$ where E is a Polish space, and $c\colon (-\mathbb{N}) \to (0, \infty)$ a uniform entropy scaling. Then $h_c(X; \mathcal{F}) = h_c(\mathcal{F}^X)$, where $\mathcal{F}^X$ is the filtration generated by the Markov process $(\pi_n X)_{n \leqslant 0}$.*

Note that $h_c(\mathcal{F}^X)$ is the entropy of the filtration $\mathcal{F}^X$ as well as the entropy of the σ-field $\sigma(\pi_n X; n \leqslant 0)$ when we consider $\mathcal{F}$ as the underlying filtration (see Remark 7.4.8). Theorem 7.4.11 has a straightforward useful consequence when the $\pi_n X$ are discrete random variables with finite entropy: it gives the upper bound $h_c(\mathcal{F}^X) \leqslant 1$ for any scaling $c(n) \sim H(\pi_n X)$, because of the increasing property of H.

This theorem will be derived from the two following lemmas and the two main propositions of this section (heredity to the generated σ-field and left-continuity).

Lemma 7.4.12 *Let $\mathcal{F}$ be a filtration. Let (E, ρ) and $(\tilde{E}, \tilde{\rho})$ be two Polish metric spaces, and $f: E \to \tilde{E}$ an isometry. Then $h_c(X) = h_c\big(f(X)\big)$ for any random variable $X \in L^1(\mathcal{F}_0; E)$.*

Proof This is a straightforward consequence of Lemma 7.2.2. □

The key lemma is the following one.

Lemma 7.4.13 *Let $\mathcal{F}$ be a filtration, $X \in L^1(\mathcal{F}_0; E)$ where E is a Polish space metrized by a distance ρ, and set $W^n = (\pi_n X, \ldots, \pi_{-1} X, X)$ for some $n \leqslant 0$. Consider the metric $\bar{\rho}_n = \frac{1}{|n|+1} \sum_{k=n}^{k=0} \rho^{(k)}$ on the state space of W^n. Then $\pi_n W^n = \phi(\pi_n X)$ where ϕ is an isometry.*

Proof For the proof we consider the distance $\tilde{\rho}_n = \sum_{k=n}^{k=0} \rho^{(k)}$ instead of $\bar{\rho}_n$ on the state space of W_n. For each $n \leqslant 0$ and $k \in \{n, \ldots, 0\}$, one has $\pi_k W_n = g_k^n(\pi_n X, \ldots, \pi_k X)$ for some functions g_k^n related by the fact that $g_{k-1}^n(\mu_n, \ldots, \mu_{k-1})$ is the distribution of $g_k^n(\mu_n, \ldots, \mu_{k-1}, M_k)$ where $M_k \sim \mu_{k-1}$. Therefore

$$\begin{aligned}&\tilde{\rho}_n^{(k-1)}\big(g_{k-1}^n(\mu_n, \ldots, \mu_{k-1}), g_{k-1}^n(\mu'_n, \ldots, \mu'_{k-1})\big)\\ &\quad = \inf_{(M_k, M'_k)} \mathbb{E}\Big[\tilde{\rho}_n^{(k)}\big(g_k^n(\mu_n, \ldots, \mu_{k-1}, M_k), g_k^n(\mu'_n, \ldots, \mu'_{k-1}, M'_k)\big)\Big]\end{aligned} \tag{\#}$$

where the infimum is take over all joinings (M_k, M'_k) of μ_{k-1} and μ'_{k-1}. Using this relation, the equality

$$\begin{aligned}&\tilde{\rho}_n^{(k)}\big(g_k^n(\mu_n, \ldots, \mu_k), g_k^n(\mu'_n, \ldots, \mu'_k)\big)\\ &\qquad = \rho^{(n)}(\mu_n, \mu'_n) + \tilde{\rho}_{n+1}^{(k)}\big(g_k^{n+1}(\mu_{n+1}, \ldots, \mu_k), g_k^{n+1}(\mu'_{n+1}, \ldots, \mu'_k)\big)\end{aligned}$$

is easy to derive. Indeed, denoting by $H(n, k)$ this equality, then $H(n, 0)$ is nothing but the equality $\tilde{\rho}_n = \rho^{(n)} + \tilde{\rho}_{n+1}$ and the implication from $H(n, k)$ to $H(n, k-1)$ is easy to derive from relation (#).

Now, by (#),

$$\begin{aligned}&\tilde{\rho}_n^{(n)}\big(g_n^n(\mu_n), g_n^n(\mu'_n)\big)\\ &\quad = \inf_{(M_{n+1}, M'_{n+1})} \mathbb{E}\Big[\tilde{\rho}_n^{(n+1)}\big(g_{n+1}^n(\mu_n, M_{n+1}), g_{n+1}^n(\mu'_n, M'_{n+1})\big)\Big]\end{aligned}$$

where the infimum is taken over all joinings (M_{n+1}, M'_{n+1}) of μ_n and μ'_n. Hence, by relation $H(n, n+1)$,

$$\begin{aligned}
&\tilde{\rho}_n^{(n)}\big(g_n^n(\mu_n), g_n^n(\mu'_n)\big) \\
&\quad = \rho^{(n)}(\mu_n, \mu'_n) + \inf_{(M_{n+1}, M'_{n+1})} \mathbb{E}\Big[\tilde{\rho}_{n+1}^{(n+1)}\big(g_{n+1}^{n+1}(M_{n+1}), g_{n+1}^{n+1}(M'_{n+1})\big)\Big],
\end{aligned}$$

and using this equality we can prove by recursion (starting at $n = 0$) the equality

$$\tilde{\rho}_n^{(n)}\big(g_n^n(\mu_n), g_n^n(\mu'_n)\big) = \big(|n| + 1\big)\rho^{(n)}(\mu_n, \mu'_n)$$

which is obviously equivalent to the statement of the lemma. □

It is interesting to note that Theorem 7.2.5 is an easy corollary of the previous lemma, Lemma 7.2.2, and Proposition 7.2.4. Indeed, this provides a new proof of this theorem, cleaner than the one given in [11].

The following lemma is a continuity-like property of $X \mapsto h_c(X; \mathcal{F})$. A cleaner continuity property of the scaled entropy, relying on this lemma, is the content of Proposition 7.4.15.

Lemma 7.4.14 *Let $\mathcal{F}$ be a filtration, and E a Polish space. Let $(X_k)_{k\geqslant 1}$ be a sequence in $L^1(\mathcal{F}_0; E)$ such that $X_k \to X$ in L^1 for some random variable $X \in L^1(\mathcal{F}_0; E)$, and such that $\sigma(X_k) \subset \sigma(X)$ for every $k \geqslant 1$. If, for a given scaling $c\colon (-\mathbb{N}) \to (0, \infty)$, there exists $\ell \geqslant 0$ such that $h_c(X_k, \mathcal{F}) \leqslant \ell$ for every k sufficiently large, then $h_c(X; \mathcal{F}) \leqslant \ell$.*

Proof We denote by $k(\epsilon)$ an integer such that $\mathbb{E}\big[\rho(X_k, X)\big] \leqslant \epsilon$ for every $k \geqslant k(\epsilon)$. Hence, for $k \geqslant k(\epsilon)$, the inequality $H_n^{\epsilon}(X_k) \geqslant H_n^{2\epsilon}(X)$ holds for every n by definition of $H_n^{\epsilon}(\cdot)$ and Lemma 7.2.1.

We write the proof for the upper scaled entropy. The proof for the lower scaled entropy is similar. Set $a = h_c(X; \mathcal{F})$. We firstly check that $a < \infty$. Assuming $a = \infty$, there exists $\epsilon_0 > 0$ such that $\limsup_{n\to-\infty} \frac{H_n^{2\epsilon_0}(X)}{c(n)} > \ell + 1$. Therefore one can take $k_0 \geqslant k(\epsilon_0)$ sufficiently large in order that $h_c(X_{k_0}, \mathcal{F}) \leqslant \ell$ and such that $\limsup_{n\to-\infty} \frac{H_n^{\epsilon_0}(X_{k_0})}{c(n)} > \ell + 1$. But $\epsilon \mapsto H_n^{\epsilon}(X_{k_0})$ is decreasing, therefore the inequality $\limsup_{n\to-\infty} \frac{H_n^{\epsilon}(X_{k_0})}{c(n)} > \ell + 1$ holds for every $\epsilon \leqslant \epsilon_0$, a contradiction of the assumption of the lemma.

Knowing now that $a < \infty$, we check that $\ell \geqslant a$. Given $\delta > 0$, there exists $\epsilon_0 > 0$ such that $\limsup_{n\to-\infty} \frac{H_n^{2\epsilon_0}(X)}{c(n)} > a - \delta$. Taking $k \geqslant k(\epsilon_0)$, one gets $\limsup_{n\to-\infty} \frac{H_n^{\epsilon}(X_k)}{c(n)} > a - \delta$ for every $\epsilon \leqslant \epsilon_0$ because $\epsilon \mapsto H_n^{\epsilon}(X_k)$ is decreasing. Taking k sufficiently large in order that $h_c(X_k, \mathcal{F}) \leqslant \ell$, one finally gets $\ell \geqslant a$. □

The property given in the following proposition will be called the *left-continuity* of the scaled entropy.

Proposition 7.4.15 (Left-Continuity) *Let $\mathcal{F}$ be a filtration, $\mathcal{B} \subset \mathcal{F}_0$ a σ-field, and $(\mathcal{B}_k)_{k\geqslant 1}$ an increasing sequence of σ-fields such that $\mathcal{B}_k \nearrow \mathcal{B}$. Then $h_c(\mathcal{B}) = \lim h_c(\mathcal{B}_k)$ for any scaling c.*

Proof The inequality $h_c(\mathcal{B}) \geqslant h_c(\mathcal{B}_k)$ is an obvious consequence of the inclusion $\mathcal{B}_k \subset \mathcal{B}$. Passing to the limit, we get $h_c(\mathcal{B}) \geqslant \lim h_c(\mathcal{B}_k)$. To show the converse inequality, consider an integrable $\mathcal{B}$-measurable random variable X taking its values in $[0, 1]$. Then $X = \lim X_k$ in L^1 where X_k is a $\mathcal{B}_k$-measurable random variable. By Lemma 7.4.14, $h_c(X) \leqslant \sup h_c(X_k)$, therefore $h_c(X) \leqslant \sup h_c(\mathcal{B}_k)$. But $h_c(\mathcal{B}_k)$ is increasing in k, hence $\sup h_c(\mathcal{B}_k) = \lim h_c(\mathcal{B}_k)$. That shows that $h_c(\mathcal{B}) \leqslant \lim h_c(\mathcal{B}_k)$. □

Proposition 7.4.16 (Heredity to the Generated σ- Field) *Let $\mathcal{F}$ be a filtration and $X \in L^1(\mathcal{F}_0; E)$ where E is a Polish space. Then $h_c(\sigma(X)) = h_c(X)$ for any scaling $c\colon (-\mathbb{N}) \to (0, \infty)$.*

Proof If $Y = f(X)$ for some Lipschitz function f, then it is easy to check that $h_c(Y) \leqslant h_c(X)$ with the help of Lemma 7.2.2. The set of random variables $f(X)$, $f\colon E \to \mathbb{R}^m$ Lipschitzian, is dense in $L^1\big(\sigma(X); \mathbb{R}^m\big)$. Indeed, this is the content of lemma 2.15 in [10] when $m = 1$, and the case $m \geqslant 2$ obviously follows from the case $m = 1$. Therefore, by Lemma 7.4.14, we know that $h_c(Y) \leqslant h_c(X)$ for every random variable $Y \in L^1\big(\sigma(X); \mathbb{R}^m\big)$. The case $m = 1$ yields the inequality $h_c(\sigma(X)) \leqslant h_c(X)$. It remains to show the converse inequality.

Firstly, consider a random variable U taking its values in $[0, 1]$ and such that $\sigma(X) = \sigma(U)$. From what we have seen above, we know that $h_c(Y) \leqslant h_c(U)$, hence $h_c(Y) \leqslant h_c\big(\sigma(X)\big)$, for every random variable $Y \in L^1\big(\sigma(X); \mathbb{R}^m\big)$.

Now, consider a sequence $(X_k)_{k\geqslant 1}$ in $L^1\big(\sigma(X); E\big)$ such that $X_k \to X$ in L^1 and each X_k takes only finitely many values. We know that $h_c(X) \leqslant \sup h_c(X_k)$ by Lemma 7.4.14. For a given $k \geqslant 1$, denote by F the finite subset of E in which X_k takes its values. Note that $h_c(X_k)$ is the same either we look at X_k as a E-valued random variable or a F-valued random variable. Considering the sup-norm on $\mathbb{R}^m$, where $m = \#F$, it is well-known that there exists a distance-preserving function $f\colon F \to \mathbb{R}^m$. By Lemma 7.4.12, $h_c(X_k) = h_c\big(f(X_k)\big)$. But we have seen that $h_c\big(f(X_k)\big) \leqslant h_c\big(\sigma(X)\big)$, therefore we finally get the inequality $h_c(X) \leqslant h_c\big(\sigma(X)\big)$. □

Now we can quickly prove Theorem 7.4.11.

Proof (Proof of Theorem 7.4.11) Let $\mathcal{B}_n = \sigma(\pi_n X, \dots, \pi_{-1}X, X)$. Firstly, we get the equality $h_c(\mathcal{B}_n; \mathcal{F}^X) = h_c(\pi_n X; \mathcal{F}^X)$ by Proposition 7.4.16 (heredity to the generated σ-field), Lemma 7.4.13 and Lemma 7.4.12. Secondly, it is not difficult to see that $h_c(\pi_n X; \mathcal{F}^X) = h_c(X; \mathcal{F}^X)$. Thus, $h_c(\mathcal{B}_n; \mathcal{F}^X) = h_c(X; \mathcal{F}^X) = h_c(X; \mathcal{F})$. But $h_c(\mathcal{B}_n; \mathcal{F}^X) \to h_c(\mathcal{F}^X)$ by Proposition 7.4.15, and then the theorem follows. □

We provide the following corollary as an easy consequence of Proposition 7.4.15 and Proposition 7.4.16.

Corollary 7.4.17 *When the final σ-field $\mathcal{F}_0$ of a filtration $\mathcal{F}$ is essentially separable, then $h_c(\mathcal{F}) = \sup_X h_c(X)$ for any scaling c where the supremum is taken over all simple $\mathcal{F}_0$-measurable random variables X.*

Proof This supremum is obviously lower than $h_c(\mathcal{F})$. To show the converse inequality, take an increasing sequence $(\mathcal{B}_k)_{k\geqslant 1}$ of finitely generated sub-σ-fields of $\mathcal{F}_0$ such that $\mathcal{B}_k \nearrow \mathcal{F}_0$, and take a simple random variable X_k generating $\mathcal{B}_k$ for each $k \geqslant 1$. By Proposition 7.4.15 and Proposition 7.4.16,

$$h_c(\mathcal{F}) = \lim_{k\to\infty} h_c(\mathcal{B}_k) = \lim_{k\to\infty} h_c(X_k),$$

thereby showing the desired converse inequality. □

The next corollary is a consequence of Lemma 7.4.13 and Proposition 7.4.16. It generalizes Proposition 7.2.3 (standardness is an asymptotic property).

Corollary 7.4.18 (Entropy is an Asymptotic Quantity) *Let $\mathcal{F}$ be a filtration, $n_0 \leqslant 0$ be an integer, and denote by $\mathcal{F}^{n_0]} = (\mathcal{F}_{n_0+n})_{n\leqslant 0}$ the filtration $\mathcal{F}$ truncated at n_0. Let $c\colon (-\mathbb{N}) \to (0,\infty)$ be a scaling and denote $c^{n_0]} = (c_{n_0+n})_{n\leqslant 0}$ its truncation at n_0. Then $h_{c^{n_0]}}(\mathcal{F}^{n_0]}) = h_c(\mathcal{F})$.*

Proof It is not difficult to derive the equality

$$H_n^{\epsilon}(X_{n_0}, \mathcal{F}^{n_0]}) = H_{n_0+n}^{\epsilon}(X_{n_0}; \mathcal{F}) \tag{7.5}$$

for every integrable $\mathcal{F}_{n_0}$-measurable random variable X_{n_0}, every $n \leqslant 0$ and every $\epsilon > 0$. This provides the inequality $h_{c^{n_0]}}(\mathcal{F}^{n_0]}) \leqslant h_c(\mathcal{F})$.

Conversely, if $W^{n_0} = (\pi_{n_0}X_0, \dots, X_0)$ where X_0 is an integrable $\mathcal{F}_0$-measurable random variable, then $h_c(X_0; \mathcal{F}) \leqslant h_c(W^{n_0}; \mathcal{F})$ by Proposition 7.4.16 and because H is increasing (Definition 7.4.1). But Lemma 7.4.13 provides the equality

$$H_{n_0+n}^{\epsilon}(W^{n_0}; \mathcal{F}) = H_{n_0+n}^{\epsilon}(\pi_{n_0}X_0; \mathcal{F})$$

for every $n \leqslant 0$. Hence equality (7.5) gives

$$H_{n_0+n}^{\epsilon}(W^{n_0}; \mathcal{F}) = H_n^{\epsilon}(\pi_{n_0}X_0; \mathcal{F}^{n_0]}),$$

therefore $h_c(W^{n_0}; \mathcal{F}) = h_{c^{n_0]}}(\pi_{n_0}X_0; \mathcal{F}^{n_0]})$ and finally we get the inequality $h_c(X_0; \mathcal{F}) \leqslant h_{c^{n_0]}}(\pi_{n_0}X_0; \mathcal{F}^{n_0]})$. This provides the inequality $h_c(\mathcal{F}) \leqslant h_{c^{n_0]}}(\mathcal{F}^{n_0]})$. □

7.4.3 Extensions that do Not Increase Entropy

Say that a filtration $\mathcal{G}$ is an *extension* of a filtration $\mathcal{F}$ if the filtration $\mathcal{F}$ is immersible in $\mathcal{G}$ (see [4] or [10] if needed). An independent enlargement of a filtration $\mathcal{F}$ is an extension of $\mathcal{F}$. A *parametric extension* is another important kind of extension, defined as follows. A *superinnovation* of a filtration $\mathcal{F}$ is a sequence $(V_n)_{n\leqslant 0}$ of random variables such that each V_n is independent of $\mathcal{F}_{n-1}$ and satisfies $\mathcal{F}_n \subset \mathcal{F}_{n-1} \vee \sigma(V_n)$. It is also called a *parameterization* or a *governing process* in some published papers. Once that is given, the filtration $\mathcal{F}$ is immersed in the filtration $\mathcal{G}$ defined by $\mathcal{G}_n = \mathcal{F}_n \vee \sigma(V_m; m \leqslant n)$ (this follows from lemma 1.6 in [10]), and $\mathcal{G}$ is called a *parametric extension* of $\mathcal{F}$.

When $\mathcal{G}$ is a parametric extension or an independent enlargement of $\mathcal{F}$ with a Vershikian filtration, it is known that $\mathcal{G}$ is Vershikian if $\mathcal{F}$ is Vershikian. We generalize these two results by showing that in such cases, $\mathcal{G}$ has the same scaled entropy as $\mathcal{F}$. We will apply these two theorems on some examples in Sect. 7.6.

The next theorem is a generalization of proposition 6.1 in [11], which says that a parametric extension of a Vershikian filtration is a Vershikian filtration.

Theorem 7.4.19 (Parametric Extensions Do Not Increase Entropy) *Let $\mathcal{F}$ be a filtration and $(V_n)_{n\leqslant 0}$ be a superinnovation of $\mathcal{F}$. Denote by $\mathcal{G}$ the parametric extension of $\mathcal{F}$ defined by $\mathcal{G}_n = \mathcal{F}_n \vee \sigma(V_m; m \leqslant n)$. Then $h_c(\mathcal{G}) = h_c(\mathcal{F})$ for any scaling c.*

Proof By the left-continuity of the scaled entropy (Proposition 7.4.15), we know that $h_c(\mathcal{G}) = \lim h_c(\mathcal{B}_m)$ where $\mathcal{B}_m = \mathcal{F}_0 \vee \sigma(V_{m+1}, \dots, V_0)$. Thanks to the equality $\mathcal{B}_m = \mathcal{F}_m \vee \sigma(V_{m+1}, \dots, V_0)$, we get the inequality $h_c(\mathcal{B}_m; \mathcal{G}) \geqslant h_c(\mathcal{F}_m; \mathcal{G})$. Using the same equality, it is not difficult to check that $\pi_m X$ is measurable with respect to $\mathcal{F}_m$ for any $\mathcal{B}_m$-measurable integrable random variable X. Since $h_c(X) = h_c(\pi_m X)$, one gets $h_c(\mathcal{B}_m; \mathcal{G}) \leqslant h_c(\mathcal{F}_m; \mathcal{G})$ and finally $h_c(\mathcal{B}_m; \mathcal{G}) = h_c(\mathcal{F}_m; \mathcal{G})$. Now, $\mathcal{F}$ is immersed in $\mathcal{G}$, therefore $h_c(\mathcal{F}_m; \mathcal{G}) = h_c(\mathcal{F}_m; \mathcal{F})$. But $h_c(\mathcal{F}_m; \mathcal{F}) = h_c(\mathcal{F}_0; \mathcal{F})$ because of the equality $h_c(X) = h_c(\pi_m X)$ holding for every $\mathcal{F}_0$-measurable integrable random variable X. □

The corollary below is an application of the previous theorem. It generalizes corollary 6.1 of [11]. In the proof, we use the fact that for every Markov process $(X_n)_{n\leqslant 0}$, there exists a superinnovation $(V_n)_{n\leqslant 0}$ of the filtration generated by $(X_n)_{n\leqslant 0}$ satisfying the additional property $\sigma(X_{n-1}, V_n) \supset \sigma(X_n)$. This stems from lemma 3.41 in [10]. Note that the filtration $\mathcal{E}$ in the corollary is immersed in the filtration $\mathcal{F}_{\phi(\cdot)}$; that means that the process $(X_{\phi(n)})_{n\leqslant 0}$ is Markovian with respect to $\mathcal{F}_{\phi(\cdot)}$.

Corollary 7.4.20 *Let $(X_n)_{n\leqslant 0}$ be a Markov process and let $\mathcal{F}$ be the filtration it generates. Let $\phi\colon (-\mathbb{N}) \to (-\mathbb{N})$ be a strictly increasing map, denote by $\mathcal{F}_{\phi(\cdot)}$ the filtration $(\mathcal{F}_{\phi(n)})_{n\leqslant 0}$, and denote by $\mathcal{E}$ the filtration generated by the Markov process $(X_{\phi(n)})_{n\leqslant 0}$. Then, for any scaling c,*

$$h_c(X_{\phi(n)}; \mathcal{F}) \xrightarrow[n\to-\infty]{} h_c(\mathcal{F}),$$

$$h_c(X_{\phi(n)}; \mathcal{E}) \xrightarrow[n\to-\infty]{} h_c(\mathcal{E})$$

and $h_c(\mathcal{F}_{\phi(\cdot)}) = h_c(\mathcal{E})$.

Proof Take a superinnovation $(V_n)_{n\leqslant 0}$ of $\mathcal{F}$ such that $\sigma(X_{n-1}, V_n) \supset \sigma(X_n)$ for every $n \leqslant 0$, and let $\mathcal{G}$ be the parametric extension of $\mathcal{F}$ with $(V_n)_{n\leqslant 0}$. By Theorem 7.4.19, we know that $h_c(\mathcal{F}) = h_c(\mathcal{G})$. By Proposition 7.4.15 (left-continuity of the scaled entropy), we know that $h_c(\mathcal{G}) = \lim h_c(\mathcal{B}_n)$ where $\mathcal{B}_n = \sigma(X_{\phi(n)}, V_{\phi(n)+1}, \ldots, V_0)$. Now, by noting that the Vershik prediction $\pi_{\phi(n)}X$ is $\sigma(X_{\phi(n)})$-measurable for every $\mathcal{B}_n$-measurable random variable X, one gets $h_c(\mathcal{B}_n) = h_c(X_{\phi(n)})$. More precisely, we should write $h_c(\mathcal{B}_n; \mathcal{G}) = h_c(X_{\phi(n)}; \mathcal{G})$, but $h_c(X_{\phi(n)}; \mathcal{G}) = h_c(X_{\phi(n)}; \mathcal{F})$ because $\mathcal{F}$ is immersed in $\mathcal{G}$. Thus we have shown $h_c(\mathcal{F}) = \lim h_c(X_{\phi(n)})$. This result applied to the Markov process $(X_{\phi(n)})_{n\leqslant 0}$ and with the identity map instead of ϕ shows that $h_c(\mathcal{E}) = \lim h_c(X_{\phi(n)}; \mathcal{E})$, and note that $h_c(X_{\phi(n)}; \mathcal{E}) = h_c(X_{\phi(n)}; \mathcal{F}_{\phi(\cdot)})$ because $\mathcal{E}$ is immersed in $\mathcal{F}_{\phi(\cdot)}$. To finish, the equality $h_c(\mathcal{E}) = h_c(\mathcal{F}_{\phi(\cdot)})$ follows from an application of Theorem 7.4.19 to the filtration $\mathcal{E}$ and its superinnovation $\big((V_{\phi(n-1)+1}, \ldots, V_{\phi(n)})\big)_{n\leqslant 0}$. □

The next theorem is a generalization of lemma 19 in [4], which says that the supremum of two independent Vershikian filtrations is a Vershikian filtration (saying that two filtrations are independent when they are defined on the same probability space and their final σ-fields are independent). To prove it, we use the following lemma, whose proof is given in the proof of lemma 19 of [4]. Note that $\pi_n(R_1, R_2)$ in this lemma is defined with respect to the filtration $\mathcal{G} = \mathcal{F}^1 \vee \mathcal{F}^2$. On the other hand, each of the filtrations $\mathcal{F}^1$ and $\mathcal{F}^2$ is immersed in $\mathcal{G}$, therefore there is no possible ambiguity about the underlying filtration with respect to which $\pi_n R_1$ and $\pi_n R_2$ are defined.

Lemma 7.4.21 *Let $\mathcal{F}^1$ and $\mathcal{F}^2$ be two independent filtrations. Let (E_1, ρ_1) and (E_2, ρ_2) be two Polish metric spaces. We take the metric $\rho\big((x_1, x_2), (x_1', x_2')\big) = \rho_1(x_1, x_1') + \rho_2(x_2, x_2')$ on the product Polish space $E = E_1 \times E_2$. Then there are some Borelian maps $i_n : E_1^{(n)} \times E_2^{(n)} \to E^{(n)}$ such that*

$$\rho^{(n)}(i_n u, i_n v) \leqslant \rho_1^{(n)}(u_1, v_1) + \rho_2^{(n)}(u_2, v_2)$$

for every $u = (u_1, u_2)$ and $v = (v_1, v_2)$ in $E_1^{(n)} \times E_2^{(n)}$, and they are such that

$$\pi_n(R_1, R_2) = i_n(\pi_n R_1, \pi_n R_2)$$

for every $\mathcal{F}_0^1$-measurable random variable R_1 taking its values in (E_1, ρ_1) and every $\mathcal{F}_0^2$-measurable random variable R_2 taking its values in (E_2, ρ_2).

Theorem 7.4.22 *Let $\mathcal{F}^1$ and $\mathcal{F}^2$ be two independent filtrations. If $\mathcal{F}^2$ is Vershikian, then the equality $h_c(\mathcal{F}^1 \vee \mathcal{F}^2) = h_c(\mathcal{F}^1)$ holds for any scaling c.*

Proof We know that $h_c(\mathcal{F}^1 \vee \mathcal{F}^2) \geqslant h_c(\mathcal{F}^1)$ because $\mathcal{F}_1$ is immersed in $\mathcal{F}_1 \vee \mathcal{F}_2$. It remains to show $h_c(\mathcal{F}^1 \vee \mathcal{F}^2) \leqslant h_c(\mathcal{F}^1)$. Let $R = (R_1, R_2)$ where R_1 and R_2 are random variables as in the previous lemma. Take $\epsilon > 0$. Thanks to the

Vershik property of $\mathcal{F}^2$, for every small enough n there exists $s_2 \in E_2^{(n)}$ such that $\mathbb{E}\big[\rho_2^{(n)}(\pi_n R_2, s_2)\big] < \epsilon$. Take this n small enough in order to get a simple $\sigma(\pi_n R_1)$-measurable random variable S_1 such that $\mathbb{E}\big[\rho_1^{(n)}(\pi_n R_1, S_1)\big] < \epsilon$. The simple random variable $S = (S_1, s_2)$ has the same entropy as S_1, and by the previous lemma,

$$\rho^{(n)}(\pi_n R, i_n S) \leqslant \rho_1^{(n)}(\pi_n R_1, S_1) + \rho_2^{(n)}(\pi_n R_2, s_2),$$

therefore $\mathbb{E}\big[\rho^{(n)}(\pi_n R, i_n S)\big] < 2\epsilon$. Since $H(i_n S) \leqslant H(S)$, that shows that $H^{2\epsilon}(\pi_n R) \leqslant H^{\epsilon}(\pi_n R_1)$. Therefore $h_c(R; \mathcal{F}^1 \vee \mathcal{F}^2) \leqslant h_c(R_1; \mathcal{F}_1)$. Consequently, the inequality $h_c(\mathcal{F}^1 \vee \mathcal{F}^2) \leqslant h_c(\mathcal{F}^1)$ holds, and the proof is over. □

7.5 Entropy of Next-Jump Time Filtrations

In this section, we consider, for a given sequence $(p_n)_{n \leqslant 0}$ of jumping probabilities, the next-jump time process $(V_n)_{n \leqslant 0}$ introduced in Sect. 7.3. Its filtration is denoted, as before, by $\mathcal{F}$. Using the Shannon entropy as the measure of entropy, we study the scaled entropy of $\mathcal{F}$ in the Kolmogorovian non-standard case, that is, in view of Proposition 7.3.1 and Theorem 7.3.7, the case when $\sum p_n = \infty$ and $\sum p_n^2 < \infty$. It is understood that we consider this situation throughout this section. Two non-standard next-jump time filtrations defined by two distinct jumping probabilities sequences are not locally isomorphic, therefore there is no real interest to compare them with the scaled entropy. But this case-study is a nice example because it illustrates Theorem 7.4.11 and because of its simplicity.

We assume without loss of generality that $p_n < 1$ for every $n < 0$. Indeed, as seen in Sect. 7.3, $p_n = 1$ only for finitely many values (Lemma 7.3.2). Therefore, we can make this assumption by applying point (2) of Lemma 7.3.3 with the largest integer n_0 such that $p_{n_0} = 1$, and by knowing that the scaled entropy is an asymptotic quantity (Corollary 7.4.18).

Thus, by point (1) of Lemma 7.3.3, the identifiability condition ($\star$) holds for the next-jump time process when one takes $X_0 = V_{-1}$, and by point (3) of Lemma 7.2.6, we know that the filtration $(\mathcal{F}_n)_{n \leqslant -1}$ is generated by the process of Vershik progressive predictions $(\pi_n V_{-1})_{n \leqslant -1}$. Therefore the calculation of the scaled entropy $h_c(\mathcal{F})$ is greatly simplified by Theorem 7.4.11 because $h_c(\mathcal{F}) = h_c(V_{-1})$ by virtue of this theorem. Moreover, by point (3) of Lemma 7.2.6, one can replace the ϵ-entropy $H^{\epsilon}(\pi_n V_{-1})$ with the ϵ-entropy $H^{\epsilon}(V_n)$, considering that V_n takes its values in the set $\{0, \dots, |n|\}$ equipped with the iterated Kantorovich metric ρ_n.

But one can even replace the ρ_n with the discrete $0{-}1$ metric. Indeed, as seen in Sect. 7.3.3, the iterated Kantorovich metrics ρ_n satisfy $\rho_n(x, x') \geqslant \epsilon_0$ for every

$n \leqslant 0$ and every $x \neq x'$, where $\epsilon_0 > 0$ does not depend on n, x and x'. Therefore, one can replace $\mathbb{E}\big[\rho_n(V_n, S)\big]$ with $\Pr(V_n \neq S)$ in the definition of $H^\epsilon(V_n)$, because $h_c(V_{-1})$ pertains on $H^\epsilon(V_n)$ only for small ϵ.

To sum up, the lower and upper scaled entropies of $\mathcal{F}$ are

$$h_c^-(\mathcal{F}) = \lim_{\epsilon \to 0} \liminf_{n \to -\infty} \frac{H^\epsilon(V_n)}{c(n)} \quad \text{and} \quad h_c^+(\mathcal{F}) = \lim_{\epsilon \to 0} \limsup_{n \to -\infty} \frac{H^\epsilon(V_n)}{c(n)},$$

where V_n is viewed as a random variable taking its values in the set $\{0, \ldots, |n|\}$ equipped with the discrete 0−1 metric. Thus, the calculation of $h_c(\mathcal{F})$ comes down to a quite elementary problem. That does not mean its calculation is an easy task. This problem is investigated in [14], where the right member is denoted by $h_c(\mu)$, where μ is the measure on $\mathbb{N}$ we introduced before Proposition 7.3.1. We noted that this measure characterizes the sequence of jumping probabilities $(p_n)_{n \leqslant 0}$, which corresponds to the *reversed hazard rate* of μ in [14]. The next-jump time process $(V_n)_{n \leqslant 0}$ is the process $(X_{-n})_{n \leqslant 0}$ with the notations of [14].

It is clear that the inequality $h_c(\mathcal{F}) \leqslant 1$ always holds for the scaling $c(n) = H(V_n)$ because $H^\epsilon(V_n) \leqslant H(V_n)$. In [14], it is shown that $H(V_n) \to \infty$ in the Kolmogorovian non-standard situation. With this scaling, using the examples studied in [14], one has some examples such that $h_c(\mathcal{F}) = r$ for any $r \in [0, 1]$. For instance, $h_c(\mathcal{F}) = 1$ in the uniform case $p_n = (|n| + 1)^{-1}$. The jumping probabilities of the examples in [14] yielding $h_c(\mathcal{F}) < 1$ are shown on Fig. 7.3. For the examples giving $h_c(\mathcal{F}) = 0$ for the scaling $c(n) = H(V_n)$, a scaling c' such that $h_{c'}(\mathcal{F}) = 1$ is derived in [14].

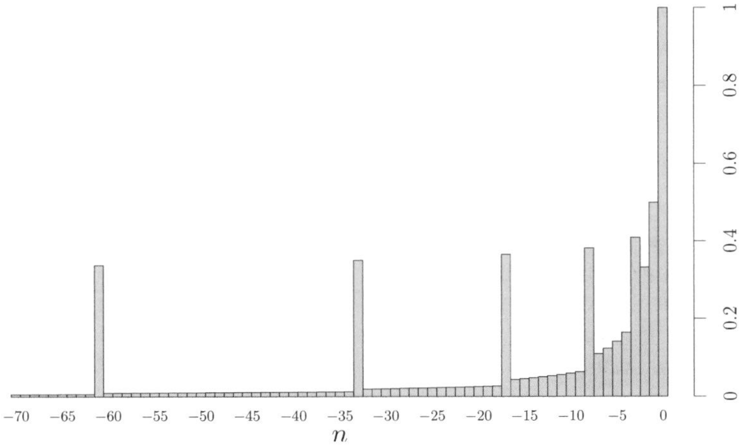

Fig. 7.3 Jumping probabilities yielding $h_c(\mathcal{F}) < 1$

7.6 Entropy of Poly-Adic Filtrations

The pioneering works of Vershik focused on poly-adic filtrations, that is to say, filtrations $\mathcal{F}$ such that for every $n \leqslant 0$, there exists a random variable η_n uniformly distributed on a finite set, independent of $\mathcal{F}_{n-1}$, and such that $\mathcal{F}_n = \mathcal{F}_{n-1} \vee \sigma(\eta_n)$. Such a random variable η_n is called an *innovation* of $\mathcal{F}$ (at time n), and denoting by r_n the size of the set on which it is uniformly distributed, $\mathcal{F}$ is said to be $(r_n)_{n\leqslant 0}$-adic (this makes sense because any other innovation of $\mathcal{F}$ at time n is uniformly distributed on r_n values). For such filtrations, Vershik defined the *exponential entropy*, originally in [18]. We will give this definition below.

In spite of the equality $\mathcal{F}_n = \mathcal{F}_m \vee \sigma(\eta_{m+1}, \dots, \eta_n)$ holding for every $m < n \leqslant 0$, the Kolmogorov property $\mathcal{F}_{-\infty} = \{\varnothing, \Omega\}$ does not ensure that $\mathcal{F}_n = \sigma(\eta_m; m \leqslant n)$. In other words, it does not ensure that $\mathcal{F}$ is generated by the process of innovations $(\eta_n)_{n\leqslant 0}$ In fact, standardness of a such a filtration is known to be equivalent to the existence of a process of innovations $(\eta'_n)_{n\leqslant 0}$ generating this filtration. This is one of the main results of Vershik's theory of filtrations. The difficult point to prove in this result is the existence of $(\eta'_n)_{n\leqslant 0}$ assuming standardness, whereas the converse is easy to prove with the help of Proposition 7.2.4.

Throughout this section, when a $(r_n)_{n\leqslant 0}$-adic filtration is under consideration, we denote by $(\ell_n)_{n\leqslant 0}$ the integer sequence associated to $(r_n)_{n\leqslant 0}$ by setting $\ell_n = \prod_{i=n+1}^{0} r_i$ (agreeing with $\ell_0 = 1$).

In Sect. 7.6.1, with the help of Lemma 7.6.3, we will see that the definition below makes sense. In particular, we will notice that the $\pi_n X$ take only finitely many values.

Definition 7.6.1 Let $\mathcal{F}$ be a $(r_n)_{n\leqslant 0}$-adic filtration and X a $\mathcal{F}_0$-measurable random variable taking only finitely many values.

1. The exponential entropy of X with respect to $\mathcal{F}$ is the number
$$h(X; \mathcal{F}) = \lim_{n\to-\infty} \frac{H(\pi_n X)}{\ell_n} = \inf_{n\leqslant 0} \frac{H(\pi_n X)}{\ell_n} \in [0, \infty)$$
where $\ell_n = \prod_{i=n+1}^{0} r_n$ and H, unless something else is said, is the Shannon entropy (in a given logarithmic base).
2. The exponential entropy, of $\mathcal{F}$ is $h(\mathcal{F}) = \sup h(X; \mathcal{F}) \in [0, \infty]$ where the supremum is taken over all $\mathcal{F}_0$-measurable random variables X taking only finitely many values.

Note the obvious inequality $h_c(\mathcal{F}) \leqslant h(\mathcal{F})$ when c is the scaling $c(n) = \ell_n$.

It is shown in [20] that $h(\mathcal{F}) = 0$ when $\mathcal{F}$ is a standard poly-adic filtration.[2] We will not use this result in the present paper. But in the case of the slowness condition (Δ) about the poly-adicity sequence $(r_n)_{n\leqslant 0}$, it follows from Theorem 7.6.4, and in the case of the opposite condition $(\neg\Delta)$, it follows from Theorem 7.6.14 (see below).

The main results of this section are listed below, where the two conditions (Δ) and (∇) about the speed of the poly-adicity sequence $(r_n)_{n\leqslant 0}$, respectively a slowness condition and a fastness condition, are

$$(\Delta): \sum_{n=-\infty}^{0} \frac{r_n \log r_n}{\ell_{n-1}} < \infty \qquad \text{and} \qquad (\nabla): \frac{\log r_n}{\ell_n} \to \infty.$$

1. Theorem 7.6.4, whose credit is given to Gorbulsky ([6]), is about the equality $h_c(\mathcal{F}) = h(\mathcal{F})$ between the exponential entropy and the scaled entropy with scaling $c(n) = \ell_n$ under the slowness condition (Δ).
2. Theorem 7.6.10, due to Vershik ([18, 20]), states that poly-adic filtrations have a zero exponential entropy under the fastness condition (∇), which is stronger than the negation $(\neg\Delta)$ of the slowness condition (Δ).
3. Theorem 7.6.14 includes Theorem 7.6.10 when the Kolmogorovian assumption holds. It states that Kolmogorovian poly-adic filtrations have a zero exponential entropy when the slowness condition (Δ) is not fulfilled. Its proof relies on a highly non-trivial theorem by Heicklen [7], which belongs to ergodic theory rather than probability theory.
4. Theorem 7.6.15, due to Vershik [18, 20], gives the value of the exponential entropy for the split-word filtrations under the slowness condition (Δ).

Gorbulsky showed Theorem 7.6.4 in the dyadic case $r_n \equiv 2$ only. Our proof of the generalization to condition (Δ) essentially uses the same mathematics. In addition to the points listed above, we will investigate the entropy of some filtrations called *filtrations of unordered pairs* in [22], which are more or less the filtrations $\mathcal{F}^{X_0}$ generated by the process $(\pi_n X_0)_{n\leqslant 0}$ of the Vershik progressive predictions of the final letter X_0 of a split-word process $(X_n)_{n\leqslant 0}$.

7.6.1 The $\pi_n X$ in Poly-Adic Filtrations and the Exponential Entropy

For poly-adic filtrations, the Vershik progressive predictions $\pi_n X$ and the iterated Kantorovich distances $\rho^{(n)}$ as defined in Sect. 7.2 have a convenient representation,

[2] It is clear that $h(X; \mathcal{F}) = 0$ when X is measurable with respect to $\sigma(\eta_n, \ldots, \eta_0)$ for any process of innovations $(\eta_n)_{n\leqslant 0}$. Thus, knowing that standardness of a poly-adic filtration means the existence of a generating process of innovations, $h(\mathcal{F}) = 0$ follows from a result similar to the left-continuity property of the scaled entropy (Proposition 7.4.15).

the one given in the following lemma which is a consequence of lemma 4.6 and lemma 4.7 in [13]. For our purposes, we only state this result for a countable state space A equipped with the $0-1$ metric. In this lemma and hereafter, it is understood that G_n is the group of automorphisms of the $(r_{n+1}, \dots, r_0)$-ary tree. If needed, the reader is referred to [13] for details about the group G_n of tree automorphisms and its action on the set of ℓ_n-words A^{ℓ_n}.

Lemma 7.6.2 *Let $\mathcal{F}$ be a $(r_n)_{n\leqslant 0}$-adic filtration and X a $\mathcal{F}_0$-measurable random variable taking its values in a countable set A. Then $\pi_n X$ can be identified to the G_n-orbit of a random word X_n on A having length ℓ_n. Using this identification and starting with the $0-1$ metric ρ on A, the n-th iterated Kantorovich metric $\rho^{(n)}$ on $A^{(n)}$ is transported to the metric $\bar{\rho}_n$ on the quotient set A^{ℓ_n}/G_n, given by*

$$\bar{\rho}_n(\Gamma, \Gamma') = \min_{w\in\Gamma, w'\in\Gamma'} \delta_n(w, w')$$

for every pair of orbits Γ and Γ', where $\delta_n(w, w')$ is the Hamming distance between the ℓ_n-words w and w' (the proportion of positions at which the letters of w and w' differ).

We will use this lemma throughout this section. Though we do not provide its proof, it is easy to derive it from the first part of the following lemma.

Lemma 7.6.3 *Let $\mathcal{B}_{-1}$ be a σ-field and ϵ_0 a random variable independent of $\mathcal{B}_{-1}$ uniformly taking its values in a set with finite size $r_0 \geqslant 2$, which we assume to be $\{1, \dots, r_0\}$ without loss of generality. Define the σ-field $\mathcal{B}_0 = \mathcal{B}_{-1} \vee \sigma(\epsilon_0)$.*

Let X_0 be a $\mathcal{B}_0$-measurable random variable taking its values in a Polish space A.

1. *There exist r_0 random variables $X_{-1}(1), \dots, X_{-1}(r_0)$, taking their values in A and measurable with respect to $\mathcal{B}_{-1}$, and such that $X_0 = X_{-1}(\epsilon_0)$.*
2. *For such random variables and when A is countable, one has $H(X_{-1}) \leqslant r_0 H(X_0)$, where $X_{-1} = \big(X_{-1}(1), \dots, X_{-1}(r_0)\big)$ and H is the Shannon entropy.*

Proof For the first point, write $X_0 = f(B_{-1}, \epsilon_0)$ for some $\mathcal{B}_{-1}$-measurable random variable B_{-1} and some measurable function f, and set $X_{-1}(i) = f(B_{-1}, i)$. For the second point, check that the law of X_0 is the average law of the $X_{-1}(i)$, hence $H(X_{-1}) \leqslant H\big(X_{-1}(1)\big) + \cdots + H\big(X_{-1}(r_0)\big) \leqslant r_0 H(X_0)$ by concavity of the Shannon entropy. □

This lemma justifies that the exponential entropy is well defined (Definition 7.6.1). Indeed, apply it with $\mathcal{B}_{-1} = \mathcal{F}_n$, $\epsilon_0 = \eta_{n+1}$, $X_{-1} = \pi_n X$ and $X_0 = \pi_{n+1} X$. Then the first part shows that $\pi_n X$ takes only finitely many values, and the second part implies $H(\pi_n X) \leqslant r_{n+1} H(\pi_{n+1} X)$ and then, by recursively applying this inequality one gets that the quantity $\frac{H(\pi_n X)}{\ell_n}$ is increasing, thereby justifying the equality $\lim_{n\to-\infty} \frac{H(\pi_n X)}{\ell_n} = \inf_{n\leqslant 0} \frac{H(\pi_n X)}{\ell_n}$.

7.6.2 Gorbulsky's Theorem

In [6], Gorbulsky proved Theorem 7.6.4 below in the case when $r_n \equiv 2$. We show that this result more generally holds for poly-adicity sequences $(r_n)_{n\leqslant 0}$ satisfying the slowness condition

$$(\Delta): \qquad \sum_{n=-\infty}^{0} \frac{\log r_n!}{\ell_{n-1}} < \infty, \quad \text{equivalent to} \quad \sum_{n=-\infty}^{0} \frac{r_n \log r_n}{\ell_{n-1}} < \infty.$$

For example, all bounded sequences $(r_n)_{n\leqslant 0}$ satisfy condition (Δ), and it is also fulfilled in the case when $r_n = |n| + 1$.

Theorem 7.6.4 *Let $\mathcal{F}$ be a $(r_n)_{n\leqslant 0}$-adic filtration. Assume that condition (Δ) is fulfilled by the poly-adicity sequence $(r_n)_{n\leqslant 0}$ and consider the scaling $c(n) = \ell_n$. Then the scaled entropy of X equals its exponential entropy:*

$$h_c(X; \mathcal{F}) = \lim_{n\to-\infty} \frac{H(\pi_n X)}{\ell_n}$$

for every $\mathcal{F}_0$-measurable random variable X taking only finitely many values. Consequently the scaled entropy $h_c(\mathcal{F})$ of $\mathcal{F}$ equals its exponential entropy $h(\mathcal{F})$.

In this theorem, it is understood that we use the Shannon entropy as the measure of entropy in the scaled entropy. The last claim of the theorem is derived from Corollary 7.4.17.

This section is devoted to the proof of this theorem. Only Lemma 7.6.8 below will be involved in the proof. It is a consequence of Lemma 7.6.6 and Lemma 7.6.7, and the result of elementary analysis stated in the following lemma will be used in Lemma 7.6.6.

Lemma 7.6.5 *Let $(u_n)_{n\geqslant 0}$ and $(v_n)_{n\geqslant 0}$ be two sequences of positive numbers. Assume that $v_n \searrow 0$ and $\sum u_n v_n < \infty$. Then $\epsilon \sum_{i=0}^{n(\epsilon)} u_i \to 0$ when $\epsilon \to 0^+$, where $n(\epsilon) = \min\{n \mid v_{n+1} < \epsilon\}$.*

Proof Let $\delta > 0$. Take M_1 such that $\sum_{i=m+1}^{n} u_i v_i < \delta$ whenever $n > m \geqslant M_1$. Now take M_2 such that $v_m \sum_{i=0}^{M_1} u_i < \delta$ whenever $m \geqslant M_2$. Set $N = \max\{M_1, M_2\}$. If $\epsilon \leqslant v_{N+1}$, then $n(\epsilon) > N$ and

$$\epsilon \sum_{i=0}^{n(\epsilon)} u_i \leqslant v_{n(\epsilon)} \sum_{i=0}^{n(\epsilon)} u_i \leqslant v_{n(\epsilon)} \sum_{i=0}^{M_1} u_i + \sum_{i=M_1+1}^{n(\epsilon)} u_i v_i < 2\delta,$$

thereby showing the desired result. □

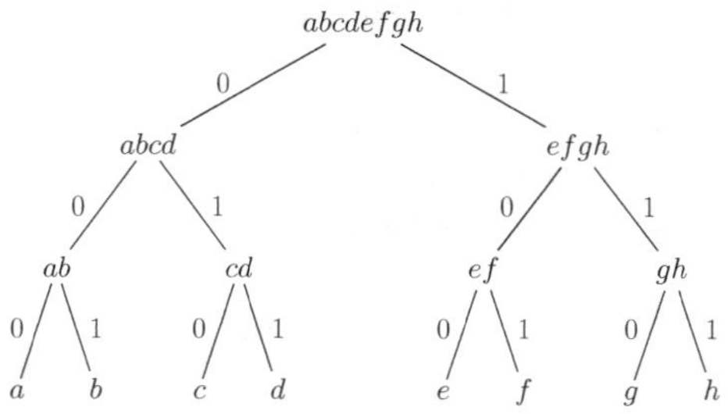

Fig. 7.4 The tree structure of the word $abcdefgh$

In the next lemmas, we denote by $|B|$ the number of words contained in a subset $B \subset A^{\ell_n}$. Recall that $\bar{\rho}_n$ in the following lemma is the quotient distance introduced in Lemma 7.6.2.

Lemma 7.6.6 *Assume that condition* (Δ) *is fulfilled by the poly-adicity sequence* $(r_n)_{n\leqslant 0}$. *For any pair of orbits* $\Gamma, \Gamma' \in A^{\ell_n}/G_n$,

$$\frac{\big|\log|\Gamma| - \log|\Gamma'|\big|}{\ell_n} \leqslant d\big(\bar{\rho}_n(\Gamma, \Gamma')\big),$$

where d *is a function satisfying* $\lim_{\epsilon\to 0} d(\epsilon) = 0$.

Proof Consider a word w of length ℓ_n with its tree structure as shown by Fig. 7.4. We denote by $\Gamma(w)$ its G_n-orbit.

The word w at level n is the concatenation of the r_{n+1} words w_i of length ℓ_{n+1} at level $n+1$. If the G_{n+1}-orbits $\Gamma(w_i)$ of the subwords w_i are pairwise distinct, then $|\Gamma(w)| = r_n!|\Gamma(w_1)|\ldots|\Gamma(w_{r_n})|$. If they are all equal, then $|\Gamma(w)| = |\Gamma(w_1)|\ldots|\Gamma(w_{r_n})|$. Generally, $|\Gamma(w)| = M_{n,1}|\Gamma(w_1)|\ldots|\Gamma(w_{r_n})|$ where $M_{n,1}$ is a multinomial coefficient lying between 1 and $r_{n+1}!$. Continuing so on, we find

$$\log|\Gamma(w)| = \sum_{j=n}^{-1}\sum_{i=1}^{\ell_n/\ell_j} \log M_{j,i}$$

where $M_{j,i}$ is a multinomial coefficient lying between 1 and $r_{j+1}!$. Writing another word w' of length ℓ_n in the same way, we get

$$\big|\log|\Gamma(w)| - \log|\Gamma(w')|\big| \leqslant \sum_{j=n}^{-1}\sum_{i=1}^{\ell_n/\ell_j} |\log M_{j,i} - \log M'_{j,i}|.$$

Each deviation $|\log M_{j,i} - \log M'_{j,i}|$ is bounded by $\log r_{j+1}!$. If the letters of w and w' differ at $\epsilon\ell_n$ positions, then at each level j there are at most $\min\left(\frac{\ell_n}{\ell_j}, \epsilon\ell_n\right)$

non-zero deviations $|\log M_{j,i} - \log M'_{j,i}|$. Therefore,

$$\begin{aligned}\frac{\big|\log|\Gamma(w)| - \log|\Gamma(w')|\big|}{\ell_n} &\leqslant \sum_{j=n}^{k(\epsilon)-1} \frac{\log r_{j+1}!}{\ell_j} + \epsilon \sum_{j=k(\epsilon)}^{-1} \log r_{j+1}! \\ &\leqslant \sum_{j=n}^{k(\epsilon)-1} \frac{r_{j+1}\log r_{j+1}}{\ell_j} + \epsilon \sum_{j=k(\epsilon)}^{-1} r_{j+1}\log r_{j+1}\end{aligned}$$

where $k(\epsilon) = \max\{k \mid \ell_{k-1}^{-1} < \epsilon\}$.

Under the (Δ) condition, the first sum in the right member goes to $\sum_{j=-\infty}^{k(\epsilon)-1} \frac{r_{j+1}\log r_{j+1}}{\ell_j}$ when $n \to -\infty$, and this goes to 0 when $\epsilon \to 0$ because $k(\epsilon)$ goes to $-\infty$. The second sum goes to 0 too because of Lemma 7.6.5. □

Lemma 7.6.7 *For any subset $B \subset A^{\ell_n}$, the log-number of words in an ϵ-neighbourhood of B does not exceed*

$$\log|B| + h(\epsilon)\ell_n + \epsilon\ell_n \log(\#A - 1)$$

where $h(\epsilon) = \epsilon \log \frac{1}{\epsilon} + (1-\epsilon)\log \frac{1}{1-\epsilon}$.

Proof The words in an ϵ-ball around a ℓ_n-word w are obtained by taking $k = 0, \ldots, \lfloor \epsilon\ell_n \rfloor$ positions in w and changing the letters at these positions. Then the number of such words is bounded above by

$$(\#A - 1)^{\lfloor \epsilon\ell_n \rfloor} \times \sum_{k=0}^{\lfloor \epsilon\ell_n \rfloor} \binom{\ell_n}{k}.$$

The lemma follows from the inequality

$$\log \sum_{k=0}^{\lfloor \epsilon\ell_n \rfloor} \binom{\ell_n}{k} \leqslant h(\epsilon)\ell_n,$$

which is derived from the classical large deviations inequality for independent symmetric Bernoulli variables (corollary 2.20 in [16]). □

Lemma 7.6.8 *Under condition (Δ), for any orbit $\Gamma \in A^{\ell_n}/G_n$, the log-number of orbits in an ϵ-neighborhood of Γ (for the $\bar{\rho}_n$ distance) does not exceed a value L_n^ϵ satisfying*

$$\lim_{\epsilon\to 0} \limsup_{n\to-\infty} \frac{L_n^\epsilon}{\ell_n} = 0.$$

Proof The number of orbits in an ϵ-neighbourhood $V_\epsilon(\Gamma)$ of Γ is less than the number of words in $V_\epsilon(\Gamma)$ divided by the minimal length of an orbit Γ' in $V_\epsilon(\Gamma)$. Let L_n^ϵ be this ratio. Applying the two previous lemmas yields the desired result:

$$L_n^\epsilon \leqslant \big[\log|\Gamma| + h(\epsilon)\ell_n + \epsilon\log(\#A-1)\ell_n\big] - \big[\log|\Gamma| - d(\epsilon)\ell_n\big]$$
$$= \ell_n\big(h(\epsilon) + \epsilon\log(\#A-1) + d(\epsilon)\big).$$

□

Proof (Proof of Theorem 7.6.4) In the proof we use the notation

$$H(E) = -\sum_{x\in E}\mu(x)\log\mu(x)$$

where μ is the law of $\pi_n X$ and E is any set of G_n-orbits. Note that

$$H(E) \leqslant \mu(E)\log\#E - \mu(E)\log\mu(E)$$

because $H(E) = \mu(E)H(\mu') - \mu(E)\log\mu(E)$ where $\mu' = \mu(\cdot \mid E)$.

We use the ϵ-entropy as defined in Lemma 7.4.9. Let $\{A_i, B\}$ be a partition achieving $H_n^\epsilon(X)$, with $A_i \subset B(\Gamma_i, \epsilon)$ and $\mu(B) \leqslant \epsilon$. One has

$$H_n^\epsilon(X) = -\sum_i \mu(A_i)\log\mu(A_i) - \mu(B)\log(B)$$

and

$$H(\pi_n X) = \sum_i H(A_i) + H(B).$$

Firstly, $H(B) \leqslant (\log\#A)\epsilon\ell_n - \epsilon\log\epsilon$. On the other hand,

$$H(A_i) \leqslant \mu(A_i)L_n^\epsilon - \mu(A_i)\log\mu(A_i),$$

hence

$$\sum_i H(A_i) \leqslant H_n^\epsilon(X) + L_n^\epsilon.$$

Thus,

$$\frac{H_n^\epsilon(X)}{\ell_n} \leqslant \frac{H(\pi_n X)}{\ell_n} \leqslant \frac{H_n^\epsilon(X)}{\ell_n} + \frac{L_n^\epsilon - \epsilon\log\epsilon}{\ell_n} + (\log\#A)\epsilon,$$

thereby yielding the theorem by applying Lemma 7.6.8. □

The following corollary just emphasizes that the equality $h_c(X) = h(X)$ of Theorem 7.6.4 can be stated not only for poly-adic filtrations, but also for filtrations immersed in a poly-adic filtration.

Corollary 7.6.9 *Let $\mathcal{F}$ be a filtration immersed in a $(r_n)_{n\leqslant 0}$-adic filtration. Consider the scaling $c(n) = \ell_n$. If $(r_n)_{n\leqslant 0}$ fulfills the (Δ) condition, then*

$$h_c(X) = \lim_{n\to-\infty} \frac{H(\pi_n X)}{\ell_n}$$

for every simple $\mathcal{F}_0$-measurable random variable X.

Proof This is a direct consequence of Theorem 7.6.4 and the immersion property (Remark 7.4.8). □

7.6.3 (∇)-Adic Filtrations Have Zero Exponential Entropy

The (∇) condition is stronger than the negation of the (Δ) condition, because this condition is the divergence of a certain sequence whereas the (Δ) condition is the convergence of the series made up of the same sequence:

$$(\nabla): \qquad \frac{\log r_n}{\ell_n} \to \infty.$$

Vershik's following theorem is proved in [20]. Note that the hypotheses do not require $\mathcal{F}$ to be Kolmogorovian.

Theorem 7.6.10 *Let $\mathcal{F}$ be a (∇)-adic filtrations. Then it has zero exponential entropy when we use the max-entropy as the underlying measure of entropy H (see above Definition 7.4.1). Consequently it also has zero exponential entropy when we use the Shannon entropy as the underlying measure of entropy, and the same result holds for the scaled entropy with the scaling $c(n) = \ell_n$.*

The proof is based on the following combinatorial lemma.

Lemma 7.6.11 *For an alphabet A having size $\#A = p$, the number of orbits χ_n^p of the action of G_n on A^{ℓ_n} is given by $\chi_0^p = p$ and the recurrence formula*

$$\chi_{n-1}^p = \binom{r_n + \chi_n^p - 1}{r_n},$$

and one has $\log \chi_n^p = o(\ell_n)$ when condition (∇) holds.

Proof The recurrence formula obviously stems from the fact that an orbit for the action of G_{n-1} is obtained by choosing a list of r_n orbits for the action of G_n, with possible repetitions. Then

$$\log \chi_{n-1}^{p} = \log\left[(r_n + \chi_n^p - 1)\cdots(r_n+1)\right] - \log\left[(\chi_n^p - 1)!\right] \leqslant 2\chi_n^p \log r_n$$

by the inequality $\log(r_n + 1) \leqslant 2\log r_n$. Set $t_n = \frac{\log \chi_n^p}{\ell_n}$. Note that $r_n = \exp(\beta_n \ell_n)$ where β_n is the quantity going to ∞ under the (∇) condition. Thus

$$t_{n-1} \leqslant \frac{2\chi_n^p \log r_n}{\ell_{n-1}} = \frac{2\exp(t_n\ell_n)\beta_n\ell_n}{\ell_n \exp(\beta_n\ell_n)} = 2\beta_n \exp\big((t_n - \beta_n)\ell_n\big).$$

The number of orbits cannot exceed the number of words, and this yields the inequality $t_n \leqslant \log p$. Therefore the right member of the last inequality goes to 0 under the (∇) condition. □

Proof (Proof of Theorem 7.6.10) Given any $\mathcal{F}_0$-measurable random variable X taking its values in a finite set A, one has $H_0(\pi_n X) \leqslant \log \chi_n^{\#A}$ with the notations of the previous lemma, where H_0 is the max-entropy. Then the result for the max-entropy follows from this lemma, and the result for the Shannon entropy H follows because of $H \leqslant H_0$. The result for the scaled entropy follows from the obvious inequality $h_c(\mathcal{F}) \leqslant h(\mathcal{F})$. □

7.6.4 $(\neg\Delta)$-Adic Filtrations Have Zero Exponential Entropy

Theorem 7.6.14 given in this section implies Theorem 7.6.10 in the case of a Kolmogorovian filtration, but its proof relies on Heicklen's theorem, rephrased below in Theorem 7.6.13, whose proof is far to be trivial and belongs to ergodic theory rather than probability theory.

Heicklen's theorem derived in [7] deals with the ergodic free actions of the group $\Gamma = \bigoplus_{n=-\infty}^{0} \mathbb{Z}/r_n\mathbb{Z}$ on a Lebesgue space, where $(r_n)_{n\leqslant 0}$ is a sequence of integers as before. One can write $\Gamma = \cup_{n=0}^{-\infty}\Gamma_n$ where $\Gamma_n = \sum_{k=n+1}^{0} \mathbb{Z}/r_k\mathbb{Z}$. Then, when a Γ-action on a Lebesgue space is given, one can associate to it a filtration $(\mathcal{F}_n)_{n\leqslant 0}$ on the Lebesgue space by defining $\mathcal{F}_n$ as the σ-field of Γ_n-invariant sets. This filtration is Kolmogorovian when the Γ-action is ergodic, and it is $(r_n)_{n\leqslant 0}$-adic when the Γ-action is free. Conversely, any Kolmogorovian $(r_n)_{n\leqslant 0}$-adic filtration on a Lebesgue space can be derived in this way from a free ergodic Γ-action. We refer the reader to [7] or [5] for these claims.

In the present paper we never assume that the filtrations $\mathcal{F}$ are defined on a Lebesgue space. But when the final σ-field $\mathcal{F}_0$ is essentially separable, then $\mathcal{F}$ is isomorphic to a Lebesgue space (see [1]). Thus, when restricting oneself to filtrations $\mathcal{F}$ whose final σ-field $\mathcal{F}_0$ is essentially separable, one can say that every Kolmogorovian $(r_n)_{n\leqslant 0}$-adic filtration comes from a free ergodic Γ-action in the above way.

The exponential entropy $h(\mathcal{F})$ of the filtration $\mathcal{F}$ corresponding to a Γ-action T and the entropy $h(T)$ of T are related, as shown by the next lemma. We refer to [5] for the definition of $h(T)$.

Lemma 7.6.12 *Let T be a free ergodic Γ-action on a Lebesgue space, and $\mathcal{F}$ be its corresponding $(r_n)_{n\leqslant 0}$-adic filtration. Then $h(\mathcal{F}) \leqslant h(T)$.*

Proof Let X_0 be a simple $\mathcal{F}_0$-measurable taking its values in a finite set A. Let P_0 be a finite partition of the Lebesgue space generating the same σ- field as X_0. Define the partition $P_n = \bigvee_{g\in\Gamma_n} T^g P_0$. The σ-field generated by the partition P_n is the same as the σ-field generated by a $\mathcal{F}_n$-measurable random word X_n having ℓ_n letters on the alphabet A, and which is one of the r_n subwords of X_{n-1}, uniformly taken at random, when one splits X_{n-1} into r_n subwords each having ℓ_n letters. Therefore, the n-th Vershik progressive prediction $\pi_n X_0$ of X_0 is the orbit of X_n under the action of the group G_n of tree automorphims. We refer to [13] for this claim, as we referred for Lemma 7.6.2. Hence, the exponential entropy $h(X_0; \mathcal{F})$ of X_0,

$$h(X_0; \mathcal{F}) := \lim_{n\to-\infty} \frac{H(\pi_n X_0)}{\ell_n},$$

satisfies

$$h(X_0; \mathcal{F}) \leqslant \lim_{n\to-\infty} \frac{H(X_n)}{\ell_n} =: h(T, P_0) \leqslant h(T) := \sup_{P \text{ finite partition}} h(T, P).$$

Therefore $h(\mathcal{F}) \leqslant h(T)$. □

Heicklen proved the following theorem.

Theorem 7.6.13 (Heicklen [7]) *Let T be a free ergodic Γ-action on a Lebesgue space, and $\mathcal{F}$ be its corresponding filtration. When the (Δ) condition about the sequence $(r_n)_{n\leqslant 0}$ does not hold, there exists an action S with zero entropy and whose corresponding filtration is isomorphic to $\mathcal{F}$.*

From this theorem and the previous lemma, one easily gets the following theorem.

Theorem 7.6.14 *Let $\mathcal{F}$ be a Kolmogorovian $(\neg\Delta)$-adic filtration, whose final σ-field $\mathcal{F}_0$ is essentially separable. Then it has a zero exponential entropy, and consequently it also has a zero scaled entropy $h_c(\mathcal{F})$ with the scaling $c(n) = \ell_n$.*

Proof Up to isomorphism, $\mathcal{F}$ is a filtration on a Lebesgue space, and it is the filtration corresponding by the way explained above to a free ergodic Γ-action on this Lebesgue space. The result follows from Heicklen's theorem and Lemma 7.6.12. □

7.6.5 (Δ)-Adic Split-Word Filtrations

The poly-adic filtrations of the split-word processes with i.i.d. letters were studied in [3, 4, 10, 15]. In the more general case of stationary letters, standardness of these filtrations is closely connected, as shown in [13], to the notion of scale of an automorphism introduced by Vershik in [19]. Theorem 7.6.15 below, which is a rephrasing of theorem 4.1 in [20], provides the exponential entropy of these filtrations under condition (Δ). The hypotheses of the theorem do not require $\mathcal{F}$ to be Kolmogorovian.

Given a sequence of integers $(r_n)_{n\leqslant 0}$, setting as before $\ell_n = r_{n+1}\dots r_0$, and given an alphabet A, a $(r_n)_{n\leqslant 0}$*-adic split-word process on* A is a Markov process $(X_n, \eta_n)_{n\leqslant 0}$ satisfying conditions below for each $n \leqslant 0$, where we denote by $\mathcal{F}$ the filtration it generates:

- X_n is a random word on A of length ℓ_n;
- η_n is a random variable uniformly distributed on $\{1, 2, \dots, r_n\}$ and is independent of $\mathcal{F}_{n-1}$, and the word X_n is the η_n-th letter of of X_{n-1} treated as a r_n-word on A^{ℓ_n}.

Obviously the filtration $\mathcal{F}$ generated by $(X_n, \eta_n)_{n\leqslant 0}$ is a $(r_n)_{n\leqslant 0}$-adic filtration for which $(\eta_n)_{n\leqslant 0}$ is a process of innovations.

For example, one can define such a process by taking a stationary probability measure on $A^{\mathbb{Z}}$ and then taking for the law of X_n the projection of this measure on ℓ_n consecutive coordinates. In this case and when A is finite, by standard ergodic theory, the Kolmogorov entropy of this stationary probability measure can be written

$$\theta = \lim_{n\to-\infty} \frac{H(X_n)}{\ell_n} \in [0, +\infty) \tag{7.6}$$

when we use the Shannon entropy H (with a given logarithmic base). More generally, it follows from Lemma 7.6.3 that this limit θ exists for any split-word process on a finite alphabet A and

$$\theta = \inf_{n\leqslant 0} \frac{H(X_n)}{\ell_n}.$$

With the terminology of [8], $\mathcal{F}$ is an adic filtration on the Bratteli graph shown on Fig. 7.5, called the graph of the ordered pairs by Vershik (in contrast with the graph of unordered pairs that we will see in Sect. 7.6.6).

The proof of the theorem involves the cardinal of the group of tree automorphisms G_n given by

$$\log \#G_n = \ell_n \sum_{m=n+1}^{0} \frac{\log r_m!}{\ell_{m-1}}.$$

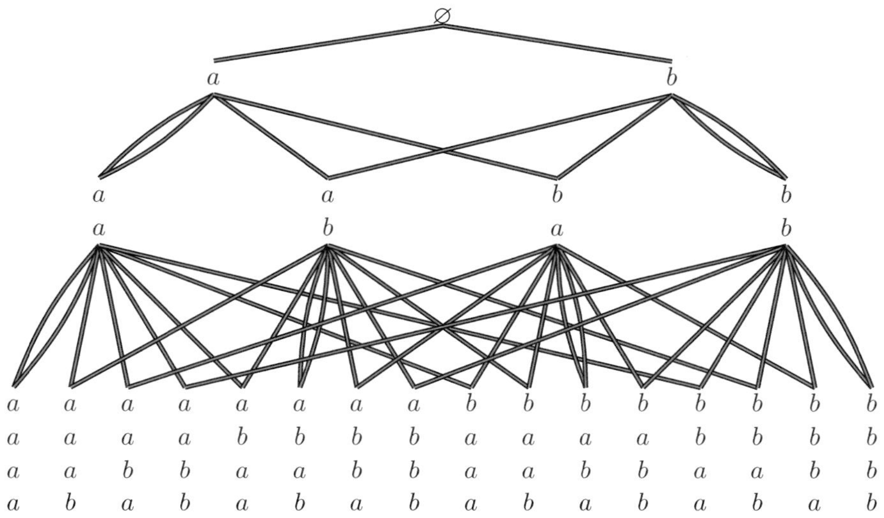

Fig. 7.5 The Bratteli graph of the ordered pairs

Theorem 7.6.15 *Assume A is finite. For the scaling $c(n) = \ell_n$ and under condition (Δ), the scaled entropy of $\mathcal{F}$ is $h_c(\mathcal{F}) = \theta$ when we use the Shannon entropy as the underlying measure of entropy. It is also the exponential entropy of $\mathcal{F}$ because of Theorem 7.6.4.*

Proof By Theorem 7.4.19, the c-scaled entropy $h_c(\mathcal{F})$ is equal to the c-scaled entropy of the filtration $\mathcal{E}$ generated by the Markov process $(X_n)_{n\leqslant 0}$. By Corollary 7.4.20 applied with $\phi(n) = n$, we know that $h_c(\mathcal{E}) = \lim_{m\to-\infty} h_c(X_m; \mathcal{E})$. Furthermore, $h_c(X_m; \mathcal{E}) = h_c(X_m; \mathcal{F})$ because $\mathcal{E}$ is immersed in $\mathcal{F}$ by lemma 1.6 of [10].

Thus we know that $h_c(\mathcal{F}) = \lim_{m\to-\infty} h_c(X_m; \mathcal{F})$. And we know that $h_c(X_m; \mathcal{F}) = \lim_{n\to-\infty} \frac{H(\pi_n X_m)}{\ell_n}$ by Theorem 7.6.4.

We firstly compare $H(\pi_n X_0)/\ell_n$ with $H(X_n)/\ell_n$. The n-th Vershik progressive prediction $\pi_n X_0$ of X_0 is identified with the orbit of X_n under the action of the group of tree automorphisms G_n. Below we denote the group of tree automorphisms by $G_n\big(\{r_i\}_{i=n+1}^0\big)$ to show its dependence on the r_n. Therefore $H(\pi_n X_0) = H(X_n) - H(X_n \mid \pi_n X_0)$, and the conditional entropy $H(X_n \mid \pi_n X_0)$ is less than the logarithm of the length of the orbit $\pi_n X_0$, and a fortiori it is less than the logarithm of the number of tree automorphisms. Thus,

$$H(X_n) \geqslant H(\pi_n X_0) \geqslant H(X_n) - \log \#G_n\big(\{r_i\}_{i=n+1}^0\big).$$

In the same way, for $n < m$,

$$H(X_n) \geqslant H(\pi_n X_m) \geqslant H(X_n) - \log \#G_{n-m}\big(\{r_i\}_{i=n+1}^m\big),$$

therefore

$$\frac{H(X_n)}{\ell_n} \geqslant \frac{H(\pi_n X_m)}{\ell_n} \geqslant \frac{H(X_n)}{\ell_n} - \frac{\log \#G_{n-m}\left(\{r_i\}_{i=n+1}^m\right)}{\ell_n}$$

$$= \frac{H(X_n)}{\ell_n} - \frac{1}{\ell_n}\frac{\ell_n}{\ell_m}\sum_{k=n}^{m}\frac{\log r_{k+1}!}{\ell_k/\ell_m}$$

$$= \frac{H(X_n)}{\ell_n} - \sum_{k=n}^{m}\frac{\log r_{k+1}!}{\ell_k},$$

hence

$$\theta \geqslant \lim_{n\to-\infty}\frac{H(\pi_n X_m)}{\ell_n} \geqslant \theta - \sum_{k=-\infty}^{m}\frac{\log r_{k+1}!}{\ell_k}.$$

The (Δ) condition being $\lim_{m\to-\infty}\sum_{k=-\infty}^{m}\frac{\log r_{k+1}!}{\ell_k} = 0$, the proof is over. □

To illustrate Theorem 7.4.19 and Theorem 7.4.22, consider the filtration $\mathcal{G}$ of the split-word process $(X_n, \eta_n)_{n\leqslant 0}$, with uniform letters on an alphabet $A = \{a, b, c, d\}$ having four letters. We know that $h_c(\mathcal{G}) = \log 4$ by the previous theorem. Now take the function $f\colon \{a, b, c, d\} \to \{0, 1\}$ sending a and b to 0, and sending c and d to 1. Then the process $\big(f(X_n), \eta_n\big)_{n\leqslant 0}$, where $f(X_n)$ denotes the word obtained by applying f to the letters of X_n, is the split-word process with uniform letters on the alphabet $\{0, 1\}$. The scaled entropy of its filtration $\mathcal{F}$ is $h_c(\mathcal{F}) = \log 2$. The filtration $\mathcal{F}$ is immersed in $\mathcal{G}$. However, Theorem 7.4.19 shows that $\mathcal{G}$ is not immersible in a parametric extension of $\mathcal{F}$ and Theorem 7.4.22 shows that $\mathcal{G}$ is not immersible in an independent enlargement of $\mathcal{F}$ with a standard filtration.

As a side note, let us mention that, for some reasons beyond the scope of this paper, every Kolmogorovian $(r_n)_{n\leqslant 0}$-adic filtration whose final σ-field is essentially separable is generated by a split-word process on a countable alphabet as long as its final σ-field is essentially separable. This follows from the ergodic theoretic fact that such a filtration can always be derived from a free ergodic Γ-action on a Lebesgue space, as said in Sect. 7.6.4, and from the existence of a countable generator for such an action.

7.6.6 Dyadic Filtrations of Unordered Pairs

Let $(X_n, \epsilon_n)_{n\leqslant 0}$ be a dyadic split-word process and $\mathcal{F}$ the filtration it generates. Consider the scaling $c(n) = \ell_n = 2^{|n|}$. We know the scaling entropy of $\mathcal{F}$ by Theorem 7.6.15. It is interesting to wonder about the scaled entropy $h_c(X_0)$ of the final letter X_0, that is to say, in view of Theorem 7.4.11, the scaled entropy $h_c(\mathcal{F}^{X_0})$

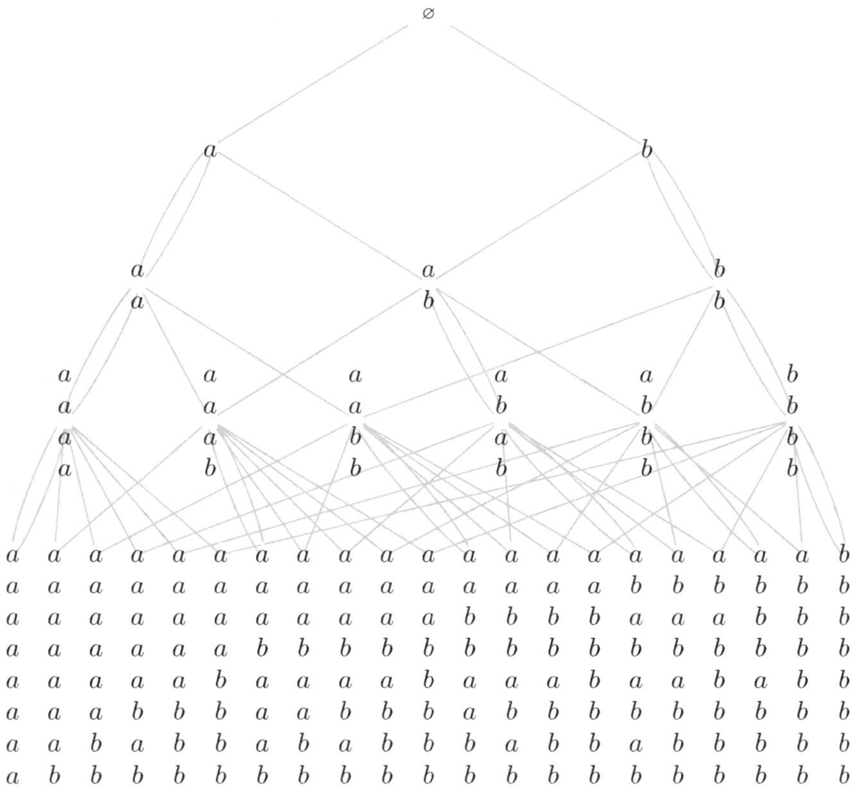

Fig. 7.6 The Bratteli graph of the unordered pairs [22]

of the filtration $\mathcal{F}^{X_0}$ generated by the Markov process $(\pi_n X_0)_{n \leqslant 0}$. Here we provide a result for the case of an alphabet $A = \{a, b\}$ having only two letters.

The Markov process $(\pi_n X_0)_{n \leqslant 0}$ can be seen as a random walk on the vertices of the *graph of the unordered pairs* shown on Fig. 7.6, and which can be found in [22].

The filtration $\mathcal{F}^{X_0}$ generated by $(\pi_n X_0)_{n \leqslant 0}$ is not dyadic, because $\pi_{n+1} X_0$ is deterministic given $\mathcal{F}_n$ for certain values of $\pi_n X_0$. This is shown on Fig. 7.6 by the double edges. Nevertheless, one has $h_c(X_0) = h(X_0)$ by Corollary 7.6.9. Indeed, there is a dyadic superinnovation of $\mathcal{F}^{X_0}$, that is to say a sequence $(\epsilon'_n)_{n \leqslant 0}$ of independent symmetric Bernoulli random variables which is a superinnovation (see Theorem 7.4.19) of $\mathcal{F}^{X_0}$ and which is a process of innovations of the enlarged filtration $\mathcal{G}^{X_0}$ defined by $\mathcal{G}_n^{X_0} = \mathcal{F}_n^{X_0} \vee \sigma(\epsilon'_m; m \leqslant n)$. To construct such a dyadic superinnovation, consider for every $n \leqslant 0$ an arbitrary but fixed order on the set of G_n-orbits, and set

$$\epsilon'_n = \begin{cases} \epsilon_n & \text{if } \pi_{n-1}X_0 \text{ is symmetric;} \\ 1 & \text{if } \pi_{n-1}X_0 = \{\Gamma_1, \Gamma_2\} \text{ with } \Gamma_1 < \Gamma_2 \text{ and } \pi_n X_0 = \Gamma_1; \\ 2 & \text{if } \pi_{n-1}X_0 = \{\Gamma_1, \Gamma_2\} \text{ with } \Gamma_1 < \Gamma_2 \text{ and } \pi_n X_0 = \Gamma_2. \end{cases}$$

As we previously recalled, $h_c(X_0) = h_c(\mathcal{F}^{X_0})$ because of Theorem 7.4.11. Moreover we know that $h_c(\mathcal{F}^{X_0}) = h_c(\mathcal{G}^{X_0})$ by Theorem 7.4.19. In fact it can be shown, with the help of Theorem 7.4.19 and lemma 5.3 in [8], that one can always "drop" the multiple edges when we are interested in the scaling entropy of the filtration associated (in the way explained in [8]) to a Bratteli graph endowed with a central probability measure.

The bounds for $h_c(X_0)$ we give in Proposition 7.6.18 are derived from the two following lemmas. The first one, giving the maximal length of a G_n-orbit, is a copy of lemma 3.6 in [20], to which we refer for the proof. The second one gives an asymptotic equivalent of the number of G_n-orbits (the number of vertices at level n of the graph of the unordered pairs). In [20], it is stated in lemma 3.7 but the given value of γ is not correct.

Lemma 7.6.16 *For an alphabet with two letters, the maximal length of a G_n-orbit is $2^{\frac{3}{4}2^{|n|}-1}$ for every $n \leqslant -2$.*

Lemma 7.6.17 *For an alphabet with two letters, the base 2 logarithm of the number χ_n of G_n-orbits is equivalent to $\gamma 2^{|n|}$ where $0.428 < \gamma < 0.429$.*

Proof It is easy to see that the number of orbits χ_n is given by $\chi_0 = 2$ and $\chi_{n-1} = \frac{\chi_n(\chi_n+1)}{2}$ (this is a particular case of Lemma 7.6.11). By the equality

$$\frac{\log_2 \chi_{n-1}}{2^{|n-1|}} = \frac{\log_2 \chi_n}{2^{|n|}} + \frac{\log_2\left(1 + \frac{1}{\chi_n}\right) - 1}{2^{|n-1|}},$$

the sequence $\frac{\log_2 \chi_n}{2^{|n|}}$ is decreasing, and for every $n < n_0 < 0$,

$$\frac{\log_2 \chi_{n_0}}{2^{|n_0|}} - \frac{1}{2^{|n_0|}} \leqslant \frac{\log_2 \chi_n}{2^{|n|}} \leqslant \frac{\log_2 \chi_{n_0}}{2^{|n_0|}}.$$

Taking $n_0 = -11$ gives the bounds on the limit γ. □

Proposition 7.6.18 *For an alphabet with two letters,*

$$\theta - \frac{3}{4} \leqslant h(X_0) \leqslant \min(\theta, \gamma)$$

where γ is given in Lemma 7.6.17 and θ is defined in Eq. (7.6). *Here we use the Shannon entropy with logarithmic base 2 as the underlying measure of entropy and $h(X_0)$ is the exponential entropy of X_0 (Definition 7.6.1), but by Theorem 7.6.4 it is the same as the scaled entropy $h_c(X_0)$ with the scaling $c(n) = \ell_n = 2^{|n|}$.*

Proof We start, as in the proof of Theorem 7.6.15, with the equality $H(\pi_n X_0) = H(X_n) - H(X_n \mid \pi_n X_0)$. But this time we bound from above the conditional entropy $H(X_n \mid \pi_n X_0)$ by the logarithm of the maximal length of a G_n-orbit. In addition we bound from above $H(\pi_n X_0)$ by the logarithm of the number of G_n-orbits. Then, using Lemma 7.6.16 and Lemma 7.6.17, we get

$$\frac{H(X_n)}{2^{|n|}} - \frac{3}{4} \leqslant \frac{H(\pi_n X_0)}{2^{|n|}} \leqslant \min\left\{\frac{H(X_n)}{2^{|n|}}, \frac{\log_2 \chi_n}{2^{|n|}}\right\},$$

and then the result follows by taking the limit. □

For example, this result shows that the filtration $\mathcal{F}^{X_0}$ in the uniform case $\theta = 1$ is not isomorphic to the filtration $\mathcal{F}^{X_0}$ in a case when $\theta < \frac{1}{4}$.

As an application of Theorem 7.4.19 and Theorem 7.4.22, consider the uniform case $\theta = 1$. Then we know that $h_c(\mathcal{F}) = \theta$ by Theorem 7.6.15, whereas $h_c(\mathcal{F}^{X_0}) \leqslant \gamma < \theta$ by Proposition 7.6.18. Thus by Theorem 7.4.19 we know that $\mathcal{F}$ is not immersible in a parametric extension of $\mathcal{F}^{X_0}$ and by Theorem 7.4.22, we know that $\mathcal{F}$ is not immersible in an independent enlargement of $\mathcal{F}^{X_0}$ with a standard filtration.

References

1. M. Barlow, M. Émery, F. Knight, S. Song, M. Yor, Autour d'un théorème de Tsirelson sur des filtrations browniennes et non-browniennes, in *Séminaire de Probabilités XXXII*. Lecture Notes in Mathematics, vol. 1686 (Springer, Berlin, 1998), pp. 264–305
2. F. Bolley, Separability and completeness for the Wasserstein distance, in *Séminaire de Probabilités XLI*. Lecture Notes in Mathematics, vol. 1934 (Springer, Berlin, 2008), pp. 371–377
3. G. Ceillier, The filtration of the split-words process. Probab. Theory Relat. Fields **153**(1–2), 269–292 (2012)
4. M. Émery, W. Schachermayer, On Vershik's standardness criterion and Tsirelson's notion of cosiness, in *Séminaire de Probabilités XXXV*. Lectures Notes in Mathematics, vol. 1755 (Springer, Berlin, 2001), pp. 265–305
5. A. Fieldsteel, J.R. Hasfura-Buenaga, Dyadic equivalence to completely positive entropy. Trans. Am. Math. Soc. **350**(3), 1143–1166 (1998)
6. A.D. Gorbulsky, Interrelations between various definitions of the entropy of decreasing sequences of partitions; scaling. J. Math. Sci. **121**(3), 2319–2325 (2004)
7. D. Heicklen, Entropy and r-equivalence. Ergod. Theory Dyn. Syst. **18**(5), 1139–1157 (1998)
8. É. Janvresse, S. Laurent, T. de la Rue, Standardness of monotonic Markov filtrations. Markov Process. Related Fields **22**(4), 697–736 (2016)
9. S. Laurent, Further comments on the representation problem for stationary processes. Statist. Probab. Lett. **80**, 592–596 (2010)

10. S. Laurent, On standardness and I-cosiness, in *Séminaire de Probabilités XLIII*. Lecture Notes in Mathematics, vol. 2006 (Springer, Berlin, 2010), pp. 127–186
11. S. Laurent, On Vershikian and I-cosy random variables and filtrations. Teoriya Veroyatnostei i ee Primeneniya **55**, 104–132 (2010). Also published in: Theory Probab. Appl. **55**, 54–76 (2011)
12. S. Laurent, Standardness and non-standardness of next-jump time filtrations. Electron. Commun. Probab. **18**(56), 1–11 (2013)
13. S. Laurent, Vershik's intermediate level standardness criterion and the scale of an automorphism, in *Séminaire de Probabilités XLV*. Lecture Notes in Mathematics, vol. 2078 (Springer, Berlin, 2013), pp. 123–139
14. S. Laurent, The reversed discrete hazard rate and the scaled entropy of a discrete measure. Preprint (2016). https://hal.archives-ouvertes.fr/hal-01251793
15. M. Smorodinsky, Processes with no standard extension. Israel J. Math. **107**, 327–331 (1998)
16. R. van der Hofstad, in *Random Graphs and Complex Networks*, vol. 1 (Cambridge University Press, Cambridge, 2016)
17. A.M. Vershik, Decreasing sequences of measurable partitions, and their applications, in *Doklady Akademii Nauk SSSR*, vol. 193 (1970), pp. 748–751. English translation: Soviet Math. Dokl. 11, 1007–1011 (1970)
18. A.M. Vershik, Continuum of pairwise nonisomorphic diadic sequences. Funktsional'nyi Analiz i Ego Prilozheniya **5**(3), 16–18 (1971). English translation: Funct. Anal. Appl. **5**(3), 182–184 (1971)
19. A.M. Vershik, Four definitions of the scale of an automorphism. Funktsional'nyi Analiz i Ego Prilozheniya **7**(3), 1–17 (1973). English translation: Funct. Anal. Appl. **7**(3), 169–181 (1973)
20. A.M. Vershik, The theory of decreasing sequences of measurable partitions (in Russian). Algebra i Analiz **6**(4), 1–68 (1994). English translation: St. Petersburg Math. J. **6**(4), 705–761 (1995)
21. A.M. Vershik, A.D. Gorbulsky, Scaled entropy of filtrations of σ-fields. Theory of Probab. Appl. **52**(3), 493–508 (2008)
22. A.M. Vershik, The problem of describing central measures on the path spaces of graded graphs. Funct. Anal. Appl. **48**(4), 26–46 (2014)

Chapter 8
Solving Rough Differential Equations with the Theory of Regularity Structures

Antoine Brault

Abstract The purpose of this article is to solve rough differential equations with the theory of regularity structures. These new tools recently developed by Martin Hairer for solving semi-linear partial differential stochastic equations were inspired by the rough path theory. We take a pedagogical approach to facilitate the understanding of this new theory. We recover results of the rough path theory with the regularity structure framework. Hence, we show how to formulate a fixed point problem in the abstract space of modelled distributions to solve the rough differential equations. We also give a proof of the existence of a rough path lift with the theory of regularity structure.

8.1 Introduction

Let $T > 0$ be a finite time horizon. Suppose that we want to solve the following ordinary differential equation

$$\forall t \in [0, T],\ \mathrm{d}y_t = F(y_t)\mathrm{d}W_t, \qquad y_0 = \xi, \tag{8.1}$$

where $W : [0, T] \to \mathbb{R}^n$ and $F : \mathbb{R}^d \to \mathcal{L}(\mathbb{R}^n, \mathbb{R}^d)$ are regular functions. Equation (8.1) can be reformulated as

$$\forall t \in [0, T], \quad y_t = \xi + \int_0^t F(y_u)\mathrm{d}W_u. \tag{8.2}$$

A. Brault (✉)
Institut de Mathématiques de Toulouse, CNRS UMR 5219, Université Paul Sabatier, Toulouse Cedex, France
e-mail: abrault@dim.uchile.cl

C. Donati-Martin et al. (eds.), *Séminaire de Probabilités L*, Lecture Notes in Mathematics 2252, https://doi.org/10.1007/978-3-030-28535-7_8

When W is smooth, Eq. (8.1) is well-defined as

$$y_t = \xi + \int_0^t F(y_u)\dot{W}_u \mathrm{d}u,$$

where $\dot{W}$ represents the derivative of W. Therefore, it becomes an ordinary differential equation that can be solved by a fixed-point argument.

Unfortunately, there are many natural situations in which we would like to consider the equation of type (8.2) for an irregular path W. This is notably the case when dealing with stochastic processes. For example the paths of the Brownian motion are almost surely nowhere differentiable [11]. It is then impossible to interpret (8.1) in a classical sense. Indeed, even if $\dot{W}$ is understood as a distribution, it is not possible in general to define a natural product between distributions, as y is itself to be thought as a distribution.

On the one hand, to overcome this issue, Itô's theory [11] was built to define properly an integral against a martingale M (for example the Brownian motion): $\int_0^t Z_u \mathrm{d}M_u$, where Z must have some good properties. The definition is not pathwise as it involves a limit in probability. Moreover, this theory is successful to develop a stochastic calculus with martingales but fails when this property vanishes. This is the case for the fractional Brownian motion, a natural process in modelling. Another bad property is that the map $W \mapsto y$ is not continuous in general with the associated uniform topology [13].

On the other hand, Young proved in [19] that we can define the integral of f against g if f is α-Hölder, g is β-Hölder with $\alpha + \beta > 1$ as

$$\int_0^t f \mathrm{d}g = \lim_{|P| \to 0} \sum_{u,v \text{ successive points in } P} f(u)(g(v) - g(u)),$$

where P is a subdivision of $[0, t]$ and $|P|$ denotes its mesh. This result is sharp, so that it is not possible to extend it to the case $\alpha + \beta \leq 1$ [19]. If W is α-Hölder it seems natural to think that y is α-Hölder, too. So assuming $\alpha < 1/2$ then $2\alpha < 1$, and Young's integral fails to give a meaning to (8.1). The fractional Brownian motion which depends on a parameter H giving its Hölder regularity cannot be dealt with Young's integral as soon as $H \leq 1/2$.

Lyons introduced in [15] the rough path theory which overcomes Young's limitation. The main idea is to construct for $0 \leq s \leq t \leq T$ an object $\mathbb{W}_{s,t}$ which "looks like" $\int_s^t (W_u - W_s)\mathrm{d}W_u$ and then define an integral against $(W, \mathbb{W})$. This is done with the sewing lemma (Theorem 8.27). This theory enabled to solve (8.1) in most of the cases and to define a topology such that the Itô's map $(W, \mathbb{W}) \mapsto y$ is continuous. Here, the rough path $(W, \mathbb{W})$ "encodes" the path W with algebraic operations. It is an extension of the Chen series developed in [3] and [14] to solve controlled differential equations. Since the original article of T. Lyons, other approaches of the rough paths theory were developed in [5, 8] and [1]. The article

[4] deals with the linear rough equations with a bounded operators. For monographs about the rough path theory, the reader can refer to [16] or [7].

Recently, Hairer developed in [10] the theory of regularity structures which can be viewed as a generalisation of the rough path theory. It allows to give a good meaning and to solve singular stochastic partial differential equations (SPDE). One of the main ideas is to build solutions from approximations at different scales. This is done with the reconstruction theorem (Theorem 8.46). Another fruitful theory was introduced to solve SPDE in [9] and also studied in [2].

The main goal of this article is to make this new theory understandable to people who are familiar with rough differential equations or ordinary differential equations.

Thus, we propose to solve (8.1) with the theory of regularity structures, when the Hölder regularity of W is in $(1/3, 1/2]$. In particular, we build the rough integral (Theorem 8.27) and the tensor of order 2: $\mathbb{W}$ (Theorem 8.18) with the reconstruction theorem.

Our approach is very related to [6, Chapter 13] where is established the link between rough differential equations and the theory of regularity structures. However, we give here the detailed proofs of Theorems 8.18 and 8.27 with the reconstruction theorem. It seems important to make the link between the two theories but is skipped in [6].

This article can be read without knowing about rough path or regularity structure theories.

After introducing notation in Sect. 8.2, we introduce in Sect. 8.3 the Hölder spaces which allow us to "measure" the regularity of a function. Then, we present the rough path theory in Sect. 8.4. In Sects. 8.5 and 8.6 we give the framework of the theory of regularity structures and the modelled distributions for solving (8.1). We prove in Sects. 8.7 and 8.8 the existence of the controlled rough path integral and the existence of a rough path lift. Finally, after having defined the composition of a function with a modelled distribution in Sect. 8.9, we solve the rough differential equation (8.1) in Sect. 8.10.

8.2 Notations

We denote by $\mathcal{L}(A, B)$, the set of linear continuous maps between two vector spaces A and B. Throughout the article, C denotes a positive constant whose value may change. For two functions f and g, we write $f \lesssim g$ if there is a constant C such that $f \leq Cg$. The symbol $:=$ means that the right hand side of the equality defines the left hand side. For a function Z from $[0, T]$ to a vector space, its increment is denoted by $Z_{s,t} := Z_t - Z_s$. If $X_1, \ldots, X_k$ are k vectors of a linear space, we denote by $\mathrm{Vect}\langle X_1, \ldots, X_k\rangle$ the subspace generated by the linear combinations of these vectors. Let T be a non-negative real, we denote by $[0, T]$ a compact interval of $\mathbb{R}$. For a continuous function $f : [0, T] \to E$, where $(E, \|\cdot\|)$ is a Banach space, we denote by $\|f\|_{\infty,T}$ the supremum of $\|f(t)\|$ for $t \in [0, T]$. The tensor product is denoted by $\otimes$. We denote $\lfloor\cdot\rfloor$ the floor function.

8.3 Hölder Spaces

8.3.1 Classical Hölder Spaces with a Positive Exponent

We introduce Hölder spaces which allow us to characterize the regularity of a non-differentiable function.

Definition 8.1 For $0 \le \alpha < 1$ and $T > 0$, the function $f : [0, T] \to E$ is α-Hölder if

$$\sup_{s \neq t \in [0,T]} \frac{\|f(t) - f(s)\|}{|t - s|^\alpha} < +\infty.$$

We denote by $C^\alpha(E)$ the *space of α-Hölder functions* equipped with the semi-norm

$$\|f\|_{\alpha,T} := \sup_{s \neq t \in [0,T]} \frac{\|f(t) - f(s)\|}{|t - s|^\alpha}.$$

If $\alpha \ge 1$ such that $\alpha = q + \beta$ where $q \in \mathbb{N}$ and $\beta \in [0, 1)$, we set $f \in C^\alpha(E)$ if f has q derivatives and $f^{(q)}$ is β-Hölder, where $f^{(q)}$ denotes the derivative of order k ($f^{(0)} := f$).

We denote by $C^\alpha = C^\alpha(\mathbb{R}^n)$. For $q \in \mathbb{N}$, we denote by C_b^q the set of all functions $f \in C^q$ such that

$$\|f\|_{C_b^q} := \sum_{k=0}^{q} \left\| f^{(k)} \right\|_\infty < +\infty. \tag{8.3}$$

Finally, for $q \in \mathbb{N}$, we define C_0^q the set of functions in C_b^q with a compact support.

Remark 8.2 The linear space of α-Hölder functions $C^\alpha(E)$ is a non separable Banach space endowed with one of the two equivalent norms $\|f(0)\| + \|f\|_{\alpha,T}$ or $\|f\|_{\infty,T} + \|f\|_{\alpha,T}$.

Remark 8.3 If f is α-Hölder on $[0, T]$, then f is β-Hölder for $\beta < \alpha$, *i.e.* $C^\alpha(E) \subset C^\beta(E)$.

8.3.2 Localised Test Functions and Hölder Spaces with a Negative Exponent

In Eq. (8.1), typically W is in C^α with $\alpha \in (0, 1)$. We need to deal with the derivative of W is the sense of distribution which should be of negative Hölder regularity $\alpha - 1 < 0$. We give in this section the definition of the space C^α with $\alpha < 0$. We show in Lemma 8.10 that an Hölder function is α-Hölder if and only if the derivative in the sense of distribution is $\alpha - 1$-Hölder with $\alpha \in (0, 1)$.

For $r > 0$, we denote by B_r the space of all functions in $\eta \in C_b^r$ compactly supported on $[-1, 1]$, such that $\|\eta\|_{C_b^r} \leq 1$.

Definition 8.4 For $\lambda > 0$, $s \in \mathbb{R}$ and a test function $\eta \in B_r$, we define the *test function localised* at s by

$$\eta_s^\lambda(t) := \frac{1}{\lambda}\eta\left(\frac{t-s}{\lambda}\right),$$

for all $t \in \mathbb{R}$.

Remark 8.5 The lower is λ, the more η_s^λ is localised at s, as can be seen in Fig. 8.1.

Remark 8.6 We work here with $t, s \in \mathbb{R}$, because we want to solve stochastic ordinary differential equations. But in the case of stochastic partial differential equations, the parameters t and s belong to $\mathbb{R}^e$ where e is an integer, see [10].

Definition 8.7 For $\alpha < 0$, we define the *Hölder space* C^α as elements in the dual of C_0^r where r is an integer strictly greater than $-\alpha$ and such that for any $\xi \in C^\alpha$ the following estimate holds

$$|\xi(\eta_s^\lambda)| \leq C(T)\lambda^\alpha, \tag{8.4}$$

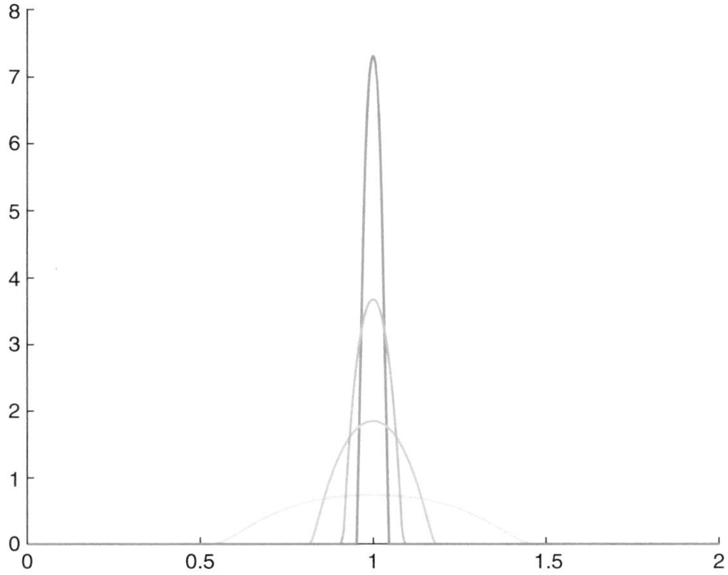

Fig. 8.1 Representation of η_s^λ for $s = 1$, $\lambda \in \{0.5, 0.2, 0.1, 0.05\}$ and with $\eta(s) = \exp(-1/(1 - s^2))\mathbb{1}_{(-1,1)}(s)$

where $C(T) \geq 0$ is a constant uniform over all $s \in [0, T]$, $\lambda \in (0, 1]$ and $\eta \in B_r$.

We define the semi-norm on C^α as the lowest constant $C(T)$ for a fixed compact $[0, T]$, i.e

$$\xi_{\alpha,T} := \sup_{s\in[0,T]} \sup_{\eta\in B_r} \sup_{\lambda\in(0,1]} \left| \frac{\xi(\eta_s^\lambda)}{\lambda^\alpha} \right|.$$

Remark 8.8 The space C^α does not depend on the choice of r, see for example [6] Exercise 13.31, p. 209.

Remark 8.9 With Definition 8.1, we can give a meaning of an α-Hölder function for $\alpha \in \mathbb{R}$. Moreover it is possible to show that if f is a function in C^α with $\alpha = q + \beta > 0$ where q is an integer and $\beta \in (0, 1)$, then for every $x \in [0, T]$ and localised functions η_x^λ,

$$|(f - P_x)(\eta_x^\lambda)| \leq C\lambda^\beta,$$

where C is uniform over $x \in [0, T]$, $\lambda \in (0, 1]$ and $\eta \in B_r$ (r a positive integer), P_x is the Taylor expansion of f of the order q in x, and $f - P_x$ is view as the canonical function associated.

Now, when we say that $f \in C^\alpha$ we should distinguish two cases:

- if $\alpha \geq 0$, f is an α-Hölder *function* in the sense of Definition 8.1
- if $\alpha < 0$, f is an α-Hölder *distribution* in the sense of Definition 8.7.

We give here a characterization of the space C^α for $\alpha \in (-1, 0)$ which is useful to make a link between the rough path and the regularity structures theories.

Lemma 8.10 *For any $\beta \in (0, 1)$, the distribution $\xi \in C^{\beta-1}$ if and only if there exist a function $z \in C^\beta$ such that $z(0) = 0$ and*

$$\forall \eta \in C_0^1,\ \xi(\eta) = -\langle z, \eta' \rangle. \tag{8.5}$$

Which means that $z' = \xi$ in the sense of distribution. Moreover, for all $t \in [0, 1]$,

$$z(t) = \sum_{k\in I_l} \langle \xi, \phi_k^l \rangle \int_0^t \phi_k^l + \sum_{j\geq l} \sum_{k\in I_j} \langle \xi, \psi_k^j \rangle \int_0^t \psi_k^j,$$

where ϕ, ψ are defined in Theorem 8.11 with a compact support in $[-c, c]$ ($c \geq 0$), l is an integer such that $2^{-l}c \leq 1$ and $I_j := [-\lfloor c \rfloor, 2^j + \lfloor c \rfloor] \bigcap \mathbb{Z}$.

The proof of Lemma 8.10 requires to introduce elements of the wavelet theory. The proof of the following theorem can be found in [18].

Theorem 8.11 *There exist $\phi, \psi \in C_0^1(\mathbb{R})$ such that for all $n \in \mathbb{N}$*

$$\{\phi_k^i := 2^{i/2}\phi(2^i \cdot -k),\ k \in \mathbb{Z}\} \cup \{\psi_k^j := 2^{j/2}\psi(2^j \cdot -k),\ k \in \mathbb{Z},\ j \geq i\} \tag{8.6}$$

is an orthonormal basis of $L^2(\mathbb{R})$. *This means that for all* $f \in L^2(\mathbb{R})$, $i \in \mathbb{N}$ *we can write*

$$f(t) = \sum_{j \geq i}^{+\infty} \sum_{k \in \mathbb{Z}} \langle f, \psi_k^j \rangle \psi_k^j(t) + \sum_{k \in \mathbb{Z}} \langle f, \phi_k^i \rangle \phi_k^i(t), \tag{8.7}$$

where the convergence is in $L^2(\mathbb{R})$. *Moreover, we have the very useful property,*

$$\int \psi(t) t^k \mathrm{d}t = 0, \tag{8.8}$$

for $k \in \{0, 1\}$.

Remark 8.12 The notation in Definition 8.4 for η_s^λ and in Theorem 8.11 for ϕ_k^i, ψ_k^j are similar but the meaning are slightly different.

We now proceed to the proof of Lemma 8.10.

Proof (Lemma 8.10) The first implication is trivial and does not require the wavelet analysis. If there exists $z \in C^\alpha$ such that for any $\eta \in C_0^1$, $\xi(\eta) = -\langle z, \dot{\eta} \rangle$, then for $\lambda \in (0, 1)$ and $s \in \mathbb{R}$,

$$\begin{aligned} \xi(\eta_s^\lambda) &= -\frac{1}{\lambda^2} \int_{\mathbb{R}} z(u) \dot{\eta} \left(\frac{u-s}{\lambda} \right) \mathrm{d}u \\ &= -\frac{1}{\lambda^2} \int_{\mathbb{R}} (z(u) - z(-\lambda + s)) \dot{\eta} \left(\frac{u-s}{\lambda} \right) \mathrm{d}u, \end{aligned}$$

where the last equality holds because η is compactly supported.

With the condition $\eta \in B_1$, $u \mapsto \dot{\eta}((u-s)/\lambda)$ is supported on $[-\lambda + s, \lambda + s]$, which yields to the bound

$$|\xi(\eta_s^\lambda)| \leq 2 \, \|\eta\|_{C^1} \, \|z\|_\alpha \, \lambda^{\alpha - 1}, \tag{8.9}$$

and proves that $\xi \in C^{\alpha-1}$.

Now, we prove the converse. Let $\phi, \psi \in C_0^1$ be defined in Theorem 8.11. Let $c \geq 0$ be a constant such that supports of ϕ and ψ are in $[-c, c]$. We denote l an integer such that $2^{-l} c \leq 1$. Thus, the support of ϕ_0^l is in $[-1, 1]$ and the support of ψ_k^j is smaller than 2 for $j \geq l$.

For $\xi \in C^{\alpha-1}$ for $\alpha \in (0, 1)$ we define for $t \in [0, 1]$,

$$z(t) := \sum_{k \in \mathbb{Z}} \langle \xi, \phi_k^l \rangle \int_0^t \phi_k^l + \sum_{j \geq l} \sum_{k \in \mathbb{Z}} \langle \xi, \psi_k^j \rangle \int_0^t \psi_k^j. \tag{8.10}$$

Noting that for $j \geq l$ and $k \in \mathbb{Z}$, ϕ_k^l and ψ_k^l are compactly supported in $[2^{-j}(k-c), 2^{-j}(k+c)]$, the terms $\int_0^1 \phi_k^j$ and $\int_0^1 \psi_k^j$ vanish when $2^{-j}(k+c) \leq 0$ and $1 \leq 2^{-j}(k-c)$. Thus, we can rewrite (8.10) as

$$z(t) = \sum_{k \in I_l} \langle \xi, \phi_k^l \rangle \int_0^t \phi_k^l + \sum_{j \geq l} \sum_{k \in I_j} \langle \xi, \psi_k^j \rangle \int_0^t \psi_k^j, \tag{8.11}$$

where $I_j := [-\lfloor c \rfloor, 2^j + \lfloor c \rfloor] \bigcap \mathbb{Z}$. The series on the right hand side of (8.10) converges in the sense of distributions. We need to justify that the limit z is in C^α.

We denote for any integer $N \in \mathbb{N}$,

$$S_N^z := \sum_{j=l}^{N} S_j, \tag{8.12}$$

where $S_j(t) := \sum_{k \in I_l} \langle \xi, \psi_k^j \rangle \int_0^t \psi_k^j$. According to (8.4), for all $j \geq l$ and $k \in I_j$

$$|\langle \xi, \psi_k^j \rangle| \leq C 2^{j/2 - j\alpha}.$$

For $|t-s| \leq 1$, let $j_0 \leq N$ be an integer such that $2^{-j_0} \leq |t-s| < 2^{-j_0+1}$. This is always possible for N large enough. On the one hand, if $l \leq j_0$, for $l \leq j \leq j_0$,

$$\begin{aligned} |S_j(t) - S_j(s)| &\leq \left\| S_j' \right\|_\infty |t-s| \\ &\leq |t-s| \sup_{u \in [0,1]} \sum_{I_j} |\langle \xi, \psi_k^j \rangle| \cdot |\psi_k^j(u)| \\ &\leq C 2^{j(1-\alpha)} |t-s|, \end{aligned} \tag{8.13}$$

where we use the fact that $\sum_{k \in I_j} |\psi(2^j t - k)| \leq C$ for a constant $C \geq 0$, because ψ is compactly supported. On the other hand, for $j > \max(j_0, l)$,

$$|S_j(t) - S_j(s)| \leq 2 \left\| S_j \right\|_\infty \tag{8.14}$$

$$\leq 2 \sup_{u \in [0,1]} \sum_{k=0}^{2^j - 1} |\langle \xi, \psi_k^j \rangle| \cdot |2^{-j} \hat{\psi}_k^j(u)|, \tag{8.15}$$

$$\leq 2C 2^{-j\alpha} \sup_{u \in [0,1]} \sum_{k=0}^{2^j - 1} |\hat{\psi}(2^j u - k)|, \tag{8.16}$$

where $\hat{\psi} := \int_0^t \psi$. Because $\int_{\mathbb{R}} \psi = 0$, there is a constant $C' \geq 0$ independent of j such that $\sum_{k=0}^{2^j-1} |\hat{\psi}(2^j u - k) - \hat{\psi}(-k)| < C'$. So finally, for $j > \max(j_0, l)$,

$$|S_j(t) - S_j(s)| \leq C 2^{-j\alpha}. \tag{8.17}$$

Thus, combining (8.13) and (8.17), for $N \geq l$,

$$\begin{aligned}\sum_{j=l}^{N} |S_j(t) - S_j(s)| &\leq C|t-s| \sum_{j=l}^{j_0} 2^{j(1-\alpha)} + C \sum_{j=j_0+1}^{\infty} 2^{-j\alpha} \\ &\leq C'|t-s|^\alpha,\end{aligned}$$

where C' is a new constant ($\sum_{j=l}^{j_0} 2^{j(1-\alpha)} = 0$ if $j_0 < l$). It follows that $\left\| S_N^z \right\|_{\alpha,1}$ is uniformly bounded in N and thus that $z \in C^\alpha$.

Now, we want to check that $\xi = \dot{z}$ in the distribution framework. For any $\eta \in C_0^2$,

$$\begin{aligned}\langle z, \dot{\eta}\rangle &= \sum_{k\in\mathbb{Z}} \langle \xi, \phi_k^l\rangle \Big\langle \int_0^t \phi_k^l, \dot{\eta}\Big\rangle + \sum_{j\geq l} \sum_{k\in\mathbb{Z}} \langle \xi, \psi_k^j\rangle \Big\langle \int_0^t \psi_k^j, \dot{\eta}\Big\rangle \\ &= -\sum_{k\in\mathbb{Z}} \langle \xi, \phi_k^l\rangle \langle \phi_k^l, \eta\rangle - \sum_{j\geq l} \sum_{k\in\mathbb{Z}} \langle \xi, \psi_k^j\rangle \langle \psi_k^j, \eta\rangle \\ &= -\langle \xi, \sum_{k\in\mathbb{Z}} \phi_k^l \langle \phi_k^l, \eta\rangle + \sum_{j\geq l} \sum_{k\in\mathbb{Z}} \psi_k^j \langle \psi_k^j, \eta\rangle\rangle \\ &= -\langle \xi, \eta\rangle,\end{aligned}$$

where the commuting of the serie and ξ is justified by the continuity of ξ in C_0^1 and the convergence of the following serie in C_0^1,

$$S_N^\eta := \sum_{j=l}^{N} \sum_{k\in\mathbb{Z}} \psi_k^j \langle \psi_k^j, \eta\rangle. \tag{8.18}$$

Indeed, we have

$$\begin{aligned}|\langle \psi_k^j, \eta\rangle| &\leq 2^{-j/2} \left| \int \psi(x) \eta(2^{-j}(x+k)) \mathrm{d}x \right| \\ &\leq 2^{-j/2} \int |\psi(x)| |\eta(2^{-j}(x+k)) - \eta(2^{-j}k) - \eta'(2^{-j}k)(2^j x)| \mathrm{d}x \\ &\leq \|\psi\|_\infty \left\|\eta'\right\|_\infty 2^{-j/2} 2^{-2j},\end{aligned}$$

where we use the fact that $\int \psi(t)t^k = 0$ for an integer $k \leq 1$. This implies that

$$\sum_{j=l}^{N} \left\| \sum_{k\in\mathbb{Z}} \psi_k^j \langle \psi_k^j, \eta \rangle \right\|_{C^1} \leq \sum_{j=l}^{N} \left\| \sum_{k\in\mathbb{Z}} \psi_k^j \langle \psi_k^j, \eta \rangle \right\|_{\infty} + \sum_{j=l}^{N} \left\| \sum_{k\in\mathbb{Z}} 2^j \psi_k'^j \langle \psi_k^j, \eta \rangle \right\|_{\infty}$$
$$\leq \left(\sum_{j=0}^{N} 2^{-2j} + \sum_{j=0}^{N} 2^{-j} \right) \|\psi\|_\infty \left\|\eta'\right\|_\infty ,$$

which proves that S_N^η is absolutely convergent in C_0^1.

Now by density of C_0^2 in C_0^1 and the continuity of ξ on C_0^1 we conclude that $\langle z, \dot{\eta} \rangle = -\langle \xi, \eta \rangle$ holds for $\eta \in C_0^1$. □

8.4 Elements of Rough Path Theory

We introduce here the elements of the rough path theory for solving Eq. (8.2). The notions discussed are reformulated in the regularity structure framework in the following sections. For an extensive introduction the reader can refer to [6], and for complete monographs to [7, 16].

8.4.1 The Space of Rough Paths

Let W be a continuous function from $[0, T]$ to $\mathbb{R}^n$.

We set $\alpha \in (1/3, 1/2]$. Then, (8.2) has not meaning, because the integral term is not defined. The main idea of the rough path theory is to define an object $\mathbb{W}_{s,t}$ which has the same algebraic and analytical properties as $\int_s^t W_{s,u} \otimes \mathrm{d}W_u$, the integral of the increment of the path against itself.

The importance of the iterated integrals can be understood with the classical linear differential equations where the solutions are provided with the exponential function. Indeed, if $W : [0, T] \to \mathbb{R}$ is smooth, the solutions of

$$\mathrm{d}y_t = y_t \mathrm{d}W_t \tag{8.19}$$

are

$$y_{s,t} = \exp(W_{s,t}) = 1 + \int_s^t \mathrm{d}W_{t_1,s} + \int_s^t \int_s^{t_1} \mathrm{d}W_{t_2,s} \mathrm{d}W_{t_1,s} + \cdots . \tag{8.20}$$

Definition 8.13 An *α-Hölder rough path* with $\alpha \in (1/3, 1/2]$ is an ordered pair $\mathbf{W} := (W, \mathbb{W})$ of functions, where $W : [0, T] \to \mathbb{R}^n$ and $\mathbb{W} : [0, T]^2 \to \mathbb{R}^n \otimes \mathbb{R}^n$ such that

1. For $s, u, t \in [0, T]$, $\mathbb{W}_{s,t} - \mathbb{W}_{s,u} - \mathbb{W}_{u,t} = W_{s,u} \otimes W_{u,t}$ (Chen's relation), i.e., for every $1 \le i, j \le n$, $\mathbb{W}^{i,j}_{s,t} - \mathbb{W}^{i,j}_{s,u} - \mathbb{W}^{i,j}_{u,t} = W^i_{s,u} W^j_{u,t}$.
2. The function W is α-Hölder and $\mathbb{W}$ is 2α-Hölder in the sense

$$\|\mathbb{W}\|_{2\alpha,T} := \sup_{s \neq t \in [0,T]} \frac{\|\mathbb{W}_{s,t}\|}{|t-s|^{2\alpha}} < +\infty.$$

One calls $\mathbb{W}$ the *second order process*. We denote by $\mathscr{C}^\alpha$ the *space of α-Hölder rough paths* endowed with the semi-norm

$$\|\mathbf{W}\|_{\alpha,T} = \|W\|_{\alpha,T} + \|\mathbb{W}\|_{2\alpha,T} \,.$$

Remark 8.14 The second order process $\mathbb{W}_{s,t}$ can be thought of as $\int_s^t W_{s,u} \otimes \mathrm{d}W_u$.

Remark 8.15 The first condition which is called Chen's relation represents the algebraic property of $\int_s^t W_{s,u} \otimes \mathrm{d}W_u$. Indeed, if W is smooth,

$$\int_s^t W^i_{s,v} \dot{W}^j_v \mathrm{d}v - \int_s^u W^i_{s,v} \dot{W}^j_v \mathrm{d}v - \int_u^t W^i_{u,v} \dot{W}^j_v \mathrm{d}v = W^i_{s,u} W^j_{u,t}$$

for all $1 \le i, j \le n$ and $0 \le s \le u \le t$.

Remark 8.16 The second condition is also an extension of the analytic property of the smooth case.

Remark 8.17 If $\mathbb{W}$ is a second order process of W, for any 2α-Hölder function F taking values in $\mathbb{R}^n \otimes \mathbb{R}^n$, $(s, t) \mapsto \mathbb{W}_{s,t} + F_t - F_s$ satisfies also the two properties of Definition 8.13. So if $\mathbb{W}$ exists, it is not unique at all.

Building $\mathbb{W}$ from W is non-trivial as soon as $n \ge 2$.

Theorem 8.18 *For any $W \in C^\alpha$ with $\alpha \in (1/3, 1/2]$ there exists a rough path lift $\mathbb{W}$, i.e. $\mathbf{W} = (W, \mathbb{W}) \in \mathscr{C}^\alpha$ in a way that the map $W \mapsto \mathbf{W}$ is continuous for the topology defined in Definition 8.13.*

Proof This result was proved in [17]. We prove of this result in the case $\alpha \in (1/3, 1/2]$ in Sect. 8.8 as an application of the reconstruction theorem (Theorem 8.46). □

8.4.2 Controlled Rough Paths

The aim of this section is to define an integrand against $\mathbf{W}$, called a controlled rough path by W. This approach was developed by Gubinelli in [8]. We introduce a function with the same regularity as W which is not differentiable with respect to time but with respect to W itself. This is the concept of the Gubinelli's derivative.

Definition 8.19 Let W be in C^α, we call a *controlled rough path by* W the pair $(y, y') \in C^\alpha(\mathbb{R}^d) \times C^\alpha(\mathbb{R}^{d\times n})$ such that

$$y_{s,t} = y'_s W_{s,t} + R^y_{s,t}, \tag{8.21}$$

with $\|R^y\|_{2\alpha,T} < +\infty$. The function y' is the *Gubinelli's derivative* of y with respect to W.

We denote $\mathscr{D}^{2\alpha}_W$ the *space of the controlled rough paths* (y, y') driven by W endowed with the semi-norm

$$\left\|(y, y')\right\|^W_{2\alpha,T} := \left\|y'\right\|_{\alpha,T} + \left\|R^y\right\|_{2\alpha,T}. \tag{8.22}$$

Remark 8.20 The identity (8.21) looks like a Taylor expansion of first order

$$f_t = f_s + f'_s(t-s) + O(|t-s|^2),$$

but $(W_t - W_s)$ substitutes the usual polynomial expression $(t-s)$, y'_s the normal derivative and the remainder term is of order 2α whereas order 2. The theory of regularity structures is a deep generalization of this analogy.

Remark 8.21 The Gubinelli's derivative y' is matrix-valued which depends on y and W.

Remark 8.22 Unlike the rough path space $\mathscr{C}^\alpha$ (see Definition 8.13) which is not a linear space, $\mathscr{D}^{2\alpha}_W$ is a Banach space with the norm $\|y_0\| + \left\|y'_0\right\| + \left\|(y, y')\right\|^W_{2\alpha,T}$ or the norm $\|y\|_{\infty,T} + \left\|y'\right\|_{\infty,T} + \left\|(y, y')\right\|^W_{2\alpha,T}$. These two norms are equivalent.

Remark 8.23 The uniqueness of y' depends on the regularity of W. If W is too smooth, for example in $C^{2\alpha}$, then y is in $C^{2\alpha}$, and every continuous function y' matches with the definition of the Gubinelli's derivative, particularly $y' = 0$. But we can prove that y' is uniquely determined by y when W is irregular enough. The reader can refer to the Chapter 4 of [6] for detailed explanations.

8.4.3 Integration Against Rough Paths

If F is a linear operator A, the differential equation (8.1) can be restated on an integral form as

$$y_t = \xi + A\int_0^t y_u \mathrm{d}W_u. \tag{8.23}$$

To give a meaning to (8.23) we must define an integral term $\int_0^t y_u \mathrm{d}W_u$.

When $W \in C^\alpha$, $y \in C^\beta$ with $\alpha + \beta > 1$, we are able to define (8.23) with Young's integral. Unfortunately, the solution y of (8.23) inherits of the regularity of W. Hence, Young's theory allows us to solve (8.23) only when $\alpha > 1/2$.

When $\alpha \in (1/3, 1/2]$, we need to "improve" the path W in taking into account of $\mathbb{W}$ in the definition of the integral.

8.4.4 Young's Integration

Young's integral was developed by Young in [19] and then used by Lyons in [14] to deal with differential equations driven by a stochastic process.

The integral is defined with a Riemann sum. Let $\mathcal{P}$ be a subdivision of $[s, t]$, we denote by $|\mathcal{P}|$ the mesh of $\mathcal{P}$. We want to define the integral as follows:

$$\int_s^t y_u \mathrm{d}W_u = \lim_{|\mathcal{P}| \to 0} \sum_{u,v \in \mathcal{P}} y_u W_{u,v},$$

where $u, v \in \mathcal{P}$ denotes successive points of the subdivision.

Theorem 8.24 *If $W \in C^\alpha$ and $y \in C^\beta$ with $\alpha + \beta > 1$, $\sum_{u,v\in\mathcal{P}} y_u W_{u,v}$ converges when $|\mathcal{P}| \to 0$. The limit is independent of the choice of $\mathcal{P}$, and it is denoted as $\int_s^t y_u \mathrm{d}W_u$. Moreover the bilinear map $(W, y) \to \int_s^t y_u \mathrm{d}W_u$ is continuous from $C^\alpha \times C^\beta$ to C^α.*

Proof For the original proof cf. [19]. □

Some important properties of the classical Riemann integration holds.

Proposition 8.25

1. *Chasles' relation holds.*
2. *When $t \to s$ we have the following approximation*

$$\int_s^t y_u \mathrm{d}W_u = y_s W_{s,t} + O(|t-s|^{\alpha+\beta}). \tag{8.24}$$

3. *The map $t \mapsto \int_s^t y_u \mathrm{d}W_u$ is α-Hölder continuous.*
4. *If F is C^1, $F(y)$ is C^β-Hölder and the Young integral $\int_s^t F(y_u)\mathrm{d}W_u$ is well-defined as above.*

Remark 8.26 Unfortunately with Young's construction, when $\alpha \le 1/2$, we can find two sequences of smooth functions $W^{1,n}$ and $W^{2,n}$ converging to W in C^α but such that $\int_s^t F(W^{1,n})\mathrm{d}W^{1,n}$ and $\int_s^t F(W^{2,n})_n \mathrm{d}W^{2,n}$ converge to two different limits for a smooth function F. See for an example the Lejay's area bubbles in [12].

8.4.5 Controlled Rough Path Integration

The rough integral relies on the controlled rough paths introduced previously. Remark 8.26 shows that if $y, W \in C^\alpha$, we cannot define a continuous integral such as $\int_s^t y_u \mathrm{d}W_u$ looks like $y_s W_{s,t}$ when $t \to s$. We must use the structure of controlled rough paths to define a "good" integral of y against W. Then, given a rough path $\mathbf{W} \in \mathscr{C}^\alpha$ and considering a controlled rough path $(y, y') \in \mathscr{D}_W^{2\alpha}$ we would like to build an integral $\int_s^t y_u \mathrm{d}W_u$ as a good approximation of $y_s W_{s,t} + y'_s \mathbb{W}_{s,t}$ when $t \to s$.

Theorem 8.27 *For $\alpha \in (1/3, 1/2]$, let $\mathbf{W} = (W, \mathbb{W}) \in \mathscr{C}^\alpha$ be an α-Hölder rough path. Given a controlled rough path driven by W : $(y, y') \in \mathscr{D}_W^{2\alpha}$ we consider the sum $\sum_{u,v \in \mathcal{P}} y_u W_{u,v} + y'_u \mathbb{W}_{u,v}$ where $\mathcal{P}$ is a subdivision of $[s, t]$ ($s \le t \in [0, T]$). This sum converges when the mesh of $\mathcal{P}$ goes to 0. We define the integral of y against $\mathbf{W}$ as*

$$\int_s^t y_u \mathrm{d}\mathbf{W}_u := \lim_{|\mathcal{P}| \to 0} \sum_{u,v \in \mathcal{P}} y_u W_{u,v} + y'_u \mathbb{W}_{u,v}.$$

The limit exists and does not depend on the choice of the subdivision. Moreover, the map $(y, y') \to (t \in [0, T] \mapsto \int_0^t y_u \mathrm{d}\mathbf{W}_u, y)$ from $\mathscr{D}_W^{2\alpha}$ into itself is continuous.

Proof The classical proof uses the sewing lemma [6, Lemma 4.2]. We give a proof with the reconstruction theorem (Theorem 8.46) in Sect. 8.7. □

To solve (8.1), we need to show that if F is a smooth function, then $F(y_t)$ remains a controlled rough path. The following proposition shows that $(F(y), (F(y))')$ defined by:

$$F(y)_t = F(y_t), \qquad F(y)'_t = F'(y_t)y_t, \tag{8.25}$$

is a controlled rough path.

Proposition 8.28 *Let $F : \mathbb{R}^d \to \mathcal{L}(\mathbb{R}^n, \mathbb{R}^d)$ be a function twice continuously differentiable such that F and its derivative are bounded. Given $(y, y') \in \mathscr{D}_W^{2\alpha}$ let $(F(y), F(y)') \in \mathscr{D}_W^{2\alpha}$ defined as above (8.25). Then, there is a constant $C_{\alpha,T}$ depending only on α and T such as*

$$\left\| F(y), F(y)' \right\|_{2\alpha,T}^{W} \le C_{\alpha,T} \left\| F \right\|_{C_b^2} (1 + \left\| W \right\|_\alpha)^2 (\left\| y'_0 \right\| + \left\| y, y' \right\|_{2\alpha,T}^{W})^2,$$

where $\|F\|_{C_b^2} = \|F\|_\infty + \left\|F'\right\|_\infty + \left\|F''\right\|_\infty$.

Proof We can find the proof in [6]. This proposition is equivalent to Theorem 8.58, which is formulated in the regularity structure framework. □

8.5 Regularity Structures

8.5.1 Definition of a Regularity Structure

The theory of regularity structures was introduced by Hairer in [10]. The tools developed in this theory allow us to solve a very wide range of semi-linear partial differential equations driven by an irregular noise.

This theory can be viewed as a generalisation of the Taylor expansion theory to irregular functions. The main idea is to describe the regularity of a function at small scales and then to reconstruct this function with the reconstruction operator of Theorem 8.46.

First we give the definition of a regularity structure.

Definition 8.29 A *regularity structure* is a 3-tuple $\mathscr{T} = (\mathcal{A}, \mathcal{T}, \mathcal{G})$ where

- The *index set* $\mathcal{A} \subset \mathbb{R}$ is bounded from below, locally finite and such that $0 \in \mathcal{A}$.
- The *model space* $\mathcal{T}$ is a graded linear space indexed by $\mathcal{A}$: $\mathcal{T} = \bigoplus_{\alpha \in \mathcal{A}} \mathcal{T}_\alpha$, where each $\mathcal{T}_\alpha$ is a non empty Banach space. The elements of $\mathcal{T}_\alpha$ are said of *homogeneity* α. For $\underline{\tau} \in \mathcal{T}$, we denote $\|\underline{\tau}\|_\alpha$ the norm of the component of $\underline{\tau}$ in $\mathcal{T}_\alpha$. Furthermore, $\mathcal{T}_0 = \mathrm{Vect}\langle \underline{1} \rangle$ is isomorphic to $\mathbb{R}$.
- The set $\mathcal{G}$ is a set of continuous linear operators acting on $\mathcal{T}$ such as for $\Gamma \in \mathcal{G}$, $\Gamma(\underline{1}) = \underline{1}$ and $\underline{\tau} \in \mathcal{T}_\alpha$, $\Gamma\underline{\tau} - \underline{\tau} \in \bigoplus_{\beta < \alpha} \mathcal{T}_\beta$. The set $\mathcal{G}$ is called *structure group*.

Remark 8.30 We underline the elements of the model space for the sake of clarity.

Remark 8.31 We set $m := \min \mathcal{A}$, $\Gamma\underline{\tau} = \underline{\tau}$ for every $\underline{\tau} \in \mathcal{T}_m$.

Let us explain the motivations of this definition. The classic polynomial Taylor expansion of order $m \in \mathbb{N}$ is given, between 0 and $t \in \mathbb{R}$, where t converges to 0 by

$$f(t) = P(t) + o(t^m), \quad \text{where } P(t) = \sum_{k=0}^{m} \frac{f^{(k)}(0)}{k!} t^k.$$

In this case the approximation P of f is indexed by integers and the space $\mathcal{T}$ is the polynomial space. For all $h \in \mathbb{R}$, the operator Γ_h associates a Taylor expansion at point t with a Taylor expansion at a point $t + h$. The polynomial $\Gamma_h(P(t)) - P(t)$ is of order less than $m - 1$:

$$\Gamma_h(P(t)) - P(t) := P(t+h) - P(t) = \sum_{k=0}^{m} \frac{f^{(k)}(0)}{k!} ((t+h)^k - t^k).$$

Moreover we have the structure of group on $(\Gamma_h, h \in \mathbb{R})$:

$$\Gamma_{h'} \circ \Gamma_h P(t) = \Gamma_{h'}\left(\sum_{k=0}^{d-1} \frac{f^k(0)}{k!}(t+h)^k\right) = \sum_{k=0}^{d-1} \frac{f^k(0)}{k!}\Gamma_{h'}((t+h)^k)$$
$$= \sum_{k=0}^{d-1} \frac{f^k(0)}{k!}(t+h+h')^k$$
$$= \Gamma_{h+h'} P(t).$$

Hence, we can define the polynomial regularity structure as following.

Definition 8.32 We define $\mathscr{T}^1 = (\mathcal{A}^1, \mathcal{T}^1, \mathcal{G}^1)$ the *canonical polynomial regularity structure* as

- $\mathcal{A}^1 = \mathbb{N}$ is the index set.
- For $k \in \mathcal{A}^1$ we define $\mathcal{T}^1_k = \text{Vect}\langle \underline{X}^k \rangle$. The subspace $\mathcal{T}^1_k$ contains the monomial of order k. The polynomial model space is $\mathcal{T}^1 = \bigoplus_{k \in \mathcal{A}} \mathcal{T}^1_k$.
- For $h \in \mathbb{R}$, $\Gamma^1_h \in \mathcal{G}^1$ is given by

$$\Gamma^1_h(\underline{X}^k) = (\underline{X} + h\underline{1})^k.$$

For $P_k \in \mathcal{T}^1_k$, there is $a_k \in \mathbb{R}$ such that $P_k = a_k \underline{X}^k$. We define the norm on $\mathcal{T}^1_k$ by $\| P_k \|_k = |a_k|$.

With the same arguments we define the polynomial regularity structure and its model associated in $\mathbb{R}^n$.

Definition 8.33 We define $\mathscr{T}^p = (\mathcal{A}^p, \mathcal{T}^p, \mathcal{G}^p)$ the *canonical polynomial regularity structure* on $\mathbb{R}^n$ as

- $\mathcal{A}^p = \mathbb{N}$ is the index set.
- For $\delta \in \mathcal{A}^p$, and k a multi-index of $\mathbb{N}^n$ such that $|k| := k_1 + \cdots + k_n = \delta$, we define $T^p_\delta = \text{Vect}\langle \underline{X}^k := \prod_{i=1}^n \underline{X}_i^{k_i}, |k| = \delta \rangle$. This space T^p_k is a linear space of homogeneous polynomial with n variables and of order δ. For $P_\delta \in T^p_\delta$, there are real coefficients $(a_k)_{|k|=\delta}$ such that $P_\delta = \sum_{|k|=\delta} a_k \underline{X}^k$. We chose the norm on T^p_δ such that $\| P_\delta \|_\delta := \sum_{|k|=\delta} |a_k|$.
  We define $\mathcal{T}^p = \bigoplus_{\delta \in A} T^p_\delta$ as the polynomial model space.
- For $h \in \mathbb{R}^n$, $\Gamma^p_h \in G^p$ is given by

$$\Gamma^p_h(\underline{X}^k) = \prod_{i=1}^n (\underline{X}_i + h_i \underline{1})^{k_i}.$$

Remark 8.34 The polynomial regularity structure is a trivial example of regularity structure which we introduce for a better understanding. But the strength of this theory is to deal with negative degree of homogeneity.

8.5.2 Definition of a Model

Definition 8.35 Given a regularity structure $\mathscr{T} = (\mathcal{A}, \mathcal{T}, \mathcal{G})$, a *model* $M = (\Pi, \Gamma)$ is two sets of functions such that for any $s, t, u \in \mathbb{R}$

- The operator Π_s is continuous and linear from $\mathcal{T}$ to $\mathcal{D}'(\mathbb{R}, \mathbb{R}^n)$.
- $\Gamma_{t,s}$ belongs to $\mathcal{G}$, so it is a linear operator acting on $\mathcal{T}$.
- The following algebraic relations hold: $\Pi_s \Gamma_{s,t} = \Pi_t$ and $\Gamma_{s,t}\Gamma_{t,u} = \Gamma_{s,u}$.
- The following analytic relations hold: for every $\gamma > 0$, $\beta < \alpha \leq \gamma$ with $\alpha, \beta \in \mathcal{A}$ and $\underline{\tau} \in \mathcal{T}_\alpha$, there is a constant $C(T, \gamma)$ uniform over $s, t \in [0, T]$, $\lambda \in (0, 1]$, $\phi \in B_r$ such that

$$
\begin{aligned}
|\Pi_s(\underline{\tau})(\phi_s^\lambda)| &\leq C(T,\gamma)\lambda^\alpha \left\|\underline{\tau}\right\|_\alpha \\
\text{and } \left\|\Gamma_{s,t}(\underline{\tau})\right\|_\beta &\leq C(T,\gamma)|t-s|^{\alpha-\beta} \left\|\underline{\tau}\right\|_\alpha .
\end{aligned}
\tag{8.26}
$$

We denote respectively by $\|\Pi\|_{\gamma,T}$ and $\|\Gamma\|_{\gamma,T}$ the smallest constants such that the bounds (8.26) hold. Namely,

$$
\|\Pi\|_{\gamma,T} := \sup_{s\in[0,T]} \sup_{\phi\in B_r} \sup_{\lambda\in(0,1]} \sup_{\alpha<\gamma} \sup_{\underline{\tau}\in T_\alpha} \frac{|\Pi_s(\underline{\tau})(\phi_s^\lambda)|_\alpha}{\lambda^\alpha \left\|\underline{\tau}\right\|_\alpha}
$$

$$
\text{and } \|\Gamma\|_{\gamma,T} := \sup_{s\neq t\in[0,T]} \sup_{\beta<\alpha<\gamma} \sup_{\underline{\tau}\in T_\alpha} \frac{\left\|\Gamma_{s,t}(\underline{\tau})\right\|_\beta}{|t-s|^{\alpha-\beta}\left\|\underline{\tau}\right\|_\alpha}.
$$

The two operators $\|\cdot\|_{\gamma,T}$ define semi-norms.

The easiest regularity structure which we can describe is the polynomial one (see Definition 8.33). We can now define the model associated to this regularity structure.

Definition 8.36 Given that $\mathscr{T}^p = (\mathcal{A}^p, \mathcal{T}^p, \mathcal{G}^p)$ the canonical polynomial regularity structure on $\mathbb{R}^n$ defined in the Definition 8.33, we define the *model of the polynomial regularity structure* $M^p = (\Pi^p, \Gamma^p)$ such that for all $x, y \in \mathbb{R}^n$ and k a multi-index of order n,

$$
\begin{aligned}
\Pi_x^p(\underline{X}^k)(y) &:= ((y_1 - x_1)^{k_1}, \ldots, (y_n - x_n)^{k_n}), \\
\Gamma_{x,y}^p(\underline{X}^k) &:= \Gamma_{x-y}(\underline{X}^k).
\end{aligned}
$$

Proof It is straightforward to check that this definition is in accordance with the one of a model (Definition 8.42 below). □

Remark 8.37 The operator Π_s which associates to an element of the abstract space a distribution which approximates this element in s. Typically for polynomial regularity structure on $\mathbb{R}$,

$$\Pi_s^p(\underline{X}^k) = (t \to (t-s)^k).$$

Remark 8.38 In the model space, the operator $\Gamma_{s,t}$ gives an expansion in a point s, given an expansion in a point t. For example

$$\Gamma_{s,t}^p(\underline{X}^k) = \Gamma_{s-t}^p(\underline{X}^k) = (\underline{X} + (s-t)\underline{1})^k. \tag{8.27}$$

Remark 8.39 The first algebraic relation means that if a distribution looks like $\underline{\tau}$ near t, the same distribution looks like $\Gamma_{s,t}(\underline{\tau})$ near s. In practice, we use this relation to find the suitable operator $\Gamma_{t,s}$. The second algebraic relation is natural. It says that moving an expansion from u to s is the same as moving an expansion from u to t and then from t to s.

Remark 8.40 The first analytic relation has to be understood as Π_s approximating $\underline{\tau} \in T_\alpha$ in s with the precision λ^α. The relation (8.27) shows that the second analytic relation is natural. Indeed,

$$(\underline{X} + (t-s)\underline{1})^k = \sum_{i=0}^{k} \binom{k}{i} (t-s)^{k-i} \underline{X}^i,$$

so for $\ell \le k$, $\left\|\Gamma_{s,t}^p(\underline{X}^k)\right\|_\ell = \binom{k}{i}|t-s|^{k-\ell}$, where $\binom{k}{i} = \frac{k!}{i!(k-i)!}$ are the binomial coefficients.

8.5.3 The Rough Path Regularity Structure

We now reformulate the results of Sects. 8.4.1 and 8.4.2 to build up a regularity structure.

In order to find the regularity structure of rough paths, we make some computations for $n = 1$. Then, we give the proof in the general case after Definition 8.41.

We fix $\alpha \in (1/3, 1/2]$ and a rough path $\mathbf{W} = (W, \mathbb{W}) \in \mathscr{C}^\alpha$. We show how to build the regularity structure of rough paths.

Let $(y, y') \in \mathscr{D}_W^{2\alpha}$ be a controlled rough path. According to Definition 8.19, $y_t = y_s + y'_s W_{s,t} + O(|t-s|^{2\alpha})$. To describe the expansion of y with the regularity structure framework, we set the symbol $\underline{1}$ constant of homogeneity 0 and the symbol

$\underline{W}$ of homogeneity α. This leads us to define the elements of the regularity structure of the controlled rough path (y, y') evaluated at time t by

$$\underline{Y}(t) = y_t \underline{1} + y'_t \underline{W}.$$

Moreover, we would like to build the rough path integral $\int y\mathrm{d}\mathbf{W}$ in the regularity structure context. So we introduce abstract elements $\underline{\dot{W}}$ and $\underline{\dot{\mathbb{W}}}$ which "represent" $\mathrm{d}\mathbf{W} = \mathrm{d}(W, \mathbb{W})$. The function W is α-Hölder, so we define the homogeneity of $\underline{\dot{W}}$ as $\alpha - 1$. The second order process $\mathbb{W}$ is 2α-Hölder, which leads us to define the homogeneity of $\underline{\dot{\mathbb{W}}}$ as $2\alpha - 1$.

Finally, with the notation of Definition 8.29, $A = \{\alpha - 1, 2\alpha - 1, 0, \alpha\}$, $\mathcal{T} = \mathrm{Vect}\langle \underline{\dot{W}}, \underline{\dot{\mathbb{W}}}, \underline{1}, \underline{W}\rangle$. Besides, we order the elements in $\mathrm{Vect}\langle\cdot\rangle$ by homogeneity.

It remains to define $\mathcal{G}$ and an associated model. We start by building the model (Π, Γ). For $s \in [0, T]$, Π_s should transform the elements of $\mathcal{T}$ to distributions (or functions when it is possible) which approximate this elements at the point s. On the one hand we define

$$\Pi_s(\underline{\dot{W}})(\phi) := \int \phi(t)\mathrm{d}W_t, \qquad \Pi_s(\underline{\dot{\mathbb{W}}})(\phi) := \int \phi(t)\mathrm{d}\mathbb{W}_{s,t},$$

where ϕ is a test function. Both integrals are well-defined because ϕ is smooth. The homogeneities of $\underline{\dot{W}}$ and $\underline{\dot{\mathbb{W}}}$ are negative, so they are mapped with distributions. On the other hand, $\underline{1}$ and $\underline{W}$ have positive homogeneities, so we can approximate them in s with functions as

$$\Pi_s(\underline{1})(t) := 1, \qquad \Pi_s(\underline{W})(t) := W_{s,t}.$$

Now, we define $\Gamma_{s,t}(\underline{\tau})$ for every $\beta \in A$ and $s, t \in [0, T]$ and $\underline{\tau} \in \mathcal{T}_\beta$. According to Definition 8.35: $\Pi_s\Gamma_{s,t}(\underline{\tau})(\phi) = \Pi_t(\underline{\tau})(\phi)$. Moreover, following Definition 8.29, $\Gamma_{s,t}$ should be a linear combination of elements of homogeneity lower than $\underline{\tau}$ and with the coefficient 1 in front of $\underline{\tau}$. First, it seems obvious to set $\Gamma_{s,t}(\underline{1}) = \underline{1}$, because $\underline{1}$ represents a constant. Then we look for $\Gamma_{s,t}(\underline{W}) = \underline{W} + a_{s,t}\underline{1}$ as a function where $a_{s,t}$ has to be determined. If it is not enough, we would look for $\Gamma_{s,t}(\underline{W})$ with more elements of our structure T. By linearity

$$\Pi_s(\underline{W} + a_{s,t}\underline{1})(u) = W_{s,u} + a_{s,t},$$

so we want that $W_{s,u} + a_{s,t} = \Pi_t(\underline{W})(u) = W_{t,u}$. Finally, we have to choose $a_{s,t} = W_{t,s}$. Given that $\underline{\dot{W}}$ has the lowest homogeneity of our structure, we set $\Gamma_{s,t}(\underline{\dot{W}}) = \underline{\dot{W}}$ in order to respect the last item of Definition 8.29. With the same reason as for $\underline{W}$ and using the Chen's relation of Defintion 8.13, we find that $\Gamma_{s,t}(\underline{\dot{\mathbb{W}}}) = \underline{\dot{\mathbb{W}}} + W_{t,s}\underline{\dot{W}}$ (see the proof of Definition 8.41).

All we did here is in one dimension. With the same arguments we can find the regularity structure of a rough path in $\mathbb{R}^n$.

Definition 8.41 For $\alpha \in (1/3, 1/2]$, given a rough path $\mathbf{W} = (W, \mathbb{W}) \in \mathscr{C}^\alpha$ which take value in $\mathbb{R}^n \bigoplus (\mathbb{R}^n \otimes \mathbb{R}^n)$. We define the *regularity structure of rough paths* $\mathscr{T}^r = (\mathcal{A}^r, \mathcal{T}^r, \mathcal{G}^r)$ and the model associated $M^r = (\Pi^r, \Gamma^r)$ as

(i) Index set $\mathcal{A}^r := \{\alpha - 1, 2\alpha - 1, 0, \alpha\}$.
(ii) Model space $\mathcal{T}^r := \mathcal{T}^r_{\alpha-1} \bigoplus \mathcal{T}^r_{2\alpha-1} \bigoplus \mathcal{T}^r_0 \bigoplus \mathcal{T}^r_\alpha$, with

$$\mathcal{T}^r_{\alpha-1} := \mathrm{Vect}\langle \underline{\dot{W}}^i, i = 1, \cdots, n\rangle, \quad \mathcal{T}^r_{2\alpha-1} := \mathrm{Vect}\langle \underline{\dot{\mathbb{W}}}^{i,j}, i, j = 1, \cdots, n\rangle,$$

$$\mathcal{T}^r_0 := \mathrm{Vect}\langle \underline{1}\rangle, \quad \mathcal{T}^r_\alpha := \mathrm{Vect}\langle \underline{W}^i, i = 1, \cdots, n\rangle.$$

(iii) For i, j integers between 1 and n, $h \in \mathbb{R}^n$ and Γ^r_h in the structure group $\mathcal{G}^r$, the following relations hold

$$\Gamma^r_h(\underline{\dot{W}^i}) := \underline{\dot{W}^i}, \qquad \Gamma^r_h(\underline{\dot{\mathbb{W}}^{i,j}}) := \underline{\dot{\mathbb{W}}^{i,j}} + h^i \underline{\dot{W}^j},$$

$$\Gamma^r_h(\underline{1}) := \underline{1}, \text{ and } \Gamma^r_h(\underline{W^i}) := \underline{W^i} + h^i \underline{1}.$$

(iv) For i, j two integers between 1 and n, for $s, t \in [0, T]$,

$$\Pi^r_s(\underline{\dot{W}^i})(\phi) := \int \phi(t) \mathrm{d}W^i_t, \qquad \Pi^r_s(\underline{\dot{\mathbb{W}}^{i,j}})(\phi) := \int \phi(t) \mathrm{d}\mathbb{W}^{i,j}_{s,t},$$

$$\Pi^r_s(\underline{1})(t) := 1, \qquad \Pi^r_s(\underline{W^i})(t) := W^i_{s,t},$$

where ϕ is a test function.
(v) For $s, t \in \mathbb{R}$, $\Gamma^r_{s,t} := \Gamma^r_{|h=W_{t,s}}$.

Proof Checking that this definition respects the definitions of a regularity structure (Definition 8.29) and of a model (Definition 8.35) is straightforward.

Here we only show where Chen's relation of Definition 8.13 is fundamental to show that the algebraic condition of Definition 8.35: $\Pi^r_s \Gamma^r_{s,t} \underline{\dot{\mathbb{W}}} = \Pi^r_t \underline{\dot{\mathbb{W}}}$ holds.

According to the definition above $\Gamma^r_{s,t} \underline{\dot{\mathbb{W}}^{i,j}} = \underline{\dot{\mathbb{W}}^{i,j}} + h^i \underline{\dot{W}^j}$. So we have

$$\Pi^r_s(\Gamma^r_{s,t} \underline{\dot{\mathbb{W}}^{i,j}})(\phi) = \int \phi(u) \mathrm{d}\mathbb{W}^{i,j}_{s,u} + W^i_{t,s} \int \phi(u) \mathrm{d}W^j_{s,u}. \tag{8.28}$$

In differentiating Chen's relation $\mathbb{W}^{i,j}_{s,u} = \mathbb{W}^{i,j}_{s,t} + \mathbb{W}^{i,j}_{t,u} + W^i_{s,t} W^j_{t,u}$ with respect to u we get $\mathrm{d}\mathbb{W}^{i,j}_{s,u} = \mathrm{d}\mathbb{W}^{i,j}_{t,u} + W^i_{s,t} \mathrm{d}W^j_{t,u}$. It follows that

$$\Pi^r_s(\Gamma^r_{s,t} \underline{\dot{\mathbb{W}}^{i,j}})(\phi) = \int \phi(u) \mathrm{d}\mathbb{W}^{i,j}_{t,u} + W^i_{s,t} \int \phi(u) \mathrm{d}W^j_{s,u} + W^i_{t,s} \int \phi(u) \mathrm{d}W^j_{s,u}. \tag{8.29}$$

Finally $\Pi_s^r(\Gamma_{s,t}^r \underline{\dot{\mathbb{W}}}^{i,j})(\phi) = \int \phi(u) \mathrm{d}\mathbb{W}_{t,u}^{i,j} = \Pi_t^r \underline{\dot{\mathbb{W}}}$, which is the algebraic condition required. □

8.6 Modelled Distributions

8.6.1 Definition and the Reconstruction Operator

We have defined a regularity structure. We now introduce the space of functions from $[0, T]$ to $\mathcal{T}$, the *model space* of a regularity structure. These abstract functions should represent at each point of $[0, T]$, a "Taylor expansion" of a real function.

We showed in Sect. 8.5.3 how to build an abstract function $\underline{Y}(t) = y_t \underline{1} + y_t' \underline{W}$ which represents the expansion of a real controlled rough path (y, y') at a point t. The most important result of the theory of regularity structures is to show how to build a real function or distribution from an abstract function. Namely, given an approximation of a function at each time, how to reconstruct "continuously" the function. This is given by the *reconstruction map* theorem.

Definition 8.42 Given a regularity structure $(\mathcal{A}, \mathcal{T}, \mathcal{G})$ and a model $M = (\Pi, \Gamma)$, for $\gamma \in \mathbb{R}$ we define the space D_M^γ of *modelled distributions* as functions $\underline{f} : [0, T] \to \mathcal{T}_{<\gamma} := \bigoplus_{\beta<\gamma} \mathcal{T}_\beta$ such that for all $s, t \in [0, T]$ and for all $\beta < \gamma$,

$$\left\| \underline{f}(s) - \Gamma_{s,t}(\underline{f}(t)) \right\|_\beta \leq C(T)|t - s|^{\gamma-\beta},$$

where $C(T)$ is a constant which depends only on T.

Recalling that $\|\cdot\|_\beta$ is the norm of the component in $\mathcal{T}_\beta$, we define by

$$\left\| \underline{f} \right\|_{\gamma,T} := \sup_{s \neq t \in [0,T]} \sup_{\beta<\gamma} \frac{\left\| \underline{f}(t) - \Gamma_{t,s}(\underline{f}(s)) \right\|_\beta}{|t - s|^{\gamma-\beta}}$$

a semi-norm on the space D_M^γ. It is also possible to consider the norm

$$\left\| \underline{f} \right\|_{\gamma,T}^* := \sup_{t \in [0,T]} \sup_{\beta<\gamma} \left\| \underline{f}(t) \right\|_\beta + \left\| \underline{f} \right\|_{\gamma,T}.$$

Moreover $\|\cdot\|_{\gamma,T}^*$ is equivalent to

$$\sup_{\beta<\gamma} \left\| \underline{f}(0) \right\|_\beta + \left\| \underline{f} \right\|_{\gamma,T},$$

so from now we use these two norms without distinction.

Remark 8.43 For a fixed model M, the modelled distributions space D_M^γ is a Banach space with the norm $\| \ \|_{\gamma,T}^*$.

Remark 8.44 We choose the same notation for the semi-norm on D_M^γ as on $\mathscr{D}_W^\gamma$ (the space of modelled distributions and on C^α (the space of Hölder functions or distributions).

So when $\underline{f} \in D_M^\gamma$, we have to understand $\left\| \underline{f} \right\|_{\gamma,T}$ with Definition 8.42 but when $f \in C^\alpha$, $\|f\|_{\alpha,T}$ is the Hölder norm of Definition 8.1 (for functions $\alpha > 0$) or 8.7 (for distributions $\alpha < 0$).

Remark 8.45 The modelled distribution space D_M^γ can be thought of as abstract γ-Hölder functions. Indeed, for an integer p and $\delta \in [0, 1)$ such that $\gamma = p + \delta$, if f is a smooth function

$$\left| f(x) - \sum_{k=0}^{p} \frac{f^{(k)}(y)}{k!}(y-x)^k \right| \leq C|t-s|^\delta,$$

according to the Taylor's inequality. Hence, Definition 8.42 of modelled distributions has to be seen as an extension of the Taylor inequality in a no classical way.

Now we are able to outline the main theorem of the theory of regularity structures which given a modelled distribution allows us to build a "real" distribution approximated at each point by the modelled distribution.

Theorem 8.46 (Reconstruction Map) *Given a regularity structure $\mathscr{T} = (\mathcal{A}, \mathcal{T}, \mathcal{G})$ and a model $M = (\Pi, \Gamma)$, for a real $\gamma > \alpha_* = \min \mathcal{A}$ and an integer $r > |\alpha_*|$ there is a linear continuous map $\mathcal{R} : D_M^\gamma \to C^{\alpha_*}$ such that for all $\underline{f} \in D_M^\gamma$,*

$$| \left[\mathcal{R}(\underline{f}) - \Pi_s(\underline{f}(s)) \right] (\phi_s^\lambda)| \leq C \, \|\Pi\|_{\gamma,T} \left\| \underline{f} \right\|_{\gamma,T}^* \lambda^\gamma, \tag{8.30}$$

where C depends uniformly over $\phi \in B_r$, $\lambda \in (0, 1]$, $s \in [0, T]$.

Moreover if $\gamma > 0$, the bound (8.30) *defined $\mathcal{R}(\underline{f})$ uniquely.*

If $(\tilde{\Pi},)$ is an other model for $\mathscr{T}$ and $\tilde{\mathcal{R}}$ the reconstruction map associated to the model, we have the bound

$$\begin{aligned} &|\mathcal{R}(\underline{f}) - \tilde{\mathcal{R}}(\underline{\tilde{f}}) - \Pi_s(\underline{f}(s)) + \tilde{\Pi}_s(\underline{\tilde{f}}(s))](\eta_s^\lambda)| \\ &\leq C \left(\left\| \tilde{\Pi} \right\|_{\gamma,T} \left\| \underline{f} - \underline{\tilde{f}} \right\|_{\gamma,T}^* + \left\| \Pi - \tilde{\Pi} \right\|_{\gamma,T} \left\| \underline{f} \right\|_{\gamma,T}^* \right) \lambda^\gamma, \end{aligned} \tag{8.31}$$

where C depends uniformly over $\phi \in B_r$, $\lambda \in (0, 1]$, $s \in [0, T]$, as above.

Proof The proof uses the wavelet analysis in decomposing the function $\underline{f}$ in a smooth wavelet basis. The proof requires many computation. A complete one can be found in [10] and a less exhaustive one is in [6]. The construction of $\mathcal{R}(\underline{f})$ is the following. We define a sequence $(\mathcal{R}^j(\underline{f}))_{j\in\mathbb{N}}$ such that

$$\mathcal{R}^j(\underline{f}) := \sum_{k\in\mathbb{Z}} \Pi_{k/2^j}(\underline{f}(k/2^j))(\phi_k^j)\phi_k^j, \tag{8.32}$$

where ϕ_k^j is defined in Definition 8.11 with a regularity at almost r. Then, we show that $\mathcal{R}^j(\underline{f})$ converges weakly to a distribution $\mathcal{R}(\underline{f})$ which means that $\mathcal{R}^j(\underline{f})(\eta)$ converges to $\mathcal{R}(\underline{f})(\eta)$ for all $\eta \in C_0^r$. And we show that the bound (8.30) holds. □

Remark 8.47 It can be proved that if for all $s \in [0, T]$ and $\underline{\tau} \in \mathcal{T}$, $\Pi_s\underline{\tau}$ is a continuous function then $\mathcal{R}(\underline{f})$ is also a continuous function such that

$$\mathcal{R}(\underline{f})(s) = \Pi_s(\underline{f}(s))(s). \tag{8.33}$$

Corollary 8.48 *With the same notation as in Theorem 8.46, for every $\gamma > 0$, there is a constant C such as*

$$\left\|\mathcal{R}(\underline{f})\right\|_{\alpha,T} \le C \left\|\Pi\right\|_{\gamma,T} \left\|\underline{f}\right\|^*_{\gamma,T}.$$

Proof According to Theorem 8.46, for $\phi \in B_r$,

$$\frac{|\mathcal{R}(\underline{f})(\phi_s^\lambda)|}{\lambda^\alpha} \le \frac{|\Pi_s(\underline{f}(s))(\phi_s^\lambda)|}{\lambda^\alpha} + C \left\|\Pi\right\|_{\gamma,T} \left\|\underline{f}\right\|^*_{\gamma,T} \lambda^{\gamma-\alpha},$$

and according to the Definition 8.35,

$$\frac{|\Pi_s(\underline{f}(s))(\phi_x^\lambda)|}{\lambda^\alpha} \le \left\|\Pi\right\|_{\gamma,T} \left\|\underline{f}\right\|_{\gamma,T}.$$

So finally

$$\begin{aligned}\left\|\mathcal{R}(\underline{f})\right\|_{\alpha,T} &\le \left\|\Pi\right\|_{\gamma,T} \left\|\underline{f}\right\|_{\gamma,T} + C \left\|\Pi\right\|_{\gamma,T} \left\|\underline{f}\right\|^*_{\gamma,T} \lambda^{\gamma-\alpha} \\ &\le C \left\|\Pi\right\|_{\gamma,T} \left\|\underline{f}\right\|^*_{\gamma,T},\end{aligned}$$

which, by letting λ going to 0 proves the inequality. □

8.6.2 Modelled Distribution of Controlled Rough Paths

We reformulate the definition of a controlled rough path in the regularity structures framework.

Definition 8.49 Given $(W, \mathbb{W}) \in \mathscr{C}^\alpha$, $(y, y') \in \mathscr{D}_W^{2\alpha}$, the rough path regularity structure $(\mathcal{A}^r, \mathcal{T}^r, \mathcal{G}^r)$ and $M^r = (\Pi^r, \Gamma^r)$ the model associated (cf. Definition 8.41), we define a modelled distribution $\underline{Y} \in D_{M^r}^{2\alpha}$ such that

$$\underline{Y}(t) = y_t \underline{1} + y'_t \underline{W}, \quad \forall t \in [0, T].$$

The space $D_{M^r}^{2\alpha}$ is the space of the *modelled distributions of the controlled rough paths.*

Remark 8.50 This definition is a particular case of modelled distributions of Definition 8.42.

Proof Let check that $\underline{Y}$ is in $D_{M^r}^{2\alpha}$. For every $s, t \in [0, T]$,

$$\begin{aligned}\underline{Y}(t) - \Gamma_{t,s}^r(\underline{Y}(s)) &= \underline{Y}(t) - \Gamma_{t,s}^r(y_s \underline{1} + y'_s \underline{W}) \\ &= \underline{Y}(t) - (y_s \underline{1} + y'_s \underline{W} + y'_s W_{s,t} \underline{1}),\end{aligned}$$

using the Definition 8.41. Then, we have

$$\left\| \underline{Y}(t) - \Gamma_{t,s}^r(\underline{Y}(s)) \right\|_0 = \left\| y(t) - y(s) - y'(s) W_{s,t} \right\| \le C|t - s|^{2\alpha},$$

according to the definition 8.19 of controlled rough paths. Besides,

$$\left\| \underline{Y}(t) - \Gamma_{t,s}^r(\underline{Y}(s)) \right\|_\alpha = \left\| y'(t) - y'(s) \right\| \le C|t - s|^\alpha,$$

which proves that $\underline{Y} \in D_{M^r}^{2\alpha}$. □

Proposition 8.51 *With the notations of Definition 8.49, the application* $(y, y') \in \mathscr{D}_W^{2\alpha} \mapsto \underline{Y} \in D_{M^r}^{2\alpha}$ *is an isomorphism and the norms* $\|y\|_{\infty,T} + \left\|y'\right\|_{\infty,T} + \left\|(y, y')\right\|_{2\alpha,T}^W$ *and* $\left\|\underline{Y}\right\|_{2\alpha,T}^*$ *are equivalent.*

Proof We prove the only equivalence between the two norms.

With the notation of Definition 8.19, we recall that

$$y_{s,t} = y'_s W_{s,t} + R_{s,t}^y, \tag{8.34}$$

and that $\left\|(y, y')\right\|_{2\alpha,T}^W = \left\|y'\right\|_{\alpha,T} + \|R^y\|_{2\alpha,T}$. Then according to the previous proof and Definition 8.42,

$$\left\|\underline{Y}\right\|_{2\alpha,T} = \sup\left\{\left\|y'\right\|_{\alpha,T}, \left\|R^y\right\|_{2\alpha,T}\right\}. \tag{8.35}$$

So we have $\left\|\underline{Y}\right\|_{2\alpha,T} \le \left\|(y, y')\right\|_{2\alpha,T}^W$ and $\left\|(y, y')\right\|_{2\alpha,T}^W \le 2\left\|\underline{Y}\right\|_{2\alpha,T}$. In adding the terms $\|y\|_{\infty,T} + \left\|y'\right\|_{\infty,T}$ to each semi-norms, we obtain the result. □

8.7 Rough Path Integral with the Reconstruction Map

The power of the theory of regularity structures is to give a sense in some cases of a product of distributions. Indeed, it is not possible in general to extend the natural product between functions to the space of distributions.

To build the controlled rough path integral of Theorem 8.27, with the theory of regularity structures we need to give a meaning to the product between y and $\dot{W}$, where $\dot{W}$ is a distribution. We start by giving a meaning to the abstract product between $\underline{Y}$ and $\underline{\dot{W}}$. When the product has good properties, we use the reconstruction map (Theorem 8.46) to define a "real" multiplication.

Definition 8.52 (Multiplication in the Model Space) Given a regularity structure $(\mathcal{A}, \mathcal{T}, \mathcal{G})$, we say that the continuous bilinear map $\star : \mathcal{T}^2 \to \mathcal{T}$ defines a *multiplication* (*product*) on the model space $\mathcal{T}$ if

- For all $\underline{\tau} \in \mathcal{T}$, on has $\underline{1} \star \underline{\tau} = \underline{1}$,
- For every $\underline{\tau} \in T_\alpha$ and $\underline{\sigma} \in T_\beta$, on has $\underline{\tau} \star \underline{\sigma} \in T_{\alpha+\beta}$, if $\alpha + \beta \in \mathcal{A}$ and $\underline{\tau} \star \underline{\sigma} = 0$ if $\alpha + \beta \notin \mathcal{A}$.
- For every $\underline{\tau} \in T_\alpha$, $\underline{\sigma} \in T_\beta$ and $\Gamma \in \mathcal{G}$, $\Gamma(\underline{\tau} \star \underline{\sigma}) = \Gamma(\underline{\tau}) \star \Gamma(\underline{\sigma})$.

We denote by $|\underline{\tau}|$ the homogeneity α of the symbol $\underline{\tau}$. The last item of the definition can be rephrased as $|\underline{\tau} \star \underline{\sigma}| = |\underline{\tau}| + |\underline{\sigma}|$.

Remark 8.53 For example in the following Theorem 8.54, we define within the regularity structure of rough paths the multiplication described in the table below:

$\star$	$\underline{\dot{W}}$	$\underline{\dot{\mathbb{W}}}$	$\underline{1}$	$\underline{W}$
$\underline{\dot{W}}$			$\underline{W}$	$\underline{\dot{\mathbb{W}}}$
$\underline{\dot{\mathbb{W}}}$			$\underline{\dot{\mathbb{W}}}$	
$\underline{1}$	$\underline{\dot{W}}$	$\underline{\dot{\mathbb{W}}}$	$\underline{1}$	$\underline{W}$
$\underline{W}$	$\underline{\dot{\mathbb{W}}}$		$\underline{W}$	

We are now able to build the rough integral with the reconstruction theorem (Theorem 8.46). The operator I corresponding to the integral of a controlled rough path against a rough path.

Theorem 8.54 *We set $\alpha \in (1/3, 1/2]$. There is a linear map $I : D^{2\alpha}_{M^r} \to C^\alpha$ such that for all $\underline{Y} \in D^{2\alpha}_{M^r}$, $I(\underline{Y})(0) = 0$ and such that the map L defined by*

$$\forall t \in [0, T],\ L(\underline{Y})(t) := I(\underline{Y})(t)\underline{1} + \langle \underline{Y}(t), \underline{1} \rangle \underline{W}$$

is linear and continuous from $D^{2\alpha}_{M^r}$ into itself. The symbol $\langle \cdot, \underline{1} \rangle$ denotes the coordinate along $\underline{1}$.

Remark 8.55 Recalling that if $\underline{Y} \in D^{2\alpha}_{M^r}$, according to the Definition 8.49 there is $(y, y') \in \mathscr{D}^{2\alpha}_W$ such that

$$\underline{Y}(t) = y_t \underline{1} + y'_t \underline{W}, \tag{8.36}$$

we show in the proof of the Theorem 8.54 that

$$I(\underline{Y})(t) = \int_0^t y_s \mathrm{d}\mathbf{W}_s, \tag{8.37}$$

where $\int_0^t y_s \mathrm{d}\mathbf{W}_s$ is defined in Theorem 8.27. Thus L is the equivalent in the modeled distribution space of the map

$$(y, y') \in \mathscr{D}^{2\alpha}_W \mapsto \left(\int_0^{\cdot} y_s \mathrm{d}\mathbf{W}_s, y_{\cdot} \right) \in \mathscr{D}^{2\alpha}_W. \tag{8.38}$$

Remark 8.56 The proof of the existence of I is the same as in Theorem 8.27 (classical sewing lemma). But we show how Theorem 8.46 (reconstruction map) can be adapted to recover the result.

Proof For $\underline{Y}$ in $D^{2\alpha}_{M^r}$, we define the point-wise product between $\underline{Y}$ and $\underline{\dot{W}}$ as in Remark 8.53, i.e $\underline{Y}(t) \star \underline{\dot{W}} := y_t \underline{\dot{W}} + y'_t \underline{W\dot{W}}$, where $\underline{W\dot{W}} := \underline{W} \star \underline{\dot{W}} := \underline{\dot{\mathbb{W}}}$. We denote this product $\underline{Y}\underline{\dot{W}}(t)$, to simplify the notation. Using the fact that $|\underline{W}|+|\underline{\dot{W}}| = 2\alpha - 1 = |\underline{\dot{\mathbb{W}}}|$ it is straightforward to check that the product is consistent with the Definition 8.52.

We check now that $\underline{Y}\underline{\dot{W}}$ is in $D^{3\alpha-1}_{M^r}$. According to Definition 8.41 item (v), we compute

$$\Gamma^r_{t,s}\left(\underline{Y}\underline{\dot{W}}(s)\right) = (y_s + y'_s W_{s,t})\underline{\dot{W}} + y'_s \underline{\dot{\mathbb{W}}},$$

since $\underline{Y} \in D^{2\alpha}_{M^r}$ with Definition 8.49,

$$\left\|\underline{Y}\underline{\dot{W}}(t) - \Gamma^r_{t,s}\left(\underline{Y}\underline{\dot{W}}(s)\right)\right\|_{\alpha-1} = \left\|y_{s,t} - y'_s W_{s,t}\right\| \lesssim |t-s|^{2\alpha}, \tag{8.39}$$

$$\left\|\underline{Y}\underline{\dot{W}}(t) - \Gamma^r_{t,s}\left(\underline{Y}\underline{\dot{W}}(s)\right)\right\|_{2\alpha-1} = \left\|y'_{s,t}\right\| \lesssim |t-s|^{\alpha}. \tag{8.40}$$

Thus, by Definition 8.42, we get that $\underline{Y}\underline{\dot{W}} \in D^{3\alpha-1}_{M^r}$.

Thus, given that $3\alpha - 1 > 0$, we can apply the reconstruction theorem in the positive case.

So there is a unique distribution $\mathcal{R}(\underline{Y}\underline{\dot{W}})$ in $C^{\alpha-1}$ such that for every $s \in [0, T]$, $\lambda > 0$ and every localized test function η^λ_s of Definition 8.4,

$$\left|\mathcal{R}(\underline{Y}\underline{\dot{W}})(\eta^\lambda_s) - y_s \int \eta^\lambda_s(u)\mathrm{d}W_u - y'_s \int \eta^\lambda_s(u)\mathrm{d}\mathbb{W}_{s,u}\right| \leq C\,\|\eta\|_{C^1}\,\lambda^{3\alpha-1}, \tag{8.41}$$

where we use relations of the item (iv) of Definition 8.41.

We define with Lemma 8.10 the operator $I : D^{2\alpha}_{M^r} \to C^\alpha$ such that $I(\underline{Y}) \in C^\alpha$ is associated to $\mathcal{R}(\underline{Y}\,\underline{\dot{W}})$. It means that $I\left(\underline{Y}\right)(0) := 0$ and $\langle I\left(\underline{Y}\right), \eta'\rangle := -\langle \mathcal{R}(\underline{Y}\,\underline{\dot{W}}), \eta\rangle$. More precisely, we have for $|t-s| \leq 1$,

$$I(\underline{Y})_{s,t} = \sum_{k\in I_l} \langle \mathcal{R}(\underline{Y}\,\underline{\dot{W}}), \phi^l_k\rangle \int_s^t \phi^l_k + \sum_{j\geq l}\sum_{k\in I_l} \langle \mathcal{R}(\underline{Y}\,\underline{\dot{W}}), \psi^j_k\rangle \int_s^t \psi^j_k. \tag{8.42}$$

Moreover, according to Theorem 8.11, we can choose the integer l such that $2^{-l} \leq |t-s| < 2^{-l+1}$.

We have

$$I(\underline{Y})_{s,t} - y_s W_{s,t} - y'_s \mathbb{W}_{s,t} = \sum_{k\in I_l} \langle \mathcal{R}(\underline{Y}\,\underline{\dot{W}}) - \Pi_s(\underline{Y}\,\underline{\dot{W}}(s)), \phi^l_k\rangle \int_s^t \phi^l_k \tag{8.43}$$

$$+ \sum_{j\geq l}\sum_{k\in I_j} \langle \mathcal{R}(\underline{Y}\,\underline{\dot{W}}) - \Pi_s(\underline{Y}\,\underline{\dot{W}}(s)), \psi^j_k\rangle \int_s^t \psi^j_k. \tag{8.44}$$

We have

$$\langle \mathcal{R}(\underline{Y}\,\underline{\dot{W}}) - \Pi_s(\underline{Y}\,\underline{\dot{W}}(s)), \psi^j_k\rangle = \langle \mathcal{R}(\underline{Y}\,\underline{\dot{W}}) - \Pi_{k/2^j}(\underline{Y}\,\underline{\dot{W}}(k/2^j)), \psi^j_k\rangle$$
$$+ \langle \Pi_{k/2^j}(\underline{Y}\,\underline{\dot{W}}(k/2^j)) - \Pi_s(\underline{Y}\,\underline{\dot{W}}(s)), \psi^j_k\rangle. \tag{8.45}$$

The first term of the right side of (8.45) is bounded by (8.30),

$$|\langle \mathcal{R}(\underline{Y}\,\underline{\dot{W}}) - \Pi_{k/2^j}(\underline{Y}\,\underline{\dot{W}}(k/2^j)), \psi^j_k\rangle| \leq C 2^{-j/2} 2^{j(1-3\alpha)}. \tag{8.46}$$

For bounding the second term of the right side of (8.45) we use the algebraic relations between Π and Γ as well as the relations (8.26),

$$\langle \Pi_{k/2^j}(\underline{Y}\,\underline{\dot{W}}(k/2^j)) - \Pi_s(\underline{Y}\,\underline{\dot{W}}(s)), \psi^j_k\rangle$$
$$= \langle \Pi_{k/2^j}\left(\underline{Y}\,\underline{\dot{W}}(k/2^j) - \Gamma_{k/2^j,s}\underline{Y}\,\underline{\dot{W}}(s)\right), \psi^j_k\rangle.$$

Yet $\underline{Y}\,\underline{\dot{W}} \in D^{3\alpha-1}_{M^r}$, so with (8.39) and (8.40), we have

$$\left\| \underline{Y}\,\underline{\dot{W}}(k/2^j) - \Gamma_{k/2^j,s}\underline{Y}\,\underline{\dot{W}}(s)\right\|_\beta \leq C|k/2^j - s|^{3\alpha-1-\beta},$$

for $\beta \in \{2\alpha - 1, \alpha - 1\}$. Finally, we obtain with the bounds (8.26),

$$\left| \langle \Pi_{k/2^j}(\underline{Y}\,\underline{\dot{W}}(k/2^j)) - \Pi_s(\underline{Y}\,\underline{\dot{W}}(s)), \psi_k^j \rangle \right| \tag{8.47}$$

$$\leq \sum_{\beta \in \{2\alpha-1, \alpha-1\}} 2^{-j\beta - j/2} \left| \frac{k}{2^j} - s \right|^{3\alpha - 1 - \beta}. \tag{8.48}$$

Moreover, we have $k/2^j \in [-c/2^j - s, c/2^j + t]$ for all terms that are non-vanishing in (8.43) and (8.44). Since $j \geq l$ in the sums and that we assume $2^{-j} \leq 2^{-l} \leq |t - s| < 2^{-l+1}$, we have

$$\left| \frac{k}{2^j} - s \right| \leq C|t - s|, \tag{8.49}$$

for all non-vanishing terms in the sums (8.43) and (8.44).

Firstly we bound (8.43). On the one hand, using (8.46), (8.48), and (8.49) and the fact that $|t - s| < 2^{-l+1}$, we obtain

$$\left| \langle \mathcal{R}(\underline{Y}\,\underline{\dot{W}}) - \Pi_s(\underline{Y}\,\underline{\dot{W}}(s)), \psi_k^j \rangle \right| \leq C 2^{-l/2} 2^{-l(3\alpha - 1)}. \tag{8.50}$$

On another hand, we have

$$\begin{aligned} \left| \int_s^t \phi_k^l \right| &\leq 2^{l/2} |t - s| \sup_{t \in \mathbb{R}} \|\phi(t)\| \\ &\leq C 2^{-l/2}. \end{aligned} \tag{8.51}$$

Thus, because there is only a finite number of terms independent on l that contribute to the sum (8.43), we obtain with (8.50) and (8.51) the following bound on (8.43):

$$\left| \sum_{k \in I_l} \langle \mathcal{R}(\underline{Y}\,\underline{\dot{W}}) - \Pi_s(\underline{Y}\,\underline{\dot{W}}(s)), \phi_k^l \rangle \int_s^t \phi_k^l \right| \leq C 2^{-l3\alpha} \leq C|t - s|^{3\alpha}, \tag{8.52}$$

where C does not depends on l.

Now, we bound (8.44). On the one hand, using (8.46), (8.48), and (8.49), we have for $j \geq l$,

$$\begin{aligned} &|\langle \mathcal{R}(\underline{Y}\,\underline{\dot{W}}) - \Pi_s(\underline{Y}\,\underline{\dot{W}}(s)), \psi_k^j \rangle| \\ &\quad \leq C 2^{-j/2} \left[2^{j(1-3\alpha)} + |t - s|^{2\alpha} 2^{-j(\alpha-1)} + |t - s|^{\alpha} 2^{-j(2\alpha-1)} \right]. \end{aligned} \tag{8.53}$$

On an other hand, we observe that

$$\left| \sum_{k \in I_j} \int_s^t \psi_k^j \right| \leq C 2^{-j/2}, \tag{8.54}$$

because a primitive of ψ has a compact support and the fact that $\int \psi = 0$. Then, combining (8.53) and (8.54) we obtain,

$$\begin{aligned}
&\left| \sum_{j \geq l} \sum_{k \in I_j} \langle \mathcal{R}(\underline{Y} \underline{\dot{W}}) - \Pi_s(\underline{Y} \underline{\dot{W}}(s)), \psi_k^j \rangle \int_s^t \psi_k^j \right| \\
&\quad \leq \sum_{j \geq l} 2^{-3j\alpha} + |t-s|^{2\alpha} 2^{-j\alpha} + |t-s|^{\alpha} 2^{-j2\alpha} \\
&\quad \leq C 2^{-3l\alpha} + |t-s|^{2\alpha} 2^{-l\alpha} + |t-s|^{\alpha} 2^{-l2\alpha} \\
&\quad \leq C|t-s|^{3\alpha}.
\end{aligned} \tag{8.55}$$

With (8.52) and (8.55) we obtain the bound of the left hand side of (8.43),

$$|I(\underline{Y})_{s,t} - y_s W_{s,t} - y_s' \mathbb{W}_{s,t}| \leq C|t-s|^{3\alpha}. \tag{8.56}$$

To show that $L(\underline{Y})$ is in $D_{M^r}^{2\alpha}$, we compute $\Gamma_{t,s}^r\big(L(\underline{Y})(s)\big) = (I(\underline{Y})(s) + y_s W_{s,t})\underline{1} + y_s \underline{W}$ and we use the estimation (8.56). Thus, we have

$$\begin{aligned}
\left\| L(\underline{Y})(t) - \Gamma_{t,s}^r\big(L(\underline{Y})(s)\big) \right\|_0 &= \left\| I(\underline{Y})(t) - I(\underline{Y})(s) - y_s W_{s,t} \right\| \\
&\leq \left\| y' \right\|_{\infty,T} \|\mathbb{W}\|_{2\alpha,T} |t-s|^{2\alpha} + C|t-s|^{3\alpha},
\end{aligned} \tag{8.57}$$

$$\text{and } \left\| L(\underline{Y})(t) - \Gamma_{t,s}\big(L(\underline{Y})(s)\big) \right\|_\alpha = \left\| y_{s,t} \right\| \leq \|y\|_\alpha |t-s|^\alpha, \tag{8.58}$$

which proves that $L(\underline{Y})$ is in $D_{M^r}^{2\alpha}$.

It remains to prove the continuity of L. According to (8.30), the constant C in (8.56) is proportional to $\left\| \underline{Y} \right\|_{\gamma,T}^*$. So we have,

$$|I(\underline{Y})_{s,t} - y_s W_{s,t}| \leq \left\| y' \right\|_\infty \|\mathbb{W}\|_{2\alpha,T} |t-s|^{2\alpha} + C \left\| Y \right\|_{2\alpha,T}^* |t-s|^{3\alpha},$$

which allows with the previous computation (8.57) and (8.58) to bound

$$\left\| L(\underline{Y}) \right\|_{2\alpha,T}^* \leq C \left\| \underline{Y} \right\|_{2\alpha,T}^* . \tag{8.59}$$

This concludes the proof. □

8.8 Existence of a Rough Path Lift

As an application of the reconstruction operator in the case $\gamma \leq 0$, we prove Theorem 8.18 which states that for any $W \in C^\alpha$ ($\alpha \in (1/3, 1/2]$) with values in $\mathbb{R}^n$, it exists a rough path lift $\mathbb{W}$ and that the map $W \mapsto \mathbf{W}$ is continuous from C^α to $\mathscr{C}^\alpha$.

Proof (Theorem 8.18) We consider the regularity structure $(\mathcal{A}^e, \mathcal{T}^e, \mathcal{G}^e)$ such that $\mathcal{A}^e = \{\alpha - 1, 0\}$, $\mathcal{T}^e = \mathrm{Vect}\langle \underline{\dot{W}}^i, i = 1, \dots, n\rangle \bigoplus \mathrm{Vect}\langle \underline{1}\rangle$ and for $\Gamma_h^e \in \mathcal{G}$, $\Gamma_h^e(\underline{\dot{W}}) = \underline{\dot{W}}$, $\Gamma_h^e(\underline{1}) = \underline{1}$. We associate the model $M^e = (\Pi^e, \Gamma^e)$ such that for every $s, t \in [0, T]$, $\eta \in B_1$

$$\Pi_s^e(\underline{\dot{W}})(\eta) := \int \eta(t)\mathrm{d}W_t, \qquad \Pi_s^e(\underline{1})(t) := 1,$$

and $\Gamma_{s,t}^e := \Gamma_{W_{t,s}}^e$.

For $0 \leq s \leq t \leq 1$, and integers $0 \leq i, j \leq n$, the modelled distribution $\underline{\dot{\mathbb{W}}}$ given by $\underline{\dot{\mathbb{W}}^{i,j}}(s) := W_s^i \underline{\dot{W}}^j$ is in $D_M^{2\alpha-1}$. Indeed $\underline{\dot{\mathbb{W}}^{i,j}}(t) - \Gamma_{t,s}^e\left(\underline{\dot{\mathbb{W}}^{i,j}}(s)\right) = W_t^i \underline{\dot{W}}^j - W_s^i \underline{\dot{W}}^j = W_{s,t}^i \underline{\dot{W}}^j$, then

$$\left\| \underline{\dot{\mathbb{W}}^{i,j}}(t) - \Gamma_{t,s}^e\left(\underline{\dot{\mathbb{W}}^{i,j}}(s)\right)\right\|_{\alpha-1} \leq |t-s|^\alpha.$$

So, $\gamma - (\alpha - 1) = \alpha$, we have $\gamma = 2\alpha - 1$. We conclude using the Definition 8.42.

Given that $\alpha \in (1/3, 1/2]$, we have $2\alpha - 1 \leq 0$. Thus, the uniqueness of the reconstruction map does not hold. But, according to Theorem 8.46, there exists $\mathcal{R}(\underline{\dot{\mathbb{W}}}) \in C^{\alpha-1}$ such that

$$|[\mathcal{R}(\underline{\dot{\mathbb{W}}}) - \Pi_s^e(\underline{\dot{\mathbb{W}}})](\eta_s^\lambda)| \leq C\lambda^{2\alpha-1}, \tag{8.60}$$

where $\eta \in B_1$. With Lemma 8.10, we define $z \in C^\alpha$ as the primitive of $\mathcal{R}(\underline{\dot{\mathbb{W}}})$ such that $z(0) = 0$. Moreover, we have for all $s, t \in [0, 1]$,

$$z_{s,t} = \sum_{k \in I_l} \langle \mathcal{R}(\underline{\dot{\mathbb{W}}}), \phi_k^l\rangle \int_s^t \phi_k^l + \sum_{j \geq l}\sum_{k \in I_j} \langle \mathcal{R}(\underline{\dot{\mathbb{W}}}), \psi_k^j\rangle \int_s^t \psi_k^j, \tag{8.61}$$

and

$$W_{s,t} = \sum_{k \in I_l} \langle \Pi_s^e(\underline{\dot{W}}), \phi_k^l\rangle \int_s^t \phi_k^l + \sum_{j \geq l}\sum_{k \in I_j} \langle \Pi_s^e(\underline{\dot{W}}), \psi_k^j\rangle \int_s^t \psi_k^j, \tag{8.62}$$

which yields to

$$W_s \otimes W_{s,t} = \sum_{k\in I_l} \langle \Pi_s^e(\dot{\underline{\mathbb{W}}}(s)), \phi_k^l \rangle \int_s^t \phi_k^l + \sum_{j\geq l} \sum_{k\in I_j} \langle \Pi_s^e(\dot{\underline{\mathbb{W}}}(s)), \psi_k^j \rangle \int_s^t \psi_k^j. \tag{8.63}$$

If there is a constant $C \geq 0$ such that,

$$|z_{s,t} - W_s \otimes W_{s,t}| \leq C|t-s|^{2\alpha}, \tag{8.64}$$

then setting $\mathbb{W}_{s,t} := z_{s,t} - W_s \otimes W_{s,t}$, the pair $(W, \mathbb{W})$ belongs to $\mathscr{C}^\alpha$ according to the Definition 8.13. Let us prove (8.64). We have

$$\begin{aligned} z_{s,t} - W_s \otimes W_{s,t} &= \sum_{k\in I_l} \langle \mathcal{R}(\dot{\underline{\mathbb{W}}}) - \Pi_s^e(\dot{\underline{\mathbb{W}}}(s)), \phi_k^l \rangle \int_s^t \phi_k^l \\ &\quad + \sum_{j\geq l} \sum_{k\in I_j} \langle \mathcal{R}(\dot{\underline{\mathbb{W}}}) - \Pi_s^e(\dot{\underline{\mathbb{W}}}(s)), \psi_k^j \rangle \int_s^t \psi_k^j. \end{aligned} \tag{8.65}$$

From (8.30), we have the bounds

$$|\langle \mathcal{R}(\dot{\underline{\mathbb{W}}}) - \Pi_s^e(\dot{\underline{\mathbb{W}}}(s)), \phi_k^j \rangle| \leq C 2^{-j/2 - j(2\alpha-1)}, \tag{8.66}$$

and

$$|\langle \mathcal{R}(\dot{\underline{\mathbb{W}}}) - \Pi_s^e(\dot{\underline{\mathbb{W}}}(s)), \psi_k^j \rangle| \leq C 2^{-j/2 - j(2\alpha-1)}. \tag{8.67}$$

Then, combining (8.65), (8.66), and (8.67), we proceed as in the proof of Lemma 8.10 to show (8.64).

It remains to show the continuity. If there is another path $\tilde{W} \in C^\alpha$, we define as for W, a model $(\tilde{\Pi}, \tilde{\Gamma})$, a modelled distribution $\tilde{\dot{\underline{\mathbb{W}}}}$, a reconstruction map $\tilde{\mathcal{R}}$ and then $\tilde{\mathbb{W}}$. By denoting

$$\Delta\Pi_{s,k/2^j} := [\Pi(\dot{\underline{\mathbb{W}}}(k/2^j)) - \Pi(\dot{\underline{\mathbb{W}}}(s)) - \tilde{\Pi}(\tilde{\dot{\underline{\mathbb{W}}}}(k/2^j)) + \tilde{\Pi}(\tilde{\dot{\underline{\mathbb{W}}}}(s))](\psi_k^j), \tag{8.68}$$

we have

$$\begin{aligned} &|\Delta\Pi_{s,k/2^j}| \\ &\quad \leq \left\| W - \tilde{W} \right\|_{\alpha,T} \left(\|W\|_{\alpha,T} + \left\| \tilde{W} \right\|_{\alpha,T} \right) |s - k/2^j| 2^{j/2(1-\alpha)} 2^{-j/2}. \end{aligned} \tag{8.69}$$

According to the bounds (8.31) and (8.69) and in writing

$$[\mathcal{R}(\underline{\dot{\mathbb{W}}}) - \Pi_s(\underline{\dot{\mathbb{W}}}(s)) - \tilde{\mathcal{R}}(\underline{\dot{\tilde{\mathbb{W}}}}) + \Pi_s(\underline{\dot{\tilde{\mathbb{W}}}}(s))](\psi_k^j)$$
$$= \mathcal{R}(\underline{\dot{\mathbb{W}}}) - \Pi_{k/2^j}(\underline{\dot{\mathbb{W}}}(k/2^j)) - \tilde{\mathcal{R}}(\underline{\dot{\tilde{\mathbb{W}}}}) + \tilde{\Pi}_{k/2^j}(\underline{\dot{\tilde{\mathbb{W}}}}(k/2^j)) + \Delta\Pi_{s,k/2^j},$$

we get

$$|\mathbb{W}_{s,t} - \tilde{\mathbb{W}}_{s,t}| \leq C\Big[\left\|\tilde{\Pi}\right\|_{2\alpha-1,T}\left\|\underline{\dot{\mathbb{W}}} - \underline{\dot{\tilde{\mathbb{W}}}}\right\|^*_{2\alpha-1,T} + \left\|\Pi - \tilde{\Pi}\right\|_{2\alpha-1,T}\left\|\underline{\dot{\mathbb{W}}}\right\|^*_{2\alpha-1,T}$$
$$+ \left\|W - \tilde{W}\right\|_{\alpha,T}\left(\|W\|_{\alpha,T} + \left\|\tilde{W}\right\|_{\alpha,T}\right)\Big]|t-s|^{2\alpha}.$$

Yet we have, $\left\|\underline{\dot{\mathbb{W}}} - \underline{\dot{\tilde{\mathbb{W}}}}\right\|^*_{2\alpha-1,T} = \left\|W - \tilde{W}\right\|_{\infty,T} + \left\|W - \tilde{W}\right\|_{\alpha,T}$, and

$$\left\|\Pi - \tilde{\Pi}\right\|_{2\alpha-1,T} \leq C\left\|W - \tilde{W}\right\|_{\alpha,T}. \tag{8.70}$$

So finally,

$$\left\|\mathbf{W} - \tilde{\mathbf{W}}\right\|_{\alpha,T} \leq C\left\|W - \tilde{W}\right\|_{\alpha,T}, \tag{8.71}$$

which proves the continuity. □

Remark 8.57 Given that $2\alpha - 1$ is negative, the uniqueness of $\mathbb{W}$ does not hold, which is in accordance with Remark 8.17.

8.9 Composition with a Smooth Function

Before solving the general rough differential equation (8.1) with the theory of regularity structures, we should give a sense of the composition of a modelled distribution with a function. Then we will be able to consider (8.1) in the space of the modelled distributions.

The composition of a modelled distribution $\underline{f} \in D_M^\gamma$ with a smooth function F is developed in [10]. The author gives a general theorem which allows the composition with an arbitrary smooth function F when $\underline{f}$ takes its values in a model space $\mathcal{T}$ such that the smallest index of homogeneity is equal to 0, *i.e.* $\forall t \in \mathbb{R},\ \underline{f}(t) \in \mathrm{Vect}\langle \underline{1}, \ldots \rangle$. Thus, it is possible to define the composition as a Taylor expansion

$$\hat{F} \circ \underline{f}(t) = \sum_k \frac{F^{(k)}(\bar{\underline{f}}(t))}{k!}(\underline{f}(t) - \bar{\underline{f}}(t)\underline{1})^k, \tag{8.72}$$

where $\bar{f}$ is the coordinate of $\underline{f}$ onto $\underline{1}$. The definition above makes no sense if the product between elements of the regularity structure is not defined. We can also find the general definition in [10]. This is not useful here. The idea of the decomposition (8.72) is to compute a Taylor expansion of F in $\bar{f}$ the part of $\underline{f}$ which is the first approximation of $\mathcal{R}\underline{f}$.

Here we just prove (what is needed for solving (8.1)) that $\hat{F} \circ \underline{f}$ lives in the same space as $\underline{f}$ and that $\hat{F}$ is Lipschitz in the particular case of modelled distribution of controlled rough paths.

Theorem 8.58 *Let $F \in C_b^2(\mathbb{R}^d, \mathcal{L}(\mathbb{R}^n, \mathbb{R}^d))$. For $\alpha \in (1/3, 1/2]$, given a rough path $\mathbf{W} = (W, \mathbb{W}) \in \mathscr{C}^\alpha$, the controlled rough path $(y, y') \in \mathscr{D}_W^{2\alpha}$, for all $\underline{Y} \in D_{M^r}^{2\alpha}$ defined by $\underline{Y}(t) = y_t \underline{1} + y_t' \underline{W}$, the map $\hat{F}$ such that*

$$\hat{F} \circ \underline{Y}(t) := F(y_t)\underline{1} + F'(y_t)y_t'\underline{W}, \tag{8.73}$$

is in $D_{M^r}^{2\alpha}$. Moreover if $F \in C_b^3$ the function associated $\hat{F}$ is Lipschitz, i.e. *for all $\underline{Y}, \widetilde{\underline{Y}} \in D_{M^r}^{2\alpha}$*

$$\left\| \hat{F}(\underline{Y}) - \hat{F}(\widetilde{\underline{Y}}) \right\|_{2\alpha,T}^* \leq C \left\| \underline{Y} - \widetilde{\underline{Y}} \right\|_{2\alpha,T}^*, \tag{8.74}$$

where C is a constant.

Remark 8.59 This theorem shows that the space $D_{M^r}^{2\alpha}$ is stable by a non linear composition $\hat{F}$, provided that $\hat{F}$ is regular enough. So with Theorem 8.54, we can build the integral

$$I(\hat{F}(\underline{Y})) = \int_0^\cdot F(y_s) \mathrm{d}\mathbf{W}_s.$$

Proof Firstly, let us show that $\hat{F}$ is a map from $D_{M^r}^{2\alpha}$ to $D_{M^r}^{2\alpha}$. A straightforward computation leads us to the two following expressions

$$\left\| \hat{F}(\underline{Y})(t) - \Gamma_{t,s}^r \left(\hat{F}(\underline{Y})(s) \right) \right\|_0 = \left\| F'(y_t)y_t' - F'(y_s)y_s' \right\|,$$

$$\left\| \hat{F}(\underline{Y})(t) - \Gamma_{t,s}^r \left(\hat{F}(\underline{Y})(s) \right) \right\|_\alpha = \left\| F(y_t) - F(y_s) - F'(y_s)y_s' W_{s,t} \right\|.$$

Let us denote the left-hand of the first equality $\Delta_{s,t}^0$ and of the second one $\Delta_{s,t}^\alpha$. We obtain

$$\begin{aligned}
\Delta_{s,t}^0 &\leq \left\| F'(y_t) \right\| \left\| y_t' - y_s' \right\| + \left\| y_s' \right\| \left\| F'(y_t) - F'(y_s) \right\| \\
&\leq \left\| F' \right\|_{\infty,T} \left\| y' \right\|_\alpha |t-s|^\alpha + \left\| y' \right\|_{\infty,T} \left\| F'' \right\|_{\infty,T} \|Y\|_{\alpha,T} |t-s|^\alpha
\end{aligned}$$

and

$$\begin{aligned}\Delta^{\alpha}_{s,t} &= \left\|F(y_t) - F(y_s) - F'(y_s)(y_{s,t} - R^y_{s,t})\right\| \\ &\leq \left\|F(y_t) - F(y_s) - F'(y_s)y_{s,t}\right\| + \left\|F'(y_s)R^y_{s,t}\right\| \\ &\leq \frac{1}{2}\left\|F''\right\|_{\infty,T}\left\|y_{s,t}\right\|^2 + \left\|F'\right\|_{\infty,T}\left\|R^y\right\|_{2\alpha,T}|t-s|^{2\alpha} \\ &\leq \frac{1}{2}\left\|F''\right\|_{\infty,T}\|y\|^2_{\alpha,T}|t-s|^{2\alpha} + \left\|F'\right\|_{\infty,T}\left\|R^y\right\|_{2\alpha,T}|t-s|^{2\alpha}.\end{aligned}$$

This proves that $\hat{F}(\underline{Y}) \in D^{2\alpha}_{M^r}$.

We now prove the inequality (8.74). A more general proof can be found in [10]. We define $\widetilde{\underline{Z}} = \underline{Y} - \widetilde{\underline{Y}}$, which is in $D^{2\alpha}_{M^r}$ by linearity. We denote by $Q_{<2\alpha}$ the projection onto $\mathcal{T}_{<2\alpha}$. Using the integration by parts formula, one can check that

$$\hat{F}(\underline{Y}(s)) - \hat{F}(\widetilde{\underline{Y}}(s)) = \sum_{k=0}^{1}\int_0^1 F^{(k)}(\widetilde{y}_s + uz_s)Q_{<2\alpha}\left[\left[(\widetilde{y}'_s + uz'_s)\underline{W}\right]^k \widetilde{\underline{Z}}(s)\right]\mathrm{d}u.$$

Then, we compute the expansion between s and t of $\underline{\Delta}(s) := \hat{F}(\underline{Y}(s)) - \hat{F}(\widetilde{\underline{Y}}(s))$. We denote $\underline{A}_u(s) := \widetilde{\underline{Y}}(s) + u\widetilde{\underline{Z}}(s)$. When u is fixed, $\underline{A}_u$ is in $D^{2\alpha}_{M^r}$. We have

$$\begin{aligned}\Gamma_{t,s}\underline{\Delta}(s) &= \sum_{k=0}^{1}\int_0^1 F^{(k)}(\underline{A}_u(s))\Gamma_{t,s}Q_{<2\alpha}\left([A'_u(s)\underline{W}]^k\widetilde{\underline{Z}}(s)\right)\mathrm{d}u \\ &= \sum_{k=0}^{1}\int_0^1 F^{(k)}(\underline{A}_u(s))[\Gamma_{t,s}(A'_u(s)\underline{W})]^k\Gamma_{t,s}\widetilde{\underline{Z}}(s)\mathrm{d}u + \underline{R}(s,t),\end{aligned}$$

where $\underline{R}$ is a remainder such that $\left\|\underline{R}(s,t)\right\|_{\beta} \lesssim |t-s|^{2\alpha-\beta}$ for $\beta \in \{0,\alpha\}$. From now, we denote by $\underline{R}$ all the remainder terms which satisfy this property.

We now shift the last expression from s to t. On the one hand

$$\Gamma^r_{t,s}(A'_u(s)\underline{W}) = \Gamma^r_{t,s}\underline{A}_u(s) - A_u(s)\underline{1} = \underline{A}_u(t) - A_u(s)\underline{1} + \underline{R}(s,t).$$

On the other hand

$$\Gamma^r_{t,s}\widetilde{\underline{Z}}(s) = \widetilde{\underline{Z}}(t) + \underline{R}(s,t).$$

This yields

$$\Gamma^r_{t,s}\underline{\Delta}(s) = \sum_{k=0}^{1}\int_0^1 F^{(k)}(\underline{A}_u(s))[A'_u(t)\underline{W} + (A_u(t) - A_u(s))\underline{1}]^k\widetilde{\underline{Z}}(s)\mathrm{d}u + \underline{R}(s,t).$$

It remains to shift $F^{(k)}$ from s to t. With the classical Taylor expansion formula,

$$F^{(k)}(A_u(s)) = \sum_{0\le l+k\le 1} F^{(k+l)}(A_u(t))(A_u(s) - A_u(t))^l + O(|t-s|^{2\alpha-k\alpha}),$$

because $\|A_u(t) - A_u(s)\| \le |t-s|^\alpha$. The bound

$$\left\| [A_u'(t)\underline{W} + (A_u(t) - A_u(s))\underline{1}]^k \right\|_\beta \lesssim |t-s|^{k\alpha-\beta}$$

holds. Finally, with the two previous expressions,

$$\begin{aligned}
\Gamma^r_{t,s}\underline{\Delta}(s) &= \sum_{0\le l+k\le 1} F^{(k+l)}(A_u(t))(A_u(s) - A_u(t))^l \\
&\quad \times [A_u'(t)\underline{W} + (A_u(t) - A_u(s))\underline{1})]^k \widetilde{\underline{Z}}(t) + \|\widetilde{\underline{Z}}\|^*_{2\alpha,T}\, O(|t-s|^{2\alpha-\beta}) \\
&= \sum_{0\le k\le 1} F^{(k)}(A_u(t))[A_u'(t)\underline{W}]^k \widetilde{\underline{Z}}(t) + \|\widetilde{\underline{Z}}\|^*_{2\alpha,T}\, O(|t-s|^{2\alpha-\beta}) \\
&= \underline{\Delta}(t) + \|\widetilde{\underline{Z}}\|^*_{2\alpha,T}\, O|t-s|^{2\alpha-\beta},
\end{aligned}$$

which proves the inequality. □

8.10 Solving the Rough Differential Equations

Theorem 8.54 combined with Theorem 8.58 allow us to solve the rough differential equations in the modelled distribution space $D^{2\alpha}_{M^r}$.

Theorem 8.60 *Given $\xi \in \mathbb{R}^d$, $F \in C_b^3(\mathbb{R}^d, \mathcal{L}(\mathbb{R}^n, \mathbb{R}^d))$, a rough path $\mathbf{W} = (W, \mathbb{W}) \in \mathscr{C}^\beta$ with $\beta \in (1/3, 1/2)$, there is a unique modelled distribution $\underline{Y} \in D^{2\beta}_{M^r}$ such that for all $t \in [0, T]$,*

$$\underline{Y}(t) = \xi\underline{1} + L(\hat{F}(\underline{Y}))(t), \tag{8.75}$$

where L is defined in Theorem 8.54.

Proof We prove that the operator $N(\underline{Y}) := \xi\underline{1} + L(\hat{F}(\underline{Y}))$ where L is defined in Theorem 8.54, has a unique fixed point. For this we show that the unit ball of $D^{2\alpha}_{M^r}$ is invariant under the action of N, and then that N is a strict contraction.

These two properties can be obtained by choosing a wise time interval $[0, T]$. We take a rough path $\mathbf{W} = (W, \mathbb{W}) \in \mathscr{C}^\beta \subset \mathscr{C}^\alpha$ with $1/3 < \alpha < \beta < 1/2$ and $\underline{Y} \in D^{2\alpha}_{M^r}$. This trick allows us to have a $T^{\beta-\alpha}$ in our estimates. Thus, with a T small enough we prove the fixed point property. We start by choosing $T \le 1$.

According to Theorem 8.58 $\hat{F}(\underline{Y}) \in D^{2\alpha}_{M^r}$, thus Theorem 8.54 shows that $N(\underline{Y}) \in D^{2\alpha}_{M^r}$. If $\underline{Y}$ is a fixed point of N then $\underline{Y} \in D^{2\beta}_{M^r}$, thanks to the fact that $\mathbf{W} \in \mathscr{C}^{\beta}$. Indeed,

$$\left\|\underline{Y}(t) - \Gamma_{t,s}\underline{Y}(s)\right\|_{\beta} = \left\|y_{s,t}\right\| \leq \left\|y'\right\|_{\infty,T} \|W\|_{2\beta,T} |t-s|^{2\beta} + \left\|R^{y}\right\|_{2\alpha,T} |t-s|^{2\alpha},$$

and

$$\left\|\underline{Y}(t) - \Gamma_{t,s}\underline{Y}(s)\right\|_{0} = \left\|y_{s,t} - y'_s W_{s,t}\right\| \leq \left\|y'\right\|_{\infty,T} \left\|\mathbb{W}_{s,t}\right\| + O(|t-s|^{3\alpha}).$$

As a result of the fixed point property $y' = F(y)$. This proves that $\underline{Y} \in D^{2\beta}_{M^r}$.

We recall that $\left\|\underline{Y}\right\|^{*}_{2\alpha,T} = \sup_{\epsilon\in\{0,\alpha\}} \left\|\underline{Y}(0)\right\|_{\epsilon} + \left\|\underline{Y}\right\|_{2\alpha,T}$, where

$$\left\|\underline{Y}\right\|_{2\alpha,T} = \sup_{t,s\in[0,T],\epsilon\in\{0,\alpha\}} \frac{\left\|\underline{Y}(t) - \Gamma_{t,s}\underline{Y}(s)\right\|_{\epsilon}}{|t-s|^{\epsilon}}.$$

It is more convenient to work with the semi-norm $\|\cdot\|_{2\alpha,T}$, so we define the affine ball unit on $[0, T]$

$$B_T = \{\underline{Y} \in D^{2\alpha}_{M^r},\ \underline{Y}(0) = \xi\underline{1} + f(\xi)\underline{W},\ \left\|\underline{Y}\right\|_{2\alpha,T} \leq 1\}.$$

Invariance
For $\underline{Y} \in B_T$, on has $\left\|\hat{F}(\underline{Y})\right\|_{2\alpha,T} \leq \left\|\underline{Y}\right\|_{2\alpha,T}$ and

$$N(\underline{Y}) = \left\|L(\hat{F}(\underline{Y}))\right\|_{2\alpha,T}.$$

On the on hand, according to the reconstruction map,

$$\begin{aligned}
&\left\|(I\hat{F}(\underline{Y}))_{s,t} - F(y_s)W_{s,t}\right\| \\
&\leq \left\|F'(y)y'\right\|_{\infty,T} \|\mathbb{W}\|_{2\alpha,T} |t-s|^{2\alpha} + C\left\|\hat{F}(\underline{Y})\right\|^{*}_{2\alpha,T} |t-s|^{3\alpha} \\
&\leq \left\|\hat{F}(\underline{Y})\right\|^{*}_{2\alpha,T} \|\mathbb{W}\|_{2\alpha} |t-s|^{2\alpha} + C\left\|\hat{F}(\underline{Y})\right\|^{*}_{2\alpha,T} |t-s|^{3\alpha} \\
&\leq \left\|F'\right\|_{\infty}\left[\left\|(\underline{Y})\right\|^{*}_{2\alpha,T} \|\mathbb{W}\|_{2\beta,T} T^{\beta-\alpha}|t-s|^{2\alpha} + C\left\|\underline{Y}\right\|^{*}_{2\alpha,T} |t-s|^{2\alpha}T^{\alpha}\right],
\end{aligned}$$

because $\|\cdot\|_{\beta} \leq \|\cdot\|_{\alpha} T^{\beta-\alpha}$. Using the fact that $T^{\alpha} \leq T^{\beta-\alpha}$ and that $\underline{Y} \in B_T$ we obtain

$$\left\|N(\underline{Y})\right\|_{0} \leq CT^{\beta-\alpha},$$

where C is independent of $\underline{Y}$. On the other hand,

$$\begin{aligned}\left\|y_{s,t}\right\| &\leq \left\|y'\right\|_{\infty,T} \|W\|_{\alpha,T} |t-s|^{\alpha} + \left\|R^{y}\right\|_{2\alpha,T} |t-s|^{2\alpha} \\ &\leq \left\|\underline{Y}\right\|^{*}_{2\alpha,T} \|W\|_{\beta,T} T^{\beta-\alpha}|t-s|^{\alpha} + \left\|R^{y}\right\|_{2\alpha,T} T^{\alpha}|t-s|^{\alpha} \\ &\leq \left\|\underline{Y}\right\|^{*}_{2\alpha,T} \|W\|_{\beta,T} T^{\beta-\alpha}|t-s|^{\alpha} + \left\|\underline{Y}\right\|^{*}_{2\alpha,T} T^{\alpha}|t-s|^{\alpha}.\end{aligned}$$

Using the last inequality

$$\begin{aligned}\|F(y)\|_{\alpha,T} &\leq \left\|F'\right\|_{\infty} \|y\|_{\alpha,T} \\ &\leq \left\|\underline{Y}\right\|^{*}_{2\alpha,T} \|W\|_{\beta,T} T^{\beta-\alpha}|t-s|^{\alpha} + \left\|\underline{Y}\right\|^{*}_{2\alpha,T} T^{\alpha}|t-s|^{\alpha},\end{aligned}$$

which leads to $\left\|N(\underline{Y})\right\|_{\alpha} \leq CT^{\beta-\alpha}$. Finally, we obtain the following estimate $\left\|N(\underline{Y})\right\|_{2\alpha,T} \leq CT^{\beta-\alpha}$, where C does not depend on $\underline{Y}$. By choosing $T = T_0$ small enough, we show that $N(B_{T_0}) \subset B_{T_0}$.

Contraction
For $\underline{Y}, \widetilde{\underline{Y}} \in D^{2\alpha}_{M^r}$,

$$\begin{aligned}\left\|N(\underline{Y}) - N(\widetilde{\underline{Y}})\right\|_{2\alpha,T} &\leq \left\|N(\underline{Y}) - N(\widetilde{\underline{Y}})\right\|_{0} + \left\|N(\underline{Y}) - N(\widetilde{\underline{Y}})\right\|_{\alpha} \\ &\leq C\left\|\hat{F}(Y) - \hat{F}(\widetilde{\underline{Y}})\right\|^{*}_{2\alpha,T} T^{\beta-\alpha} + \|F(y) - F(\widetilde{y})\|_{\alpha} \\ &\leq C\left\|\underline{Y} - \widetilde{\underline{Y}}\right\|^{*}_{2\alpha,T} T^{\beta-\alpha} + \left\|F'\right\|_{\infty} \|y - \widetilde{y}\|_{\alpha},\end{aligned}$$

according to (8.74). Then it is easy to show that

$$\|y - \widetilde{y}\|_{\alpha} \leq CT^{\beta-\alpha} \left\|\underline{Y} - \widetilde{\underline{Y}}\right\|_{2\alpha,T}.$$

Finally, $\left\|N(\underline{Y}) - N(\widetilde{\underline{Y}})\right\|_{2\alpha,T} \leq CT^{\beta-\alpha} \left\|\underline{Y} - \widetilde{\underline{Y}}\right\|_{2\alpha,T}$ where C does not depend on neither $\underline{Y}$ nor $\widetilde{\underline{Y}}$. So with T small enough, $N(B_T) \subset B_T$ and N is a strict contraction. So, there is a unique solution $\underline{Y} \in D^{2\alpha}_{M^r}$ to (8.75) on $[0, T]$. As mentioned at the beginning of the proof, $\underline{Y}$ is in$D^{2\beta}_{M^r}$.

□

Corollary 8.61 *Given $\xi \in \mathbb{R}^d$, $F \in C^3_b(\mathbb{R}^d, \mathcal{L}(\mathbb{R}^n, \mathbb{R}^d))$, a rough path $\mathbf{W} = (W, \mathbb{W}) \in \mathscr{C}^{\beta}$ with $\beta \in (1/3, 1/2)$, there is a unique controlled rough path $(y, y') \in \mathscr{D}^{2\beta}_{W}$ such that for all $t \in [0, T]$*

$$y(t) = \xi + \int_0^t F(y_u)\mathrm{d}\mathbf{W}_u, \tag{8.76}$$

where the integral has to be understood as the controlled rough path integral (Theorem 8.27).

Remark 8.62 Actually, we can extend this result to $T = +\infty$, because T is chosen uniformly with respect to parameters of the problem.

Proof It suffices to project Eq. (8.75) onto $\underline{1}$ and onto $\underline{W}$. □

Acknowledgements I am very grateful to Laure Coutin and Antoine Lejay for their availability, help and their careful rereading.

I deeply thank Peter Friz for suggesting me this topic during my master thesis and for welcoming me at the Technical University of Berlin for 4 months.

References

1. I. Bailleul, Flows driven by rough paths. Rev. Mat. Iberoam. **31**, 901–934 (2015)
2. I. Bailleul, F. Bernicot, D. Frey, Higher order paracontrolled calculus, 3d-PAM and multiplicative burgers equations (2015). Preprint arXiv:1506.08773
3. K.-T. Chen, Integration of paths, geometric invariants and a generalized Baker–Hausdorff formula. Ann. Math. **65**, 163–178 (1957)
4. L. Coutin, A. Lejay, Perturbed linear rough differential equations. Ann. Math. Blaise Pascal **21**(1), 103–150 (2014)
5. A.M. Davie, Differential equations driven by rough paths: an approach via discrete approximation. Appl. Math. Res. eXpress **2008**, abm009 (2008)
6. P. Friz, M. Hairer, *A Course of Rough Paths* (Springer, Berlin, 2014)
7. P.K. Friz, N.B. Victoir, *Multidimensional Stochastic Processes as Rough Paths: Theory and Applications*, vol. 120 (Cambridge University Press, Cambridge, 2010)
8. M. Gubinelli, Controlling rough paths. J. Funct. Anal. **216**(1), 86–140 (2004)
9. M. Gubinelli, P. Imkeller, N. Perkowski, Paracontrolled distributions and singular PDEs. Preprint arXiv:1210.2684 (2012)
10. M. Hairer, A theory of regularity structures. Invent. Math. **198**(2), 269–504 (2014)
11. I. Karatzas, S. Shreve, *Brownian Motion and Stochastic Calculus*, vol. 113 (Springer, Berlin, 2012)
12. A. Lejay, Trajectoires rugueuses. Matapli **98**, 119–134 (2012)
13. T. Lyons, On the nonexistence of path integrals. Proc. R. Soc. Lond. A **432**(1885), 281–290 (1991)
14. T. Lyons, Differential equations driven by rough signals (I): an extension of an inequality of L.C. Young. Math. Res. Lett **1**(4), 451–464 (1994)
15. T.J. Lyons, Differential equations driven by rough signals. Rev. Mat. Iberoamericana **14**(2), 215–310 (1998)
16. T. Lyons, Z. Qian, *System Control and Rough Paths* (Oxford Science Publications, Oxford, 2002)
17. T. Lyons, N. Victoir, An extension theorem to rough paths. Ann. Inst. H. Poincaré Anal. Non Linéaire **24**(5), 835–847 (2007)
18. Y. Meyer, *Wavelets and Operators*, vol. 1 (Cambridge University Press, Cambridge, 1995)
19. L.C. Young, An inequality of the Hölder type, connected with Stieltjes integration. Acta Math. **67**(1), 251–282 (1936)

Chapter 9
On the Euler–Maruyama Scheme for Degenerate Stochastic Differential Equations with Non-sticky Condition

Dai Taguchi and Akihiro Tanaka

Abstract The aim of this paper is to study weak and strong convergence of the Euler–Maruyama scheme for a solution of one-dimensional degenerate stochastic differential equation $\mathrm{d}X_t = \sigma(X_t)\mathrm{d}W_t$ with non-sticky condition. For proving this, we first prove that the Euler–Maruyama scheme also satisfies non-sticky condition. As an example, we consider stochastic differential equation $\mathrm{d}X_t = |X_t|^\alpha \mathrm{d}W_t$, $\alpha \in (0, 1/2)$ with non-sticky boundary condition and we give some remarks on CEV models in mathematical finance.

Keywords Stochastic differential equations · Non-sticky condition · Euler–Maruyama scheme · Hölder continuous diffusion coefficient · Mathematical finance · CEV models

9.1 Introduction

Let $X = (X_t)_{t\in[0,T]}$ be a solution of one-dimensional stochastic differential equations (SDEs)

$$\mathrm{d}X_t = \sigma(X_t)\mathrm{d}W_t,\ t \in [0, T],\ X_0 = x_0 \in \mathbb{R}, \tag{9.1}$$

where $W = (W_t)_{t\in[0,T]}$ is a one-dimensional standard Brownian motion on a probability space $(\Omega, \mathcal{F}, \mathbb{P})$ with a filtration $(\mathcal{F}_t)_{t\geq 0}$ satisfying the usual conditions. It is well-known that if the coefficient σ is Lipschitz continuous then a solution of Eq. (9.1) can be constructed by a limit of Picard's successive approximation, and the solution satisfies the pathwise uniqueness.

D. Taguchi (✉)
Graduate School of Engineering Science, Osaka University, Toyonaka, Osaka, Japan

A. Tanaka
Graduate School of Engineering Science, Osaka University, Toyonaka, Osaka, Japan

Sumitomo Mitsui Banking Corporation, Chiyoda-ku, Tokyo, Japan

C. Donati-Martin et al. (eds.), *Séminaire de Probabilités L*, Lecture Notes in Mathematics 2252, https://doi.org/10.1007/978-3-030-28535-7_9

In the one-dimensional setting, Engelbert and Schmidt [11] provided an equivalent condition on σ for the existence of a weak solution and uniqueness in law for SDE (9.1), by using time change of a Brownian motion (see also [10]). More precisely, they proved that Eq. (9.1) has a non-exploding weak solution for every initial condition $X_0 = x_0 \in \mathbb{R}$ if and only if $I(\sigma) \subset Z(\sigma)$, and the solution is unique in the sense of probability law if and only if $I(\sigma) = Z(\sigma)$. Here the sets $I(\sigma)$ and $Z(\sigma)$ are defined by $I(\sigma) := \{x \in \mathbb{R}; \int_{-\varepsilon}^{\varepsilon} \sigma(x+y)^{-2}\mathrm{d}y = +\infty, \ \forall \varepsilon > 0\}$ and $Z(\sigma) := \{x \in \mathbb{R} \ ; \ \sigma(x) = 0\}$. However, there exists a function σ such that $I(\sigma) \subsetneq Z(\sigma)$, so in this setting, the uniqueness in law does not hold. For example, if $\sigma(x) := |x|^{\alpha}$ for $\alpha \in (0, 1/2)$ then $I(\sigma) = \emptyset$, $Z(\sigma) = \{0\}$, and if $x_0 = 0$ then $X_t = 0$ and the time change a Brownian motion are solutions of the SDE, and moreover, if $x_0 \neq 0$, there is a solution which spends at zero. Therefore, as a concept of a solution of SDE, Engelbert and Schmidt [12] introduced a *fundamental solution* of Eq. (9.1), which is a solution of SDE (9.1) with the following *non-sticky condition*:

$$\mathbb{E}\left[\int_0^T \mathbf{1}_{Z(\sigma)}(X_s)\mathrm{d}s\right] = 0, \tag{9.2}$$

that is, in other words, $\sigma^2(X_s(\omega)) > 0$, Leb $\otimes$ $\mathbb{P}$-a.e., and proved that there exists a weak solution for a fundamental solution of SDE (9.1) and uniqueness in law holds (see, Theorem 5.4 in [12]).

On the other hand, the pathwise uniqueness for a solution of SDE is an important concept of a uniqueness for a solution of SDE. Yamada and Watanabe [32] proved that the pathwise uniqueness implies uniqueness in the sense of probability law, and weak existence and pathwise uniqueness imply the solution is a strong solution. Moreover, they also showed that under one-dimensional setting, if the diffusion coefficient σ is α-Hölder continuous with exponent $\alpha \in [1/2, 1]$, then the pathwise uniqueness holds (see also [23] and [26] for discontinuous setting of σ). Besides, Girsanov [13] and Barlow [3] provided some examples of α-Hölder continuous function σ with $\alpha \in (0, 1/2)$ such that the pathwise uniqueness fails for SDE (9.1), and thus the Hölder exponent $\alpha = 1/2$ is sharp.

Under such a background on the pathwise uniqueness, Manabe and Shiga [24] studied a solution of SDEs with non-sticky boundary condition $\mathbb{E}[\int_0^T \mathbf{1}_{\{0\}}(X_s)\mathrm{d}s] = 0$ (see also page 221 of [16]). They proved that if the diffusion coefficient σ is bounded, continuous and odd function and continuously differentiable on $\mathbb{R} \setminus \{0\}$ such that (i) $\int_0^{\delta} \sigma(y)^{-2}\mathrm{d}y < \infty$ for some $\delta > 0$ and (ii) the limit $\lim_{x \searrow 0} xa'(x)a(x)^{-1}$ exists and is not $1/2$, then two solutions X^1 and X^2 of SDE (9.1) with non-sticky boundary condition $\mathbb{E}[\int_0^T \mathbf{1}_{\{0\}}(X_s)\mathrm{d}s] = 0$ and the same initial value and driven by the same Brownian motion, satisfy $\mathbb{P}(|X_t^1| = |X_t^2|, \ \forall t \geq 0) = 1$. However, sign of solutions does not know from the information of driving Brownian motion. Moreover, additionally if $\sigma(0) = 0$ then the pathwise uniqueness holds for the following reflected SDE with non-sticky boundary condition

$$X_t = x_0 + \int_0^t \sigma(X_s)\mathrm{d}W_s + L_t^0(X) \geq 0, \ \mathbb{E}\left[\int_0^t \mathbf{1}_{\{0\}}(X_s)\mathrm{d}s\right] = 0, \ t \in [0,T], \ x_0 \geq 0, \tag{9.3}$$

where $L^0(X)$ is a local time of X at the origin. Recently, these results were extended by Bass and Chen [4] and Bass et al. [5]. It was shown in [5] (resp. [4]) that if $\sigma(x) = |x|^\alpha$ with $\alpha \in (0, 1/2)$ then a strong solution of SDE (9.1) with non-sticky boundary condition $\mathbb{E}[\int_0^T \mathbf{1}_{\{0\}}(X_s)\mathrm{d}s] = 0$ (resp. reflected SDE (9.3)) exists and the pathwise uniqueness holds by using excursion theory and pseudo-strong Markov property (resp. approximation argument). Note that in the case of $\alpha = 0$, that is, $\sigma(x) = \mathbf{1}(x \neq 0)$, Pascu and Pascu [27] studies sticky and non-sticky solutions.

Under the viewpoint of numerical analysis, we often use the Euler–Maruyama scheme $X^{(n)} = (X_t^{(n)})_{t\in[0,T]}$ which is a discrete approximation for a solution of SDE (9.1) defined by $\mathrm{d}X_t^{(n)} = \sigma(X_{\eta_n(t)}^{(n)})\mathrm{d}W_t$, $X_0^{(n)} = x_0$, $t \in [0,T]$, where $\eta_n(s) := t_k^{(n)} = kT/n$, if $s \in [t_k^{(n)}, t_{k+1}^{(n)})$. It is well-known that if the coefficient σ is Lipschitz continuous, then $X^{(n)}$ has strong (L^p-sup) rate of convergence 1/2, that is, $\mathbb{E}[\sup_{0\leq t\leq T}|X_t - X_t^{(n)}|^p]^{1/p} \leq Cn^{-1/2}$ for any $p \geq 1$, (see, e.g. [21]). On the other hand, the Euler–Maruyama scheme can be applied to many directions not only numerical analysis. Indeed, Maruyama [25] used the scheme for proving Girsanov's theorem for one-dimensional SDE $\mathrm{d}X_t = b(X_t)\mathrm{d}t + \mathrm{d}W_t$. Moreover, Skorokhod constructed a (weak) solution of SDE with continuous and linear growth coefficients as a limit of the Euler–Maruyama scheme (see, chapter 3, section 3 in [29]). Skorokhod's arguments can be also applied to a construction of a solution, which is based on the approximation argument of the coefficients, (see, e.g. chapter 3 in [30]). On the other hand, Yamada [33] proved that if the diffusion coefficient σ is α-Hölder continuous with $\alpha \in [1/2, 1]$, then the Euler–Maruyama scheme $X^{(n)}$ converges to the unique strong solution of SDE in L^2-sup sense. Recently, in the same setting, the rate of convergence was provided (see, [15] and [34]), by using Yamada and Watanabe approximation arguments or Itô–Tanaka formula. The result of Yamada [33] also extended by Kaneko and Nakao [19]. They showed that by using the similar arguments of Skorokhod [29], if the pathwise uniqueness holds for SDE with continuous and linear growth coefficients, then the Euler–Maruyama scheme and a solution of SDE with smooth approximation of the coefficients converge to the solution of corresponding SDE in L^2-sup sense. For results on weak convergence, when the uniqueness in law holds for SDE with discontinuous coefficients, then Yan [34] provided some equivalent conditions for the weak convergence of the Euler–Maruyama scheme, by using a limit theorem of stochastic integrals. Moreover, recently, Ankirchner et al. [1] proved that the weak convergence of the Euler–Maruyama scheme with continuous diffusion coefficient σ such that $I(\sigma) = Z(\sigma) = \emptyset$.

Inspired by the above previous works, in this paper, we study weak and strong convergence of the Euler–Maruyama scheme for a solution of SDE (9.1) with non-sticky condition (9.2). We first prove that the Euler–Maruyama scheme $X^{(n)}$ defined below (see, (9.4)) also satisfies the non-sticky condition (9.2). As an application of this fact, we prove that the Euler–Maruyama scheme converges weakly to a unique non-sticky weak solution of SDE, and if the pathwise uniqueness holds then it converges to a unique non-sticky strong solution of SDE in L^p-sup sense for any $p \geq 1$. The idea of proof is also based on arguments of Yan [34] and Skorokhod [29], and prove that by using occupation time formula if the limit of sub-sequence of the Euler–Maruyama scheme exists, then the limit satisfies the non-sticky condition (see, Lemma 9.5 below). As an example, the unique strong solution of SDE $\mathrm{d}X_t = |X_t|^\alpha \mathrm{d}W_t$ with non-sticky boundary condition $\mathbb{E}[\int_0^T \mathbf{1}_{\{0\}}(X_s)\mathrm{d}s] = 0$ for $\alpha \in (0, 1/2)$ can be approximated by the Euler–Maruyama scheme.

This paper is structured as follows. In Sect. 9.2, we prove the weak (resp. strong) convergence for the Euler–Maruyama scheme to a solution of SDE with non-sticky condition by using the uniqueness in law (resp. pathwise uniqueness). In Sect. 9.2.1, we provide the definition of the Euler–Maruyama scheme and prove that it satisfies the non-sticky condition. In Sect. 9.2.2, we state the main theorems of this present paper. We prove some auxiliary estimates in Sect. 9.2.4 and provide the proof of main theorems in Sect. 9.2.5.

Notations

We give some basic notations and definitions used throughout this paper. For a Lipschitz continuous function $f : \mathbb{R} \to \mathbb{R}$, we define $\|f\|_{\mathrm{Lip}} := \sup_{x \neq y} \frac{|f(x)-f(y)|}{|x-y|}$. For a given $T > 0$, we denote by $C[0, T]$ the space of continuous functions $w : [0, T] \to \mathbb{R}$ with metric ρ defined by $\rho(w, w') = \sup_{0 \leq t \leq T} |w_t - w'_t|$, and by $C_b(C[0, T]^k; \mathbb{R})$, $k \in \mathbb{N}$, a continuous function $f : C[0, T]^k \to \mathbb{R}$ such that $\sup_{w \in C[0,T]^k} |f(w)|$ is finite. We denote the sign function by $\mathrm{sgn}(x) := -\mathbf{1}_{(-\infty,0]}(x) + \mathbf{1}_{(0,\infty)}(x)$ for $x \in \mathbb{R}$. For a measurable function $\sigma : \mathbb{R} \to \mathbb{R}$, we define $I(\sigma) := \{x \in \mathbb{R}; \int_{-\varepsilon}^{\varepsilon} \sigma(x + y)^{-2}\mathrm{d}y = +\infty,\ \forall \varepsilon > 0\}$ and $Z(\sigma) := \{x \in \mathbb{R}\ ;\ \sigma(x) = 0\}$, and we denote by $D(\sigma)$ the set of all discontinuous points of σ. For a continuous semi-martingale $Y = (Y_t)_{t \geq 0}$, we denote $L^x(Y) = (L_t^x(Y))_{t \geq 0}$ the symmetric local time of Y at the level $x \in \mathbb{R}$. We may write a solution of SDE (9.1) by expressing (X, W).

9.2 Weak and Strong Convergence for the Euler–Maruyama Scheme

Throughout this paper, we suppose the following assumptions for the diffusion coefficient σ.

Assumption 9.1 $\sigma : \mathbb{R} \to \mathbb{R}$ is a measurable function and $Z(\sigma)$ is not the empty set and is a countable set, that is, σ is degenerate.

9.2.1 Euler–Maruyama Scheme

We define the Euler–Maruyama scheme $X^{(n)} = (X_t^{(n)})_{t\in[0,T]}$ for SDE (9.1) by

$$X_t^{(n)} = x_n + \int_0^t \sigma(X_{\eta_n(s)}^{(n)})\mathrm{d}W_s, \tag{9.4}$$

where the sequence $\{x_n\}_{n\in\mathbb{N}} \subset \mathbb{R}\backslash Z(\sigma)$ satisfies $\lim_{n\to\infty} x_n = x_0 \in \mathbb{R}$ and $\eta_n(s) := t_k^{(n)} = kT/n$, if $s \in [t_k^{(n)}, t_{k+1}^{(n)})$. Note that since $Z(\sigma)$ is a countable set, there exists such a sequence $\{x_n\}_{n\in\mathbb{N}}$. From here, we fix the sequence $\{x_n\}_{n\in\mathbb{N}}$.

Remark 9.1 Usually the initial value of the Euler–Maruyama scheme $X_0^{(n)}$ is defined by x_0. However, if $Z(\sigma) \neq \emptyset$ and $X_0^{(n)} = x_0 \in Z(\sigma)$, then $X_t^{(n)} = x_0$ for all $t \in [0, T]$. Therefore, in order to approximate a solution of SDE (9.1) with non-sticky condition (9.2), we need to take an approximate sequence $\{x_n\}_{n\in\mathbb{N}}$ from $\mathbb{R} \setminus Z(\sigma)$.

Now we prove that the Euler–Maruyama scheme (9.4) satisfies the non-sticky condition.

Lemma 9.1 *For any $n \in \mathbb{N}$, $X^{(n)}$ satisfies the non-sticky condition*

$$\mathbb{E}\left[\int_0^T \mathbf{1}_{Z(\sigma)}(X_s^{(n)})\mathrm{d}s\right] = 0.$$

Proof We first prove by induction that for each $k = 0, \ldots, n-1$, it holds that

$$\mathbb{P}(X_s^{(n)} \notin Z(\sigma)) = 1, \text{ for any } s \in (t_k^{(n)}, t_{k+1}^{(n)}].$$

Since $X_s^{(n)} = x_n + \sigma(x_n)W_s$ for any $s \in (0, t_1^{(n)}]$ and $\sigma(x_n) \neq 0$, we have $\mathbb{P}(X_s^{(n)} \in Z(\sigma)) = 0$, that is, $\mathbb{P}(X_s^{(n)} \notin Z(\sigma)) = 1$. Thus the statement holds for $k = 0$.

Now we assume that the statement holds for $\ell = 1, \ldots, k-1$. Then since $X_s^{(n)} = X_{t_k^{(n)}}^{(n)} + \sigma(X_{t_k^{(n)}}^{(n)})(W_s - W_{t_k^{(n)}})$ for any $s \in (t_k^{(n)}, t_{k+1}^{(n)}]$, by the assumption $\mathbb{P}(X_{t_k^{(n)}}^{(n)} \notin Z(\sigma)) = 1$, we have

$$\begin{aligned}
&\mathbb{P}(X_s^{(n)} \in Z(\sigma)) \\
&= \mathbb{E}\left[\mathbb{P}\left(X_{t_k^{(n)}}^{(n)} + \sigma(X_{t_k^{(n)}}^{(n)})(W_s - W_{t_k^{(n)}}) \in Z(\sigma) \,\middle|\, X_{t_k^{(n)}}^{(n)}\right)\mathbf{1}(X_{t_k^{(n)}}^{(n)} \notin Z(\sigma))\right].
\end{aligned}$$

Note that random variables $X^{(n)}_{t^{(n)}_k}$ and $W_s - W_{t^{(n)}_k}$ are independent, thus we have

$$\begin{aligned}
&\mathbb{P}(X^{(n)}_s \in Z(\sigma)) \\
&= \mathbb{E}\left[\mathbb{P}\left(x + \sigma(x)(W_s - W_{t^{(n)}_k}) \in Z(\sigma) \right) \Big|_{x = X^{(n)}_{t^{(n)}_k}} \mathbf{1}(X^{(n)}_{t^{(n)}_k} \notin Z(\sigma)) \right] = 0.
\end{aligned}$$

This concludes the case for k. Hence we have for each $k = 0, \ldots, n-1$, it holds that $\mathbb{P}(X^{(n)}_s \notin Z(\sigma)) = 1$ for any $s \in (t^{(n)}_k, t^{(n)}_{k+1}]$.

Using this fact, we have

$$\mathbb{E}\left[\int_0^T \mathbf{1}_{Z(\sigma)}(X^{(n)}_s) \mathrm{d}s \right] = \sum_{k=0}^{n-1} \int_{t^{(n)}_k}^{t^{(n)}_{k+1}} \mathbb{P}(X^{(n)}_s \in Z(\sigma)) \mathrm{d}s = 0,$$

which concludes the statement. □

9.2.2 Main Results

In this subsection, we provide a weak and strong convergence for the Euler–Maruyama scheme.

We need the following assumptions on the diffusion coefficient σ.

Assumption 9.2

(i) For any $z \in Z(\sigma)$,

$$\lim_{\varepsilon \searrow 0} \int_{-\varepsilon}^{\varepsilon} \frac{1}{\sigma(z+y)^2} \mathrm{d}y = 0.$$

(ii) The diffusion coefficient σ is of linear growth, (i.e., there exists $K > 0$ such that for any $x \in \mathbb{R}$, $|\sigma(x)| \leq K(1 + |x|)$), continuous almost everywhere with respect to Lebesgue measure and $\sigma_1(y)^2 > 0$ for any $y \in D(\sigma)$, where $\sigma_1(y)^2 := \liminf_{x \to y} \sigma(x)^2$ for $y \in \mathbb{R}$.

Remark 9.2

(i) Assumption 9.2 (i) implies that for any $z \in Z(\sigma)$, there exists $\varepsilon(z) > 0$ such that $\int_{-\varepsilon(z)}^{\varepsilon(z)} \frac{1}{\sigma(z+y)^2} \mathrm{d}y < \infty$, thus $I(\sigma) = \emptyset \neq Z(\sigma)$. Therefore from the result of Engelbert and Schmidt (see, e.g. Theorem 5.5.7 in [20]), the uniqueness in law does not hold for SDE (9.1). However, it follows from Theorem 5.4 in [12] that a solution of SDE (9.1) with non-sticky condition (9.2) exists and uniqueness in law holds by using time change of a Brownian motion.

(ii) It follows from Assumption 9.2 (i) that if the Euler–Maruyama scheme converges to some stochastic process, almost surely, then the limit satisfies the non-sticky condition, (see, Lemma 9.5 (iv)).

We obtain the following result on the weak convergence of the Euler–Maruyama scheme.

Theorem 9.1 *Suppose that Assumption 9.2 holds. Let* $X = (X_t)_{0 \le t \le T}$ *be a solution of SDE* (9.1) *with non-sticky condition* (9.2) *and* $\{X^{(n)}\}_{n \in \mathbb{N}}$ *be the Euler–Maruyama scheme for* X *defined by* (9.4). *Then for any* $f \in C_b(C[0,T];\mathbb{R})$,

$$\lim_{n \to \infty} \mathbb{E}[f(X^{(n)})] = \mathbb{E}[f(X)].$$

If σ is continuous and the pathwise uniqueness holds for X, then we have the strong convergence for the Euler–Maruyama scheme.

Theorem 9.2 *Suppose that Assumption 9.2 holds and* σ *is continuous. Let* $X = (X_t)_{t \in [0,T]}$ *be a solution of SDE* (9.1) *with non-sticky condition* (9.2) *and* $\{X^{(n)}\}_{n \in \mathbb{N}}$ *be the Euler–Maruyama scheme for* X *defined by* (9.4).

(i) If the pathwise uniqueness holds for X*, then for any* $p \in [1,\infty)$,

$$\lim_{n \to \infty} \mathbb{E}\left[\sup_{0 \le t \le T} \left|X_t - X_t^{(n)}\right|^p\right] = 0.$$

(ii) Suppose that $\mathbb{P}(|X_t| = |X'_t|, \forall t \ge 0) = 1$*, for any the other solution* X' *of SDE* (9.1) *driven by the same Brownian motion, with non-sticky condition* (9.2). *Then for any* $p \in [1,\infty)$,

$$\lim_{n \to \infty} \mathbb{E}\left[\sup_{0 \le t \le T} \left||X_t| - |X_t^{(n)}|\right|^p\right] = 0.$$

9.2.3 Examples and Applications

As examples of Theorem 9.2, we have two corollaries.

The first example is an application of a result in [5].

Corollary 9.1 *Let* $\sigma(x) = |x|^\alpha$, $\alpha \in (0, 1/2)$ *and* $X = (X_t)_{0 \le t \le T}$ *be a solution of SDE* (9.1) *with non-sticky boundary condition* $\mathbb{E}[\int_0^T \mathbf{1}_{\{0\}}(X_s)\mathrm{d}s] = 0$*, and* $\{X^{(n)}\}_{n \in \mathbb{N}}$ *be the Euler–Maruyama scheme for* X *defined by* (9.4). *Then for any* $p \in [1,\infty)$,

$$\lim_{n \to \infty} \mathbb{E}\left[\sup_{0 \le t \le T} \left|X_t - X_t^{(n)}\right|^p\right] = 0.$$

Proof From Theorem 1.2 in [5], the pathwise uniqueness holds for SDE $\mathrm{d}X_t = |X_t|^{\alpha}\mathrm{d}W_t$, $X_0 = x_0 \in \mathbb{R}$ with non-sticky boundary condition. On the other hand, since $Z(\sigma) = \{0\}$ and $\alpha \in (0, 1/2)$, it holds that

$$\lim_{\varepsilon \searrow 0} \int_{-\varepsilon}^{\varepsilon} \frac{1}{|y|^{2\alpha}} \mathrm{d}y = \lim_{\varepsilon \searrow 0} \frac{2\varepsilon^{1-2\alpha}}{1-2\alpha} = 0.$$

Hence $\sigma(x) = |x|^{\alpha}$ satisfies Assumption 9.2. From Theorem 9.2, we conclude the statement. □

We give a financial application of SDE considered in Corollary 9.1. In mathematical finance, constant elasticity of variance (CEV) models introduced by Cox [7]

$$\mathrm{d}X_t = (X_t)^{\alpha}\mathrm{d}W_t, \ X_0 = x_0 > 0, \ \alpha \in (0, 1],$$

have been studied by many authors (see, e.g. [2, 8, 14, 17] and [18]). If $\alpha \in [1/2, 1]$, then as mentioned in the introduction, pathwise uniqueness holds (see, Theorem 1 in [32] or Proposition 5.2.13 in [20]). Moreover, the boundary point zero is absorbing, that is, the process remains at zero after it reaches zero (see, e.g. Proposition 6.1.3.1 in [18]).

On the other hand, if $\alpha \in (0, 1/2)$, then the pathwise uniqueness does not hold for CEV models, and the boundary zero is regular, that is, the solution can get in to the boundary zero and can get out from the boundary zero, (see, e.g. [6] or Example 5.4 in [9]). Therefore one may consider CEV models with absorbing boundary (see, [8]), or reflecting boundary by setting $X_t := (1-\alpha)^{\frac{1}{1-\alpha}} (\rho_t)^{\frac{1}{2(1-\alpha)}}$, where $\rho = (\rho_t)_{t \geq 0}$ be a $(1-2\alpha)/(1-\alpha)$-dimensional squared Bessel process (see, the explicit form of the density function given in [18], page 367, case $\beta (= \alpha - 1) < 0$).

Recently, there are some studies on CEV models $\mathrm{d}X_t = |X_t|^{\alpha}\mathrm{d}W_t$ with "free boundary condition" (see, e.g. subsection 2.2 in [2]) to extend them as $\mathbb{R}$-valued processes. However, as mentioned in the introduction, the uniqueness in law and pathwise uniqueness do not hold (in particular, there is no density function) without some boundary conditions. Therefore, if one would like to extend CEV models as $\mathbb{R}$-valued processes, then as one approach, the non-sticky boundary is useful.

Finally, we give a relation between CEV model with non-sticky boundary condition and squared Bessel process. Let $X = (X_t)_{t \in [0,T]}$ be a solution of SDE $\mathrm{d}X_t = |X_t|^{\alpha}\mathrm{d}W_t$, $X_0 = x_0 \in \mathbb{R}$, for $\alpha \in (0, 1/2)$. We first do not assume any boundary condition for X. Let $g(x) := |x|^{2(1-\alpha)}/(1-\alpha)^2$. Then it is easy to see that

$$g'(x) = \frac{2\mathrm{sgn}(x)}{1-\alpha}|x|^{1-2\alpha}, \ \forall x \in \mathbb{R} \quad \text{and} \quad g''(x) = \frac{2(1-2\alpha)}{1-\alpha} \frac{1}{|x|^{2\alpha}}, \ \forall x \in \mathbb{R} \setminus \{0\}.$$

So we cannot apply Itô's formula for g, but since g is convex and X is a continuous martingale, we can apply Itô–Tanaka formula (see, e.g. Theorem 1.5 in chapter VI of [28]) to obtain

$$Y_t := g(X_t) = g(x_0) + \int_0^t g'(X_s)\mathrm{d}X_s + \frac{1}{2}\int_{\mathbb{R}} L_t^x(X)g''(\mathrm{d}x),$$

where $g''(\mathrm{d}x)$ is the second derivative measure of g, and is given by

$$g''(\mathrm{d}x) = \frac{2(1-2\alpha)}{1-\alpha}\frac{\mathbf{1}_{\mathbb{R}\setminus\{0\}}(x)}{|x|^{2\alpha}}\mathrm{d}x.$$

Therefore, using the occupation time formula (see, e.g. Corollary 1.6 in chapter VI of [28]), we have

$$\begin{aligned} Y_t &= g(x_0) + 2\int_0^t \sqrt{Y_s}\,\mathrm{sgn}(X_s)\mathrm{d}W_s + \frac{1-2\alpha}{1-\alpha}\int_0^t \frac{\mathbf{1}_{\mathbb{R}\setminus\{0\}}(X_s)}{|X_s|^{2\alpha}}\mathrm{d}\langle X\rangle_s \\ &= g(x_0) + 2\int_0^t \sqrt{Y_s}\,\mathrm{sgn}(X_s)\mathrm{d}W_s + \frac{1-2\alpha}{1-\alpha}\int_0^t \mathbf{1}_{\mathbb{R}\setminus\{0\}}(X_s)\mathrm{d}s. \end{aligned}$$

We now assume non-sticky boundary condition for X, then $\mathbf{1}_{\mathbb{R}\setminus\{0\}}(X_s) = 1$ for all $s \in [0,T]$, almost surely and thus

$$Y_t = g(x_0) + 2\int_0^t \sqrt{Y_s}\mathrm{d}\widetilde{W}_s + \frac{1-2\alpha}{1-\alpha}t,$$

where $\widetilde{W} = (\widetilde{W}_t)_{t\in[0,T]}$ is a Brownian motion defined by $\mathrm{d}\widetilde{W}_t := \mathrm{sgn}(X_t)\mathrm{d}W_t$. Therefore, the law of Y is a $(1-2\alpha)/(1-\alpha)$-dimensional squared Bessel process.

Remark 9.3 Note that one may use Itô's formula for $Y_t = |X_t|^{2(1-\alpha)}/(1-\alpha)^2$ for "some" $t \geq 0$, in order to prove Y satisfies the equation $\mathrm{d}Y_t = 2\sqrt{Y_t}\mathrm{d}W_t + \frac{1-2\alpha}{1-\alpha}\mathrm{d}t$, (see, e.g. [17] and [14]). The above computation shows that this is true for $t < \inf\{s > 0;\, X_s = 0\}$.

The second example is an application of a result in [24].

Corollary 9.2 *Let* $Z(\sigma) = \{0\}$, $X = (X_t)_{t\in[0,T]}$ *be a solution of SDE* (9.1) *with non-sticky boundary condition* $\mathbb{E}[\int_0^T \mathbf{1}_{\{0\}}(X_s)\mathrm{d}s] = 0$, *and* $\{X^{(n)}\}_{n\in\mathbb{N}}$ *be the Euler–Maruyama scheme for* X *defined by* (9.1). *Suppose that* $\sigma : \mathbb{R} \to \mathbb{R}$ *satisfies Assumption 9.2, and is a bounded, continuous and odd function and continuously differentiable on* $\mathbb{R}\setminus\{0\}$ *such that the limit* $\lim_{x\searrow 0} xa'(x)a(x)^{-1}$ *exists and is not* $1/2$. *Then for any* $p \in [1,\infty)$,

$$\lim_{n\to\infty}\mathbb{E}\left[\sup_{0\leq t\leq T}\left||X_t| - |X_t^{(n)}|\right|^p\right] = 0.$$

Proof From Assumption 9.2 (ii), there exists $\delta > 0$ such that $\int_0^\delta \sigma(y)^{-2}\mathrm{d}y < \infty$. Hence it follows from Theorem 1 in [24] that the assumptions on Theorem 9.2 hold. Thus we conclude the proof. □

9.2.4 Auxiliary Estimates

In this subsection, we introduce some useful estimates for proving Theorems 9.1 and 9.2.

We first prove the following standard inequalities on a solution of SDE (9.1), the Euler–Maruyama scheme defined by (9.4) and their local times.

Lemma 9.2 *Let X be a solution of SDE* (9.1) *and $\{X^{(n)}\}_{n\in\mathbb{N}}$ be the Euler–Maruyama scheme for X defined by* (9.4). *Suppose that σ is of linear growth. Then for any $p \geq 1$, there exists a positive constant $C_p > 0$ such that*

$$\mathbb{E}\left[\sup_{0\le t\le T}|X_t|^p\right] + \sup_{n\in\mathbb{N}}\mathbb{E}\left[\sup_{0\le t\le T}|X_t^{(n)}|^p\right] \le C_p, \tag{9.5}$$

$$\mathbb{E}[|X_t - X_s|^p]^{1/p} + \sup_{n\in\mathbb{N}}\mathbb{E}[|X_t^{(n)} - X_s^{(n)}|^p]^{1/p} \le C_p|t-s|^{1/2}, \tag{9.6}$$

for any $t, s \in [0, T]$. Moreover, there exists $C_0 > 0$ such that

$$\sup_{y\in\mathbb{R}}\mathbb{E}[L_T^y(X)] + \sup_{y\in\mathbb{R},\, n\in\mathbb{N}}\mathbb{E}[L_T^y(X^{(n)})] \le C_0. \tag{9.7}$$

Proof Since $\{x_n\}_{n\in\mathbb{N}}$ is bounded and σ is of linear growth, the estimates (9.5) and (9.6) can be shown by applying Gronwall's inequality and Burkholder–Davis–Gundy's inequality, thus it will be omitted.

We prove (9.7). By Itô–Tanaka formula, we have for any $y \in \mathbb{R}$,

$$\begin{aligned} L_T^y(X) &= |X_T - y| - |x_0 - y| - \int_0^T \mathrm{sgn}(X_s - y)\mathrm{d}X_s \\ &\le |x_0| + |X_T| + \left|\int_0^T \mathrm{sgn}(X_s - y)\sigma(X_s)\mathrm{d}W_s\right| \end{aligned}$$

and by the same way

$$L_T^y(X^{(n)}) \le |x_n| + |X_T^{(n)}| + \left|\int_0^T \mathrm{sgn}(X_s^{(n)} - y)\sigma(X_{\eta_n(s)}^{(n)})\mathrm{d}W_s\right|.$$

Hence by using Burkholder–Davis–Gundy's inequality and (9.5) with $p = 1, 2$, we conclude (9.7). □

The following lemma is a key estimate for the non-sticky condition.

Lemma 9.3 *Suppose that Assumption 9.2 hold. Let X be a solution of SDE* (9.1) *with non-sticky condition* (9.2) *and $\{X^{(n)}\}_{n\in\mathbb{N}}$ be the Euler–Maruyama scheme for X defined by* (9.4)*. Let $z \in Z(\sigma)$ and $f_z : \mathbb{R} \to [0,1]$ be a Lipschitz continuous function with $\operatorname{supp} f_z \subset [z-\varepsilon, z+\varepsilon]$ for some $\varepsilon > 0$. Then there exists $C > 0$ which does not depend on n, z and ε such that*

$$\mathbb{E}\left[\int_0^T f_z(X_s)\mathrm{d}s\right] \le C\int_{-\varepsilon}^{\varepsilon} \frac{1}{\sigma(z+y)^2}\mathrm{d}y \tag{9.8}$$

and

$$\mathbb{E}\left[\int_0^T f_z(X_s^{(n)})\mathrm{d}s\right] \le C\left\{\int_{-\varepsilon}^{\varepsilon} \frac{1}{\sigma(z+y)^2}\mathrm{d}y + \frac{\|f_z\|_{\mathrm{Lip}}}{n^{1/2}}\right\}. \tag{9.9}$$

Proof We first prove (9.8). Since X satisfies the non-sticky condition (9.2), we have $\sigma(X_s(\omega))^2 > 0$, Leb$\otimes\mathbb{P}$-a.e. Thus by using Fatou's lemma and the occupation time formula (see, e.g. Corollary 1.6 in chapter VI of [28]) and Lemma 9.2, we have

$$\begin{aligned}
\mathbb{E}\left[\int_0^T f_z(X_s)\mathrm{d}s\right] &= \mathbb{E}\left[\int_0^T \frac{f_z(X_s)\mathbf{1}_{\{\sigma(X_s)^2>0\}}}{\sigma(X_s)^2}\mathrm{d}\langle X\rangle_s\right] \\
&\le \liminf_{N\to\infty}\mathbb{E}\left[\int_{\mathbb{R}} \frac{f_z(y)\mathbf{1}_{\{\sigma(y)^2>1/N\}}}{\sigma(y)^2}L_T^y(X)\mathrm{d}y\right] \\
&\le \sup_{y\in\mathbb{R}}\mathbb{E}\left[L_T^y(X)\right]\int_{z-\varepsilon}^{z+\varepsilon}\frac{1}{\sigma(y)^2}\mathrm{d}y \\
&\le C_0\int_{-\varepsilon}^{\varepsilon}\frac{1}{\sigma(z+y)^2}\mathrm{d}y,
\end{aligned}$$

which implies (9.8).

Now we prove (9.9). Since, from Lemma 9.1, we have $\sigma(X_s^{(n)}(\omega))^2 > 0$, Leb $\otimes$ $\mathbb{P}$-a.e. Hence by using Lipschitz continuity of f_z, Fatou's lemma, the occupation time formula and Lemma 9.2, we have

$$\begin{aligned}
\mathbb{E}\left[\int_0^T f_z(X_s^{(n)})\mathrm{d}s\right] &\le \mathbb{E}\left[\int_0^T f_z(X_{\eta_n(s)}^{(n)})\mathrm{d}s\right] + \int_0^T \mathbb{E}\left[\left|f_z(X_s^{(n)}) - f_z(X_{\eta_n(s)}^{(n)})\right|\right]\mathrm{d}s \\
&\le \mathbb{E}\left[\int_0^T \frac{f_z(X_{\eta_n(s)}^{(n)})\mathbf{1}_{\{\sigma(X_{\eta_n(s)}^{(n)})^2>0\}}}{\sigma(X_{\eta_n(s)}^{(n)})^2}\mathrm{d}\langle X^{(n)}\rangle_s\right] \\
&\quad + \|f_z\|_{\mathrm{Lip}}\int_0^T \mathbb{E}\left[\left|X_s^{(n)} - X_{\eta_n(s)}^{(n)}\right|\right]\mathrm{d}s
\end{aligned}$$

$$
\begin{aligned}
&\leq \liminf_{N\to\infty} \mathbb{E}\left[\int_{\mathbb{R}} \frac{f_z(y)\mathbf{1}_{\{\sigma(y)^2>1/N\}}}{\sigma(y)^2} L_T^y(X^{(n)})\mathrm{d}y\right] + \frac{C_1 T^{3/2}\|f_z\|_{\mathrm{Lip}}}{n^{1/2}} \\
&\leq \sup_{n,\in\mathbb{N},\ y\in\mathbb{R}} \mathbb{E}\left[L_T^y(X^{(n)})\right] \int_{z-\varepsilon}^{z+\varepsilon} \frac{1}{\sigma(y)^2}\mathrm{d}y + \frac{C_1 T^{3/2}\|f_z\|_{\mathrm{Lip}}}{n^{1/2}} \\
&\leq \max\{C_0, C_1 T^{3/2}\}\left\{\int_{-\varepsilon}^{\varepsilon} \frac{1}{\sigma(z+y)^2}\mathrm{d}y + \frac{\|f_z\|_{\mathrm{Lip}}}{n^{1/2}}\right\},
\end{aligned}
$$

which implies (9.9). □

Now we introduce the following key lemma which is proved by Skorokhod (see, e.g. Theorem in [29], Chapter 3, section 3, page 32), and which shows convergence in probability for a sequence of stochastic integrals.

Lemma 9.4 (Skorokhod [29]) *Let $(W, \mathcal{F}^W)$ and $(W^n, \mathcal{F}^{W^n})$, $n \in \mathbb{N}$ be Brownian motions and $f_n = (f_n(t))_{t\in[0,T]}$ be a $\mathcal{F}^{W^n}$-adapted stochastic processes such that $\int_0^t f_n(s)\mathrm{d}W_s^n$ is well-defined, for all $n \in \mathbb{N}$. Suppose that for any $t \in [0, T]$, W_t^n and $f_n(t)$ converges to W_t and a $\mathcal{F}^W$-adapted process $f(t)$ in probability, respectively, and the stochastic integral $\int_0^t f(s)\mathrm{d}W_s$ is well-defined. Suppose further that the following conditions are satisfied for $\{f_n\}_{n\in\mathbb{N}}$:*

(a) For any $\varepsilon > 0$, there exists $K > 0$ such that for any $n \in \mathbb{N}$,

$$\mathbb{P}\left(\sup_{0\leq t\leq T} |f_n(t)| > K\right) \leq \varepsilon.$$

(b) For any $\varepsilon > 0$,

$$\lim_{h\searrow 0}\lim_{n\to\infty} \sup_{|t_1-t_2|\leq h} \mathbb{P}\left(|f_n(t_2) - f_n(t_1)| > \varepsilon\right) = 0.$$

Then it holds that for any $t \in [0, T]$

$$\lim_{n\to\infty} \int_0^t f_n(s)\mathrm{d}W_s^n = \int_0^t f(s)\mathrm{d}W_s,$$

in probability.

Finally, we prove the following key lemma in order to show main theorems and in particular to deal with non-sticky condition.

Lemma 9.5 *Suppose that Assumption 9.2 holds. Let (X, W) be a solution of SDE (9.1) with non-sticky condition (9.2) and $\{X^{(k)}\}_{k\in\mathbb{N}}$ be a sub-sequence of the Euler–Maruyama scheme defined by (9.4). Then there exists a probability space $(\widehat{\Omega}, \widehat{\mathcal{F}}, \widehat{\mathbb{P}})$, a sub-sub-sequence $\{k_\ell\}_{\ell\in\mathbb{N}}$ and three-dimensional continuous processes $\widehat{Y}^{k_\ell} = (\widehat{X}^{k_\ell}, \widehat{X}^{(k_\ell)}, \widehat{W}^{k_\ell})$ and $\widehat{Y} = (\widehat{X}, \widehat{X}^{(*)}, \widehat{W})$ defined on the probability space*

$(\widehat{\Omega}, \widehat{\mathcal{F}}, \widehat{\mathbb{P}})$ *such that the following properties are satisfied:*

(i) The law of stochastic processes $(X, X^{(k_\ell)}, W)$ *and* $(\widehat{X}^{k_\ell}, \widehat{X}^{(k_\ell)}, \widehat{W}^{k_\ell})$ *coincide for each* $\ell \in \mathbb{N}$. *In particular,* $(\widehat{X}^{k_\ell}, \widehat{X}^{(k_\ell)}, \widehat{W}^{k_\ell})$ *can be chosen as follows: there exist measurable maps* $\phi_{k_\ell} : \widehat{\Omega} \to \Omega$, $\ell \in \mathbb{N}$ *such that*

$$(\widehat{X}^{k_\ell}, \widehat{X}^{(k_\ell)}, \widehat{W}^{k_\ell}) = (X \circ \phi_{k_\ell}, X^{(k_\ell)} \circ \phi_{k_\ell}, W \circ \phi_{k_\ell}).$$

(ii) $\widehat{\mathbb{P}}(\lim_{\ell\to\infty} \sup_{0\le t\le T} |\widehat{Y}_t^{k_\ell} - \widehat{Y}_t| = 0) = 1.$

(iii) $\widehat{W}$ *is a Brownian motion and* $\widehat{X}$, $\widehat{X}^{(*)}$ *are continuous martingales on* $(\widehat{\Omega}, \widehat{\mathcal{F}}, \widehat{\mathbb{P}})$.

(iv) $\widehat{X}$ *and* $\widehat{X}^{(*)}$ *satisfy non-sticky condition*

$$\widehat{\mathbb{E}}\left[\int_0^T \mathbf{1}_{Z(\sigma)}(\widehat{X}_s)\mathrm{d}s\right] = \widehat{\mathbb{E}}\left[\int_0^T \mathbf{1}_{Z(\sigma)}(\widehat{X}_s^{(*)})\mathrm{d}s\right] = 0. \tag{9.10}$$

(v) There exist an extension $(\widetilde{\Omega}, \widetilde{\mathcal{F}}, \widetilde{\mathbb{P}})$ *of* $(\widehat{\Omega}, \widehat{\mathcal{F}}, \widehat{\mathbb{P}})$ *and Brownian motions* $\widetilde{B} = (\widetilde{B}_t)_{t\in[0,T]}$, $\widetilde{B}^{(*)} = (\widetilde{B}_t^{(*)})_{t\in[0,T]}$ *such that* $(\widehat{X}, \widetilde{B})$ *and* $(\widehat{X}^{(*)}, \widetilde{B}^{(*)})$ *are solutions of SDE* (9.1) *with non-sticky condition* (9.10).

(vi) If σ *is continuous then* $(\widehat{X}, \widehat{W})$ *and* $(\widehat{X}^{(*)}, \widehat{W})$ *are solutions of SDE* (9.1) *with non-sticky condition* (9.10).

Proof Proof of (i) and (ii). We first note that since the diffusion coefficient σ is of linear growth, the estimates in Lemma 9.2 hold. Hence it follows from Theorem 4.3 and the proof of Theorem 4.2 in [16] that the family of three-dimensional stochastic process $\{(X, X^{(k)}, W)\}_{k\in\mathbb{N}}$ is tight in $C[0, T]^3$, and thus is relatively compact in $C[0, T]^3$ by Prohorov's Theorem (see, e.g. Theorem 2.4.7 in [20]). Hence there exist a sub-sequence $\{k_\ell\}_{\ell\in\mathbb{N}}$ and $X^{(*)}$ such that $\lim_{\ell\to\infty} \mathbb{E}[f(X, X^{(k_\ell)}, W)] = \mathbb{E}[f(X, X^{(*)}, W)]$, for any $f \in C_b(C[0, T]^3; \mathbb{R})$. Therefore, by using Skorohod's representation theorem (see, e.g. Theorem 1.2.7 in [16] or Theorem 1.10.4 in [31]) and Addendum 1.10.5 in [31], there exists a probability space $(\widehat{\Omega}, \widehat{\mathcal{F}}, \widehat{\mathbb{P}})$, three-dimensional continuous processes $\widehat{Y}^{k_\ell} = (\widehat{X}^{k_\ell}, \widehat{X}^{(k_\ell)}, \widehat{W}^{k_\ell})$ and $\widehat{Y} = (\widehat{X}, \widehat{X}^{(*)}, \widehat{W})$ defined on the probability space $(\widehat{\Omega}, \widehat{\mathcal{F}}, \widehat{\mathbb{P}})$ and measurable maps $\phi_{k_\ell} : \widehat{\Omega} \to \Omega$, $\ell \in \mathbb{N}$ such that the properties (i) and (ii) are satisfied.

Proof of (iii). We first prove $\widehat{W}$ is a Brownian motion on $(\widehat{\Omega}, \widehat{\mathcal{F}}, \widehat{\mathbb{P}})$. From the property (i), $\widehat{W}^{k_\ell}$ is a Brownian motion, so $\widehat{W}^{k_\ell}$ and $(|\widehat{W}_t^{k_\ell}|^2 - t)_{t\in[0,T]}$ are martingales. Therefore it follows from Lemma A.1 in [34] and the above property (ii) that $\widehat{W}$ and $(|\widehat{W}|_t^2 - t)_{t\in[0,T]}$ are martingales, thus the quadratic variation of $\widehat{W}_t$ is t for all $t \in [0, T]$. Lévy's Theorem (e.g. Theorem 3.3.16 in [20]) implies that $\widehat{W}$ is a Brownian motion.

Next, we prove $\widehat{X}$ and $\widehat{X}^{(*)}$ are continuous martingales. By using the above property (i), it holds that $(\widehat{X}^{k_\ell}, \widehat{W}^{k_\ell})$ satisfies the following equations

$$\widehat{X}_t^{k_\ell} = x_0 + \int_0^t \sigma(\widehat{X}_s^{k_\ell})\mathrm{d}\widehat{W}_s^{k_\ell} \quad \text{and} \quad \widehat{\mathbb{E}}\left[\int_0^T \mathbf{1}_{Z(\sigma)}(\widehat{X}_s^{k_\ell})\mathrm{d}s\right] = 0 \tag{9.11}$$

and by using Lemma 9.1, $(\widehat{X}^{(k_\ell)}, \widehat{W}^{k_\ell})$ satisfies the following equations

$$\widehat{X}_t^{(k_\ell)} = x_{k_\ell} + \int_0^t \sigma(\widehat{X}_{\eta_{k_\ell}(s)}^{(k_\ell)})\mathrm{d}\widehat{W}_s^{k_\ell} \quad \text{and} \quad \widehat{\mathbb{E}}\left[\int_0^T \mathbf{1}_{Z(\sigma)}(\widehat{X}_s^{(k_\ell)})\mathrm{d}s\right] = 0. \tag{9.12}$$

Thus from Lemma 9.2, sequences of stochastic process $\{\widehat{X}^{k_\ell}\}_{\ell\in\mathbb{N}}$ and $\{\widehat{X}^{(k_\ell)}\}_{\ell\in\mathbb{N}}$ are uniformly integrable martingales, which uniformly converge to $\widehat{X}$ and $\widehat{X}^{(*)}$, respectively. Hence from Lemma A.1 in [34], we conclude $\widehat{X}$ and $\widehat{X}^{(*)}$ are continuous martingales.

Proof of (iv). For $\varepsilon > 0$ and $z \in Z(\sigma)$, we define a continuous function $f_{\varepsilon,z} : \mathbb{R} \to [0, 1]$ by

$$f_{\varepsilon,z}(x) := \begin{cases} -\dfrac{x-z}{\varepsilon} + 1 & \text{if } 0 \leq x - z < \varepsilon, \\ \dfrac{x-z}{\varepsilon} + 1 & \text{if } -\varepsilon < x - z < 0, \\ 0 & \text{if } |x-z| \geq \varepsilon. \end{cases}$$

Then it is easy to see that $\lim_{\varepsilon\searrow 0} f_{\varepsilon,z}(x) = \mathbf{1}_{\{z\}}(x)$ for each $x \in \mathbb{R}$, and $f_{\varepsilon,z}$ is Lipschitz continuous with $\|f_{\varepsilon,z}\|_{\mathrm{Lip}} = 2/\varepsilon$. Recall that $\widehat{X}^{k_\ell}$ and $\widehat{X}^{(k_\ell)}$ satisfy the equations (9.11) and (9.12), respectively. Hence from the dominated convergence theorem, Lemma 9.3 and Assumption 9.2 (i), we have

$$\begin{aligned}
&\widehat{\mathbb{E}}\left[\int_0^T \mathbf{1}_{Z(\sigma)}(\widehat{X}_s)\mathrm{d}s\right] + \widehat{\mathbb{E}}\left[\int_0^T \mathbf{1}_{Z(\sigma)}(\widehat{X}_s^{(*)})\mathrm{d}s\right] \\
&= \sum_{z\in Z(\sigma)} \left\{\widehat{\mathbb{E}}\left[\int_0^T \mathbf{1}_{\{z\}}(\widehat{X}_s)\mathrm{d}s\right] + \widehat{\mathbb{E}}\left[\int_0^T \mathbf{1}_{\{z\}}(\widehat{X}_s^{(*)})\mathrm{d}s\right]\right\} \\
&= \sum_{z\in Z(\sigma)} \lim_{\varepsilon\searrow 0} \left\{\widehat{\mathbb{E}}\left[\int_0^T f_{\varepsilon,z}(\widehat{X}_s)\mathrm{d}s\right] + \widehat{\mathbb{E}}\left[\int_0^T f_{\varepsilon,z}(\widehat{X}_s^{(*)})\mathrm{d}s\right]\right\} \\
&= \sum_{z\in Z(\sigma)} \lim_{\varepsilon\searrow 0} \lim_{\ell\to\infty} \left\{\widehat{\mathbb{E}}\left[\int_0^T f_{\varepsilon,z}(\widehat{X}_s^{k_\ell})\mathrm{d}s\right] + \widehat{\mathbb{E}}\left[\int_0^T f_{\varepsilon,z}(\widehat{X}_s^{(k_\ell)})\mathrm{d}s\right]\right\} \\
&\leq 2C \sum_{z\in Z(\sigma)} \lim_{\varepsilon\searrow 0} \lim_{\ell\to\infty} \left\{\int_{-\varepsilon}^{\varepsilon} \frac{1}{\sigma(z+y)^2}\mathrm{d}y + \frac{1}{\varepsilon k_\ell^{1/2}}\right\} \\
&= 2C \sum_{z\in Z(\sigma)} \lim_{\varepsilon\searrow 0} \int_{-\varepsilon}^{\varepsilon} \frac{1}{\sigma(z+y)^2}\mathrm{d}y = 0,
\end{aligned}$$

which concludes (iv).

Proof of (v). The proof is almost the same as Lemma 2.3 and Theorem 2.1 in [34]. We first prove that for each $t \in [0, T]$,

$$\lim_{\ell\to\infty} \langle \widehat{X}^{k_\ell}\rangle_t = \langle \widehat{X}\rangle_t \quad \text{and} \quad \lim_{\ell\to\infty} \langle \widehat{X}^{(k_\ell)}\rangle_t = \langle \widehat{X}^{(*)}\rangle_t \tag{9.13}$$

in $L^1(\widehat{\Omega}, \widehat{\mathcal{F}}, \widehat{\mathbb{P}})$ and

$$\lim_{\ell\to\infty} \int_0^t \sigma(\widehat{X}_s^{k_\ell})^2 \mathbf{1}(\widehat{X}_s \notin D(\sigma)) \mathrm{d}s = \int_0^t \sigma(\widehat{X}_s)^2 \mathbf{1}(\widehat{X}_s \notin D(\sigma)) \mathrm{d}s, \tag{9.14}$$

$$\lim_{\ell\to\infty} \int_0^t \sigma(\widehat{X}_{\eta_{k_\ell}(s)}^{(k_\ell)})^2 \mathbf{1}(\widehat{X}_s^{(*)} \notin D(\sigma)) \mathrm{d}s = \int_0^t \sigma(\widehat{X}_s^{(*)})^2 \mathbf{1}(\widehat{X}_s^{(*)} \notin D(\sigma)) \mathrm{d}s, \tag{9.15}$$

in $L^1(\widehat{\Omega}, \widehat{\mathcal{F}}, \widehat{\mathbb{P}})$. From the property (ii), continuous martingales $\widehat{X}^{k_\ell}$ and $\widehat{X}^{(k_\ell)}$ converge to $\widehat{X}$ and $\widehat{X}^{(*)}$ almost surely in $C[0, T]$, respectively. Hence it follows from Theorem 2.2 in [22] that $\int_0^\cdot \widehat{X}_s^{k_\ell} \mathrm{d}\widehat{X}_s^{k_\ell}$ and $\int_0^\cdot \widehat{X}_s^{(k_\ell)} \mathrm{d}\widehat{X}_s^{(k_\ell)}$ converge to $\int_0^\cdot \widehat{X}_s \mathrm{d}\widehat{X}_s$ and $\int_0^\cdot \widehat{X}_s^{(*)} \mathrm{d}\widehat{X}_s^{(*)}$ in probability as $\ell \to \infty$, respectively. Since for any squared integrable continuous martingale M, $\langle M\rangle_t = M_t^2 - M_0^2 - 2\int_0^t M_s \mathrm{d}M_s$, we have

$$\lim_{\ell\to\infty} \langle \widehat{X}^{k_\ell}\rangle_t = \langle \widehat{X}\rangle_t \quad \text{and} \quad \lim_{\ell\to\infty} \langle \widehat{X}^{(k_\ell)}\rangle_t = \langle \widehat{X}^{(*)}\rangle_t,$$

in probability. On the other hand, since σ is of linear growth, from Lemma 9.2, the classes

$$\left\{\langle \widehat{X}^{k_\ell}\rangle_t,\ \langle \widehat{X}^{(k_\ell)}\rangle_t\ ;\ t \in [0, T],\ \ell \in \mathbb{N}\right\} \text{ and}$$
$$\left\{\sigma(\widehat{X}_t^{k_\ell})^2,\ \sigma(\widehat{X}_{\eta_{k_\ell}(t)}^{(k_\ell)})^2\ ;\ t \in [0, T],\ \ell \in \mathbb{N}\right\}$$

are uniformly integrable, thus we conclude (9.13), (9.14), and (9.15).

Recall that $\sigma_1(y)^2 := \liminf_{x\to y} \sigma(x)^2$ for $y \in \mathbb{R}$. Using Fatou's lemma and (9.13), we have for any $0 \le r < t \le T$,

$$\begin{aligned}\widehat{\mathbb{E}}\left[\int_r^t \sigma_1(\widehat{X}_s)^2 \mathrm{d}s\right] &= \widehat{\mathbb{E}}\left[\int_r^t \liminf_{\ell\to\infty} \sigma(\widehat{X}_s^{k_\ell})^2 \mathrm{d}s\right] \le \liminf_{\ell\to\infty} \widehat{\mathbb{E}}\left[\int_r^t \sigma(\widehat{X}_s^{k_\ell})^2 \mathrm{d}s\right] \\ &= \widehat{\mathbb{E}}\left[\langle \widehat{X}\rangle_t - \langle \widehat{X}\rangle_r\right]\end{aligned}$$

and by the same way,

$$\widehat{\mathbb{E}}\left[\int_r^t \sigma_1(\widehat{X}_s^{(*)})^2 \mathrm{d}s\right] \le \widehat{\mathbb{E}}\left[\langle \widehat{X}^{(*)}\rangle_t - \langle \widehat{X}^{(*)}\rangle_r\right].$$

Therefore, using the occupation time formula and Lemma 9.2, we have

$$\widehat{\mathbb{E}}\left[\int_0^T \mathbf{1}_{D(\sigma)}(\widehat{X}_s)\sigma_1(\widehat{X}_s)^2 \mathrm{d}s\right] \leq \widehat{\mathbb{E}}\left[\int_0^T \mathbf{1}_{D(\sigma)}(\widehat{X}_s)\mathrm{d}\langle \widehat{X}\rangle_s\right] = \widehat{\mathbb{E}}\left[\int_{D(\sigma)} L_T^y(\widehat{X})\mathrm{d}y\right]$$
$$\leq \mathrm{Leb}(D(\sigma)) \sup_{y\in\mathbb{R}} \widehat{\mathbb{E}}[L_T^y(\widehat{X})] = 0$$

and by the same way

$$\widehat{\mathbb{E}}\left[\int_0^T \mathbf{1}_{D(\sigma)}(\widehat{X}_s^{(*)})\sigma_1(\widehat{X}_s^{(*)})^2 \mathrm{d}s\right] = 0.$$

Recall that from the assumption, $\sigma_1(y)^2 = \liminf_{x\to y} \sigma(x)^2 > 0$ for $y \in D(\sigma)$, so we obtain

$$\int_0^T \mathbf{1}_{D(\sigma)}(\widehat{X}_s)\mathrm{d}s = \int_0^T \mathbf{1}_{D(\sigma)}(\widehat{X}_s^{(*)})\mathrm{d}s = 0, \tag{9.16}$$

$\widehat{\mathbb{P}}$-almost surely. Therefore, it hold from (9.14), (9.16), and (9.15) that for any $t \in [0, T]$,

$$\lim_{\ell\to\infty} \widehat{\mathbb{E}}\left[\left|\langle \widehat{X}^{k_\ell}\rangle_t - \int_0^t \sigma(\widehat{X}_s)^2 \mathrm{d}s\right|\right]$$
$$= \lim_{\ell\to\infty} \widehat{\mathbb{E}}\left[\left|\int_0^t \left\{\sigma(\widehat{X}_s^{k_\ell})^2 - \sigma(\widehat{X}_s)^2\right\} \mathbf{1}(\widehat{X}_s \notin D(\sigma))\mathrm{d}s\right|\right] = 0 \tag{9.17}$$

and

$$\lim_{\ell\to\infty} \widehat{\mathbb{E}}\left[\left|\langle \widehat{X}^{(k_\ell)}\rangle_t - \int_0^t \sigma(\widehat{X}_s^{(*)})^2 \mathrm{d}s\right|\right] = 0. \tag{9.18}$$

Therefore, from (9.13), (9.17), and (9.18), we obtain $\langle \widehat{X}\rangle_t = \int_0^t \sigma(\widehat{X}_s)^2 \mathrm{d}s$ and $\langle \widehat{X}^{(*)}\rangle_t = \int_0^t \sigma(\widehat{X}_s^{(*)})^2 \mathrm{d}s$, $\widehat{\mathbb{P}}$-almost surely. Since $\widehat{X}$ and $\widehat{X}^{(*)}$ are square integrable continuous martingales, by using martingale representation theorem (see, e.g. chapter 2, Theorem 7.1' in [16]), there exist an extension $(\widetilde{\Omega}, \widetilde{\mathcal{F}}, \widetilde{\mathbb{P}})$ of $(\widehat{\Omega}, \widehat{\mathcal{F}}, \widehat{\mathbb{P}})$ and Brownian motions $\widetilde{B} = (\widetilde{B}_t)_{t\in[0,T]}$, $\widetilde{B}^{(*)} = (\widetilde{B}_t^{(*)})_{t\in[0,T]}$ such that

$$\widehat{X}_t = x_0 + \int_0^t \sigma(\widehat{X}_s)\mathrm{d}\widetilde{B}_s \quad \text{and} \quad \widehat{X}_t^{(*)} = x_0 + \int_0^t \sigma(\widehat{X}_s^{(*)})\mathrm{d}\widetilde{B}_s^{(*)},$$

$\widetilde{\mathbb{P}}$-almost surely, thus from the property (iv), we conclude the statement of (v).

Proof of (vi). We first prove that $\{\widehat{X}^{k_\ell}\}_{\ell\in\mathbb{N}}$ and $\{\widehat{X}^{(k_\ell)}\}_{\ell\in\mathbb{N}}$ satisfy the following two properties:

(a) For any $\varepsilon > 0$ and a measurable, polynomial growth function $f : \mathbb{R} \to \mathbb{R}$, there exists $K \equiv K(\varepsilon, f) > 0$ such that

$$\sup_{\ell\in\mathbb{N}} \max\left\{\widehat{\mathbb{P}}\left(\sup_{0\le s\le T} |f(\widehat{X}_s^{k_\ell})| \ge K\right), \widehat{\mathbb{P}}\left(\sup_{0\le s\le T} |f(\widehat{X}_s^{(k_\ell)})| \ge K\right)\right\} < \varepsilon.$$

(b) For any $\widetilde{\varepsilon} > 0$ and a continuous function $g : \mathbb{R} \to \mathbb{R}$,

$$\lim_{h\to 0}\lim_{\ell\to\infty}\sup_{|t_1-t_2|\le h}\max\left\{\widehat{\mathbb{P}}\left(\left|g(\widehat{X}_{t_1}^{k_\ell}) - g(\widehat{X}_{t_2}^{k_\ell})\right| > \widetilde{\varepsilon}\right),\right.$$
$$\left.\widehat{\mathbb{P}}\left(\left|g(\widehat{X}_{t_1}^{(k_\ell)}) - g(\widehat{X}_{t_2}^{(k_\ell)})\right| > \widetilde{\varepsilon}\right)\right\} = 0$$

Indeed, the property (a) follows from Markov's inequality and Lemma 9.2. In order to prove the property (b), we use the property (a) with $f(x) = x$. Then since g is uniformly continuous on the interval $[-K, K]$, there exists $\delta \equiv \delta(\widetilde{\varepsilon}, K) > 0$ such that for any $x, y \in [-K, K]$, if $|x - y| < \delta$ then $|g(x) - g(y)| < \widetilde{\varepsilon}$. Therefore, it follows from Markov's inequality and Lemma 9.2 that

$$\sup_{\ell\in\mathbb{N}}\widehat{\mathbb{P}}\left(\left|g(\widehat{X}_{t_1}^{k_\ell}) - g(\widehat{X}_{t_2}^{k_\ell})\right| \ge \widetilde{\varepsilon}\right)$$
$$\le \sup_{\ell\in\mathbb{N}}\widehat{\mathbb{P}}\left(\left|g(\widehat{X}_{t_1}^{k_\ell}) - g(\widehat{X}_{t_2}^{k_\ell})\right| \ge \widetilde{\varepsilon},\ \sup_{s\in[0,T]} |\widehat{X}_s^{k_\ell}| \le K\right) + \sup_{\ell\in\mathbb{N}}\widehat{\mathbb{P}}\left(\sup_{s\in[0,T]} |\widehat{X}_s^{k_\ell}| \ge K\right)$$
$$\le \sup_{\ell\in\mathbb{N}}\widehat{\mathbb{P}}\left(\left|\widehat{X}_{t_1}^{k_\ell} - \widehat{X}_{t_2}^{k_\ell}\right| \ge \delta\right) + \varepsilon \le \frac{C_1|t_1 - t_2|^{1/2}}{\delta} + \varepsilon$$

and by the same way,

$$\sup_{\ell\in\mathbb{N}}\widehat{\mathbb{P}}\left(\left|g(\widehat{X}_{t_1}^{(k_\ell)}) - g(\widehat{X}_{t_2}^{(k_\ell)})\right| \ge \widetilde{\varepsilon}\right) \le \frac{C_1|t_1 - t_2|^{1/2}}{\delta} + \varepsilon.$$

By taking $h \to 0$, since ε is arbitrary, the property (b) follows.

Recall that σ is continuous, $\lim_{n\to\infty} x_n = x_0$ and $\widehat{W}$ is a Brownian motion. It follows from Lemma 9.4 and the above properties (a), (b) with $f = g = \sigma$ that, by letting $\ell \to \infty$, the limits $\widehat{X}$ and $\widehat{X}^{(*)}$ are satisfies the equation

$$\widehat{X}_t = x_0 + \int_0^t \sigma(\widehat{X}_s)\mathrm{d}\widehat{W}_s \quad \text{and} \quad \widehat{X}_t^{(*)} = x_0 + \int_0^t \sigma(\widehat{X}_s^{(*)})\mathrm{d}\widehat{W}_s$$

and thus from the property (iv), we conclude the statement of (vi). □

9.2.5 *Proof of Main Theorems*

Before proving Theorem 9.1, we recall the following elementally fact on calculus. Let $\{a_n\}_{n\in\mathbb{N}}$ be a sequence on $\mathbb{R}$ and $a \in \mathbb{R}$. If for any sub-sequence $\{a_{n_k}\}_{k\in\mathbb{N}}$ of $\{a_n\}_{n\in\mathbb{N}}$, there exists a sub-sub-sequence $\{a_{n_{k_\ell}}\}_{\ell\in\mathbb{N}}$ such that $\lim_{\ell\to\infty} a_{n_{k_\ell}} = a$, then the sequence $\{a_n\}_{n\in\mathbb{N}}$ converges to a. By using the this fact and Lemma 9.5, we prove Theorem 9.1.

Proof of Theorem 9.1
It is enough to prove that for any sub-sequence $\{X^{(k)}\}_{k\in\mathbb{N}}$ of the Euler–Maruyama scheme $\{X^{(n)}\}_{n\in\mathbb{N}}$ defined by (9.4), there is a sub-sub-sequence $\{X^{(k_\ell)}\}_{\ell\in\mathbb{N}}$ such that for any $f \in C_b(C[0,T];\mathbb{R})$,

$$\lim_{\ell\to\infty} \mathbb{E}[f(X^{(k_\ell)})] = \mathbb{E}[f(X)].$$

Let $\{X^{(k)}\}_{k\in\mathbb{N}}$ be a sub-sequence of the Euler–Maruyama scheme $\{X^{(n)}\}_{n\in\mathbb{N}}$. From Lemma 9.5, there exists a probability space $(\widehat{\Omega}, \widehat{\mathcal{F}}, \widehat{\mathbb{P}})$, a sub-sequence $\{k_\ell\}_{\ell\in\mathbb{N}}$ and 3-dimensional continuous processes $\widehat{Y}^{k_\ell} = (\widehat{X}^{k_\ell}, \widehat{X}^{(k_\ell)}, \widehat{W}^{k_\ell})$ and $\widehat{Y} = (\widehat{X}, \widehat{X}^{(*)}, \widehat{W})$ defined on the probability space $(\widehat{\Omega}, \widehat{\mathcal{F}}, \widehat{\mathbb{P}})$ such that the properties (i)–(v) are satisfied.

For the proof of this theorem, we only use $\widehat{X}^{(k_\ell)}$ and $\widehat{X}^{(*)}$, do not use $(\widehat{X}^{k_\ell}, \widehat{W}^{k_\ell})$ and $(\widehat{X}, \widehat{W})$. From the property (i), (ii) in Lemma 9.5 and using the dominated convergence theorem, we have

$$\lim_{\ell\to\infty} \mathbb{E}[f(X^{(k_\ell)})] = \lim_{\ell\to\infty} \widehat{\mathbb{E}}[f(\widehat{X}^{(k_\ell)})] = \widehat{\mathbb{E}}[f(\widehat{X}^{(*)})], \tag{9.19}$$

for any $f \in C_b(C[0,T];\mathbb{R})$.

On the other hand, the property (iv) and (v) imply that there exist an extension $(\widetilde{\Omega}, \widetilde{\mathcal{F}}, \widetilde{\mathbb{P}})$ of $(\widehat{\Omega}, \widehat{\mathcal{F}}, \widehat{\mathbb{P}})$ and Brownian motion $\widetilde{B}^{(*)} = (\widetilde{B}_t^{(*)})_{t\in[0,T]}$ such that $(\widehat{X}^{(*)}, \widetilde{B}^{(*)})$ is a solution of SDE (9.1) with non-sticky condition (9.10). Hence from the uniqueness in law for SDE (9.1) with non-sticky condition (9.10) (see, Remark 9.2 (i)) and (9.19), we have

$$\lim_{\ell\to\infty} \mathbb{E}[f(X^{(k_\ell)})] = \widehat{\mathbb{E}}[f(\widehat{X}^{(*)})] = \widetilde{\mathbb{E}}[f(\widehat{X}^{(*)})] = \mathbb{E}[f(X)]$$

for any $f \in C_b(C[0,T];\mathbb{R})$. This concludes the statement.

Proof of Theorem 9.2
The proof for the statement (ii) is similar to (i), thus we only prove the statement (i).

The proof is based on [19], that is, we prove the statement by contradiction. We suppose that the statement (i) is not true, that is, there exist $\varepsilon_0 > 0$ and a sub-sequence $\{n_k\}_{k\in\mathbb{N}}$ such that

$$\mathbb{E}\left[\sup_{0\leq t\leq T}\left|X_t - X_t^{(n_k)}\right|^p\right] \geq \varepsilon_0, \text{ for any } k \in \mathbb{N}. \tag{9.20}$$

We now denote $X^{(k)}$ by $X^{(n_k)}$ to simplify. Then from Lemma 9.5, there exist a probability space $(\widehat{\Omega}, \widehat{\mathcal{F}}, \widehat{\mathbb{P}})$, a sub-sequence $\{k_\ell\}_{\ell\in\mathbb{N}}$ and 3-dimensional continuous processes $\widehat{Y}^{k_\ell} = (\widehat{X}^{k_\ell}, \widehat{X}^{(k_\ell)}, \widehat{W}^{k_\ell})$ and $\widehat{Y} = (\widehat{X}, \widehat{X}^{(*)}, \widehat{W})$ defined on the probability space $(\widehat{\Omega}, \widehat{\mathcal{F}}, \widehat{\mathbb{P}})$ such that the properties (i)–(vi) are satisfied.

Note that from Lemma 9.2, the family of random variables $\{\sup_{0\leq t\leq T}|\widehat{X}_t^{k_\ell} - \widehat{X}_t^{(k_\ell)}|^p\}_{\ell\in\mathbb{N}}$ is uniformly integrable. Therefore, from the assumption (9.20) and the property (i), (ii) in Lemma 9.5, we have

$$\begin{aligned}\varepsilon_0 &\leq \liminf_{\ell\to\infty}\mathbb{E}\left[\sup_{0\leq t\leq T}\left|X_t - X_t^{(k_\ell)}\right|^p\right] \\ &= \liminf_{\ell\to\infty}\widehat{\mathbb{E}}\left[\sup_{0\leq t\leq T}\left|\widehat{X}_t^{k_\ell} - \widehat{X}_t^{(k_\ell)}\right|^p\right] \\ &= \widehat{\mathbb{E}}\left[\sup_{0\leq t\leq T}\left|\widehat{X}_t - \widehat{X}_t^{(*)}\right|^p\right].\end{aligned} \tag{9.21}$$

On the other hand, the property (iv) and (vi) imply that $\widehat{X}$ and $\widehat{X}^{(*)}$ are solutions of SDE (9.1) driven by the same Brownian motion $\widehat{W}$, with non-sticky condition (9.10) on the probability space $(\widehat{\Omega}, \widehat{\mathcal{F}}, \widehat{\mathbb{P}})$. Hence from the assumption on the pathwise uniqueness, (9.20) and (9.21), we conclude $0 < \varepsilon_0 \leq 0$. This is the contradiction.

Acknowledgements The authors would like to thank Professor Masatoshi Fukushima for his valuable comments. The authors would also like to thank an anonymous referee for his/her careful readings and advices. The first author was supported by JSPS KAKENHI Grant Number 17H06833. The second author was supported by Sumitomo Mitsui Banking Corporation.

References

1. S. Ankirchner, T. Kruse, M. Urusov, A functional limit theorem for coin tossing Markov chains (2018). Preprint arXiv:1902.06249v1
2. A. Antonov, M. Konikov, M. Spector, The free boundary SABR: natural extension to negative rates. SSRN 2557046 (2015)

3. M.T. Barlow, One dimensional stochastic differential equations with no strong solution. Lond. Math. Soc. **2**(2), 335–347 (1982)
4. R.F. Bass, Z.Q. Chen, One-dimensional stochastic differential equations with singular and degenerate coefficients. Sankhyā **67**(1), 19–45 (2005)
5. R.F. Bass, K. Burdzy, Z.Q. Chen, Pathwise uniqueness for a degenerate stochastic differential equation. Ann. Probab. **35**(6), 2385–2418 (2007)
6. A.N. Borodin, P. Salminen, *Handbook of Brownian Motion-Facts and Formulae* (Birkhäuser, Basel, 2012)
7. J.C. Cox, The constant elasticity of variance option pricing model. J. Portf. Manag. **23**(5), 15–17 (1996)
8. F. Delbaen, H. Shirakawa, A note on option pricing for the constant elasticity of variance model. Asia-Pacific Finan. Markets **9**(2), 85–99 (2002)
9. R. Durrett, *Stochastic Calculus: A Practical Introduction* (CRC Press, Boca Raton, 1996)
10. H.J. Engelbert, J. Hess, Stochastic integrals of continuous local martingales, II. Math. Nachr. **100**(1), 249–269 (1981)
11. H.J. Engelbert, W. Schmidt, On one-dimensional stochastic differential equations with generalized drift. Lecture Notes Control Inform. Sci. **69**, 143–155 (1985)
12. H.J. Engelbert, W. Schmidt, On solutions of one-dimensional stochastic differential equations without drift. Wahrscheinlichkeitstheorie verw. Gebiete **68**(3), 287–314 (1985)
13. I.V. Girsanov, An example of non-uniqueness of the solution of the stochastic equation of K. Ito. Theory Probab. Appl. **7**(3), 325–331 (1962)
14. A. Gulisashvili, B. Horvath, A. Jacquier, Mass at zero in the uncorrelated SABR model and implied volatility asymptotics. Quant. Finan. **18**(10), 1753–1765 (2018)
15. I. Gyöngy, M. Rásonyi, A note on Euler approximations for SDEs with Hölder continuous diffusion coefficients. Stoch. Process. Appl. **121**, 2189–2200 (2011)
16. N. Ikeda, S. Watanabe, *Stochastic Differential Equations and Diffusion Processes*, 2nd edn. North-Holland Mathematical Library, vol. 24 (North-Holland, Amsterdam; Kodansha, Tokyo 1981)
17. O. Islah, O. Solving SABR in exact form and unifying it with LIBOR market model. SSRN 1489428 (2009)
18. M. Jeanblanc, M. Yor, M. Chesney, *Mathematical Methods for Financial Markets* (Springer, Berlin, 2009)
19. H. Kaneko, S. Nakao, A note on approximation for stochasitc differential equations. Sémin. Probab. **22**, 155–162 (1988)
20. I. Karatzas, S.E. Shreve, *Brownian Motion and Stochastic Calculus*, 2nd edn. (Springer, Berlin, 1991)
21. P. Kloeden, E. Platen, *Numerical Solution of Stochastic Differential Equations* (Springer, Berlin, 1995)
22. T.G. Kurtz, P. Protter, Weak limit theorems for stochastic integrals and stochastic differential equations. Ann. Probab. **19**(3), 1035–1070 (1991)
23. J.F. Le Gall, One-dimensional stochastic differential equations involving the local times of the unknown process, in *Stochastic Analysis and Applications* (Springer, Berlin, 1984), pp. 51–82
24. S. Manabe, T. Shiga, On one–dimensional stochastic differential equations with non-sticky boundary condition. J. Math. Kyoto Univ. **13**(3), 595–603 (1973)
25. G. Maruyama, On the transition probability functions of the Markov process. Nat. Sci. Rep. Ochanomizu Univ. **5**, 10–20 (1954)
26. S. Nakao, On the pathwise uniqueness of solutions of one-dimensional stochastic differential equations. Osaka J. Math. **9**, 513–518 (1972)
27. M.N. Pascu, N.R. Pascu, A note on the sticky Brownian motion on $\mathbb{R}$. Bull. Transilv. Univ. Brasov Ser. III **4**(2), 57–62 (2011)
28. D. Revuz, M. Yor, *Continuous Martingales and Brownian Motion*, 3rd edn. (Springer, Berlin, 1999)
29. A.V. Skorokhod, *Studies in the Theory of Random Processes* (Addison-Wesley, Washington, 1965)

30. H. Tanaka, M. Hasegawa, Stochastic differential equations. Semin. Probab. **19** (1964) (in Japanese)
31. A.W. Van Der Vaart, J.A. Wellner, Weak convergence and empirical processes (Springer, New York, 1996)
32. T. Yamada, S. Watanabe, On the uniqueness of solutions of stochastic differential equations. J. Math. Kyoto Univ. **11**, 155–167 (1971)
33. T. Yamada, Sur une construction des solutions déquations différentielles stochastiques dans le cas non-Lipschitzien. Sémin. Probab. **12**, 114–131 (1978)
34. B.L. Yan, The Euler scheme with irregular coefficients. Ann. Probab. **30**(3), 1172–1194 (2002)

Chapter 10
On a Construction of Strong Solutions for Stochastic Differential Equations with Non-Lipschitz Coefficients: A Priori Estimates Approach

Toshiki Okumura

Abstract Given a stochastic differential equation of which coefficients satisfy Yamada–Watanabe condition or Nakao-Le Gall condition. We prove that its strong solution can be constructed on any probability space using a priori estimates and also using Ito theory based on Picard's approximation scheme.

10.1 Introduction

The present paper proposes concrete and direct constructions of strong solutions for stochastic differential equations (SDEs) with non-Lipschitz coefficients. It is well known that in Ito classical theory on SDEs (see Ito [3]) under the global Lipschitz condition for coefficients, the existence and the uniqueness hold for their strong solutions. The theory is based on the Picard's iteration method and then the existence and the uniqueness follow naturally by Picard's successive approximation procedure of strong solutions. Although Ito theory is beautifully established, the global Lipschitz condition imposed on coefficients is too strict and too restricted for the purpose of discussing various SDEs raised both in the theory of stochastic calculus and in its applications. Consider for examples, the SDE associated to square Bessel processes, to Wright–Fischer model in population genetics, to Cox–Ingersoll–Ross (CIR) model in mathematical finance, and also to skewed symmetric Brownian motions. The classical Ito theory covers none of these examples.

As is well known that the frame of the weak existence theory [10, 11] is wide enough to cover the all examples cited in the above. We know also that

---

The views expressed in this paper are those of the author and do not necessarily represent the views of The Dai-ichi Life Insurance Company, Limited.

T. Okumura (✉)
The Dai-ichi Life Insurance Company, Limited, Chiyoda-ku, Tokyo, Japan

C. Donati-Martin et al. (eds.), *Séminaire de Probabilités L*, Lecture Notes in Mathematics 2252, https://doi.org/10.1007/978-3-030-28535-7_10

the pathwise uniqueness holds for solutions of SDEs in the above by Yamada–Watanabe condition [13], or by that of Nakao-Le Gall [6, 7]. Then the existence of strong solutions for them follows immediately by Yamada–Watanabe Theorem [13]. However, we would like to point out that the proof of the existence by Yamada–Watanabe involves no construction procedure of strong solutions.

In this situation, the investigation on concrete construction of strong solutions under non Lipschitz conditions appears to be interesting. Our paper is motivated by a construction of strong solutions given by Stefan Ebenfeld [1]. His method covers CIR model in mathematical finance. The method is based on a priori estimates and also on Ito classical theory. The benefit of this approach is that the argument only requires some fundamental knowledge about stochastic and functional analysis.

The first part of the paper (Sects. 10.2–10.5) is devoted to the improvement of the result by Stefan Ebenfeld [1]. We show a concrete construction procedure of strong solutions under Yamada–Watanabe condition. Although our construction method based on a priori estimates is new, the Euler Maruyama approximation method gives an another construction of strong solutions under the same condition (see Yamada [12], Kaneko and Nakao [5]).

In the final part of the paper, we discuss the existence of strong solutions under Nakao-Le Gall condition. Since coefficients are allowed to be discontinuous, the Euler–Maruyama method based on the continuity of coefficients does not cover this case. Discontinuous points of coefficient raise various difficulties to be discussed carefully in the proof. The stochastic calculus based on local times and their occupation formulas plays important roles to overcome these difficulties.

10.2 The Main Result Under Yamada–Watanabe Condition

10.2.1 Assumptions

We discuss under the following assumptions.

(1) Let $(\Omega, \mathcal{F}, \{\mathcal{F}_t\}, \mathbb{P})$ be a filtered probability space where the filtration satisfies the usual conditions.
(2) Let W be a Brownian motion with respect to $(\Omega, \mathcal{F}, \{\mathcal{F}_t\}, \mathbb{P})$.
(3) Let $T > 0$, and let $X_0 \in \mathbb{R}$.
(4) Let $b \in C([0, T] \times \mathbb{R}, \mathbb{R})$, and let $\sigma \in C([0, T] \times \mathbb{R}, \mathbb{R})$.
(5) Let b and σ satisfy the following linear growth condition

$$\exists C > 0 \quad \forall t \in [0, T] \quad \forall x \in \mathbb{R},$$

$$|b(t, x)| + |\sigma(t, x)| \leq C(1 + |x|).$$

(6) Let b satisfy the following Lipschitz continuity condition

$$\exists C > 0 \quad \forall t \in [0, T] \quad \forall x, y \in \mathbb{R},$$
$$|b(t, x) - b(t, y)| \leq C|x - y|.$$

(7) Let σ satisfy the following continuity condition

$$\forall \epsilon > 0 \quad \forall t \in [0, T] \quad \forall x, y \in \mathbb{R},$$
$$|\sigma(t, x) - \sigma(t, y)| \leq h(|x - y|),$$

where h is a strictly increasing continuous function defined on $[0, \infty)$ with $h(0) = 0$ such that

(a)

$$h(x) \leq C(1 + x); \quad \exists C > 0 \quad \forall x \in [0, \infty),$$

(b)

$$\int_{(0,\epsilon)} h^{-2}(u) du = \infty; \quad \forall \epsilon > 0.$$

We consider the following SDE

$$X_t = X_0 + \int_0^t b(s, X_s) ds + \int_0^t \sigma(s, X_s) dW_s. \tag{10.1}$$

The following result shows that the assumptions are sufficient to guarantee pathwise uniqueness of solutions.

Proposition 10.1 (Yamada and Watanabe [13]) *Under the assumptions (1)–(7), pathwise uniqueness holds for SDE (10.1).*

A detailed proof of Proposition 10.1 can be found for example in the book of Karatzas and Shreve [4], Proposition 5.3.20.

10.2.2 Main Theorem

Here, we present the main theorem. The proof of the main theorem is discussed along the lines of Stefan Ebenfeld [1].

Theorem 10.1 (Strong Existence) *Under the assumptions (1–7), the SDE (10.1) has a strong solution.*

The proof of the theorem is based on a particular approximation of the SDE (10.1). Let $\chi_\epsilon(x)$ be a function in $C^\infty(\mathbb{R})$, such that

$$\chi_\epsilon(x) := \begin{cases} 0 & ; x \le -\epsilon, \\ \text{positive} & ; -\epsilon < x < \epsilon, \\ 0 & ; x \ge \epsilon, \end{cases}$$

and

$$\int_{-\infty}^{\infty} \chi_\epsilon(x)dx = 1.$$

Moreover, we define that

$$\sigma_\epsilon(t, x) := \int_{-\infty}^{\infty} \sigma(t, x - y)\chi_\epsilon(y)dy = (\sigma * \chi_\epsilon)(x),$$

where the symbol $*$ stands for the convolution operator. Consider the following approximated SDE

$$X_t^{(\epsilon)} = X_0 + \int_0^t b(s, X_s^{(\epsilon)})ds + \int_0^t \sigma_\epsilon(s, X_s^{(\epsilon)})dW_s. \tag{10.2}$$

First, we mention that σ_ϵ belongs to $C^\infty(\mathbb{R})$ and satisfies following properties.

Lemma 10.1 *Under the assumption (7) for* σ*,*

(i) for any $0 < \delta < 1$*, there exists a constant* $\epsilon(\delta) > 0$ *such that*

$$|\sigma_\epsilon(t, x) - \sigma(t, x)| \le \delta \quad \textit{holds}; \quad \forall \epsilon \in (0, \epsilon(\delta)] \quad \forall t \in [0, T] \quad \forall x \in \mathbb{R},$$

(ii) for any $\epsilon > 0$*,*

$$|\sigma_\epsilon(t, x) - \sigma(t, x)| \le h(\epsilon) \quad \textit{holds}; \quad \forall t \in [0, T] \quad \forall x \in \mathbb{R},$$

(iii) for σ_ϵ*,*

$$|\sigma_\epsilon(t, x) - \sigma_\epsilon(t, y)| \le h(|x - y|) \quad \textit{holds}; \quad \forall t \in [0, T] \quad \forall x, y \in \mathbb{R}.$$

Proof (Proof of (i)) Since $h(t)$ is continuous and $h(0) = 0$, there exists $\epsilon(\delta) > 0$ such that

$$h(|u|) \le \delta; \quad 0 \le |u| \le \epsilon(\delta).$$

Therefore, we have for $0 \le \epsilon \le \epsilon(\delta)$,

$$\begin{aligned}
|\sigma_\epsilon(t,x) - \sigma(t,x)| &= |\int_{-\infty}^{\infty} \sigma(t, x-y)\chi_\epsilon(y)dy - \int_{-\infty}^{\infty} \sigma(t,x)\chi_\epsilon(y)dy| \\
&\le \int_{-\infty}^{\infty} h(|y|)\chi_\epsilon(y)dy \\
&= \int_{-\epsilon}^{\epsilon} h(|y|)\chi_\epsilon(y)dy \\
&\le \delta \int_{-\epsilon}^{\epsilon} \chi_\epsilon(y)dy \\
&= \delta.
\end{aligned}$$

□

Proof (Proof of (ii)) (ii) can be proved in the same way as (i). □

Proof (Proof of (iii)) From the assumption (7) and the definition of χ_ϵ, we have

$$\begin{aligned}
|\sigma_\epsilon(t,x) - \sigma_\epsilon(t,y)| &= |\int_{-\infty}^{\infty} \sigma(t, x-u)\chi_\epsilon(u)du - \int_{-\infty}^{\infty} \sigma(t, y-u)\chi_\epsilon(u)du| \\
&= |\int_{-\infty}^{\infty} \chi_\epsilon(u)[\sigma(t, x-u) - \sigma(t, y-u)]du| \\
&\le \int_{-\infty}^{\infty} \chi_\epsilon(u)|\sigma(t, x-u) - \sigma(t, y-u)|du \\
&\le \int_{-\infty}^{\infty} \chi_\epsilon(u)h(|(x-u) - (y-u)|)du \\
&= \int_{-\infty}^{\infty} \chi_\epsilon(u)h(|x-y|)du \\
&= h(|x-y|)\int_{-\infty}^{\infty} \chi_\epsilon(u)du \\
&= h(|x-y|).
\end{aligned}$$

□

Remark 10.1 From Lemma 10.1, there exists $C > 0$ such that

$$|b(t,x)| + |\sigma_\epsilon(t,x)| \le C(1 + |x|)$$

holds for any $t \in [0, T]$, $x \in \mathbb{R}$, $\epsilon > 0$.

Since σ_ϵ belongs to $C^\infty(\mathbb{R})$ and satisfies obviously local Lipschitz condition, it is shown that the approximated SDE (10.2) has a unique strong solution. See, for example, Theorem 5.12.1 in the book of Rogers and Williams [9].

10.3 A Priori Estimates

10.3.1 The High Norm

We will mention the well known result on the boundedness of solutions for approximated SDE (10.2) in the sense of the high norm.

Lemma 10.2 (A Priori Estimate in the High Norm) *Solutions of the approximated SDE (10.2) satisfy the following estimate*

$$\begin{aligned} &\exists C > 0 \quad \forall \epsilon \leq 1, \\ &\sup_{t\in[0,T]} \mathbb{E}[|X_t^{(\epsilon)}|^4] \leq C. \end{aligned} \tag{10.3}$$

A detailed proof of the lemma can be found, for example, in the book of Karatzas and Shreve [4], Problem 5.3.15.

10.3.2 The Low Norm

The next lemma on a priori estimate in the low norm for the approximated SDE (10.2) will play essential roles in the proof of our main theorem. The lemma requires a smooth approximation of the function $|x|$. Therefore, we introduce a sequence $(a_n)_{n\in\mathbb{N}}$ of positive numbers by

$$a_0 := 1; \quad \int_{a_n}^{a_{n-1}} \frac{1}{nh^2(x)} dx = 1.$$

Next, we choose a sequence $(\rho_n)_{n\in\mathbb{N}}$ of smooth mollifiers with the following properties

$$supp(\rho_n) \subset [a_n, a_{n-1}]; \quad 0 \leq \rho_n(x) \leq \frac{2}{nh^2(x)}; \quad \int_{a_n}^{a_{n-1}} \rho_n(x)dx = 1.$$

Finally, we define a sequence $(\varphi_n)_{n\in\mathbb{N}}$ of smooth functions by

$$\varphi_n(x) := \int_0^{|x|} \int_0^y \rho_n(z)dzdy + a_{n-1}.$$

Then, $(\varphi_n)_{n\in\mathbb{N}}$ has the following properties

$$\varphi_n(x) \geq |x|; \quad |\varphi_n'(x)| \leq 1; \quad \varphi_n''(x) = \rho_n(|x|).$$

In other words, φ_n is a smooth approximation of the function $|x|$ from above with a bounded first-order derivative and a second-order derivative having support in the interval $[a_n, a_{n-1}]$.

Lemma 10.3 (A Priori Estimate in the Low Norm) *Solutions of the approximated SDE (10.2) satisfy the following a priori estimate*

$$\begin{aligned}&\forall\alpha > 0 \quad \exists 0 < \beta \leq 1 \quad \forall 0 < \epsilon_1, \epsilon_2 \leq \beta,\\ &\sup_{t\in[0,T]} \mathbb{E}[|X_t^{(\epsilon_1)} - X_t^{(\epsilon_2)}|] \leq \alpha.\end{aligned} \tag{10.4}$$

Proof Put

$$\begin{aligned}\Delta_t^{(\epsilon_1,\epsilon_2)} &:= X_t^{(\epsilon_1)} - X_t^{(\epsilon_2)}\\ &= \int_0^t [b(s, X_s^{(\epsilon_1)}) - b(s, X_s^{(\epsilon_2)})]ds + \int_0^t [\sigma_{\epsilon_1}(s, X_s^{(\epsilon_1)}) - \sigma_{\epsilon_2}(s, X_s^{(\epsilon_2)})]dW_s.\end{aligned}$$

Applying Ito formula to the approximated SDE (10.2), we obtain the following representation

$$\begin{aligned}\varphi_n(\Delta_t^{(\epsilon_1,\epsilon_2)}) = \varphi_n(0) &+ \int_0^t \varphi_n'(\Delta_s^{(\epsilon_1,\epsilon_2)})[b(s, X_s^{(\epsilon_1)}) - b(s, X_s^{(\epsilon_2)})]ds\\ &+ \frac{1}{2}\int_0^t \varphi_n''(\Delta_s^{(\epsilon_1,\epsilon_2)})[\sigma_{\epsilon_1}(s, X_s^{(\epsilon_1)}) - \sigma_{\epsilon_2}(s, X_s^{(\epsilon_2)})]^2 ds\\ &+ \int_0^t \varphi_n'(\Delta_s^{(\epsilon_1,\epsilon_2)})[\sigma_{\epsilon_1}(s, X_s^{(\epsilon_1)}) - \sigma_{\epsilon_2}(s, X_s^{(\epsilon_2)})]dW_s.\end{aligned}$$

We note that due to the uniform boundedness of φ_n' and the linear growth condition (see Remark 10.1), the Ito integral in the above is a martingale with mean 0. Let $0 < \beta \leq 1$, $0 < \epsilon_1, \epsilon_2 \leq \beta$. By Ito formula, we have

$$\begin{aligned}\mathbb{E}[|\Delta_t^{(\epsilon_1,\epsilon_2)}|] &\leq \mathbb{E}[\varphi_n(\Delta_t^{(\epsilon_1,\epsilon_2)})]\\ &= \mathbb{E}[\varphi_n(0)] + \mathbb{E}[\int_0^t \varphi_n'(\Delta_s^{(\epsilon_1,\epsilon_2)})[b(s, X_s^{(\epsilon_1)}) - b(s, X_s^{(\epsilon_2)})]ds]\\ &\quad + \frac{1}{2}\mathbb{E}[\int_0^t \varphi_n''(\Delta_s^{(\epsilon_1,\epsilon_2)})[\sigma_{\epsilon_1}(s, X_s^{(\epsilon_1)}) - \sigma_{\epsilon_2}(s, X_s^{(\epsilon_2)})]^2 ds]\\ &\quad + \mathbb{E}[\int_0^t \varphi_n'(\Delta_s^{(\epsilon_1,\epsilon_2)})[\sigma_{\epsilon_1}(s, X_s^{(\epsilon_1)}) - \sigma_{\epsilon_2}(s, X_s^{(\epsilon_2)})]dW_s]\\ &= a_{n-1} + \mathbb{E}[\int_0^t \varphi_n'(\Delta_s^{(\epsilon_1,\epsilon_2)})[b(s, X_s^{(\epsilon_1)}) - b(s, X_s^{(\epsilon_2)})]ds]\\ &\quad + \frac{1}{2}\mathbb{E}[\int_0^t \varphi_n''(\Delta_s^{(\epsilon_1,\epsilon_2)})[\sigma_{\epsilon_1}(s, X_s^{(\epsilon_1)}) - \sigma_{\epsilon_2}(s, X_s^{(\epsilon_2)})]^2 ds],\end{aligned}$$

using $\varphi_n''(x) = \rho_n(|x|)$;

$$\le a_{n-1} + \mathbb{E}[\int_0^t |\varphi_n'(\Delta_s^{(\epsilon_1,\epsilon_2)})[b(s, X_s^{(\epsilon_1)}) - b(s, X_s^{(\epsilon_2)})]|ds]$$
$$+ \frac{1}{2}\mathbb{E}[\int_0^t \rho_n(|\Delta_s^{(\epsilon_1,\epsilon_2)}|)[\sigma_{\epsilon_1}(s, X_s^{(\epsilon_1)}) - \sigma_{\epsilon_2}(s, X_s^{(\epsilon_2)})]^2 ds],$$

using $|\varphi_n'(x)| \le 1$;

$$\le a_{n-1} + \mathbb{E}[\int_0^t |[b(s, X_s^{(\epsilon_1)}) - b(s, X_s^{(\epsilon_2)})]|ds]$$
$$+ \frac{1}{2}\mathbb{E}[\int_0^t \rho_n(|\Delta_s^{(\epsilon_1,\epsilon_2)}|)|\sigma_{\epsilon_1}(s, X_s^{(\epsilon_1)}) - \sigma_{\epsilon_2}(s, X_s^{(\epsilon_2)})|^2 ds]$$
$$= a_{n-1} + \mathbb{E}[\int_0^t |[b(s, X_s^{(\epsilon_1)}) - b(s, X_s^{(\epsilon_2)})]|ds]$$
$$+ \frac{1}{2}\mathbb{E}[\int_0^t \rho_n(|\Delta_s^{(\epsilon_1,\epsilon_2)}|)|\sigma_{\epsilon_1}(s, X_s^{(\epsilon_1)}) - \sigma_{\epsilon_2}(s, X_s^{(\epsilon_2)})$$
$$- \sigma(s, X_s^{(\epsilon_1)}) + \sigma(s, X_s^{(\epsilon_1)}) + \sigma(s, X_s^{(\epsilon_2)}) - \sigma(s, X_s^{(\epsilon_2)})|^2 ds]$$
$$\le a_{n-1} + \mathbb{E}[\int_0^t |[b(s, X_s^{(\epsilon_1)}) - b(s, X_s^{(\epsilon_2)})]|ds]$$
$$+ \frac{1}{2}\mathbb{E}[\int_0^t \rho_n(|\Delta_s^{(\epsilon_1,\epsilon_2)}|)(3|\sigma(s, X_s^{(\epsilon_1)}) - \sigma(s, X_s^{(\epsilon_2)})|^2$$
$$+ 3|\sigma_{\epsilon_1}(s, X_s^{(\epsilon_1)}) - \sigma(s, X_s^{(\epsilon_1)})|^2 + 3|\sigma(s, X_s^{(\epsilon_2)}) - \sigma_{\epsilon_2}(s, X_s^{(\epsilon_2)})|^2)ds].$$

Here, we give $0 < \delta < 1$. Then, by the assumption (7) and Lemma 10.1, we have for any $0 < \epsilon_1, \epsilon_2 \le \beta \le \epsilon(\delta)$,

$$\mathbb{E}[|\Delta_t^{(\epsilon_1,\epsilon_2)}|] \le a_{n-1} + \mathbb{E}[\int_0^t |[b(s, X_s^{(\epsilon_1)}) - b(s, X_s^{(\epsilon_2)})]|ds]$$
$$+ \frac{1}{2}\mathbb{E}[\int_0^t \rho_n(|\Delta_s^{(\epsilon_1,\epsilon_2)}|)(3[h(|\Delta_s^{(\epsilon_1,\epsilon_2)}|)]^2 + 3\delta^2 + 3\delta^2)ds]$$
$$\le a_{n-1} + \mathbb{E}[\int_0^t |[b(s, X_s^{(\epsilon_1)}) - b(s, X_s^{(\epsilon_2)})]|ds]$$
$$+ \frac{1}{2}\mathbb{E}[\int_0^t \rho_n(|\Delta_s^{(\epsilon_1,\epsilon_2)}|)(3[h(|\Delta_s^{(\epsilon_1,\epsilon_2)}|)]^2 + 3\delta + 3\delta)ds],$$

using $0 \le \rho_n(x) \le \frac{2}{nh^2(x)}$;

$$\le a_{n-1} + \mathbb{E}[\int_0^t |[b(s, X_s^{(\epsilon_1)}) - b(s, X_s^{(\epsilon_2)})]|ds]$$

$$+ \frac{1}{2}\mathbb{E}[\int_0^t \frac{2I_{[a_n,a_{n-1}]}(|\Delta_s^{(\epsilon_1,\epsilon_2)}|)}{nh^2(|\Delta_s^{(\epsilon_1,\epsilon_2)}|)}(3[h(|\Delta_s^{(\epsilon_1,\epsilon_2)}|)]^2 + 6\delta)ds]$$

$$\le a_{n-1} + \mathbb{E}[\int_0^t |[b(s, X_s^{(\epsilon_1)}) - b(s, X_s^{(\epsilon_2)})]|ds]$$

$$+ \frac{1}{2}T\frac{2\cdot 6\delta}{nh^2(a_n)} + \frac{1}{2}\mathbb{E}[\int_0^t \frac{2}{nh^2(|\Delta_s^{(\epsilon_1,\epsilon_2)}|)} \cdot 3|h(|\Delta_s^{(\epsilon_1,\epsilon_2)}|)|^2 ds]$$

$$= a_{n-1} + \mathbb{E}[\int_0^t |[b(s, X_s^{(\epsilon_1)}) - b(s, X_s^{(\epsilon_2)})]|ds]$$

$$+ T\frac{6\delta}{nh^2(a_n)} + \frac{3}{n}\mathbb{E}[\int_0^t 1ds],$$

with $t \in [0, T]$;

$$\le a_{n-1} + \mathbb{E}[\int_0^t |[b(s, X_s^{(\epsilon_1)}) - b(s, X_s^{(\epsilon_2)})]|ds]$$

$$+ T\frac{6\delta}{nh^2(a_n)} + \frac{3}{n}\mathbb{E}[\int_0^T 1ds]$$

$$= a_{n-1} + \mathbb{E}[\int_0^t |[b(s, X_s^{(\epsilon_1)}) - b(s, X_s^{(\epsilon_2)})]|ds]$$

$$+ T\frac{6\delta}{nh^2(a_n)} + \frac{3}{n}T,$$

by the assumption (6);

$$\le a_{n-1} + \mathbb{E}[\int_0^t C|\Delta_s^{(\epsilon_1,\epsilon_2)}|ds] + T\frac{6\delta}{nh^2(a_n)} + \frac{3}{n}T$$

$$= a_{n-1} + \frac{T}{n}(\frac{6\delta}{h^2(a_n)} + 3) + C\int_0^t \mathbb{E}[|\Delta_s^{(\epsilon_1,\epsilon_2)}|]ds.$$

Combining all the estimates stated earlier we obtain our final estimate

$$\mathbb{E}[|\Delta_t^{(\epsilon_1,\epsilon_2)}|] \le \gamma(n,\delta) + C\int_0^t \mathbb{E}[|\Delta_s^{(\epsilon_1,\epsilon_2)}|]ds, \tag{10.5}$$

$$\gamma(n,\delta) := a_{n-1} + \frac{T}{n}(\frac{6\delta}{h^2(a_n)} + 3).$$

Choosing n sufficiently large first and then choosing δ sufficiently small, we can take $\gamma(n,\delta)$ arbitrarily small. By Gronwall's inequality, Eq. (10.5) implies

$$\mathbb{E}[|\Delta_t^{(\epsilon_1,\epsilon_2)}|] \leq \gamma(n,\delta)e^{Ct} \leq \gamma(n,\delta)e^{CT}. \tag{10.6}$$

Thus, for any $\alpha > 0$, choosing n and δ such that $\gamma(n,\delta)e^{CT} < \alpha$, we have for any $0 < \epsilon_1, \epsilon_2 \leq \beta \leq \epsilon(\delta)$,

$$\mathbb{E}[|\Delta_t^{(\epsilon_1,\epsilon_2)}|] \leq \alpha. \tag{10.7}$$

□

10.4 Uniform Integrability

Consider the following Banach spaces ($1 \leq p < \infty$)

$$H^p := C([0,T], L^p(\Omega, \mathcal{F}, \mathbb{P})),$$

$$||X||_p := \sup_{t\in[0,T]} (\mathbb{E}[|X_t|^p])^{\frac{1}{p}}.$$

Moreover, we consider the following subsets ($1 \leq p < \infty$)

$$N^p := \{X \in H^p | X \text{ is adapted with respect to the filtration} \{\mathcal{F}_t\}\}.$$

Since N^p is a closed subspace of H^p, it is also a Banach space with respect to the norm $||\cdot||_p$. We define $\epsilon_n := \frac{1}{n}$ and write $X^{(n)}$ instead of $X^{(\epsilon_n)}$. By Lemma 10.2, the sequence $(X^{(n)})_{n\in\mathbb{N}}$ is bounded in N^4. Moreover, according to Lemma 10.3, the sequence $(X^{(n)})_{n\in\mathbb{N}}$ is a Cauchy sequence in N^1. We will show that the sequence $(X^{(n)})_{n\in\mathbb{N}}$ is also a Cauchy sequence in N^2.

Lemma 10.4 *$(X^{(n)})_{n\in\mathbb{N}}$ is a Cauchy sequence in N^2.*

Proof Lemma 10.2 implies immediately

$$\sup_{n\in\mathbb{N}} ||X^{(n)}||_2 < \infty. \tag{10.8}$$

If $(X^{(n)})$ is not a Cauchy sequence in N^2, there exist a positive constant C and some subsequences $(p_n)_{n\in\mathbb{N}}$ and $(q_n)_{n\in\mathbb{N}}$ such that

$$\lim_{n\to\infty} \sup_{0\leq s\leq T} \mathbb{E}[|X_s^{(p_n)} - X_s^{(q_n)}|^2] = C > 0. \tag{10.9}$$

Note that

$$\begin{aligned}\mathbb{E}[\int_0^T |X_s^{(p_n)} - X_s^{(q_n)}|ds] &= \int_0^T \mathbb{E}[|X_s^{(p_n)} - X_s^{(q_n)}|]ds \\ &\le \int_0^T (\sup_{0\le s\le T} \mathbb{E}[|X_s^{(p_n)} - X_s^{(q_n)}|])ds \\ &= T||X^{(p_n)} - X^{(q_n)}||_1.\end{aligned}$$

Since $(X^{(n)})$ is a Cauchy sequence in N^1,

$$\lim_{n\to\infty} \mathbb{E}[\int_0^T |X_s^{(p_n)} - X_s^{(q_n)}|ds] \le \lim_{n\to\infty} T||X^{(p_n)} - X^{(q_n)}||_1 = 0. \tag{10.10}$$

We can choose subsequences (p_n') and (q_n') such that

$$\lim_{n\to\infty} |X_s^{(p_n')} - X_s^{(q_n')}| = 0 \tag{10.11}$$

almost everywhere on $[0, T] \times \Omega$ w.r.t. $dt \times d\mathbb{P}$. Assume that for some subsequences $(p_n)_{n\in\mathbb{N}}$ and $(q_n)_{n\in\mathbb{N}}$,

$$\lim_{n\to\infty} \sup_{0\le s\le T} \mathbb{E}[|X_s^{(p_n)} - X_s^{(q_n)}|^2] = C > 0. \tag{10.12}$$

Since

$$\lim_{n\to\infty} ||X^{(p_n)} - X^{(q_n)}||_1 = 0,$$

we can choose subsequences $(p_n')_{n\in\mathbb{N}}$, $(q_n')_{n\in\mathbb{N}}$ such that

$$\lim_{n\to\infty} |X_s^{(p_n')} - X_s^{(q_n')}| = 0 \tag{10.13}$$

almost surely on $[0, T] \times \Omega$ w.r.t. $dt \times d\mathbb{P}$. Let $\epsilon_n' = \frac{1}{p_n'}$ and $\epsilon_n'' = \frac{1}{q_n'}$. We have

$$\begin{aligned}||X^{(p_n')} - X^{(q_n')}||_2^2 &= \sup_{0\le s\le T} \mathbb{E}[|X_s^{(p_n')} - X_s^{(q_n')}|^2] \\ &\le 2\mathbb{E}[\int_0^T (\sigma_{\epsilon_n'}(s, X_s^{(p_n')}) - \sigma_{\epsilon_n''}(s, X_s^{(q_n')}))^2 ds] \\ &\quad + 2\mathbb{E}[\int_0^T (b(s, X_s^{(p_n')}) - b(s, X_s^{(q_n')}))^2 ds]\end{aligned}$$

Now, we define as follows;

$$L_1^{(n)}(T) := \mathbb{E}[\int_0^T (\sigma_{\epsilon_n'}(s, X_s^{(p_n')}) - \sigma_{\epsilon_n''}(s, X_s^{(q_n')}))^2 ds], \tag{10.14}$$

$$L_2^{(n)}(T) := \mathbb{E}[\int_0^T (b(s, X_s^{(p_n')}) - b(s, X_s^{(q_n')}))^2 ds]. \tag{10.15}$$

For $L_1^{(n)}(T)$, we observe that

$$\begin{aligned} L_1^{(n)}(T) \le\ & 3\mathbb{E}[\int_0^T (\sigma_{\epsilon_n'}(s, X_s^{(p_n')}) - \sigma(s, X_s^{(p_n')}))^2 ds] \\ & + 3\mathbb{E}[\int_0^T (\sigma(s, X_s^{(p_n')}) - \sigma(s, X_s^{(q_n')}))^2 ds] \\ & + 3\mathbb{E}[\int_0^T (\sigma_{\epsilon_n''}(s, X_s^{(q_n')}) - \sigma(s, X_s^{(q_n')}))^2 ds]. \end{aligned}$$

Using Lemma 10.1 (ii) and the assumption (7), we have

$$\begin{aligned} L_1^{(n)}(T) \le\ & 3T(h^2(\epsilon_n') + h^2(\epsilon_n'')) \\ & + 3\mathbb{E}[\int_0^T h^2(|X_s^{(p_n')} - X_s^{(q_n')}|) ds]. \end{aligned}$$

Note that

$$h^2(|X_s^{(p_n')} - X_s^{(q_n')}|) \le 2C^2 + 2C^2 |X_s^{(p_n')} - X_s^{(q_n')}|^2.$$

Since

$$\sup_n ||X^{(p_n')} - X^{(q_n')}||_4 < \infty$$

holds, the family of processes

$$h^2(|X_s^{(p_n')} - X_s^{(q_n')}|)$$

is uniformly integrable on $[0, T] \times \Omega$ w.r.t. $dt \times d\mathbb{P}$. Since $h(\epsilon)$ tends to 0 $(\epsilon \downarrow 0)$, we have by Eq. (10.13),

$$\lim_{n \to \infty} L_1^{(n)}(T) = 0.$$

For $L_2^{(n)}(T)$, we have by the assumption (6) that

$$L_2^{(n)}(T) \leq \mathbb{E}[\int_0^T C^2|X_s^{(p_n')} - X_s^{(q_n')}|^2 ds].$$

Since the family of processes $|X_s^{(p_n')} - X_s^{(q_n')}|^2$ is uniformly integrable, Eq. (10.13) implies

$$\lim_{n\to\infty} L_2^{(n)}(T) = 0.$$

Thus we observe

$$\lim_{n\to\infty} ||X_s^{(p_n')} - X_s^{(q_n')}||_2^2 = 0.$$

This fact contradicts Eq. (10.12). □

Therefore, the sequence $(X^{(n)})_{n\in\mathbb{N}}$ converges to some $\widetilde{X} \in N^2$. Since the convergence in N^2 implies in N^1, we have

$$\lim_{n\to\infty} (||X^{(n)} - \widetilde{X}||_1 + ||X^{(n)} - \widetilde{X}||_2) = 0. \tag{10.16}$$

10.5 Proof of The Main Result

Now, we are in a position to prove our main theorem.

Proof (Proof of Theorem) We use the following notation for the right-hand sides of the SDEs under consideration

$$RHS_t^{(n)} := X_0 + \int_0^t b(s, X_s^{(n)})ds + \int_0^t \sigma_{\epsilon_n}(s, X_s^{(n)})dW_s,$$

$$\widetilde{RHS}_t := X_0 + \int_0^t b(s, \widetilde{X}_s)ds + \int_0^t \sigma(s, \widetilde{X}_s)dW_s.$$

Fix $N > 0$. Since σ_ϵ is C^∞-function, it satisfies Lipschitz condition on $(t, x) \in [0, T] \times [-N, N]$. Let

$$\tau_N := \inf\{s : X_s^{(n)} \notin [-N, N]\}. \tag{10.17}$$

Note that

$$|\sigma_\epsilon(t, x)| + |b(t, x)| \leq C(1 + |x|), \tag{10.18}$$

it is well known that

$$\lim_{N\to\infty} T \wedge \tau_N = T \quad (a.s.), \tag{10.19}$$

see, for example, Theorem 5.12.1 in Rogers and Williams [9]. We know also by [9] that for the strong solutions $X^{(n)}$ satisfy their respective SDEs in the following sense

$$\mathbb{E}[\sup_{0\le t\le T\wedge\tau_N} |X_t^{(n)} - RHS_t^{(n)}|^2] \le 4\mathbb{E}[|X_{T\wedge\tau_N}^{(n)} - RHS_{T\wedge\tau_N}^{(n)}|^2] = 0. \tag{10.20}$$

By Lemma 10.2, we know that the family of variables

$$|X_t^{(n)} - RHS_t^{(n)}|^2, \quad t \in [0, T]$$

is uniformly integrable. Thus, letting $N \to \infty$, we have

$$\mathbb{E}[\sup_{0\le t\le T} |X_t^{(n)} - RHS_t^{(n)}|^2] = 0. \tag{10.21}$$

This implies the following weaker condition

$$||X^{(n)} - RHS^{(n)}||_2 = 0. \tag{10.22}$$

In the following, $C_1, C_2 > 0$ denotes some generic constants independent of n. With the help of the linear growth condition (see (5) in 2.1 Assumptions), we obtain the following estimate

$$\mathbb{E}[\int_0^T (|b(s, \widetilde{X}_s)|^2 + |\sigma(s, \widetilde{X}_s)|^2)ds] \le C^2T(1 + ||\widetilde{X}||_2^2) < \infty.$$

This implies that $\widetilde{RHS}$ has continuous paths $\mathbb{P}$-a.s. and satisfies the following regularity condition

$$\mathbb{E}[\sup_{t\in[0,T]} |\widetilde{RHS_t}|^2] < \infty.$$

With the help of the Lipschitz continuity condition for b and the modulus of continuity condition for σ and σ_ϵ (see Lemma 10.1), we obtain the following statement of convergence

$$\lim_{n\to\infty} ||RHS^{(n)} - \widetilde{RHS}||_2^2$$

$$\leq 2 \lim_{n\to\infty} (\sup_{[0,T]} \mathbb{E}[|\int_0^t (b(s, X_s^{(n)}) - b(s, \widetilde{X}_s))ds|^2]$$

$$+ \sup_{[0,T]} \mathbb{E}[|\int_0^t (\sigma_{\epsilon_n}(s, X_s^{(n)}) - \sigma(s, \widetilde{X}_s))dW_s|^2])$$

$$\leq C_1(\lim_{n\to\infty} \mathbb{E}[\int_0^T |b(s, X_s^{(n)}) - b(s, \widetilde{X}_s)|^2 ds]$$

$$+ \lim_{n\to\infty} \mathbb{E}[\int_0^T |\sigma_{\epsilon_n}(s, X_s^{(n)}) - \sigma(s, \widetilde{X}_s)|^2 ds])$$

$$= C_1(\lim_{n\to\infty} \mathbb{E}[\int_0^T |b(s, X_s^{(n)}) - b(s, \widetilde{X}_s)|^2 ds]$$

$$+ \lim_{n\to\infty} \mathbb{E}[\int_0^T |\sigma_{\epsilon_n}(s, X_s^{(n)}) - \sigma(s, X_s^{(n)}) + \sigma(s, X_s^{(n)}) - \sigma(s, \widetilde{X}_s)|^2 ds])$$

$$\leq C_1(\lim_{n\to\infty} \mathbb{E}[\int_0^T |b(s, X_s^{(n)}) - b(s, \widetilde{X}_s)|^2 ds]$$

$$+ \lim_{n\to\infty} \mathbb{E}[\int_0^T |\sigma_{\epsilon_n}(s, X_s^{(n)}) - \sigma(s, X_s^{(n)})|^2 + |\sigma(s, X_s^{(n)}) - \sigma(s, \widetilde{X}_s)|^2 ds]),$$

using Lemma 10.1;

$$\leq C_2(\lim_{n\to\infty} \mathbb{E}[\int_0^T (|X_s^{(n)} - \widetilde{X}_s|)^2 ds]$$

$$+ \lim_{n\to\infty} \mathbb{E}[\int_0^T |\delta|^2 + |h^2(|X_s^{(n)} - \widetilde{X}_s|)|ds]),$$

using Lemma 10.4 and noticing that $\delta > 0$ is arbitrary;

$$= 0.$$

Combining the estimates stated earlier, we see that $\widetilde{X}$ satisfies the SDE (10.1) in the following sense

$$||\widetilde{X} - \widetilde{RHS}||_2 = 0. \tag{10.23}$$

Although $\widetilde{RHS}$ is a modification of $\widetilde{X}$ having continuous path $\mathbb{P}$-a.s., the same is generally not true for $\widetilde{X}$. Therefore, we consider $\widetilde{RHS}$ instead of $\widetilde{X}$ using the following notation

$$X := \widetilde{RHS},$$
$$RHS_t := X_0 + \int_0^t b(s, X_s)ds + \int_0^t \sigma(s, X_s)dW_s.$$

Since X and $\widetilde{X}$ coincide as elements of N^2, the linear growth condition (see (5) in 2.1 Assumptions) yields the following estimate

$$\mathbb{E}[\int_0^T (|b(s, X_s)|^2 + |\sigma(s, |X_s|)|^2)ds] \leq C^2(1 + ||X||_2^2) < \infty. \tag{10.24}$$

This implies that RHS has continuous paths $\mathbb{P}$-a.s. and satisfies the following regularity condition

$$\mathbb{E}[\sup_{t\in[0,T]} |RHS_t|^2] < \infty. \tag{10.25}$$

Finally, with the help of the Lipschitz continuity condition for b (see (6) in 2.1 Assumptions), the modulus of continuity condition for σ (see (7) in 2.1 Assumptions), and Doob's maximal inequality, we see that X satisfies the SDE (10.1) in the sense of Ito theory

$$\begin{aligned}
\mathbb{E}[\sup_{t\in[0,T]} |X_t - RHS_t|^2] &= \mathbb{E}[\sup_{t\in[0,T]} |\widetilde{RHS_t} - RHS_t|^2] \\
&\leq 2(\mathbb{E}[\sup_{t\in[0,T]} |\int_0^t (b(s, \widetilde{X}_s) - b(s, X_s))ds|^2] \\
&\quad + \mathbb{E}[\sup_{t\in[0,T]} |\int_0^t (\sigma(s, \widetilde{X}_s) - \sigma(s, X_s))dW_s|^2]) \\
&\leq C_1(\mathbb{E}[\int_0^T |b(s, \widetilde{X}_s) - b(s, X_s)|^2 ds] \\
&\quad + \mathbb{E}[\int_0^T |\sigma(s, \widetilde{X}_s) - \sigma(s, X_s)|^2 ds]) \\
&\leq C_1(||\widetilde{X} - X||_2 + \int_0^T \mathbb{E}[h^2(|\widetilde{X}_s - X_s|)]ds) \\
&= 0.
\end{aligned}$$

Thus, X is the desired strong solution. This concludes the proof. □

10.6 The Main Result under Nakao-Le Gall Condition

In the present section, we construct concretely a strong solution of SDE under Nakao-Le Gall condition. We consider the following SDE

$$X_t = X_0 + \int_0^t \sigma(X_s) dW_s. \tag{10.26}$$

We assume that σ satisfies Nakao-Le Gall condition.

Definition 10.1 (Nakao-Le Gall Condition) σ be $\mathbb{R} \to \mathbb{R}$, Borel measurable. There exist two positive constants $0 < k < K < \infty$ such that

$$0 < k \leq \sigma(x) \leq K < \infty \quad \forall x \in \mathbb{R}.$$

And, there exists bounded increasing function f such that

$$|\sigma(x) - \sigma(y)|^2 \leq |f(x) - f(y)| \quad \forall x, y \in \mathbb{R}, \tag{10.27}$$

where f is not necessarily continuous.

The main result in this section is the following theorem. Although the result of the theorem is known, our proof of the theorem proposes a concrete construction of strong solution.

Theorem 10.2 *Under Nakao-Le Gall condition, the SDE (10.26) has a strong solution.*

To prove the theorem, we prepare some approximation techniques. Here, let $f(-\infty)$ and $f(\infty)$ be

$$f(-\infty) := \lim_{x \to -\infty} f(x), \tag{10.28}$$

$$f(\infty) := \lim_{x \to \infty} f(x), \tag{10.29}$$

then we obtain $-\infty < f(-\infty) < f(\infty) < \infty$. Let $v(f)$ be

$$v(f) := f(\infty) - f(-\infty), \tag{10.30}$$

$v(f)$ is called the total variation of f.

Remark 10.2 Let D be a set of the discontinuous points of f. Since f is a bounded increasing function, it is well known that D is a countable set. Let $(f_l)_{l \in \mathbb{N}}$ be a sequence of C^∞-functions such that

$$f_l \leq f, \tag{10.31}$$

$$\lim_{l \to \infty} f_l(x) = f(x) \quad for\ x \notin D, \tag{10.32}$$

and

$$v(f_l) \le v(f). \tag{10.33}$$

We will construct an example of such sequence (f_l). Let $(g_l)_{l\in\mathbb{N}}$ be a sequence of C^∞-functions such that

$$g_l(u) = \begin{cases} 0 & ; u \le 0, \\ g(u) > 0 & ; 0 < u < \frac{1}{l}, \\ 0 & ; u \ge \frac{1}{l}, \end{cases}$$

and

$$\int_{-\infty}^{\infty} g_l(u)du = 1.$$

Put

$$f_l(x) := \int_{-\infty}^{\infty} f(x-u)g_l(u)du. \tag{10.34}$$

Note that

$$\begin{aligned} f_l(x) &= \int_{-\infty}^{\infty} f(x-u)g_l(u)du \\ &\le \int_{-\infty}^{\infty} f(x)g_l(u)du \\ &= f(x). \end{aligned}$$

This implies (10.31). Let $x \notin D$. For any $\epsilon > 0$, there exists l such that

$$f(x) - \epsilon \le f(x-u) \le f(x) \tag{10.35}$$

holds for $0 \le u \le \frac{1}{l}$. We have

$$\begin{aligned} f(x) - \epsilon &= \int_{-\infty}^{\infty} (f(x) - \epsilon)g_l(u)du \\ &\le \int_{-\infty}^{\infty} f(x-u)g_l(u)du \\ &= f_l(x) \\ &\le f(x). \end{aligned}$$

Thus,

$$\lim_{l\to\infty} f_l(x) = f(x) \tag{10.36}$$

holds for $x \notin D$. By the definition of $f_l(x)$, we observe that

$$f(x - \frac{1}{l}) \le f_l(x) \le f(x). \tag{10.37}$$

This implies $f_l(-\infty) = f(-\infty)$, and also $f_l(\infty) \le f(\infty)$. Therefore, we have

$$v(f_l) = f_l(\infty) - f_l(-\infty) \le v(f). \tag{10.38}$$

Let $\sigma_\epsilon(x)$ be

$$\sigma_\epsilon(x) := \int_{-\infty}^{\infty} \sigma(x-y)\chi_\epsilon(y)dy = (\sigma * \chi_\epsilon)(x), \tag{10.39}$$

where the function $\chi_\epsilon(x)$ is given in Sect. 10.2.2. Then, $\sigma_\epsilon(x)$ is $C^\infty(\mathbb{R})$ function. Consider the following approximated SDE

$$X_t^{(\epsilon)} = X_0 + \int_0^t \sigma_\epsilon(X_s^{(\epsilon)})dW_s. \tag{10.40}$$

Lemma 10.5

(i) *σ_ϵ is a function in C^∞ and $0 < k \le \sigma_\epsilon(x) \le K < \infty$,*
(ii) *let x be continuous point of σ, then $\lim_{\epsilon\downarrow 0} \sigma_\epsilon(x) = \sigma$,*
(iii) *$|\sigma_\epsilon(x) - \sigma_\epsilon(y)|^2 \le |f(x\vee y+\epsilon) - f(x\wedge y-\epsilon)|$, where $x \vee y := max(x, y)$ and $x \wedge y := min(x, y)$.*

Proof (Proof of (i))

$$\sigma_\epsilon(x) = \int_{-\infty}^{\infty} \sigma(x-y)\chi_\epsilon(y)dy \ge \int_{-\infty}^{\infty} k\chi_\epsilon(y)dy = k, \tag{10.41}$$

and

$$\sigma_\epsilon(x) = \int_{-\infty}^{\infty} \sigma(x-y)\chi_\epsilon(y)dy \le \int_{-\infty}^{\infty} K\chi_\epsilon(y)dy = K. \tag{10.42}$$

□

Proof (Proof of (ii)) Let x be a continuous point of σ. $\forall\eta > 0$, $\exists\delta > 0$ such that

$$\sigma(x) - \eta \le \sigma(y) \le \sigma(x) + \eta, \tag{10.43}$$

for any y such that $|x-y|<\delta$. For $0<\epsilon<\delta$,

$$
\begin{aligned}
|\sigma(x)-\sigma_\epsilon(x)| &= |\int_{-\infty}^{\infty}[\sigma(x)-\sigma(x-y)]\chi_\epsilon(y)dy| \\
&\le \int_{-\infty}^{\infty}|[\sigma(x)-\sigma(x-y)]|\chi_\epsilon(y)dy \\
&\le \eta.
\end{aligned}
$$

□

Proof (Proof of (iii)) Assume that $x>y$. By Schwarz inequality, we have

$$
\begin{aligned}
|\sigma_\epsilon(x)-\sigma_\epsilon(y)|^2 &= |\int_{-\infty}^{\infty}\sigma(x-u)\chi_\epsilon(u)du-\int_{-\infty}^{\infty}\sigma(y-u)\chi_\epsilon(u)du|^2 \\
&\le \int_{-\infty}^{\infty}|\sigma(x-u)-\sigma(y-u)|^2\chi_\epsilon(u)du \\
&\le \int_{-\infty}^{\infty}f(x-u)\chi_\epsilon(u)du-\int_{-\infty}^{\infty}f(y-u)\chi_\epsilon(u)du \\
&\le f(x+\epsilon)-f(y-\epsilon).
\end{aligned}
$$

By similar arguments for $y\ge x$, we have (iii). □

Here we introduce some local times which play important roles in the proof of Lemma 10.5. Let $L_t^a(X_\bullet^{(\epsilon)})$ be the local time at a of the process $X_\bullet^{(\epsilon)}$ such that

$$
L_t^a(X_\bullet^{(\epsilon)}) := |X_t^{(\epsilon)}-a|-|X_0^{(\epsilon)}-a|-\int_0^t \operatorname{sgn}(X_s^{(\epsilon)}-a)dX_s^{(\epsilon)}, \tag{10.44}
$$

(see Revuz and Yor [8] Chapter 6).

Let $Z_t^{(\epsilon_1,\epsilon_2,\theta)}$ be

$$
Z_t^{(\epsilon_1,\epsilon_2,\theta)} := X_t^{(\epsilon_1)}+\theta(X_t^{(\epsilon_2)}-X_t^{(\epsilon_1)}), \quad 0<\epsilon_1,\epsilon_2\le 1, \quad 0\le\theta\le 1. \tag{10.45}
$$

Let $L_t^a(Z_\bullet^{(\epsilon_1,\epsilon_2,\theta)})$ be the local time at a of the process $Z_\bullet^{(\epsilon_1,\epsilon_2,\theta)}$ such that

$$
\begin{aligned}
L_t^a(Z_\bullet^{(\epsilon_1,\epsilon_2,\theta)}) := |Z_t^{(\epsilon_1,\epsilon_2,\theta)}-a|-|Z_0^{(\epsilon_1,\epsilon_2,\theta)}-a| \\
-\int_0^t \operatorname{sgn}(Z_s^{(\epsilon_1,\epsilon_2,\theta)}-a)dZ_s^{(\epsilon_1,\epsilon_2,\theta)}.
\end{aligned} \tag{10.46}
$$

We have the next lemma.

Lemma 10.6

(i) Let C_L be the constant such that

$$C_L := \sup_{\epsilon\in(0,1]} \sup_{a\in\mathbb{R}} \mathbb{E}[L_T^a(X_\bullet^{(\epsilon)})]. \tag{10.47}$$

Then, $C_L < \infty$ holds and it is independent of ϵ.

(ii) Let $\widetilde{C}_L$ be the constant such that

$$\widetilde{C}_L := \sup_{(\epsilon_1,\epsilon_2,\theta)\in(0,1]\times(0,1]\times[0,1]} \sup_{a\in\mathbb{R}} \mathbb{E}[L_T^a(Z_\bullet^{(\epsilon_1,\epsilon_2,\theta)})]. \tag{10.48}$$

Then, $\widetilde{C}_L < \infty$ holds and it is independent of $(\epsilon_1, \epsilon_2, \theta)$.

Proof (Proof of (ii)) By the definition of $L_t^a(Z_\bullet^{(\epsilon_1,\epsilon_2,\theta)})$, we have

$$\begin{aligned}
0 &\le L_t^a(Z_\bullet^{(\epsilon_1,\epsilon_2,\theta)}) \\
&\le L_T^a(Z_\bullet^{(\epsilon_1,\epsilon_2,\theta)}) \\
&\le |Z_T^{(\epsilon_1,\epsilon_2,\theta)} - Z_0^{(\epsilon_1,\epsilon_2,\theta)}| - \int_0^T \operatorname{sgn}(Z_s^{(\epsilon_1,\epsilon_2,\theta)}) dZ_s^{(\epsilon_1,\epsilon_2,\theta)}.
\end{aligned}$$

Then,

$$\begin{aligned}
(L_T^a(Z_\bullet^{(\epsilon_1,\epsilon_2,\theta)}))^2 &\le 2\Big(\int_0^T (\sigma_{\epsilon_1}(X_s^{(\epsilon_1)}) + \theta(\sigma_{\epsilon_2}(X_s^{(\epsilon_2)}) - \sigma_{\epsilon_1}(X_s^{(\epsilon_1)}))dW_s\Big)^2 \\
&\quad + 2\Big(\int_0^T \operatorname{sgn}(Z_s^{(\epsilon_1,\epsilon_2,\theta)} - a)(\sigma_{\epsilon_1}(X_s^{(\epsilon_1)}) \\
&\quad + \theta(\sigma_{\epsilon_2}(X_s^{(\epsilon_2)}) - \sigma_{\epsilon_1}(X_s^{(\epsilon_1)}))dW_s\Big)^2.
\end{aligned} \tag{10.49}$$

Therefore, we have

$$\begin{aligned}
\mathbb{E}[(L_T^a(Z_\bullet^{(\epsilon_1,\epsilon_2,\theta)}))^2] &\le 2\mathbb{E}\Big[\int_0^T (\sigma_{\epsilon_1}(X_s^{(\epsilon_1)}) + \theta(\sigma_{\epsilon_2}(X_s^{(\epsilon_2)}) - \sigma_{\epsilon_1}(X_s^{(\epsilon_1)})))^2 ds\Big] \\
&\quad + 2\mathbb{E}\Big[\int_0^T (\operatorname{sgn}(Z_s^{(\epsilon_1,\epsilon_2,\theta)}))^2(\sigma_{\epsilon_1}(X_s^{(\epsilon_1)}) \\
&\quad + \theta(\sigma_{\epsilon_2}(X_s^{(\epsilon_2)}) - \sigma_{\epsilon_1}(X_s^{(\epsilon_1)})))^2 ds\Big].
\end{aligned} \tag{10.50}$$

Using the assumption on σ, we have

$$\mathbb{E}[(L_T^a(Z_\bullet^{(\epsilon_1,\epsilon_2,\theta)}))^2] \le 36K^2T < \infty, \tag{10.51}$$

where K is independent of $(a, \epsilon_1, \epsilon_2, \theta)$. This implies immediately (ii). □

Now, the proof of (i) is similar. It is well known that we have following occupation formulas. Let g be a non-negative Borel function. We have

$$\int_0^t g(X_s^{(\epsilon)})d < X_\bullet^{(\epsilon)} >_s = \int_{-\infty}^{\infty} g(a)L_t^a(X_\bullet^{(\epsilon)})da, \tag{10.52}$$

and also

$$\int_0^t g(Z_s^{(\epsilon_1,\epsilon_2,\theta)})d < Z_\bullet^{(\epsilon_1,\epsilon_2,\theta)} >_s = \int_{-\infty}^{\infty} g(a)L_t^a(Z_\bullet^{(\epsilon_1,\epsilon_2,\theta)})da, \tag{10.53}$$

where $< X_\bullet^{(\epsilon)} >$ is the quadratic variation of the process $X_\bullet^{(\epsilon)}$ such that

$$< X_\bullet^{(\epsilon)} >_t := \int_0^t (\sigma_\epsilon(X_s^{(\epsilon)}))^2 ds. \tag{10.54}$$

And also, $< Z_\bullet^{(\epsilon_1,\epsilon_2,\theta)} >_t$ is the quadratic variation of the process $Z_\bullet^{(\epsilon_1,\epsilon_2,\theta)}$ such that

$$< Z_\bullet^{(\epsilon_1,\epsilon_2,\theta)} >_t := \int_0^t (\sigma_{\epsilon_1}(X_s^{(\epsilon_1)}) + \theta(\sigma_{\epsilon_2}(X_s^{(\epsilon_2)}) - \sigma_{\epsilon_1}(X_s^{(\epsilon_1)})))^2 ds, \tag{10.55}$$

(see Revuz and Yor [8] Chapter 6).

Here, we state a lemma which is very useful in the rest of the paper. Let $B \subset [0, T]$ be a Borel set. Leb.B means the Lebesgue measure of the set B.

Lemma 10.7 *We have*

(i) Leb. $\{s\ ;\ 0 \le s \le T,\ X_s^{(\epsilon)} \in D\} = 0$ *(a.s.),*
(ii) Leb. $\{s\ ;\ 0 \le s \le T,\ Z_s^{(\epsilon_1,\epsilon_2,\theta)} \in D\} = 0$ *(a.s.).*

Proof (Proof of (i)) Note that by Lemma 10.5,

$$< X_\bullet^{(\epsilon)} >_t = \int_0^t \sigma_{\epsilon_1}(X_s^{(\epsilon_1)})^2 ds \ge k^2 t, \quad k > 0. \tag{10.56}$$

Since D is a countable set and $a \to L_T^a$ is non-negative continuous,

$$\begin{aligned} k^2 \int_0^T I_D(X_s^{(\epsilon)})ds &\le \int_0^T I_D(X_s^{(\epsilon)})d < X^{(\epsilon)} >_s \\ &= \int_{-\infty}^{\infty} I_D(a)L_T^a(X_\bullet^{(\epsilon)})da \\ &= 0 \quad (a.s.). \end{aligned}$$

This implies (i). □

Proof (Proof of (ii)) Note that for $k > 0$,

$$< Z_\bullet^{(\epsilon_1,\epsilon_2,\theta)} >_t = \int_0^t ((1-\theta)\sigma_{\epsilon_1}(X_s^{(\epsilon_1)}) + \theta\sigma_{\epsilon_2}(X_s^{(\epsilon_2)}))^2 ds \geq k^2 t. \tag{10.57}$$

The similar argument as in the proof of (i) implies (ii), see, for example, Exercise 1.32 p. 237 in Revuz and Yor [8]. Closely related technique to Lemma 10.7 is employed in Hashimoto and Tsuchiya [2]. □

The next lemma is crucial in the proof of Theorem 10.2.

Lemma 10.8 (A Priori Estimates) *For any* α*, there exits* $0 < \beta \leq 1$ *such that* $\forall 0 < \epsilon_1, \epsilon_2 \leq \beta$,

$$\sup_{t\in[0,T]} \mathbb{E}[|X_t^{(\epsilon_1)} - X_t^{(\epsilon_2)}|] \leq \alpha. \tag{10.58}$$

Proof Put

$$\Delta_t^{(\epsilon_1,\epsilon_2)} := X_t^{(\epsilon_1)} - X_t^{(\epsilon_2)}.$$

Let $a_0 = 1 > a_1 > \cdots > a_{n-1} > a_n \cdots$, such that

$$\int_{a_n}^{a_{n-1}} \frac{dx}{x} = n.$$

We choose a sequence $(\rho_n)_{n\in\mathbb{N}}$ of smooth functions such that

$$supp(\rho_n) \subset [a_n, a_{n-1}]; \quad 0 \leq \rho_n(x) \leq \frac{2}{nx}; \quad \int_{a_n}^{a_{n-1}} \rho_n(x)dx = 1.$$

We define a sequence $(\psi_n)_{n\in\mathbb{N}}$ of smooth functions by

$$\psi_n(x) := \int_0^{|x|}\int_0^y \rho_n(u)dudy + a_{n-1}.$$

Then, $(\psi_n)_{n\in\mathbb{N}}$ has the following properties

$$\psi_n(x) \geq |x|; \quad |\psi_n'(x)| \leq 1; \quad \psi_n''(x) = \rho_n(|x|).$$

Moreover, we have

$$
\begin{aligned}
|\Delta_t^{(\epsilon_1,\epsilon_2)}| &\le \psi_n(\Delta_t^{(\epsilon_1,\epsilon_2)}) \\
&= a_{n-1} + \int_0^t \psi_n'(\Delta_s^{(\epsilon_1,\epsilon_2)})[\sigma_{\epsilon_1}(X_s^{(\epsilon_1)}) - \sigma_{\epsilon_2}(X_s^{(\epsilon_2)})]dW_s \\
&\quad + \frac{1}{2}\int_0^t \psi_n''(\Delta_s^{(\epsilon_1,\epsilon_2)})[\sigma_{\epsilon_1}(X_s^{(\epsilon_1)}) - \sigma_{\epsilon_2}(X_s^{(\epsilon_2)})]^2 ds.
\end{aligned}
$$

Since $|\psi_n'(x)| \le 1$, σ_{ϵ_1} and σ_{ϵ_2} are bounded, then

$$
\int_0^t \psi_n'(\Delta_s^{(\epsilon_1,\epsilon_2)})[\sigma_{\epsilon_1}(X_s^{(\epsilon_1)}) - \sigma_{\epsilon_2}(X_s^{(\epsilon_2)})]dW_s \tag{10.59}
$$

is a martingale with mean 0. Therefore, we have

$$
\begin{aligned}
\mathbb{E}[|\Delta_t^{(\epsilon_1,\epsilon_2)}|] &\le a_{n-1} + \frac{1}{2}\mathbb{E}[\psi_n''(\Delta_s^{(\epsilon_1,\epsilon_2)})[\sigma_{\epsilon_1}(X_s^{(\epsilon_1)}) - \sigma_{\epsilon_2}(X_s^{(\epsilon_2)})]^2 ds] \\
&\le a_{n-1} + \frac{3}{2}\mathbb{E}[\int_0^t \rho_n(|\Delta_s^{(\epsilon_1,\epsilon_2)}|)][\sigma_{\epsilon_1}(X_s^{(\epsilon_1)}) - \sigma(X_s^{(\epsilon_1)})]^2 ds] \\
&\quad + \frac{3}{2}\mathbb{E}[\int_0^t \rho_n(|\Delta_s^{(\epsilon_1,\epsilon_2)}|)[\sigma_{\epsilon_2}(X_s^{(\epsilon_2)}) - \sigma(X_s^{(\epsilon_2)})]^2 ds] \\
&\quad + \frac{3}{2}\mathbb{E}[\int_0^t \rho_n(|\Delta_s^{(\epsilon_1,\epsilon_2)}|)[\sigma(X_s^{(\epsilon_1)}) - \sigma(X_s^{(\epsilon_2)})]^2 ds].
\end{aligned}
$$

Here, we define

$$
J_t^{(\epsilon_1,\epsilon_2)}(1) := \mathbb{E}[\int_0^t \rho_n(|\Delta_s^{(\epsilon_1,\epsilon_2)}|)][\sigma_{\epsilon_1}(X_s^{(\epsilon_1)}) - \sigma(X_s^{(\epsilon_1)})]^2 ds], \tag{10.60}
$$

$$
J_t^{(\epsilon_1,\epsilon_2)}(2) := \mathbb{E}[\int_0^t \rho_n(|\Delta_s^{(\epsilon_1,\epsilon_2)}|)[\sigma_{\epsilon_2}(X_s^{(\epsilon_2)}) - \sigma(X_s^{(\epsilon_2)})]^2 ds], \tag{10.61}
$$

and

$$
J_t^{(\epsilon_1,\epsilon_2)}(3) := \mathbb{E}[\int_0^t \rho_n(|\Delta_s^{(\epsilon_1,\epsilon_2)}|)[\sigma(X_s^{(\epsilon_1)}) - \sigma(X_s^{(\epsilon_2)})]^2 ds]. \tag{10.62}
$$

Now, we remember that

$$
\rho_n(|x|) \le I_{[a_n,a_{n-1}]}(|x|)\frac{2}{n|x|},
$$

and

$$\int_{a_n}^{a_{n-1}} \rho_n(u)du = 1.$$

Consider $J_t^{(\epsilon_1,\epsilon_2)}(1)$ and $J_t^{(\epsilon_1,\epsilon_2)}(2)$, we have

$$J_t^{(\epsilon_1,\epsilon_2)}(1) \leq \frac{2}{na_n}\mathbb{E}[\int_0^t [\sigma_{\epsilon_1}(X_s^{(\epsilon_1)}) - \sigma(X_s^{(\epsilon_1)})]^2 ds], \tag{10.63}$$

and also

$$J_t^{(\epsilon_1,\epsilon_2)}(2) \leq \frac{2}{na_n}\mathbb{E}[\int_0^t [\sigma_{\epsilon_2}(X_s^{(\epsilon_2)}) - \sigma(X_s^{(\epsilon_2)})]^2 ds]. \tag{10.64}$$

Now, we consider the term $J_\bullet^{(\epsilon_1,\epsilon_2)}(3)$. By Eq. (10.27), we note that

$$J_t^{(\epsilon_1,\epsilon_2)}(3) \leq \mathbb{E}[\int_0^t \rho_n(|\Delta_s^{(\epsilon_1,\epsilon_2)}|)|f(X_s^{(\epsilon_1)}) - f(X_s^{(\epsilon_2)})|ds]. \tag{10.65}$$

Let $\widetilde{J}_t^l = J_t^{(\epsilon_1,\epsilon_2,l)}(3)$ be

$$\widetilde{J}_t^l = J_t^{(\epsilon_1,\epsilon_2,l)}(3) := \mathbb{E}[\int_0^t \rho_n(|\Delta_s^{(\epsilon_1,\epsilon_2)}|)|f_l(X_s^{(\epsilon_1)}) - f_l(X_s^{(\epsilon_2)})|ds].$$

By Hadamard formula;

$$f_l(x) - f_l(y) = (x-y)\int_0^1 f_l'(x+\theta(y-x))d\theta, \tag{10.66}$$

we have

$$\begin{aligned}
\widetilde{J}_t^l &= \mathbb{E}[\int_0^t \rho_n(|\Delta_s^{(\epsilon_1,\epsilon_2)}|)|f_l(X_s^{(\epsilon_1)}) - f_l(X_s^{(\epsilon_2)})|ds] \\
&\leq 2\mathbb{E}[\int_0^t I_{[a_n,a_{n-1}]}(|\Delta_s^{(\epsilon_1,\epsilon_2)}|)\frac{|f_l(X_s^{(\epsilon_1)}) - f_l(X_s^{(\epsilon_2)})|}{n|\Delta_s^{(\epsilon_1,\epsilon_2)}|}ds] \\
&= \frac{2}{n}\mathbb{E}[\int_0^t \int_0^1 f_l'(X_s^{(\epsilon_1)} + \theta(X_s^{(\epsilon_2)} - X_s^{(\epsilon_1)}))d\theta ds].
\end{aligned}$$

Let g be a non-negative Borel function. We have the occupation formula

$$\int_0^t g(Z_s^{(\epsilon_1,\epsilon_2,\theta)})d < Z_\bullet^{(\epsilon_1,\epsilon_2,\theta)} >_s = \int_{-\infty}^{\infty} g(a)L_t^a(Z_\bullet^{(\epsilon_1,\epsilon_2,\theta)})da. \tag{10.67}$$

Since we know that

$$< Z_\bullet^{(\epsilon_1,\epsilon_2,\theta)} >_t \geq k^2 t,$$

we obtain

$$\begin{aligned}\widetilde{J}_t^l &\leq \frac{2}{n}\mathbb{E}[\int_0^1 d\theta \int_0^t f_l'(Z_s^{(\epsilon_1,\epsilon_2,\theta)})ds] \\ &\leq \frac{2}{nk^2}\mathbb{E}[\int_0^1 d\theta \int_{-\infty}^{\infty} L_t^a(Z_\bullet^{(\epsilon_1,\epsilon_2,\theta)}) f_l'(a)da].\end{aligned}$$

For $\theta \in [0, 1]$, $0 < \epsilon_1, \epsilon_2 \leq 1$, we have

$$\begin{aligned}\widetilde{J}_t^l &\leq \frac{2}{n} \cdot \frac{\widetilde{C}_L}{k^2} \int_{-\infty}^{\infty} f_l'(a)da \\ &\leq \frac{2}{n} \cdot \frac{\widetilde{C}_L}{k^2} v(f_l) \\ &\leq \frac{2}{n} \cdot \frac{\widetilde{C}_L}{k^2} v(f).\end{aligned} \tag{10.68}$$

Since

$$\lim_{l\to\infty} f_l = f(x), \quad x \notin D \tag{10.69}$$

and

$$Leb.\{s\,;\ 0 \leq s \leq T,\ X_s^{(\epsilon_1)} \in D \ or\ X_s^{(\epsilon_2)} \in D\} = 0 \quad (a.s.), \tag{10.70}$$

we have

$$\lim_{l\to\infty} |f_l(X_s^{(\epsilon_1)}) - f_l(X_s^{(\epsilon_2)})| = |f(X_s^{(\epsilon_1)}) - f(X_s^{(\epsilon_2)})| \tag{10.71}$$

almost surely on $[0, T] \times \Omega$, w.r.t. $dt \times d\mathbb{P}$. Note that f and f_l are uniformly bounded. We have

$$\lim_{l\to\infty} \widetilde{J}^{(\epsilon_1,\epsilon_2,l)} = \mathbb{E}[\int_0^t \rho_n(|\Delta_s^{(\epsilon_1,\epsilon_2)}|)|f(X_s^{(\epsilon_1)}) - f(X_s^{(\epsilon_2)})|ds]. \tag{10.72}$$

By the inequalities (10.68), we obtain

$$J_t^{(\epsilon_1,\epsilon_2)}(3) \leq \frac{2}{n} \cdot \frac{\widetilde{C}_L}{k^2} v(f). \tag{10.73}$$

Finally, we will come back to estimate $\mathbb{E}[|\Delta_t^{(\epsilon_1,\epsilon_2)}|]$. We know that

$$\mathbb{E}[|\Delta_t^{(\epsilon_1,\epsilon_2)}|] \leq a_{n-1} + \frac{3}{2}J_t^{(\epsilon_1,\epsilon_2)}(1) + \frac{3}{2}J_t^{(\epsilon_1,\epsilon_2)}(2) + \frac{3}{2}J_t^{(\epsilon_1,\epsilon_2)}(3).$$

By (10.63) and also by (10.64) we obtain that

$$\begin{aligned}\frac{3}{2}J_t^{(\epsilon_1,\epsilon_2)}(1) + \frac{3}{2}J_t^{(\epsilon_1,\epsilon_2)}(2) &\leq \frac{3}{na_n}\mathbb{E}[\int_0^t [\sigma_{\epsilon_1}(X_s^{(\epsilon_1)}) - \sigma(X_s^{(\epsilon_1)})]^2 ds] \\ &+ \frac{3}{na_n}\mathbb{E}[\int_0^t [\sigma_{\epsilon_2}(X_s^{(\epsilon_2)}) - \sigma(X_s^{(\epsilon_2)})]^2 ds].\end{aligned} \tag{10.74}$$

Let $\alpha > 0$ be given, choose n such that

$$a_{n-1} < \frac{\alpha}{3},$$

and also

$$\frac{3}{2}J_t^{(\epsilon_1,\epsilon_2)}(3) \leq \frac{3}{n} \cdot \frac{\widetilde{C}_L}{k^2}v(f) < \frac{\alpha}{3}. \tag{10.75}$$

For this n, we have

$$\begin{aligned}\frac{3}{2}J_t^{(\epsilon_1,\epsilon_2)}(1) + \frac{3}{2}J_t^{(\epsilon_1,\epsilon_2)}(2) &\leq \frac{3}{na_n}(\mathbb{E}[\int_0^t (\sigma_{\epsilon_1}(X_s^{(\epsilon_1)}) - \sigma(X_s^{(\epsilon_1)}))^2 ds] \\ &+ \mathbb{E}[\int_0^t (\sigma_{\epsilon_2}(X_s^{(\epsilon_2)}) - \sigma(X_s^{(\epsilon_2)}))^2 ds]) \\ &\leq \frac{3}{na_nk^2}(\mathbb{E}[\int_{-\infty}^{\infty} (\sigma_{\epsilon_1}(a) - \sigma(a))^2 L_t^a(X_\cdot^{(\epsilon_1)}) da] \\ &+ \mathbb{E}[\int_{-\infty}^{\infty} (\sigma_{\epsilon_2}(a) - \sigma(a))^2 L_t^a(X_\cdot^{(\epsilon_2)}) da]).\end{aligned}$$

As is well known, the local time $L_t^a(X_\cdot^{(\epsilon_1)})$ can be written as $L_t^a(X_\cdot^{(\epsilon_1)}) = L^a_{<X^{(\epsilon_1)}>_t}(B_\cdot)$, where B is called the Dambis–Dubins–Schwarz Brownian motion. See, Chapter 5 and Chapter 6 in Revuz and Yor [8]. Note that

$$< X^{(\epsilon_1)} >_t \leq k^2 t$$

and

$$0 \leq L^a_{<X^{(\epsilon_1)}>_t}(B_\cdot) \leq L^a_{k^2 t}(B_\cdot),$$

hold. Then we obtain

$$\mathbb{E}[\int_{-\infty}^{\infty}(\sigma_{\epsilon_1}(a)-\sigma(a))^2 L_t^a(X_\cdot^{(\epsilon_1)})da] \leq \mathbb{E}[\int_{-\infty}^{\infty}(\sigma_{\epsilon_1}(a)-\sigma(a))^2 L_{k^2 t}^a(B_\cdot)da]$$
$$\leq 4k^2 \cdot k^2 t.$$

Since $a \mapsto L_{k^2 t}^a$ is a continuous function with a compact support a.s.,

$$\lim_{\epsilon_1 \to 0} \int_{-\infty}^{\infty}(\sigma_{\epsilon_1}(a)-\sigma(a))^2 L_{k^2 t}^a(B_\cdot)da = 0, \tag{10.76}$$

holds a.s.. By Lebesgue convergence theorem, we obtain

$$\lim_{\epsilon_1 \to 0} \mathbb{E}[\int_{-\infty}^{\infty}(\sigma_{\epsilon_1}(a)-\sigma(a))^2 L_t^a(X_\cdot^{(\epsilon_1)})da]$$
$$\leq \lim_{\epsilon_1 \to 0} \mathbb{E}[\int_{-\infty}^{\infty}(\sigma_{\epsilon_1}(a)-\sigma(a))^2 L_{k^2 t}^a(B_\cdot)da]$$
$$= 0.$$

Thus we have proved Lemma 10.8. □

Proof (Proof of Theorem 10.2) In this part, we use the Notation and some basic arguments on functional analysis employed in the Sect. 10.4. We define $\epsilon_n := 1/n$ and write $X^{(n)}$ instead of $X^{(\epsilon_n)}$. Since $0 < k \leq \sigma_{\epsilon_n} \leq K < \infty$, there exists $C > 0$ such that for $n \in \mathbb{N}$

$$\sup_{t \in [0,T]} \mathbb{E}[|X_t^{(n)}|^4] \leq C. \tag{10.77}$$

This result is called a priori estimate in the High Norm.

Lemma 10.9 *$(X^{(n)})_{n \in \mathbb{N}}$ is a Cauchy sequence in N^2.*

□

Proof Let for some subsequences $(p_n)_{n \in \mathbb{N}}$, $(q_n)_{n \in \mathbb{N}}$,

$$\lim_{n \to \infty} \sup_{0 \leq t \leq T} \mathbb{E}[|X_t^{(p_n)} - X_t^{(q_n)}|^2] = C > 0 \tag{10.78}$$

holds. Since

$$\lim_{n \to \infty} ||X^{(p_n)} - X^{(q_n)}||_1 = 0,$$

we can choose subsequences $(p'_n)_{n\in\mathbb{N}}$, $(q'_n)_{n\in\mathbb{N}}$ such that

$$\lim_{n\to\infty} |X_T^{(p'_n)} - X_T^{(q'_n)}| = 0 \quad (a.s.).$$

Using

$$\sup_n \mathbb{E}[|X_T^{(p'_n)} - X_T^{(q'_n)}|^4] < \infty,$$

the family of variables $|X_T^{(p'_n)} - X_T^{(q'_n)}|^2$ is uniformly integrable. Note that $X_t^{(p'_n)} - X_t^{(q'_n)}$ $(0 \le t \le T)$ is a martingale, we have by Doob's maximal inequality

$$\begin{aligned} ||X^{(p'_n)} - X^{(q'_n)}||_2^2 &\le \mathbb{E}[\sup_{0\le t\le T} |X_t^{(p'_n)} - X_t^{(q'_n)}|^2] \\ &\le 4\mathbb{E}[|X_T^{(p'_n)} - X_T^{(q'_n)}|^2]. \end{aligned} \tag{10.79}$$

Thus, we observe that

$$\lim_{n\to\infty} ||X^{(p'_n)} - X^{(q'_n)}||_2 = 0. \tag{10.80}$$

This fact contradicts Eq. (10.78). □

Therefore, the sequence $(X^{(n)})_{n\in\mathbb{N}}$ converges to some $\widetilde{X} \in N^2$. Moreover, we observe that

$$\lim_{n\to\infty} (\| X^{(n)} - \widetilde{X} \|_1 + \| X^{(n)} - \widetilde{X} \|_2) = 0. \tag{10.81}$$

Let

$$RHS_t^{(n)} := X_0 + \int_0^t \sigma_{\epsilon_n}(X_s^{(n)}) dW_s. \tag{10.82}$$

From Ito theory, we know that the strong solutions $X^{(n)}$ satisfy their respective SDEs in the following sense

$$\mathbb{E}[\sup_{0\le t\le T} |X_t^{(n)} - RHS_t^{(n)}|^2] = 0. \tag{10.83}$$

This implies

$$\| X_t^{(n)} - RHS^{(n)} \|_2^2 = 0. \tag{10.84}$$

Since $X^{(n)} \in N^2$ $(n = 1, 2, \cdots)$ is a sequence of martingales, there exists a martingale version of the process $\widetilde{X}$. Let X be a martingale version of $\widetilde{X}$. Here, let

$$RHS_t := X_0 + \int_0^t \sigma(X_s) dW_s. \tag{10.85}$$

Lemma 10.10

$$Leb.\{s\,;\ 0 \le s \le T,\ X_s \in D\} = 0 \quad (a.s.) \tag{10.86}$$

holds.

□

Proof Since the sequence of $< X^{(n)} >_t$

$$k^2 t \le < X^{(n)} >_t, \quad n = 1, 2, \cdots, \quad (a.s.)$$

converges to

$$< X >_t, \quad 0 \le t \le T, \quad (a.s.),$$

we have

$$k^2 t \le < X >_t, \quad 0 \le t \le T, \quad (a.s.). \tag{10.87}$$

Let $L_t^a(X)$ be the local time at a of X. We have

$$\begin{aligned} k^2 \int_0^T I_D(X_s) ds &\le \int_0^T I_D(X_s) d < X >_s \\ &= \int_{-\infty}^{\infty} L_T^a(X) I_D(a) da. \end{aligned} \tag{10.88}$$

Let

$$\widehat{C}_L := \sup_{a \in \mathbb{R}} \mathbb{E}[L_T^a(X)]. \tag{10.89}$$

By the argument employed in the proof of Lemma 10.6, we can prove that $\widehat{C}_L < \infty$ holds. Thus, we have

$$\begin{aligned} \mathbb{E}[\int_0^T I_D(X_s) ds] &\le \frac{1}{k^2} \mathbb{E}[\int_{-\infty}^{\infty} L_T^a(X) I_D(a) da] \\ &\le \frac{\widehat{C}_L}{k^2} \int_{-\infty}^{\infty} I_D(a) da \\ &= 0. \end{aligned}$$

This implies

$$Leb.\{s\,;\ 0 \le s \le T,\ X_s \in D\} = 0 \quad (a.s.). \tag{10.90}$$

Now, we will show that $RHS^{(n)}$ converges RHS in N^2. Observe using Lemma 10.5 (iii) that

$$\begin{aligned}
&\lim_{n\to\infty} \parallel RHS^{(n)} - RHS \parallel_2^2 \\
&= \lim_{n\to\infty} \sup_{t\in[0,T]} \mathbb{E}[|\int_0^t \sigma_{\epsilon_n}(X_s^{(n)}) - \sigma(X_s)dW_s|^2] \\
&= \lim_{n\to\infty} \mathbb{E}[\int_0^T \left\{\sigma_{\epsilon_n}(X_s^{(n)}) - \sigma(X_s)\right\}^2 ds] \\
&= \lim_{n\to\infty} \mathbb{E}[\int_0^T \left\{\sigma_{\epsilon_n}(X_s^{(n)}) - \sigma_{\epsilon_n}(X_s) + \sigma_{\epsilon_n}(X_s) - \sigma(X_s)\right\}^2 ds] \\
&\le 2 \lim_{n\to\infty} \mathbb{E}[\int_0^T |\sigma_{\epsilon_n}(X_s^{(n)}) - \sigma_{\epsilon_n}(X_s)|^2 ds] \\
&\quad + 2 \lim_{n\to\infty} \mathbb{E}[\int_0^T |\sigma_{\epsilon_n}(X_s) - \sigma(X_s)|^2 ds] \\
&\le 2 \lim_{n\to\infty} \mathbb{E}[\int_0^T |f(X_s^{(n)} \vee X_s + \epsilon_n) - f(X_s^{(n)} \wedge X_s - \epsilon_n)| ds] \\
&\quad + 2 \lim_{n\to\infty} \mathbb{E}[\int_0^T (\sigma_{\epsilon_n}(X_s) - \sigma(X_s))^2 ds].
\end{aligned}$$

Let

$$S_1^{(n)}(T) := \mathbb{E}[\int_0^T |f(X_s^{(n)} \vee X_s + \epsilon_n) - f(X_s^{(n)} \wedge X_s - \epsilon_n)| ds], \tag{10.91}$$

and also

$$S_2^{(n)}(T) := \mathbb{E}[\int_0^T (\sigma_{\epsilon_n}(X_s) - \sigma(X_s))^2 ds]. \tag{10.92}$$

By Doob's maximal inequality, we have

$$\begin{aligned}
\mathbb{E}[\sup_{0\le s\le T} |X_s^{(n)} - X_s|^2] &\le \mathbb{E}[|X_T^{(n)} - X_T|^2] \\
&\le 4||X^{(n)} - X||_2^2.
\end{aligned} \tag{10.93}$$

This implies

$$\lim_{n\to\infty} \sup_{0\le s\le T} |X_s^{(n)} - X_s| = 0 \quad (a.s.). \tag{10.94}$$

By Lemma 10.10, we have

$$Leb.\{s\,;\ 0 \le s \le T,\ X_s \in D\} = 0 \quad (a.s.). \tag{10.95}$$

Note that $X_s^{(n)} \vee X_s + \epsilon_n$ and $X_s^{(n)} \wedge X_s - \epsilon_n$ converge to X_s. We observe that

$$Leb.\{s\,;\ 0 \le s \le T,\ \lim_{n\to\infty} |f(X_s^{(n)} \vee X_s + \epsilon_n) - f(X_s^{(n)} \wedge X_s - \epsilon_n)|$$
$$\textit{does not converge to } 0\} = 0 \quad (a.s.).$$

Since f is a bounded function, we have

$$\lim_{n\to\infty} S_1^{(n)}(T) = 0. \tag{10.96}$$

For $S_2^{(n)}(T)$, we have

$$\begin{aligned} S_2^{(n)}(T) &= \mathbb{E}[\int_0^T (\sigma_{\epsilon_n}(X_s) - \sigma(X_s))^2 ds] \\ &\le \frac{1}{k^2}\mathbb{E}[\int_0^T (\sigma_{\epsilon_n}(X_s) - \sigma(X_s))^2 d < X >_s] \\ &\le \frac{1}{k^2}\mathbb{E}[\int_{-\infty}^{\infty} (\sigma_{\epsilon_n}(a) - \sigma(a))^2 L_T^a(X.)da]. \end{aligned}$$

Since $(\sigma_{\epsilon_n}(a) - \sigma(a))^2$ is uniformly bounded by $4k^2$, and $a \mapsto L_T^a(X.)$ is a continuous function with a compact support a.s., we have

$$\lim_{n\to\infty} \int_{-\infty}^{\infty} (\sigma_{\epsilon_n}(a) - \sigma(a))^2 L_T^a(X.)da = 0 \quad (a.s.).$$

By Lebesgue convergence theorem, we can conclude

$$\lim_{n\to\infty} S_2^{(n)}(T) = 0. \tag{10.97}$$

Therefore, we have

$$\lim_{n\to\infty} ||RHS^{(n)} - RHS||_2^2 = 0. \tag{10.98}$$

Since $X^{(n)} = RHS^{(n)}$ converges to $\widetilde{X}$ in N^2, $X^{(n)} = RHS^{(n)}$ converges to X in N^2. We have

$$||X - RHS||_2^2 = 0. \tag{10.99}$$

Note that X is a martingale having continuous paths,

$$\begin{aligned} \mathbb{E}[\sup_{0\le s\le T} |X_t - RHS_t|^2] &\le 4\mathbb{E}[|X_T - RHS_T|^2] \\ &\le 4||X - RHS||_2^2 \\ &= 0. \end{aligned} \tag{10.100}$$

Thus, X is the desired strong solution. This concludes the proof of Theorem 10.2. □

Acknowledgements This work is motivated by stimulating discussions with Toshio Yamada. Also, we received valuable comments from Tomoyuki Ichiba. The author would like to thank them very much. Finally, we are grateful to the anonymous referee who made significant suggestions.

References

1. S. Ebenfeld, Energy methods for stochastic differential equations. Int. J. Probab. Stoch. Proces. **82**(3), 231–239 (2010)
2. H. Hashimoto, T. Tsuchiya, Stability problems for Cantor stochastic differential equations. Stoch. Proc. Appl. **128**(1), 211–232 (2018)
3. K. Ito, On stochastic differential equations. Mem. Am. Math. Sor. **4**, 1–51 (1951)
4. I. Karatzas, S.E. Shreve, *Brownian Motion and Stochastic Calculus* (Springer, New York, 1991)
5. H. Kaneko, S. Nakao, A note on approximation for stochastic differential equations. Sémin. Probab. Strasbourg **22**, 155–162 (1988)
6. J.F. Le Gall, Applications du temps local aux équations différentielles stochastiques unidimensionnelles, in *Séminaire de probabilités XVII*. Lecture Notes in Mathematics, vol. 986 (Springer, Berlin, 1983), pp. 15–31
7. S. Nakao, On the pathwise uniqueness of solutions of one-dimensional stochastic differential equations. Osaka J. Math. **9**, 513–518 (1972)
8. D. Revuz, M. Yor, *Continuous Martingales and Brownian Motion* (Springer, New York, 1991)
9. L.C.G. Rogers, D. Williams, *Diffusions, Markov Processes, and Martingales* (Cambridge University Press, Cambridge, 1986)
10. A.V. Skorohod, *Studies in the Theory of Ramdom Processes* (Addison-Wesley, Reading, 1965). Reprinted by Dover Publications, New York
11. D.W. Stroock, S.R.S. Varadhan, *Multidimensional Diffusion Processes* (Springer, New York, 1979)
12. T. Yamada, Sur une construction des solutions d'equations differentielles stochastiques dans le cas non-lipschitzien. Sémin. Probab. Strasbourg **12**, 114–131 (1978)
13. T. Yamada, S. Watanabe, On the uniqueness of solutions of stochastic differential equations. J. Math. Kyoto Univ. **11**, 155–167 (1971)

Chapter 11
Heat Kernel Coupled with Geometric Flow and Ricci Flow

Koléhè A. Coulibaly-Pasquier

Abstract We prove on-diagonal upper bound for the minimal fundamental solution of the heat equation evolving under geometric flow. In the case of Ricci flow, with non-negative Ricci curvature and a condition on the growth of volume of ball for the initial manifold, we derive Gaussian bounds for the minimal fundamental solution of the heat equation, and then for the conjugate heat equation.

11.1 Introduction

Let $(M, g(t))$ be a complete Riemannian manifold, either non compact or compact without boundary, $g(t)$ be a family of metrics on M, $\nabla^{g(t)}$ and $\Delta_{g(t)}$ the corresponding gradient and Laplace-Beltrami operator, $\mathrm{Ric}_{g(t)}$ the corresponding Ricci curvature, $\mu_{g(t)}$ the corresponding Riemannian volume, $d_{g(t)}(x, y)$ be the distance function, and $B_{g(t)}(x, r)$ the geodesic ball of radius r for the distance $d_{g(t)}$. Sometimes to reduce the notation, when there are no risk of confusion concerning the family of metric we simply write ∇^t, Δ_t, μ_t, ...

Let $\alpha_{i,j}(t)$ be a family of symmetric 2-tensors on M. We consider the following heat equation coupled with a geometric flow.

$$\begin{cases} \partial_t g_{i,j}(t) = \alpha_{i,j}(t), \\ \partial_t f(t,x) = \frac{1}{2}\Delta_t f(t,x), \\ f(0,x) = f_0(x). \end{cases} \tag{11.1}$$

We are interested in estimating the minimal fundamental solution of (11.1). For the existence of minimal fundamental solution in non compact case we refer to Chapter 24 of [9]. An estimate of this fundamental solution, already give an estimate

K. A. Coulibaly-Pasquier (✉)
Institut Élie Cartan de Lorraine, UMR 7502, Université de Lorraine and CNRS, Villers-lès-Nancy, France
e-mail: kolehe.coulibaly@univ-lorraine.fr

C. Donati-Martin et al. (eds.), *Séminaire de Probabilités L*, Lecture Notes in Mathematics 2252, https://doi.org/10.1007/978-3-030-28535-7_11

of the conjugate heat equation, which is the density of the $g(t)$-Brownian motion introduced in [1, 10] see also [11]. Moreover estimate of the fundamental solution of the heat equation have many geometric applications, both in constant metric case and geometric flows for instance in [5, 8, 21, 23].

Such a flow have been investigated in the literature. We mention the following situations.

- The most famous case is when $\alpha_{i,j}(t) := 0$. This is the case of constant metric and Eq. (11.1) is the usual heat equation in M.
- The Ricci flow corresponds to $\alpha(t) := -\operatorname{Ric}_{g(t)}$.
- We can also consider $\alpha_{i,j}(t) := -2hH_{i,j}(g(t))$, where $H_{i,j}(g(t))$ is the second fundamental form according to the metric $g(t)$, and h is the mean curvature, when the family of metric derives from the mean curvature flow.

The existence of the Ricci flow $\partial_t g(t) = -\operatorname{Ric}_{g(t)}$ for compact manifold was proved in [18]. Under additional assumptions, the existence of the Ricci flow for a complete manifold was proved in [24]. For the last example, the existence result of the mean curvature flow for a compact manifold could be found in [12].

Using stochastic calculus we prove on-diagonal upper bound for the minimal fundamental solution of the heat equation (11.1), for general geometric flow. As far as we know this result is new. We derive a Gaussian upper bound for the minimal heat kernel coupled with the Ricci flow, in the case of positive Ricci curvature and condition on the growth of volume of ball for the initial manifold (i.e Hypothesis H1 in Theorem 11.3).

Related Results Previous stochastic proof of Harnack inequality with power appear in [3] for the constant metric case. Note that our coupling is different from the coupling in [3], and simplify the argument since we do not need to take care of different cutlocus. The Harnack inequality with power we obtain also appears in [7], and is obtain by different way.

For the Ricci flow, Gaussian upper bounds could be found as example in [23] where the author use Harnack inequality and doubling volume property. An over one by Zhang and Cao [5] uses Sobolev type inequality that is conserved along Ricci flow.

Outline The paper is organized as follows. In Sect. 11.2 we define a horizontal coupling. We use this coupling and Girsanov's Theorem in order to generalize Harnack inequality with power—for inhomogeneous heat equation—introduced by Wang [25] see also [3, 14]. We also use this coupling to give some isoperimetric-type Harnack inequality in Lemma 11.1 and ultracontractivity of the heat kernel in Corollary 11.3.

In Sect. 11.3, since the heat kernel of (11.1) is in general non symmetric, the Gaussian bound is not a direct consequence of Harnack inequality with power as in [14]. To overcome this difficulty we use the dual process and derive on-diagonal upper estimate of the heat kernel of (11.1) in Theorem 11.2.

Section 11.4 is devoted to the case of Ricci flow. We use modification of Grigor'yan trick to derive Gaussian Heat kernel bounds from the on-diagonal upper bound. The principal result of this section is Theorem 11.3 and Corollary 11.8.

11.2 Coupling and Harnack Inequality with Power

11.2.1 Coupling

In the first part of this section, we focus on the operator of type $L_t := \frac{1}{2}\Delta_{g(t)}$, where $\Delta_{g(t)}$ is the Laplace operator associated to a time dependent family of metrics $g(t)_{t\in[0,T_c[}$. We suppose that $(M, g(t))$ is complete for all $t \in [0, T_c[$. Let $x \in M$ and $t \mapsto X_t(x)$ be the $g(t)$-Brownian motion started at x. The notion of $g(t)$-Brownian motion, i.e. a L_t diffusion, parallel transport, and damped parallel transport has been given in [1, 10]. We also suppose in this section that all $g(t)$-Brownian motion is non-explosive (i.e. stochastically complete).

Since we use different family of metrics all construction depends on the family of metrics.

Let $//_t^{g(t),X_\cdot(x)}$ be the $g(t)$ parallel transport above $t \mapsto X_t(x)$, which is a linear isometry

$$
\begin{aligned}
//_t^{g(t),X_\cdot(x)} &: (T_xM, g(0)) \longrightarrow (T_{X_t(x)}M, g(t)) \\
//_0^{g(t),X_\cdot(x)} &= Id_{T_xM}
\end{aligned}
$$

Let $\mathbf{W}_t^{g(t),X_\cdot(x)}$ be the damped parallel transport that satisfies the following Stratonovich covariant equation:

$$
\begin{aligned}
&* d(//_t^{g(t),X_\cdot(x)})^{-1}(\mathbf{W}_t^{g(t),X_\cdot(x)}) \\
&= -\frac{1}{2}(//_t^{g(t),X_\cdot(x)})^{-1}\Big(\mathrm{Ric}_{g(t)} - \partial_t g(t)\Big)^{\#g(t)}(\mathbf{W}_{0,t}^{g(t),X_\cdot(x)})\, dt.
\end{aligned}
$$

It is a linear operator between:

$$
\begin{aligned}
\mathbf{W}_t^{g(t),X_\cdot(x)} &: T_xM \longrightarrow T_{X_t(x)}M \\
\mathbf{W}_0^{g(t),X_\cdot(x)} &= \mathrm{Id}_{T_xM}\,.
\end{aligned}
$$

In [2] we give a construction of a process with value in a space of curves. Since we sometimes change the underlying family of metrics, we incorporate this family of metrics in the notation.

Let $x, y \in M$, $u \mapsto \gamma(u)$ be a $g(0)$ geodesic curve such that $\gamma(0) = x$ and $\gamma(1) = y$ and $t \mapsto (X_t(u)_{u\in[0,1]})$ be the horizontal L_t-diffusion in C^1 path space

$C^1([0,1], M)$ over $X_t(x)$ that starts at γ, where $X_t(x)$ is a $g(t)$-Brownian motion that starts at x. By assumption it is defined for all $t \in [0, T]$ with $T < T_c$.

We will recall the usual properties satisfied by the horizontal L_t-diffusion in C^1 path space Theorem 3.1 [2]:

The family

$$u \mapsto (X_t(u))_{t\in[0,T]}$$

is a family of L_t-diffusions. It is a.s. continuous in (t, u) and C^1 in u, satisfies

$$X_t(0) = X_t(x) \text{ and } X_0(u) = \gamma(u),$$

and solves the equation

$$\partial_u X_t(u) = \mathbf{W}_t^{g(t),X(u)}(\dot{\gamma}(u)). \tag{11.2}$$

Furthermore, $X_.(u)$ satisfies the following Itô stochastic differential equation

$$d^{\nabla_t} X_t(u) = P_{0,u}^{g(t),X_t(.)} d^{\nabla_t} X_t(0), \tag{11.3}$$

where

$$P_{0,u}^{g(t),X_t(.)} : T_{X_t(0)}M \to T_{X_t(u)}M$$

denotes usual parallel transport along the C^1-curve

$$[0, u] \to M, \quad v \mapsto X_t(v),$$

with respect to the metric $g(t)$.

We often use the notation $//_t$ for $//_t^{g(t),X_.(x)}$ when there no risk of confusion of the underling process and the family of metrics.

Proposition 11.1 *Suppose that the $g(t)$-Brownian motion starting at x is non-explosive. The diagonal process $t \mapsto X_t(\frac{t}{T})$ satisfies the following stochastic differential equation:*

$$d^{\nabla_t}(X_.(\frac{\cdot}{T}))_t = P_{0,\frac{t}{T}}^{g(t),X_t(.)} d^{\nabla_t} X_t(0) + \frac{1}{T}\mathbf{W}_t^{g(t),X_.(\frac{t}{T})}\dot{\gamma}(\frac{t}{T})\, dt$$

Proof We pass to the Stratonovich differential to obtain the following chain rule formula at time t_0:

$$*d(X_.(\frac{\cdot}{T}))_{t_0} = *d(X_.(\frac{t_0}{T}))_{t_0} + \frac{dX_{t_0}(\frac{t}{T})}{dt}\Big|_{t=t_0} dt_0.$$

We use (11.2) to identify the last term of the right hand side:

$$\frac{dX_{t_0}(\frac{t}{T})}{dt}|_{t=t_0} = \frac{1}{T}\mathbf{W}_{t_0}^{g(t),X_.(\frac{t_0}{T})}(\dot{\gamma}(\frac{t_0}{T})).$$

Now we come back to the Itô differential equation using the following relation:

$$d^{\nabla_t} Y_t = //_t^{Y_.}\left(d\int_0^t (//_s^{Y_.})^{-1} * dY_s\right),$$

and we obtain

$$\begin{aligned}
&d^{\nabla_{t_0}}(X_.(\frac{\dot{}}{T}))_{t_0}\\
&= //_{t_0}\left(d\int_0^{t_0} //_s^{-1} * d\left(X_.\left(\frac{t_0}{T}\right)\right)_s + \frac{1}{T}//_s^{-1}\mathbf{W}_s^{g(s),X_.(\frac{s}{T})}\left(\dot{\gamma}\left(\frac{s}{T}\right)\right)ds\right)\\
&= d^{\nabla_{t_0}}\left(X_.\left(\frac{t_0}{T}\right)\right)_{t_0} + \frac{1}{T}\mathbf{W}_{t_0}^{g(t_0),X_.(\frac{t_0}{T})}\dot{\gamma}\left(\frac{t_0}{T}\right).
\end{aligned}$$

We then use (11.3) to identify

$$d^{\nabla_{t_0}}(X_.(\frac{t_0}{T}))_{t_0} = P_{0,\frac{t_0}{T}}^{g(t_0),X_{t_0}(.)} d^{\nabla_{t_0}} X_{t_0}(0).$$

Thus concludes the proof. □

Let

$$\begin{aligned}
N_t &:= -\frac{1}{T}\int_0^t \langle P_{0,\frac{s}{T}}^{g(s),X_s(.)} d^{\nabla_s} X_s(0), \mathbf{W}_s^{g(s),X_.(\frac{s}{T})}\dot{\gamma}(\frac{s}{T})\rangle_{g(s)},\\
R_t &:= \exp\left(N_t - \frac{1}{2}\langle N\rangle_t\right).
\end{aligned}$$

In many situations Novikov's criterion is satisfied. Therefore we could expect R_t to be a martingale. Define the new probability measure $\mathbb{Q}$ as:

$$\mathbb{Q} := R_T\mathbb{P}.$$

Proposition 11.2 *Suppose that the $g(t)$-Brownian motion starting at x is non-explosive and suppose that Novikov's criterion is satisfied for N_t. Then under $\mathbb{Q}$, the process $X_t(\frac{t}{T})$ is a L_t-diffusion that starts at x, and finishes at $X_T(1) = X_T(y)$, i.e. under $\mathbb{Q}$, $X_T(1)$ have the same distribution as the $g(t)$-Brownian motion at time T that start at y.*

Proof One could directly apply Girsanov's theorem. We prefer here to give a direct proof. Let $f \in C_b^2(M, \mathbb{R})$, since N_t satisfy Novikov's condition R_t is a $\mathbb{P}$-martingale. We use Itô formula to compute:

$$df(X_t(\frac{t}{T})) = \langle \nabla^t f(X_t(\frac{t}{T})), d^{\nabla_t}(X_.(\frac{\cdot}{T}))_t \rangle_{g(t)} + \frac{1}{2} \mathrm{Hess}_t\, f(X_t(\frac{t}{T}))(d^{\nabla_t}(X_.(\frac{\cdot}{T}))_t, d^{\nabla_t}(X_.(\frac{\cdot}{T}))_t).$$

Since $P_{0,\frac{t}{T}}^{g(t),X_t(.)}$ is an isometry for the metric $g(t)$

$$df(X_t(\frac{t}{T})) = \langle \nabla^t f(X_t(\frac{t}{T})), P_{0,\frac{t}{T}}^{g(t),X_t(.)} d^{\nabla_t} X_t(0) \rangle_{g(t)} + \frac{1}{T} \langle \nabla^t f(X_t(\frac{t}{T})), \mathbf{W}_t^{g(t),X_.(\frac{t}{T})} \dot{\gamma}(\frac{t}{T}) \rangle_{g(t)}\, dt + \frac{1}{2} \Delta_t f(X_t(\frac{t}{T}))dt.$$

Moreover

$$dR_t d(f(X_t(\frac{t}{T}))) = -\frac{1}{T} R_t \langle \nabla^t f(X_t(\frac{t}{T})), \mathbf{W}_t^{g(t),X_.(\frac{t}{T})} \dot{\gamma}(\frac{t}{T}) \rangle_{g(t)}\, dt.$$

This implies

$$d(R_t f(X_t(\frac{t}{T}))) = \frac{1}{2} R_t \Delta_t f(X_t(\frac{t}{T}))dt + dM_t^{\mathbb{P}},$$

where $M_t^{\mathbb{P}}$ is a martingale for $\mathbb{P}$. On the other hand, since R_t is a $\mathbb{P}$-martingale, we have

$$R_t \int_0^t \Delta_s f(X_s(\frac{s}{T}))ds = \int_0^t R_s \Delta_s f(X_s(\frac{s}{T}))ds + \tilde{M}^{\mathbb{P}},$$

where $\tilde{M}_s^{\mathbb{P}}$ is a $\mathbb{P}$-martingale.

Thus

$$R_t \Big(f(X_t(\frac{t}{T})) - \frac{1}{2} \int_0^t \Delta_s f(X_s(\frac{s}{T}))ds \Big)$$

is a martingale.

Since U_t is a $\mathbb{Q}$-martingale if and only if $R_t U_t$ is a $\mathbb{P}$-martingale, $f(X_t(\frac{t}{T})) - \frac{1}{2} \int_0^t \Delta_s f(X_s(\frac{s}{T}))ds$ is then a $\mathbb{Q}$ martingale i.e. $X_t(\frac{t}{T})$ is a L_t diffusion under the probability $\mathbb{Q}$. It is clear that it finishes at $X_T(y)$. Thus $X_t(\frac{t}{T})$ can be seen as a coupling between two L_t diffusions that starts at different points up to changing probability. □

11.2.2 Harnack Inequality with Power and Some Semigroup Property

Let T_c be the maximal life time of geometric flow $g(t)_{t\in[0,T_c[}$. For all $T < T_c$, let X_t^T be a $g(T-t)$-Brownian motion and $//_{0,t}^T := //_{0,t}^{g(T-t),X^T_\cdot}$ be the associated parallel transport. In this case, for a solution $f(t,.)$ of (11.1), $f(T-t, X_t^T(x))$ is a local martingale for any $x \in M$. Hence the following representation holds for the solution:

$$P_{0,T} f_0(x) := f(T,x) = \mathbb{E}_x[f_0(X_T^T)].$$

The subscript T refers to the fact that a time reversal step is involved.

Let $\mathbf{W}_{0,t}^T := \mathbf{W}_{0,t}^{g(T-t),X^T_\cdot}$ be the damped parallel transport along the $g(T-t)$-Brownian motion. We recall the covariant differential equation satisfied by this damped parallel transport (11.2.1):

$$*d((//_{0,t}^T)^{-1}(\mathbf{W}_{0,t}^T)) = -\frac{1}{2}(//_{0,t}^T)^{-1}(\mathrm{Ric}_{g(T-t)} - \partial_t(g(T-t)))^{\#g(T-t)}(\mathbf{W}_{0,t}^T)\, dt$$

with

$$\mathbf{W}_{0,t}^T : T_xM \longrightarrow T_{X_t^T(x)}M, \mathbf{W}_{0,0}^T = \mathrm{Id}_{T_xM}.$$

By the over subscript T we mean that the family of metrics is $g(T-t)$.

Proposition 11.3 *Suppose that there exist $\overline{\alpha}, \underline{\alpha} \geq 0$, $\underline{K} \geq 0$ such that for all $t \in [0,T]$:*

$$-\underline{\alpha} g(t) \leq \alpha(t) \leq \overline{\alpha} g(t),$$

$$-(d-1)\underline{K}^2 g(t) \leq \mathrm{Ric}(t)$$

then the $g(t)$-Brownian motion, and the $g(T-t)$-Brownian motion does not explode before the time T.

Proof This is a sufficient condition but it is far from being necessary one, for the process to do not explode. Let $x, y \in M$ and let $d_t(x,y)$ be the Riemannian distance from x to y computed with the metric $g(t)$. Let $Cut_t(x)$ be the set of cutlocus of x for the metric $g(t)$. Consider a fixed point $x_0 \in M$, and X_t a $g(t)$-Brownian motion starting at X_0. Using the Itô-Tanaka formula for $d_t(x_0, X_t)$ that have been proved for constant metric by Kendall, and generalized to $g(t)$-Brownian motion in [20] Theorem 2, we have:

$$\begin{aligned} d_t(x_0, X_t) = d_0(x_0, X_0) + \int_0^t \mathbb{1}_{X_s \notin Cut_s(x_0)} \left(\tfrac{1}{2}\Delta_{g(s)} d_s(x_0, .) + \tfrac{\partial d_s(x_0,.)}{\partial s}\right)(X_s)\, ds \\ + \int_0^t \mathbb{1}_{X_s \notin Cut_s(x_0)} \langle \nabla^s d_s(x_0, X_s), d^{\nabla_s} X_s \rangle_{g(s)} - L_t, \end{aligned} \tag{11.4}$$

where L_t is the local time at $Cut_t(x_0)$. The local time is non-decreasing non negative process that increase only when X_t touches $Cut_t(x_0)$. Moreover the distance $d_t(x_0, x)$ is smooth if $x \notin Cut_t(x_0) \cup x_0$. Let $x \notin Cut_t(x_0)$ and $\gamma : [0, d_t(x_0, x)] \to M$ be the $g(t)$-geodesic from x_0 to x. We have:

$$\frac{\partial d_s(x_0, x)}{\partial s} = \frac{1}{2} \int_0^{d_s(x_0,x)} \alpha(s)(\dot{\gamma}(u), \dot{\gamma}(u))\, du \leq \frac{\overline{\alpha}}{2} d_s(x_0, x).$$

Recall the Laplacian comparison Theorem:

$$\Delta_{g(s)} d_s(x_0, x) \leq (d-1)\underline{K} \coth(\underline{K} d_s(x_0, x)).$$

We then get the following control of the drift term (using $x \coth(x) \leq 1 + x$ for $x \geq 0$), and $F(x) := (d-1)(\frac{\overline{\alpha} x}{(d-1)} + \frac{1}{x} + \underline{K})$:

$$\begin{aligned} &\left(\frac{1}{2}\Delta_{g(s)} d_s(x_0, .) + \frac{\partial d_s(x_0, .)}{\partial s}\right)(x) \\ &\leq \frac{1}{2}\Big((d-1)\underline{K} \coth(\underline{K} d_s(x_0, x) + \overline{\alpha} d_s(x_0, x))\Big) \\ &\leq F(d_s(x_0, x)). \end{aligned}$$

Since $\|\nabla^{g(t)} d_t(X_0, .)\|_{g(t)} = 1$ and $Cut_t(x_0)$ have 0 as $g(t)$ volume, the martingale part of $d_t(x_0, X_t)$ is a real Brownian motion. We finish the proof using the comparison theorem of stochastic differential equation, and the usual criterion of non-explosion of a one dimensional diffusion. For the $g(T-t)$-Brownian motion, we simply to change $\overline{\alpha}$ by $\underline{\alpha}$ in the above formula. □

Remark 11.1 For the backward Ricci flow, it is shown in [20] without any assumption as in the above proposition that the $g(t)$-Brownian motion does not explode. But the sufficient condition for the existence of the forward Ricci flow in complete Riemannian manifolds as given by Shi in [24, Theorem 1.1], that is the boundedness of the initial Riemannian tensor (for the metric $g(0)$) also gives a bound of the Ric tensor along the flow (for bounded time). Hence the conditions for non explosion of the $g(t)$-Brownian motion given in the above proposition is satisfied, at least for small time, if the initial metric satisfies Shi's condition for the complete manifolds.

In the following proposition R_t^T is defined as R_t but according to the family of metrics $g(T-t)$ instead of $g(t)$.

Proposition 11.4 *Suppose that the $g(t)$-Brownian motion started at x is non-explosive for the first point, and the $g(T-t)$-Brownian motion started at x is non-explosive for the second point.*

1. *If there exists $C \in \mathbb{R}$ such that $\mathrm{Ric}_{g(t)} - \alpha(t) \geq Cg(t)$, then R_t is a martingale, and for $\beta \geq 1$*

$$\mathbb{E}[R_t^\beta] \leq e^{\frac{1}{2}\beta(\beta-1)\frac{d_0^2(x,y)}{T^2}\frac{1-e^{-Ct}}{C}}.$$

2. *If there exists $\tilde{C} \in \mathbb{R}$ such that $\mathrm{Ric}_{g(t)} + \alpha(t) \geq \tilde{C}g(t)$ then R_t^T is a martingale and for $\beta \geq 1$*

$$\mathbb{E}[(R_t^T)^\beta] \leq e^{\frac{1}{2}\beta(\beta-1)\frac{d_T^2(x,y)}{T^2}\frac{1-e^{-\tilde{C}t}}{\tilde{C}}}. \tag{11.5}$$

If $C = 0$ then we take for convention that for all t, $\frac{1-e^{-Ct}}{C} = t$.

Proof Without loss of generality, we just make the proof for R_t, the computation is the same as for R_t^T. Let $X_t(x)$ be a $g(t)$-Brownian motion and let $v \in T_xM$. We use short notation for the $g(t)$ parallel transport and the damped parallel transport along $X_t(x)$, $//_t := //_t^{g(t),X_.(x)}$ and $\mathbf{W}(X_.(x))_t := \mathbf{W}_t^{g(t),X_.(x)}$. Then we use the isometry property of the parallel transport, i.e., $//_s : (T_xM, g(0)) \mapsto (T_{X_s(x)}M, g(s))$, to deduce

$$\begin{aligned}
&* d\langle \mathbf{W}(X_.(x))_s v, \mathbf{W}(X_.(x))_s v\rangle_{g(s)} \\
&= *d\langle //_s^{-1}\mathbf{W}(X_.(x))_s v, //_s^{-1}\mathbf{W}(X_.(x))_s v\rangle_{g(0)} \\
&= 2\langle *d//_s^{-1}\mathbf{W}(X_.(x))_s v, //_s^{-1}\mathbf{W}(X_.(x))_s v\rangle_{g(0)} \\
&= 2\langle //_s * d//_s^{-1}\mathbf{W}(X_.(x))_s v, \mathbf{W}(X_.(x))_s v\rangle_{g(s)} \\
&= -\langle (\mathrm{Ric}_{g(s)} - \partial_s(g(s)))^{\#g(s)}(\mathbf{W}(X_.(x))_s v), \mathbf{W}(X_.(x))_s v\rangle_{g(s)}\, ds \\
&\leq -C \parallel \mathbf{W}(X_.(x))_s v \parallel^2 ds.
\end{aligned}$$

By Gronwall's lemma we get

$$\parallel \mathbf{W}(X_.(x))_s v \parallel_{g(s)} \leq e^{-\frac{1}{2}Cs} \parallel v \parallel_{g(0)}.$$

Recall that $N_t := -\frac{1}{T}\int_0^t \langle P_{0,\frac{s}{T}}^{g(s),X_s(.)} d^{\nabla_s} X_s(0), \mathbf{W}(X_.(\frac{s}{T}))_s \dot{\gamma}(\frac{s}{T})\rangle_{g(s)}$, and $P_{0,\frac{s}{T}}^{g(s),X_s(.)}$ is a $g(s)$ isometry and $d^{\nabla_s} X_s(0) = //_s e_i dw^i$ where w is a $\mathbb{R}^n$-Brownian motion, and $(e_i)_{i=1..n}$ is an orthonormal basis of $T_x M$. Then

$$\begin{aligned}\langle N\rangle_t &= \frac{1}{T^2}\int_0^t \parallel \mathbf{W}(X_.(\frac{s}{T}))_s \dot{\gamma}(\frac{s}{T}) \parallel_{g(s)}^2 \, ds \\ &\leq \frac{1}{T^2}\int_0^t e^{-Cs} \parallel \dot{\gamma}(\frac{s}{T}) \parallel_{g(0)}^2 \, ds \\ &\leq \frac{1}{T^2} d_0^2(x,y) \int_0^t e^{-Cs}\, ds.\end{aligned}$$

So by Nokinov's criterion, R_t is a martingale. Let $\beta \geq 1$,

$$\begin{aligned}\mathbb{E}[R_t^\beta] &= \mathbb{E}[e^{\beta N_t - \frac{\beta}{2}\langle N\rangle_t}] \\ &= \mathbb{E}[e^{\beta N_t - \frac{\beta^2}{2}\langle N\rangle_t} e^{\frac{\beta(\beta-1)}{2}\langle N\rangle_t}] \\ &\leq e^{\frac{1}{2}\beta(\beta-1)\frac{d_0^2(x,y)}{T^2}\frac{1-e^{-Ct}}{C}}.\end{aligned}$$

By the same computation we have

$$\begin{aligned}\langle N^T\rangle_t &= \frac{1}{T^2}\int_0^t \parallel \mathbf{W}^T(X_.(\frac{s}{T}))_s \dot{\gamma}(\frac{s}{T}) \parallel_{g(T-s)}^2 \, ds \\ &\leq \frac{1}{T^2}\int_0^t e^{-\tilde{C}s} \parallel \dot{\gamma}(\frac{s}{T}) \parallel_{g(T)}^2 \, ds \\ &\leq \frac{1}{T^2} d_T^2(x,y) \int_0^t e^{-\tilde{C}s}\, ds.\end{aligned}$$

Thus R_t^T is a martingale. Given $\beta \geq 1$ we have similarly (11.5). □

Remark 11.2 In the case of Ricci flow, $\partial_t g(t) = -\operatorname{Ric}_{g(t)}$, then $\partial_t g(T-t) = \operatorname{Ric}_{g(T-t)}$ so the process $X_t^T(x)$ does not explode (we do not need proposition 11.3, but [20]) and the condition of the above proposition is satisfied with $\tilde{C} = 0$ and

$$\mathbb{E}[(R_T^T)^\beta] \leq e^{\frac{1}{2}\beta(\beta-1)\frac{d_T^2(x,y)}{T}}.$$

Using the horizontal L_t-diffusion, we could give a alternative proof of Theorem 3.2 in [4] (isoperimetric-type Harnack inequality) for the constant metric case, and also a generalisation for inhomogeneous diffusions.

Lemma 11.1 *If there exists* $\tilde{C} \in \mathbb{R}$ *such that* $\mathrm{Ric}_{g(t)} + \alpha(t) \geq \tilde{C} g(t)$ *and if the* $g(T-t)$*-Brownian motion does not explode then for every measurable set A,*

$$P_{0,T}(\mathbb{1}_A)(x) \leq P_{0,T}(\mathbb{1}_{A_0^{\rho_T}})(y).$$

Where $\rho_T := e^{-\frac{\tilde{C}T}{2}} d_T(x, y)$ *and* $A_0^\epsilon := \{z \in M \, s.t. \, d_0(z, A) \leq \epsilon\}$

Proof We could give a proof with the usual Kendall coupling, but we have to manage the different cutlocus. We prefer here give a proof using the horizontal L_t diffusion in C^1 path space. Since the $g(T-t)$- Brownian motion does not explode, it is the same for the L_{T-t}-horizontal diffusion. Let γ be a $g(T)$ geodesic such that $\gamma(0) = x$ and $\gamma(1) = y$. By 11.2,

$$\partial_u X_t(u) = \mathbf{W}^T(X_\cdot^T(\gamma(u)))_t(\dot{\gamma}(u))$$

and

$$\| \mathbf{W}^T(X_\cdot^T(x))_s v \|_{g(T-s)} \leq e^{-\frac{1}{2}\tilde{C}s} \| v \|_{g(T)} .$$

We then get

$$d_0(X_T^T(x), X_T^T(y)) \leq \int_0^1 \|\partial_u X_t(u)\|_{g(0)} \, du \leq e^{-\frac{1}{2}\tilde{C}T} d_T(x, y) = \rho_T.$$

Hence $\{X_T^T(x) \in A\} \subset \{X_T^T(y) \in A_0^{\rho_T}\}$ and

$$P_{0,T}(\mathbb{1}_A)(x) = \mathbb{E}[\mathbb{1}_A(X_T^T(x))] \leq \mathbb{E}[\mathbb{1}_{A_0^{\rho_T}}(X_T^T(y))] = P_{0,T}(\mathbb{1}_{A_0^{\rho_T}})(y).$$

Thus concludes the proof. □

Corollary 11.1

1. *If* $\dot{g} = 0$ *and* $\mathrm{Ric}_g \geq K$ *then we can take* $\rho_T = e^{-\frac{KT}{2}} d(x, y)$*. This as actually Theorem 3.2 in [4] for the Riemannian case.*
2. *If* $g(t)$ *satisfies the Ricci flow,* $\partial_t g(t) = -\mathrm{Ric}_{g(t)}$*, so* $X_t^T(x)$ *does not explode ([20]) and since the damped parallel transport is an isometry we could take* $\rho_T = d_T(x, y)$

We are now ready to give the Harnack inequality with power. Let f be a solution of (11.1) and let $P_{0,T}$ be the inhomogeneous heat kernel associated to (11.1), i.e.

$$P_{0,T} f_0(x) := f(T, x) = \mathbb{E}_x[f_0(X_T^T)].$$

Theorem 11.1 *Suppose that the $g(T-t)$-Brownian motion X_t^T does not explode, and that the process R_t^T is a martingale. Then for all $\alpha > 1$ and $f_0 \in C_b(M)$ we have:*

$$| P_{0,T} f_0 |^\alpha (x) \leq \mathbb{E}[(R_T^T)^{\frac{\alpha}{\alpha-1}}]^{\alpha-1} P_{0,T} | f_0 |^\alpha (y).$$

Moreover if there exists $\tilde{C} \in \mathbb{R}$ such that

$$\mathrm{Ric}_{g(t)} + \alpha(t) \geq \tilde{C} g(t)$$

then we have:

$$| P_{0,T} f_0 |^\alpha (x) \leq e^{\frac{\alpha}{2(\alpha-1)} \frac{d_T^2(x,y)}{T^2} \frac{1-e^{-\tilde{C}T}}{\tilde{C}}} P_{0,T} | f_0 |^\alpha (y).$$

Proof We write $\tilde{X}_t^T := X_t^T(\frac{t}{T})$ the diagonal process associated to the family of metrics $g(T-t)$, and use Proposition 11.2, and Hölder inequality:

$$\begin{aligned} | P_{0,T} f_0 |^\alpha (x) &= | \mathbb{E}^{\mathbb{Q}}[f_0(\tilde{X}_T^T)] |^\alpha \\ &= | \mathbb{E}^{\mathbb{P}}[R_T^T f_0(\tilde{X}_T^T)] |^\alpha \\ &\leq \mathbb{E}^{\mathbb{P}}[(R_T^T)^{\frac{\alpha}{\alpha-1}}]^{\alpha-1} \mathbb{E}^{\mathbb{P}}[| f_0 |^\alpha (\tilde{X}_T^T)] \\ &= \mathbb{E}^{\mathbb{P}}[(R_T^T)^{\frac{\alpha}{\alpha-1}}]^{\alpha-1} \mathbb{E}_y^{\mathbb{P}}[| f_0 |^\alpha (X_T^T(y))] \\ &= \mathbb{E}^{\mathbb{P}}[(R_T^T)^{\frac{\alpha}{\alpha-1}}]^{\alpha-1} P_{0,T} | f_0 |^\alpha (y). \end{aligned}$$

The last part in the theorem is an application of Proposition 11.4. □

We will denote by μ_t the volume measure associated to the metric $g(t)$, and for A a measurable set, $\mu_t(A) := \int_A 1 \, d\mu_t$, and $B_t(x, r)$ the ball for the metric $g(t)$ of center x and radius r.

Corollary 11.2 *Suppose that the $g(T-t)$-Brownian motion X_t^T does not explode, and there exists $\tilde{C} \in \mathbb{R}$ such that $\mathrm{Ric}_{g(t)} + \alpha(t) \geq \tilde{C} g(t)$. Moreover suppose that there exists a function $\tau : [0, T] \mapsto \mathbb{R}$ such that:*

$$\frac{1}{2} \mathrm{trace}_{g(t)}(\alpha(t))(y) \leq \tau(t), \quad \forall (t, y) \in [0, T] \times M$$

then for $f_0 \in L^\alpha(\mu_0)$

$$| P_{0,T} f_0 | (x) \leq \frac{e^{\frac{\int_0^T \tau(s)\, ds + 1}{\alpha}}}{\left(\mu_T(B_T(x, \sqrt{\frac{2(\alpha-1)T^2}{\alpha(\frac{1-e^{-\tilde{C}T}}{\tilde{C}})}}))\right)^{\frac{1}{\alpha}}} \| f_0 \|_{L^\alpha(\mu_0)} .$$

Proof By Proposition 11.4 R_t^T is a martingale. If $f_0 \in C_b(M) \cap L^\alpha(\mu_0)$ we apply Theorem 11.1 and get:

$$| P_{0,T} f_0 |^\alpha (x) \le e^{\frac{\alpha}{2(\alpha-1)} \frac{d_T^2(x,y)}{T^2} \frac{1-e^{-\tilde{C}T}}{\tilde{C}}} P_{0,T} | f_0 |^\alpha (y).$$

We integrate both sides along the ball $B_T\big(x, \sqrt{\frac{2(\alpha-1)T^2}{\alpha(\frac{1-e^{-\tilde{C}T}}{\tilde{C}})}}\big)$, with respect to the measure μ_T, in y and obtain:

$$\begin{aligned}
&\mu_T(B_T\big(x, \sqrt{\frac{2(\alpha-1)T^2}{\alpha(\frac{1-e^{-\tilde{C}T}}{\tilde{C}})}})) \mid P_{0,T} f_0 |^\alpha (x) \\
&\le e \int_{B_T\big(x, \sqrt{\frac{2(\alpha-1)T^2}{\alpha(\frac{1-e^{-\tilde{C}T}}{\tilde{C}})}}\big)} P_{0,T} \mid f_0 |^\alpha (y)\, d\mu_T(y) \\
&\le e \int_M P_{0,T} \mid f_0 |^\alpha (y)\, d\mu_T(y).
\end{aligned}$$

We have that $\frac{d}{dt}\mu_t(y) = \frac{1}{2}\operatorname{trace}_{g(t)}(\alpha(t))(y) d\mu_t(y)$, and by the Stokes theorem we have:

$$\begin{aligned}
\frac{d}{dt}\int_M P_{0,t} \mid f_0 |^\alpha (y)\, d\mu_t(y) &= \int_M P_{0,t} \mid f_0 |^\alpha (y) \frac{d}{dt} d\mu_t(y) \\
&\le \tau(t) \int_M P_{0,t} \mid f_0 |^\alpha (y)\, d\mu_t(y).
\end{aligned}$$

We deduce that:

$$\int_M P_{0,t} \mid f_0 |^\alpha (y)\, d\mu_t(y) \le e^{\int_0^t \tau(s)\, ds} \parallel f_0 \parallel^\alpha_{L^\alpha(\mu_0)} .$$

Hence for $f_0 \in C_b(M) \cap L^\alpha(\mu_0)$

$$| P_{0,T} f_0 | (x) \le \frac{e^{\frac{\int_0^T \tau(s)\, ds + 1}{\alpha}}}{\Big(\mu_T(B_T(x, \sqrt{\frac{2(\alpha-1)T^2}{\alpha(\frac{1-e^{-\tilde{C}T}}{\tilde{C}})}}))\Big)^{\frac{1}{\alpha}}} \parallel f_0 \parallel_{L^\alpha(\mu_0)} .$$

We conclude by a classical density argument that the same inequality is true for $f_0 \in L^\alpha(\mu_0)$. □

Corollary 11.3 *If the family of metric comes from the Ricci flow and if*

$$(\tau(t)) = -\frac{1}{2} \inf_{y \in M} R(t, y) < \infty, \forall t \in [0, T]$$

where $R(t, y)$ is the scalar curvature at y for the metric $g(t)$ then we have

$$| P_{0,T} f_0 | (x) \leq \frac{e^{\frac{\int_0^T \tau(s)\, ds + 1}{\alpha}}}{\left(\mu_T(B_T(x, \sqrt{\frac{2(\alpha-1)T}{\alpha}}))\right)^{\frac{1}{\alpha}}} \parallel f_0 \parallel_{L^\alpha(\mu_0)} .$$

If $\inf_{x \in M} \left(\mu_T(B_T(x, \sqrt{\frac{2(\alpha-1)T}{\alpha}}))\right) =: C_T > 0$ *then as a linear operator:*

$$\parallel P_{0,T} \parallel_{L^\alpha(\mu_0) \mapsto L^\infty(\mu_0)} \leq \frac{e^{\frac{\int_0^T \tau(s)\, ds + 1}{\alpha}}}{C_T^{\frac{1}{\alpha}}}.$$

Proof If $g(t)$ comes from Ricci flow then $g(T - t)$ satisfies a backward Ricci flow. Then the process $X_t^T(x)$ does not explode before T [20]. Moreover we have $\tilde{C} = 0$ in Proposition 11.4, then the process R_t^T is a martingale and we could apply the above corollary. □

11.3 Non Symmetry of the Inhomogeneous Heat Kernel, and Heat Kernel Estimate

Unfortunately the non homogeneous heat kernel is non symmetric in general. The goal of this section is to by-pass this difficulty. This will be achieved by the study of the dual process and time reverse.

Let $\frac{\partial}{\partial_t} g(t) := \alpha(t)$ where α is a time-dependent symmetric 2-tensor. We suppose that there exist functions $\tau(t)$ and $\underline{\tau}(t)$ such that:

$$\begin{cases} \frac{1}{2} \sup_{y \in M} \operatorname{trace}_{g(t)}(\alpha(t))(y) \leq \tau(t) \\ \frac{1}{2} \inf_{y \in M} \operatorname{trace}_{g(t)}(\alpha(t))(y) \geq \underline{\tau}(t). \end{cases} \quad (11.6)$$

Consider the following heat operator where the subscript mean the variable in which we differentiate: $L_{t,x} := -\frac{\partial}{\partial_t} + \frac{1}{2}\Delta_{g(t)}$. Let $x, y \in M$ and $0 < \tau < \sigma \leq t$. Denote by $P(x, t, y, \tau)$ the fundamental solution of

$$\begin{cases} L_{t,x} P(x, t, y, \tau) = 0 \\ \lim_{t \searrow \tau} P(., t, y, \tau) = \delta_y(.) \end{cases} \quad (11.7)$$

Using Itô's formula we obtain as in [10]:

$$X_{t-\tau}^{t-}(x) \stackrel{\mathscr{L}}{=} P(x,t,y,\tau)\,d\mu_\tau(y)$$

Let $v, u \in C^{1,2}(\mathbb{R}, M)$, the space of functions that are differentiable in time, and differentiable twice in space. Consider the adjoint operator L^* of L with respect to $\langle Lu, v\rangle := \int_0^T \int_M (Lu) v d\mu_t\, dt$. As in Guenther [17], it satisfies

$$L_{t,x}^* = \frac{1}{2}\Delta_t + \frac{\partial}{\partial_t} + \frac{1}{2}\operatorname{trace}_{g(t)}(\alpha(t)).$$

The fundamental solution $P^*(y,\tau,x,t)$ of L^*, satisfies:

$$\begin{cases} L_{\tau,y}^* P^*(y,\tau,x,t) = 0 \\ \lim_{\tau\nearrow t} P^*(.,\tau,x,t) = \delta_x(.). \end{cases} \tag{11.8}$$

Using Duhamel's principle the adjoint property yields:

$$P(x,t,y,\tau) = P^*(y,\tau,x,t).$$

After a time reversal, $P^*(y,t-s,x,t)$ satisfies the following heat equation:

$$\begin{cases} \partial_s P^*(y,t-s,x,t) = \frac{1}{2}\Delta_{g(t-s),y} P^* + \frac{1}{2}\operatorname{trace}_{g(t-s)}(\alpha(t-s))(y) P^* \\ \lim_{s\searrow 0} P^*(y,t-s,x,t) = \delta_y. \end{cases} \tag{11.9}$$

Using the Feynman-Kac formula, we conclude that:

$$P^*(y,t-s,x,t) \le e^{\frac{1}{2}\int_0^s \tau(t-u)du}\overline{P}(y,s,x,t),$$

where $\overline{P}(y,s,x,t)$ be the fundamental solution of

$$\begin{cases} \partial_s f(s,x) = \frac{1}{2}\Delta_{g(t-s)} f(s,x) \\ f(0,x) = f_0(x); \end{cases} \tag{11.10}$$

i.e., $\overline{P}(y,s,x,t)$ satisfies:

$$\begin{cases} \partial_s \overline{P}(y,s,x,t) = \frac{1}{2}\Delta_{g(t-s),y}\overline{P}(y,s,x,t) \\ \lim_{s\searrow 0}\overline{P}(.,s,x,t) = \delta_x(.). \end{cases} \tag{11.11}$$

Theorem 11.2 *Suppose that* (11.6) *is satisfied and that:*

- *the $g(s)$-Brownian motion does not explode before the time $\frac{t}{2}$ and there exists $C \in \mathbb{R}$ such that $\forall s \in [0, \frac{t}{2}]$:*

$$\mathrm{Ric}_{g(s)} - \alpha(s) \geq C g(s);$$

- *the $g(t-s)$-Brownian motion does not explode before the time $\frac{t}{2}$ and there exist $\tilde{C} \in \mathbb{R}$ such that $\forall s \in [0, \frac{t}{2}]$:*

$$\mathrm{Ric}_{g(t-s)} + \alpha(t-s) \geq \tilde{C} g(t-s).$$

Then the fundamental solution of (11.1) *that we note $P(x,t,y,0)$ satisfies for all $0 < t < T_c$:*

$$P(x,t,y,0) \leq e \frac{e^{\frac{1}{2}\int_0^t \overline{\tau}(s)\,ds}}{\left(\mu_t(B_t(x, \sqrt{\frac{(\frac{t}{2})^2}{(\frac{1-e^{-\tilde{C}\frac{t}{2}}}{\tilde{C}})}}))\right)^{\frac{1}{2}}} \frac{e^{-\frac{1}{2}\int_0^{\frac{t}{2}} \underline{\tau}(s)\,ds}}{\left(\mu_0(B_0(y, \sqrt{\frac{(\frac{t}{2})^2}{(\frac{1-e^{-C\frac{t}{2}}}{C})}}))\right)^{\frac{1}{2}}}.$$

Proof By the Chapman-Kolmogorov formula we have:

$$\begin{aligned} P(x,t,y,0) &= \int_M P(x,t,z,\frac{t}{2}) P(z,\frac{t}{2},y,0)\, d\mu_{\frac{t}{2}}(z) \\ &= \int_M P(x,t,z,\frac{t}{2}) P^*(y,0,z,\frac{t}{2})\, d\mu_{\frac{t}{2}}(z) \\ &\leq \Big(\int_M (P(x,t,z,\frac{t}{2}))^2 d\mu_{\frac{t}{2}}(z)\Big)^{\frac{1}{2}} \Big(\int_M (P^*(y,0,z,\frac{t}{2}))^2 d\mu_{\frac{t}{2}}(z)\Big)^{\frac{1}{2}}. \end{aligned}$$

Recall that $P(x, \frac{t}{2}+s, z, \frac{t}{2})$ is the fundamental solution, which starts at δ_x at time $s=0$, of:

$$\begin{cases} \partial_s f(s,x) = \frac{1}{2}\Delta_{g(\frac{t}{2}+s)} f(s,x) \\ f(0,x) = f_0(x) \end{cases} \tag{11.12}$$

Then we have:

$$P_{0,\frac{t}{2}} f_0(x) := f(\frac{t}{2}, x) = \mathbb{E}[f_0(X^{t-\cdot}_{\frac{t}{2}}(x))].$$

According to the proof of Corollary 11.2, for $f_0 \in C_b(M) \cap L^2(\mu_{\frac{t}{2}})$:

$$| P_{0,\frac{t}{2}} f_0 | (x) \leq \frac{e^{\frac{\int_0^{\frac{t}{2}} \tau(\frac{t}{2}+s)\,ds+1}{2}}}{\left(\mu_t(B_t(x,\sqrt{\frac{(\frac{t}{2})^2}{(\frac{1-e^{-\tilde{C}(\frac{t}{2})}}{\tilde{C}})}}))\right)^{\frac{1}{2}}} \parallel f_0 \parallel_{L^2(\mu_{\frac{t}{2}})} .$$

Given $x_0 \in M$ and $n \in \mathbb{N}$, we apply the above inequality to $f_0(y) := P(x,t,y,\frac{t}{2}) \wedge (n\mathbb{1}_{B(x_0,n)}(y))$ to obtain:

$$\begin{aligned}
&\int_M \left(P(x,t,z,\frac{t}{2}) \wedge (n\mathbb{1}_{B(x_0,n)}(z))\right)^2 d\mu_{\frac{t}{2}}(z) \\
&\leq \int_M P(x,t,z,\frac{t}{2})\left(P(x,t,z,\frac{t}{2}) \wedge n\mathbb{1}_{B(x_0,n)}(z)\right) d\mu_{\frac{t}{2}}(z) \\
&\leq \frac{e^{\frac{\int_0^{\frac{t}{2}} \tau(\frac{t}{2}+s)\,ds+1}{2}}}{\left(\mu_t(B_t(x,\sqrt{\frac{(\frac{t}{2})^2}{(\frac{1-e^{-\tilde{C}(\frac{t}{2})}}{\tilde{C}})}}))\right)^{\frac{1}{2}}} \left(\int_M \left(P(x,t,z,\frac{t}{2}) \wedge (n\mathbb{1}_{B(x_0,n)}(z))\right)^2 d\mu_{\frac{t}{2}}(z)\right)^{\frac{1}{2}}.
\end{aligned}$$

Letting n goes to infinity, we obtain that $z \to P(x,t,z,\frac{t}{2})$ is in $L^2(\mu_{\frac{t}{2}})$ for $t > 0$, and that:

$$\left(\int_M \left(P(x,t,z,\frac{t}{2})\right)^2 d\mu_{\frac{t}{2}}(z)\right)^{\frac{1}{2}} \leq \frac{e^{\frac{\int_0^{\frac{t}{2}} \tau(\frac{t}{2}+s)\,ds+1}{2}}}{\left(\mu_t(B_t(x,\sqrt{\frac{(\frac{t}{2})^2}{(\frac{1-e^{-\tilde{C}(\frac{t}{2})}}{\tilde{C}})}}))\right)^{\frac{1}{2}}}.$$

Recall that:

$$P^*\left(y,0,x,\frac{t}{2}\right) \leq e^{\frac{1}{2}\int_0^{\frac{t}{2}} \tau(u)du}\,\overline{P}\left(y,\frac{t}{2},x,\frac{t}{2}\right),$$

where $\overline{P}(y,\frac{t}{2},x,\frac{t}{2})$ is the heat kernel at time $\frac{t}{2}$, which starts at time 0 at δ_y, of the following equation:

$$\begin{cases} \partial_s f(s,x) = \frac{1}{2}\Delta_{g(\frac{t}{2}-s)} f(s,x) \\ f(0,x) = f_0(x). \end{cases} \tag{11.13}$$

We also have:

$$\overline{P}_{0,\frac{t}{2}} f_0(x) := f(\frac{t}{2},x) = \mathbb{E}[f_0(X^{g(\cdot)}_{\frac{t}{2}}(x))]$$

To make a direct link with 11.2, we could think that the family of metrics is $s \mapsto g(\frac{t}{2} - s)$, so many changes of signs are involved. However, the proof of the following is the same as the one of Corollary 11.2. We get for $f_0 \in B_b(M) \cap L^2(\mu_{\frac{t}{2}})$:

$$| \overline{P}_{0,\frac{t}{2}} f_0 | (y) \leq \frac{e^{\frac{-\int_0^{\frac{t}{2}} \underline{\tau}(s)\, ds + 1}{2}}}{\left(\mu_0(B_0(y, \sqrt{\frac{(\frac{t}{2})^2}{(\frac{1-e^{-C(\frac{t}{2})}}{C})}}))\right)^{\frac{1}{2}}} \parallel f_0 \parallel_{L^2(\mu_{\frac{t}{2}})} .$$

Similarly $z \to \overline{P}(y, \frac{t}{2}, z, \frac{t}{2})$ is in $L^2(\mu_{\frac{t}{2}})$ and

$$\left(\int_M (\overline{P}(y, \frac{t}{2}, z, \frac{t}{2}))^2 \, d\mu_{\frac{t}{2}}(z) \right)^{\frac{1}{2}} \leq \frac{e^{\frac{-\int_0^{\frac{t}{2}} \underline{\tau}(s)\, ds + 1}{2}}}{\left(\mu_0(B_0(y, \sqrt{\frac{(\frac{t}{2})^2}{(\frac{1-e^{-C(\frac{t}{2})}}{C})}}))\right)^{\frac{1}{2}}} .$$

We obtain:

$$\left(\int_M (P^*(y, 0, z, \frac{t}{2}))^2 d\mu_{\frac{t}{2}}(z) \right)^{\frac{1}{2}} \leq e^{\frac{1}{2}\int_0^{\frac{t}{2}} \tau(u)du} \left(\int_M \overline{P}^2(y, \frac{t}{2}, z, \frac{t}{2}) \, d\mu_{\frac{t}{2}}(z) \right)^{\frac{1}{2}}$$

$$\leq e^{\frac{1}{2}\int_0^{\frac{t}{2}} \tau(u)du} \frac{e^{\frac{-\int_0^{\frac{t}{2}} \underline{\tau}(s)\, ds + 1}{2}}}{\left(\mu_0(B_0(y, \sqrt{\frac{(\frac{t}{2})^2}{(\frac{1-e^{-C(\frac{t}{2})}}{C})}}))\right)^{\frac{1}{2}}} .$$

□

Remark 11.3 Having a heat kernel estimate for the heat equation we have simultaneously a kernel estimate of conjugate equation.

Remark 11.4 The hypothesis $\mathrm{Ric}_{g(t-s)} + \alpha(t-s) \geq \tilde{C} g(t-s)$, for $s \in [0, \frac{t}{2}]$ is a kind of quantitative super Ricci flow as defined in [22] (if $\tilde{C} = 0$ this is exactly the definition of super Ricci flow). This quantitative version of super Ricci flow allow us to control the rate of expansion of the damped parallel transport along the $g(t - .)$-Brownian motion.
The hypothesis $\mathrm{Ric}_{g(s)} - \alpha(s) \geq C g(s)$ for $s \in [0, \frac{t}{2}]$ allow us to control the rate of expansion of the damped parallel transport along the dual process, namely the process associated to $\overline{P}$ i.e. $g(.)$-Brownian motion.

Remark 11.5 If $g(t) = g(0)$ is constant, and $\mathrm{Ric}_{g(0)} \geq 0$ we have $\tau(t) = \underline{\tau}(t) = 0$, $C = \tilde{C} = 0$ and we deduce a Li-Yau on-diagonal estimate of the usual heat equation on complete manifolds as in [21] (up to some constant):

$$P_t(x, y) \leq e \frac{1}{\left(\mu_0(B_0(x, \sqrt{\tfrac{t}{2}}))\right)^{\frac{1}{2}}} \frac{1}{\left(\mu_0(B_0(y, \sqrt{\tfrac{t}{2}}))\right)^{\frac{1}{2}}}.$$

Using the symmetry of the heat kernel in the constant metric case we do not need to consider the dual as in the above theorem.

11.4 Grigor'yan Tricks, On-diagonal Estimate to Gaussian Estimate, the Ricci Flow Case

In this section we use the on-diagonal estimate of the previous section to derive a Gaussian type estimate of the minimal heat kernel coupled with Ricci flow (for complete manifold with non negative Ricci curvature). The proof involves several steps. In particular, we use a modification of Grigor'yan tricks [15, 16] to control integrability of the square of the heat kernel outside some ball, combined to an adapted version of Hamilton entropy estimate to control the difference of the heat kernel at two points. This type of strategy, is a modification of different arguments which appears in the literature on the Ricci flow (e.g. Cao-Zhang [5]).

Proposition 11.5 *Let $g(t)_{t\in[0,T_c[}$ be a family of metrics that satisfy the Ricci flow i.e. $\dot{g}(t) = -\mathrm{Ric}_{g(t)}$. Suppose that for all $t \in [0, T_c[$, $\mathrm{Ric}_{g(t)} \geq 0$ then we have the following on-diagonal estimate for the heat kernel and so for the conjugate heat kernel:*

$$P(x, t, y, 0) \leq e \frac{e^{-\frac{1}{4}\int_{\frac{t}{2}}^{t} \inf_M R(s,.)\, ds}}{\left(\mu_t(B_t(x, \sqrt{\tfrac{t}{2}}))\right)^{\frac{1}{2}}} \frac{e^{\frac{1}{4}\int_0^{\frac{t}{2}} \left(\sup_M R(s,.) - \inf_M R(s,.)\right) ds}}{\left(\mu_0(B_0(y, \sqrt{\tfrac{t}{2}}))\right)^{\frac{1}{2}}}.$$

Proof We could use Proposition 11.3 with $\overline{K} = 0$ and $\overline{\alpha} = 0$ to get the non-explosion of the $g(t)$-Brownian motion, and as in [20] the $g(T-t)$-Brownian motion does not explode. We then use Theorem 11.2 with $C = 0$ and $\tilde{C} = 0$ to get the following on-diagonal estimate, for all $t \in]0, T_c[$:

$$P(x, t, y, 0) \leq e \frac{e^{\frac{1}{2}\int_0^t \tau(s)\, ds}}{\left(\mu_t(B_t(x, \sqrt{\tfrac{t}{2}}))\right)^{\frac{1}{2}}} \frac{e^{-\frac{1}{2}\int_0^{\frac{t}{2}} \underline{\tau}(s)\, ds}}{\left(\mu_0(B_0(y, \sqrt{\tfrac{t}{2}}))\right)^{\frac{1}{2}}}.$$

Recall that in the case of Ricci flow: $\tau(s) = -\frac{1}{2} \inf_M R(s, .) \leq 0$. □

Remark 11.6 Note that in the proof of Theorem 11.2 the time $\frac{t}{2}$ is arbitrary and so in Proposition 11.5. Thus if we have only a control of $\sup_M R$ for small time, we also have a on-diagonal estimate.

We start this section by the following Hamilton estimate.

Lemma 11.2 *Let f be a positive solution of* (11.1)*, where $\alpha_{i,j}(t) = -(\mathrm{Ric}_{g(t)})_{i,j}$, $t \in]0, T_c[$ and $M_{\frac{t}{2}} := \sup_{x\in M} f(\frac{t}{2}, x)$ then for all $x, y \in M$,*

$$f(t,x) \le \sqrt{f(t,y)}\sqrt{M_{\frac{t}{2}} e^{\frac{d_t^2(x,y)}{t}}}.$$

Proof By the homogeneity of the desired inequality under multiplication by a constant, and the linearity of the heat equation, we can suppose that $f > 1$, by taking for $\epsilon > 0$, $f_\epsilon = \frac{f+2\epsilon}{\inf_M f+\epsilon}$ and take the limit in ϵ.
Since no confusion could arise, we will simply write without subscript ∇, $\|.\|$, and $\|.\|_{HS}$, for $\nabla^{g(t)}$, $\|.\|_{g(t)}, \ldots$ Recall the Hilbert-Schmidt norm is defined as $\|\alpha\|^2_{HS} = g^{il}g^{jm}\alpha_{ij}\alpha_{lm}$, for a 2-tensor $\alpha := \alpha_{ij}dx_i \otimes dx_j$.
Using an orthonormal frame and Weitzenbock's formula, we obtain the following equation

$$\begin{aligned}
&(-\partial_t + \frac{1}{2}\Delta_{g(t)})\left(\frac{\| \nabla f \|^2}{f}\right)(t,x)\\
&= \frac{1}{f}\Big(\| \operatorname{Hess} f - \frac{\nabla f \otimes \nabla f}{f} \|^2_{HS} + (\mathrm{Ric}_{g(t)} + \dot{g})(\nabla f, \nabla f)\Big)(t,x).
\end{aligned}$$

Thus, in the case of Ricci flow

$$(-\partial_t + \frac{1}{2}\Delta_{g(t)})\Big(\frac{\| \nabla f \|^2}{f}\Big) \ge 0.$$

By a direct computation

$$(-\partial_t + \frac{1}{2}\Delta_{g(t)})(f\log f)(t,x) = \frac{1}{2}\frac{\| \nabla f \|^2}{f}(t,x).$$

Let

$$N_s := h(s)\frac{\| \nabla f \|^2}{f}(t-s, X_s^t(x)) + (f\log f)(t-s, X_s^t(x)),$$

where $X_s^t(x)$ is a $g(t-s)$-Brownian motion started at x. If $h(s) := \frac{t/2-s}{2}$ then by Itô formula, it is easy to see that N_s is a super-martingale. So we have:

$$\mathbb{E}[N_0] \le \mathbb{E}[N_{\frac{t}{2}}],$$

that is:

$$\begin{aligned}\frac{t}{4}\frac{\| \nabla f \|^2}{f}(t,x)+(f\log f)(t,x) &\leq \mathbb{E}[(f\log f)(\frac{t}{2},X^t_{\frac{t}{2}}(x))]\\ &\leq \mathbb{E}[f(\frac{t}{2},X^t_{\frac{t}{2}}(x))]\log(M_{\frac{t}{2}})\\ &= f(t,x)\log(M_{\frac{t}{2}}),\end{aligned}$$

where we use $f>1$ and that $f(t-s,X^t_s(x))$ is a martingale. The above computation yields

$$\frac{\| \nabla f \|}{f}(t,x)\leq \frac{2}{\sqrt{t}}\sqrt{\log(\frac{M_{\frac{t}{2}}}{f(t,x)})},$$

and consequently

$$\left\| \nabla\sqrt{\log(\frac{M_{\frac{t}{2}}}{f(x,t)})}\right\| \leq \frac{1}{\sqrt{t}}.$$

After integrating this inequality along a $g(t)$-geodesic between x and y, we get

$$\sqrt{\log\left(\frac{M_{\frac{t}{2}}}{f(y,t)}\right)}\leq \sqrt{\log\left(\frac{M_{\frac{t}{2}}}{f(x,t)}\right)}+\frac{d_t(x,y)}{\sqrt{t}},$$

that yields to

$$f(t,x)\leq \sqrt{f(t,y)}\sqrt{M_{\frac{t}{2}}e^{\frac{d_t^2(x,y)}{t}}}.$$

Now, we adapt the argument of Grigor'yan to the situation of Ricci flow (with non negative Ricci curvature). □

Lemma 11.3 *Suppose that* (11.6) *is satisfied, the family of metrics* $g(t)$ *comes from the Ricci flow, and let* B *be a measurable set, then:*

$$\frac{e^{-\frac{1}{2}\int_0^t \tau(s)\,ds}}{\mu_0(B)^{\frac{1}{2}}}\leq \frac{1}{\mu_t(B)^{\frac{1}{2}}}\leq \frac{e^{-\frac{1}{2}\int_0^t \underline{\tau}(s)\,ds}}{\mu_0(B)^{\frac{1}{2}}}.$$

Moreover if $\mathrm{Ric}_{g(t)}\geq 0$ *for all* $t\in[0,T_c[$ *then for all* $x\in M$ *and* $r>0$ *we have:*

$$\frac{1}{\mu_0(B_t(x,r))^{\frac{1}{2}}}\leq \frac{1}{\mu_0(B_0(x,r))^{\frac{1}{2}}}.$$

Proof Recall that:

$$\frac{d}{dt}\mu_t = \frac{1}{2}\,\mathrm{trace}_{g(t)}(\dot{g}(t))\mu_t.$$

In the case of a Ricci flow this becomes $\frac{d}{dt}\mu_t(dx) = -\frac{1}{2}R(x,t)\mu_t(dx)$. Thus, the first inequality of the lemma follows from an integration. For the second point, it is clear that $\mathrm{Ric} \geq 0$ yields that $d_t(x, y)$ is non increasing in time. Then $B_0(x, r) \subset B_t(x, r)$, which clearly gives $\frac{1}{\mu_0(B_t(x,r))^{\frac{1}{2}}} \leq \frac{1}{\mu_0(B_0(x,r))^{\frac{1}{2}}}$. □

The above lemma immediately yields the following remark.

Remark 11.7 Suppose $\dot{g}(t) = -\mathrm{Ric}_{g(t)}$ and $\mathrm{Ric}_{g(.)} \geq 0$. Using the estimate in Lemma 11.5 and Lemma 11.3 we have the following estimate:

$$P(x,t,y,0) \leq e\frac{e^{\frac{1}{4}\int_0^t\left(\sup_M R(s,.)-\inf_M R(s,.)\right)ds}}{\left(\mu_0(B_0(x,\sqrt{\frac{t}{2}}))\right)^{\frac{1}{2}}}\frac{e^{\frac{1}{4}\int_0^{\frac{t}{2}}\sup_M R(s,.)\,ds}}{\left(\mu_0(B_0(y,\sqrt{\frac{t}{2}}))\right)^{\frac{1}{2}}}.$$

Proposition 11.6 *Let* $g(t)$ *be a solution of Ricci flow such that* $\mathrm{Ric}_g(t) \geq 0$*, and let* $r > 0$, $t_0 > t \geq 0$*, and define:*

$$\xi(y,t) = \begin{cases} \frac{-(r-d_t(x,y))^2}{(t_0-t)} & if \quad d_t(x,y) \leq r \\ 0 & if \quad d_t(x,y) \geq r \end{cases}$$

and $\Lambda(t) = \frac{1}{2}\int_0^t \inf_{x\in M}(R(s,x))ds$*. If* f *is a solution of* (11.1) *then for* $t_2 < t_1 < t_0$*:*

$$\int_M f^2(t_1,y)e^{\xi(y,t_1)}\mu_{t_1}(dy) \leq e^{-(\Lambda(t_1)-\Lambda(t_2))}\int_M f^2(t_2,y)e^{\xi(y,t_2)}\mu_{t_2}(dy).$$

Proof Let $\gamma(s)$ be the $g(t)$-geodesic between x and y. Using Remark 6 in [22] and the fact that $\mathrm{Ric}_{g(t)} \geq 0$ we get

$$\begin{aligned}\frac{d}{dt}d_t^2(x,y) &= \frac{d}{dt}\int_0^1 \parallel \dot{\gamma}(s) \parallel_{g(t)}^2\, ds \\ &= -\int_0^1 \mathrm{Ric}_{g(t)}(\dot{\gamma}(s),\dot{\gamma}(s))\,ds \leq 0.\end{aligned}$$

So for $y \in B_t(x, r)$ we have

$$\frac{d}{dt}\xi(y,t) \leq \frac{-(r-d_t(x,y))^2}{(t_0-t)^2}.$$

We also have

$$\| \nabla^t \xi(y,t) \|^2_{g(t)} = \frac{4(r - d_t(x,y))^2}{(t_0 - t)^2}.$$

Then

$$\frac{d}{dt}\xi(y,t) \le -\frac{\| \nabla^t \xi(y,t) \|^2_{g(t)}}{4},$$

and the above inequality is clear if $y \notin B_t(x,r)$. Let $f(t,x)$ a solution of (11.1) then we have:

$$\begin{aligned}
&\frac{d}{dt}\int_M f^2(t,y)e^{\xi(y,t)}\mu_t(dy)\\
&= \int_M \Big(f(t,y)\Delta_t f(t,y) + f^2(t,y)\frac{d}{dt}\xi(y,t) - \frac{R(y,t)}{2} f^2(t,y)\Big)e^{\xi(y,t)}\,\mu_t(dy)\\
&= \int_M -\langle \nabla^t f, \nabla^t(fe^{\xi})\rangle_{g(t)} + f^2\frac{d}{dt}\xi e^{\xi} - \frac{R}{2} f^2 e^{\xi}\,\mu_t(dy)\\
&= \int_M \Big(-\langle \nabla^t f, \nabla^t f\rangle_{g(t)} - \langle \nabla^t f, f\nabla^t \xi\rangle_{g(t)} + f^2\frac{d}{dt}\xi\Big)e^{\xi} - \frac{R}{2} f^2 e^{\xi}\,\mu_t(dy)\\
&\le -\int_M \Big(\langle \nabla^t f, \nabla^t f\rangle + 2\langle \nabla^t f, f\nabla^t \frac{\xi}{2}\rangle + f^2\frac{\| \nabla^t \xi \|^2}{4}\Big)e^{\xi} - \frac{R}{2} f^2 e^{\xi}\,\mu_t(dy)\\
&= -\int_M \Big(\langle \nabla^t f, \nabla^t f\rangle + 2\langle \nabla^t f, f\nabla^t \frac{\xi}{2}\rangle + f^2\frac{\| \nabla^t \xi \|^2}{4}\Big)e^{\xi} - \frac{R}{2} f^2 e^{\xi}\,\mu_t(dy)\\
&= -\int_M \| \nabla^t f + f\nabla^t \frac{\xi}{2} \|^2_{g(t)}\,\mu_t(dy) - \int_M \frac{R}{2} f^2 e^{\xi}\,\mu_t(dy)\\
&\le -\int_M \frac{R}{2} f^2 e^{\xi}\,\mu_t(dy).
\end{aligned}$$

The result follows. □

We define

$$I_r(t) := \int_{M\setminus B_t(x,r)} f^2(t,y)\mu_t(dy).$$

Proposition 11.7 *Under the same assumptions as in Proposition 11.6. Let* $\rho < r$ *and* f *be a solution of* (11.1). *We have:*

$$I_r(t_1) \le e^{-(\Lambda(t_1)-\Lambda(t_2))}\Big(I_\rho(t_2) + e^{\frac{-(r-\rho)^2}{(t_1-t_2)}}\int_M f^2(t_2,y)\mu_{t_2}(dy)\Big)$$

for $t_2 < t_1$.

Proof For $t_2 < t_1$,

$$\begin{aligned}
I_r(t_1) &= \int_{M\setminus B_{t_1}(x,r)} f^2(t_1,y)\mu_{t_1}(dy) \le \int_M f^2(t_1,y)e^{\xi(y,t_1)}\mu_{t_1}(dy) \\
&\le e^{-(\Lambda(t_1)-\Lambda(t_2))}\int_M f^2(t_2,y)e^{\xi(y,t_2)}\mu_{t_2}(dy) \\
&\le e^{-(\Lambda(t_1)-\Lambda(t_2))}\Big(\int_{B_{t_2}(x,\rho)} f^2(t_2,y)e^{\xi(y,t_2)}\mu_{t_2}(dy) \\
&\quad + \int_{M\setminus B_{t_2}(x,\rho)} f^2(t_2,y)e^{\xi(y,t_2)}\mu_{t_2}(dy)\Big) \\
&\le e^{-(\Lambda(t_1)-\Lambda(t_2))}\Big(I_\rho(t_2) + \int_{B_{t_2}(x,\rho)} f^2(t_2,y)e^{\xi(y,t_2)}\mu_{t_2}(dy)\Big) \\
&\le e^{-(\Lambda(t_1)-\Lambda(t_2))}\Big(I_\rho(t_2) + e^{\frac{-(r-\rho)^2}{A(t_0-t_2)}}\int_M f^2(t_2,y)\mu_{t_2}(dy)
\end{aligned}$$

Then remark that the definition of $I_r(t)$ is independent of t_0 and of the corresponding ξ, so we can pass to the limit when $t_0 \searrow t_1$ to obtain the desired result. □

We apply the above proposition to the heat kernel $P(x,t,y,0)$ of the Eq. (11.7) which also satisfies (11.1).

Theorem 11.3 *If $\dot g(t) = -\operatorname{Ric}_{g(t)}$ for all $t \in [0,T_c[$ and the following assumptions are satisfied:*

- *H1 : if M is not compact, we suppose that there exists a uniform constant $c_n > 0$ such that for all $x \in M$ we have $\mu_0(B_{g(0)}(x,r)) \ge c_n r^n$ (that is a non collapsing condition).*
- *H2 : $\operatorname{Ric}_{g(t)} \ge 0$ for all $t \in [0,T_c[$.*

Then for all $a > 1$ there exist two positive explicit constants q_a, m_a depending only on a, c_n and the dimension, such that we have the following heat kernel estimate for all $t \in]0,T_c[$ and $x_0, y_0 \in M$:

$$P(y_0,t,x_0,0) \le q_a \frac{e^{\int_0^t \frac{1}{2}\sup_M R(u,.)-\frac{1}{4}\inf_M R(u,.)du}}{\big(\mu_0(B_0(x_0,\sqrt{t}))\big)^{\frac{1}{2}}\mu_0(B_0(y_0,\sqrt{t}))^{\frac{1}{2}}} e^{-\frac{m_a d_t(x_0,y_0)^2}{16t}}.$$

The values of q_a and m_a are given by (11.16) *in the proof. Moreover we could optimize m_a, in terms of $a > 1$, to get a better control for points which are far.*

Proof We could suppose that $\int_0^t \sup_M R(s,.)\,ds < \infty$ for all $t \in [0,T_c[$, else the conclusion is satisfied. Let $f(t,x) := P(x,t,y,0)$ be the heat kernel of (11.1) that is the solution of Eq. (11.7). Note that H2 gives also a condition for non-explosion of the $g(t)$-Brownian motion by Proposition 11.3. Then we have by the proof of Theorem 11.2:

$$\begin{aligned}\int_M f^2(t,x)\mu_t(dx) &= \int_M P^2(x,t,y,0)\mu_t(dx)\\ &= \int_M P^{*2}(y,0,x,t)\mu_t(dx)\\ &\le e\frac{e^{\int_0^t \tau(u)-\underline{\tau}(u)\,du}}{\left(\mu_0(B_0(y,\sqrt{t}))\right)}\\ &= e\frac{e^{\frac{1}{2}\int_0^t \sup_M R(u,.)-\inf_M R(u,.)du}}{\left(\mu_0(B_0(y,\sqrt{t}))\right)}.\end{aligned}$$

Let $0 < \rho < r$, and $t_2 < t_1 < t_0$ then apply Proposition 11.7 to $f(t,x) := P(x,t,y,0)$ to get:

$$\begin{aligned}I_r(t_1) &\le e^{-(\Lambda(t_1)-\Lambda(t_2))}\left(I_\rho(t_2) + e^{\frac{-(r-\rho)^2}{(t_1-t_2)}}\int_M f^2(t_2,y)\mu_{t_2}(dy)\right)\\ &\le e^{-(\Lambda(t_1)-\Lambda(t_2))}\left(I_\rho(t_2) + e\cdot e^{\frac{-(r-\rho)^2}{(t_1-t_2)}}\frac{e^{\frac{1}{2}\int_0^{t_2}\sup_M R(u,.)-\inf_M R(u,.)du}}{\left(\mu_0(B_0(y,\sqrt{t_2}))\right)}\right).\end{aligned}$$

Let $a > 1$ be a constant. Let us define $r_k := (\frac{1}{2} + \frac{1}{k+2})r$ and $t_k := \frac{t}{a^k}$.

Thus Proposition 11.7 can be applied to $r_{k+1} < r_k$ and $t_{k+1} < t_k$, yielding to the same estimate as before:

$$\begin{aligned}&I_{r_k}(t_k)\\ &\le e^{-(\Lambda(t_k)-\Lambda(t_{k+1}))}\left(I_{r_{k+1}}(t_{k+1}) + e\cdot e^{\frac{-(r_k-r_{k+1})^2}{(t_k-t_{k+1})}}\frac{e^{\frac{1}{2}\int_0^{t_{k+1}}\sup_M R(u,.)-\inf_M R(u,.)du}}{\left(\mu_0(B_0(y,\sqrt{t_{k+1}}))\right)}\right)\\ &\le e^{-(\Lambda(t_k)-\Lambda(t_{k+1}))}\left(I_{r_{k+1}}(t_{k+1}) + e\cdot e^{\frac{-(r_k-r_{k+1})^2}{(t_k-t_{k+1})}}\frac{e^{\frac{1}{2}\int_0^{t_0}\sup_M R(u,.)-\inf_M R(u,.)du}}{\left(\mu_0(B_0(y,\sqrt{t_{k+1}}))\right)}\right).\end{aligned}$$

Applying recursively this inequality, and use H2 to see that Λ is non decreasing, we have for all k:

$$I_{r_0}(t_0) \le e^{-(\Lambda(t_0)-\Lambda(t_{k+1}))}I_{r_{k+1}}(t_{k+1}) + e\sum_{i=0}^{k} e^{-\Lambda(t_0)}e^{\frac{-(r_i-r_{i+1})^2}{(t_i-t_{i+1})}}\frac{e^{\frac{1}{2}\int_0^{t_0}\sup_M R(u,.)du}}{\left(\mu_0(B_0(y,\sqrt{t_{i+1}}))\right)} \tag{11.14}$$

We also have $\lim_{k\longrightarrow\infty} I_{r_k}(t_k) = 0$ (see Lemma 11.4 in Appendix).

So we can pass to the limit when k goes to infinity in Eq. (11.14) to get:

$$I_{r_0}(t_0) \le e \cdot e^{-\Lambda(t_0)} e^{\frac{1}{2}\int_0^{t_0} \sup_M R(u,.)du} \sum_{i=0}^{\infty} e^{\frac{-(r_i - r_{i+1})^2}{(t_i - t_{i+1})}} \frac{1}{\left(\mu_0(B_0(y, \sqrt{t_{i+1}}))\right)}.$$

Recall that $r_i - r_{i+1} = \frac{r}{(i+3)(i+2)}$ and $t_i - t_{i+1} = \frac{t}{a^i}(1 - \frac{1}{a})$. Moreover, by Bishop-Gromov theorem, Theorem 4.19 [13] in the case $\mathrm{Ric} \ge 0$ we have

$$\frac{\mu_0(B_0(y, \sqrt{t_i}))}{\mu_0(B_0(y, \sqrt{t_{i+1}}))} \le a^{\frac{n}{2}} := c_a.$$

Iterating the above inequality we get:

$$\frac{\mu_0(B_0(y, \sqrt{t_0}))}{\mu_0(B_0(y, \sqrt{t_{i+1}}))} \le (c_a)^{i+1}.$$

So we have:

$$\begin{aligned}
I_{r_0}(t_0) &\le e \frac{e^{-\Lambda(t_0)} e^{\frac{1}{2}\int_0^{t_0} \sup_M R(u,.)du}}{\mu_0(B_0(y, \sqrt{t_0}))} \sum_{i=0}^{\infty} e^{\frac{-(r_i - r_{i+1})^2}{(t_i - t_{i+1})}} (c_a)^{i+1} \\
&\le e \frac{e^{-\Lambda(t_0)} e^{\frac{1}{2}\int_0^{t_0} \sup_M R(u,.)du}}{\mu_0(B_0(y, \sqrt{t_0}))} \sum_{i=0}^{\infty} e^{\frac{-(\frac{r}{(i+3)(i+2)})^2}{(\frac{t}{a^i}(1-\frac{1}{a}))} + (i+1)\log(c_a)} \\
&\le e \frac{e^{-\Lambda(t_0)} e^{\frac{1}{2}\int_0^{t_0} \sup_M R(u,.)du}}{\mu_0(B_0(y, \sqrt{t_0}))} \sum_{i=0}^{\infty} e^{\frac{-a^{i+1} r^2}{t_0(a-1)(i+3)^4} + (i+1)\log(c_a)}.
\end{aligned}$$

There exists a constant m_a such that $\frac{a^{i+1}}{(a-1)(i+3)^4} \ge m_a(i+2)$, and thus we get:

$$\begin{aligned}
I_{r_0}(t_0) &\le e \frac{e^{-\Lambda(t_0)} e^{\frac{1}{2}\int_0^{t_0} \sup_M R(u,.)du}}{\mu_0(B_0(y, \sqrt{t_0}))} \sum_{i=0}^{\infty} e^{\frac{-m_a r^2}{t_0}(i+2) + (i+1)\log(c_a)} \\
&\le e \frac{e^{-\Lambda(t_0)} e^{\frac{1}{2}\int_0^{t_0} \sup_M R(u,.)du}}{\mu_0(B_0(y, \sqrt{t_0}))} e^{\frac{-m_a r^2}{t_0}} \sum_{i=0}^{\infty} e^{-(i+1)(\frac{m_a r^2}{t_0} - \log(c_a))}.
\end{aligned}$$

If $\frac{m_a r^2}{t_0} - \log(c_a) \ge \log(2)$ then

$$I_{r_0}(t_0) \le e \frac{e^{-\Lambda(t_0)} e^{\frac{1}{2}\int_0^{t_0} \sup_M R(u,.)du}}{\mu_0(B_0(y, \sqrt{t_0}))} e^{\frac{-m_a r^2}{t_0}},$$

If $\frac{m_a r^2}{t_0} - \log(c_a) < \log(2)$ then

$$\begin{aligned} I_{r_0}(t_0) &\leq \int_M P^2(x, t_0, y, 0)\mu_{t_0}(dx) \\ &= \int_M P^{*2}(y, 0, x, t_0)\mu_{t_0}(dx) \\ &\leq e \frac{e^{\frac{1}{2}\int_0^{t_0} \sup_M R(u,.) - \inf_M R(u,.) du}}{\left(\mu_0(B_0(y, \sqrt{t_0}))\right)} \\ &\leq e \frac{e^{\frac{1}{2}\int_0^{t_0} \sup_M R(u,.) - \inf_M R(u,.) du}}{\left(\mu_0(B_0(y, \sqrt{t_0}))\right)} e^{\log(2)+\log(c_a)-\frac{m_a r^2}{t_0}}. \end{aligned}$$

We have that for all $a > 1$ there exists a constant $q_a := 2ea^{\frac{n}{2}}$ and $\frac{(e \ln a)^5}{a^2(a-1)5^5} \leq m_a$, so we could take in the following $m_a = \frac{(e \ln a)^5}{a^2(a-1)5^5}$ such that:

$$I_r(t) \leq q_a \frac{e^{\frac{1}{2}\int_0^{t} \sup_M R(u,.) - \inf_M R(u,.) du}}{\left(\mu_0(B_0(y, \sqrt{t}))\right)} e^{-\frac{m_a r^2}{t}}. \tag{11.15}$$

Case 1 points which are far.

Let $x_0, y_0 \in M$ such that $d_t(x_0, y_0) \geq \sqrt{t}$, let $r := \frac{d_t(x_0, y_0)}{2}$, then by (11.15) (with $I_r(t)$ defined with $f(t, x) = P(x, t, x_0, 0)$, there exists $z_0 \in B_t(y_0, \sqrt{\frac{t}{4}}) \subset M \backslash B_t(x_0, r)$ such that:

$$\begin{aligned} \mu_t(B_t(y_0, \sqrt{\frac{t}{4}}) P^2(z_0, t, x_0, 0) &\leq I_r(t) \\ &\leq q_a \frac{e^{\frac{1}{2}\int_0^{t} \sup_M R(u,.) - \inf_M R(u,.) du}}{\left(\mu_0(B_0(x_0, \sqrt{t}))\right)} e^{-\frac{m_a d_t(x_0, y_0)^2}{4t}}. \end{aligned}$$

Then there exists $z_0 \in B_t(y_0, \sqrt{\frac{t}{4}})$ such that:

$$P^2(z_0, t, x_0, 0) \leq q_a \frac{e^{\frac{1}{2}\int_0^{t} \sup_M R(u,.) - \inf_M R(u,.) du}}{\left(\mu_0(B_0(x_0, \sqrt{t})) \mu_t(B_t(y_0, \sqrt{\frac{t}{4}})\right)} e^{-\frac{m_a d_t(x_0, y_0)^2}{4t}}$$

By Lemma 11.3 (comparison of volume)

$$P(z_0,t,x_0,0) \le (q_a)^{\frac{1}{2}}(\psi(t))^{\frac{1}{2}}\frac{e^{\frac{1}{4}\int_0^t \sup_M R(u,.)-\inf_M R(u,.)du}}{\sqrt{(\mu_0(B_0(x_0,\sqrt{t}))\mu_0(B_0(y_0,\sqrt{\frac{t}{4}}))}}e^{-\frac{m_a d_t(x_0,y_0)^2}{8t}},$$

where $\psi(t) = e^{\frac{1}{2}\int_0^t \sup_M R(u,.)du}$.

We conclude the proof by using Lemma 11.2 (for $f(t,x) := P(x,t,x_0,0)$) to compare the solution of the heat equation at different points. We have:

$$\begin{aligned} P(y_0,t,x_0,0) &\le \sqrt{P(z_0,t,x_0,0)}\sqrt{\sup_M P(.,\frac{t}{2},x_0,0)e^{\frac{d_t(z_0,y_0)^2}{t}}} \\ &\le \sqrt{P(z_0,t,x_0,0)}\sqrt{\sup_M P(.,\frac{t}{2},x_0,0)e^{\frac{1}{4}}} \end{aligned}$$

Note that using Remark 11.7 and H1 we have, where c_n is the constant coming from H1:

$$\begin{aligned} \sup_M P(.,\frac{t}{2},x_0,0) &\le \frac{e\psi(\frac{t}{4})^{\frac{1}{2}}e^{\frac{1}{4}\int_0^{\frac{t}{2}}\left(\sup_M R(s,.)-\inf_M R(s,.)\right)ds}}{\left(\mu_0(B_0(x_0,\sqrt{\frac{t}{4}}))\right)^{\frac{1}{2}}c_n^{\frac{1}{2}}\left(\sqrt{\frac{t}{4}}\right)^{\frac{n}{2}}} \\ &\le \frac{\tilde{c}_n^{\frac{1}{2}}e\psi(\frac{t}{4})^{\frac{1}{2}}e^{\frac{1}{4}\int_0^{\frac{t}{2}}\left(\sup_M R(s,.)-\inf_M R(s,.)\right)ds}}{\left(\mu_0(B_0(x_0,\sqrt{\frac{t}{4}}))\right)^{\frac{1}{2}}c_n^{\frac{1}{2}}\left(\mu_0(B_0(y_0,\sqrt{\frac{t}{4}}))\right)^{\frac{1}{2}}} \\ &\le q_n\frac{\psi(t)^{\frac{1}{2}}e^{\frac{1}{4}\int_0^t\left(\sup_M R(s,.)-\inf_M R(s,.)\right)ds}}{\left(\mu_0(B_0(x_0,\sqrt{t}))\right)^{\frac{1}{2}}\left(\mu_0(B_0(y_0,\sqrt{t}))\right)^{\frac{1}{2}}}, \end{aligned}$$

where in second and last inequality we use Bishop-Gromov theorem to compare volume of ball in positive Ricci curvature case to the corresponding Euclidean volume, i.e. ($r \mapsto \frac{\mu(B(x,r))}{\tilde{c}_n r^n}$ is non increasing and smaller than 1) and $\tilde{c}_n$ is the volume of the Euclidean unit ball in dimension n, we have $q_n = \frac{\tilde{c}_n^{\frac{1}{2}}e2^n}{c_n^{\frac{1}{2}}}$. Hence we have for $\tilde{q}_n = (q_n 2^{\frac{n}{2}})^{\frac{1}{2}}$:

$$P(y_0,t,x_0,0) \le \tilde{q}_n(q_a)^{\frac{1}{4}}\frac{\psi(t)^{\frac{1}{2}}e^{\frac{1}{4}\int_0^t\left(\sup_M R(s,.)-\inf_M R(s,.)\right)ds}}{\left(\mu_0(B_0(x_0,\sqrt{t}))\right)^{\frac{1}{2}}\left(\mu_0(B_0(y_0,\sqrt{t}))\right)^{\frac{1}{2}}}e^{-\frac{m_a d_t(x_0,y_0)^2}{16t}}.$$

Case 2 points are close.

For points x_0, y_0 which are closed that is $d_t(x_0, y_0) \leq \sqrt{t}$ the above inequality is a consequence of 11.7 since in this case $e^{\frac{-m_a}{16}} \leq e^{\frac{-m_a d_t^2(x_0,y_0)}{16t}}$.

Then after changing the function ψ we get:

$$P(y_0, t, x_0, 0) \leq \tilde{q}_a \frac{e^{\int_0^t \frac{1}{2} \sup_M R(u,.) - \frac{1}{4} \inf_M R(u,.) du}}{\left(\mu_0(B_0(x_0, \sqrt{t}))\right)^{\frac{1}{2}} \mu_0(B_0(y_0, \sqrt{t}))^{\frac{1}{2}}} e^{-\frac{m_a d_t(x_0,y_0)^2}{16t}}$$

where

$$\tilde{q}_a = (\tilde{q}_n(2ea^{\frac{n}{2}})^{\frac{1}{4}}) \vee (e2^{\frac{n}{2}} e^{\frac{m_a}{16}}), m_a = \frac{(e \ln a)^5}{a^2(a-1)5^5}. \tag{11.16}$$

□

Proposition 11.8 *With the same Hypothesis as in the Theorem 11.3, there exist* $c, \lambda > 0$ *that depend on* n, c_n, m_a *such that we get the following lower bound estimate:*

$$c \frac{e^{\int_0^t -\lambda \sup_M R(u,.) + \frac{1}{4} \inf_M R(u,.)\, du}}{\mu_0(B_0(z_0, \sqrt{t})^{\frac{1}{2}} \mu_0(B_0(y, \sqrt{t})^{\frac{1}{2}}} e^{\frac{-4d_t(z_0,y)^2}{t}} \leq P(z_0, t, y, 0).$$

Proof We will get the lower bound from the upper bound. Since

$$\begin{aligned}\frac{d}{dt} \int_M P(x, t, y, 0)\mu_t(dx) &= -\frac{1}{2} \int_M P(x, t, y, 0) R(t, x)\mu_t(dx) \\ &\geq -\frac{1}{2} \sup_M R(t, .) \int_M P(x, t, y, 0)\mu_t(dx)\end{aligned}$$

we have

$$\int_M P(x, t, y, 0)\mu_t(dx) \geq e^{-\int_0^t \frac{1}{2} \sup_M R(t,.)}.$$

Let $\beta > 0$ large enough, we use Schwarz's inequality:

$$\begin{aligned}&\int_{B_t(y,\sqrt{\beta t})} P^2(x, t, y, 0)\mu_t(dx) \\ &\geq \frac{1}{\mu_t(B_t(y, \sqrt{\beta t}))} \left(\int_{B_t(y,\sqrt{\beta t})} P(x, t, y, 0)\mu_t(dx)\right)^2 = \frac{1}{\mu_t(B_t(y, \sqrt{\beta t}))} \\ &\times \left(\int_M P(x, t, y, 0)\mu_t(dx) - \int_{M \setminus B_t(y,\sqrt{\beta t})} P(x, t, y, 0)\mu_t(dx)\right)^2.\end{aligned}$$

We will use the upper bound in Theorem 11.3 and H1 to get an estimate of the last term in the above equation. Let $I(t) := e^{\int_0^t \frac{1}{2}\sup_M R(u,.)-\frac{1}{4}\inf_M R(u,.)du}$

$$\begin{aligned}
\int_{M\setminus B_t(y,\sqrt{\beta t})} P(x,t,y,0)\mu_t(dx) &\leq \frac{q_a I(t)}{c_n t^{\frac{n}{2}}}\int_{M\setminus B_t(y,\sqrt{\beta t})} e^{-\frac{m_a d_t(x,y)^2}{16t}}\mu_t(dx)\\
&\leq \frac{q_a I(t)}{c_n t^{\frac{n}{2}}}\int_{M\setminus B_t(y,\sqrt{\beta t})} e^{-\frac{m_a d_t(x,y)^2}{32t}}\mu_t(dx)e^{-\frac{m_a\beta}{32}}\\
&\leq \frac{q_a I(t)e^{-\frac{m_a\beta}{32}}}{c_n t^{\frac{n}{2}}}\sum_{k=1}^{\infty}\int_{B_t(y,2^k\sqrt{\beta t})\setminus B_t(y,2^{k-1}\sqrt{\beta t})} e^{-\frac{m_a d_t(x,y)^2}{32t}}\mu_t(dx)\\
&\leq \frac{q_a I(t)e^{-\frac{m_a\beta}{32}}}{c_n t^{\frac{n}{2}}}\sum_{k=1}^{\infty} e^{-\frac{m_a 2^{2(k-1)}\beta}{32}}\mu_t(B_t(y,2^k\sqrt{\beta t}))\\
&\leq \frac{q_a I(t)e^{-\frac{m_a\beta}{32}}}{c_n}\sum_{k=1}^{\infty} e^{-\frac{m_a 2^{2(k-1)}\beta}{32}}\tilde{c}_n(2^k\sqrt{\beta})^n,
\end{aligned}$$

where use H2 and the volume comparison theorem in the last inequality. Let $\beta = 2(\frac{1}{2}\int_0^t \sup_M R(u,.)\,du + C)/(m_a/32)$, since for all $k \geq 0$, $2^{2(k)} \geq k+1$ and $R(u,.) \geq 0$:

$$\begin{aligned}
I(t)\sum_{k=1}^{\infty} e^{-\frac{m_a 2^{2(k-1)}\beta}{32}}(2^k\sqrt{\beta})^n &\leq I(t)2^n\beta^{\frac{n}{2}}\sum_{k=0}^{\infty} e^{-\frac{m_a(k+1)\beta}{32}}2^{nk}\\
&\leq I(t)\beta^{\frac{n}{2}}e^{-\frac{m_a\beta}{32}}\frac{2^n}{1-2^n(e^{-\frac{m_a\beta}{32}})}\\
&\leq e^{-\frac{1}{2}\int_0^t \sup_M R(u,.)\,du - C}\beta^{\frac{n}{2}}\frac{2^n}{1-2^n(e^{-\frac{m_a\beta}{32}})}.
\end{aligned}$$

The last term goes to 0 as C tends to infinity. Hence for a constant C large enough (that only depends on n, c_n, q_a and m_a) and independent on t, we have:

$$\int_{M\setminus B_t(y,\sqrt{\beta t})} P(x,t,y,0)\mu_t(dx) \leq \frac{1}{2}e^{-\frac{m_a\beta}{32}} \leq \frac{1}{2}e^{-\frac{1}{2}\int_0^t \sup_M R(u,.)\,du}.$$

Thus

$$\int_{B_t(y,\sqrt{\beta t})} P^2(x,t,y,0)\mu_t(dx) \geq \frac{1}{4\mu_t(B_t(y,\sqrt{\beta t}))}(e^{-\frac{1}{2}\int_0^t \sup_M R(u,.)\,du})^2.$$

Hence there exist $x_1 \in B_t(y, \sqrt{\beta t})$ such that:

$$P(x_1, t, y, 0) \geq \frac{1}{2\mu_t(B_t(y, \sqrt{\beta t}))}(e^{-\frac{1}{2}\int_0^t \sup_M R(u,.)\,du})$$

Using the volume comparison theorem, Hypotheses H2 and H1, we have:

$$\mu_t(B_t(y, \sqrt{\beta t})) \leq \tilde{c}_n \beta^{\frac{n}{2}} t^{\frac{n}{2}} \leq \frac{\tilde{c}_n}{c_n} \beta^{\frac{n}{2}} \mu_0(B_0(y, \sqrt{t})),$$

so there exists a constant cst that depends on the dimension and the constant c_n of H1

$$P(x_1, t, y, 0) \geq cst \frac{1}{\mu_0(B_0(y, \sqrt{t}))} \frac{e^{-\frac{1}{2}\int_0^t \sup_M R(u,.)\,du}}{(\int_0^t \sup_M R(u, .)\,du + 2C)/(m_a/32))^{\frac{n}{2}}}.$$

Since for all $x \geq 0$ we have

$$\frac{e^{-\frac{x}{2}}}{(x + 2C)^{\frac{n}{2}}} \geq \frac{e^{-(\frac{1}{2}+\frac{n}{4C})x}}{(2C)^{\frac{n}{2}}}$$

we get for a $cst(n, c_n, m_a)$ that can change from a line to line and using H2:

$$P(x_1, t, y, 0) \geq cst \frac{e^{-(\frac{1}{2}+\frac{n}{4C})\int_0^t \sup_M R(u,.)\,du}}{\mu_0(B_0(y, \sqrt{t}))} \geq cst \frac{e^{-(\frac{1}{2}+\frac{n}{4C})\int_0^t \sup_M R(u,.)\,du}}{t^{\frac{n}{2}}}.$$

We conclude the proof by using lemma 11.2 (for $f(t, x) := P(x, t, y, 0)$) to compare the solution of the heat equation at different points. We have for all z_0:

$$P(x_1, t, y, 0) \leq \sqrt{P(z_0, t, y, 0)} \sqrt{\sup_M P(., \frac{t}{2}, y, 0) e^{\frac{d_t(z_0,x_1)^2}{t}}}$$

With the triangle inequality

$$d_t(z_0, x_1)^2 \leq 2d_t(z_0, y)^2 + 2\beta t$$

and so

$$P(x_1, t, y, 0) \leq \sqrt{P(z_0, t, y, 0)} \sqrt{\sup_M P(., \frac{t}{2}, y, 0) e^{\frac{2d_t(z_0,y)^2}{t}} e^{2\beta}}.$$

As in the proof of Theorem 11.3 and using H1 we get:

$$\sup_M P(.,\frac{t}{2},y,0) \le cst\frac{e^{\int_0^t \left(\frac{1}{2}\sup_M R(s,.)-\frac{1}{4}\inf_M R(s,.)\right)ds}}{\left(\mu_0(B_0(z_0,\sqrt{t}))\right)^{\frac{1}{2}}\left(\mu_0(B_0(y,\sqrt{t}))\right)^{\frac{1}{2}}}$$
$$\le cst\frac{e^{\int_0^t \left(\frac{1}{2}\sup_M R(s,.)-\frac{1}{4}\inf_M R(s,.)\right)ds}}{t^{\frac{n}{2}}}.$$

Hence:

$$cst\frac{e^{-(\frac{1}{2}+\frac{n}{4C})\int_0^t \sup_M R(u,.)\,du}}{t^{\frac{n}{2}}} \le P(x_1,t,y,0) \tag{11.17}$$
$$\le \sqrt{P(z_0,t,y,0)}\sqrt{\sup_M P(.,\frac{t}{2},y,0)e^{\frac{2d_t(z_0,y)^2}{t}}}\,e^{2\beta}.$$

Thus

$$cst\frac{e^{-(\frac{1}{2}+\frac{n}{4C})\int_0^t \sup_M R(u,.)\,du}}{t^{\frac{n}{2}}}$$
$$\le \sqrt{P(z_0,t,y,0)}\frac{e^{\int_0^t \left(\frac{1}{4}\sup_M R(s,.)-\frac{1}{8}\inf_M R(s,.)\right)ds}}{t^{\frac{n}{4}}}e^{\frac{2d_t(z_0,y)^2}{t}}e^{2\beta}$$

We recall the definition of β there exist some constants $c, \lambda > 0$ that depend on n, c_n, m_a such that:

$$c\frac{e^{\int_0^t -\lambda\sup_M R(u,.)+\frac{1}{8}\inf_M R(u,.)\,du}}{t^{\frac{n}{4}}}e^{\frac{-2d_t(z_0,y)^2}{t}} \le \sqrt{P(z_0,t,y,0)}.$$

After using H1 again for a constant that could change from line to line, we have:

$$c\frac{e^{\int_0^t -\lambda\sup_M R(u,.)+\frac{1}{4}\inf_M R(u,.)\,du}}{\mu_0(B_0(z_0,\sqrt{t})^{\frac{1}{2}}\mu_0(B_0(y,\sqrt{t})^{\frac{1}{2}}}e^{\frac{-4d_t(z_0,y)^2}{t}} \le P(z_0,t,y,0),$$

which is the desired lower bound. □

Remark 11.8 The constants $\frac{1}{2}$ and $\frac{1}{4}$ are far from being optimal in Theorem 11.3. Hypothesis H1 and H2 are used many time to compare the volume of some ball and the Euclidean one's.

Note that for manifold of dimension three, H2 reduce to $\text{Ric}_{g(0)} \ge 0$ e.g. in Corollary 9.2 [18] and page 193 in [8].

The existence of Ricci flow for a complete non-compact manifold at least for short time have been proved in [24] Theorem 2.1, under boundedness conditions for the Riemannian tensor of $(M, g(0))$. There also have some uniqueness results in this direction in [6] Theorem 1.1.

Appendix

Lemma 11.4 *With the same hypothesis as in Theorem 11.3 and suppose* $\int_0^t \sup_M R(s,.)\,ds < \infty$ *for all* $t \in [0, T_c[$. *Let* $a > 1$ *be a constant,* $r_k := (\frac{1}{2} + \frac{1}{k+2})r$ *and* $t_k := \frac{t}{a^k}$. *We have*

$$\lim_{k \longrightarrow \infty} I_{r_k}(t_k) = 0.$$

Proof We use H1 and $\int_0^t \sup_M R(s,.)\,ds < \infty$ to get a global polynomial bound of the heat kernel i.e. Remark 11.7. We obtain that for all $t \leq \frac{T_c}{2}$:

$$\begin{aligned} P(x,t,y,0) &\leq \frac{Cst}{\left(\mu_0(B_0(x,\sqrt{\frac{t}{2}}))\right)^{\frac{1}{2}} \mu_0(B_0(y,\sqrt{\frac{t}{2}}))\big)^{\frac{1}{2}}} \\ &\leq \frac{cst}{t^{\frac{n}{2}}}. \end{aligned}$$

Then for a constant which could change from line to line:

$$\begin{aligned} I_r(t) &= \int_{(M\setminus B_t(y,r))} P^2(x,t,y,0)\mu_t(dx) \leq \frac{cst}{t^{\frac{n}{2}}} \int_{(M\setminus B_t(y,r))} P(x,t,y,0)\mu_t(dx) \\ &\leq \frac{cst}{t^{\frac{n}{2}}} \int_{(M\setminus B_t(y,r))} P^*(y,0,x,t)\mu_t(dx) \\ &\leq \frac{cst}{t^{\frac{n}{2}}} \int_{(M\setminus B_t(y,r))} \overline{P}(y,t,x,t)\mu_t(dx) \\ &\leq cst \frac{\mathbb{P}_y(\tau_r < t)}{t^{\frac{n}{2}}} \end{aligned}$$

where in $\tau_r := \inf\{t > 0, d_t(X_t(y), y) = r\}$, and $X_t(y)$ is a $g(t)$-Brownian motion started at y, note that $X_t(y)$ does not explode using H2 and Proposition 11.3.

Let $\rho_t := d_t(X_t(y), y)$, use (11.4) and Itô's formula we get for the real Brownian motion $b_t := \int_0^t \mathbb{1}_{X_s \notin Cut_s(y)} \langle \nabla^s d_s(y, X_s), d^{\nabla_s} X_s \rangle_{g(s)}$

$$\begin{aligned} d\rho_t^2 = 2\rho_t d\rho_t + dt &\leq 2\rho_t \big(\mathbb{1}_{X_t \notin Cut_t(x_0)} (\frac{1}{2}\Delta_{g(t)} d_t(y,.) \\ &+ \frac{\partial d_t(y,.)}{\partial t})(X_t)\big)\,dt + dt + 2\rho_t db_t. \end{aligned}$$

By the Laplacian comparison theorem (Theorem 3.4.2 in [19]) we also have, since by H2, $\mathrm{Ric}_{g(t)} \geq 0$ for all $t \in [0, T_c[$, within the cutlocus:

$$\Delta_{g(t)} d_t(y, .) \leq \frac{n-1}{d_t(y, .)} \quad \text{and} \quad \frac{\partial d_t(y, .)}{\partial t} \leq 0.$$

Hence

$$d\rho_t^2 \leq n dt + 2\rho_t db_t,$$

at time $t = \tau_r$,

$$r^2 \leq n\tau_r + \int_0^{\tau_r} 2\rho_t db_t.$$

On the event $\{\tau_r \leq t\}$,

$$\frac{r^2 - nt}{2} \leq \int_0^{\tau_r} \rho_t db_t.$$

Note that $\int_0^{\tau_r} \rho_t db_t = W_T$, where W is a Brownian motion and $T = \int_0^{\tau_r} \rho_s^2 \, ds$. On the event $\{\tau_r \leq t\}$ we also have $T \leq r^2 t$. Then the event $\{\tau_r \leq t\}$ implies

$$\frac{r^2 - nt}{2} \leq W_T \leq \sup_{s \in [0, r^2 t]} W_s \sim r\sqrt{t}|W_1|,$$

hence for $r^2 - nt > 0$,

$$\begin{aligned} \mathbb{P}_y(\tau_r < t) &\leq \mathbb{P}_0\left(\frac{r^2 - nt}{2r\sqrt{t}} \leq |W_1|\right) \\ &\leq \frac{2}{\sqrt{2\pi}\left(\frac{r^2-nt}{2r\sqrt{t}}\right)} e^{-\left(\frac{r^2-nt}{2r\sqrt{t}}\right)^2}. \end{aligned}$$

Since $a > 1$ be a constant $r_k := (\frac{1}{2} + \frac{1}{k+2})r$ and $t_k := \frac{t}{a^k}$, we have

$$\frac{\mathbb{P}_y(\tau_{r_k} < t_k)}{t_k^{\frac{n}{2}}} \xrightarrow[k \to \infty]{} 0$$

and then

$$\lim_{k \longrightarrow \infty} I_{r_k}(t_k) = 0.$$

□

References

1. M. Arnaudon, K.A. Coulibaly, A. Thalmaier, Brownian motion with respect to a metric depending on time: definition, existence and applications to Ricci flow. C.R. Math. **346**(13–14), 773–778 (2008)
2. M. Arnaudon, K.A. Coulibaly, A. Thalmaier, Horizontal diffusion in C^1 path space, in *Séminaire de Probabilités XLIII*, Lecture Notes in Mathematics, vol. 2006 (Springer, Berlin, 2011), pp. 73–94
3. M. Arnaudon, A. Thalmaier, F.-Y. Wang, Harnack inequality and heat kernel estimates on manifolds with curvature unbounded below. Bull. Sci. Math. **130**(3), 223–233 (2006)
4. D. Bakry, I. Gentil, M. Ledoux. On Harnack inequalities and optimal transportation. Ann. Sc. Norm. Super. Pisa Cl. Sci. (5), **14**(3), 705–727 (2015)
5. X. Cao, Q.S. Zhang, The conjugate heat equation and ancient solutions of the Ricci flow. Adv. Math. **228**(5), 2891–2919 (2011)
6. B.-L. Chen, X.-P. Zhu, Uniqueness of the Ricci flow on complete noncompact manifolds. J. Differ. Geom. **74**(1), 119–154 (2006)
7. L.-J. Cheng, Diffusion semigroup on manifolds with time-dependent metrics. Forum Math. **29**(4), 775–798 (2017)
8. B. Chow, S.-C. Chu, D. Glickenstein, C. Guenther, J. Isenberg, T. Ivey, D. Knopf, P. Lu, F. Luo, L. Ni, The Ricci flow: techniques and applications. Part II, in *Mathematical Surveys and Monographs*, vol. 144 (American Mathematical Society, Providence, 2008). Analytic aspects
9. B. Chow, S.-C. Chu, D. Glickenstein, C. Guenther, J. Isenberg, T. Ivey, D. Knopf, P. Lu, F. Luo, L. Ni, The Ricci flow: techniques and applications. Part III. Geometric-analytic aspects, in *Mathematical Surveys and Monographs*, vol. 163 (American Mathematical Society, Providence, 2010)
10. K.A. Coulibaly-Pasquier, Brownian motion with respect to time-changing Riemannian metrics, applications to Ricci flow. Ann. Inst. Henri Poincaré Probab. Stat. **47**(2), 515–538 (2011)
11. K.A. Coulibaly-Pasquier, Some stochastic process without birth, linked to the mean curvature flow. Ann. Probab. **39**(4), 1305–1331 (2011)
12. M. Gage, R.S. Hamilton, The heat equation shrinking convex plane curves. J. Differ. Geom. **23**(1), 69–96 (1986)
13. S. Gallot, D. Hulin, J. Lafontaine, *Riemannian Geometry*, 3rd edn. Universitext (Springer, Berlin, 2004)
14. F.-Z. Gong, F.-Y. Wang, Heat kernel estimates with application to compactness of manifolds. Q. J. Math. **52**(2), 171–180 (2001)
15. A. Grigor´ yan, Integral maximum principle and its applications. Proc. Roy. Soc. Edinburgh Sect. A **124**(2), 353–362 (1994)
16. A. Grigor´ yan, Gaussian upper bounds for the heat kernel on arbitrary manifolds. J. Differ. Geom. **45**(1), 33–52 (1997)
17. C.M. Guenther, The fundamental solution on manifolds with time-dependent metrics. J. Geom. Anal. **12**(3), 425–436 (2002)
18. R.S. Hamilton, Three-manifolds with positive Ricci curvature. J. Differ. Geom. 17(2), 255–306 (1982)
19. E.P. Hsu, Stochastic analysis on manifolds, in *Graduate Studies in Mathematics*, vol. 38 (American Mathematical Society, Providence, 2002)
20. K. Kuwada, R. Philipowski, Non-explosion of diffusion processes on manifolds with time-dependent metric. Math. Z. **268**(3–4), 979–991 (2011)
21. P. Li, S.-T. Yau, On the parabolic kernel of the Schrödinger operator. Acta Math. **156**(3–4), 153–201 (1986)
22. R.J. McCann, P.M. Topping, Ricci flow, entropy and optimal transportation. Am. J. Math. **132**(3), 711–730 (2010)

23. L. Ni, Ricci flow and nonnegativity of sectional curvature. Math. Res. Lett. **11**(5–6), 883–904 (2004)
24. W.-X. Shi, Deforming the metric on complete Riemannian manifolds. J. Differ. Geom. **30**(1), 223–301 (1989)
25. F.-Y. Wang, Logarithmic Sobolev inequalities on noncompact Riemannian manifolds. Probab. Theory Relat. Fields **109**(3), 417–424 (1997)

Chapter 12
Scaled Penalization of Brownian Motion with Drift and the Brownian Ascent

Hugo Panzo

Abstract We study a scaled version of a two-parameter Brownian penalization model introduced by Roynette-Vallois-Yor (Period Math Hungar 50:247–280, 2005). The original model penalizes Brownian motion with drift $h \in \mathbb{R}$ by the weight process $\big(\exp(\nu S_t) : t \geq 0\big)$ where $\nu \in \mathbb{R}$ and $\big(S_t : t \geq 0\big)$ is the running maximum of the Brownian motion. It was shown there that the resulting penalized process exhibits three distinct phases corresponding to different regions of the (ν, h)-plane. In this paper, we investigate the effect of penalizing the Brownian motion concurrently with scaling and identify the limit process. This extends an existing result for the $\nu < 0,\ h = 0$ case to the whole parameter plane and reveals two additional "critical" phases occurring at the boundaries between the parameter regions. One of these novel phases is Brownian motion conditioned to end at its maximum, a process we call the *Brownian ascent*. We then relate the Brownian ascent to some well-known Brownian path fragments and to a random scaling transformation of Brownian motion that has attracted recent interest.

12.1 Introduction

Brownian penalization was introduced by Roynette, Vallois, and Yor in a series of papers where they considered limit laws of the Wiener measure perturbed by various weight processes, see the monographs [15, 28] for a complete list of the early works. One motivation for studying Brownian penalizations is that they can be seen as a way to condition Wiener measure by an event of probability 0. Another reason penalizations are interesting is that they often exhibit phase transitions typical of statistical mechanics models. Let $C\big(\mathbb{R}_+; \mathbb{R}\big)$ denote the space of continuous functions from $[0, \infty)$ to $\mathbb{R}$ and let $X = (X_t : t \geq 0)$ denote the canonical process on this space. For each $t \geq 0$, let $\mathcal{F}_t$ denote the σ-algebra generated by

H. Panzo (✉)
Department of Mathematics, University of Connecticut, Storrs, CT, USA
e-mail: hugo.panzo@uconn.edu

C. Donati-Martin et al. (eds.), *Séminaire de Probabilités L*, Lecture Notes in Mathematics 2252, https://doi.org/10.1007/978-3-030-28535-7_12

$(X_s : s \le t)$, and let $\mathcal{F}_\infty$ denote the σ-algebra generated by $\cup_{t\ge 0}\mathcal{F}_t$. We write $\mathcal{F}$ for the filtration $(\mathcal{F}_t : t \ge 0)$. Let P_0 denote the Wiener measure on $\mathcal{F}_\infty$, that is the unique measure under which X is standard Brownian motion. As usual, we write E_0 for the corresponding expectation. Our starting point is the following definition of Brownian penalization adapted from [21].

Definition 12.1 Suppose the $\mathcal{F}$-adapted weight process $\Gamma = (\Gamma_t : t \ge 0)$ takes non-negative values and that $0 < E_0[\Gamma_t] < \infty$ for $t \ge 0$. For each $t \ge 0$, consider the Gibbs probability measure on $\mathcal{F}_\infty$ defined by

$$Q_t^\Gamma(\Lambda) := \frac{E_0[1_\Lambda \Gamma_t]}{E_0[\Gamma_t]}, \quad \Lambda \in \mathcal{F}_\infty. \tag{12.1}$$

We say that the weight process Γ satisfies the *penalization principle for Brownian motion* if there exists a probability measure Q^Γ on $\mathcal{F}_\infty$ such that:

$$\forall s \ge 0,\ \forall \Lambda_s \in \mathcal{F}_s,\quad \lim_{t\to\infty} Q_t^\Gamma(\Lambda_s) = Q^\Gamma(\Lambda_s). \tag{12.2}$$

In this case Q^Γ is called *Wiener measure penalized by* Γ or simply *penalized Wiener measure* when there is no ambiguity. Similarly, $(X_t : t \ge 0)$ under Q^Γ is called *Brownian motion penalized by* Γ or *penalized Brownian motion*.

Remark 12.1 Let P_x denote the distribution of Brownian motion starting at $x \in \mathbb{R}$ and E_x its corresponding expectation. By replacing E_0 with E_x in (12.1), it is straightforward to modify the above definition of Q_t^Γ and Q^Γ to yield the analogous measures $Q_{x,t}^\Gamma$ and Q_x^Γ which account for a general starting point. Since the main focus of our work is on computing scaling limits, and any fixed starting point would be scaled to 0, we restrict our attention to penalizing P_0 for the sake of clarity.

In many cases of interest, a one-parameter family of weight processes $(\Gamma^\nu : \nu \in \mathbb{R})$ is considered. The parameter ν allows us to adjust or "tune" the strength of penalization and plays a role similar to that of the inverse temperature in statistical mechanics models. In this case we write Q_t^ν and Q^ν for the corresponding Gibbs and penalized measures, respectively. This notational practice is modified accordingly when dealing with a two-parameter weight process.

A natural choice for Γ are non-negative functions of the running maximum $S_t = \sup_{s\le t} X_s$ and these *supremum penalizations* were considered by the aforementioned authors in [30]. Their results include as a special case the one-parameter weight process $\Gamma_t = \exp(\nu S_t)$ with $\nu < 0$ which we briefly describe here.

Theorem 12.1 (Roynette-Vallois-Yor [30] Theorem 3.6) *Let* $\Gamma_t = \exp(\nu S_t)$ *with* $\nu < 0$. *Then* Γ *satisfies the penalization principle for Brownian motion and the penalized Wiener measure* Q^ν *has the representation*

$$Q^\nu(\Lambda_t) = E_0\left[1_{\Lambda_t} M_t^\nu\right],\ \forall t \ge 0,\ \forall \Lambda_t \in \mathcal{F}_t$$

where $M^\nu = (M_t^\nu : t \geq 0)$ is a positive $(\mathcal{F}, P_0)$-martingale starting at 1 which has the form

$$M_t^\nu = \exp(\nu S_t) - \nu \exp(\nu S_t)(S_t - X_t).$$

Remark 12.2 M^ν is an example of an *Azéma-Yor martingale*. More generally,

$$F(S_t) - f(S_t)(S_t - X_t),\ t \geq 0$$

is an $(\mathcal{F}, P_0)$-local martingale whenever f is locally integrable and $F(y) = \int_0^y f(x)dx$, see Theorem 3 of [16].

Let $S_\infty = \lim_{t\to\infty} S_t$ which exists in the extended sense due to monotonicity. Then the measure Q^ν can be further described in terms of a path decomposition at S_∞.

Theorem 12.2 (Roynette-Vallois-Yor [30] Theorem 4.6) *Let $\nu < 0$ and Q^ν be as in Theorem 12.1.*

1. *Under Q^ν, the random variable S_∞ is exponential with parameter $-\nu$.*
2. *Let $g = \sup\{t \geq 0 : X_t = S_\infty\}$. Then $Q^\nu(0 < g < \infty) = 1$, and under Q^ν the following hold:*
   - *(a) the processes $(X_t : t \leq g)$ and $(X_g - X_{g+t} : t \geq 0)$ are independent,*
   - *(b) $(X_g - X_{g+t} : t \geq 0)$ is distributed as a Bessel(3) process starting at 0,*
   - *(c) conditionally on $S_\infty = y > 0$, the process $(X_t : t \leq g)$ is distributed as a Brownian motion started at 0 and stopped when it first hits the level y.*

More recently, penalization has been studied for processes other than Brownian motion. In this case, the measure of the underlying processes is referred to as the *reference measure*. Similar supremum penalization results have been attained for simple random walk [6, 7], stable processes [33, 34], and integrated Brownian motion [21]. This paper builds upon the work of Roynette-Vallois-Yor that appeared in [29] where they penalized Brownian motion using the two-parameter weight process $\Gamma_t = \exp(\nu S_t + hX_t)$ with $\nu, h \in \mathbb{R}$. An easy application of Girsanov's theorem shows that this is equivalent to using the weight process $\Gamma_t = \exp(\nu S_t)$ with $\nu \in \mathbb{R}$ and replacing the reference measure P_0 by the distribution of Brownian motion with drift h, henceforth referred to as P_0^h. Now their two-parameter penalization can also be seen as supremum penalization of Brownian motion with drift. In this model, the two parameters ν (for penalization) and h (for drift) can have competing effects, leading to interesting phase transitions. That is to say the penalized Wiener measure is qualitatively different depending on where (ν, h) lies in the parameter plane. To describe the resulting phases, we first partition the parameter plane into six disjoint regions (the origin is excluded to avoid trivialities).

Theorem 12.3 (Roynette-Vallois-Yor [29] Theorem 1.7) *Let $\Gamma_t = \exp(\nu S_t + hX_t)$ with $\nu, h \in \mathbb{R}$. For $\nu < 0$, let Q^ν be as in Theorem 12.1. Then Γ satisfies*

Fig. 12.1
$L_1 = \{(\nu, h) : \nu < 0,\ h = 0\}$
$R_1 = \{(\nu, h) : h < -\nu,\ h > 0\}$
$L_2 = \{(\nu, h) : h = -\nu,\ \nu < 0\}$
$R_2 = \{(\nu, h) : h > -\nu,\ h > -\frac{1}{2}\nu\}$
$L_3 = \{(\nu, h) : h = -\frac{1}{2}\nu,\ \nu > 0\}$
$R_3 = \{(\nu, h) : h < 0,\ h < -\frac{1}{2}\nu\}$

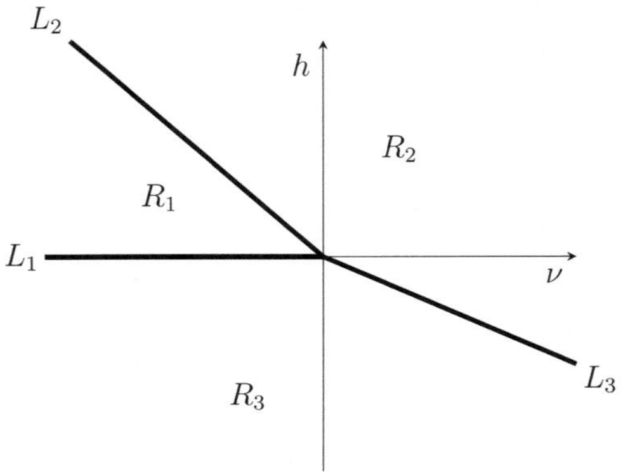

the penalization principle for Brownian motion and the penalized Wiener measure $Q^{\nu,h}$ has three phases which are given by

$$dQ^{\nu,h} = \begin{cases} dQ^{\nu+h} & : (\nu, h) \in L_1 \cup R_1 \\ dP_0^{\nu+h} & : (\nu, h) \in L_2 \cup R_2 \cup L_3 \\ \frac{\nu+2h}{2h} \exp(\nu S_\infty) dP_0^h & : (\nu, h) \in R_3. \end{cases}$$

Remark 12.3 While Theorem 12.3 had only three phases hence needing only three regions in the parameter plane, the rationale behind our choice of six regions in the phase diagram depicted in Fig. 12.1 will become clear when we state our main results.

One topic that has attracted interest is the following: to what extent can sets $\Lambda \in \mathcal{F}_\infty$ replace the sets $\Lambda_s \in \mathcal{F}_s$ in the limit (12.2) which defines penalization? This can't be done in complete generality, e.g. the case of supremum penalization from Theorem 12.2. Here we saw that S_∞ is exponentially distributed under Q^ν, yet it is easy to see that for each $t \geq 0$, $S_\infty = \infty$ almost surely under Q_t^ν. The last chapter of [28] is devoted to studying this question. Intuitively speaking, the effect of penalizing by Γ_t on the probability of an event $\Lambda_s \in \mathcal{F}_s$ should be more pronounced the closer s is to t. Hence letting s keep pace with t as $t \to \infty$ instead of having s remain fixed might result in a different outcome. This leads to the notion of *scaled Brownian penalizations*. Roughly speaking, scaled penalization amounts to penalizing and scaling simultaneously. To be more specific, we first introduce the family of scaled processes $X^{\alpha,t} = (X_s^{\alpha,t} : 0 \leq s \leq 1)$ indexed by $t > 0$ with scaling exponent $\alpha \geq 0$ where $X_s^{\alpha,t} := X_{st}/t^\alpha$. Now with a weight process Γ that satisfies the penalization principle for Brownian motion and the right choice of α, we then compare the weak limits of $X^{\alpha,t}$ under $Q^\Gamma|_{\mathcal{F}_t}$ (penalizing then scaling) and under $Q_t^\Gamma|_{\mathcal{F}_t}$ (scaling and penalizing simultaneously) as $t \to \infty$. Next we give an example of such a result for supremum penalization.

Theorem 12.4 (Roynette-Yor [28] Theorem 4.18) *Let* $(R_s : 0 \leq s \leq 1)$ *and* $(\mathfrak{m}_s : 0 \leq s \leq 1)$ *denote a Bessel(3) process and Brownian meander, respectively. Let* $\Gamma_t = \exp(\nu S_t)$ *with* $\nu < 0$ *and* Q^ν *be the penalized Wiener measure. Then the distribution of* $X^{\frac{1}{2},t}$ *under* $Q^\nu|_{\mathcal{F}_t}$ *and* $Q^\nu_t|_{\mathcal{F}_t}$ *converges weakly to* $(-R_s : 0 \leq s \leq 1)$ *and* $(-\mathfrak{m}_s : 0 \leq s \leq 1)$*, respectively, as* $t \to \infty$.

Notice how the scaling limits are similar (their path measures are mutually absolutely continuous by the Imhof relation (12.48)) yet at the same time quite different (the Bessel(3) process is a time-homogenous Markov process while the Brownian meander is a time-inhomogeneous Markov process). Since Q^ν is just a special case of $Q^{\nu,h}$, a natural question is whether results similar to Theorem 12.4 can be proven for the two-parameter model from Theorem 12.3. This is the primary goal of this paper.

12.1.1 Main Results

Our main results lie in computing the weak limits of $X^{\alpha,t}$ under $Q^{\nu,h}|_{\mathcal{F}_t}$ and $Q^{\nu,h}_t|_{\mathcal{F}_t}$ as $t \to \infty$ for all $(\nu, h) \in \mathbb{R}^2$ with appropriate α. This extends the scaled penalization result of Theorem 12.4 to the whole parameter plane of the two-parameter model from Theorem 12.3. In doing so, we reveal two additional "critical" phases. These new phases correspond to the parameter rays L_2 and L_3 which occur at the interfaces of the other regions, see Fig. 12.1. We call the two novel processes corresponding to these critical phases the *Brownian ascent* and the *up-down process*.

- Loosely speaking, the Brownian ascent $(\mathfrak{a}_s : 0 \leq s \leq 1)$ is a Brownian path of duration 1 conditioned to end at its maximum, i.e. conditioned on the event $\{X_1 = S_1\}$. It can be represented as a path transformation of the Brownian meander, see Sect. 12.6. While the sobriquet *ascent* is introduced in the present paper, this transformed meander process first appeared in [2] in the context of Brownian *first passage bridge*. More recently, the same transformed meander has appeared in [26], although no connection is made in that paper to [2] or Brownian motion conditioned to end at its maximum.
- The up-down process is a random mixture of deterministic up-down paths that we now describe. For $0 \leq \Theta \leq 1$, let $\mathfrak{u}^\Theta = (\mathfrak{u}^\Theta_s : 0 \leq s \leq 1)$ be the continuous path defined by

$$\mathfrak{u}^\Theta_s = \Theta - |s - \Theta|, \ 0 \leq s \leq 1.$$

It is easy to see that $\mathfrak{u}^\Theta$ linearly interpolates between the points $(0, 0)$, (Θ, Θ), and $(1, 2\Theta - 1)$ when $0 < \Theta < 1$, while $\mathfrak{u}^0$ or $\mathfrak{u}^1$ is simply the path of constant slope -1 or 1, respectively. To simplify notation, we write $\mathfrak{u}$ instead of $\mathfrak{u}^1$. The up-down process with slope h is defined as the process $h\mathfrak{u}^U$ where U is a Uniform[0, 1] random variable. Roughly speaking, the up-down process of slope h starts at 0 and moves with slope h until a random Uniform[0, 1] time U then it moves with slope $-h$ for the remaining time $1-U$. In our results the initial slope is always positive, hence the name up-down.

Theorem 12.5 *Let $W=(W_s:0\le s\le 1)$ and $R=(R_s:0\le s\le 1)$ be Brownian motion and Bessel(3) processes, each starting at* 0. *Let $\mathfrak{a}$ be a Brownian ascent, $\mathfrak{m}$ be a Brownian meander, and $\mathfrak{e}$ be a normalized Brownian excursion. Let $\mathfrak{u}^U$ be the up-down process with U a Uniform*[0, 1] *random variable. Then the weak limits of $X^{\alpha,t}$ as $t\to\infty$ are as follows:*

Region	α	Limit under $Q^{\nu,h}\vert_{\mathcal{F}_t}$	Limit under $Q_t^{\nu,h}\vert_{\mathcal{F}_t}$	Proof
L_1	1/2	$-R$	$-\mathfrak{m}$	Theorem 12.4
R_1			$-\mathfrak{e}$	Section 12.3.4
L_2		W	$\mathfrak{a}$	Section 12.3.1
R_2	1	$(\nu+h)\mathfrak{u}$		Section 12.3.2
L_3			$-h\mathfrak{u}^U$	Section 12.3.3
R_3		$h\mathfrak{u}$		Section 12.3.2

Remark 12.4 The L_1 row is a restatement of Theorem 12.4. All remaining rows are new.

Remark 12.5 Note the abrupt change in behavior when going from L_3 to R_3 in the $Q^{\nu,h}\vert_{\mathcal{F}_t}$ column, namely going from slope $\nu+h=-h$ to slope h. However, in the $Q_t^{\nu,h}\vert_{\mathcal{F}_t}$ column, i.e. the scaled penalization, the change from R_2 to L_3 to R_3 is in some sense less abrupt because the up-down process switches from R_2 behavior (slope $\nu+h=-h$) to R_3 behavior (slope h) at a random Uniform[0, 1] time.

With the exception of the up-down process, the limits in the $\alpha=1$ regions (ballistic scaling) are all deterministic. This suggests that we try subtracting the deterministic drift from the canonical process and then scale this centered process. This leads to a functional central limit theorem which is proved in Sect. 12.4.

Theorem 12.6 *Let $W=(W_s:0\le s\le 1)$ be Brownian motion starting at* 0. *Then the weak limits of the centered and scaled canonical process as $t\to\infty$ are as follows:*

Region	Process	Limit under $Q^{\nu,h}\vert_{\mathcal{F}_t}$	Limit under $Q_t^{\nu,h}\vert_{\mathcal{F}_t}$
R_2	$\frac{X_{\bullet t}-(\nu+h)t\mathfrak{u}_\bullet}{\sqrt{t}}$	W	
L_3			see Theorem 12.7
R_3	$\frac{X_{\bullet t}-ht\mathfrak{u}_\bullet}{\sqrt{t}}$		

Here the $\bullet$ stands for a time parameter that ranges between 0 and 1. For example, this notation allows us to write $X_{\bullet t}-ht\mathfrak{u}_\bullet$ rather than the unwieldy $(X_{st}-ht\mathfrak{u}_s:0\le s\le 1)$. The center-right entry in the above table is exceptional because in that case we don't have a deterministic drift that can be subtracted in order to center the process. In light of the L_3 row of Theorem 12.5, we need to subtract a random drift that switches from $-h$ to h at an appropriate time. More specifically, define $\Theta_t = \inf\{s : X_s = S_t\}$. Then for each $t > 0$, we want to subtract from $X_{\bullet t}$ a continuous process starting at 0 that has drift $-ht$ until time Θ_t/t and has drift ht thereafter. A candidate for this random centering process is $2S_{\bullet t}+ht\mathfrak{u}_\bullet$ which is seen to have the desired up-down drift. Indeed, this allows us to fill in the remaining entry of the table with a functional central limit theorem which is proved in Sect. 12.5.

Theorem 12.7 *Let* $W = (W_s : 0 \leq s \leq 1)$ *be Brownian motion starting at* 0. *If* $(\nu, h) \in L_3$ *then*

$$\frac{X_{\bullet t} - (2S_{\bullet t} + ht\mathfrak{u}_\bullet)}{\sqrt{t}}$$

under $Q_t^{\nu,h}|_{\mathcal{F}_t}$ *converges weakly to* W *as* $t \to \infty$.

12.1.2 Notation

For the remainder of this paper, the non-negative real numbers are denoted by $\mathbb{R}_+$ and we use X for the $C(\mathbb{R}_+; \mathbb{R})$ canonical process under P_x or P_x^h. Abusing notation, we also use X for the $C([0, 1]; \mathbb{R})$ canonical process under the analogous measures P_x or P_x^h. We use a subscript when restricting to paths starting at a particular point, say $C_0([0, 1]; \mathbb{R})$ for paths starting at 0. As explained below Theorem 12.6, we often use the bullet point $\bullet$ to stand for a time parameter that ranges between 0 and 1. Expectation under P_x and P_x^h is denoted by E_x and E_x^h, respectively. Expectation under Q^ν and Q_t^ν is denoted by $Q^\nu[\cdot]$ and $Q_t^\nu[\cdot]$, respectively. We always have S_t and I_t denoting the running extrema of the canonical process, namely $S_t = \sup_{s \leq t} X_s$ and $I_t = \inf_{s \leq t} X_s$. When dealing with the canonical process, we use Θ_t and θ_t to refer to the first time at which the maximum and minimum, respectively, is attained over the time interval $[0, t]$. Also, τ_x and γ_x refer to the first and last hitting times of x, respectively. More specifically, $\Theta_t = \inf\{s : X_s = S_t\}$, $\theta_t = \inf\{s : X_s = I_t\}$, $\tau_x = \inf\{t : X_t = x\}$, and $\gamma_x = \sup\{t : X_t = x\}$. For the Bessel(3) process we use R. A Brownian path of duration 1 starting at 0 is written as W. We use Fraktur letters for the various time-inhomogeneous processes that appear in this paper: Brownian ascent $\mathfrak{a}$, standard Brownian bridge $\mathfrak{b}$, pseudo-Brownian bridge $\tilde{\mathfrak{b}}$, normalized Brownian excursion $\mathfrak{e}$, Brownian meander $\mathfrak{m}$ and co-meander $\tilde{\mathfrak{m}}$, up-down process $\mathfrak{u}^U$, and also for the deterministic path with slope 1 which we write as $\mathfrak{u}$. The path transformations ϕ and Π_δ are described in Definition 12.2 and Definition 12.3, respectively. We use $\|F\|$ to denote the supremum norm of a bounded path functional $F : C([0, 1]; \mathbb{R}) \to \mathbb{R}$.

12.1.3 Organization of Paper

In Sect. 12.2, we introduce a duality relation between regions of the parameter plane and use it to compute the partition function asymptotics. Section 12.3 is devoted to the proof of Theorem 12.5. In Sect. 12.4, we describe a modification of path functionals that allows us to decouple them from an otherwise dependent factor. This is used to prove Theorem 12.6 in the remainder of that section. Section 12.5 contains the proof of Theorem 12.7. In Sect. 12.6, we provide several constructions of the Brownian ascent and give a connection to some recent literature. In Sect. 12.7, we comment on work in preparation and suggest some directions for future research. We gather various known results used throughout the paper and include them along with references in the Appendices.

12.2 Duality and Partition Function Asymptotics

The normalization constant $E_0[\Gamma_t]$ appearing in the Gibbs measure (12.1) is known as the *partition function*. The first step in proving a scaled penalization result is to obtain an asymptotic for the partition function as $t \to \infty$. Indeed, differing asymptotics may indicate different phases and the asymptotic often suggests the right scaling exponent, e.g. diffusive scaling for power law behavior and ballistic scaling for exponential behavior. In this section we compute partition function asymptotics for our two-parameter model in each of the parameter phases. While some of these asymptotics appeared in [29], we derive them all for the sake of completeness. Most of the cases boil down to an application of Watson's lemma or Laplace's method. Refer to Appendix 5 for precise statements of these tools from asymptotic analysis.

To reduce the number of computations required, we make use of a duality relation between the regions R_2 and R_3 and critical lines L_1 and L_2. Recall that P_0 is invariant under the path transformation $X_\bullet \mapsto X_{1-\bullet} - X_1$, for example, see Lemma 2.9.4 in [13]. So the joint distribution of X_1 and S_1 under P_0 coincides with that of $-X_1$ and $\sup_{s \le 1}(X_{1-s} - X_1) = S_1 - X_1$ under P_0. Consequently,

$$\begin{aligned} E_0\left[\exp\left(\nu S_t + hX_t\right)\right] &= E_0\left[\exp\left(\sqrt{t}(\nu S_1 + hX_1)\right)\right] \\ &= E_0\left[\exp\left(\sqrt{t}\big(\nu S_1 - (\nu + h)X_1\big)\right)\right] \\ &= E_0\left[\exp\left(\nu S_t - (\nu + h)X_t\right)\right] \end{aligned}$$

follows from Brownian scaling. An easy calculation while referring to Fig. 12.1 shows that if $(\nu, h) \in L_1$, then $\big(\nu, -(\nu + h)\big) \in L_2$ and vice versa. The same holds for R_2 and R_3. In fact, the map $(\nu, h) \mapsto \big(\nu, -(\nu + h)\big)$ is an involution from R_2 onto R_3 and from L_1 onto L_2. Hence the partition function asymptotics for L_2

and R_2 can be obtained from those of L_1 and R_3 by substituting $\big(\nu, -(\nu+h)\big)$ for (ν, h).

We can push this idea further by applying it to functionals of the entire path. First we need some new definitions.

Definition 12.2 For $X_\bullet \in C\big([0,1];\mathbb{R}\big)$, define $\phi : C\big([0,1];\mathbb{R}\big) \to C\big([0,1];\mathbb{R}\big)$ by

$$\phi X_s = X_{1-s} - (X_1 - X_0),\ 0 \le s \le 1.$$

For $F : C\big([0,1];\mathbb{R}\big) \to \mathbb{R}$, let F_ϕ denote $F \circ \phi$.

This path transformation reverses time and shifts the resulting path so that it starts at the same place. It is clear that ϕ is a linear transformation from $C\big([0,1];\mathbb{R}\big)$ onto $C\big([0,1];\mathbb{R}\big)$ that is continuous and an involution. Additionally, F_ϕ is bounded continuous whenever F is and $\|F_\phi\| = \|F\|$. As mentioned above, P_0 is invariant under ϕ so we have $E_0[F_\phi(X_\bullet)] = E_0[F(X_\bullet)]$. This duality will be exploited again in Sects. 12.3.1, 12.3.2, and 12.4.

Proposition 12.1 *The partition function has the following asymptotics as $t \to \infty$:*

$$E_0[\exp(\nu S_t + hX_t)] \sim \begin{cases} -\frac{1}{\nu}\sqrt{\frac{2}{\pi t}} & : L_1 = \{(\nu,h) : \nu < 0,\ h = 0\} \\ -\frac{\nu}{h^2(\nu+h)^2}\sqrt{\frac{2}{\pi t^3}} & : R_1 = \{(\nu,h) : h < -\nu,\ h > 0\} \\ -\frac{1}{\nu}\sqrt{\frac{2}{\pi t}} & : L_2 = \{(\nu,h) : h = -\nu,\ \nu < 0\} \\ 2\frac{\nu+h}{\nu+2h}\exp\left(\frac{1}{2}(\nu+h)^2 t\right) & : R_2 = \{(\nu,h) : h > -\nu,\ h > -\frac{1}{2}\nu\} \\ 2h^2 t\exp\left(\frac{1}{2}h^2 t\right) & : L_3 = \{(\nu,h) : h = -\frac{1}{2}\nu,\ \nu > 0\} \\ \frac{2h}{\nu+2h}\exp\left(\frac{1}{2}h^2 t\right) & : R_3 = \{(\nu,h) : h < 0,\ h < -\frac{1}{2}\nu\}. \end{cases}$$

Proof We divide the proof into four cases.

L_1 and L_2 case

The L_1 asymptotic follows from Watson's lemma being applied to

$$E_0\left[\exp\left(\nu\sqrt{t}S_1\right)\right] = \int_0^\infty \exp\left(\nu\sqrt{t}y\right)\frac{2}{\sqrt{2\pi}}\exp\left(-\frac{y^2}{2}\right)dy$$

where $\nu < 0$. For $(\nu,h) \in L_2$, we appeal to duality and substitute $\big(\nu, -(\nu+h)\big) = (\nu, 0)$ for (ν,h) in the L_1 asymptotic.

R_2 and R_3 case
Let $(\nu, h) \in R_3$. Using a Girsanov change of measure, we can write

$$E_0\left[\exp(\nu S_t + hX_t)\right] = E_0^h\left[\exp(\nu S_t)\right]\exp\left(\frac{1}{2}h^2 t\right).$$

In R_3 we have $h < 0$ so S_∞ is almost surely finite under P_0^h. In fact, S_∞ has the Exponential$(-2h)$ distribution under P_0^h. Since $\nu < -2h$, dominated convergence implies

$$\lim_{t\to\infty} E_0^h\left[\exp(\nu S_t)\right] = E_0^h\left[\exp(\nu S_\infty)\right] = \frac{2h}{\nu + 2h}.$$

This gives the R_3 asymptotic. Once again, we can use duality to get the R_2 asymptotic by substituting $(\nu, -(\nu + h))$ for (ν, h) in the R_3 asymptotic.

L_3 case
In L_3 we have $(\nu, h) = (-2h, h)$, hence Pitman's $2S - X$ theorem implies

$$\begin{aligned} E_0\left[\exp\left(\sqrt{t}(\nu S_1 + hX_1)\right)\right] &= E_0\left[\exp\left(\sqrt{t}(-2hS_1 + hX_1)\right)\right] \\ &= E_0\left[\exp\left(-h\sqrt{t}R_1\right)\right] \end{aligned}$$

where $(R_t : t \geq 0)$ is a Bessel(3) process. So using the Bessel(3) transition density (12.51), we can write the L_3 partition function as

$$\int_0^\infty \exp\left(-h\sqrt{t}y\right)\sqrt{\frac{2}{\pi}}y^2\exp\left(-\frac{y^2}{2}\right)dy. \tag{12.3}$$

After the change of variables $y \mapsto y\sqrt{t}$, we arrive at

$$\sqrt{\frac{2}{\pi}}t^{\frac{3}{2}}\int_0^\infty y^2\exp\left(-\left(hy + \frac{y^2}{2}\right)t\right)dy$$

whose asymptotic can be ascertained by a direct application of Laplace's method.

R_1 case
Let $(R_t : t \geq 0)$ denote a Bessel(3) process starting at 0 and U an independent Uniform[0, 1] random variable. For t fixed, the identity in law

$$(S_t, S_t - X_t) \overset{\mathcal{L}}{=} (UR_t, (1 - U)R_t)$$

follows from Pitman's $2S - X$ theorem, see Item C in Chapter 1 of [28]. When $\nu + 2h \neq 0$, this identity leads to

$$\begin{aligned} E_0\left[\exp(\nu S_t + hX_t)\right] &= E_0\left[\exp((\nu+2h)R_tU - hR_t)\right] \\ &= E_0\left[\exp\left(-h\sqrt{t}R_1\right)\int_0^1 \exp\left((\nu+2h)\sqrt{t}R_1u\right)du\right] \\ &= \frac{1}{(\nu+2h)\sqrt{t}}\left(E_0\left[\frac{\exp((\nu+h)\sqrt{t}R_1)}{R_1}\right]\right. \\ &\quad \left. -E_0\left[\frac{\exp(-h\sqrt{t}R_1)}{R_1}\right]\right) = \frac{1}{(\nu+2h)}\sqrt{\frac{2}{\pi t}} \\ &\quad \int_0^\infty \exp\left((\nu+h)\sqrt{t}y\right) y\exp\left(-\frac{y^2}{2}\right)dy \\ &\quad -\frac{1}{(\nu+2h)}\sqrt{\frac{2}{\pi t}}\int_0^\infty \exp\left(-h\sqrt{t}y\right) y\exp\left(-\frac{y^2}{2}\right)dy. \end{aligned}$$

Since $h > 0$ and $\nu + h < 0$, Watson's lemma can be applied to both integrals and their asymptotics combined. If $\nu + 2h = 0$ instead, we can use the reasoning from the L_3 *case* to show that the partition function is equal to (12.3). However, unlike that case, now we have $h > 0$ so Watson's lemma can be applied to yield the desired asymptotic.

□

12.3 Proof of Theorem 12.5

We divide the proof into four cases.

12.3.1 L_2 Case

In this section we prove the L_2 row in the table from Theorem 12.5. That the limit under $Q^{\nu,h}|_{\mathcal{F}_t}$ is W follows trivially from Brownian scaling and Theorem 12.3 since $Q^{\nu,h} = P_0^{\nu+h} = P_0$ when $(\nu, h) \in L_2$. To prove the limit under $Q_t^{\nu,h}|_{\mathcal{F}_t}$ is $\mathfrak{a}$, we show that

$$\lim_{t\to\infty} Q_t^{\nu,h}\left[F\left(\frac{X_{\bullet t}}{\sqrt{t}}\right)\right] = E\left[F(\mathfrak{m}_1 - \mathfrak{m}_{1-\bullet})\right] \qquad (12.4)$$

for any bounded continuous $F : C([0,1];\mathbb{R}) \to \mathbb{R}$ and $(\nu,h) \in L_2$. Then the desired result follows from Proposition 12.2.

Proof (Proof of (12.4)*)* The idea behind the proof is to use duality to transfer the result of Theorem 12.4 from the L_1 phase to the L_2 phase. Recall that in L_2 we have $\nu < 0$ and $\nu + h = 0$. Then the invariance property of ϕ and Theorem 12.4 imply that

$$\begin{aligned}\lim_{t\to\infty} Q_t^{\nu,h}\left[F\left(\frac{X_{\bullet t}}{\sqrt{t}}\right)\right] &= \lim_{t\to\infty}\frac{E_0\left[F_\phi\left(\frac{X_{\bullet t}}{\sqrt{t}}\right)\exp\left(\nu S_t-(\nu+h)X_t\right)\right]}{E_0\left[\exp\left(\nu S_t-(\nu+h)X_t\right)\right]}\\ &= \lim_{t\to\infty}\frac{E_0\left[F_\phi\left(\frac{X_{\bullet t}}{\sqrt{t}}\right)\exp\left(\nu S_t\right)\right]}{E_0\left[\exp\left(\nu S_t\right)\right]}\\ &= E\left[F_\phi(-\mathfrak{m}_\bullet)\right]\\ &= E\left[F(\mathfrak{m}_1-\mathfrak{m}_{1-\bullet})\right].\end{aligned}$$

□

12.3.2 R_2 and R_3 Case

In this section we prove the R_2 and R_3 rows in the table from Theorem 12.5. This is done by showing that the limits

$$\lim_{t\to\infty} Q^{\nu,h}\left[F\left(\frac{X_{\bullet t}}{t}\right)\right] = \begin{cases} F\big((\nu+h)\mathfrak{u}_\bullet\big) & : (\nu,h)\in R_2\\ F\big(h\mathfrak{u}_\bullet\big) & : (\nu,h)\in R_3\end{cases} \tag{12.5}$$

and

$$\lim_{t\to\infty} Q_t^{\nu,h}\left[F\left(\frac{X_{\bullet t}}{t}\right)\right] = \begin{cases} F\big((\nu+h)\mathfrak{u}_\bullet\big) & : (\nu,h)\in R_2\\ F\big(h\mathfrak{u}_\bullet\big) & : (\nu,h)\in R_3\end{cases} \tag{12.6}$$

hold for any bounded continuous $F : C([0,1];\mathbb{R}) \to \mathbb{R}$. First we need a lemma which asserts that ballistic scaling of Brownian motion with drift h results in a deterministic path of slope h.

Lemma 12.1 *Let $h \in \mathbb{R}$. Then $X_{\bullet t}/t$ converges to $h\mathfrak{u}_\bullet$ in probability under P_0^h as $t \to \infty$.*

Proof For any $\epsilon > 0$ we have

$$\begin{aligned}\lim_{t\to\infty} P_0^h\left(\left\|\frac{X_{\bullet t}}{t} - h\mathrm{u}_\bullet\right\| > \epsilon\right) &= \lim_{t\to\infty} P_0\left(\left\|\frac{X_{\bullet t} + ht\mathrm{u}_\bullet}{t} - h\mathrm{u}_\bullet\right\| > \epsilon\right)\\ &= \lim_{t\to\infty} P_0\left(\|X_\bullet\| > \epsilon\sqrt{t}\right)\\ &= 0\end{aligned}$$

since $\|X_\bullet\|$ is almost surely finite under P_0. □

Proof (Proof of (12.5)*)* When $(\nu, h) \in R_2$, we can use Theorem 12.3, Lemma 12.1, and bounded convergence to get

$$\begin{aligned}\lim_{t\to\infty} Q^{\nu,h}\left[F\left(\frac{X_{\bullet t}}{t}\right)\right] &= \lim_{t\to\infty} E_0^{\nu+h}\left[F\left(\frac{X_{\bullet t}}{t}\right)\right]\\ &= F\left((\nu+h)\mathrm{u}_\bullet\right).\end{aligned}$$

When $(\nu, h) \in R_3$, we can use Theorem 12.3, Lemma 12.1, and dominated convergence to get

$$\begin{aligned}\lim_{t\to\infty} Q^{\nu,h}\left[F\left(\frac{X_{\bullet t}}{t}\right)\right] &= \frac{\nu+2h}{2h}\lim_{t\to\infty} E_0^h\left[F\left(\frac{X_{\bullet t}}{t}\right)\exp(\nu S_\infty)\right]\\ &= \frac{\nu+2h}{2h} F\left(h\mathrm{u}_\bullet\right) E_0^h\left[\exp(\nu S_\infty)\right]\\ &= F\left(h\mathrm{u}_\bullet\right).\end{aligned}$$

Here we used the fact that S_∞ has the Exponential$(-2h)$ distribution under P_0^h. □

Proof (Proof of (12.6)*)* We first show that the limit holds in the R_3 case and then use duality to transfer this result to the R_2 case. Accordingly, suppose $(\nu, h) \in R_3$. By a Girsanov change of measure, we can write

$$E_0\left[F\left(\frac{X_{\bullet t}}{t}\right)\exp\left(\nu S_t + hX_t\right)\right] = \exp\left(\frac{1}{2}h^2 t\right) E_0^h\left[F\left(\frac{X_{\bullet t}}{t}\right)\exp\left(\nu S_t\right)\right].$$

Dividing this by the partition function asymptotic from Proposition 12.1 and using Lemma 12.1 with dominated convergence gives us

$$\begin{aligned}\lim_{t\to\infty} Q_t^{\nu,h}\left[F\left(\frac{X_{\bullet t}}{t}\right)\right] &= \frac{\nu+2h}{2h}\lim_{t\to\infty} E_0^h\left[F\left(\frac{X_{\bullet t}}{t}\right)\exp\left(\nu S_t\right)\right]\\ &= \frac{\nu+2h}{2h} F\left(h\mathrm{u}_\bullet\right) E_0^h\left[\exp(\nu S_\infty)\right]\\ &= F\left(h\mathrm{u}_\bullet\right).\end{aligned}$$

Now suppose $(\nu, h) \in R_2$. Then $\big(\nu, -(\nu + h)\big) \in R_3$. Hence the invariance property of ϕ and the above result imply

$$\begin{aligned}\lim_{t\to\infty} Q_t^{\nu,h}\left[F\left(\frac{X_{\bullet t}}{t}\right)\right] &= \lim_{t\to\infty} Q_t^{\nu,-(\nu+h)}\left[F_\phi\left(\frac{X_{\bullet t}}{t}\right)\right] \\ &= F_\phi\big(-(\nu+h)\mathrm{u}_\bullet\big) \\ &= F\big((\nu+h)\mathrm{u}_\bullet\big).\end{aligned}$$

□

12.3.3 L_3 Case

In this section we prove the L_3 row in the table from Theorem 12.5. The proof that the limit under $Q^{\nu,h}|_{\mathcal{F}_t}$ is $(\nu + h)\mathrm{u}$ is identical to that of the R_2 case of (12.5) since $Q^{\nu,h} = P_0^{\nu+h}$ when $(\nu, h) \in L_3$ by Theorem 12.3. To prove the limit under $Q_t^{\nu,h}|_{\mathcal{F}_t}$ is $-h\mathrm{u}^U$, we show that

$$\lim_{t\to\infty} Q_t^{\nu,h}\left[F\left(\frac{X_{\bullet t}}{t}\right)\right] = E\left[F\left(-h\mathrm{u}_\bullet^U\right)\right] \tag{12.7}$$

for any bounded Lipschitz continuous $F : C\big([0, 1]; \mathbb{R}\big) \to \mathbb{R}$ and $(\nu, h) \in L_3$.

First we need some preliminary results on Bessel(3) bridges and related path decompositions. Let $u > 0$ and $\rho, r \geq 0$ with $\rho = 0$ or $r = 0$. Then the Bessel(3) bridge of length u from ρ to r can be represented by

$$\sqrt{\left(\rho + (r-\rho)\frac{s}{u} + \mathfrak{b}_s^{(1)}\right)^2 + \left(\mathfrak{b}_s^{(2)}\right)^2 + \left(\mathfrak{b}_s^{(3)}\right)^2}, \quad 0 \leq s \leq u \tag{12.8}$$

where $\mathfrak{b}^{(i)}$, $i = 1, 2, 3$ are independent Brownian bridges of length u from 0 to 0, see [18]. Note that this representation does not hold when both $\rho > 0$ and $r > 0$ as discussed in [36].

If $x, y > 0$ and $0 < u < 1$, then the path $(X_s : 0 \leq s \leq 1)$ under P_x conditionally given $\{(I_1, X_1, \theta_1) = (0, y, u)\}$ can be decomposed into a concatenation of two Bessel(3) bridges, both of which have 0 as either a starting or ending point. More precisely, the path fragments

$$(X_s : 0 \leq s \leq u) \text{ and } (X_s : u \leq s \leq 1)$$

are independent and distributed respectively like

$$(R_s : 0 \leq s \leq u) \text{ given } \{(R_0, R_u) = (x, 0)\}$$

and

$$(R_{s-u} : u \leq s \leq 1) \text{ given } \{(R_0, R_{1-u}) = (0, y)\}.$$

This follows from Theorem 2.1.(ii) in [19] and the discussion in the Introduction of [2].

Lemma 12.2 *Let $x, y > 0$ and $0 < u < 1$. Consider the path $\omega^u_{x,y}$ in $C_0([0,1];\mathbb{R})$ that linearly interpolates between the points $(0,0)$, (u, x) and $(1, x-y)$. Specifically, $\omega^u_{x,y}$ is given by*

$$\omega^u_{x,y}(s) = \begin{cases} x\frac{s}{u} & : 0 \leq s \leq u \\ x - y\frac{s-u}{1-u} & : u < s \leq 1. \end{cases}$$

Suppose $f(t) > 0$ for $t > 0$ and $\lim_{t\to\infty} f(t) = \infty$. If $F : C([0,1];\mathbb{R}) \to \mathbb{R}$ is bounded Lipschitz continuous, then

$$\lim_{t\to\infty} E_0\left[F\left(\frac{X_\bullet}{f(t)}\right)\middle| S_1 = xf(t), S_1 - X_1 = yf(t), \Theta_1 = u\right] = F\left(\omega^u_{x,y}(\bullet)\right)$$

and the convergence is uniform on $\{(x, y, u) \in \mathbb{R}^3 : x > 0, y > 0, 0 < u < 1\}$.

Proof Using the translation and reflection symmetries of Wiener measure, we can write

$$\begin{aligned} &E_0\left[F\left(\frac{X_\bullet}{f(t)}\right)\middle| S_1 = xf(t), S_1 - X_1 = yf(t), \Theta_1 = u\right] \\ &= E_{xf(t)}\left[F\left(x - \frac{X_\bullet}{f(t)}\right)\middle| I_1 = 0, X_1 = yf(t), \theta_1 = u\right]. \end{aligned}$$

Together with (12.8) and the path decomposition noted above, this implies

$$E_0\left[F\left(\frac{X_\bullet}{f(t)}\right)\middle| S_1 = xf(t), S_1 - X_1 = yf(t), \Theta_1 = u\right] = E\left[F\left(x - \frac{Y^{(t)}_\bullet}{f(t)}\right)\right] \tag{12.9}$$

where $Y^{(t)}$ is defined by

$$Y^{(t)}_s := \begin{cases} \sqrt{\left(xf(t)\frac{u-s}{u} + \mathfrak{b}^{(1)}_s\right)^2 + \left(\mathfrak{b}^{(2)}_s\right)^2 + \left(\mathfrak{b}^{(3)}_s\right)^2} & : 0 \leq s \leq u \\ \sqrt{\left(yf(t)\frac{s-u}{1-u} + \mathfrak{b}^{(4)}_{s-u}\right)^2 + \left(\mathfrak{b}^{(5)}_{s-u}\right)^2 + \left(\mathfrak{b}^{(6)}_{s-u}\right)^2} & : u < s \leq 1. \end{cases}$$

Here $\mathfrak{b}^{(i)}$, $1 \le i \le 6$ are independent Brownian bridges from 0 to 0 of length u or $1-u$ as applicable. Let $\|\cdot\|_2$ denote the Euclidean norm on $\mathbb{R}^3$. Then we have

$$\left|\omega_{x,y}^u(s) - \left(x - \frac{Y_s^{(t)}}{f(t)}\right)\right|$$

$$= \begin{cases} \left| \frac{1}{f(t)} \left\| \left(xf(t)\frac{u-s}{u} + \mathfrak{b}_s^{(1)}, \mathfrak{b}_s^{(2)}, \mathfrak{b}_s^{(3)}\right)\right\|_2 - \left\|\left(x\frac{u-s}{u}, 0, 0\right)\right\|_2 \right| & : 0 \le s \le u \\ \left| \frac{1}{f(t)} \left\| \left(yf(t)\frac{s-u}{1-u} + \mathfrak{b}_{s-u}^{(4)}, \mathfrak{b}_{s-u}^{(5)}, \mathfrak{b}_{s-u}^{(6)}\right)\right\|_2 - \left\|\left(y\frac{s-u}{1-u}, 0, 0\right)\right\|_2 \right| & : u < s \le 1. \end{cases}$$

Now notice that the reverse triangle inequality implies

$$\left|\omega_{x,y}^u(s) - \left(x - \frac{Y_s^{(t)}}{f(t)}\right)\right| \le \begin{cases} \frac{1}{f(t)} \left\|\left(\mathfrak{b}_s^{(1)}, \mathfrak{b}_s^{(2)}, \mathfrak{b}_s^{(3)}\right)\right\|_2 & : 0 \le s \le u \\ \frac{1}{f(t)} \left\|\left(\mathfrak{b}_{s-u}^{(4)}, \mathfrak{b}_{s-u}^{(5)}, \mathfrak{b}_{s-u}^{(6)}\right)\right\|_2 & : u < s \le 1. \end{cases} \tag{12.10}$$

Suppose F has Lipschitz constant K. Then it follows from (12.10) and subadditivity of the square root function that

$$\left| F\left(\omega_{x,y}^u(\bullet)\right) - E\left[F\left(x - \frac{Y_\bullet^{(t)}}{f(t)}\right)\right]\right| \le \frac{K}{f(t)} \sum_{i=1}^{6} E\left[\left\|\mathfrak{b}_\bullet^{(i)}\right\|\right]. \tag{12.11}$$

This bound is uniform in x and y but has an implicit dependence on u. We can easily remedy this situation by noting that Brownian scaling implies that the expected value of the uniform norm of a Brownian bridge from 0 to 0 of length u is an increasing function of u. Noting that $u \le 1$, we can write

$$\sum_{i=1}^{6} E\left[\left\|\mathfrak{b}_\bullet^{(i)}\right\|\right] \le 6E\left[\|\mathfrak{b}_\bullet\|\right]$$

where $\mathfrak{b}$ is a standard Brownian bridge from 0 to 0 of length 1. This leads to a version of (12.11) which is uniform on $\{(x, y, u) \in \mathbb{R}^3 : x > 0, y > 0, 0 < u < 1\}$, namely

$$\left| F\left(\omega_{x,y}^u(\bullet)\right) - E\left[F\left(x - \frac{Y_\bullet^{(t)}}{f(t)}\right)\right]\right| \le \frac{6K}{f(t)} E\left[\|\mathfrak{b}_\bullet\|\right].$$

Together with (12.9) this proves the lemma since $f(t) \to \infty$ as $t \to \infty$. □

Proof (Proof of (12.7)*)* Recalling that $\nu = -2h$ on L_3, Brownian scaling implies

$$E_0\left[F\left(\frac{X_{\bullet t}}{t}\right)\exp\left(\nu S_t + hX_t\right)\right] = E_0\left[F\left(\frac{X_\bullet}{\sqrt{t}}\right)\exp\left(-h\sqrt{t}(2S_1 - X_1)\right)\right]. \tag{12.12}$$

For $0 < u < 1$, define

$$f_t(x, y, u) := \begin{cases} E_0\left[F\left(\frac{X_\bullet}{\sqrt{t}}\right)\Big| S_1 = x\sqrt{t}, S_1 - X_1 = y\sqrt{t}, \Theta_1 = u\right] & : x, y > 0 \\ 0 & : \text{otherwise.} \end{cases}$$

Using the tri-variate density (12.55), the right-hand side of (12.12) can be written as

$$\int_0^1\int_0^\infty\int_0^\infty \frac{xy\ f_t\left(\frac{x}{\sqrt{t}}, \frac{y}{\sqrt{t}}, u\right)}{\pi\sqrt{u^3(1-u)^3}} \exp\left(-h\sqrt{t}(x+y) - \frac{x^2}{2u} - \frac{y^2}{2(1-u)}\right) dxdydu.$$

Changing variables $x \mapsto x\sqrt{u} - uh\sqrt{t}$ and $y \mapsto y\sqrt{1-u} - (1-u)h\sqrt{t}$ gives

$$\frac{h^2 t}{\pi} e^{\frac{1}{2}h^2 t} \int_0^1 \int_{h\sqrt{(1-u)t}}^\infty \int_{h\sqrt{ut}}^\infty \tilde{f}_t(x, y, u)\ g_t(x, y, u) \exp\left(-\frac{x^2 + y^2}{2}\right) dxdydu \tag{12.13}$$

where we defined

$$\tilde{f}_t(x, y, u) := f_t\left(\frac{x\sqrt{u}}{\sqrt{t}} - uh, \frac{y\sqrt{1-u}}{\sqrt{t}} - (1-u)h, u\right)$$

and

$$g_t(x, y, u) := \frac{\left(x - h\sqrt{ut}\right)\left(y - h\sqrt{(1-u)t}\right)}{h^2 t\sqrt{u(1-u)}}.$$

Now we divide (12.13) by the L_3 partition function asymptotic from Proposition 12.1 which gives

$$\int_0^1 \int_{-\infty}^\infty \int_{-\infty}^\infty \tilde{f}_t(x, y, u)\ g_t(x, y, u)\ 1_{A_t}(x, y)\ \frac{1}{2\pi} \exp\left(-\frac{x^2 + y^2}{2}\right) dxdydu \tag{12.14}$$

where we defined

$$A_t := \left\{(x, y) : x > h\sqrt{ut}, y > h\sqrt{(1-u)t}\right\}.$$

At this stage we want to find the limit of (12.14) as $t \to \infty$ by appealing to Lemma 12.5 with μ being the probability measure on $\mathbb{R} \times \mathbb{R} \times [0, 1]$ having density $\frac{1}{2\pi} \exp\left(-\frac{x^2+y^2}{2}\right)$. In this direction, note that $\tilde{f}_t$ is bounded and the fact that the convergence in Lemma 12.2 is uniform and $(x, y, u) \mapsto F\left(\omega^u_{x,y}(\bullet)\right)$ is continuous on $\{(x, y, u) \in \mathbb{R}^3 : x > 0, y > 0, 0 < u < 1\}$ implies that

$$\begin{aligned}\lim_{t\to\infty} \tilde{f}_t(x, y, u) &= F\left(\omega^u_{-hu,-h(1-u)}(\bullet)\right)\\ &= F\left(-h\mathfrak{u}^u_\bullet\right)\end{aligned}$$

μ-almost surely. Additionally, $g_t 1_{A_t}$ is non-negative and converges μ-almost surely to 1 as $t \to \infty$. Lastly, by reversing the steps that led from (12.12) to (12.14), we see that

$$\begin{aligned}&\lim_{t\to\infty} \int_0^1 \int_{-\infty}^{\infty} \int_{-\infty}^{\infty} g_t(x, y, u)\, 1_{A_t}(x, y)\, \frac{1}{2\pi} \exp\left(-\frac{x^2+y^2}{2}\right) dxdydu\\ &= \lim_{t\to\infty} \frac{E_0\left[\exp\left(\nu S_t + hX_t\right)\right]}{2h^2 t \exp\left(\frac{1}{2}h^2 t\right)}\\ &= 1.\end{aligned}$$

Hence by Lemma 12.5 we can conclude that

$$\begin{aligned}\lim_{t\to\infty} Q_t^{\nu,h}\left[F\left(\frac{X_{\bullet t}}{t}\right)\right] &= \int_0^1 \int_{-\infty}^{\infty} \int_{-\infty}^{\infty} F\left(-h\mathfrak{u}^u_\bullet\right) \frac{1}{2\pi} \exp\left(-\frac{x^2+y^2}{2}\right) dxdydu\\ &= \int_0^1 F\left(-h\mathfrak{u}^u_\bullet\right) du\\ &= E\left[F\left(-h\mathfrak{u}^U_\bullet\right)\right]\end{aligned}$$

as desired. □

12.3.4 R_1 Case

In this section we prove the R_1 row in the table from Theorem 12.5. That the limit under $Q^{\nu,h}|_{\mathcal{F}_t}$ is $-R$ follows from Theorem 12.3 and Theorem 12.4 since $Q^{\nu,h} = Q^{\nu+h}$ with $\nu + h < 0$ when $(\nu, h) \in R_1$. To prove the limit under $Q_t^{\nu,h}|_{\mathcal{F}_t}$ is $-\mathfrak{e}$, we show that

$$\lim_{t\to\infty} Q_t^{\nu,h}\left[F\left(\frac{X_{\bullet t}}{\sqrt{t}}\right)\right] = E\left[F(-\mathfrak{e}_\bullet)\right] \tag{12.15}$$

for any bounded continuous $F : C([0,1]; \mathbb{R}) \to \mathbb{R}$ and $(\nu, h) \in R_1$.

Proof (Proof of (12.15)*)* From Brownian scaling, we have

$$E_0\left[F\left(\frac{X_{\bullet t}}{\sqrt{t}}\right)\exp\left(\nu S_t + hX_t\right)\right] = E_0\left[F\left(X_\bullet\right)\exp\left(\nu\sqrt{t}S_1 + h\sqrt{t}X_1\right)\right]. \tag{12.16}$$

Since $\nu < 0$, we can write

$$\exp\left(\nu\sqrt{t}S_1\right) = -\nu\sqrt{t}\int_0^\infty e^{\nu\sqrt{t}x}1_{S_1<x}dx.$$

Hence the right-hand side of (12.16) can be written as

$$\begin{aligned} &-\nu\sqrt{t}\int_0^\infty e^{\nu\sqrt{t}x}E_0\left[F\left(X_\bullet\right)\exp\left(h\sqrt{t}X_1\right); S_1 < x\right]dx \\ &= -\nu\sqrt{t}\int_0^\infty e^{\nu\sqrt{t}x}E_0\left[F\left(-X_\bullet\right)\exp\left(-h\sqrt{t}X_1\right); I_1 > -x\right]dx \\ &= -\nu\sqrt{t}\int_0^\infty e^{(\nu+h)\sqrt{t}x}E_x\left[F\left(x - X_\bullet\right)\exp\left(-h\sqrt{t}X_1\right); I_1 > 0\right]dx \end{aligned} \tag{12.17}$$

where the two equalities follow from the reflection and translation symmetries of Wiener measure. The h-transform representation of the Bessel(3) path measure from Proposition 12.8 can be used to rewrite the expectation appearing in (12.17) in terms of a Bessel(3) process $(R_s : 0 \leq s \leq 1)$. This leads to

$$-\nu\sqrt{t}\int_0^\infty e^{(\nu+h)\sqrt{t}x}E_x\left[F\left(x - R_\bullet\right)\exp\left(-h\sqrt{t}R_1\right)\frac{x}{R_1}\right]dx.$$

Next we disintegrate the Bessel(3) path measure into a mixture of Bessel(3) bridge measures by conditioning on the endpoint of the path. Refer to Proposition 1 in [11] for a precise statement of a more general result. See also Theorem 1 in [5] where weak continuity of the bridge measures with respect to their starting and ending points is established. This results in

$$-\nu\sqrt{t}\int_0^\infty\int_0^\infty e^{(\nu+h)\sqrt{t}x - h\sqrt{t}y}\frac{x}{y}E_x\left[F\left(x - R_\bullet\right)|R_1 = y\right]P_x(R_1 \in dy)dx. \tag{12.18}$$

Using the Bessel(3) transition density formula (12.52), we can now write (12.18) as

$$-\nu\sqrt{\frac{2t}{\pi}}\int_0^\infty\int_0^\infty e^{(\nu+h)\sqrt{t}x - h\sqrt{t}y}f(x,y)\sinh(xy)e^{-\frac{x^2+y^2}{2}}dydx$$

where we defined

$$f(x, y) := E_x\left[F\left(x - R_\bullet\right) | R_1 = y\right].$$

Applying the change of variables $x \mapsto x/\sqrt{t}$ and $y \mapsto y/\sqrt{t}$ gives

$$-\frac{\nu}{h^2(\nu+h)^2}\sqrt{\frac{2}{\pi t}}\int_0^\infty\int_0^\infty g(x,y) f\left(\frac{x}{\sqrt{t}}, \frac{y}{\sqrt{t}}\right)\frac{1}{xy}\sinh\left(\frac{xy}{t}\right)e^{-\frac{x^2+y^2}{2t}}\,dydx \tag{12.19}$$

where we defined

$$g(x, y) := h^2(\nu+h)^2 xy \exp\big((\nu+h)x - hy\big).$$

After dividing (12.19) by the R_1 partition function asymptotic from Proposition 12.1, we see that showing

$$\lim_{t\to\infty}\int_0^\infty\int_0^\infty g(x,y) f\left(\frac{x}{\sqrt{t}}, \frac{y}{\sqrt{t}}\right)\frac{t}{xy}\sinh\left(\frac{xy}{t}\right)e^{-\frac{x^2+y^2}{2t}}\,dydx = E\left[F(-\mathfrak{e}_\bullet)\right] \tag{12.20}$$

will prove (12.15). Notice that for all $x, y > 0$, the limit

$$\lim_{t\to\infty} f\left(\frac{x}{\sqrt{t}}, \frac{y}{\sqrt{t}}\right)\frac{t}{xy}\sinh\left(\frac{xy}{t}\right)e^{-\frac{x^2+y^2}{2t}} = E\left[F(-\mathfrak{e}_\bullet)\right]$$

follows from the weak continuity of the bridge measures with respect to their starting and ending points which was noted above and the fact that a normalized Brownian excursion is simply a Bessel(3) bridge from 0 to 0 of unit length. Additionally, the convexity of sinh on [0, 1] along with the inequality $2xy \le x^2+y^2$ leads to the bound

$$\left| f\left(\frac{x}{\sqrt{t}}, \frac{y}{\sqrt{t}}\right)\frac{t}{xy}\sinh\left(\frac{xy}{t}\right)e^{-\frac{x^2+y^2}{2t}}\right| \le \|F\|\sinh(1)$$

which holds for all $x, y, t > 0$. Noting that g is a probability density, (12.20) now follows from bounded convergence. □

12.4 Proof of Theorem 12.6

In this section we prove Theorem 12.6 by showing that the following limits hold for any bounded continuous $F : C([0,1];\mathbb{R}) \to \mathbb{R}$:

$$\text{if } (\nu,h) \in R_2 \cup L_3, \text{ then } \lim_{t\to\infty} Q^{\nu,h}\left[F\left(\frac{X_{\bullet t} - (\nu+h)t\mathrm{u}_\bullet}{\sqrt{t}}\right)\right] = E_0\left[F(X_\bullet)\right], \tag{12.21}$$

$$\text{if } (\nu,h) \in R_3, \text{ then } \lim_{t\to\infty} Q^{\nu,h}\left[F\left(\frac{X_{\bullet t} - ht\mathrm{u}_\bullet}{\sqrt{t}}\right)\right] = E_0\left[F(X_\bullet)\right], \tag{12.22}$$

$$\left.\begin{array}{l} \text{if } (\nu,h) \in R_2, \text{ then } \lim_{t\to\infty} Q_t^{\nu,h}\left[F\left(\dfrac{X_{\bullet t} - (\nu+h)t\mathrm{u}_\bullet}{\sqrt{t}}\right)\right] \\ \\ \text{if } (\nu,h) \in R_3, \text{ then } \lim_{t\to\infty} Q_t^{\nu,h}\left[F\left(\dfrac{X_{\bullet t} - ht\mathrm{u}_\bullet}{\sqrt{t}}\right)\right] \end{array}\right\} = E_0\left[F(X_\bullet)\right]. \tag{12.23}$$

Proof (Proof of (12.21)*)* If $(\nu,h) \in R_2 \cup L_3$, we can use Theorem 12.3, a path transformation that adds drift $\nu+h$, and Brownian scaling to write for all $t > 0$

$$\begin{aligned} Q^{\nu,h}\left[F\left(\frac{X_{\bullet t} - (\nu+h)t\mathrm{u}_\bullet}{\sqrt{t}}\right)\right] &= E_0^{\nu+h}\left[F\left(\frac{X_{\bullet t} - (\nu+h)t\mathrm{u}_\bullet}{\sqrt{t}}\right)\right] \\ &= E_0\left[F\left(\frac{X_{\bullet t}}{\sqrt{t}}\right)\right] \\ &= E_0\left[F\left(X_\bullet\right)\right] \end{aligned}$$

from which (12.21) follows. □

An idea that will be helpful for the proof of (12.22) and also in the next section is to modify the path functional F in such a way so that it "ignores" the beginning of the path. After proving a limit theorem for the modified path functional, we lift this result to the original functional by controlling the error arising from the modification. Here we state some definitions and notation that make this procedure precise.

Definition 12.3 For $0 < \delta \le 1$ and $X_\bullet \in C([0,1];\mathbb{R})$, define $\Pi_\delta : C([0,1];\mathbb{R}) \to C_0([0,1];\mathbb{R})$ by

$$\Pi_\delta X_s = \begin{cases} \frac{X_\delta}{\delta} s : 0 \le s < \delta \\ \\ X_s \;\; : \delta \le s \le 1. \end{cases}$$

For $F : C([0, 1]; \mathbb{R}) \to \mathbb{R}$, let F_δ denote $F \circ \Pi_\delta$ and define $\Delta_\delta^F : C([0, 1]; \mathbb{R}) \to \mathbb{R}_+$ by

$$\Delta_\delta^F(X_\bullet) = |F_\delta(X_\bullet) - F(X_\bullet)|.$$

So Π_δ replaces the initial $[0, \delta]$ segment of the path $X_\bullet$ with a straight line which interpolates between the points $(0, 0)$ and (δ, X_δ) while Δ_δ^F is the absolute error that results from using F_δ instead of F. The Markov property implies that under P_0, the random variable $F_\delta(X_\bullet)$ is independent of the initial $[0, \delta]$ part of the path $X_\bullet$ after conditioning on X_δ. Note that if F is bounded continuous then so are F_δ and Δ_δ^F with $\|F_\delta\| \leq \|F\|$ and $\|\Delta_\delta^F\| \leq 2\|F\|$. Also notice that $\lim_{\delta \searrow 0} \Delta_\delta^F(X_\bullet) = 0$ for any $X_\bullet \in C_0([0, 1]; \mathbb{R})$ whenever F is continuous.

Proof (Proof of (12.22)*)* We divide the proof into two stages, the first for F_δ and the second for F.

Stage 1: convergence for F_δ

Suppose $(\nu, h) \in R_3$ and fix $0 < \delta \leq 1$. In this case Theorem 12.3 implies

$$Q^{\nu,h}\left[F_\delta\left(\frac{X_{\bullet t} - ht\mathrm{u}_\bullet}{\sqrt{t}}\right)\right] = \frac{\nu + 2h}{2h} E_0^h\left[F_\delta\left(\frac{X_{\bullet t} - ht\mathrm{u}_\bullet}{\sqrt{t}}\right) e^{\nu S_\infty}\right]. \tag{12.24}$$

We can rewrite the expectation appearing on the right-hand side of (12.24) as

$$\underbrace{E_0^h\left[F_\delta\left(\frac{X_{\bullet t} - ht\mathrm{u}_\bullet}{\sqrt{t}}\right) e^{\nu S_{\delta t}}; \Theta_\infty \leq \delta t\right]}_{A_t} + \underbrace{E_0^h\left[F_\delta\left(\frac{X_{\bullet t} - ht\mathrm{u}_\bullet}{\sqrt{t}}\right) e^{\nu S_\infty}; \Theta_\infty > \delta t\right]}_{B_t}.$$

Since $(\nu, h) \in R_3$, we know that $\exp(\nu S_\infty)$ is integrable and $\Theta_\infty < \infty$ almost surely under P_0^h. Hence by dominated convergence we have $B_t = o(1)$ as $t \to \infty$ and consequently

$$A_t = E_0^h\left[F_\delta\left(\frac{X_{\bullet t} - ht\mathrm{u}_\bullet}{\sqrt{t}}\right) e^{\nu S_\infty}\right] + o(1) \text{ as } t \to \infty.$$

Similarly, we can show that

$$A_t = E_0^h\left[F_\delta\left(\frac{X_{\bullet t} - ht\mathrm{u}_\bullet}{\sqrt{t}}\right) e^{\nu S_{\delta t}}\right] + o(1) \text{ as } t \to \infty.$$

Together, these imply

$$E_0^h\left[F_\delta\left(\frac{X_{\bullet t} - ht\mathrm{u}_\bullet}{\sqrt{t}}\right) e^{\nu S_\infty}\right] = E_0^h\left[F_\delta\left(\frac{X_{\bullet t} - ht\mathrm{u}_\bullet}{\sqrt{t}}\right) e^{\nu S_{\delta t}}\right] + o(1) \text{ as } t \to \infty. \tag{12.25}$$

Applying a path transformation that adds drift h on the right-hand side of (12.25) and using Brownian scaling results in

$$E_0^h\left[F_\delta\left(\frac{X_{\bullet t}-ht\mathrm{u}_\bullet}{\sqrt{t}}\right)\exp(\nu S_\infty)\right]$$
$$=E_0\left[F_\delta\left(X_\bullet\right)\exp\left(\nu\sqrt{t}\sup_{0\le s\le\delta}\left\{X_s+h\sqrt{t}s\right\}\right)\right]+o(1)\text{ as }t\to\infty.$$

Combining this with (12.24), we have established that

$$Q^{\nu,h}\left[F_\delta\left(\frac{X_{\bullet t}-ht\mathrm{u}_\bullet}{\sqrt{t}}\right)\right]$$
$$=\frac{\nu+2h}{2h}E_0\left[F_\delta\left(X_\bullet\right)\exp\left(\nu\sqrt{t}\sup_{0\le s\le\delta}\left\{X_s+h\sqrt{t}s\right\}\right)\right]+o(1)\text{ as }t\to\infty. \tag{12.26}$$

Now notice that $F_\delta\left(X_\bullet\right)$ and $\sup_{0\le s\le\delta}\left\{X_s+h\sqrt{t}s\right\}$ are independent after conditioning on X_δ. So with $p_\delta(\cdot,\cdot)$ denoting the transition density of Brownian motion at time δ, we see that the expectation appearing on the right-hand side of (12.26) is equal to

$$\int_{-\infty}^{\infty}E_0\left[F_\delta(X_\bullet)|X_\delta=x\right]E_0\left[\exp\left(\nu\sqrt{t}\sup_{0\le s\le\delta}\left\{X_s+h\sqrt{t}s\right\}\right)\middle|X_\delta=x\right]p_\delta(0,x)dx. \tag{12.27}$$

From the particular pathwise construction of Brownian bridge given in (5.6.29) of [13], it follows that the distribution of Brownian bridge plus a constant drift is the same as Brownian bridge with an appropriately shifted endpoint. Hence the second expectation appearing inside the integral in (12.27) is seen to equal

$$E_0\left[\exp\left(\nu\sqrt{t}\sup_{0\le s\le\delta}X_s\right)\middle|X_\delta=x+h\sqrt{t}\delta\right].$$

By using the distribution of the maximum of a Brownian bridge from (12.56), this expectation has the integral representation

$$\int_0^\infty\frac{4y-2x-2h\sqrt{t}\delta}{\delta}\exp\left(\nu\sqrt{t}y-\frac{2y(y-x-h\sqrt{t}\delta)}{\delta}\right)dy$$

when t is large enough such that $x + h\sqrt{t}\delta \leq 0$. After some manipulations and recalling that $\nu + 2h < 0$, we can use Watson's lemma to compute the limit of this integral which holds for all $x \in \mathbb{R}$:

$$\lim_{t\to\infty} \int_0^\infty \left(\frac{4y-2x}{\delta} - 2h\sqrt{t}\right) \exp\left(\frac{2yx-2y^2}{\delta}\right) e^{(\nu+2h)\sqrt{t}y} dy = \frac{2h}{\nu+2h}.$$

Now we want to invoke Lemma 12.5 to find the limit of (12.27) hence we need to verify that

$$\lim_{t\to\infty} \int_{-\infty}^{\infty} E_0\left[\exp\left(\nu\sqrt{t} \sup_{0\leq s\leq\delta} \left\{X_s + h\sqrt{t}s\right\}\right) \middle| X_\delta = x\right] p_\delta(0,x)dx = \frac{2h}{\nu+2h}. \tag{12.28}$$

By working backwards starting from the left-hand side of (12.28) and reversing the conditioning, scaling, and path transformation, this can be reduced to checking

$$\lim_{t\to\infty} E_0^h\left[\exp\left(\nu S_{\delta t}\right)\right] = \frac{2h}{\nu+2h}$$

which follows from dominated convergence since $(\nu, h) \in R_3$ and S_∞ has the Exponential$(-2h)$ distribution under P_0^h. Now we can evaluate the limit of (12.27) as

$$\lim_{t\to\infty} \int_{-\infty}^{\infty} E_0\left[F_\delta(X_\bullet)|X_\delta = x\right] E_0\left[\exp\left(\nu\sqrt{t} \sup_{0\leq s\leq\delta} \left\{X_s + h\sqrt{t}s\right\}\right) \middle| X_\delta = x\right]$$

$$p_\delta(0,x)dx = \frac{2h}{\nu+2h} \int_{-\infty}^{\infty} E_0\left[F_\delta(X_\bullet)|X_\delta = x\right] p_\delta(0,x)dx = \frac{2h}{\nu+2h} E_0\left[F_\delta(X_\bullet)\right].$$

Combining this with (12.26) leads to

$$\lim_{t\to\infty} Q^{\nu,h}\left[F_\delta\left(\frac{X_{\bullet t} - ht\mathrm{u}_\bullet}{\sqrt{t}}\right)\right] = E_0\left[F_\delta(X_\bullet)\right]$$

as desired.

Stage 2: convergence for F
Recall the notation Δ_δ^F from Definition 12.3. From *Stage 1*, we know that

$$\lim_{t\to\infty} \left| Q^{\nu,h}\left[F_\delta\left(\frac{X_{\bullet t} - ht\mathrm{u}_\bullet}{\sqrt{t}}\right)\right] - E_0\left[F_\delta(X_\bullet)\right] \right| = 0.$$

Hence the triangle inequality implies that

$$\limsup_{t\to\infty}\left|Q^{\nu,h}\left[F\left(\frac{X_{\bullet t}-ht\mathrm{u}_\bullet}{\sqrt{t}}\right)\right]-E_0\left[F(X_\bullet)\right]\right|$$
$$\le \limsup_{t\to\infty} Q^{\nu,h}\left[\Delta_\delta^F\left(\frac{X_{\bullet t}-ht\mathrm{u}_\bullet}{\sqrt{t}}\right)\right]+E_0\left[\Delta_\delta^F(X_\bullet)\right].$$

Since F is bounded continuous, we know that the last term on the right-hand side of this inequality vanishes as $\delta \searrow 0$ by bounded convergence. This leads to

$$\limsup_{t\to\infty}\left|Q^{\nu,h}\left[F\left(\frac{X_{\bullet t}-ht\mathrm{u}_\bullet}{\sqrt{t}}\right)\right]-E_0\left[F(X_\bullet)\right]\right|$$
$$\le \lim_{\delta\searrow 0}\limsup_{t\to\infty} Q^{\nu,h}\left[\Delta_\delta^F\left(\frac{X_{\bullet t}-ht\mathrm{u}_\bullet}{\sqrt{t}}\right)\right]. \tag{12.29}$$

Using Theorem 12.3, we can express the right-hand side of (12.29) as

$$\frac{\nu+2h}{2h}\lim_{\delta\searrow 0}\limsup_{t\to\infty} E_0^h\left[\Delta_\delta^F\left(\frac{X_{\bullet t}-ht\mathrm{u}_\bullet}{\sqrt{t}}\right)\exp(\nu S_\infty)\right]. \tag{12.30}$$

Since $h < -\frac{1}{2}\nu$ when $(\nu, h) \in R_3$, we can find $p > 1$ such that $-2h > p\nu$. Let q be the Hölder conjugate of p. Using Hölder's inequality, we can upper bound (12.30) by

$$\frac{\nu+2h}{2h}\lim_{\delta\searrow 0}\limsup_{t\to\infty} E_0^h\left[\left(\Delta_\delta^F\left(\frac{X_{\bullet t}-ht\mathrm{u}_\bullet}{\sqrt{t}}\right)\right)^q\right]^{\frac{1}{q}} E_0^h[\exp(p\nu S_\infty)]^{\frac{1}{p}}$$
$$=\frac{\nu+2h}{2h}\lim_{\delta\searrow 0} E_0\left[\left(\Delta_\delta^F(X_\bullet)\right)^q\right]^{\frac{1}{q}}\left(\frac{2h}{p\nu+2h}\right)^{\frac{1}{p}}$$
$$=0.$$

Here we used a path transformation that adds drift h along with Brownian scaling to eliminate t from the first expectation and used the fact that S_∞ has the Exponential$(-2h)$ distribution under P_0^h to compute the second expectation. Now it follows that

$$\limsup_{t\to\infty}\left|Q^{\nu,h}\left[F\left(\frac{X_{\bullet t}-ht\mathrm{u}_\bullet}{\sqrt{t}}\right)\right]-E_0\left[F(X_\bullet)\right]\right|=0$$

which proves (12.22).

□

Proof (Proof of (12.23)*)* We first show that the limit holds in the R_3 case and then use duality to transfer this result to the R_2 case. Accordingly, suppose $(\nu, h) \in R_3$. Using a Girsanov change of measure, we can write

$$E_0\left[F\left(\frac{X_{\bullet t} - ht\mathrm{u}_\bullet}{\sqrt{t}}\right)e^{\nu S_t + hX_t}\right] = e^{\frac{1}{2}h^2 t}E_0^h\left[F\left(\frac{X_{\bullet t} - ht\mathrm{u}_\bullet}{\sqrt{t}}\right)e^{\nu S_t}\right]. \tag{12.31}$$

By repeating the same argument that led to (12.25) while using F instead of F_δ, we can establish that

$$E_0^h\left[F\left(\frac{X_{\bullet t} - ht\mathrm{u}_\bullet}{\sqrt{t}}\right)e^{\nu S_t}\right] = E_0^h\left[F\left(\frac{X_{\bullet t} - ht\mathrm{u}_\bullet}{\sqrt{t}}\right)e^{\nu S_\infty}\right] + o(1) \text{ as } t \to \infty.$$

Combining this with (12.31) and using the partition function asymptotic from Proposition 12.1 results in

$$\begin{aligned}\lim_{t\to\infty} Q_t^{\nu,h}\left[F\left(\frac{X_{\bullet t} - ht\mathrm{u}_\bullet}{\sqrt{t}}\right)\right] &= \lim_{t\to\infty}\frac{\nu+2h}{2h}E_0^h\left[F\left(\frac{X_{\bullet t} - ht\mathrm{u}_\bullet}{\sqrt{t}}\right)e^{\nu S_\infty}\right]\\ &= \lim_{t\to\infty} Q^{\nu,h}\left[F\left(\frac{X_{\bullet t} - ht\mathrm{u}_\bullet}{\sqrt{t}}\right)\right]\\ &= E_0[F(X_\bullet)]\end{aligned}$$

where the last two equalities follow from Theorem 12.3 and (12.22), respectively.

Now suppose $(\nu, h) \in R_2$. Then $\big(\nu, -(\nu+h)\big) \in R_3$. Hence the invariance property of ϕ and the above result imply

$$\begin{aligned}\lim_{t\to\infty} Q_t^{\nu,h}\left[F\left(\frac{X_{\bullet t} - ht\mathrm{u}_\bullet}{\sqrt{t}}\right)\right] &= \lim_{t\to\infty} Q_t^{\nu,-(\nu+h)}\left[F_\phi\left(\frac{X_{\bullet t} - ht\mathrm{u}_\bullet}{\sqrt{t}}\right)\right]\\ &= E_0[F_\phi(X_\bullet)]\\ &= E_0[F(X_\bullet)].\end{aligned}$$

□

12.5 Proof of Theorem 12.7

In this section we prove Theorem 12.7 by showing that

$$\lim_{t\to\infty} Q_t^{\nu,h}\left[F\left(\frac{X_{\bullet t} - (2S_{\bullet t} + ht\mathrm{u}_\bullet)}{\sqrt{t}}\right)\right] = E_0\left[F(X_\bullet)\right] \tag{12.32}$$

for any bounded Lipschitz continuous $F : C\big([0,1];\mathbb{R}\big) \to \mathbb{R}$ and $(\nu, h) \in L_3$.

Proof (Proof of (12.32)*)* We divide the proof into two stages, the first for F_δ and the second for F.

Stage 1: convergence for $\boldsymbol{F_\delta}$
Suppose $(\nu, h) \in L_3$ and fix $0 < \delta \le 1$. Recalling that $\nu = -2h$, we can use Brownian scaling and Pitman's $2S - X$ theorem to write

$$\begin{aligned} & E_0\left[F_\delta\left(\frac{X_{\bullet t} - (2S_{\bullet t} + ht\mathrm{u}_\bullet)}{\sqrt{t}}\right)\exp\left(\nu S_t + hX_t\right)\right] \\ &\quad = E_0\left[F_\delta\left(-R_\bullet - h\sqrt{t}\mathrm{u}_\bullet\right)\exp\left(-h\sqrt{t}R_1\right)\right]. \end{aligned} \tag{12.33}$$

Fix $x > 0$. Then the absolute continuity relation from Lemma 12.3 implies that the right-hand side of (12.33) is equal to

$$E_x\left[F_\delta\left(-R_\bullet - h\sqrt{t}\mathrm{u}_\bullet\right)\exp\left(-h\sqrt{t}R_1\right)\frac{xR_\delta\exp\left(\frac{x^2}{2\delta}\right)}{\delta\sinh\left(\frac{xR_\delta}{\delta}\right)}\right].$$

Now we can use Proposition 12.8 to switch from the Bessel(3) process to Brownian motion. Hence the above expectation is equal to

$$E_x\left[F_\delta\left(-X_\bullet - h\sqrt{t}\mathrm{u}_\bullet\right)\exp\left(-h\sqrt{t}X_1\right)\frac{X_\delta X_1\exp\left(\frac{x^2}{2\delta}\right)}{\delta\sinh\left(\frac{xX_\delta}{\delta}\right)}; I_1 > 0\right].$$

Next we use a Girsanov change of measure to add drift $-h\sqrt{t}$. This results in

$$E_x^{-h\sqrt{t}}\left[F_\delta\left(-X_\bullet - h\sqrt{t}\mathrm{u}_\bullet\right)\exp\left(-h\sqrt{t}x + \frac{1}{2}h^2t\right)\frac{X_\delta X_1\exp\left(\frac{x^2}{2\delta}\right)}{\delta\sinh\left(\frac{xX_\delta}{\delta}\right)}; I_1 > 0\right].$$

Applying a path transformation that adds drift $-h\sqrt{t}$ while changing the measure back to that of Brownian motion without drift yields

$$E_x\left[F_\delta\left(-X_\bullet\right)\frac{(X_\delta - h\sqrt{t}\delta)(X_1 - h\sqrt{t})e^{\frac{x^2}{2\delta} - h\sqrt{t}x + \frac{1}{2}h^2t}}{\delta\sinh\left(\frac{x}{\delta}(X_\delta - h\sqrt{t}\delta)\right)}; \inf_{0\le s\le 1}\{X_s - h\sqrt{t}s\} > 0\right].$$

Now we divide this by the L_3 partition function asymptotic from Proposition 12.1 which gives

$$E_x\left[F_\delta\left(-X_\bullet\right)\frac{(X_\delta - h\sqrt{t}\delta)(X_1 - h\sqrt{t})e^{\frac{x^2}{2\delta} - h\sqrt{t}x}}{2h^2t\delta\sinh\left(\frac{x}{\delta}(X_\delta - h\sqrt{t}\delta)\right)}1_{A_t}\right] \tag{12.34}$$

where we defined

$$A_t := \left\{ \inf_{0 \le s \le 1} \{X_s - h\sqrt{t}s\} > 0 \right\}.$$

At this point we want to use Lemma 12.5 to find the limit of (12.34) as $t \to \infty$. Toward this end, note that

$$\frac{(X_\delta - h\sqrt{t}\delta)(X_1 - h\sqrt{t})e^{\frac{x^2}{2\delta} - h\sqrt{t}x}}{2h^2 t\delta \sinh\left(\frac{x}{\delta}(X_\delta - h\sqrt{t}\delta)\right)} 1_{A_t}$$

is non-negative for all $t > 0$. Recalling that $x > 0$ and $h < 0$, we see that almost surely under P_x

$$\lim_{t \to \infty} \frac{(X_\delta - h\sqrt{t}\delta)(X_1 - h\sqrt{t})e^{\frac{x^2}{2\delta} - h\sqrt{t}x}}{2h^2 t\delta \sinh\left(\frac{x}{\delta}(X_\delta - h\sqrt{t}\delta)\right)} 1_{A_t} = \exp\left(\frac{x^2 - 2xX_\delta}{2\delta}\right).$$

Additionally, by reversing the steps that led from (12.33) to (12.34), we see that

$$\begin{aligned}\lim_{t \to \infty} E_x\left[\frac{(X_\delta - h\sqrt{t}\delta)(X_1 - h\sqrt{t})e^{\frac{x^2}{2\delta} - h\sqrt{t}x}}{2h^2 t\delta \sinh\left(\frac{x}{\delta}(X_\delta - h\sqrt{t}\delta)\right)} 1_{A_t}\right] &= \lim_{t \to \infty} \frac{E_0\left[\exp\left(\nu S_t + hX_t\right)\right]}{2h^2 t \exp\left(\frac{1}{2}h^2 t\right)} \\ &= 1.\end{aligned}$$

This agrees with

$$E_x\left[\exp\left(\frac{x^2 - 2xX_\delta}{2\delta}\right)\right] = 1$$

which follows from a routine calculation. Hence we can conclude from Lemma 12.5, reflection symmetry of Wiener measure, and Lemma 12.3 that

$$\begin{aligned}&\lim_{t \to \infty} E_x\left[F_\delta(-X_\bullet) \frac{(X_\delta - h\sqrt{t}\delta)(X_1 - h\sqrt{t})e^{\frac{x^2}{2\delta} - h\sqrt{t}x}}{2h^2 t\delta \sinh\left(\frac{x}{\delta}(X_\delta - h\sqrt{t}\delta)\right)}; \inf_{0 \le s \le 1} \{X_s - h\sqrt{t}s\} > 0\right] \\ &= E_x\left[F_\delta(-X_\bullet) \exp\left(\frac{x^2 - 2xX_\delta}{2\delta}\right)\right] \\ &= E_{-x}\left[F_\delta(X_\bullet) \exp\left(\frac{x^2 + 2xX_\delta}{2\delta}\right)\right] \\ &= E_0\left[F_\delta(X_\bullet)\right].\end{aligned}$$

This shows that for any $0 < \delta \le 1$ we have

$$\lim_{t\to\infty} Q_t^{\nu,h}\left[F_\delta\left(\frac{X_{\bullet t} - (2S_{\bullet t} + ht\mathrm{u}_\bullet)}{\sqrt{t}}\right)\right] = E_0\left[F_\delta(X_\bullet)\right].$$

Stage 2: convergence for F

We proceed as in the beginning of *Stage 2* in the proof of (12.22). Similarly to (12.29) we have

$$\limsup_{t\to\infty}\left|Q_t^{\nu,h}\left[F\left(\frac{X_{\bullet t} - (2S_{\bullet t} + ht\mathrm{u}_\bullet)}{\sqrt{t}}\right)\right] - E_0[F(X_\bullet)]\right|$$
$$\le \lim_{\delta\searrow 0}\limsup_{t\to\infty} Q_t^{\nu,h}\left[\Delta_\delta^F\left(\frac{X_{\bullet t} - (2S_{\bullet t} + ht\mathrm{u}_\bullet)}{\sqrt{t}}\right)\right]. \tag{12.35}$$

Unlike (12.30) however, we can't use Hölder's inequality to get a useful bound for (12.35) since (ν, h) is on the critical line L_3. Instead, we make use of the Lipschitz continuity of F. Suppose F has Lipschitz constant K. Then for any $X_\bullet \in C\big([0,1];\mathbb{R}\big)$ we have

$$\begin{aligned}\Delta_\delta^F(X_\bullet) = |F(\Pi_\delta X_\bullet) - F(X_\bullet)| &\le K\|\Pi_\delta X_\bullet - X_\bullet\| \\ &\le 2K\sup_{0\le s\le\delta}|X_s|.\end{aligned}$$

Along with (12.33), this implies that (12.35) is bounded above by

$$2K\lim_{\delta\searrow 0}\limsup_{t\to\infty}\frac{E_0\left[\sup\limits_{0\le s\le\delta}\left|R_s + h\sqrt{t}s\right|\exp\left(-h\sqrt{t}R_1\right)\right]}{E_0\left[\exp\left(-h\sqrt{t}R_1\right)\right]}. \tag{12.36}$$

With $p_1(\cdot,\cdot)$ denoting the Bessel(3) transition density at time 1, we can write the expectation appearing in the numerator of (12.36) as a mixture of Bessel(3) bridges by conditioning on the endpoint

$$\int_0^\infty E_0\left[\sup_{0\le s\le\delta}\left|R_s + h\sqrt{t}s\right|\,\middle|\, R_1 = y\right]e^{-h\sqrt{t}y}p_1(0,y)dy. \tag{12.37}$$

Let $\|\cdot\|_2$ denote the Euclidean norm on $\mathbb{R}^3$. Using (12.8) and recalling that $h < 0$, we can write the expectation appearing in (12.37) as

$$E\left[\sup_{0\le s\le\delta}\left|\left\|\left(\mathfrak{b}_s^{(1)} + ys, \mathfrak{b}_s^{(2)}, \mathfrak{b}_s^{(3)}\right)\right\|_2 - \left\|\left(|h|\sqrt{t}s, 0, 0\right)\right\|_2\right|\right]$$

where $\mathfrak{b}^{(i)}$, $i = 1, 2, 3$ are independent Brownian bridges of length 1 from 0 to 0. Now notice that the reverse triangle inequality implies this is bounded above by

$$E\left[\sup_{0\le s\le \delta}\left\|\left(\mathfrak{b}_s^{(1)} + (y + h\sqrt{t})s,\, \mathfrak{b}_s^{(2)},\, \mathfrak{b}_s^{(3)}\right)\right\|_2\right].$$

Using subadditivity of the square root function and the triangle inequality, this is bounded above by

$$3E\left[\sup_{0\le s\le \delta}|\mathfrak{b}_s|\right] + \left|y + h\sqrt{t}\right|\delta. \tag{12.38}$$

By substituting (12.38) for the expectation appearing in (12.37), we see that the latter expression is bounded above by

$$3E\left[\sup_{0\le s\le \delta}|\mathfrak{b}_s|\right] E_0\left[\exp\left(-h\sqrt{t}R_1\right)\right] + \delta\int_0^\infty \left|y + h\sqrt{t}\right| e^{-h\sqrt{t}y} p_1(0, y)dy. \tag{12.39}$$

Now substituting (12.39) for the expectation appearing in the numerator of (12.36) leads to the upper bound

$$2K\lim_{\delta\searrow 0}\left(3E\left[\sup_{0\le s\le \delta}|\mathfrak{b}_s|\right] + \delta\limsup_{t\to\infty}\frac{\int_0^\infty \left|y + h\sqrt{t}\right| e^{-h\sqrt{t}y} p_1(0, y)dy}{E_0\left[\exp\left(-h\sqrt{t}R_1\right)\right]}\right). \tag{12.40}$$

We can evaluate the lim sup term appearing in (12.40) explicitly by using the L_3 partition function asymptotic from Proposition 12.1 and the Bessel(3) transition density formula (12.51) to write

$$\begin{aligned}
&\lim_{t\to\infty}\frac{\int_0^\infty \left|y + h\sqrt{t}\right| e^{-h\sqrt{t}y} p_1(0, y)dy}{E_0\left[\exp\left(-h\sqrt{t}R_1\right)\right]} \\
&= \lim_{t\to\infty}\int_0^\infty \sqrt{\frac{2}{\pi}}\frac{\left|y + h\sqrt{t}\right| y^2}{2h^2 t} e^{-h\sqrt{t}y - \frac{y^2}{2} - \frac{1}{2}h^2 t} dy.
\end{aligned}$$

Applying the change of variables $y \mapsto y - h\sqrt{t}$ and using dominated convergence results in

$$\lim_{t\to\infty}\int_{h\sqrt{t}}^\infty \sqrt{\frac{2}{\pi}}\frac{|y|\left(y - h\sqrt{t}\right)^2}{2h^2 t} e^{-\frac{y^2}{2}} dy = \int_{-\infty}^\infty \frac{1}{\sqrt{2\pi}}|y| e^{-\frac{y^2}{2}} dy = \sqrt{\frac{2}{\pi}}.$$

Hence (12.40) equals

$$2K \lim_{\delta \searrow 0} \left(3E \left[\sup_{0 \le s \le \delta} |\mathfrak{b}_s| \right] + \delta \sqrt{\frac{2}{\pi}} \right) = 0.$$

Here we used dominated convergence and the fact that $\sup_{0 \le s \le 1} |\mathfrak{b}_s|$ is integrable and $\mathfrak{b}$ is continuous with $\mathfrak{b}_0 = 0$. Now it follows that

$$\limsup_{t \to \infty} \left| Q_t^{\nu,h} \left[F \left(\frac{X_{\bullet t} - (2S_{\bullet t} + ht\mathfrak{u}_\bullet)}{\sqrt{t}} \right) \right] - E_0[F(X_\bullet)] \right| = 0$$

which proves (12.32).

□

12.6 Brownian Ascent

We informally defined the Brownian ascent as a Brownian path of duration 1 conditioned on the event $\{X_1 = S_1\}$. Since this is a null event, some care is needed to make the conditioning precise. Accordingly, we condition on the event $\{S_1 - X_1 < \epsilon\}$ and let $\epsilon \searrow 0$. This leads to an equality in law between the Brownian ascent and a path transformation of the Brownian meander. While this result along with the other Propositions in this section are likely obvious to those familiar with Brownian path fragments, we include proofs for the convenience of non-experts. The reader can refer to Appendix 1 for some basic information on the Brownian meander.

Proposition 12.2

$$(\mathfrak{a}_s : 0 \le s \le 1) \stackrel{\mathcal{L}}{=} (\mathfrak{m}_1 - \mathfrak{m}_{1-s} : 0 \le s \le 1)$$

Proof The idea behind the proof is to use the invariance property of ϕ from Definition 12.2 together with a known limit theorem for the meander, similarly to proving (12.4). Let $F : C([0, 1]; \mathbb{R}) \to \mathbb{R}$ be bounded and continuous. Then we have

$$\begin{aligned} E[F(\mathfrak{a}_\bullet)] &= \lim_{\epsilon \searrow 0} E_0[F(X_\bullet)|S_1 - X_1 < \epsilon] = \lim_{\epsilon \searrow 0} E_0[F(-X_\bullet)|X_1 - I_1 < \epsilon] \\ &= \lim_{\epsilon \searrow 0} E_0[F_\phi(-X_\bullet)|I_1 > -\epsilon] = E[F_\phi(-\mathfrak{m}_\bullet)] \\ &= E[F(\mathfrak{m}_1 - \mathfrak{m}_{1-\bullet})] \end{aligned}$$

where weak convergence to Brownian meander in the last limit follows from Theorem 2.1 in [9]. □

Recall Lévy's equivalence

$$\Big((S_t - X_t, S_t) : t \geq 0\Big) \stackrel{\mathcal{L}}{=} \Big((|X_t|, L_t^0(X)) : t \geq 0\Big) \tag{12.41}$$

where $L_t^0(X)$ denotes the local time of X at the level 0 up to time t. This equality in law holds under P_0, see Item B in Chapter 1 of [28]. Since conditioning $(X_s : 0 \leq s \leq 1)$ on the event $\{|X_1| < \epsilon\}$ and letting $\epsilon \searrow 0$ results in a standard Brownian bridge, we can apply the same argument of Proposition 12.2 to both sides of (12.41) and get the following result which can also be seen to follow from a combination of Proposition 12.2 and Théorème 8 of [3].

Proposition 12.3 *Let $\big(L_s^0(\mathfrak{b}) : 0 \leq s \leq 1\big)$ denote the local time process at the level 0 of a Brownian bridge of length 1 from 0 to 0. Then we have the equality in law*

$$\left(\sup_{0\leq u\leq s} \mathfrak{a}_u : 0 \leq s \leq 1\right) \stackrel{\mathcal{L}}{=} \big(L_s^0(\mathfrak{b}) : 0 \leq s \leq 1\big).$$

We can also construct the Brownian ascent from a Brownian path by scaling the pre-maximum part of the path so that it has duration 1.

Proposition 12.4 *Let Θ denote the almost surely unique time at which the standard Brownian motion W attains its maximum over the time interval $[0, 1]$. Then we have the equality in law*

$$(\mathfrak{a}_s : 0 \leq s \leq 1) \stackrel{\mathcal{L}}{=} \left(\frac{W_{s\Theta}}{\sqrt{\Theta}} : 0 \leq s \leq 1\right).$$

Proof From Denisov's path decomposition Theorem 12.9 we have

$$(\mathfrak{m}_s : 0 \leq s \leq 1) \stackrel{\mathcal{L}}{=} \left(\frac{W_\Theta - W_{\Theta - s\Theta}}{\sqrt{\Theta}} : 0 \leq s \leq 1\right).$$

Reflecting both processes about 0 gives

$$(-\mathfrak{m}_s : 0 \leq s \leq 1) \stackrel{\mathcal{L}}{=} \left(\frac{W_{\Theta - s\Theta} - W_\Theta}{\sqrt{\Theta}} : 0 \leq s \leq 1\right).$$

Applying ϕ to both processes results in

$$(\mathfrak{m}_1 - \mathfrak{m}_{1-s} : 0 \leq s \leq 1) \stackrel{\mathcal{L}}{=} \left(\frac{W_{s\Theta}}{\sqrt{\Theta}} : 0 \leq s \leq 1\right).$$

Now the desired result follows from Proposition 12.2. □

There is an absolute continuity relation between the path measures of the Brownian ascent and Brownian motion run up to the first hitting time of 1 and then rescaled to have duration 1. This random scaling construction is reminiscent of Pitman and Yor's *agreement formula* for Bessel bridges, see [19].

Proposition 12.5 *Let* τ_1 *be the first hitting time of* 1 *by* X. *For any measurable* $F : C\big([0,1];\mathbb{R}\big) \to \mathbb{R}_+$ *we have*

$$E[F(\mathfrak{a}_\bullet)] = \sqrt{\frac{\pi}{2}} E_0\left[F\left(\frac{X_{\bullet\tau_1}}{\sqrt{\tau_1}}\right)\frac{1}{\sqrt{\tau_1}}\right].$$

Proof Proposition 12.2 and the Imhof relation (12.48) imply that

$$E[F(\mathfrak{a}_\bullet)] = \sqrt{\frac{\pi}{2}} E_0\left[F\left(R_1 - R_{1-\bullet}\right)\frac{1}{R_1}\right] \tag{12.42}$$

where R is a Bessel(3) process. Let γ_1 denote the last hitting time of 1 by R. Since $R_\bullet \mapsto F(R_1 - R_{1-\bullet})R_1$ is also a non-negative measurable path functional, we can use Theorem 12.8 to rewrite the right-hand side of (12.42) as

$$\begin{aligned} \sqrt{\frac{\pi}{2}} E_0\left[F\left(R_1 - R_{1-\bullet}\right) R_1 \frac{1}{R_1^2}\right] &= \sqrt{\frac{\pi}{2}} E_0\left[F\left(\frac{R_{\gamma_1} - R_{(1-\bullet)\gamma_1}}{\sqrt{\gamma_1}}\right)\frac{R_{\gamma_1}}{\sqrt{\gamma_1}}\right] \\ &= \sqrt{\frac{\pi}{2}} E_0\left[F\left(\frac{1 - R_{(1-\bullet)\gamma_1}}{\sqrt{\gamma_1}}\right)\frac{1}{\sqrt{\gamma_1}}\right]. \end{aligned} \tag{12.43}$$

Now Williams' time reversal Theorem 12.11 can be used to conclude that (12.43) is equal to

$$\sqrt{\frac{\pi}{2}} E_0\left[F\left(\frac{X_{\bullet\tau_1}}{\sqrt{\tau_1}}\right)\frac{1}{\sqrt{\tau_1}}\right]$$

which completes the proof. □

The process

$$\left(\frac{X_{s\tau_1}}{\sqrt{\tau_1}} : 0 \le s \le 1\right) \tag{12.44}$$

under P_0 which appears in Proposition 12.5 has recently been studied by Elie, Rosenbaum, and Yor in [10, 25–27]. Among other results, they derive the density of the random variable α defined by

$$\alpha = \frac{X_{U\tau_1}}{\sqrt{\tau_1}}$$

where U is a Uniform[0, 1] random variable independent of X. The non-obvious fact that $E_0[\alpha] = 0$ leads to an interesting corollary of Proposition 12.5.

Corollary 12.1 *Let U be a Uniform*[0, 1] *random variable independent of* $\mathfrak{a}$. *Then*

$$E\left[\int_0^1 \frac{\mathfrak{a}_s}{\mathfrak{a}_1} ds\right] = E\left[\frac{\mathfrak{a}_U}{\mathfrak{a}_1}\right] = 0.$$

12.6.1 Brownian Co-ascent

In this section we show how the process (12.44) is related to the Brownian co-meander. The reader unfamiliar with the co-meander can refer to Appendix 1 for some basic information. The following proposition suggests that a suitable name for the process (12.44) is the *Brownian co-ascent* since it is constructed from the co-meander in the same manner that the ascent is constructed from the meander, viz Proposition 12.2.

Proposition 12.6 *If X has distribution* P_0 *then*

$$\left(\frac{X_{s\tau_1}}{\sqrt{\tau_1}} : 0 \leq s \leq 1\right) \stackrel{\mathcal{L}}{=} (\tilde{\mathfrak{m}}_1 - \tilde{\mathfrak{m}}_{1-s} : 0 \leq s \leq 1).$$

Proof Let $F : C([0, 1]; \mathbb{R}) \to \mathbb{R}$ be bounded and continuous. Then by Theorem 2.1. in [26] we have

$$E_0\left[F\left(\frac{X_{\bullet\tau_1}}{\sqrt{\tau_1}}\right)\right] = E_0\left[F(R_1 - R_{1-\bullet})\frac{1}{R_1^2}\right]$$

and by (12.49) we have

$$E_0\left[F(R_1 - R_{1-\bullet})\frac{1}{R_1^2}\right] = E[F\left(\tilde{\mathfrak{m}}_1 - \tilde{\mathfrak{m}}_{1-\bullet}\right)].$$

The proposition follows from combining these two identities. □

From now on we refer to the process (12.44) as the Brownian co-ascent and denote it by $\left(\tilde{\mathfrak{a}}_s : 0 \leq s \leq 1\right)$. This allows us to state as an immediate corollary of Proposition 12.5 the following absolute continuity relation between the ascent and co-ascent which can also be seen as a counterpart of (12.50).

Corollary 12.2 *For any measurable* $F : C([0, 1]; \mathbb{R}) \to \mathbb{R}_+$ *we have*

$$E[F(\mathfrak{a}_\bullet)] = \sqrt{\frac{\pi}{2}} E\left[F(\tilde{\mathfrak{a}}_\bullet)\tilde{\mathfrak{a}}_1\right].$$

Next we give an analogue of Proposition 12.3 for the Brownian co-ascent. Let $(\ell_t : t \geq 0)$ denote the inverse local time of X at the level 0, that is, $\ell_t = \inf\{s : L_s^0(X) > t\}$. The *pseudo-Brownian bridge* $\tilde{\mathfrak{b}}$ was introduced in [4] and has representation

$$\left(\tilde{\mathfrak{b}}_s : 0 \leq s \leq 1\right) \stackrel{\mathcal{L}}{=} \left(\frac{X_{s\ell_1}}{\sqrt{\ell_1}} : 0 \leq s \leq 1\right)$$

where the right-hand side is under P_0, see also [25].

Proposition 12.7 *Let* $\left(L_s^0(\tilde{\mathfrak{b}}) : 0 \leq s \leq 1\right)$ *denote the local time process at the level* 0 *of a pseudo-Brownian bridge. Then we have the equality in law*

$$\left(\sup_{0 \leq u \leq s} \tilde{\mathfrak{a}}_u : 0 \leq s \leq 1\right) \stackrel{\mathcal{L}}{=} \left(L_s^0(\tilde{\mathfrak{b}}) : 0 \leq s \leq 1\right).$$

Proof Define $T_1 = \inf\{t : S_t = 1\}$. Notice that $T_1 = \tau_1$ almost surely under P_0. Hence

$$\left(\sup_{0 \leq u \leq s} \frac{X_{u\tau_1}}{\sqrt{\tau_1}} : 0 \leq s \leq 1\right) \stackrel{\mathcal{L}}{=} \left(\frac{1}{\sqrt{T_1}} S_{sT_1} : 0 \leq s \leq 1\right) \tag{12.45}$$

under P_0. Additionally, by Lévy's equivalence (12.41) we have

$$\left(\frac{1}{\sqrt{T_1}} S_{sT_1} : 0 \leq s \leq 1\right) \stackrel{\mathcal{L}}{=} \left(\frac{1}{\sqrt{\ell_1}} L_{s\ell_1}^0(X) : 0 \leq s \leq 1\right) \tag{12.46}$$

under P_0. Moreover, the representation of $\tilde{\mathfrak{b}}$ along with the scaling relation between Brownian motion and its local time implies

$$\left(\frac{1}{\sqrt{\ell_1}} L_{s\ell_1}^0(X) : 0 \leq s \leq 1\right) \stackrel{\mathcal{L}}{=} \left(L_s^0(\tilde{\mathfrak{b}}) : 0 \leq s \leq 1\right). \tag{12.47}$$

The proposition follows from combining (12.45), (12.46) and (12.47) along with the representation (12.44) of $\tilde{\mathfrak{a}}$. □

12.7 Concluding Remarks

Two natural directions for generalizing the main results of this paper are to change the weight process or the reference measure. Scaled penalization of Brownian motion with drift $h \in \mathbb{R}$ by the weight process $\Gamma_t = \exp\left(-\nu(S_t - I_t)\right)$ is one such possibility. This *range penalization* with $\nu > 0$ has been investigated in [31] for $h = 0$, in [20] for $0 < |h| < \nu$, and recently in [14] for $|h| = \nu$. While the first two papers identify the corresponding scaling limit, only partial results are known in the critical case $|h| = \nu$. A related model replaces the Brownian motion with drift by reflecting Brownian motion with drift and penalizes by the supremum instead of the range. The asymmetry imposed by the reflecting barrier at 0 now makes the sign of h relevant. Work in preparation by the current author describes the scaling limit in the critical case for both of these models.

Another interesting question is to what extent can the absolute continuity relation Proposition 12.5 and the path constructions Propositions 12.2 and 12.4 be generalized to processes other than Brownian motion? While all three of these can be nominally applied to many processes, it's not obvious if they yield a genuine ascent, that is, the process conditioned to end at its maximum. Scale invariance is an underlying theme in all of these results so it makes sense to first consider self-similar processes such as strictly stable Lévy processes and Bessel processes. In this direction, existing work on stable meanders and the stable analogue of Denisov's decomposition found in Chapter VIII of [1] would be a good starting point.

Acknowledgements The author would like to thank Iddo Ben-Ari for his helpful suggestions and encouragement and also Jim Pitman and Ju-Yi Yen for their tips on the history of the Brownian meander and co-meander as well as pointers to the literature.

Appendix 1: Normalized Brownian Excursion, Meander and Co-meander

The normalized Brownian excursion, meander and co-meander can be constructed from the excursion of Brownian motion which straddles time 1. In fact, this is usually how these processes are defined, see Chapter 7 in [35]. Define $g_1 = \sup\{t < 1 : X_t = 0\}$ as the last zero before time 1 and $d_1 = \inf\{t > 1 : X_t = 0\}$ as the first zero after time 1. Then the normalized excursion, meander and co-meander have representation

$$(\mathfrak{e}_s : 0 \le s \le 1) \overset{\mathcal{L}}{=} \left(\frac{|X_{g_1 + s(d_1 - g_1)}|}{\sqrt{d_1 - g_1}} : 0 \le s \le 1 \right),$$

$$(\mathfrak{m}_s : 0 \le s \le 1) \overset{\mathcal{L}}{=} \left(\frac{|X_{g_1 + s(1 - g_1)}|}{\sqrt{1 - g_1}} : 0 \le s \le 1 \right)$$

and

$$\left(\tilde{\mathfrak{m}}_s : 0 \le s \le 1\right) \overset{\mathcal{L}}{=} \left(\frac{|X_{d_1 + s(1 - d_1)}|}{\sqrt{d_1 - 1}} : 0 \le s \le 1 \right),$$

respectively, where the right-hand sides are under P_0.

The laws of the meander and co-meander are absolutely continuous with respect to each other and to the law of the Bessel(3) process starting at 0. More specifically, for any measurable $F : C([0, 1]; \mathbb{R}) \to \mathbb{R}_+$ we have

$$E[F(\mathfrak{m}_\bullet)] = \sqrt{\frac{\pi}{2}} E_0 \left[F(R_\bullet) \frac{1}{R_1} \right] \tag{12.48}$$

$$E[F(\tilde{\mathfrak{m}}_\bullet)] = E_0 \left[F(R_\bullet) \frac{1}{R_1^2} \right] \tag{12.49}$$

$$E[F(\mathfrak{m}_\bullet)] = \sqrt{\frac{\pi}{2}} E \left[F(\tilde{\mathfrak{m}}_\bullet) \tilde{\mathfrak{m}}_1 \right]. \tag{12.50}$$

The first of these relations (12.48) is known as Imhof's relation [12, 28], while (12.49) appears as Theorem 7.4.1. in [35] and (12.50) follows from a combination of the previous two.

Appendix 2: Absolute Continuity Relations

Here we collect some useful absolute continuity relations between the laws of various processes. While the statements involve bounded measurable path functionals F, they are also valid for non-negative measurable F. The results given without proof can be found in the literature as indicated. The first two relations give us absolute continuity for Brownian motion and Bessel(3) processes starting at different points, as long as we are willing to ignore the initial $[0, \delta]$ segment of the path. See Definition 12.3 for notation that makes this precise.

Lemma 12.3 *Let $x \in \mathbb{R}$, $y > 0$, and $0 < \delta \leq 1$. Then for any bounded measurable $F : C([0, 1]; \mathbb{R}) \to \mathbb{R}$ we have*

$$E_0 \left[F_\delta(X_\bullet) \right] = E_x \left[F_\delta(X_\bullet) \exp\left(\frac{x^2 - 2x X_\delta}{2\delta} \right) \right]$$

and

$$E_0 \left[F_\delta(R_\bullet) \right] = E_y \left[F_\delta(R_\bullet) \frac{y R_\delta \exp\left(\frac{y^2}{2\delta}\right)}{\delta \sinh\left(\frac{y R_\delta}{\delta}\right)} \right].$$

Proof We only prove the first statement as the same argument applies to the second; see (12.51) and (12.52) for the Bessel(3) transition densities. First note that by the definition of F_δ and the Markov property we have for any $z \in \mathbb{R}$

$$E_0 \left[F_\delta(X_\bullet) | X_\delta = z \right] = E_x \left[F_\delta(X_\bullet) | X_\delta = z \right].$$

Now by conditioning on X_δ with $p_\delta(\cdot,\cdot)$ denoting the transition density of Brownian motion at time δ, we can write

$$\begin{aligned}
E_0\left[F_\delta(X_\bullet)\right] &= \int_{-\infty}^{\infty} E_0\left[F_\delta(X_\bullet)|X_\delta = z\right] p_\delta(0,z)dz \\
&= \int_{-\infty}^{\infty} E_x\left[F_\delta(X_\bullet)|X_\delta = z\right] \frac{p_\delta(0,z)}{p_\delta(x,z)} p_\delta(x,z)dz \\
&= \int_{-\infty}^{\infty} E_x\left[F_\delta(X_\bullet)\frac{p_\delta(0,X_\delta)}{p_\delta(x,X_\delta)}\middle| X_\delta = z\right] p_\delta(x,z)dz \\
&= E_x\left[F_\delta(X_\bullet)\frac{\exp\left(-\frac{X_\delta^2}{2\delta}\right)}{\exp\left(-\frac{(X_\delta - x)^2}{2\delta}\right)}\right] \\
&= E_x\left[F_\delta(X_\bullet)\exp\left(\frac{x^2 - 2xX_\delta}{2\delta}\right)\right].
\end{aligned}$$

□

The next relation results from an h-transform of Brownian motion by the harmonic function $h(x) = x$. See Section 1.6 of [35].

Proposition 12.8 *Let $x > 0$. Then for any bounded measurable $F : C\big([0,1];\mathbb{R}\big) \to \mathbb{R}$ we have*

$$E_x[F(R_\bullet)] = E_x\left[F(X_\bullet)\frac{X_1}{x};\, I_1 > 0\right].$$

The law of a Bessel(3) process run up to the last hitting time of $x > 0$ is, after rescaling, absolutely continuous with respect to the law of a Bessel(3) process run up to a fixed time. This is a special case of Théorème 3 in [4]; see also Theorem 8.1.1. in [35].

Theorem 12.8 *Let γ_x be the last hitting time of $x > 0$ by the Bessel(3) process R. Then for any bounded measurable $F : C\big([0,1];\mathbb{R}\big) \to \mathbb{R}$ we have*

$$E_0\left[F\left(\frac{R_{\bullet\gamma_x}}{\sqrt{\gamma_x}}\right)\right] = E_0\left[F(R_\bullet)\frac{1}{R_1^2}\right].$$

Appendix 3: Path Decompositions

Denisov's path decomposition [8] asserts that the pre and post-maximum parts of a Brownian path are rescaled independent Brownian meanders. See Corollary 17 in Chapter VIII of [1] for an extension to strictly stable Lévy processes.

Theorem 12.9 (Denisov) *Let Θ denote the almost surely unique time at which the Brownian motion W attains its maximum over the time interval $[0, 1]$. Then the transformed pre-maximum path*

$$\left(\frac{W_\Theta - W_{\Theta - s\Theta}}{\sqrt{\Theta}} : 0 \leq s \leq 1\right)$$

and the transformed post-maximum path

$$\left(\frac{W_\Theta - W_{\Theta + s(1-\Theta)}}{\sqrt{1-\Theta}} : 0 \leq s \leq 1\right)$$

are independent Brownian meanders which are independent of Θ.

Williams' path decomposition for Brownian motion with drift $h < 0$ splits the path at the time of the global maximum Θ_∞ by first picking an Exponential$(-2h)$ distributed S_∞ and then running a Brownian motion with drift $-h$ until it hits the level S_∞ for the pre-maximum path and then running Brownian motion with drift h conditioned remain below S_∞ for the post-maximum path. See Theorem 55.9 in Chapter VI of [24] for the following more precise statement.

Theorem 12.10 (Williams) *Suppose $h < 0$ and consider the following independent random elements:*

1. *$(X_t : t \geq 0)$, a Brownian motion with drift $-h$ starting at 0;*
2. *$(R_t : t \geq 0)$, a Brownian motion with drift h starting at 0 conditioned to be non-positive for all time;*
3. *and l, an Exponential$(-2h)$ random variable.*

Let $\tau_l = \inf\{t : X_t = l\}$ be the first hitting time of the level l by X. Then the process

$$\widetilde{X}_t = \begin{cases} X_t & : 0 \leq t \leq \tau_l \\ l + R_{t-\tau_l} & : \tau_l < t. \end{cases}$$

is Brownian motion with drift h starting at 0.

Williams' time reversal connects the laws of Brownian motion run until a first hitting time and a Bessel(3) process run until a last hitting time, see Theorem 49.1 in Chapter III of [23].

Theorem 12.11 (Williams) *Let* $\tau_1 = \inf\{t : X_t = 0\}$ *be the first hitting time of* 1 *by the Brownian motion* X *started at* 0. *Let* $\gamma_1 = \sup\{t : R_t = 1\}$ *be the last hitting time of* 1 *by the Bessel(3) process* R *started at* 0. *Then the following equality in law holds:*

$$(1 - X_{\tau_1 - t} : 0 \leq t \leq \tau_1) \stackrel{\mathcal{L}}{=} (R_t : 0 \leq t \leq \gamma_1).$$

Appendix 4: Density and Distribution Formulas

The Bessel(3) transition density formulas

$$p_t(0, y) = \sqrt{\frac{2}{\pi t^3}} y^2 \exp\left(-\frac{y^2}{2t}\right) dy \tag{12.51}$$

and

$$p_t(x, y) = \sqrt{\frac{2}{\pi t}} \frac{y}{x} \sinh\left(\frac{xy}{t}\right) \exp\left(-\frac{x^2 + y^2}{2t}\right) dy, \tag{12.52}$$

valid for $y \geq 0$ and $x, t > 0$, can be found in Chapter XI of [22]. The density formula for the endpoint of a Brownian meander

$$P(\mathfrak{m}_1 \in dy) = y \exp\left(-\frac{y^2}{2}\right) dy, \ y \geq 0 \tag{12.53}$$

follows from (12.51) together with the Imhof relation (12.48). The well-known arcsine law for the time of the maximum of Brownian motion states that

$$P_0(\Theta_1 \in du) = \frac{1}{\pi \sqrt{u(1-u)}} du, \ 0 < u < 1. \tag{12.54}$$

Using Denisov's path decomposition Theorem 12.9, the densities (12.53) and (12.54) can be combined to yield the joint density

$$\begin{aligned} &P_0 \left(S_1 \in dx, S_1 - X_1 \in dy, \Theta_1 \in du\right) \\ &= \frac{xy}{\pi \sqrt{u^3 (1-u)^3}} \exp\left(-\frac{x^2}{2u} - \frac{y^2}{2(1-u)}\right) dx dy du \end{aligned} \tag{12.55}$$

which holds for $x, y \geq 0$ and $0 < u < 1$.

The maximum of a Brownian bridge from 0 to a of length $T > 0$ has distribution

$$P_0\left(\sup_{0\le s\le T} X_s \ge b \,\middle|\, X_T = a\right) = \exp\left(-\frac{2b(b-a)}{T}\right) \tag{12.56}$$

where $b \ge \max\{0, a\}$, see (4.3.40) in [13].

Appendix 5: Asymptotic Analysis Tools

The following versions of these standard results in asymptotic analysis can be found in [17] and [32], respectively.

Lemma 12.4 (Watson's Lemma) *Let $q(x)$ be a function of the positive real variable x, such that*

$$q(x) \sim \sum_{n=0}^{\infty} a_n x^{\frac{n+\lambda-\mu}{\mu}} \quad \textit{as } x \to 0,$$

where λ and μ are positive constants. Then

$$\int_0^{\infty} q(x)e^{-tx}dx \sim \sum_{n=0}^{\infty} \Gamma\left(\frac{n+\lambda}{\mu}\right)\frac{a_n}{t^{\frac{n+\lambda}{\mu}}} \quad \textit{as } t \to \infty.$$

Theorem 12.12 (Laplace's Method) *Define*

$$I(t) = \int_a^b f(x)e^{-th(x)}dx$$

where $-\infty \le a < b \le \infty$ and $t > 0$. Assume that:

1. *$h(x)$ has a unique minimum on $[a, b]$ at point $x = x_0 \in (a, b)$,*
2. *$h(x)$ and $f(x)$ are continuously differentiable in a neighborhood of x_0 with $f(x_0) \neq 0$ and*

$$h(x) = h(x_0) + \frac{1}{2}h''(x_0)(x-x_0)^2 + O\big((x-x_0)^3\big) \quad \textit{as } x \to x_0,$$

3. *the integral $I(t)$ exists for sufficiently large t.*

Then

$$I(t) = f(x_0)\sqrt{\frac{2\pi}{th''(x_0)}}e^{-th(x_0)}\left(1 + O\left(\frac{1}{\sqrt{t}}\right)\right) \quad \textit{as } t \to \infty.$$

Appendix 6: Convergence Lemma

Here we give a Fatou-type lemma that helps streamline the proofs of the main theorems.

Lemma 12.5 *Suppose* $\{F_t\}_{t\geq 0}$, F, $\{X_t\}_{t\geq 0}$, *and* X *are all integrable functions defined on the same measure space* (Ω, Σ, μ) *such that* $\{|F_t|\}_{t\geq 0}$ *are bounded by* $M > 0$, $\{X_t\}_{t\geq 0}$ *are non-negative,* $\int X_t d\mu \to \int X d\mu$, *and both* $F_t \to F$ *and* $X_t \to X$ μ*-almost surely. Then we have*

$$\lim_{t\to\infty} \int F_t X_t d\mu = \int F X d\mu.$$

Proof Notice that $(M + F_t)X_t$ is non-negative for all $t \geq 0$. So by Fatou's lemma we have

$$M \int X d\mu + \liminf_{t\to\infty} \int F_t X_t d\mu = \liminf_{t\to\infty} \int (M + F_t)X_t d\mu \geq \int (M + F)X d\mu$$

which implies

$$\liminf_{t\to\infty} \int F_t X_t d\mu \geq \int F X d\mu.$$

Similarly, $(M - F_t)X_t$ is non-negative for all $t \geq 0$, hence

$$M \int X d\mu - \limsup_{t\to\infty} \int F_t X_t d\mu = \liminf_{t\to\infty} \int (M - F_t)X_t d\mu \geq \int (M - F)X d\mu$$

which implies

$$\limsup_{t\to\infty} \int F_t X_t d\mu \leq \int F X d\mu.$$

Together these inequalities imply that

$$\lim_{t\to\infty} \int F_t X_t d\mu = \int F X d\mu.$$

□

References

1. J. Bertoin, Lévy processes, in *Cambridge Tracts in Mathematics*, vol. 121 (Cambridge University Press, Cambridge, 1996)
2. J. Bertoin, L. Chaumont, J. Pitman, Path transformations of first passage bridges. Electron. Comm. Probab. **8**, 155–166 (2003) (electronic)

3. P. Biane, M. Yor, Quelques précisions sur le méandre brownien. Bull. Sci. Math. (2) **112**(1), 101–109 (1988)
4. P. Biane, J.F. Le Gall, M. Yor, Un processus qui ressemble au pont brownien, in *Séminaire de Probabilités, XXI*. Lecture Notes in Mathematics, vol. 1247 (Springer, Berlin, 1987), pp. 270–275
5. L. Chaumont, G. Uribe Bravo, Markovian bridges: weak continuity and pathwise constructions. Ann. Probab. **39**(2), 609–647 (2011)
6. P. Debs, Penalisation of the standard random walk by a function of the one-sided maximum, of the local time, or of the duration of the excursions, in *Séminaire de probabilités XLII*. Lecture Notes in Mathematics, vol. 1979 (Springer, Berlin, 2009), pp. 331–363
7. P. Debs, Penalisation of the symmetric random walk by several functions of the supremum. Markov Process. Related Fields **18**(4), 651–680 (2012)
8. I.V. Denisov, Random walk and the Wiener process considered from a maximum point. Teor. Veroyatnost. i Primenen. **28**(4), 785–788 (1983)
9. R.T. Durrett, D.L. Iglehart, D.R. Miller, Weak convergence to Brownian meander and Brownian excursion. Ann. Probab. **5**(1), 117–129 (1977)
10. R. Elie, M. Rosenbaum, M. Yor, On the expectation of normalized Brownian functionals up to first hitting times. Electron. J. Probab. **19**(37), 23 (2014)
11. P. Fitzsimmons, J. Pitman, M. Yor, Markovian bridges: construction, Palm interpretation, and splicing, in *Seminar on Stochastic Processes, 1992 (Seattle, WA, 1992)*. Progress in Probability, vol. 33 (Birkhäuser Boston, Boston, 1993), pp. 101–134
12. J.P. Imhof, Density factorizations for Brownian motion, meander and the three-dimensional Bessel process, and applications. J. Appl. Probab. **21**(3), 500–510 (1984)
13. I. Karatzas, S.E. Shreve, Brownian motion and stochastic calculus, in *Graduate Texts in Mathematics*, vol. 113 (Springer, New York, 1988)
14. M. Kolb, M. Savov, Conditional survival distributions of Brownian trajectories in a one dimensional Poissonian environment in the critical case. Electron. J. Probab. **22**(14), 29 (2017)
15. J. Najnudel, B. Roynette, M. Yor, A global view of Brownian penalisations, in *MSJ Memoirs*, vol. 19 (Mathematical Society of Japan, Tokyo, 2009)
16. J. Obłój, M. Yor, On local martingale and its supremum: harmonic functions and beyond, in *From stochastic calculus to mathematical finance* (Springer, Berlin, 2006), pp. 517–533
17. F.W.J. Olver, *Asymptotics and Special Functions* (Academic Press [A subsidiary of Harcourt Brace Jovanovich, Publishers], New York, 1974). Computer Science and Applied Mathematics
18. J. Pitman, Combinatorial stochastic processes, in *Lecture Notes in Mathematics*, vol. 1875 (Springer, Berlin, 2006). Lectures from the 32nd Summer School on Probability Theory held in Saint-Flour, July 7–24, 2002, With a foreword by Jean Picard
19. J. Pitman, M. Yor, Decomposition at the maximum for excursions and bridges of one-dimensional diffusions, in *Itô's Stochastic Calculus and Probability Theory* (Springer, Tokyo, 1996), pp. 293–310
20. T. Povel, On weak convergence of conditional survival measure of one-dimensional Brownian motion with a drift. Ann. Appl. Probab. **5**(1), 222–238 (1995)
21. C. Profeta, Some limiting laws associated with the integrated Brownian motion. ESAIM Probab. Stat. **19**, 148–171 (2015)
22. D. Revuz, M. Yor, Continuous martingales and Brownian motion, in *Grundlehren der Mathematischen Wissenschaften [Fundamental Principles of Mathematical Sciences]*, vol. 293, 2nd edn. (Springer, Berlin, 1994)
23. L.C.G. Rogers, D. Williams, Diffusions, Markov processes, and martingales, in *Cambridge Mathematical Library*, vol. 1 (Cambridge University Press, Cambridge, 2000)
24. L.C.G. Rogers, D. Williams, Diffusions, Markov processes, and martingales, in *Cambridge Mathematical Library*, Vol. 2 (Cambridge University Press, Cambridge, 2000)
25. M. Rosenbaum, M. Yor, On the law of a triplet associated with the pseudo-Brownian bridge, in *Séminaire de Probabilités XLVI*. Lecture Notes in Mathematics, vol. 2123 (Springer, Cham, 2014), pp. 359–375

26. M. Rosenbaum, M. Yor, Random scaling and sampling of Brownian motion. J. Math. Soc. Japan **67**(4), 1771–1784 (2015)
27. M. Rosenbaum, M. Yor, Some explicit formulas for the Brownian bridge, Brownian meander and Bessel process under uniform sampling. ESAIM Probab. Stat. **19**, 578–589 (2015)
28. B. Roynette, M. Yor, *Penalising Brownian Paths*. Lecture Notes in Mathematics, vol. 1969 (Springer, Berlin, 2009)
29. B. Roynette, P. Vallois, M. Yor, Limiting laws for long Brownian bridges perturbed by their one-sided maximum. III. Period. Math. Hungar. **50**(1–2), 247–280 (2005)
30. B. Roynette, P. Vallois, M. Yor, Limiting laws associated with Brownian motion perturbed by its maximum, minimum and local time. II. Studia Sci. Math. Hungar. **43**(3), 295–360 (2006)
31. U. Schmock, Convergence of the normalized one-dimensional Wiener sausage path measures to a mixture of Brownian taboo processes. Stochastics Stochastics Rep. **29**(2), 171–183 (1990)
32. W. Szpankowski, Average case analysis of algorithms on sequences, in *Wiley-Interscience Series in Discrete Mathematics and Optimization* (Wiley-Interscience, New York, 2001)
33. Y. Yano, A remarkable σ-finite measure unifying supremum penalisations for a stable Lévy process. Ann. Inst. Henri Poincaré Probab. Stat. **49**(4), 1014–1032 (2013)
34. K. Yano, Y. Yano, M. Yor, Penalisation of a stable Lévy process involving its one-sided supremum. Ann. Inst. Henri Poincaré Probab. Stat. **46**(4), 1042–1054 (2010)
35. J.Y. Yen, M. Yor, *Local Times and Excursion Theory for Brownian Motion*. Lecture Notes in Mathematics, vol. 2088 (Springer, Cham, 2013)
36. M. Yor, L. Zambotti, A remark about the norm of a Brownian bridge. Stat. Probab. Lett. **68**(3), 297–304 (2004)

Chapter 13
Interlacing Diffusions

Theodoros Assiotis, Neil O'Connell, and Jon Warren

Abstract We study in some generality intertwinings between h-transforms of Karlin–McGregor semigroups associated with one dimensional diffusion processes and those of their Siegmund duals. We obtain couplings so that the corresponding processes are interlaced and furthermore give formulae in terms of block determinants for the transition densities of these coupled processes. This allows us to build diffusion processes in the space of Gelfand–Tsetlin patterns so that the evolution of each level is Markovian. We show how known examples naturally fit into this framework and construct new processes related to minors of matrix valued diffusions. We also provide explicit formulae for the transition densities of the particle systems with one-sided collisions at either edge of such patterns.

13.1 Introduction

In this work we study in some generality intertwinings and couplings between Karlin–McGregor semigroups (see [45], also [44]) associated with one dimensional diffusion processes and their duals. Let $X(t)$ be a diffusion process with state space an interval $I \subset \mathbb{R}$ with end points $l < r$ and transition density $p_t(x, y)$. We define the Karlin–McGregor semigroup associated with X, with n particles, by its transition densities (with respect to Lebesgue measure) given by,

$$\det(p_t(x_i, y_j))_{i,j=1}^n,$$

T. Assiotis (✉)
Mathematical Institute, University of Oxford, Oxford, UK
e-mail: theo.assiotis@maths.ox.ac.uk

N. O'Connell
School of Mathematics and Statistics, University College Dublin, Dublin, Ireland
e-mail: neil.oconnell@ucd.ie

J. Warren
Department of Statistics, University of Warwick, Coventry, UK
e-mail: J.Warren@warwick.ac.uk

C. Donati-Martin et al. (eds.), *Séminaire de Probabilités L*, Lecture Notes in Mathematics 2252, https://doi.org/10.1007/978-3-030-28535-7_13

for $x, y \in W^n(I^\circ)$ where $W^n(I^\circ) = (x = (x_1, \cdots, x_n) : l < x_1 \leq \cdots \leq x_n < r)$. This sub-Markov semigroup is exactly the semigroup of n independent copies of the diffusion process X which are killed when they intersect. For such a diffusion process $X(t)$ we consider the conjugate (see [65]) or Siegmund dual (see [21] or the original paper [60]) diffusion process $\hat{X}(t)$ via a description of its generator and boundary behaviour in the next subsection. The key relation dual/conjugate diffusion processes satisfy is the following (see Lemma 13.1), with $z, z' \in I^\circ$,

$$\mathbb{P}_z(X(t) \leq z') = \mathbb{P}_{z'}(\hat{X}(t) \geq z).$$

We will obtain *couplings* of h-transforms of Karlin–McGregor semigroups associated with a diffusion process and its dual so that the corresponding processes *interlace*. We say that $y \in W^n(I^\circ)$ and $x \in W^{n+1}(I^\circ)$ interlace and denote this by $y \prec x$ if $x_1 \leq y_1 \leq x_2 \leq \cdots \leq x_{n+1}$. Note that this defines a space denoted by $W^{n,n+1}(I^\circ) = ((x, y) : l < x_1 \leq y_1 \leq x_2 \leq \cdots \leq x_{n+1} < r)$,

$$x_1 \; y_1 \; x_2 \; y_1 \; x_3 \quad x_n \; y_n \; x_{n+1}$$
$$\bullet \; \bullet \; \bullet \; \bullet \; \bullet \quad \cdots \quad \bullet \; \bullet \; \bullet \; ,$$

with the following two-level representation,

$$
\begin{array}{ccccccc}
y_1 & & y_1 & \cdots\cdots & & y_n & \\
\bullet & & \bullet & & & \bullet & \\
x_1 & x_2 & & x_3 & \cdots & x_n & x_{n+1} \\
\bullet & \bullet & & \bullet & \cdots & \bullet & \bullet
\end{array}.
$$

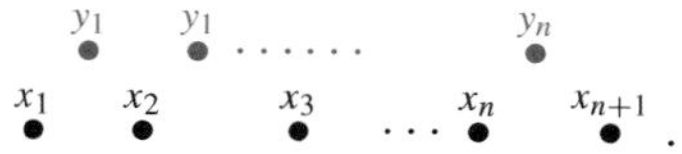

Similarly, we say that $x, y \in W^n(I^\circ)$ interlace if $l < y_1 \leq x_1 \leq y_2 \leq \cdots \leq x_n < r$ (we still denote this by $y \prec x$). Again, this defines the space $W^{n,n}(I^\circ) = ((x, y) : l < y_1 \leq x_1 \leq y_2 \leq \cdots \leq x_n < r)$,

$$y_1 \; x_1 \; y_2 \; x_2 \quad x_{n-1} \; y_n \; x_n$$
$$\bullet \; \bullet \; \bullet \; \bullet \quad \cdots \quad \bullet \; \bullet \; \bullet \; ,$$

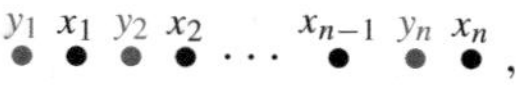

with the two-level representation,

$$
\begin{array}{ccccc}
y_1 & y_2 & \cdots & y_n & \\
\bullet & \bullet & & \bullet & \\
& x_1 & x_2 & x_{n-1} & x_n \\
& \bullet & \bullet \cdots & \bullet & \bullet
\end{array}.
$$

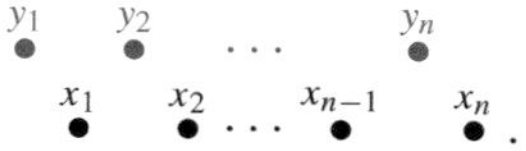

Our starting point in this paper are explicit transition kernels, actually arising from the consideration of stochastic coalescing flows. These kernels defined on $W^{n,n+1}(I^\circ)$ (or $W^{n,n}(I^\circ)$) are given in terms of block determinants and give rise to a Markov process $Z = (X, Y)$ with (sub-)Markov transition semigroup Q_t with joint dynamics described as follows. Let L and $\hat{L}$ be the generators of a pair of one dimensional diffusions in Siegmund duality. Then, after an appropriate Doob's h-transformation Y evolves *autonomously* as n $\hat{L}$-diffusions conditioned not to intersect. The X components then evolve as $n + 1$ (or n) independent L-diffusions reflected off the random Y barriers, a notion made precise in the next subsection. Our main result, Theorem 13.1 in the text, states (modulo technical assumptions) that under a special initial condition for $Z = (X, Y)$, the *non-autonomous* X component is distributed as a Markov process in its own right. Its evolution governed by an explicit Doob's h-transform of the Karlin–McGregor semigroup associated with $n+1$ (or n) L-diffusions.

At the heart of this result lie certain intertwining relations, obtained immediately from the special structure of Q_t, of the form,

$$P_t \Lambda = \Lambda Q_t \,, \tag{13.1}$$

$$\Pi \hat{P}_t = Q_t \Pi \,, \tag{13.2}$$

where Λ is an explicit positive kernel (not yet normalized), Π is the operator induced by the projection on the Y level, P_t is the Karlin–McGregor semigroup associated with the one dimensional diffusion process with transition density $p_t(x, y)$ and $\hat{P}_t$ the corresponding semigroup associated with its dual/conjugate (some conditions and more care is needed regarding boundary behaviour for which the reader is referred to the next section).

Now we move towards building a multilevel process. First, note that by concatenating $W^{1,2}(I^\circ), W^{2,3}(I^\circ), \cdots, W^{N-1,N}(I^\circ)$ we obtain the space of Gelfand–Tsetlin patterns of depth N denoted by $\mathbb{GT}(N)$,

$$\mathbb{GT}(N) = \{(X^{(1)}, \cdots, X^{(N)}) : X^{(n)} \in W^n(I^\circ),\ X^{(n)} \prec X^{(n+1)}\} \,.$$

A point $(X^{(1)}, \cdots, X^{(N)}) \in \mathbb{GT}(N)$ is typically depicted as an array as shown in the following diagram:

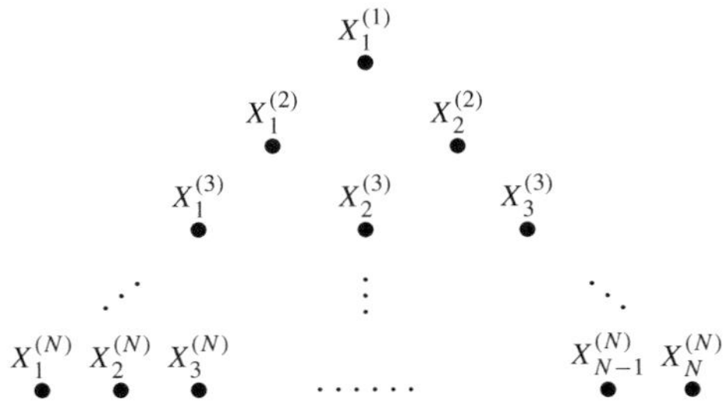

Similarly, by concatenating $W^{1,1}(I^\circ), W^{1,2}(I^\circ), W^{2,2}(I^\circ), \cdots, W^{N,N}(I^\circ)$ we obtain the space of symplectic Gelfand–Tsetlin patterns of depth N denoted by $\mathbb{GT}_{\mathbf{s}}(N)$,

$$\mathbb{GT}_{\mathbf{s}}(N) = \{(X^{(1)}, \hat{X}^{(1)} \cdots, X^{(N)}, \hat{X}^{(N)}) \\ : X^{(n)}, \hat{X}^{(n)} \in W^n(I^\circ),\ X^{(n)} \prec \hat{X}^{(n)} \prec X^{(n+1)}\} \,,$$

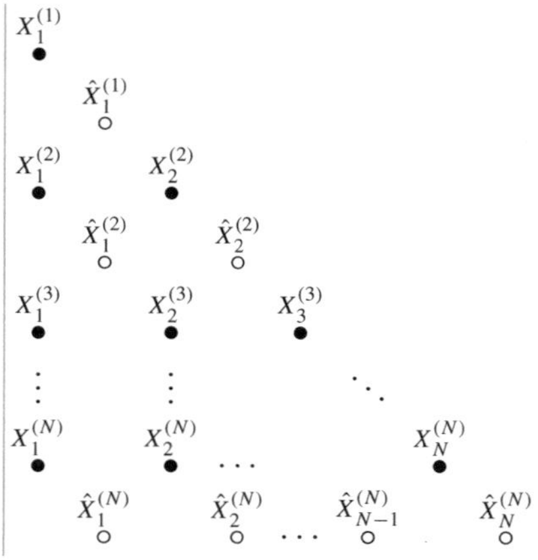

Theorem 13.1 allows us to concatenate a sequence of $W^{n,n+1}$-valued processes (or two-level processes), by a procedure described at the beginning of Sect. 13.3, in order to build diffusion processes in the space of Gelfand Tsetlin patterns so that each level is Markovian with explicit transition densities. Such examples of dynamics on *discrete* Gelfand–Tsetlin patterns have been extensively studied over the past decade as models for random surface growth, see in particular [9, 10, 68] and the more recent paper [18] and the references therein. They have also been considered in relation to building infinite dimensional Markov processes, preserving some distinguished measures of representation theoretic origin, on the boundary of these Gelfand–Tsetlin graphs via the *method of intertwiners*; see Borodin and Olshanski [11] for the type A case and more recently Cuenca [22] for the type BC. In the paper [4] we pursued these directions in some detail.

Returning to the continuum discussion both the process considered by Warren in [66] which originally provided motivation for this work and a process recently constructed by Cerenzia in [17] that involves a hard wall fit in the framework introduced here. The techniques developed in this paper also allow us to study at the process level (and not just at fixed times) the process constructed by Ferrari and Frings in [31]. The main new examples considered in this paper are:

- Interlacing diffusion processes built from non-intersecting squared Bessel processes, that are related to the LUE matrix diffusion process minors studied by König and O'Connell in [47] and a dynamical version of a model considered by Dieker and Warren in [27]. More generally, we study all diffusion processes associated with the classical orthogonal polynomials in a uniform way. This includes non-intersecting Jacobi diffusions and is related to the JUE matrix diffusion, see [28].
- Interlacing Brownian motions in an interval, related to the eigenvalue processes of Brownian motions on some classical compact groups.

- A general study of interlacing diffusion processes with discrete spectrum and connections to the classical theory of total positivity and Chebyshev systems, see for example the monograph of Karlin [44].

We now mention a couple of recent works in the literature that are related to ours. Firstly a different approach based on generators for obtaining couplings of intertwined multidimensional diffusion processes via hard reflection is investigated in Theorem 3 of [55]. This has subsequently been extended by Sun [63] to isotropic diffusion coefficients, who making use of this has independently obtained similar results to us for the specific LUE and JUE processes. Moreover, a general β extension of the intertwining relations for the random matrix related aforementioned processes was also established in the note [3] by one of us. Finally, some results from this paper have been used recently in [2] to construct an infinite dimensional Feller process on the so called *graph of spectra*, that is the continuum analogue of the Gelfand–Tsetlin graph, which leaves the celebrated Hua-Pickrell measures invariant.

We also study the interacting particle systems with one-sided collisions at either edge of such Gelfand–Tsetlin pattern valued processes and give explicit Schutz-type determinantal transition densities for them in terms of derivatives and integrals of the one dimensional kernels. This also leads to formulas for the largest and smallest eigenvalues of the LUE and JUE ensembles in analogy to the ones obtained in [66] for the GUE.

Finally, we briefly explain how this work is connected to superpositions/ decimations of random matrix ensembles (see e.g.[34]) and in a different direction to the study of strong stationary duals. This notion was considered by Fill and Lyzinski in [32] motivated in turn by the study of strong stationary times for diffusion processes (first introduced by Diaconis and Fill in [25] in the Markov chain setting).

The rest of this paper is organised as follows:

1. In Sect. 13.2 we introduce the basic setup of dual/conjugate diffusion processes, give the transition kernels on interlacing spaces and our main results on intertwinings and Markov functions.
2. In Sect. 13.3 we apply the theory developed in this paper to show how known examples easily fit into this framework and construct new ones, among others the ones alluded to above.
3. In Sect. 13.4 we study the interacting particle systems at the edges of the Gelfand–Tsetlin patterns.
4. In Sect. 13.5 we prove well-posedness of the simple systems of $SDEs$ with reflection described informally in the first paragraphs of the introduction and under assumptions that their transition kernels are given by those in Sect. 13.2.
5. In the Appendix we elaborate on and give proofs of some of the facts stated about dual diffusion processes in Sect. 13.2 and also discuss entrance laws.

13.2 Two-Level Construction

13.2.1 Set Up of Conjugate Diffusions

Since our basic building blocks will be one dimensional diffusion processes and their conjugates we introduce them here and collect a number of facts about them (for justifications and proofs see the Appendix). The majority of the facts below can be found in the seminal book of Ito and McKean [39], and also more specifically regarding the transition densities of general one dimensional diffusion processes, in the classical paper of McKean [52] and also section 4.11 of [39] which we partly follow at various places.

We consider $(X_t)_{t\geq 0}$ a time homogeneous one dimensional diffusion process with state space an interval I with endpoints $l < r$ which can be open or closed, finite or infinite (interior denoted by I°) with infinitesimal generator given by,

$$L = a(x)\frac{d^2}{dx^2} + b(x)\frac{d}{dx},$$

with domain to be specified later in this section. In order to be more concise, we will frequently refer to such a diffusion process with generator L as an L-diffusion. We make the following regularity assumption throughout the paper.

Definition 13.1 (Assumption (R)) We assume that $a(\cdot) \in C^1(I^\circ)$ with $a(x) > 0$ for $x \in I^\circ$ and $b(\cdot) \in C(I^\circ)$.

We start by giving the very convenient description of the generator L in terms of its speed measure and scale function. Define its scale function $s(x)$ by $s'(x) = \exp\left(-\int_c^x \frac{b(y)}{a(y)}dy\right)$ (the scale function is defined up to affine transformations) where c is an arbitrary point in I°, its speed measure with density $m(x) = \frac{1}{s'(x)a(x)}$ in I° with respect to the Lebesgue measure (note that it is a Radon measure in I° and also strictly positive in I°) and speed function $M(x) = \int_c^x m(y)dy$. With these definitions the formal infinitesimal generator L can be written as,

$$L = \mathcal{D}_m \mathcal{D}_s\ ,$$

where $\mathcal{D}_m = \frac{1}{m(x)}\frac{d}{dx} = \frac{d}{dM}$ and $\mathcal{D}_s = \frac{1}{s'(x)}\frac{d}{dx} = \frac{d}{ds}$.

We now define the conjugate diffusion (see [65]) or Siegmund dual (see [60]) $(\hat{X}_t)_{t\geq 0}$ of X to be a diffusion process with generator,

$$\hat{L} = a(x)\frac{d^2}{dx^2} + (a'(x) - b(x))\frac{d}{dx},$$

and domain to be given shortly.

The following relations are easy to verify and are key to us.

$$\hat{s}'(x) = m(x) \text{ and } \hat{m}(x) = s'(x).$$

So the conjugation operation swaps the scale functions and speed measures. In particular

$$\hat{L} = \mathcal{D}_{\hat{m}}\mathcal{D}_{\hat{s}} = \mathcal{D}_s\mathcal{D}_m \ .$$

Using Feller's classification of boundary points (see Appendix) we obtain the following table for the boundary behaviour of the diffusion processes with generators L and $\hat{L}$ at l or r,

Bound. class. of L	Bound. class. of $\hat{L}$
Natural	Natural
Entrance	Exit
Exit	Entrance
Regular	Regular

We briefly explain what these boundary behaviours mean. A process can neither be started at, nor reach in finite time a *natural* boundary point. It can be started from an *entrance* point but such a boundary point cannot be reached from the interior I°. Such points are called *inaccessible* and can be removed from the state space. A diffusion can reach an *exit* boundary point from I° and once it does it is absorbed there. Finally, at a *regular* (also called entrance and exit) boundary point a variety of behaviours is possible and we need to *specify* one such. We will only be concerned with the two extreme possibilities namely *instantaneous reflection* and *absorption* (sticky behaviour interpolates between the two and is not considered here). Furthermore, note that if l is *instantaneously reflecting* then (see for example Chapter 2 paragraph 7 in [12]) $Leb\{t : X_t = l\} = 0$ $a.s.$ and analogously for the upper boundary point r.

Now in order to describe the domain, $Dom(L)$, of the diffusion process with formal generator L we first define the following function spaces (with the obvious abbreviations),

$$C(\bar{I}) = \{f \in C(I^\circ) : \lim_{x\downarrow l} f(x), \lim_{x\uparrow r} f(x) \text{ exist and are finite}\} ,$$

$$\mathfrak{D} = \{f \in C(\bar{I}) \cap C^2(I^\circ) : Lf \in C(\bar{I})\} ,$$

$$\mathfrak{D}_{nat} = \mathfrak{D} ,$$

$$\mathfrak{D}_{entr} = \mathfrak{D}_{refl} = \{f \in \mathfrak{D} : (\mathcal{D}_s f)(l^+) = 0\} ,$$

$$\mathfrak{D}_{exit} = \mathfrak{D}_{abs} = \{f \in \mathfrak{D} : (Lf)(l^+) = 0\}.$$

Similarly, define $\mathfrak{D}^{nat}, \mathfrak{D}^{entr}, \mathfrak{D}^{refl}, \mathfrak{D}^{exit}, \mathfrak{D}^{abs}$ by replacing l with r in the definitions above. Then the domain of the generator of the $(X_t)_{t\geq 0}$ diffusion process (with generator L) with boundary behaviour i at l and j at r where $i, j \in \{nat, entr, refl, exit, abs\}$ is given by,

$$Dom(L) = \mathfrak{D}_i \cap \mathfrak{D}^j .$$

For justifications see for example Chapter 8 in [29] and for an entrance boundary point also Theorem 12.2 of [46] or page 122 of [52].

Coming back to conjugate diffusions note that the boundary behaviour of X_t, the L-diffusion, determines the boundary behaviour of $\hat{X}_t$, the $\hat{L}$-diffusion, except at a regular point. At such a point we define the boundary behaviour of the $\hat{L}$-diffusion to be dual to that of the L-diffusion. Namely, if l is regular reflecting for L then we define it to be regular absorbing for $\hat{L}$. Similarly, if l is regular absorbing for L we define it to be regular reflecting for $\hat{L}$. The analogous definition being enforced at the upper boundary point r. Furthermore, we denote the semigroups associated with X_t and $\hat{X}_t$ by P_t and $\hat{\mathsf{P}}_t$ respectively and note that $\mathsf{P}_t 1 = \hat{\mathsf{P}}_t 1 = 1$. We remark that at an *exit* or *regular absorbing* boundary point the transition *kernel* $p_t(x, dy)$ associated with P_t has an *atom* there with mass (depending on t and x) the probability that the diffusion has reached that point by time t started from x.

We finally arrive at the following duality relation, going back in some form to Siegmund. This is proven via an approximation by birth and death chains in Sect. 13.4 of [21]. We also give a proof in the Appendix following [67] (where the proof is given in a special case). The reader should note the restriction to the interior I°.

Lemma 13.1 *$\mathsf{P}_t \mathbf{1}_{[l,y]}(x) = \hat{\mathsf{P}}_t \mathbf{1}_{[x,r]}(y)$ for $x, y \in I^\circ$.*

Now, it is well known that, the transition density $p_t(x, y) : (0, \infty) \times I^\circ \times I^\circ \to (0, \infty)$ of any one dimensional diffusion process with a speed measure which has a continuous density with respect to the Lebesgue measure in I° (as is the case in our setting) is continuous in (t, x, y). Moreover, under our assumptions $\partial_x p_t(x, y)$ exists for $x \in I^\circ$ and as a function of (t, y) is continuous in $(0, \infty) \times I^\circ$ (see Theorem 4.3 of [52]).

This fact along with Lemma 13.1 gives the following relationships between the transition densities for $x, y \in I^\circ$,

$$p_t(x, y) = \partial_y \hat{\mathsf{P}}_t \mathbf{1}_{[x,r]}(y) = \partial_y \int_x^r \hat{p}_t(y, dz) , \tag{13.3}$$

$$\hat{p}_t(x, y) = -\partial_y \mathsf{P}_t \mathbf{1}_{[l,x]}(y) = -\partial_y \int_l^x p_t(y, dz). \tag{13.4}$$

Before closing this section, we note that the speed measure is the *symmetrizing* measure of the diffusion process and this shall be useful in what follows. In particular, for $x, y \in I^\circ$ we have,

$$\frac{m(y)}{m(x)} p_t(y, x) = p_t(x, y). \tag{13.5}$$

13.2.2 Transition Kernels for Two-Level Processes

First, we recall the definitions of the interlacing spaces our processes will take values in,

$$\begin{aligned}
W^n(I^\circ) &= ((x) : l < x_1 \le \cdots \le x_n < r) \, , \\
W^{n,n+1}(I^\circ) &= ((x, y) : l < x_1 \le y_1 \le x_2 \le \cdots \le x_{n+1} < r) \, , \\
W^{n,n}(I^\circ) &= ((x, y) : l < y_1 \le x_1 \le y_2 \le \cdots \le x_n < r) \, , \\
W^{n+1,n}(I^\circ) &= ((x, y) : l < y_1 \le x_1 \le y_2 \le \cdots \le y_{n+1} < r).
\end{aligned}$$

Note that, for $(x, y) \in W^{n,n+1}(I^\circ)$ we have $x \in W^{n+1}(I^\circ)$ and $y \in W^n(I^\circ)$, this is a minor difference in notation to the one used in [66]; in the notations of that paper $W^{n+1,n}$ is our $W^{n,n+1}$ ($\mathbb{R}$). Also define for $x \in W^n(I^\circ)$,

$$W^{\bullet,n}(x) = \{y \in W^\bullet(I^\circ) : (x, y) \in W^{\bullet,n}(I^\circ)\}.$$

We now make the following standing assumption, enforced throughout the paper, on the boundary behaviour of the one dimensional diffusion process with generator L, depending on which interlacing space our two-level process defined next takes values in. Its significance will be explained later on. Note that any possible combination is allowed between the behaviour at l and r.

Definition 13.2 (Assumption (BC)) Assume the L-diffusion has the following boundary behaviour: When considering $\boldsymbol{W^{n,n+1}(I^\circ)}$:

$$l \text{ is either } Natural \text{ or } Entrance \text{ or } Regular\ Reflecting\ , \tag{13.6}$$

$$r \text{ is either } Natural \text{ or } Entrance \text{ or } Regular\ Reflecting. \tag{13.7}$$

When considering $\boldsymbol{W^{n,n}(I^\circ)}$:

$$l \text{ is either } Natural \text{ or } Exit \text{ or } Regular\ Absorbing\ , \tag{13.8}$$

$$r \text{ is either } Natural \text{ or } Entrance \text{ or } Regular\ Reflecting. \tag{13.9}$$

When considering $\boldsymbol{W^{n+1,n}(I^\circ)}$:

$$l \text{ is either } Natural \text{ or } Exit \text{ or } Regular\ Absorbing\ , \tag{13.10}$$

$$r \text{ is either } Natural \text{ or } Exit \text{ or } Regular\ Absorbing. \tag{13.11}$$

We will need to enforce a further regularity and non-degeneracy assumption at regular boundary points for some of our results. This is a technical condition and presumably can be removed.

Definition 13.3 (Assumption (BC+)) Assume condition (**BC**) above. Moreover, if a boundary point $\mathfrak{b} \in \{l, r\}$ is regular we assume that $\lim_{x\to\mathfrak{b}} a(x) > 0$ and the limits $\lim_{x\to\mathfrak{b}} b(x)$, $\lim_{x\to\mathfrak{b}} \big(a'(x) - b(x)\big)$ exist and are finite.

We shall begin by considering the following stochastic process which we will denote by $\big(\boldsymbol{\Phi}_{0,t}(x_1), \cdots, \boldsymbol{\Phi}_{0,t}(x_n); t \geq 0\big)$. It consists of a system of n independent L-diffusions started from $x_1 \leq \cdots \leq x_n$ which *coalesce* and move together once they meet. This is a process in $W^n(I)$ which once it reaches any of the hyperplanes $\{x_i = x_{i+1}\}$ continues there forever. We have the following proposition for the finite dimensional distributions of the coalescing process:

Proposition 13.1 *For* $z, z' \in W^n(I^\circ)$,

$$\mathbb{P}\big(\boldsymbol{\Phi}_{0,t}(z_i) \leq z'_i \ \textit{for}\ 1 \leq i \leq n\big) = \det\Big(P_t \boldsymbol{1}_{[l,z'_j]}(z_i) - \boldsymbol{1}(i < j)\Big)^n_{i,j=1} .$$

Proof This is done for Brownian motions in Proposition 9 of [66] using a generic argument based on continuous non-intersecting paths. The only variation here is that there might be an atom at l which however does not alter the proof.

We now define the kernel $q_t^{n,n+1}((x, y), (x', y'))dx'dy'$ on $W^{n,n+1}(I^\circ)$ as follows:

Definition 13.4 For $(x, y), (x', y') \in W^{n,n+1}(I^\circ)$ define $q_t^{n,n+1}((x, y), (x', y'))$ by,

$$\begin{aligned} &q_t^{n,n+1}((x, y), (x', y')) \\ &= \frac{\prod_{i=1}^n \hat{m}(y'_i)}{\prod_{i=1}^n \hat{m}(y_i)} (-1)^n \frac{\partial^n}{\partial_{y_1} \cdots \partial_{y_n}} \frac{\partial^{n+1}}{\partial_{x'_1} \cdots \partial_{x'_{n+1}}} \\ &\times \mathbb{P}\big(\boldsymbol{\Phi}_{0,t}(x_i) \leq x'_i, \boldsymbol{\Phi}_{0,t}(y_j) \leq y'_j \ \text{ for all } i, j\big) . \end{aligned}$$

This density exists by virtue of the regularity of the one dimensional transition densities. It is then an elementary computation using Proposition 13.1 and Lemma 13.1, along with relation (13.4), that $q_t^{n,n+1}$ can be written out explicitly as shown below. Note that each y_i and x'_j variable appears only in a certain row or column respectively.

$$q_t^{n,n+1}((x,y),(x',y')) = \det \begin{pmatrix} A_t(x,x') & B_t(x,y') \\ C_t(y,x') & D_t(y,y') \end{pmatrix} \tag{13.12}$$

where,

$$A_t(x,x')_{ij} = \partial_{x'_j} \mathsf{P}_t \mathbf{1}_{[l,x'_j]}(x_i) = p_t(x_i, x'_j) \,,$$

$$B_t(x,y')_{ij} = \hat{m}(y'_j)(\mathsf{P}_t \mathbf{1}_{[l,y'_j]}(x_i) - \mathbf{1}(j \geq i)) \,,$$

$$C_t(y,x')_{ij} = -\hat{m}^{-1}(y_i)\partial_{y_i}\partial_{x'_j} \mathsf{P}_t \mathbf{1}_{[l,x'_j]}(y_i) = -\mathcal{D}_s^{y_i} p_t(y_i, x'_j) \,,$$

$$D_t(y,y')_{ij} = -\frac{\hat{m}(y'_j)}{\hat{m}(y_i)} \partial_{y_i} \mathsf{P}_t \mathbf{1}_{[l,y'_j]}(y_i) = \hat{p}_t(y_i, y'_j).$$

We now define for $t > 0$ the operators $Q_t^{n,n+1}$ acting on the bounded Borel functions on $W^{n,n+1}(I^\circ)$ by,

$$(Q_t^{n,n+1} f)(x,y) = \int_{W^{n,n+1}(I^\circ)} q_t^{n,n+1}((x,y),(x',y')) f(x',y') dx' dy'. \tag{13.13}$$

Then the following facts hold:

Lemma 13.2 *Assume (**R**) and (**BC**) hold for the L-diffusion. Then,*

$$Q_t^{n,n+1} 1 \leq 1,$$

$$Q_t^{n,n+1} f \geq 0 \text{ for } f \geq 0.$$

Proof The first property will follow from performing the dx' integration first in Eq. (13.13) with $f \equiv 1$. This is easily done by the very structure of the entries of $q_t^{n,n+1}$: noting that each x'_i variable appears in a single column, then using multilinearity to bring the integrals inside the determinant and the relations:

$$\int_{y'_{j-1}}^{y'_j} A_t(x,x')_{ij} dx'_j = \mathsf{P}_t \mathbf{1}_{[l,y'_j]}(x_i) - \mathsf{P}_t \mathbf{1}_{[l,y'_{j-1}]}(x_i),$$

$$\int_{y'_{j-1}}^{y'_j} C_t(y,x')_{ij} dx'_j = -\frac{1}{\hat{m}(y_i)} \partial_{y_i} \mathsf{P}_t \mathbf{1}_{[l,y'_j]}(y_i) + \frac{1}{\hat{m}(y_i)} \partial_{y_i} \mathsf{P}_t \mathbf{1}_{[l,y'_{j-1}]}(y_i),$$

and observing that by (**BC**) the boundary terms are:

$$\int_{y'_N}^{r} A_t(x,x')_{iN+1}dx'_{N+1} = 1 - \mathsf{P}_t\mathbf{1}_{[l,y'_N]}(x_i),$$

$$\int_{y'_N}^{r} C_t(y,x')_{iN+1}dx'_{N+1} = \frac{1}{\hat{m}(y_i)}\partial_{y_i}\mathsf{P}_t\mathbf{1}_{[l,y'_N]}(y_i),$$

$$\int_{l}^{y'_1} A_t(x,x')_{i1}dx'_1 = \mathsf{P}_t\mathbf{1}_{[l,y'_1]}(x_i),$$

$$\int_{l}^{y'_1} C_t(y,x')_{i1}dx'_1 = -\frac{1}{\hat{m}(y_i)}\partial_{y_i}\mathsf{P}_t\mathbf{1}_{[l,y'_1]}(y_i),$$

we are left with the integral:

$$Q_t^{n,n+1}1 = \int_{W^n(I^\circ)} \det(\hat{p}_t(y_i,y'_j))_{i,j=1}^n dy' \le 1.$$

This is just a restatement of the fact that a Karlin–McGregor semigroup, to be defined shortly in this subsection, is sub-Markov.

The *positivity* preserving property also follows immediately from the original definition, since $\mathbb{P}\big(\mathbf{\Phi}_{0,t}(x_i) \le x'_i, \mathbf{\Phi}_{0,t}(y_j) \le y'_j \text{ for all } i,j\big)$ is increasing in the x'_i and decreasing in the y_j respectively: Obviously for any k with $x'_k \le \tilde{x}'_k$ and $\tilde{y}_k \le y_k$ each of the events:

$$\big\{\mathbf{\Phi}_{0,t}(x_i) \le x'_i, i \ne k, \mathbf{\Phi}_{0,t}(x_k) \le \tilde{x}'_k, \mathbf{\Phi}_{0,t}(y_j) \le y'_j, \text{ for all } j\big\},$$
$$\big\{\mathbf{\Phi}_{0,t}(x_i) \le x'_i, \text{ for all } i, \mathbf{\Phi}_{0,t}(y_j) \le y'_j, j \ne k, \mathbf{\Phi}_{0,t}(\tilde{y}_k) \le y'_k\big\},$$

contain the event:

$$\big\{\mathbf{\Phi}_{0,t}(x_i) \le x'_i, \mathbf{\Phi}_{0,t}(y_j) \le y'_j \text{ for all } i,j\big\}.$$

Thus, the partial derivatives $\partial_{x'_i}$ and $-\partial_{y_j}$ of $\mathbb{P}\big(\mathbf{\Phi}_{0,t}(x_i) \le x'_i, \mathbf{\Phi}_{0,t}(y_j) \le y'_j \text{ for all } i,j\big)$ are positive.

In fact, $Q_t^{n,n+1}$ defined above, forms a sub-Markov semigroup, associated with a Markov process $Z = (X,Y)$, with possibly finite lifetime, described informally as follows: the X components follow independent L-diffusions reflected off the Y components. More precisely assume that the L-diffusion is given as the pathwise unique solution X to the SDE,

$$d\mathsf{X}(t) = \sqrt{2a(\mathsf{X}(t))}d\beta(t) + b(\mathsf{X}(t))dt + dK^l(t) - dK^r(t)$$

where β is a standard Brownian motion and K^l and K^r are (possibly zero) positive finite variation processes that only increase when $\mathsf{X} = l$ or $\mathsf{X} = r$, so that $\mathsf{X} \in I$ and $Leb\{t : \mathsf{X}(t) = l \text{ or } r\} = 0$ a.s. We write $\mathfrak{s}_L$ for the corresponding measurable solution map on path space, namely so that $\mathsf{X} = \mathfrak{s}_L(\beta)$.

Consider the following system of $SDEs$ with reflection in $W^{n,n+1}$ which can be described in words as follows. The Y components evolve as n autonomous $\hat{L}$-diffusions stopped when they collide or when (if) they hit l or r, and we denote this time by $T^{n,n+1}$. The X components evolve as $n+1$ L-diffusions reflected off the Y particles.

$$\begin{aligned}
dX_1(t) &= \sqrt{2a(X_1(t))}d\beta_1(t) + b(X_1(t))dt + dK^l(t) - dK_1^+(t),\\
dY_1(t) &= \sqrt{2a(Y_1(t))}d\gamma_1(t) + (a'(Y_1(t)) - b(Y_1(t)))dt,\\
dX_2(t) &= \sqrt{2a(X_2(t))}d\beta_2(t) + b(X_2(t))dt + dK_2^-(t) - dK_2^+(t),\\
&\ \vdots\\
dY_n(t) &= \sqrt{2a(Y_n(t))}d\gamma_n(t) + (a'(Y_n(t)) - b(Y_n(t)))dt,\\
dX_{n+1}(t) &= \sqrt{2a(X_{n+1}(t))}d\beta_{n+1}(t) + b(X_{n+1}(t))dt + dK_{n+1}^-(t) - dK^r(t).
\end{aligned} \tag{13.14}$$

Here $\beta_1, \cdots, \beta_{n+1}, \gamma_1, \cdots, \gamma_n$ are independent standard Brownian motions and the positive finite variation processes K^l, K^r, K_i^+, K_i^- are such that K^l (possibly zero) increases only when $X_1 = l$, K^r (possibly zero) increases only when $X_{n+1} = r$, $K_i^+(t)$ increases only when $Y_i = X_i$ and $K_i^-(t)$ only when $Y_{i-1} = X_i$, so that $(X_1(t) \le Y_1(t) \le \cdots \le X_{n+1}(t); t \ge 0) \in W^{n,n+1}(I)$ up to time $T^{n,n+1}$. Note that, X either reflects at l or r or does not visit them at all by our boundary conditions (13.6) and (13.7). The problematic possibility of an X component being trapped between a Y particle and a boundary point and pushed in opposite directions does not arise, since the whole process is then instantly stopped.

The fact that these $SDEs$ are well-posed, so that in particular (X, Y) is Markovian, is proven in Proposition 13.34 under a Yamada–Watanabe condition (incorporating a linear growth assumption), that we now define precisely and abbreviate throughout by **(YW)**. Note that, the functions $a(\cdot)$ and $b(\cdot)$ initially defined in I° can in certain circumstances be continuously extended to the boundary points l and r and this is implicit in assumption **(YW)**.

Definition 13.5 (Assumption (YW)) Let I be an interval with endpoints $l < r$ and suppose ρ is a non-decreasing function from $(0, \infty)$ to itself such that $\int_{0+} \frac{dx}{\rho(x)} = \infty$. Consider the following condition on functions $a : I \to \mathbb{R}_+$ and $b : I \to \mathbb{R}$,

$$\begin{aligned}
|\sqrt{a(x)} - \sqrt{a(y)}|^2 &\le \rho(|x-y|),\\
|b(x) - b(y)| &\le C|x-y|.
\end{aligned}$$

Moreover, we assume that the functions $\sqrt{a(\cdot)}$ and $b(\cdot)$ are of at most linear growth (for $b(\cdot)$ this is immediate by Lipschitz continuity above).

We will say that the L-diffusion satisfies (**YW**) if its diffusion and drift coefficients a and b satisfy (**YW**).

Moreover, by virtue of the following result these $SDEs$ provide a precise description of the dynamics of the two-level process $Z = (X, Y)$ associated with $Q_t^{n,n+1}$. Proposition 13.2 below will be proven in Sect. 13.5.2 as either Proposition 13.35 or Proposition 13.37, depending on the boundary conditions.

Proposition 13.2 *Assume* (**R**) *and* (**BC+**) *hold for the L-diffusion and* (**YW**) *holds for both the L and $\hat{L}$ diffusions. Then, $Q_t^{n,n+1}$ is the sub-Markov semigroup associated with the (Markovian) system of SDEs (13.14) in the sense that if $\mathbf{Q}_{x,y}^{n,n+1}$ governs the processes (X, Y) satisfying the SDEs (13.14) and with initial condition (x, y) then for any f continuous with compact support and fixed $T > 0$,*

$$\int_{W^{n,n+1}(I^\circ)} q_T^{n,n+1}((x, y), (x', y')) f(x', y') dx' dy'$$
$$= \mathbf{Q}_{x,y}^{n,n+1}\big[f(X(T), Y(T))\mathbf{1}(T < T^{n,n+1})\big].$$

For further motivation regarding the definition of $Q_t^{n,n+1}$ and moreover, a completely different argument for its semigroup property, that however does not describe explicitly the dynamics of X and Y, we refer the reader to the next Sect. 13.2.3.

We now briefly study some properties of $Q_t^{n,n+1}$, that are immediate from its algebraic structure (with no reference to the $SDEs$ above required). In order to proceed and fix notations for the rest of this section, start by defining the Karlin–McGregor semigroup P_t^n associated with n L-diffusions in I° given by the transition density, with $x, y \in W^n(I^\circ)$,

$$p_t^n(x, y)dy = \det(p_t(x_i, y_j))_{i,j=1}^n dy. \tag{13.15}$$

Note that, in the case an exit or regular absorbing boundary point exists, P_t^1 is the semigroup of the L-diffusion *killed* and not absorbed at that point. In particular it is not the same as P_t which is a Markov semigroup. Similarly, define the Karlin–McGregor semigroup $\hat{P}_t^n$ associated with n $\hat{L}$-diffusions by,

$$\hat{p}_t^n(x, y)dy = \det(\hat{p}_t(x_i, y_j))_{i,j=1}^n dy, \tag{13.16}$$

with $x, y \in W^n(I^\circ)$. The same comment regarding absorbing and exit boundary points applies here as well.

Now, define the operators $\Pi_{n,n+1}$, induced by the projections on the Y level as follows with f a bounded Borel function on $W^n(I^\circ)$,

$$(\Pi_{n,n+1}f)(x,y) = f(y).$$

The following proposition immediately follows by performing the dx' integration in the explicit formula for the block determinant (as already implied in the proof that $Q_t^{n,n+1}1 \leq 1$).

Proposition 13.3 *Assume* (**R**) *and* (**BC**) *hold for the L-diffusion. For $t > 0$ and f a bounded Borel function on $W^n(I^\circ)$ we have,*

$$\Pi_{n,n+1}\hat{P}_t^n f = Q_t^{n,n+1}\Pi_{n,n+1}f. \tag{13.17}$$

The fact that Y is distributed as n independent $\hat{L}$-diffusions killed when they collide or when they hit l or r is already implicit in the statement of Proposition 13.2. However, it is also the probabilistic consequence of the proposition above. Namely, the intertwining relation (13.17), being an instance of Dynkin's criterion (see for example Exercise 1.17 Chapter 3 of [57]), implies that the evolution of Y is Markovian with respect to the joint filtration of X and Y i.e. of the process Z and we take this as the definition of Y being autonomous. Moreover, Y is evolving according to $\hat{P}_t^n$. Thus, the Y components form an *autonomous diffusion* process. Finally, by taking $f \equiv 1$ above we get that the finite lifetime of Z exactly corresponds to the killing time of Y, which we denote by $T^{n,n+1}$.

Similarly, we define the kernel $q_t^{n,n}((x,y),(x',y'))dx'dy'$ on $W^{n,n}(I^\circ)$ as follows:

Definition 13.6 For $(x,y),(x',y') \in W^{n,n}(I^\circ)$ define $q_t^{n,n}((x,y),(x',y'))$ by,

$$\begin{aligned}
&q_t^{n,n}((x,y),(x',y'))\\
&\quad = \frac{\prod_{i=1}^n \hat{m}(y_i')}{\prod_{i=1}^n \hat{m}(y_i)}(-1)^n \frac{\partial^n}{\partial_{y_1}\cdots\partial_{y_n}}\frac{\partial^n}{\partial_{x_1'}\cdots\partial_{x_n'}}\\
&\qquad \times \mathbb{P}\big(\mathbf{\Phi}_{0,t}(x_i) \leq x_i', \mathbf{\Phi}_{0,t}(y_j) \leq y_j' \text{ for all } i,j\big).
\end{aligned}$$

We note that as before $q_t^{n,n}$ can in fact be written out explicitly,

$$q_t^{n,n}((x,y),(x',y')) = \det\begin{pmatrix} A_t(x,x') & B_t(x,y') \\ C_t(y,x') & D_t(y,y') \end{pmatrix} \tag{13.18}$$

where,

$$\begin{aligned}
A_t(x,x')_{ij} &= \partial_{x_j'}\mathsf{P}_t\mathbf{1}_{[l,x_j']}(x_i) = p_t(x_i,x_j'),\\
B_t(x,y')_{ij} &= \hat{m}(y_j')(\mathsf{P}_t\mathbf{1}_{[l,y_j']}(x_i) - \mathbf{1}(j>i)),
\end{aligned}$$

$$C_t(y,x')_{ij} = -\hat{m}^{-1}(y_i)\partial_{y_i}\partial_{x'_j}\mathsf{P}_t\mathbf{1}_{[l,x'_j]}(y_i) = -\mathcal{D}_s^{y_i}p_t(y_i,x'_j),$$

$$D_t(y,y')_{ij} = -\frac{\hat{m}(y'_j)}{\hat{m}(y_i)}\partial_{y_i}\mathsf{P}_t\mathbf{1}_{[l,y'_j]}(y_i) = \hat{p}_t(y_i,y'_j).$$

Remark 13.1 Comparing with the $q_t^{n,n+1}$ formulae everything is the same except for the indicator function being $\mathbf{1}(j > i)$ instead of $\mathbf{1}(j \geq i)$.

Define the operator $Q_t^{n,n}$ for $t > 0$ acting on bounded Borel functions on $W^{n,n}(I^\circ)$ by,

$$(Q_t^{n,n}f)(x,y) = \int_{W^{n,n}(I^\circ)} q_t^{n,n}((x,y),(x',y'))f(x',y')dx'dy'. \tag{13.19}$$

Then with the analogous considerations as for $Q_t^{n,n+1}$ (see Sect. 13.2.3 as well), we can see that $Q_t^{n,n}$ should form a sub-Markov semigroup, to which we can associate a Markov process Z, with possibly finite lifetime, taking values in $W^{n,n}(I^\circ)$, the evolution of which we now make precise.

To proceed as before, we assume that the L-diffusion is given by an SDE and we consider the following system of $SDEs$ with reflection in $W^{n,n}$ which can be described as follows. The Y components evolve as n autonomous $\hat{L}$-diffusions killed when they collide or when (if) they hit the boundary point r, a time which we denote by $T^{n,n}$. The X components evolve as n L-diffusions being kept apart by hard reflection on the Y particles.

$$\begin{aligned}
dY_1(t) &= \sqrt{2a(Y_1(t))}d\gamma_1(t) + (a'(Y_1(t)) - b(Y_1(t)))dt + dK^l(t),\\
dX_1(t) &= \sqrt{2a(X_1(t))}d\beta_1(t) + b(X_1(t))dt + dK_1^+(t) - dK_1^-(t),\\
&\vdots \\
dY_n(t) &= \sqrt{2a(Y_n(t))}d\gamma_n(t) + (a'(Y_n(t)) - b(Y_n(t)))dt,\\
dX_n(t) &= \sqrt{2a(X_n(t))}d\beta_n(t) + b(X_n(t))dt + dK_n^+(t) - dK^r(t).
\end{aligned} \tag{13.20}$$

Here $\beta_1, \cdots, \beta_n, \gamma_1, \cdots, \gamma_n$ are independent standard Brownian motions and the positive finite variation processes K^l, K^r, K_i^+, K_i^- are such that $\bar{K}^l$ (possibly zero) increases only when $Y_1 = l$, K^r (possibly zero) increases only when $X_n = r$, $K_i^+(t)$ increases only when $Y_i = X_i$ and $K_i^-(t)$ only when $Y_{i-1} = X_i$, so that $(Y_1(t) \leq X_1(t) \leq \cdots \leq X_n(t); t \geq 0) \in W^{n,n}(I)$ up to $T^{n,n}$. Note that, Y reflects at the boundary point l or does not visit it all and similarly X reflects at r or does not reach it all by our boundary assumptions (13.8) and (13.9). The intuitively problematic issue of Y_n pushing X_n upwards at r does not arise since the whole process is stopped at such instance.

That these $SDEs$ are well-posed, so that in particular (X, Y) is Markovian, again follows from the arguments of Proposition 13.34. As before, we have the following precise description of the dynamics of the two-level process $Z = (X, Y)$ associated with $Q_t^{n,n}$.

Proposition 13.4 *Assume* (**R**) *and* (**BC+**) *hold for the L-diffusion and* (**YW**) *holds for both the L and $\hat{L}$ diffusions. Then, $Q_t^{n,n}$ is the sub-Markov semigroup associated with the (Markovian) system of SDEs (13.20) in the sense that if $\mathbf{Q}_{x,y}^{n,n}$ governs the processes (X, Y) satisfying the SDEs (13.20) with initial condition (x, y) then for any f continuous with compact support and fixed $T > 0$,*

$$\int_{W^{n,n}(I^\circ)} q_T^{n,n}((x, y), (x', y'))f(x', y')dx'dy'$$
$$= \mathbf{Q}_{x,y}^{n,n}\left[f(X(T), Y(T))\mathbf{1}(T < T^{n,n})\right].$$

We also define, analogously to before, an operator $\Pi_{n,n}$, induced by the projection on the Y level by,

$$(\Pi_{n,n} f)(x, y) = f(y).$$

We have the following proposition which immediately follows by performing the dx' integration in Eq. (13.19).

Proposition 13.5 *Assume* (**R**) *and* (**BC**) *hold for the L-diffusion. For $t > 0$ and f a bounded Borel function on $W^n(I^\circ)$ we have,*

$$\Pi_{n,n}\hat{P}_t^n f = Q_t^{n,n}\Pi_{n,n} f. \tag{13.21}$$

This, again implies that the evolution of Y is Markovian with respect to the joint filtration of X and Y. Furthermore, Y is distributed as n $\hat{L}$-diffusions killed when they collide or when (if) they hit the boundary point r (note the difference here to $W^{n,n+1}$ is because of the asymmetry between X and Y and our standing assumption (13.8) and (13.9)). Hence, the Y components form a *diffusion* process and they are *autonomous*. The finite lifetime of Z analogously to before (by taking $f \equiv 1$ in the proposition above), exactly corresponds to the killing time of Y which we denote by $T^{n,n}$. As before, this is already implicit in the statement of Proposition 13.4.

Finally, we can define the kernel $q_t^{n+1,n}((x, y), (x', y'))dx'dy'$ on $W^{n+1,n}(I^\circ)$ in an analogous way and also the operator $Q_t^{n+1,n}$ for $t > 0$ acting on bounded Borel functions on $W^{n+1,n}(I^\circ)$ as well. The description of the associated process Z in $W^{n+1,n}(I^\circ)$ in words is as follows. The Y components evolve as $n + 1$ autonomous $\hat{L}$-diffusions killed when they collide (by our boundary conditions (13.10) and (13.11) if the Y particles do visit l or r they are reflecting there) and the X components evolve as n L-diffusions reflected on the Y particles. These dynamics can be described in terms of $SDEs$ with reflection under completely analogous assumptions. The details are omitted.

13.2.3 Stochastic Coalescing Flow Interpretation

The definition of $q_t^{n,n+1}$, and similarly of $q_t^{n,n}$, might look rather mysterious and surprising. It is originally motivated from considering stochastic coalescing flows. Briefly, the finite system $\big(\mathbf{\Phi}_{0,t}(x_1), \cdots, \mathbf{\Phi}_{0,t}(x_n); t \geq 0\big)$ can be extended to an infinite system of coalescing L-diffusions starting from each space time point and denoted by $(\mathbf{\Phi}_{s,t}(\cdot), s \leq t)$. This is well documented in Theorem 4.1 of [48] for example. The random family of maps $(\mathbf{\Phi}_{s,t}, s \leq t)$ from I to I enjoys among others the following natural looking and intuitive properties: the *cocycle* or *flow* property $\mathbf{\Phi}_{t_1,t_3} = \mathbf{\Phi}_{t_2,t_3} \circ \mathbf{\Phi}_{t_1,t_2}$, *independence of its increments* $\mathbf{\Phi}_{t_1,t_2} \perp \mathbf{\Phi}_{t_3,t_4}$ for $t_2 \leq t_3$ and *stationarity* $\mathbf{\Phi}_{t_1,t_2} \overset{law}{=} \mathbf{\Phi}_{0,t_2-t_1}$. Finally, we can consider its generalized inverse by $\mathbf{\Phi}_{s,t}^{-1}(x) = sup\{w : \mathbf{\Phi}_{s,t}(w) \leq x\}$ which is well defined since $\mathbf{\Phi}_{s,t}$ is non-decreasing.

With these notations in place $q_t^{n,n+1}$ can also be written as,

$$\begin{aligned} &q_t^{n,n+1}((x,y),(x',y'))dx'dy \\ &\quad = \frac{\prod_{i=1}^n \hat{m}(y_i')}{\prod_{i=1}^n \hat{m}(y_i)} \mathbb{P}\big(\mathbf{\Phi}_{0,t}(x_i) \in dx_i', \mathbf{\Phi}_{0,t}^{-1}(y_j') \in dy_j \text{ for all } i,j\big). \end{aligned} \tag{13.22}$$

We now sketch an argument that gives the semigroup property $Q_{t+s}^{n,n+1} = Q_t^{n,n+1} Q_s^{n,n+1}$. We do not try to give all the details that would render it completely rigorous, mainly because it cannot be used to precisely describe the dynamics of $Q_t^{n,n+1}$, but nevertheless all the main steps are spelled out.

All equalities below should be understood after being integrated with respect to dx'' and dy over arbitrary Borel sets. The first equality is by definition. The second equality follows from the *cocycle* property and conditioning on the values of $\mathbf{\Phi}_{0,s}(x_i)$ and $\mathbf{\Phi}_{s,s+t}^{-1}(y_j'')$. Most importantly, this is where the *boundary behaviour assumptions* (13.6) and (13.7) we made at the beginning of this subsection are used. These ensure that no possible contributions from atoms on ∂I are missed; namely the random variable $\mathbf{\Phi}_{0,s}(x_i)$ is supported (its distribution gives full mass) in I°. Moreover, it is not too hard to see from the coalescing property of the flow that, we can restrict the integration over $(x',y') \in W^{n,n+1}(I^\circ)$ for otherwise the integrand vanishes. Finally, the third equality follows from *independence* of the increments and the fourth one by *stationarity* of the flow.

$$\begin{aligned} &q_{s+t}^{n,n+1}((x,y),(x'',y''))dx''dy \\ &\quad = \frac{\prod_{i=1}^n \hat{m}(y_i'')}{\prod_{i=1}^n \hat{m}(y_i)} \\ &\qquad \times \mathbb{P}\big(\mathbf{\Phi}_{0,s+t}(x_i) \in dx_i'', \mathbf{\Phi}_{0,s+t}^{-1}(y_j'') \in dy_j \text{ for all } i,j\big) \end{aligned}$$

$$= \frac{\prod_{i=1}^n \hat{m}(y_i'')}{\prod_{i=1}^n \hat{m}(y_i)} \int_{(x',y')\in W^{n,n+1}(I^\circ)}$$

$$\times \mathbb{P}\big(\mathbf{\Phi}_{0,s}(x_i) \in dx_i', \mathbf{\Phi}_{s,s+t}(x_i') \in dx_i'', \mathbf{\Phi}_{0,s}^{-1}(y_j') \in dy_j, \mathbf{\Phi}_{s,s+t}^{-1}(y_j'') \in dy_j'\big)$$

$$= \int_{(x',y')\in W^{n,n+1}(I^\circ)} \frac{\prod_{i=1}^n \hat{m}(y_i')}{\prod_{i=1}^n \hat{m}(y_i)} \mathbb{P}\big(\mathbf{\Phi}_{0,s}(x_i) \in dx_i', \mathbf{\Phi}_{0,s}^{-1}(y_j') \in dy_j\big)$$

$$\times \frac{\prod_{i=1}^n \hat{m}(y_i'')}{\prod_{i=1}^n \hat{m}(y_i')} \mathbb{P}\big(\mathbf{\Phi}_{s,s+t}(x_i') \in dx_i'', \mathbf{\Phi}_{s,s+t}^{-1}(y_j'') \in dy_j'\big)$$

$$= \int_{(x',y')\in W^{n,n+1}(I^\circ)} q_s^{n,n+1}$$

$$\times ((x,y),(x',y'))q_t^{n,n+1}((x',y'),(x'',y''))dx'dy'dx''dy.$$

13.2.4 Intertwining and Markov Functions

In this subsection (n_1, n_2) denotes one of $\{(n, n-1), (n, n), (n, n+1)\}$. First, recall the definitions of P_t^n and $\hat{P}_t^n$ given in (13.15) and (13.16) respectively. Similarly, we record here again, the following proposition and recall that it can in principle completely describe the evolution of the Y particles and characterizes the finite lifetime of the process Z as the killing time of Y.

Proposition 13.6 *Assume* (**R**) *and* (**BC**) *hold for the L-diffusion. For $t > 0$ and f a bounded Borel function on $W^{n_1}(I^\circ)$ we have,*

$$\Pi_{n_1,n_2} \hat{P}_t^{n_1} f = Q_t^{n_1,n_2} \Pi_{n_1,n_2} f. \tag{13.23}$$

Now, we define the following integral operator Λ_{n_1,n_2} acting on Borel functions on $W^{n_1,n_2}(I^\circ)$, whenever f is integrable as,

$$(\Lambda_{n_1,n_2} f)(x) = \int_{W^{n_1,n_2}(x)} \prod_{i=1}^{n_1} \hat{m}(y_i) f(x,y)dy, \tag{13.24}$$

where we remind the reader that $\hat{m}(\cdot)$ is the density with respect to Lebesgue measure of the speed measure of the diffusion with generator $\hat{L}$.

The following intertwining relation is the fundamental ingredient needed for applying the theory of Markov functions, originating with the seminal paper of Rogers and Pitman [58]. This proposition directly follows by performing the dy integration in the explicit formula of the block determinant (or alternatively by invoking the coalescing property of the stochastic flow $\big(\mathbf{\Phi}_{s,t}(\cdot); s \le t\big)$ and the original definitions).

Proposition 13.7 *Assume* (**R**) *and* (**BC**) *hold for the L-diffusion. For* $t > 0$ *we have the following equality of positive kernels,*

$$P_t^{n_2}\Lambda_{n_1,n_2} = \Lambda_{n_1,n_2}Q_t^{n_1,n_2}. \tag{13.25}$$

Combining the two propositions above gives the following relation for the Karlin–McGregor semigroups,

$$P_t^{n_2}\Lambda_{n_1,n_2}\Pi_{n_1,n_2} = \Lambda_{n_1,n_2}\Pi_{n_1,n_2}\hat{P}_t^{n_1}. \tag{13.26}$$

Namely, the two semigroups are themselves intertwined with kernel,

$$\left(\Lambda_{n_1,n_2}\Pi_{n_1,n_2}f\right)(x) = \int_{W^{n_1,n_2}(x)}\prod_{i=1}^{n_1}\hat{m}(y_1)f(y)dy.$$

This implies the following. Suppose $\hat{h}_{n_1}$ is a strictly positive (in $\mathring{W}^{n_1}$) eigenfunction for $\hat{P}_t^{n_1}$ namely, $\hat{P}_t^{n_1}\hat{h}_{n_1} = e^{\lambda_{n_1}t}\hat{h}_{n_1}$, then (with both sides possibly being infinite),

$$(P_t^{n_2}\Lambda_{n_1,n_2}\Pi_{n_1,n_2}\hat{h}_{n_1})(x) = e^{\lambda_{n_1}t}(\Lambda_{n_1,n_2}\Pi_{n_1,n_2}\hat{h}_{n_1})(x).$$

We are interested in strictly positive eigenfunctions because they allow us to define Markov processes, however non positive eigenfunctions can be built this way as well.

We now finally arrive at our main results. We need to make precise one more notion, already referenced several times in the introduction. For a possibly sub-Markov semigroup $(\mathfrak{P}_t; t \geq 0)$ or more generally, for fixed t, a sub-Markov kernel with eigenfunction $\mathfrak{h}$ with eigenvalue e^{ct} we define the Doob's h-transform by $e^{-ct}\mathfrak{h}^{-1}\circ\mathfrak{P}_t\circ\mathfrak{h}$. Observe that, this is now an honest Markov semigroup (or Markov kernel).

If $\hat{h}_{n_1}$ is a strictly positive in $\mathring{W}^{n_1}$ eigenfunction for $\hat{P}_t^{n_1}$ then so is the function $\hat{h}_{n_1,n_2}(x, y) = \hat{h}_{n_1}(y)$ for $Q_t^{n_1,n_2}$ from Proposition 13.6. We can thus define the proper Markov kernel $Q_t^{n_1,n_2,\hat{h}_{n_1}}$ which is the h-transform of $Q_t^{n_1,n_2}$ by $\hat{h}_{n_1}$. Define $h_{n_2}(x)$, strictly positive in $\mathring{W}^{n_2}$, as follows, assuming that the integrals are finite in the case of $W^{n,n}(I^\circ)$ and $W^{n+1,n}(I^\circ)$,

$$h_{n_2}(x) = (\Lambda_{n_1,n_2}\Pi_{n_1,n_2}\hat{h}_{n_1})(x),$$

and the Markov Kernel $\Lambda_{n_1,n_2}^{\hat{h}_{n_1}}(x,\cdot)$ with $x \in \mathring{W}^{n_2}$ by,

$$(\Lambda_{n_1,n_2}^{\hat{h}_{n_1}}f)(x) = \frac{1}{h_{n_2}(x)}\int_{W^{n_1,n_2}(x)}\prod_{i=1}^{n_1}\hat{m}(y_i)\hat{h}_{n_1}(y)f(x, y)dy.$$

Finally, defining $P_t^{n_2,h_{n_2}}$ to be the Karlin–McGregor semigroup $P_t^{n_2}$ h-transformed by h_{n_2} we obtain:

Proposition 13.8 *Assume* (**R**) *and* (**BC**) *hold for the L-diffusion. Let $Q_t^{n_1,n_2}$ denote one of the operators induced by the sub-Markov kernels on $W^{n_1,n_2}(I^\circ)$ defined in the previous subsection. Let $\hat{h}_{n_1}$ be a strictly positive eigenfunction for $\hat{P}_t^{n_1}$ and assume that $h_{n_2}(x) = (\Lambda_{n_1,n_2}\Pi_{n_1,n_2}\hat{h}_{n_1})(x)$ is finite in $W^{n_2}(I^\circ)$, so that in particular $\Lambda_{n_1,n_2}^{\hat{h}_{n_1}}$ is a Markov kernel. Then, with the notations of the preceding paragraph we have the following relation for $t > 0$,*

$$P_t^{n_2,h_{n_2}} \Lambda_{n_1,n_2}^{\hat{h}_{n_1}} f = \Lambda_{n_1,n_2}^{\hat{h}_{n_1}} Q_t^{n_1,n_2,\hat{h}_{n_2}} f, \tag{13.27}$$

with f a bounded Borel function in $W^{n_1,n_2}(I^\circ)$.

This intertwining relation and the theory of Markov functions (see Section 2 of [58] for example) immediately imply the following corollary:

Corollary 13.1 *Assume $Z = (X, Y)$ is a Markov process with semigroup $Q_t^{n_1,n_2,\hat{h}_{n_2}}$, then the X component is distributed as a Markov process with semigroup $P_t^{n_2,h_{n_2}}$ started from x if (X, Y) is started from $\Lambda_{n_1,n_2}^{\hat{h}_{n_1}}(x, \cdot)$. Moreover, the conditional distribution of $Y(t)$ given $(X(s); s \le t)$ is $\Lambda_{n_1,n_2}^{\hat{h}_{n_1}}(X(t), \cdot)$.*

We give a final definition in the case of $W^{n,n+1}$ only, that has a natural analogue for $W^{n,n}$ and $W^{n+1,n}$ (we shall elaborate on the notion introduced below in Section 5.1 on well-posedness of $SDEs$ with reflection). Take $Y = (Y_1, \cdots, Y_n)$ to be an n-dimensional system of *non-intersecting* paths in $\mathring{W}^n(I^\circ)$, so that in particular $Y_1 < Y_2 < \cdots < Y_n$. Then, by *X is a system of $n+1$ L-diffusions reflected off Y* we mean processes $(X_1(t), \cdots, X_{n+1}(t); t \ge 0)$, satisfying $X_1(t) \le Y_1(t) \le X_2(t) \le \cdots \le X_{n+1}(t)$ for all $t \ge 0$, and so that the following $SDEs$ hold,

$$dX_1(t) = \sqrt{2a(X_1(t))}d\beta_1(t) + b(X_1(t))dt + dK^l(t) - dK_1^+(t),$$

$$\vdots$$

$$dX_j(t) = \sqrt{2a(X_j(t))}d\beta_j(t) + b(X_j(t))dt + dK_j^-(t) - dK_j^+(t), \tag{13.28}$$

$$\vdots$$

$$dX_{n+1}(t) = \sqrt{2a(X_{n+1}(t))}d\beta_{n+1}(t) + b(X_{n+1}(t))dt + dK_{n+1}^-(t) - dK^r(t),$$

where the positive finite variation processes K^l, K^r, K_i^+, K_i^- are such that K^l increases only when $X_1 = l$, K^r increases only when $X_{n+1} = r$, $K_i^+(t)$ increases only when $Y_i = X_i$ and $K_i^-(t)$ only when $Y_{i-1} = X_i$, so that $(X_1(t) \le Y_1(t) \le$

$\cdots \le X_{n+1}(t)) \in W^{n,n+1}(I)$ forever. Here $\beta_1, \cdots, \beta_{n+1}$ are independent standard Brownian motions which are moreover *independent* of Y. The reader should observe that the dynamics between (X, Y) are exactly the ones prescribed in the system of $SDEs$ (13.14) with the difference being that now the process has infinite lifetime. This can be achieved from (13.14) by h-transforming the Y process as explained in this section to have infinite lifetime. By pathwise uniqueness of solutions to reflecting $SDEs$, with coefficients satisfying (**YW**), in continuous time-dependent domains proven in Proposition 13.34, under any absolutely continuous change of measure for the (X, Y)-process that depends only on Y (a Doob h-transform in particular), the Eq. (13.28) still hold with the β_i independent Brownian motions which moreover remain independent of the Y process. We thus arrive at our main theorem:

Theorem 13.1 *Assume* (**R**) *and* (**BC**+) *hold for the L-diffusion and* (**YW**) *holds for both the L and $\hat{L}$ diffusions. Moreover, assume $\hat{h}_n$ is a strictly positive eigenfunction for $\hat{P}_t^n$. Suppose Y consists of n non-intersecting $\hat{L}$-diffusions h-transformed by $\hat{h}_n$, with transition semigroup $\hat{P}_t^{n,\hat{h}_n}$, and X is a system of $n+1$ L-diffusions reflected off Y started according to the distribution $\Lambda_{n,n+1}^{\hat{h}_n}(x, \cdot)$ for some $x \in \mathring{W}^{n+1}(I)$. Then X is distributed as a diffusion process with semigroup $P_t^{n+1,h_{n+1}}$ started from x, where $h_{n+1} = \Lambda_{n,n+1}\Pi_{n,n+1}\hat{h}_n$.*

Proof By Proposition 13.2 and the discussion above, the process (X,Y) evolves according to the Markov semigroup $Q_t^{n_1,n_2,\hat{h}_{n_2}}$. Then, an application of the Rogers-Pitman Markov functions criterion in [58] with the function $\phi(x, y) = x$ and the intertwining (13.27) gives that, under the initial law $\Lambda_{n,n+1}^{\hat{h}_n}(x, \cdot)$ for (X, Y), $(X(t); t \ge 0)$ is a Markov process with semigroup $P_t^{n+1,h_{n+1}}$ started from x, in particular a diffusion.

The statement and proof of the result for $W^{n,n}$ and $W^{n+1,n}$ is completely analogous.

Finally, the intertwining relation (13.27) also allows us to start the two-level process (X, Y) from a degenerate point, in particular the system of reflecting $SDEs$ when some of the Y coordinates coincide, as long as starting the process with semigroup $P_t^{n_2,h_{n_2}}$ from such a degenerate point is valid. Suppose $\left(\mu_t^{n_2,h_{n_2}}\right)_{t>0}$ is an entrance law for $P_t^{n_2,h_{n_2}}$, namely for $t, s > 0$,

$$\mu_s^{n_2,h_{n_2}} P_t^{n_2,h_{n_2}} = \mu_{t+s}^{n_2,h_{n_2}},$$

then we have the following corollary, which is obtained immediately by applying $\mu_t^{n_2,h_{n_2}}$ to both sides of (13.27):

Corollary 13.2 *Under the assumptions above, if* $\left(\mu_s^{n_2,h_{n_2}}\right)_{s>0}$ *is an entrance law for the process with semigroup* $P_t^{n_2,h_{n_2}}$ *then* $\left(\mu_s^{n_2,h_{n_2}}\Lambda_{n_1,n_2}^{\hat{h}_{n_1}}\right)_{s>0}$ *forms an entrance law for process* (X,Y) *with semigroup* $Q_t^{n_1,n_2,\hat{h}_{n_1}}$.

Hence, the statement of Theorem 13.1 generalizes, so that if X is a system of L-diffusions reflected off Y started according to an entrance law, then X is again itself distributed as a Markov process.

The entrance laws that we will be concerned with in this paper will correspond to starting the process with semigroup $P_t^{n_2,h_{n_2}}$ from a single point $(x,\cdots,x)$ for some $x \in I$. These will be given by so called time dependent *biorthogonal ensembles*, namely measures of the form,

$$\det\left(f_i(t,x_j)\right)_{i,j=1}^{n_2}\det\left(g_i(t,x_j)\right)_{i,j=1}^{n_2}. \tag{13.29}$$

Under some further assumptions on the Taylor expansion of the one dimensional transition density $p_t(x,y)$ they will be given by so called *polynomial ensembles*, where one of the determinant factors is the Vandermonde determinant,

$$\det\left(\phi_i(t,x_j)\right)_{i,j=1}^{n_2}\det\left(x_j^{i-1}\right)_{i,j=1}^{n_2}. \tag{13.30}$$

A detailed discussion is given in the Appendix.

13.3 Applications and Examples

Applying the theory developed in the previous section we will now show how some of the known examples of diffusions in Gelfand–Tsetlin patterns fit into this framework and construct new processes of this kind. In particular we will treat all the diffusions associated with Random Matrix eigenvalues, a model related to Plancherel growth that involves a wall, examples coming from Sturm-Liouville semigroups and finally point out the connection to strong stationary times and superpositions and decimations of Random Matrix ensembles.

First, recall that the space of Gelfand–Tsetlin patterns of depth N denoted by $\mathbb{GT}(N)$ is defined to be,

$$\left\{\left(x^{(1)},\cdots,x^{(N)}\right): x^{(n)} \in W^n,\ x^{(n)} \prec x^{(n+1)}\right\},$$

and also the space of symplectic Gelfand–Tsetlin patterns of depth N denoted by $\mathbb{GT}_{\mathbf{s}}(N)$ is given by,

$$\left\{\left(x^{(1)}, \hat{x}^{(1)} \cdots, x^{(N)}, \hat{x}^{(N)}\right) : x^{(n)}, \hat{x}^{(n)} \in W^n,\ x^{(n)} \prec \hat{x}^{(n)} \prec x^{(n+1)}\right\}.$$

Please note the minor discrepancy in the definition of $\mathbb{GT}(N)$ with the notation used for $W^{n,n+1}$: here for two consecutive levels $x^{(n)} \in W^n, x^{(n+1)} \in W^{n+1}$ in the Gelfand–Tsetlin pattern the pair $(x^{(n+1)}, x^{(n)}) \in W^{n,n+1}$ and not the other way round.

13.3.1 Concatenating Two-Level Processes

We will describe the construction for $\mathbb{GT}$, with the extension to $\mathbb{GT}_{\mathbf{s}}$ being analogous. Let us fix an interval I with endpoints $l < r$ and let L_n for $n = 1, \cdots, N$ be a sequence of diffusion process generators in I (satisfying (13.6) and (13.7)) given by,

$$L_n = a_n(x)\frac{d^2}{dx^2} + b_n(x)\frac{d}{dx}. \tag{13.31}$$

We will moreover denote their transition densities with respect to Lebesgue measure by $p_t^n(\cdot, \cdot)$.

We want to consider a process $(\mathbb{X}(t); t \geq 0) = \left(\left(\mathbb{X}^{(1)}(t), \cdots, \mathbb{X}^{(N)}(t)\right); t \geq 0\right)$ taking values in $\mathbb{GT}(N)$ so that, for each $2 \leq n \leq N$, $\mathbb{X}^{(n)}$ consists of n independent L_n diffusions reflected off the paths of $\mathbb{X}^{(n-1)}$. More precisely we consider the following system of reflecting $SDEs$, with $1 \leq i \leq n \leq N$, initialized in $\mathbb{GT}(N)$ and stopped at the stopping time $\tau_{\mathbb{GT}(N)}$ to be defined below,

$$d\mathbb{X}_i^{(n)}(t) = \sqrt{2a_n\left(\mathbb{X}_i^{(n)}(t)\right)}d\beta_i^{(n)}(t) + b_n\left(\mathbb{X}_i^{(n)}(t)\right)dt + dK_i^{(n),-} - dK_i^{(n),+}, \tag{13.32}$$

driven by an array $\left(\beta_i^{(n)}(t); t \geq 0, 1 \leq i \leq n \leq N\right)$ of $\frac{N(N+1)}{2}$ independent standard Brownian motions. The positive finite variation processes $K_i^{(n),-}$ and $K_i^{(n),+}$ are such that $K_i^{(n),-}$ increases only when $\mathbb{X}_i^{(n)} = \mathbb{X}_{i-1}^{(n-1)}$, $K_i^{(n),+}$ increases only when $\mathbb{X}_i^{(n)} = \mathbb{X}_i^{(n-1)}$ with $K_1^{(N),-}$ increasing when $\mathbb{X}_1^{(N)} = l$ and $K_N^{(N),+}$ increasing when $\mathbb{X}_N^{(N)} = r$, so that $\mathbb{X} = \left(\mathbb{X}^{(1)}, \cdots, \mathbb{X}^{(N)}\right)$ stays in $\mathbb{GT}(N)$ forever. The stopping $\tau_{\mathbb{GT}(N)}$ is given by,

$$\tau_{\mathbb{GT}(N)} = \inf\big\{t \geq 0 : \exists\, (n, i, j)\ 2 \leq n \leq N-1,\ 1 \leq i < j \leq n \text{ s.t. } \mathbb{X}_i^{(n)}(t) = \mathbb{X}_j^{(n)}(t)\big\}.$$

Stopping at $\tau_{\mathbb{GT}(N)}$ takes care of the problematic possibility of two of the time dependent barriers coming together. It will turn out that $\tau_{\mathbb{GT}(N)} = \infty$ almost surely under certain initial conditions of interest to us given in Proposition 13.9 below; this will be the case since then each level $\mathbb{X}^{(n)}$ will evolve according to a Doob's h-transform and thus consisting of non-intersecting paths. That the system of reflecting $SDEs$ (13.32) above is well-posed, under a Yamada–Watanabe condition on the coefficients $(\sqrt{a_n}, b_n)$ for $1 \leq n \leq N$, follows (inductively) from Proposition 13.34.

We would like Theorem 13.1 to be applicable to each pair $(\mathbb{X}^{(n-1)}, \mathbb{X}^{(n)})$, with $X = \mathbb{X}^{(n)}$ and $Y = \mathbb{X}^{(n-1)}$. To this end, for $n = 2, \cdots, N$, suppose that $\mathbb{X}^{(n-1)}$ is distributed according to the following h-transformed Karlin–McGregor semigroup by the strictly positive in $\mathring{W}^{n-1}$ eigenfunction g_{n-1} with eigenvalue $e^{c_{n-1}t}$,

$$e^{-c_{n-1}t}\frac{g_{n-1}(y_1, \cdots, y_{n-1})}{g_{n-1}(x_1, \cdots, x_{n-1})}\det\big(\widehat{p}_t^n(x_i, y_j)\big)_{i,j=1}^{n-1},$$

where $\widehat{p}_t^n(\cdot, \cdot)$ denotes the transition density associated with the dual $\widehat{L}_n$ (killed at an exit of regular absorbing boundary point) of L_n. We furthermore, denote by $\widehat{m}^n(\cdot)$ the density with respect to Lebesgue measure of the speed measure of $\widehat{L}_n$. Then, Theorem 13.1 gives that under a special initial condition (stated therein) for the joint dynamics of $(\mathbb{X}^{(n-1)}, \mathbb{X}^{(n)})$, with $X = \mathbb{X}^{(n)}$ and $Y = \mathbb{X}^{(n-1)}$, the projection on $\mathbb{X}^{(n)}$ is distributed as the G_{n-1} h-transform of n independent L_n diffusions, thus consisting of non-intersecting paths, where G_{n-1} is given by,

$$G_{n-1}(x_1, \cdots, x_n) = \int_{W^{n-1,n}(x)} \prod_{i=1}^{n-1} \widehat{m^n}(y_i) g_{n-1}(y_1, \cdots, y_{n-1}) dy_1 \cdots dy_{n-1}. \tag{13.33}$$

Consistency then demands, by comparing $(\mathbb{X}^{(n-1)}, \mathbb{X}^{(n)})$ and $(\mathbb{X}^{(n)}, \mathbb{X}^{(n+1)})$, the following condition between the transition kernels (which is also sufficient as we see below for the construction of a consistent process $(\mathbb{X}^{(1)}, \cdots, \mathbb{X}^{(N)})$), for $t > 0, x, y \in \mathring{W}^n$,

$$\begin{aligned} &e^{-c_{n-1}t}\frac{G_{n-1}(y_1, \cdots, y_n)}{G_{n-1}(x_1, \cdots, x_n)}\det\big(p_t^n(x_i, y_j)\big)_{i,j=1}^n \\ &= e^{-c_n t}\frac{g_n(y_1, \cdots, y_n)}{g_n(x_1, \cdots, x_n)}\det\left(\widehat{p}_t^{n+1}(x_i, y_j)\right)_{i,j=1}^n. \end{aligned} \tag{13.34}$$

Denote the semigroup associated with these densities by $\left(\mathfrak{P}^{(n)}(t); t > 0\right)$ and also define the Markov kernels $\mathfrak{L}^n_{n-1}(x, dy)$ for $x \in \mathring{W}^n$ by,

$$\mathfrak{L}^n_{n-1}(x, dy) = \frac{\prod_{i=1}^{n-1} \widehat{m^n}(y_i) g_{n-1}(y_1, \cdots, y_{n-1})}{G_{n-1}(x_1, \cdots, x_n)} \mathbf{1} \\ \times \left(y \in W^{n-1,n}(x)\right) dy_1 \cdots dy_{n-1}.$$

Then, by inductively applying Theorem 13.1, we easily see the following Proposition holds:

Proposition 13.9 *Assume* (**R**) *and* (**BC+**) *hold for the L_n-diffusion and* (**YW**) *holds for the pairs of $(L_n, \hat{L}_n)$-diffusions for $2 \le n \le N$. Moreover, suppose that there exist functions g_n and G_n so that the consistency relations (13.33) and (13.34) hold. Let $\nu_N(dx)$ be a measure supported in $\mathring{W}^N$. Consider the process $(\mathbb{X}(t); t \ge 0) = \left(\left(\mathbb{X}^{(1)}(t), \cdots, \mathbb{X}^{(N)}(t)\right); t \ge 0\right)$ in $\mathbb{GT}(N)$ satisfying the SDEs (13.32) and initialized according to,*

$$\nu_N(dx^{(N)}) \mathfrak{L}^N_{N-1}(x^{(N)}, dx^{(N-1)}) \cdots \mathfrak{L}^2_1(x^{(2)}, dx^{(1)}). \tag{13.35}$$

Then $\tau_{\mathbb{GT}(N)} = \infty$ almost surely, $\left(\mathbb{X}^{(n)}(t); t \ge 0\right)$ for $1 \le n \le N$ evolves according to $\mathfrak{P}^{(n)}(t)$ and for fixed $T > 0$ the law of $\left(\mathbb{X}^{(1)}(T), \cdots, \mathbb{X}^{(N)}(T)\right)$ is given by,

$$\left(\nu_N \mathfrak{P}^{(N)}_T\right)(dx^{(N)}) \mathfrak{L}^N_{N-1}(x^{(N)}, dx^{(N-1)}) \cdots \mathfrak{L}^2_1(x^{(2)}, dx^{(1)}). \tag{13.36}$$

Proof For $n = 2$ this is the statement of Theorem 13.1. Assume that the proposition is proven for $n = N - 1$. Observe that, an initial condition of the form (13.35) in $\mathbb{GT}(N)$ gives rise to an initial condition of the same form in $\mathbb{GT}(N-1)$:

$$\tilde{\nu}_{N-1}(dx^{(N-1)}) \mathfrak{L}^{N-1}_{N-2}(x^{(N-1)}, dx^{(N-2)}) \cdots \mathfrak{L}^2_1(x^{(2)}, dx^{(1)}),$$

$$\tilde{\nu}_{N-1}(dx^{(N-1)}) = \int_{\mathring{W}^N} \nu_N(dx^{(N)}) \mathfrak{L}^N_{N-1}(x^{(N)}, dx^{(N-1)}).$$

Then, by the inductive hypothesis $\left(\mathbb{X}^{(N-1)}(t); t \ge 0\right)$ evolves according to $\mathfrak{P}^{(N-1)}(t)$, with the joint evolution of $(\mathbb{X}^{(N-1)}, \mathbb{X}^{(N)})$, by (13.33) and (13.34) with $n = N - 1$, as in Theorem 13.1, with $X = \mathbb{X}^{(N)}$ and $Y = \mathbb{X}^{(N-1)}$ and with initial condition $\nu_N(dx^{(N)}) \mathfrak{L}^N_{N-1}(x^{(N)}, dx^{(N-1)})$. We thus obtain that $\left(\mathbb{X}^{(N)}(t); t \ge 0\right)$ evolves according to $\mathfrak{P}^{(N)}(t)$ and for fixed T the conditional distribution of $\mathbb{X}^{(N-1)}(T)$ given $\mathbb{X}^{(N)}(T)$ is $\mathfrak{L}^N_{N-1}\left(\mathbb{X}^{(N)}(T), dx^{(N-1)}\right)$. This, along with the inductive hypothesis on the law of $\mathbb{GT}(N-1)$ at time T yields (13.36). The fact that $\tau_{\mathbb{GT}(N)} = \infty$ is also clear since each $\left(\mathbb{X}^{(n)}(t); t \geq 0\right)$ is governed by a Doob transformed Karlin–McGregor semigroup.

Similarly, the result above holds by replacing $\nu_N(dx^{(N)})$ by an entrance law $\left(\nu_t^{(N)}(dx^{(N)})\right)_{t \geq 0}$ for $\mathfrak{P}^{(N)}(t)$, in which case $\left(\nu_N \mathfrak{P}_T^{(N)}\right)(dx^{(N)})$ gets replaced by $\nu_T^{(N)}(dx^{(N)})$.

The consistency relations (13.33) and (13.34) and the implications for which choices of $L_1, \cdots, L_N$ to make will not be studied here. These questions are worth further investigation and will be addressed in future work.

13.3.2 Brownian Motions in Full Space

The process considered here was first constructed by Warren in [66]. Suppose in our setup of the previous section we take as the L-diffusion a standard Brownian motion with generator $\frac{1}{2}\frac{d^2}{dx^2}$, speed measure with density $m(x) = 2$ and scale function $s(x) = x$. Then, its conjugate diffusion with generator $\hat{L}$ from the results of the previous section is again a standard Brownian motion, so that in particular $P_t^n = \hat{P}_t^n$. Recall that the Vandermonde determinant $h_n(x) = \prod_{1 \leq i < j \leq n}(x_j - x_i)$ is a positive harmonic function for P_t^n (see for example [66] or by iteration from the results here). Moreover, the h-transformed semigroup P_t^{n,h_n} is exactly the semigroup of n particle Dyson Brownian motion.

Proposition 13.10 *Let $x \in \mathring{W}^{n+1}(\mathbb{R})$ and consider a process $(X, Y) \in W^{n,n+1}(\mathbb{R})$ started from the distribution $\left(\delta_x, \frac{n! h_n(y)}{h_{n+1}(x)} \mathbf{1}(y \prec x) dy\right)$ with the Y particles evolving as n particle Dyson Brownian motion and the X particles as $n + 1$ standard Brownian motions reflected off the Y particles. Then, the X particles are distributed as $n + 1$ Dyson Brownian motion started from x.*

Proof We apply Theorem (13.1) with the L-diffusion being a standard Brownian motion. Observe that, (**R**), (**BC**+) and (**YW**) are easily seen to be satisfied. Finally, as recalled above the Vandermonde determinant $h_n(x) = \prod_{1 \leq i < j \leq n}(x_j - x_i)$ is a positive harmonic function for n independent Brownian motions killed when they intersect and the semigroup P_t^{n,h_n} is the one associated to n particle Dyson Brownian motion.

In fact, we can start the process from the boundary of $W^{n,n+1}(\mathbb{R})$ via an entrance law as described in the previous section. To be more concrete, an entrance law for $P_t^{n+1,h_{n+1}}$ describing the process starting from the origin, which can be obtained via a limiting procedure detailed in the Appendix is the following:

$$\mu_t^{n+1,h_{n+1}}(dx) = C_{n+1} t^{-(n+1)^2/2} \exp\Big(-\frac{1}{2t}\sum_{i=1}^{n+1} x_i^2\Big) h_{n+1}^2(x) dx.$$

Thus, from the previous section's results

$$\nu_t^{n,n+1,h_{n+1}}(dx,dy) = \mu_t^{n+1,h_{n+1}}(dx)\frac{n!h_n(y)}{h_{n+1}(x)}\mathbf{1}(y \prec x)dy,$$

forms an entrance law for the semigroup associated to the two-level process in Proposition 13.10. Hence, we obtain the following:

Proposition 13.11 *Consider a Markovian process $(X,Y) \in W^{n,n+1}(\mathbb{R})$ initialized according to the entrance law $\nu_t^{n,n+1,h_{n+1}}(dx,dy)$ with the Y particles evolving as n particle Dyson Brownian motion and the X particles as $n+1$ standard Brownian motions reflected off the Y particles. Then, the X particles are distributed as $n+1$ Dyson Brownian motion started from the origin.*

It can be seen that we are in the setting of Proposition 13.9 with the $L_k \equiv L$-diffusion a standard Brownian motion and the functions g_k, G_k being up to a multiplicative constant equal to the Vandermonde determinant $\prod_{1\le i<j\le k}(x_j - x_i)$. Thus, we can concatenate these two-level processes to build a process $\left(\mathbb{X}^n(t); t \ge 0\right) = (X_i^{(k)}(t); t \ge 0, 1 \le i \le k \le n)$ taking values in $\mathbb{GT}(n)$ recovering Proposition 6 of [66]. Being more concrete, the dynamics of $\mathbb{X}^n(t)$ are as follows: level k of this process consists of k independent standard Brownian motions reflected off the paths of level $k-1$. Then, from Proposition 13.9 we get:

Proposition 13.12 *If $\mathbb{X}^n$ is started from the origin then the kth level process $X^{(k)}$ is distributed as k particle Dyson Brownian motion started from the origin.*

Connection to Hermitian Brownian Motion We now point out the well known connection to the minor process of a Hermitian valued Brownian motion.It is a well known fact that the eigenvalues of minors of Hermitian matrices interlace. In particular, for any $n \times n$ Hermitian valued diffusion the eigenvalues of the $k \times k$ minor $\left(\lambda^{(k)}(t); t \ge 0\right)$ and of the $(k-1) \times (k-1)$ minor $\left(\lambda^{(k-1)}(t); t \ge 0\right)$ interlace: $\left(\lambda_1^{(k)}(t) \le \lambda_2^{(k-1)}(t) \le \cdots \le \lambda_k^{(k)}(t); t \ge 0\right)$. Now, let $(H(t); t \ge 0)$ be an $n \times n$ Hermitian valued Brownian motion. Then $\left(\lambda^{(k)}(t); t \ge 0\right)$ evolves as k particle Dyson Brownian motion. Also for any *fixed* time T the vector $(\lambda^{(1)}(T), \cdots, \lambda^{(n)}(T))$ has the same distribution as $\mathbb{X}(T)$, namely it is uniform on the space of $\mathbb{GT}(n)$ with bottom level $\lambda^{(n)}(T)$. However the evolution of these processes is different, in fact the interaction between two levels of the minor process $\left(\lambda^{(k-1)}(t), \lambda^{(k)}(t); t \ge 0\right)$ is quite complicated involving long range interactions and not the local reflection as in our case as shown in [1]. In fact, the evolution of $\left(\lambda^{(k-1)}(t), \lambda^{(k)}(t), \lambda^{(k+1)}(t); t \ge 0\right)$ is not even Markovian at least for some initial conditions (again see [1]).

13.3.3 Brownian Motions in Half Line and BES(3)

The process we will consider here, taking values in a symplectic Gelfand–Tsetlin pattern, was first constructed by Cerenzia in [17] as the diffusive scaling limit of the symplectic Plancherel growth model. It is built from reflecting and killed Brownian motions in the half line. We begin in the simplest possible setting:

Proposition 13.13 *Consider a process $(X, Y) \in W^{1,1}([0, \infty))$ started according to the distribution $(\delta_x, \mathbf{1}_{[0,x]}dy)$ for $x > 0$ with the Y particle evolving as a reflecting Brownian motion in $[0, \infty)$ and the X particle as a Brownian motion in $(0, \infty)$ reflected upwards off the Y particle. Then, the X particle is distributed as a $BES(3)$ process (Bessel process of dimension 3) started from x.*

Proof Take as the L-diffusion a Brownian motion absorbed when it reaches 0 and let P_t^1 be the semigroup of Brownian motion *killed* (not absorbed) at 0. Then, its dual diffusion $\hat{L}$ is a reflecting Brownian motion in the positive half line and let $\hat{P}_t^1$ be the semigroup it gives rise to. Observe that, (**R**), (**BC**+) and (**YW**) are easily seen to be satisfied. Letting, $\hat{h}_{1,1}(x) = 1$ which is clearly a positive harmonic function for $\hat{L}$, we get that $h_{1,1}(x) = \int_0^x 1dx = x$. Now, note that $P_t^{1,h_{1,1}}$ is exactly the semigroup of a $BES(3)$ process. As is well known, a Bessel process of dimension 3 is a Brownian motion conditioned to stay in $(0, \infty)$ by an h-transform with the function x. Then, from the analogue of Theorem 13.1 in $W^{n,n}$ we obtain the statement.

Now we move to the next stage of 2 particles evolving as reflecting Brownian motions being reflected off a $BES(3)$ process.

Proposition 13.14 *Consider a process $(X, Y) \in W^{1,2}([0, \infty))$ started according to the following distribution $\left(\delta_{(x_1,x_2)}, \frac{2y}{x_2^2-x_1^2}1_{[x_1,x_2]}dy\right)$ for $x_1 < x_2$ with the Y particle evolving as a BES(3) process and the X particles as reflecting Brownian motions in $[0, \infty)$ reflected off the Y particles. Then, the X particles are distributed as two non-intersecting reflecting Brownian motions started from (x_1, x_2).*

Proof We apply Theorem 13.1. We take as the L-diffusion a reflecting Brownian motion. Write P_t^2 for the Karlin–McGregor semigroup associated to 2 reflecting Brownian motions killed when they intersect. Note that, (**R**), (**BC**+) and (**YW**) are clearly satisfied. Observe that with $\hat{h}_{1,2}(x) = x$, which is a positive harmonic function for a Brownian motion killed at 0, we have:

$$h_{1,2}(x_1, x_2) = \int_{x_1}^{x_2} xdx = \frac{1}{2}(x_2^2 - x_1^2).$$

Finally note that, $P_t^{2,h_{1,2}}$ is exactly the semigroup of 2 non-intersecting reflecting Brownian motions in $[0, \infty)$.

These relations can be iterated to n and n and also n and $n+1$ particles. Define the functions:

$$\hat{h}_{n,n}(x) = \prod_{1\le i<j\le n} (x_j^2 - x_i^2),$$

$$\hat{h}_{n,n+1}(x) = \prod_{1\le i<j\le n} (x_j^2 - x_i^2) \prod_{i=1}^{n} x_i .$$

Also, consider the positive kernels Λ_{n_1,n_2}, defined in (13.24), with $\hat{m} \equiv 2$. Then, an easy calculation (after writing these functions as determinants) gives that up to a constant $h_{n,n} = \Lambda_{n,n}\hat{h}_{n,n}$ is equal to $\hat{h}_{n,n+1}$ and $h_{n,n+1} = \Lambda_{n,n+1}\hat{h}_{n,n+1}$ is equal to $\hat{h}_{n+1,n+1}$. Finally, let $\Lambda_{n,n}^{\hat{h}_{n,n}}$ and $\Lambda_{n,n+1}^{\hat{h}_{n,n+1}}$ denote the corresponding normalized Markov kernels.

Proposition 13.15 *Consider a process $(X,Y) \in W^{n,n}([0,\infty))$ started according to the distribution $(\delta_x, \Lambda_{n,n}^{\hat{h}_{n,n}}(x,\cdot))$ for $x \in \mathring{W}^n([0,\infty))$ with the Y particles evolving as n reflecting Brownian motions conditioned not to intersect in $[0,\infty)$ and the X particles as n Brownian motion in $(0,\infty)$ reflected off the Y particles. Then, the X particles are distributed as n $BES(3)$ processes conditioned never to intersect started from x.*

Proof We take as the L-diffusion a Brownian motion absorbed at 0. Then, the $\hat{L}$-diffusion is a reflecting Brownian motion. As before, (**R**), (**BC**+) and (**YW**) are clearly satisfied. Note that, $\hat{h}_{n,n}$ is a harmonic function for n reflecting Brownian motions killed when they intersect. Moreover, note that $P_t^{n,h_{n,n}}$ is exactly the semigroup of n non-intersecting $BES(3)$ processes (note that the n particle Karlin–McGregor semigroup P_t^n is that of n killed at zero Brownian motions). The statement follows from the analogue of Theorem 13.1 in $W^{n,n}$.

Proposition 13.16 *Consider a process $(X,Y) \in W^{n,n+1}([0,\infty))$ started according to the following distribution $\left(\delta_x, \Lambda_{n,n+1}^{\hat{h}_{n,n+1}}(x,\cdot)\right)$ for $x \in \mathring{W}^{n+1}([0,\infty))$ with the Y particles evolving as n BES(3) processes conditioned not to intersect and the X particles as $n+1$ reflecting Brownian motions in $[0,\infty)$ reflected off the Y particles. Then, the X particles are distributed as $n+1$ non-intersecting reflecting Brownian motions started from x.*

Proof We take as the L-diffusion a reflecting Brownian motion. Then, the $\hat{L}$-diffusion is a Brownian motion absorbed at 0. As before, the assumptions (**R**), (**BC**+) and (**YW**) are clearly satisfied. Note that, $\hat{h}_{n,n+1}$ is harmonic for the corresponding Karlin–McGregor semigroup $\hat{P}_t^n$, associated with n Brownian motions killed at zero and when they intersect. Moreover, note that the semigroup $\hat{P}_t^{n,\hat{h}_{n,n+1}}$, namely the semigroup $\hat{P}_t^n$ h-transformed by $\hat{h}_{n,n+1}$, gives the semigroup of the process Y. Finally, observe that $P_t^{n+1,h_{n,n+1}}$ is exactly the semigroup of $n+1$

non-intersecting reflecting Brownian motions. The statement follows from Theorem 13.1.

Remark 13.2 The semigroups considered above are also the semigroups of n Brownian motions conditioned to stay in a Weyl Chamber of type B and type D (after we disregard the sign of the last coordinate) respectively (see for example [42] where such a study was undertaken).

We can in fact start these processes from the origin, by using the following explicit entrance law (see for example [17] or the Appendix for the general recipe) for $P_t^{n,h_{n,n}}$ and $P_t^{n,h_{n-1,n}}$ issued from zero,

$$\mu_t^{n,h_{n,n}}(dx) = C'_{n,n} t^{-n(n+\frac{1}{2})} \exp\left(-\frac{1}{2t}\sum_{i=1}^{n} x_i^2\right) h_{n,n}^2(x)dx,$$

$$\mu_t^{n,h_{n-1,n}}(dx) = C'_{n-1,n} t^{-n(n-\frac{1}{2})} \exp\left(-\frac{1}{2t}\sum_{i=1}^{n} x_i^2\right) h_{n-1,n}^2(x)dx.$$

Concatenating these two-level processes, we construct a process $\left(\mathbb{X}_s^{(n)}(t); t \geq 0\right)$ $= (X^{(1)}(t) \prec \hat{X}^{(1)}(t) \prec \cdots \prec X^{(n)}(t) \prec \hat{X}^{(n)}(t); t \geq 0)$ in $\mathbb{GT}_s(n)$ with dynamics as follows: Firstly, $X_1^{(1)}$ is a Brownian motion reflecting at the origin. Then, for each k, the k particles corresponding to $\hat{X}^{(k)}$ perform independent Brownian motions reflecting off the $X^{(k)}$ particles to maintain interlacing. Finally, for $k \geq 2$ the k particles corresponding to $X^{(k)}$ reflect off $\hat{X}^{(k-1)}$ and also in the case of $X_1^{(k)}$ reflecting at the origin.

Then, the symplectic analogue of Proposition 13.9 (which is again proven in the same way by consistently patching together two-level processes) implies the following, recovering the results of Section 2.3 of [17]:

Proposition 13.17 *If $\mathbb{X}_s^n$ is started from the origin then the projections onto $X^{(k)}$ and $\hat{X}^{(k)}$ are distributed as k non-intersecting reflecting Brownian motions and k non-intersecting $BES(3)$ processes respectively started from the origin.*

13.3.4 Brownian Motions in an Interval

Let $I = [0, \pi]$ for concreteness and let the L-diffusion be a reflecting Brownian motion in I. Then its dual, the $\hat{L}$-diffusion is a Brownian motion absorbed at 0 or π. It will be shown in Corollary 13.3, that the minimal positive eigenfunction, is given up to a (signed) constant factor by,

$$\hat{h}_n(x) = \det(\sin(kx_j))_{k,j=1}^n. \tag{13.37}$$

This is the eigenfunction that corresponds to conditioning these Brownian motions to stay in the interval $(0,\pi)$ and not intersect forever. Also, observe that up to a constant factor $\hat{h}_n$ is given by (see the notes [20, 53] and Remark 13.4 below for the connection to classical compact groups),

$$\prod_{i=1}^{n}\sin(x_i)\prod_{1\le i<j\le n}\big(\cos(x_i)-\cos(x_j)\big).$$

Now, via the iterative procedure of producing eigenfunctions, namely by taking $\Lambda_{n,n+1}\hat{h}_n$, where $\Lambda_{n,n+1}$ is defined in (13.24), we obtain that up to a (signed) constant factor,

$$h_{n+1}(x)=\det(\cos((k-1)x_j))_{k,j=1}^{n+1}, \tag{13.38}$$

is a strictly positive eigenfunction for P_t^{n+1}. In fact, it is the minimal positive eigenfunction (again this follows from Corollary 13.3) of P_t^{n+1} and it corresponds to conditioning these reflected Brownian motions in the interval to not intersect. This is also (see [20, 53] and Remark 13.4) given up to a constant factor by,

$$\prod_{1\le i<j\le n+1}\big(\cos(x_i)-\cos(x_j)\big).$$

Define the Markov kernel:

$$(\Lambda_{n,n+1}^{\hat{h}_n}f)(x)=\frac{n!}{h_{n+1}(x)}\int_{W^{n,n+1}(x)}\hat{h}_n(y)f(x,y)dy.$$

Then we have the following result:

Proposition 13.18 *Let* $x\in\mathring{W}^{n+1}([0,\pi])$. *Consider a process* $(X,Y)\in W^{n,n+1}([0,\pi])$ *started at* $\left(\delta_x,\Lambda_{n,n+1}^{\hat{h}_n}(x,\cdot)\right)$ *with the* Y *particles evolving as* n *Brownian motions conditioned to stay in* $(0,\pi)$ *and conditioned to not intersect and the* X *particles as* $n+1$ *reflecting Brownian motions in* $[0,\pi]$ *reflected off the* Y *particles. Then the* X *particles are distributed as* $n+1$ *non-intersecting Brownian motions reflected at the boundaries of* $[0,\pi]$ *started from* x.

Proof Take as the L-diffusion a reflecting Brownian motion in $[0,\pi]$. The $\hat{L}$-diffusion is a Brownian motion absorbed at 0 or π. Observe that, the assumptions (**R**), (**BC+**) and (**YW**) are satisfied. Moreover, as noted above $\hat{h}_n$ is the ground state for n Brownian motions killed when they hit 0 or π or when they intersect. The statement of the proposition then follows from Theorem 13.1.

Remark 13.3 The dual relation, in the following sense is also true: If we reflect n Brownian motions between $n+1$ reflecting Brownian motions in $[0,\pi]$ conditioned not to intersect then we obtain n Brownian motions conditioned to stay in $(0,\pi)$ and conditioned not to intersect. This is obtained by noting that up to a constant factor $\hat{h}_n$ defined in (13.37) is given by $\Lambda_{n+1,n}h_{n+1}$, with h_{n+1} as in (13.38).

Remark 13.4 The processes studied above are related to the eigenvalue evolutions of Brownian motions on $SO(2(n+1))$ (reflecting Brownian motions in $[0,\pi]$) and $USp(2n)$ (conditioned Brownian motions in $[0,\pi]$) respectively (see e.g. [56] for skew product decompositions of Brownian motions on manifolds of matrices).

Remark 13.5 It is also possible to build the following interlacing processes with equal number of particles. Consider as the Y process n Brownian motions in $[0,\pi)$ reflecting at 0 and conditioned to stay away from π and not to intersect. In our framework $\hat{L} = \frac{1}{2}\frac{d^2}{dx^2}$ with Neumann boundary condition at 0 and Dirichlet at π. Then the minimal eigenfunction corresponding to this conditioning is given up to a sign by,

$$\det\left(\cos\left((k-\frac{1}{2})y_j\right)\right)_{k,j=1}^{n}.$$

Now let X be n Brownian motions in $(0,\pi]$ reflecting at π and reflected off the Y particles. Then the projection onto the X process (assuming the two levels (X,Y) are started appropriately) evolves as n Brownian motions in $(0,\pi]$ reflecting at π and conditioned to stay away from 0 and not to intersect. These processes are related to the eigenvalues of Brownian motions on $SO(2n+1)$ and $SO^-(2n+1)$ respectively.

13.3.5 Brownian Motions with Drifts

The processes considered here were first introduced by Ferrari and Frings in [31] (there only the *fixed time* picture was studied, namely no statement was made about the distribution of the projections on single levels as processes). They form a generalization of the process studied in the first subsection.

13.3.5.1 Hermitian Brownian with Drifts

We begin by a brief study of the matrix valued process first. Let $(Y_t; t \geq 0) = (B_t; t \geq 0)$ be an $n \times n$ Hermitian Brownian motion. We seek to add a matrix of *drifts* and study the resulting eigenvalue process. For simplicity let M be a diagonal $n \times n$ Hermitian matrix with distinct ordered eigenvalues $\mu_1 < \cdots < \mu_n$ and consider the Hermitian valued process $\left(Y_t^M; t \geq 0\right) = (B_t + tM; t \geq 0)$.

Then a computation that starts by applying Girsanov's theorem, using unitary invariance of Hermitian Brownian motion, integrating over $\mathbb{U}(n)$, the group of $n \times n$ unitary matrices, and then computing that integral using the classical Harish Chandra-Itzykson-Zuber (HCIZ) formula gives that the eigenvalues

$(\lambda_1^M(t), \cdots, \lambda_n^M(t); t \geq 0)$ of $\left(Y_t^M; t \geq 0\right)$ form a diffusion process with explicit transition density given by,

$$s_t^{n,M}(\lambda, \lambda') = \exp\left(-\frac{1}{2}\sum_{i=1}^{n} \mu_i^2 t\right) \frac{\det\left(\exp(\mu_j \lambda_i')\right)_{i,j=1}^n}{\det\left(\exp(\mu_j \lambda_i)\right)_{i,j=1}^n} \det\left(\phi_t(\lambda_i, \lambda_j')\right)_{i,j=1}^n,$$

where ϕ_t is the standard heat kernel. For a proof of this fact, which uses the theory of Markov functions, see for example [49].

Observe that, $s_t^{n,M}$ is exactly the transition density of n Brownian motions with drifts $\mu_1 < \cdots < \mu_n$ conditioned to never intersect as studied in [6]. More generally, if we look at the $k \times k$ minor of $\left(Y_t^M; t \geq 0\right)$ then its eigenvalues evolve as k Brownian motions with drifts $\mu_1 < \cdots < \mu_k$ conditioned to never intersect.

Remark 13.6 These processes also appear in the recent work of Ipsen and Schomerus [38] as the finite time Lyapunov exponents of "Isotropic Brownian motions".

Now, write $\mu^{(k)}$ for $(\mu_1, \cdots, \mu_k)$ and $P_t^{n,\mu^{(n)}}$ for the semigroup that arises from $s_t^{n,M}$. Then, $u_t^{n,\mu^{(n)}}(d\lambda)$ defined by,

$$u_t^{n,\mu^{(n)}}(d\lambda) = const_{n,t} \det(e^{-(\lambda_i - t\mu_j)^2/2t})_{i,j=1}^n \frac{\prod_{1\leq i<j\leq n}(\lambda_j - \lambda_i)}{\prod_{1\leq i<j\leq n}(\mu_j^{(n)} - \mu_i^{(n)})} d\lambda,$$

forms an entrance law for $P_t^{n,\mu^{(n)}}$ starting from the origin (see for example [31] or the Appendix).

13.3.5.2 Interlacing Construction with Drifting Brownian Motions with Reflection

Now moving on to Warren's process with drifts (as referred to in [31]). We seek to build $n+1$ Brownian motions with drifts $\mu_1 < \cdots < \mu_{n+1}$ conditioned to never intersect by reflecting off n Brownian motions with drifts $\mu_1 < \cdots < \mu_n$ conditioned to never intersect $n+1$ independent Brownian motions each with drift μ_{n+1}. We prove the following:

Proposition 13.19 *Consider a Markov process $(X, Y) \in W^{n,n+1}(\mathbb{R})$ started from the origin with the Y particles evolving as n Brownian motions with drifts $\mu_1 < \cdots < \mu_n$ conditioned to never intersect and the X particles as $n+1$ Brownian motions all with drift μ_{n+1} reflected off the Y particles. Then, the X particles are distributed as $n+1$ Brownian motions with drifts $\mu_1 < \cdots < \mu_{n+1}$ conditioned to never intersect started from the origin.*

Proof Let the L-diffusion be a Brownian motion with drift μ_{n+1}, namely with generator $L = \frac{1}{2}\frac{d^2}{dx^2} + \mu_{n+1}\frac{d}{dx}$. Then, its dual diffusion $\hat{L} = \frac{1}{2}\frac{d^2}{dx^2} - \mu_{n+1}\frac{d}{dx}$

has speed measure $\hat{m}(x) = 2e^{-2\mu_{n+1}x}$. Note that, the assumptions (**R**), (**BC+**) and (**YW**) are easily seen to be satisfied. Let $P_t^{n+1,\mu_{n+1}}$ and $\hat{P}_t^{n,\mu_{n+1}}$ denote the corresponding Karlin–McGregor semigroups. Consider the (not yet normalized) positive kernel $\Lambda_{n,n+1}^{\mu_{n+1}}$ given by,

$$(\Lambda_{n,n+1}^{\mu_{n+1}} f)(x) = \int_{W^{n,n+1}(x)} f(x,y) \prod_{i=1}^{n} 2e^{-2\mu_{n+1}y_i} dy_i.$$

and define the function

$$\hat{h}_n^{\mu_{n+1},\mu^{(n)}}(y) = \prod_{i=1}^{n} e^{\mu_{n+1}y_i} \det(e^{\mu_i y_j})_{i,j=1}^{n}.$$

Note that, $\hat{h}_n^{\mu_{n+1},\mu^{(n)}}$ is a strictly positive eigenfunction for $\hat{P}_t^{n,\mu_{n+1}}$. Moreover, the h-transform of $\hat{P}_t^{n,\mu_{n+1}}$ with $\hat{h}_n^{\mu_{n+1},\mu^{(n)}}$ is exactly the semigroup $P_t^{n,\mu^{(n)}}$ of n Brownian motions with drifts $(\mu_1, \cdots, \mu_n)$ conditioned to never intersect. By integrating the determinant we get,

$$(\Lambda_{n,n+1}^{\mu_{n+1}} \hat{h}_n^{\mu_{n+1},\mu^{(n)}})(x) = \frac{2^n}{\prod_{i=1}^{n}(\mu_{n+1} - \mu_i)} \det(e^{(\mu_i - \mu_{n+1})x_j})_{i,j=1}^{n+1},$$

and note that the h-transform of $P_t^{n+1,\mu_{n+1}}$ by $\Lambda_{n,n+1}^{\mu_{n+1}} \hat{h}_n^{\mu_{n+1},\mu^{(n)}}$ is $P_t^{n+1,\mu^{(n+1)}}$. Finally, defining the entrance law for the two-level process started from the origin by $\nu_t^{n,n+1,\mu_{n+1},\mu^{(n)}} = u_t^{n+1,\mu^{(n+1)}} \Lambda_{n,n+1}^{\mu_{n+1},\mu^{(n)}}$, we obtain the statement of the proposition from Theorem 13.1 (see also discussion after Corollary 13.2).

Remark 13.7 A 'positive temperature' version of the proposition above appears as Proposition 9.1 in [54].

We can then iteratively apply the result above to concatenate two-level processes and build a process:

$$\left(\mathbb{X}_{(\mu_1,\cdots,\mu_n)}(t); t \geq 0\right) = \left(X_{\mu_1}^{(1)}(t) \prec X_{\mu_2}^{(2)}(t) \prec \cdots \prec X_{\mu_n}^{(n)}(t); t \geq 0\right),$$

in $\mathbb{GT}(n)$ as in Proposition 13.9 whose joint dynamics are given as follows (this was also described in [31]): Level k consists of k copies of independent Brownian motions all with drifts μ_k reflected off the paths of level $k-1$. Then, from Proposition 13.9 one obtains:

Proposition 13.20 *Assume $\mu_1 < \mu_2 < \cdots < \mu_n$. Consider the process $\left(\mathbb{X}_{(\mu_1,\cdots,\mu_n)}(t); t \geq 0\right)$ defined above started from the origin. Then, the projection on $X_{\mu_k}^{(k)}$ is distributed as k Brownian motions with drifts $\mu_1 < \cdots < \mu_k$ conditioned to never intersect, issueing from the origin.*

Remark 13.8 Note that, the multilevel process whose construction is described above via the hard reflection dynamics and the minors of the Hermitian valued process $\left(Y_t^M; t \geq 0\right)$ coincide on each fixed level k (as single level processes, this is what we have proven here) and also at fixed times (this is already part of the results of [31]). However, they do not have the same law as processes. Finally, for the fixed time correlation kernel of this Gelfand–Tsetlin valued process see Theorem 1 of [31].

13.3.6 Geometric Brownian Motions and Quantum Calogero-Sutherland

A geometric Brownian motion of unit diffusivity and drift parameter α is given by the SDE,

$$ds(t) = s(t)dW(t) + \alpha s(t)dt,$$

which can be solved explicitly to give that,

$$s(t) = s(0) \exp\left(W(t) + \left(\alpha - \frac{1}{2}\right)t\right).$$

We will assume that $s(0) > 0$, so that the process lives in $(0, \infty)$. Its generator is given by,

$$L^{\alpha} = \frac{1}{2}x^2 \frac{d^2}{dx^2} + \alpha x \frac{d}{dx},$$

with both 0 and ∞ being natural boundaries. With $h_n(x) = \prod_{1 \leq i < j \leq n}(x_j - x_i)$ denoting the Vandermonde determinant it can be easily verified (although it also follows by recursively applying the results below) that h_n is a positive eigenfunction of n independent geometric Brownian motions, namely that with,

$$L_n^{\alpha} = \sum_{i=1}^{n} \frac{1}{2} x_i^2 \partial_{x_i}^2 + \alpha \sum_{i=1}^{n} x_i \partial_{x_i},$$

we have,

$$L_n^{\alpha} h_n = \frac{n(n-1)}{2}\left(\frac{n-2}{3} + \alpha\right) h_n = c_{n,\alpha} h_n.$$

The quantum Calogero-Sutherland Hamiltonian $\mathcal{H}^{\theta}_{CS}$ (see [15, 64]) is given by,

$$\mathcal{H}^{\theta}_{CS} = \frac{1}{2}\sum_{i=1}^{n}\left(x_i\partial_{x_i}\right)^2 + \theta\sum_{i=1}^{n}\sum_{j\neq i}\frac{x_i^2}{x_i - x_j}\partial_{x_i}.$$

Its relation to geometric Brownian motions lies in the following simple observation. For $\theta = 1$ this quantum Hamiltonian coincides with the infinitesimal generator of n independent geometric Brownian motions with drift parameter $\frac{1}{2}$ h-transformed by the Vandermonde determinant namely,

$$\mathcal{H}^{1}_{CS} = h_n^{-1} \circ L_n^{\frac{1}{2}} \circ h_n - c_{n,\frac{1}{2}}.$$

We now show how one can construct a $\mathbb{GT}(n)$ valued process so that the kth level consists of k geometric Brownian motions with drift parameter $n - k + \frac{1}{2}$ h-transformed by the Vandermonde determinant. The key ingredient is the following:

Proposition 13.21 *Consider a process $(X, Y) \in W^{n,n+1}((0,\infty))$ started according to the following distribution $(\delta_x, \frac{n!h_n(y)}{h_{n+1}(x)}\mathbf{1}(y \prec x)dy)$ for $x \in \mathring{W}^{n+1}((0,\infty))$ with the Y particles evolving as n non-intersecting geometric Brownian motions with drift parameter $\alpha+1$ conditioned to not intersect via an h-transform by h_n and the X particles evolving as $n+1$ geometric Brownian motions with drift parameter α being reflected off the Y particles. Then, the X particles are distributed as $n+1$ non-intersecting geometric Brownian motions with drift parameter α conditioned to not intersect via an h-transform by h_{n+1}, started form $x \in \mathring{W}^{n+1}((0,\infty))$.*

Proof Taking as the L-diffusion L^{α}, and note that its speed measure is given by $m^{\alpha}(x) = 2x^{2\alpha-2}$, the conjugate diffusion is $\widehat{L^{\alpha}} = L^{1-\alpha}$. Observe that, the assumptions (**R**), (**BC+**) and (**YW**) are clearly satisfied.

First, note that an easy calculation gives that the h-transform of $\widehat{L^{\alpha}}$ by $\widehat{m^{\alpha}}^{-1}$ is an $L^{\alpha+1}$-diffusion, namely a geometric Brownian motion with drift parameter $\alpha + 1$. Hence, an h-transform of n $\widehat{L^{\alpha}}$-diffusions by the eigenfunction $\prod_{i=1}^{n}\widehat{m^{\alpha}}^{-1}(y_i)h_n(y)$ gives n non-intersecting geometric Brownian motions with drift parameter $\alpha + 1$ conditioned to not intersect via an h-transform by h_n. The statement of the proposition is then obtained from an application of Theorem 13.1.

Remark 13.9 Observe that, under an application of the exponential map the results of Sect. 13.3.5, give a generalization of Proposition 13.21 above.

Using the proposition above it is straightforward, and we will not elaborate on, how to iterate to build the $\mathbb{GT}(n)$ valued process with the correct drift parameters on each level.

Remark 13.10 The following geometric Brownian motion,

$$ds(t) = \sqrt{2}s(t)dW(t) - (u + u' + v + v')s(t)dt,$$

also arises as a continuum scaling limit after we scale space by $1/N$ and send N to infinity of the bilateral birth and death chain with birth rates $(x-u)(x-u')$ and death rates $(x+v)(x+v')$ considered by Borodin and Olshanski in [11].

13.3.7 Squared Bessel Processes and LUE Matrix Diffusions

In this subsection we will first construct a process taking values in $\mathbb{GT}$ being the analogue of the Brownian motion model for squared Bessel processes and having close connections to the LUE matrix valued diffusion. We also build a process in $\mathbb{GT}_\mathbf{s}$ generalizing the construction of Cerenzia (after a "squaring" transformation of the state space) for all dimensions $d \geq 2$. We begin with a definition:

Definition 13.7 The squared Bessel process of dimension d, abbreviated from now on as $BESQ(d)$ process, is the one dimensional diffusion with generator in $(0,\infty)$,

$$L^{(d)} = 2x\frac{d^2}{dx^2} + d\frac{d}{dx}.$$

The origin is an entrance boundary for $d \geq 2$, a regular boundary point for $0 < d < 2$ and an exit one for $d \leq 0$. Define the index $\nu(d) = \frac{d}{2} - 1$. The density of the speed measure of $L^{(d)}$ is $m_\nu(y) = c_\nu y^\nu$ and its scale function $s_\nu(x) = \bar{c}_\nu x^{-\nu}$, $\nu \neq 0$ and $s_0(x) = logx$. Then from the results of the previous section its conjugate, the $\widehat{L^{(d)}}$ diffusion, is a $BESQ(2-d)$ process with the dual boundary condition. Moreover, the following relation will be key, see [35]: A Doob h-transform of a $BESQ(2-d)$ process by its scale function $x^{\nu+1}$ gives a $BESQ(d+2)$ process.

Note that, condition (**BC+**) only holds for dimensions $d \in (-\infty, 0] \cup [2,\infty)$; this is because for $0 < d < 2$, the origin is a regular boundary point and the diffusion coefficient degenerates (these values of the parameters will not be considered here). We use the following notation throughout, for $d \in (-\infty, 0] \cup [2,\infty)$: we write $P_t^{n,(d)}$ for the Karlin–McGregor semigroup of n $BESQ(d)$ processes killed when they intersect or when they hit the origin, in case $d \leq 0$.

We start in the simplest setting of $W^{1,1}$ and consider the situation of a single $BESQ(2-d)$ process being reflected upwards off a $BESQ(d)$ process:

Proposition 13.22 *Let $d \geq 2$. Consider a process $(X,Y) \in W^{1,1}([0,\infty))$ started according to the distribution $(\delta_x, \frac{(\nu+1)y^\nu}{x^{\nu+1}}1_{[0,x]}dy)$ for $x > 0$ with the Y particle evolving as a $BESQ(d)$ process and the X particle as a $BESQ(2-d)$ process in $(0,\infty)$ reflected off the Y particle. Then, the X particle is distributed as a $BESQ(d+2)$ process started from x.*

Proof We take as the L-diffusion a $BESQ(2-d)$ process. Then, the $\hat{L}$-diffusion is a $BESQ(d)$ process. Note that, the assumptions (**R**), (**BC+**) and (**YW**) are satisfied. Since $\hat{h}_{1,1}^{(d)}(x) = 1$ is invariant for $BESQ(d)$, the following is invariant for $BESQ(2-d)$,

$$h_{1,1}^{(d)}(x) = \int_0^x c_\nu y^\nu dy = \frac{c_\nu}{\nu+1} x^{\nu+1}.$$

Then, as already remarked above, see [35], the h-transformed process with semigroup $P_t^{1,(2-d),h_{1,1}^{(d)}}$ is exactly a $BESQ(d+2)$ process. The analogue of Theorem 13.1 in $W^{n,n}$ gives the statement of the proposition.

We expect that the restriction to $d \geq 2$ is not necessary for the result to hold (it should be true for $d > 0$). In fact, Corollary 13.13, corresponds to $d = 1$, after we perform the transformation $x \mapsto \sqrt{x}$, which in particular maps $BESQ(1)$ and $BESQ(3)$ to reflecting Brownian motion and $BES(3)$ respectively.

We now move on to an arbitrary number of particles. Define the functions,

$$\hat{h}_{n,n}^{(d)}(x) = \prod_{1\leq i<j\leq n} (x_j - x_i) = \det\left(x_i^{j-1}\right)_{i,j=1}^n,$$

$$\hat{h}_{n,n+1}^{(d)}(x) = \prod_{1\leq i<j\leq n} (x_j - x_i) \prod_{i=1}^n x_i^{\nu+1} = \det\left(x_i^{j+\nu}\right)_{i,j=1}^n.$$

Moreover, let $\Lambda_{n-1,n}$ and $\Lambda_{n,n}$ be the following positive kernels, defined as in (13.24), where we recall that $m_{\nu(d)}(\cdot)$ is the speed measure density with respect to Lebesgue measure of a $BESQ(d)$ process:

$$(\Lambda_{n-1,n} f)(x) = \int_{W^{n-1,n}(x)} \prod_{i=1}^{n-1} m_{\nu(2-d)}(y_i) f(x,y) dy,$$

$$(\Lambda_{n,n} f)(x) = \int_{W^{n,n}(x)} \prod_{i=1}^{n_1} m_{\nu(d)}(y_i) f(x,y) dy.$$

An easy calculation gives that $h_{n-1,n}^{(d)}(x) = c_{n-1,n}(\nu)(\Lambda_{n-1,n}\Pi_{n-1,n}\hat{h}_{n-1,n}^{(d)})(x)$ is equal to $\hat{h}_{n,n}^{(d)}(x)$ and $h_{n,n}^{(d)}(x) = c_{n,n}(\nu)(\Lambda_{n,n}\Pi_{n,n}\hat{h}_{n,n}^{(d)})(x)$ is equal $\hat{h}_{n,n+1}^{(d)}(x)$, where $c_{n-1,n}(\nu), c_{n,n}(\nu)$ are explicit constants whose exact values are not important in what follows. Then we have:

Proposition 13.23 *Let $d \geq 2$. Consider a process $(X, Y) \in W^{n,n+1}([0,\infty))$ started according to the following distribution $(\delta_x, \frac{n!\prod_{1\leq i<j\leq n}(y_j-y_i)}{\prod_{1\leq i<j\leq n+1}(x_j-x_i)}\mathbf{1}(y \prec x)dy)$ for $x \in \mathring{W}^{n+1}([0,\infty))$ with the Y particles evolving as n non-intersecting $BESQ(d+2)$ processes and the X particles evolving as $n+1$ $BESQ(d)$ processes being reflected off the Y particles. Then, the X particles are distributed as $n+1$ non-intersecting $BESQ(d)$ processes started form $x \in \mathring{W}^{n+1}([0,\infty))$.*

Proof Take as the L-diffusion a $BESQ(d)$ process. Then, the $\hat{L}$-diffusion is a $BESQ(2-d)$ process. Note that, the assumptions (**R**), (**BC+**) and (**YW**) are satisfied. We use the positive harmonic function $\hat{h}^{(d)}_{n,n+1}(x)$ for the semigroup $P_t^{n,(2-d)}$ of n independent $BESQ(2-d)$ processes killed when they hit 0 or when they intersect, which transforms them into n non-intersecting $BESQ(d+2)$ processes. Finally observe that, $P_t^{n+1,(d),h^{(d)}_{n,n+1}}$ is exactly the semigroup of $n+1$ $BESQ(d)$ processes conditioned to never intersect (see e.g. [47]). Then, Theorem 13.1 gives the statement of the proposition.

Proposition 13.24 *Let $d \geq 2$. Consider a process $(X, Y) \in W^{n,n}([0,\infty))$ started according to the following distribution $\left(\delta_x, \frac{c_{n,n}(\nu)\hat{h}^{(d)}_{n,n}(y)\prod_{i=1}^n m_{\nu(d)}(y_i)}{h^{(d)}_{n,n}(x)}\mathbf{1}(y \prec x)dy\right)$ for $x \in \mathring{W}^n([0,\infty))$ with the Y particles evolving as n non-intersecting $BESQ(d)$ processes and the X particles evolving as n $BESQ(2-d)$ processes being reflected off the Y particles. Then, the X particles are distributed as n non-intersecting $BESQ(d+2)$ processes started form $x \in \mathring{W}^n([0,\infty))$.*

Proof Take as the L-diffusion a $BESQ(2-d)$ process. Then, the $\hat{L}$-diffusion is a $BESQ(d)$ process. Note that, the assumptions (**R**), (**BC+**) and (**YW**) are satisfied. We use the positive harmonic function $\hat{h}^{(d)}_{n,n}(x)$ for the semigroup $P_t^{n,(d)}$ of n independent $BESQ(d)$ processes killed when they intersect. Furthermore note that, $P_t^{n,(2-d),h^{(d)}_{n,n}}$ is the semigroup of n $BESQ(d+2)$ processes conditioned to never intersect (the transformation by $h^{(d)}_{n,n}$ corresponds to transforming the $BESQ(2-d)$ processes to $BESQ(d+2)$ and then conditioning these to never intersect). Then, the analogue of Theorem 13.1 in $W^{n,n}$ gives the statement.

It is possible to start both of these processes from the origin via the following explicit entrance law for n non-intersecting $BESQ(d)$ processes (see for example [47]),

$$\mu_t^{n,(d)}(dx) = C_{n,d}t^{-n(n+\nu)} \prod_{1\leq i<j\leq n}(x_j - x_i)^2 \prod_{i=1}^n x_i^\nu e^{-\frac{1}{2t}x_i} dx.$$

Defining the two entrance laws,

$$\nu_t^{n,n,\hat{h}^{(d)}_{n,n}}(dx, dy) = \mu_t^{n,(d+2)}(dx)\frac{c_{n,n}(\nu)\hat{h}^{(d)}_{n,n}(y)\prod_{i=1}^n m_{\nu(d)}(y_i)}{h^{(d)}_{n,n}(x)}\mathbf{1}(y \prec x)dy,$$

$$\nu_t^{n,n+1,\hat{h}^{(d)}_{n,n+1}}(dx, dy) = \mu_t^{n+1,(d)}(dx)\frac{n!\prod_{1\leq i<j\leq n}(y_j - y_i)}{\prod_{1\leq i<j\leq n+1}(x_j - x_i)}\mathbf{1}(y \prec x)dy,$$

for the processes with semigroups corresponding to the pair (X, Y) described in Propositions 13.24 and 13.23 respectively, we immediately arrive at the following proposition in analogy to the case of Dyson's Brownian motion:

Proposition 13.25

(a) *Let $d \geq 2$. Consider a process $(X, Y) \in W^{n,n+1}([0,\infty))$ started according to the entrance law $\nu_t^{n,n+1,\hat{h}^{(d)}_{n,n+1}}(dx, dy)$ with the Y particles evolving as n non-intersecting $BESQ(d+2)$ processes and the X particles evolving as $n+1$ $BESQ(d)$ processes being reflected off the Y particles. Then, the X particles are distributed as $n+1$ non-intersecting $BESQ(d)$ processes issueing from the origin.*

(b) *Let $d \geq 2$. Consider a process $(X, Y) \in W^{n,n}([0,\infty))$ started according to the entrance law $\nu_t^{n,n,\hat{h}^{(d)}_{n,n}}(dx, dy)$ with the Y particles which evolve as n non-intersecting $BESQ(d)$ processes and the X particles evolving as n $BESQ(2-d)$ processes being reflected off the Y particles. Then, the X particles are distributed as n non-intersecting $BESQ(d+2)$ processes issueing from the origin.*

Making use of the proposition above we build two processes in Gelfand–Tsetlin patterns. First, the process in $\mathbb{GT}(n)$. To do this, we make repeated use of part **(a)** of Proposition 13.25 to consistently concatenate two-level processes. Note the fact that the dimension d, of the $BESQ(d)$ processes, decreases by 2 at each stage that we increase the number of particles. So we fix n the depth of the Gelfand–Tsetlin pattern and d^* the dimension of the $BESQ$ processes at the bottom of the pattern. Then, we build a consistent process,

$$\left(\mathbb{X}^{n,(d^*)}(t); t \geq 0\right) = (X_i^{(k)}(t); t \geq 0, 1 \leq i \leq k \leq n),$$

taking values in $\mathbb{GT}(n)$ with the joint dynamics described as follows: $X_1^{(1)}$ evolves as a $BESQ(d^* + 2(n-1))$ process. Moreover, for $k \geq 2$ particles at level k evolve as k independent $BESQ(d^* + 2(n-k))$ processes reflecting off the $(k-1)$ particles at the $(k-1)^{th}$ level to maintain the interlacing. Hence, from Proposition 13.9 (see discussion following it regarding the entrance laws) we obtain:

Proposition 13.26 *Let $d \geq 2$. If $\mathbb{X}^{n,(d^*)}$ is started from the origin according to the entrance law then the projection onto the kth level process $X^{(k)}$ is distributed as k $BESQ(d^* + 2(n-k))$ processes conditioned to never intersect.*

By making alternating use of parts **(a)** and **(b)** of Proposition 13.25 we construct a consistent process

$$\left(\mathbb{X}^{n,(d)}(t); t \geq 0\right) = (X^{(1)}(t) \prec \hat{X}^{(1)}(t) \prec \cdots \prec X^{(n)}(t) \prec \hat{X}^{(n)}(t); t \geq 0)$$

in $\mathbb{GT}_{\mathbf{s}}(n)$, for which Proposition 13.17 can be viewed as the $d = 1$ case, and whose joint dynamics are given as follows: $X_1^{(1)}$ evolves as a $BESQ(d)$ process. Then, for any k, the k particles corresponding to $\hat{X}^{(k)}$ evolve as k independent $BESQ(2-d)$ processes reflecting off the particles corresponding to $X^{(k)}$ in order for the interlacing to be maintained. Moreover, for $k \geq 2$ the k particles corresponding to $X^{(k)}$ evolve as k independent $BESQ(d)$ processes reflecting off the particles corresponding to $\hat{X}^{(k-1)}$ in order to maintain the interlacing.

Then, it is a consequence of the symplectic analogue of Proposition 13.9 (involving an entrance law, see the discussion following Proposition 13.9) that:

Proposition 13.27 *Let $d \geq 2$. If $\mathbb{X}^{n,(d)}$ is started from the origin then the projections onto $X^{(k)}$ and $\hat{X}^{(k)}$ are distributed as k non-intersecting $BESQ(d)$ and k non-intersecting $BESQ(d+2)$ processes respectively started from the origin.*

Connection to Wishart Processes We now spell out the connection between the processes constructed above and matrix valued diffusion processes by first considering the connection to $\mathbb{X}^{n,(d^*)}$, for d^* even. Let $d^* = 2$ for simplicity.

Take $(A(t); t \geq 0)$ to be an $n \times n$ complex Brownian matrix and consider $(H(t); t \geq 0) = (A(t)A(t)^*; t \geq 0)$. This is called the Wishart process and was first studied in the real symmetric case by Marie-France Bru in [13], see also [24] for a detailed study in the Hermitian setting and some of its properties. Then, it is well known (first proven in [47]), we have that $\left(\lambda^{(k)}(t); t \geq 0\right)$, the eigenvalues of the $k \times k$ minor of $(H(t); t \geq 0)$, evolve as k non-colliding $BESQ(2(n-k+1))$ processes. These eigenvalues then interlace with $\left(\lambda^{(k-1)}(t); t \geq 0\right)$ which evolve as $k-1$ non-colliding $BESQ(2(n-k+1)+1)$ processes with the *fixed* time T conditional density of $\lambda^{(k-1)}(T)$ given $\lambda^{(k)}(T)$ on $W^{k-1,k}(\lambda^{(k)}(T))$ being $\Lambda_{k-1,k}^{\hat{h}_{k-1,k}^{(d)}}\left(\lambda^{(k)}(T), \cdot\right)$ (see Section 3 of [31], Section 3.3 of [33]). Inductively (since for fixed T, $\lambda^{(n-k)}(T)$ is a Markov chain in k see Section 4 of [31]) this gives that the distribution at *fixed* times T of the vector $(\lambda^{(1)}(T), \cdots, \lambda^{(n)}(T))$ is uniform over the space of $\mathbb{GT}(n)$ with bottom level $\lambda^{(n)}(T)$. Moreover, by making use of this coincidence along *space-like paths* one can write down the dynamical correlation kernel (along space-like paths) of the process we constructed from Theorem 1.3 of [30].

Remark 13.11 Although $\mathbb{X}^{n,(2)}$ and the minor process described in the preceding paragraph on single levels or at fixed times coincide, the interaction between consecutive levels of the minor process should be different from local hard reflection, although the dynamics of consecutive levels of the LUE process have not been studied yet (as far as we know).

We now describe the random matrix model that parallels $\mathbb{X}_s^{n,(d)}$ for d even. Start with a row vector $\left(A^{(d)}(t); t \geq 0\right)$ of $d/2$ independent standard complex Brownian motions, then $\left(X^{(d)}(t); t \geq 0\right) = \left(A^{(d)}(t)A^{(d)}(t)^*; t \geq 0\right)$ evolves as a one dimensional $BESQ(d)$ diffusion (this is really just the definition of a $BESQ(d)$ process). Now, add another independent complex Brownian motion to make $\left(A^{(d)}(t); t \geq 0\right)$ a row vector of length $d/2+1$. Then, $\left(X^{(d)}(t); t \geq 0\right) = \left(A^{(d)}(t)A^{(d)}(t)^*; t \geq 0\right)$ evolves as a $BESQ(d+2)$ process interlacing with the aforementioned $BESQ(d)$. At fixed times, the fact that the conditional distribution of the $BESQ(d)$ process given the position x of the $BESQ(d+2)$ process is proportional to $y^{\frac{d}{2}-1}1_{[0,x]}$ follows from the conditional laws in [27] (see also [23]) and will be spelled out in a few sentences. Now, make $\left(A^{(d)}(t); t \geq 0\right)$ a $2 \times \left(\frac{d}{2}+1\right)$ matrix by adding a row of $d/2+1$ independent complex Brownian motions, the eigenvalues of $\left(X^{(d)}(t); t \geq 0\right) = \left(A^{(d)}(t)A^{(d)}(t)^*; t \geq 0\right)$ evolve as 2 $BESQ(d)$ processes which interlace with the $BESQ(d+2)$. We can continue this construction indefinitely by adding columns and rows successively of independent complex Brownian motions. As before, this eigenvalue process will coincide with $\mathbb{X}_s^{n,(d)}$ on single levels as stochastic processes but also at *fixed* times as distributions of whole interlacing arrays. We elaborate a bit on this fixed time coincidence. For simplicity, let $T = 1$. Let A be an $n \times k$ matrix of independent standard complex normal random variables. Let A' be the $n \times (k+1)$ matrix obtained from A by adding to it a column of independent standard complex normal random variables. Let λ be the n eigenvalues of AA^* and λ' be the n eigenvalues of $A'(A')^*$. We want the conditional density $\rho_{\lambda|\lambda'}(\lambda)$, of λ given λ', with respect to Lebesgue measure. From [27] (see also [23]) the conditional density $\rho_{\lambda'|\lambda}(\lambda)$ is given by,

$$\rho_{\lambda'|\lambda}(\lambda) = \frac{\prod_{1\leq i<j\leq n}(\lambda'_j - \lambda'_i)}{\prod_{1\leq i<j\leq n}(\lambda_j - \lambda_i)} e^{-\sum_{i=1}^n (\lambda'_i - \lambda_i)} \mathbf{1}(\lambda \prec \lambda') .$$

Hence, by Bayes' rule, and recalling the law of the LUE ensemble, we have,

$$\rho_{\lambda|\lambda'}(\lambda) = \left[\frac{\rho_\lambda}{\rho_{\lambda'}} \rho_{\lambda'|\lambda}\right](\lambda) = \frac{\prod_{1\leq i<j\leq n}(\lambda_j - \lambda_i)\prod_{i=1}^n \lambda_i^{\frac{d}{2}-1}}{\prod_{1\leq i<j\leq n}(\lambda'_j - \lambda'_i)\prod_{i=1}^n \lambda_i'^{\frac{d}{2}}} \mathbf{1}(\lambda \prec \lambda') .$$

Similarly to the case of $\mathbb{GT}$, by induction this gives fixed time coincidence of the two $\mathbb{GT}_{\mathbf{s}}$ valued processes.

13.3.8 Diffusions Associated with Orthogonal Polynomials

Here, we consider three diffusions in Gelfand–Tsetlin patterns associated with the classical orthogonal polynomials, Hermite, Laguerre and Jacobi. Although the one dimensional diffusion processes these are built from, the Ornstein–Uhlenbeck, the Laguerre and Jacobi are special cases of Sturm-Liouville diffusions with discrete spectrum, which we will consider in the next subsection, they are arguably the most interesting examples, with close connections to random matrices and so we consider them separately (for the classification of one dimensional diffusion operators with polynomial eigenfunctions see [51] and for a nice exposition Section 2.7 of [5]). One of the common features of the Karlin–McGregor semigroups associated with them is that they all have the Vandermonde determinant as their ground state (this follows from Corollary 13.3). At the end of this subsection we describe the connection to eigenvalue processes of minors of matrix diffusions.

Definition 13.8 The Ornstein–Uhlenbeck (OU) diffusion process in $I = \mathbb{R}$ has generator and SDE description,

$$L_{OU} = \frac{1}{2}\frac{d^2}{dx^2} - x\frac{d}{dx},$$
$$dX(t) = dB(t) - X(t)dt,$$

with $m_{OU}(x) = e^{-x^2}$ and $-\infty$ and ∞ both natural boundaries. Its conjugate diffusion process $\hat{L}_{OU}$ has generator and SDE description,

$$\hat{L}_{OU} = \frac{1}{2}\frac{d^2}{dx^2} + x\frac{d}{dx},$$
$$d\hat{X}(t) = dB(t) + \hat{X}(t)dt,$$

and again $-\infty$ and ∞ are both natural boundaries and note the drift away from the origin.

Definition 13.9 The Laguerre $Lag(\alpha)$ diffusion process in $I = [0, \infty)$ has generator and SDE description,

$$L_{Lag(\alpha)} = 2x\frac{d^2}{dx^2} + (\alpha - 2x)\frac{d}{dx},$$
$$dX(t) = 2\sqrt{X(t)}dB(t) + (\alpha - 2X(t))dt,$$

with $m_{Lag(\alpha)}(x) = x^{\alpha/2}e^{-x}$ and ∞ being natural and for $\alpha \geq 2$ the point 0 is an entrance boundary. We will only be concerned with such values of α here.

Definition 13.10 The Jacobi diffusion process $Jac(\beta, \gamma)$ in $I = [0, 1]$ has generator and SDE description,

$$L_{Jac(\beta,\gamma)} = 2x(1-x)\frac{d^2}{dx^2} + 2(\beta - (\beta+\gamma)x)\frac{d}{dx},$$
$$dX(t) = 2\sqrt{X(t)(1-X(t))}dB(t) + 2(\beta - (\beta+\gamma)X(t))dt,$$

with $m_{Jac(\beta,\gamma)}(x) = x^{\beta-1}(1-x)^{\gamma-1}$ and 0 and 1 being entrance for $\beta, \gamma \geq 1$. We will only be concerned with such values of β and γ in this section.

The restriction of parameters α, β, γ for $Lag(\alpha)$ and $Jac(\beta, \gamma)$ is so that (**BC+**) is satisfied (for a certain range of the parameters the points 0 and/or 1 are regular boundaries in which case (**BC+**) is no longer satisfied due to the fact that the diffusion coefficients degenerate at the boundary points).

We are interested in the construction of a process in $\mathbb{GT}(N)$, so that in particular at each stage the number of particles increases by one. We start in the simplest setting of $W^{1,2}$ and in particular the Ornstein–Uhlenbeck case to explain some subtleties. We will then treat all cases uniformly.

Consider a two-level process (X, Y) with the X particles evolving as two OU processes being reflected off the Y particle which evolves as an $\hat{L}_{OU}$ diffusion. Then, since this is an honest Markov process, Theorem 13.1 (whose conditions are easily seen to be satisfied) gives that if started appropriately, the projection on the X particles is Markovian with semigroup $P_t^{2,OU,\bar{h}_2}$. Here, $P_t^{2,OU,\bar{h}_2}$ is the Doob h-transformed semigroup of two independent OU processes killed when they intersect by the harmonic function $\bar{h}_2$:

$$\bar{h}_2(x_1, x_2) = \int_{x_1}^{x_2} \hat{m}_{OU}(y)dy = s_{OU}(x_2) - s_{OU}(x_1),$$

where $s_{OU}(x) = e^{x^2}F(x)$ is the scale function of the OU process and $F(x) = e^{-x^2}\int_0^x e^{y^2}dy$ is the Dawson function. We note that, although this process is built from two OU processes being kept apart (more precisely this diffusion lives in $\mathring{W}^2$), it is *not two independent OU processes conditioned to never intersect.*

However, we can initially h-transform the $\hat{L}_{OU}$ process to make it an OU process with the h-transform given by $\hat{h}_1(x) = \hat{m}_{OU}^{-1}(x)$ with eigenvalue -1. Now, note that:

$$h_2(x_1, x_2) = \int_{x_1}^{x_2} \hat{m}_{OU}(y)\hat{m}_{OU}^{-1}(y)dy = (x_2 - x_1).$$

This, as we see later in Corollary 13.3 is the ground state of the semigroup associated to two independent OU processes killed when they intersect. Thus, if we consider a two-level process (X, Y) with the X particles evolving as 2 OU processes reflected off a single OU process, we get from Theorem 13.1 that the projection on the X particles is distributed as two independent OU processes conditioned to never intersect via a Doob h-transform by h_2.

Similarly, an easy calculation gives that we can h-transform the $\hat{L}_{Lag(\alpha)}$-diffusion to make it a $Lag(\alpha + 2)$ with the h-transform being $\hat{m}_{Lag(\alpha)}^{-1}(x)$ with eigenvalue -2 and h-transform with $\hat{m}_{Jac(\beta,\gamma)}^{-1}(x)$ with eigenvalue $-2(\beta + \gamma)$ the $\hat{L}_{Jac(\beta,\gamma)}$-diffusion to make it a $Jac(\beta + 1, \gamma + 1)$ to obtain the analogous result. Furthermore, this generalizes to arbitrary n. First, let

$$h_{n+1}(x) = \frac{1}{n!} \prod_{1 \le i < j \le n+1} (x_j - x_i)$$

denote the Vandermonde determinant. By Corollary 13.3, h_{n+1} is the ground state of the semigroup associated to $n + 1$ independent copies of an OU or $Lag(\alpha)$ or $Jac(\beta, \gamma)$ diffusion killed when they intersect.

Proposition 13.28 *Assume the constants α, β, γ satisfy $\alpha \geq 2, \beta \geq 1, \gamma \geq 1$. Let (X, Y) be a two-level diffusion process in $W^{n,n+1}(I^\circ)$ started according to the distribution $(\delta_x, \frac{n! \prod_{1\leq i<j\leq n}(y_j - y_i)}{\prod_{1\leq i<j\leq n+1}(x_j - x_i)} \mathbf{1}(y \prec x)dy)$, where $x \in \mathring{W}^{n+1}(I)$, and X and Y evolving as follows:*

OU: *X as $n+1$ independent OU processes reflected off Y which evolves as n OU processes conditioned to never intersect via a Doob h-transform by h_n,*
Lag: *X as $n+1$ independent $Lag(\alpha)$ processes reflected off Y which evolves as n $Lag(\alpha+2)$ processes conditioned to never intersect via a Doob h-transform by h_n,*
Jac: *X as $n+1$ independent $Jac(\beta, \gamma)$ processes reflected off Y which evolves as n $Jac(\beta+1, \gamma+1)$ processes conditioned to never intersect via a Doob h-transform by h_n.*

Then, the X particles are distributed as,

OU: *$n+1$ OU processes conditioned to never intersect via a Doob h-transform by h_{n+1},*
Lag: *$n+1$ $Lag(\alpha)$ processes conditioned to never intersect via a Doob h-transform by h_{n+1},*
Jac: *$n+1$ $Jac(\beta, \gamma)$ processes conditioned to never intersect via a Doob h-transform by h_{n+1}, started from x.*

Proof We take as the L-diffusion an OU or $Lag(\alpha)$ or $Jac(\beta, \gamma)$ diffusion respectively. Note that, the assumptions **(R)**, **(BC+)** and **(YW)** are satisfied for $Lag(\alpha)$ for $\alpha \geq 2$ and for $Jac(\beta, \gamma)$ for $\beta \geq 1, \gamma \geq 1$ (also these assumptions are clearly satisfied for an OU process). Furthermore, observe that with,

$$\hat{h}_n(x) = \prod_{i=1}^{n} \hat{m}^{-1}(x) \prod_{1\leq i<j\leq n} (x_j - x_i),$$

we have:

$$h_{n+1}(x) = (\Lambda_{n,n+1} \Pi_{n,n+1} \hat{h}_n)(x) = \frac{1}{n!} \prod_{1\leq i<j\leq n+1} (x_j - x_i).$$

Moreover, note that the semigroup of n independent copies of an $\hat{L}$-diffusion (namely either an $\hat{L}_{OU}$ or $\hat{L}_{Lag(\alpha)}$ or $\hat{L}_{Jac(\beta,\gamma)}$ diffusion) killed when they intersect h-transformed by $\hat{h}_n$ is exactly the semigroup corresponding to Y in the statement of the proposition. Finally, making use of Theorem 13.1 we obtain the required statement.

It is rather easy to see how to iterate this construction to obtain a consistent process in a Gelfand–Tsetlin pattern. To be precise, let us fix N the depth of the pattern and constants $\alpha \geq 2$, $\beta \geq 1$ and $\gamma \geq 1$ that will be the parameters of the processes at the bottom row. Then, in the Ornstein–Uhlenbeck case level k evolves as k independent OU processes reflected off the paths at level $k-1$. In the Laguerre case level k evolves as k independent $Lag(\alpha+2(N-k))$ processes reflected off the particles at level $k-1$. Finally, in the Jacobi case level k evolves as k independent $Jac(\beta+(N-k), \gamma+(N-k))$ processes reflected off the particles at level $(k-1)$. The result giving the distribution of the projection on each level (under certain initial conditions) is completely analogous to previous sections and we omit the statement.

Remark 13.12 In the $Laguerre$ case we can build in a completely analogous way a process in $\mathbb{GT}_\mathbf{s}$ in analogy to the $BESQ(d)$ case of Proposition 13.27. In the Jacobi case (with $\beta, \gamma \geq 1$) we can build a process $(X, Y) \in W^{n,n}((0,1))$ started from the origin (according to the entrance law) with the Y particles evolving as n non-intersecting $Jac(\beta, \gamma+1)$ and the X particles as n $Jac(1-\beta, \gamma)$ in $(0,1)$ reflected off the Y particles. Then, the X particles are distributed as n non-intersecting $Jac(\beta+1, \gamma)$ processes started from the origin.

Connection to Random Matrices We now make the connection to the eigenvalues of matrix valued diffusion processes associated with orthogonal polynomials. The relation for the Ornstein–Uhlenbeck process and $Lag(d)$ processes we constructed is the same as for Brownian motions and $BESQ(d)$ processes. The only difference being, that we replace the complex Brownian motions by complex Ornstein–Uhlenbeck processes in the matrix valued diffusions (the only difference being, that this introduces a restoring $-x$ drift in both the matrix valued diffusion processes and the $SDEs$ for the eigenvalues).

We now turn to the Jacobi minor process. First, following Doumerc's PhD thesis [28] (see in particular Section 9.4.3 therein) we construct the matrix Jacobi diffusion as follows. Let $(U(t), t \geq 0)$ be a Brownian motion on $\mathbb{U}(N)$, the manifold of $N \times N$ unitary matrices and let $p+q=N$. Let n be such that $n \leq p, q$ and consider $(H(t), t \geq 0)$ the projection onto the first n rows and p columns of $(U(t), t \geq 0)$. Then $(J^{p,q}(t), t \geq 0) = (H(t)H(t)^*, t \geq 0)$ is defined to be the $n \times n$ matrix Jacobi diffusion (with parameters p, q). Its eigenvalues evolve as n non-colliding $Jac(p-(n-1), q-(n-1))$ diffusions. Its $k \times k$ minor is built by projecting onto the first k rows of $(U(t), t \geq 0)$ and it has eigenvalues $\left(\lambda^{(k)}(t), t \geq 0\right)$ that evolve as k non-colliding $Jac(p-(n-1)+n-k, q-(n-1)+n-k)$. For fixed times T, if $(U(t), t \geq 0)$ is started according to Haar measure, the distribution of $\lambda^{(k-1)}(T)$ given $\lambda^{(k)}(T)$ on $W^{k-1,k}(\lambda^{(k)}(T))$ being $\Lambda_{k-1,k}^{\hat{h}_{k-1}}\left(\lambda^{(k)}(T), \cdot\right)$ see e.g. [33]. For the connection to the process in $W^{n,n}$ described in the remark, we could have projected on the first n rows and $p+1$ columns of $(U(t), t \geq 0)$ and denoting that by $\left(H(t)', t \geq 0\right)$, then $\left(J^{p+1,q-1}(t), t \geq 0\right) = \left(H(t)'(H(t)')^*, t \geq 0\right)$ has eigenvalues evolving as n non-colliding $Jac(p-(n-1)+1, q-(n-1)-1)$ and those interlace with the eigenvalues of $(J^{p,q}(t), t \geq 0)$.

Remark 13.13 Non-colliding Jacobi diffusions have also appeared in the work of Gorin [36] as the scaling limits of some natural Markov chains on the Gelfand–Tsetlin graph in relation to the harmonic analysis of the infinite unitary group $\mathbb{U}(\infty)$.

13.3.9 Diffusions with Discrete Spectrum

13.3.9.1 Spectral Expansion and Ground State of the Karlin–McGregor Semigroup

In this subsection, we show how the diffusions associated with the classical orthogonal polynomials and the Brownian motions in an interval are special cases of a wider class of one dimensional diffusion processes with explicitly known minimal eigenfunctions for the Karlin–McGregor semigroups associated with them. We start by considering the diffusion process generator L with *discrete spectrum* $0 \geq -\lambda_1 > -\lambda_2 > \cdots$ (the absence of natural boundaries is sufficient for this, see for example Theorem 3.1 of [52]) with speed measure m and transition density given by $p_t(x, dy) = q_t(x, y)m(dy)$ where,

$$L\phi_k(x) = -\lambda_k\phi_k(x),$$

$$q_t(x, y) = \sum_{k=1}^{\infty} e^{-\lambda_k t}\phi_k(x)\phi_k(y).$$

The eigenfunctions $\{\phi_k\}_{k\geq 1}$ form an orthonormal basis of $L^2(I, m(dx))$ and the expansion $\sum_{k=1}^{\infty} e^{-\lambda_k t}\phi_k(x)\phi_k(y)$ converges uniformly on compact squares in $I^\circ \times I^\circ$. Furthermore, the Karlin–McGregor semigroup transition density with respect to $\prod_{i=1}^{n} m(dy_i)$ is given by,

$$\det(q_t(x_i, y_j))_{i,j=1}^n.$$

We now obtain an analogous spectral expansion for this. We start by expanding the determinant to get,

$$\begin{aligned}
\det(q_t(x_i, y_j))_{i,j=1}^n &= \sum_{\sigma\in\mathfrak{S}_n} sign(\sigma)\prod_{i=1}^n q_t(x_i, y_{\sigma(i)}) \\
&= \sum_{k_1,\cdots,k_n}\prod_{i=1}^n \phi_{k_i}(x_i)e^{-\lambda_{k_i}t}\sum_{\sigma\in\mathfrak{S}_n} sign(\sigma)\prod_{i=1}^n \phi_{k_i}(y_{\sigma(i)}) \\
&= \sum_{k_1,\cdots,k_n}\prod_{i=1}^n \phi_{k_i}(x_i)e^{-\lambda_{k_i}t}\det(\phi_{k_i}(y_j))_{i,j=1}^n.
\end{aligned}$$

Write $\phi_{\mathsf{k}}(y)$ for $\det(\phi_{k_i}(y_j))_{i,j=1}^n$ for an n-tuple $\mathsf{k} = (k_1, \cdots, k_n)$ and also λ_{k} for $(\lambda_{k_1}, \cdots, \lambda_{k_n})$ and note that we can restrict to $k_1, \cdots, k_n$ distinct otherwise the determinant vanishes. In fact we can restrict to $k_1, \cdots, k_n$ ordered by replacing

$k_1, \cdots, k_n$ by $k_{\tau(1)}, \cdots, k_{\tau(n)}$ and summing over $\tau \in \mathfrak{S}_n$ to obtain, with $|\lambda_{\mathsf{k}}| = \sum_{i=1}^n \lambda_{k_i}$:

$$\det(q_t(x_i, y_j))_{i,j=1}^n = \sum_{1 \le k_1 < \cdots < k_n} e^{-|\lambda_{\mathsf{k}}|t} \phi_{\mathsf{k}}(x)\phi_{\mathsf{k}}(y). \tag{13.39}$$

The expansion is converging uniformly on compacts in $W^n(I^\circ) \times W^n(I^\circ)$ for $t > 0$. Now, denoting by T the lifetime of the process we obtain, for $x = (x_1, \cdots, x_n) \in \mathring{W}^n(I)$, the following spectral expansion that converges uniformly on compacts in $x \in W^n(I^\circ)$,

$$\mathbb{P}_x(T > t) = \left[P_t^n \mathbf{1}\right](x) = \sum_{1 \le k_1 < \cdots < k_n} e^{-|\lambda_{\mathsf{k}}|t} \phi_{\mathsf{k}}(x) \langle \phi_{\mathsf{k}}, \mathbf{1} \rangle_{W^n(m)} \tag{13.40}$$

where we used the notation:

$$\langle f, g \rangle_{W^n(m)} = \int_{W^n(I^\circ)} f(x_1, \cdots, x_n) g(x_1, \cdots, x_n) \prod_{i=1}^n m(x_i) dx_i.$$

So, as $t \to \infty$ by the fact that the eigenvalues are distinct and ordered the leading exponential term is forced to be $k_i = i$ and thus:

$$\mathbb{P}_x(T > t) = \langle \phi_{(1,\cdots,n)}, \mathbf{1} \rangle_{W^n(m)} \times e^{-\sum_{i=1}^n \lambda_i t} \det(\phi_i(x_j))_{i,j=1}^n + o\left(e^{-\sum_{i=1}^n \lambda_i t}\right), \text{ as } t \to \infty.$$

Hence, we can state the following corollary.

Corollary 13.3 *The function,*

$$h_n(x) = \det(\phi_i(x_j))_{i,j=1}^n \tag{13.41}$$

is the ground state of P_t^n.

The above argument proves that $h_n(x) \ge 0$ but in fact the positivity is strict, $h_n(x) > 0$ for all $x \in W^n(I^\circ)$ which can be seen as follows. We have the eigenfunction relation, by the Andreif (or generalized Cauchy–Binet) identity:

$$\int_{W^n(I^\circ)} \det(q_t(x_i, y_j))_{i,j=1}^n \det(\phi_i(y_j))_{i,j=1}^n \prod_{i=1}^n m(y_i) dy_i = e^{-\sum_{i=1}^n \lambda_i t} \det(\phi_i(x_j))_{i,j=1}^n.$$

Assume that $\det(\phi_i(x_j))_{i,j=1}^n = 0$ for some $x \in W^n(I^\circ)$. Then, by the *strict* positivity of $\det(q_t(x_i, y_j))_{i,j}^n \prod_{i=1}^n m(y_i) > 0$ and continuity of $h_n(x)$ (see Theorem 4 of [45], also Problem 6 and its solution on pages 158–159 of [39]), the determinant $\det(\phi_i(y_j))_{i,j=1}^n$ must necessarily vanish everywhere in $W^n(I^\circ)$. Hence, we can write for all $x \in I^\circ$ $\phi_n(x) = \sum_{i=1}^{n-1} a_i\phi_i(x)$ for some constants a_i. However, this contradicts the orthonormality of the eigenfunctions and so $h_n(x) > 0$ for all $x \in W^n(I^\circ)$.

A different way to see that $h_n(x)$ is strictly positive (up to a constant) in $\mathring{W}^n(I)$ is the well known fact (see paragraph immediately after Theorem 6.2 of Chapter 1 on page 36 of [44]) that the eigenfunctions coming from Sturm-Liouville operators form a Complete T-system (CT-system) or *Chebyshev* system namely $\forall n \geq 1$,

$$h_n(x) = \det(\phi_i(x_j))_{i,j=1}^n > 0, \ x \in \mathring{W}^n(I).$$

Remark 13.14 In fact a CT-system requires that the determinant does not vanish in $W^n(I)$ so w.l.o.g multiplying by -1 if needed we can assume it is positive.

For the orthogonal polynomial diffusions and Brownian motions in an interval taking the ϕ_j's to be the Hermite, Laguerre, Jacobi polynomials (which via row and column operations give the Vandermonde determinant) and trigonometric functions (of increasing frequencies) we obtain the minimal eigenfunction.

Following this discussion, we can thus define the *conditioned semigroup* with transition kernel p_t^{n,h_n} with respect to Lebesgue measure in $W^n(I^\circ)$ as follows,

$$p_t^{n,h_n}(x, y) = e^{\sum_{i=1}^n \lambda_i t} \frac{\det(\phi_i(y_j))_{i,j=1}^n}{\det(\phi_i(x_j))_{i,j=1}^n} \det(p_t(x_i, y_j))_{i,j=1}^n.$$

13.3.9.2 Conditioning Diffusions for Non-intersection Through Local Interactions

Now, a natural question arising is the following. When is it possible to obtain n conservative (by that we mean in case l or r can be reached then they are forced to be regular reflecting) L-diffusions *conditioned via the minimal positive eigenfunction* to never intersect through the hard reflection interactions we have been studying in this work? We are able to provide an answer in Proposition 13.29 below under a certain assumption that we now explain.

First, note that L being conservative implies $\phi_1 = 1$. Furthermore, assuming that the $\phi_k \in C^{n-1}(I^\circ)$ for $1 \leq k \leq n$ and denoting by $\phi_k^{(j)}$ their jth derivative we define the Wronskian $W(\phi_1, \cdots, \phi_n)(x)$ of $\phi_1, \cdots, \phi_n$ by,

$$W(\phi_1, \cdots, \phi_n)(x) = \det\left(\phi_i^{(j-1)}(x)\right)_{i,j=1}^n.$$

Then, we say that $\{\phi_j\}_{j=1}^n$ form a (positive) Extended Complete T-system or ECT-system if for all $1 \le k \le n$,

$$W(\phi_1, \cdots, \phi_k)(x) > 0, \ \forall x \in I^\circ.$$

This is a stronger property, in particular implying that $\{\phi_j\}_{j=1}^n$ form a CT-system (see Theorem 2.3 of Chapter 2 of [44]). Assuming that the eigenfunctions in question $\{\phi_j\}_{j=1}^n$ form a (positive) ECT-system then since $\phi_1 = 1$,

$$W(\phi_2^{(1)}, \cdots, \phi_n^{(1)})(x) > 0, \ \forall x \in I^\circ,$$

and hence,

$$\hat{h}_{n-1}(x) := \det(\mathcal{D}_{\hat{m}}\phi_{i+1}(x_j))_{i,j=1}^{n-1} > 0, \ x \in \mathring{W}^{n-1}(I). \tag{13.42}$$

We then have the following positive answer for the question we stated previously:

Proposition 13.29 *Under the conditions of Theorem 13.1, furthermore assume that the generator L has discrete spectrum and its first n eigenfunctions $\{\phi_j\}_{j=1}^n$ form an ECT-system. Now assume that the X particles consist of n independent L-diffusions reflected off the Y particles which evolve as an $n-1$ dimensional diffusion with semigroup $P_t^{n-1,\hat{h}_{n-1}}$, where $\hat{h}_{n-1}$ is defined in (13.42). Then, the X particles (if the two-level process is started appropriately) are distributed as n independent L-diffusions conditioned to never intersect with semigroup P_t^{n,h_n}, where h_n is defined by (13.41).*

Proof Making use of the relations $\mathcal{D}_{\hat{m}} = \mathcal{D}_s$ and $\mathcal{D}_{\hat{s}} = \mathcal{D}_m$ between the diffusion process generator L and its dual we obtain,

$$\hat{L}\mathcal{D}_{\hat{m}}\phi_i = \mathcal{D}_{\hat{m}}\mathcal{D}_{\hat{s}}\mathcal{D}_{\hat{m}}\phi_i = \mathcal{D}_{\hat{m}}\mathcal{D}_m\mathcal{D}_s\phi_i = -\lambda_i\mathcal{D}_{\hat{m}}\phi_i.$$

Thus, $\left(e^{\lambda_i t}\mathcal{D}_{\hat{m}}\phi_i(\hat{X}(t)); t \ge 0\right)$ for each $1 \le i \le n$ is a local martingale. By virtue of boundedness (since we assume that the L-diffusion is conservative we have $\lim_{x\to l,r} \mathcal{D}_{\hat{m}}\phi_i(x) = \lim_{x\to l,r} \mathcal{D}_s\phi_i(x) = 0$) it is in fact a true martingale and so for $1 \le i \le n$,

$$\hat{P}_t^1\mathcal{D}_{\hat{m}}\phi_i = e^{-\lambda_i t}\mathcal{D}_{\hat{m}}\phi_i. \tag{13.43}$$

Then, by the well-known Andreif (or generalized Cauchy–Binet) identity we obtain,

$$\hat{P}_t^{n-1}\hat{h}_{n-1} = e^{-\sum_{i=1}^{n-1}\lambda_{i+1}t}\hat{h}_{n-1}$$

and thus $\hat{h}_{n-1}$ is a strictly positive eigenfunction for $\hat{P}_t^{n-1}$. Finally, by performing a simple integration we see that,

$$(\Lambda_{n-1,n}\Pi_{n-1,n}\hat{h}_{n-1})(x) = const_n h_n(x), \ x \in W^n(I).$$

Using Theorem 13.1 we obtain the statement of the proposition.

Obviously the diffusions associated with orthogonal polynomials and Brownian motions in an interval fall under this framework.

13.3.10 Eigenfunctions via Intertwining

In this short subsection we point out that all eigenfunctions for n copies of a diffusion process with generator L in W^n (not necessarily diffusions with discrete spectrum e.g. Brownian motions or $BESQ(d)$ processes) that are obtained by iteration of the intertwining kernels considered in this work, or equivalently from building a process in a Gelfand–Tsetlin pattern, are of the form,

$$\mathfrak{H}_n(x_1, \cdots, x_n) = \det\left(h_i^{(n)}(x_j)\right)_{i,j=1}^n, \tag{13.44}$$

for functions $\left(h_1^{(n)}, \cdots, h_n^{(n)}\right)$ (not necessarily the eigenfunctions of a one dimensional diffusion operator) given by,

$$h_i^{(n)}(x) = w_1^{(n)}(x)\int_c^x w_2^{(n)}(\xi_1)\int_c^{\xi_1} w_3^{(n)}(\xi_2)\cdots\int_c^{\xi_{i-2}} w_i^{(n)}(\xi_{i-1})d\xi_{i-1}\cdots d\xi_1, \tag{13.45}$$

for some weights $w_i^{(n)}(x) > 0$ and $c \in I^\circ$. An easy consequence of the representation above (see e.g. Theorem 1.1 of Chapter 6 of [44]) and assuming $w_i^{(n)} \in C^{n-i}(l, r)$ ($n-i$ times continuously differentiable) is that the Wronskian $W\left(h_1^{(n)}, \cdots, h_n^{(n)}\right)$ is given by for $x \in I^\circ$,

$$W\left(h_1^{(n)}, \cdots, h_n^{(n)}\right)(x) = \left[w_1^{(n)}(x)\right]^n\left[w_2^{(n)}(x)\right]^{n-1}\cdots\left[w_n^{(n)}(x)\right], \tag{13.46}$$

so that in particular $W\left(h_1^{(n)}, \cdots, h_n^{(n)}\right)(x) > 0$.

We shall restrict to the case of $\mathbb{GT}(n)$ (where the number of particles on each level increases by 1) for simplicity and prove claims (13.44) and (13.45) by induction. For $n = 1$ there is nothing to prove. We conclude by stating and proving the inductive step as a precise proposition:

Proposition 13.30 *Assume that the input, strictly positive, eigenfunction* $\mathfrak{H}_{n-1}$ *for* $n-1$ *copies of a one dimensional diffusion process is of the form (13.44) and (13.45). Then, the eigenfunction* $\mathfrak{H}_n$ *built from the intertwining relation of Karlin–McGregor semigroups (13.26) for* n *copies of its dual diffusion has the same form (13.44) and (13.45), with the weights* $\{w_i^{(n)}\}_{i=1}^n$ *satisfying an explicit recursion in terms of the* $\{w_i^{(n-1)}\}_{i=1}^{n-1}$.

Proof In order to obtain a strictly positive eigenfunction for n copies of an L-diffusion, we can in fact start more generally with n copies of an L-diffusion h-transformed by a one dimensional strictly positive eigenfunction h (denoting by L^h such a diffusion process where we assume that L^h satisfies the boundary conditions of Sect. 13.2 in order for the intertwining (13.26) to hold). It is then clear that:

$$\mathfrak{H}_n(x_1, \cdots, x_n) = \prod_{i=1}^{n} h(x_i)(\Lambda_{n-1,n}\mathfrak{H}_{n-1})(x_1, \cdots, x_n), \tag{13.47}$$

where now $\mathfrak{H}_{n-1}(x_1, \cdots, x_{n-1})$ is a strictly positive eigenfunction of $n-1$ copies of an $\widehat{L^h}$ diffusion and which by our hypothesis is given by,

$$\mathfrak{H}_{n-1}(x_1, \cdots, x_{n-1}) = \det\left(h_i^{(n-1)}(x_j)\right)_{i,j=1}^{n-1}, \tag{13.48}$$

for some functions $\left(h_1^{(n-1)}, \cdots, h_{n-1}^{(n-1)}\right)$ with a representation as in (13.45) for some weights $\{w_i^{(n-1)}\}_{i\le n-1}$. A simple integration now gives,

$$h_1^{(n)}(x) = h(x),$$

$$h_i^{(n)}(x) = h(x)\int_c^x \widehat{m^h}(y)h_{i-1}^{(n-1)}(y)dy, \quad \text{for } i \ge 2,$$

where $\widehat{m^h}(x) = h^{-2}(x)s'(x)$ is the density of the speed measure of a $\widehat{L^h}$ diffusion. We thus obtain the following recursive representation for the weights $\{w_i^{(n)}\}_{i\le n}$,

$$w_1^{(n)}(x) = h(x), \tag{13.49}$$

$$w_2^{(n)}(x) = h^{-2}(x)s'(x)w_1^{(n-1)}(x), \tag{13.50}$$

$$w_i^{(n)}(x) = w_{i-1}^{(n-1)}(x), \quad \text{for } i \ge 3. \tag{13.51}$$

13.3.11 Connection to Superpositions and Decimations

For particular entrance laws, the joint law of X and Y at a fixed time can be interpreted in terms of superpositions/decimations of random matrix ensembles (see e.g. [34]). For example, in the context of Proposition 13.11, the joint law of X and Y at time 1 agrees with the joint law of the odd (respectively even) eigenvalues in a superposition of two independent samples from the GOE_{n+1} and GOE_n ensembles, consistent with the fact that in such a superposition, the odd (respectively even) eigenvalues are distributed according to the GUE_{n+1} (respectively GUE_n) ensembles, see Theorem 5.2 in [34]. In the BESQ/Laguerre case, our Proposition 13.25 is similarly related to recent work on GOE singular values by Bornemann and La Croix [8] and Bornemann and Forrester [7].

13.3.12 Connection to Strong Stationary Duals

Strong stationary duality (SSD) first introduced by Diaconis and Fill [25] in the discrete state space setting is a fundamental notion in the study of strong stationary times which are a key tool in understanding mixing times of Markov Chains. More recently, Fill and Lyzinski [32] developed an analogous theory for diffusion processes in compact intervals. Given a conservative diffusion $\mathcal{G}$ one associates to it a SSD $\mathcal{G}^*$ such that the two semigroups are intertwined (see Definition 3.1 there). In Theorem 3.4 therein the form of the dual generator is derived and as already indicated in Remark 5.4 in the same paper this is exactly the dual diffusion $\hat{\mathcal{G}}$ h-transformed by its scale function.

In our framework, considering a two-level process in $W^{1,1}$ with $L = \hat{\mathcal{G}}$ and so $\hat{L} = \mathcal{G}$ and using the positive harmonic function $\hat{h}_1 \equiv 1$, the distribution of the projection on the X particle (under certain initial conditions) coincides with the SSD $\mathcal{G}^*$ diffusion. Hence this provides a coupling of a diffusion $\mathcal{G}$ and its strong stationary dual $\mathcal{G}^*$ respecting the intertwining between $\mathcal{G}$ and $\mathcal{G}^*$.

13.4 Edge Particle Systems

In this section we will study the autonomous particle systems at either edge of the Gelfand–Tsetlin pattern valued processes we have constructed. In the figure below, the particles we will be concerned with are denoted in •.

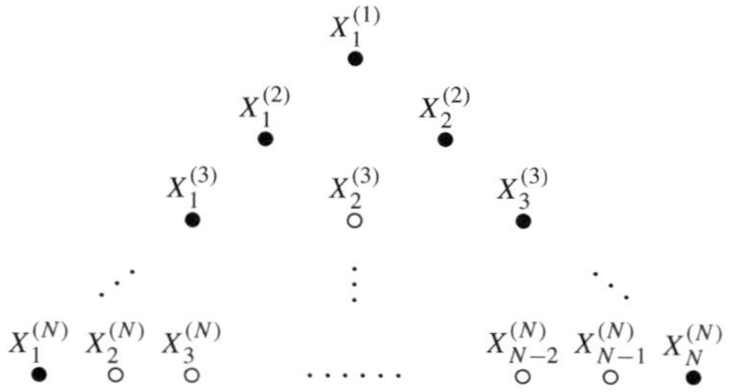

Our goal is to derive determinantal expressions for their transition densities. Such expressions were derived by Schutz for TASEP in [59] and later Warren [66] for Brownian motions. See also Johansson's work in [40], for an analogous formula for a Markov chain related to the Meixner ensemble and finally Dieker and Warren's investigation in [26], for formulae in the discrete setting based on the RSK correspondence. These so called Schutz-type formulae were the starting points for the recent complete solution of TASEP in [50] which led to the KPZ fixed point and also for the recent progress [41] in the study of the two time joint distribution in Brownian directed percolation. For a detailed investigation of the Brownian motion model the reader is referred to the book [69].

We will mainly restrict ourselves to the consideration of Brownian motions, $BESQ(d)$ processes and the diffusions associated with orthogonal polynomials. In a little bit more generality we will assume that the interacting diffusions have generators of the form,

$$L = a(x)\frac{d^2}{dx^2} + b(x)\frac{d}{dx},$$

with,

$$a(x) = a_0 + a_1 x + a_2 x^2 \quad b(x) = b_0 + b_1 x.$$

We will also make the following **standing assumption** in this section. We restrict to the case of the boundaries of the state space I being either *natural* or *entrance* thus the state space is an open interval (l, r). Under these assumptions the transition densities will be smooth in (l, r) in both the backwards and forwards variables (possibly blowing up as we approach l or r see e.g [62] and for a detailed study of the transition densities of the Wright–Fisher diffusion see [19]). This covers all the processes we built that relate to minor processes of matrix diffusions. This interacting particle system can also be seen as the solution to the following system of SDE's with one-sided collisions with $(x_1^1 \le \cdots \le x_n^n)$,

$$\begin{aligned}
X_1^{(1)}(t) &= x_1^1 + \int_0^t \sqrt{2a(X_1^{(1)}(s))}d\gamma_1^1(s) + \int_0^t b^{(1)}(X_1^{(1)}(s))ds,\\
&\vdots\\
X_m^{(m)}(t) &= x_m^m + \int_0^t \sqrt{2a(X_m^{(m)}(s))}d\gamma_m^m(s) + \int_0^t b^{(m)}(X_m^{(m)}(s))ds + K_m^{m,-}(t),
\end{aligned} \tag{13.52}$$

$$\begin{aligned}
&\vdots\\
X_n^{(n)}(t) &= x_n^n + \int_0^t \sqrt{2a(X_n^{(n)}(s))}d\gamma_n^n(s) + \int_0^t b^{(n)}(X_n^{(n)}(s))ds + K_n^{n,-}(t).
\end{aligned}$$

where γ_i^i are independent standard Brownian motions and $K_i^{i,-}$ are positive finite variation processes with the measure $dK_i^{i,-}$ supported on $\left\{t : X_i^{(i)}(t) = X_{i-1}^{(i-1)}(t)\right\}$ and

$$b^{(k)}(x) = b(x) + (n-k)a'(x) = b_0 + (n-k)a_1 + (b_1 + 2(n-k)a_2)x.$$

That these SDE's are well-posed, so that in particular the solution is Markov, follows from the same arguments as in Sect. 13.5.1. Note that, a quadratic diffusion coefficient $a(\cdot)$ and linear drift $b(\cdot)$ satisfy (**YW**). See the following figure for a description of the interaction. The arrows indicate the direction of the 'pushing force' (with magnitude the finite variation process K) applied when collisions occur between the particles so that the ordering is maintained.

$$\overset{X_1^{(1)}}{\bullet} \longrightarrow \overset{X_2^{(2)}}{\bullet} \longrightarrow \overset{X_3^{(3)}}{\bullet} \cdots \overset{X_{n-1}^{(n-1)}}{\bullet} \longrightarrow \overset{X_n^{(n)}}{\bullet}.$$

Note that our assumption that the boundary points are either *entrance* or *natural* does not always allow for an *infinite* such particle system,in particular think of the $BESQ(d)$ case where d drops down by 2 each time we add a particle. Denote by $p_t^{(k)}(x, y)$ the transition kernel associated with the $L^{(k)}$-diffusion with generator,

$$L^{(k)} = a(x)\frac{d^2}{dx^2} + b^{(k)}(x)\frac{d}{dx}.$$

Defining,

$$\mathcal{S}_t^{(k),j}(x, x') = \begin{cases} \int_l^{x'} \frac{(x'-z)^{j-1}}{(j-1)!} p_t^{(k)}(x, z)dz & j \geq 1 \\ \partial_{x'}^{-j} p_t^{(k)}(x, x') & j \leq 0 \end{cases},$$

and with $x = (x_1, \cdots, x_n)$, $x' = (x'_1, \cdots, x'_n)$,

$$s_t(x, x') = \det\left(\mathcal{S}_t^{(i),i-j}(x_i, x'_j)\right)_{i,j=1}^n, \tag{13.53}$$

we arrive at the following proposition.

Proposition 13.31 *Assume that the diffusion and drift coefficients of the generators* $L^{(k)}$ *are of the form* $a(x) = a_0 + a_1x + a_2x^2$ *and* $b^{(k)}(x) = b_0 + (n-k)a_1 + (b_1 + 2(n-k)a_2)x$ *and moreover assume that the boundaries of the state space are either natural or entrance for the* $L^{(k)}$*-diffusion; in particular this implies certain constraints on the constants* a_0, a_1, a_2, b_0, b_1*. Then, the process* $(X_1^{(1)}(t), \cdots, X_n^{(n)}(t))$ *satisfying the* $SDEs$ *(13.52), in which* $X_k^{(k)}$ *is an* $L^{(k)}$*-diffusion reflected off* $X_{k-1}^{(k-1)}$*, has transition densities* $s_t(x, x')$.

Proof First, we make the following crucial observation. Define the constant $c_{k,n} = 2(n-k-1)a_2+b_1$ and note that the $L^{(k)}$-diffusion is the h-transform of the conjugate $\widehat{L^{(k+1)}}$ with $\widehat{m^{(k+1)}}^{-1}(x)$ with eigenvalue $c_{k,n}$, so that $L^{(k)} = \left(\widehat{L^{(k+1)}}\right)^* - c_{k,n}$ which is again a bona fide diffusion process generator (with L^* denoting the formal adjoint of L with respect to Lebesgue measure). Thus, making use of (13.4) and (13.5) we obtain the following relation between the transition densities,

$$p_t^{(k)}(x,z) = -e^{c_{k,n}t}\int_l^z \partial_x p_t^{(k+1)}(x,w)dw, \tag{13.54}$$

$$\partial_z^j p_t^{(k)}(x,z) = -e^{c_{k,n}t}\partial_z^{j-1}\partial_x p_t^{(k+1)}(x,z).$$

Now, let $f : W^n(I^\circ) \mapsto \mathbb{R}$ be continuous with compact support. Then, we have the following $t=0$ boundary condition,

$$\lim_{t\to 0}\int_{W^n(I^\circ)} s_t(x,x')f(x')dx' = f(x), \tag{13.55}$$

which formally can easily be seen to hold since the transition densities along the main diagonal approximate delta functions and all other contributions vanish. We spell this out now. Let $\epsilon > 0$ and suppose f is zero in a 2ϵ neighbourhood of $\partial W^n(I^\circ)$. We consider a contribution to the Leibniz expansion of the determinant coming from a permutation ρ that is not the identity. Hence there exist $i < j$ so that $\rho(i) > i$ and $\rho(j) \le i$ and note that the factors $\mathcal{S}_t^{(i),i-\rho(i)}\left(x_i, x'_{\rho(i)}\right)$ and $\mathcal{S}_t^{(j),j-\rho(j)}\left(x_j, x'_{\rho(j)}\right)$ are contained in the contribution corresponding to ρ. Since $j-\rho(j) > 0$ and $i-\rho(i) < 0$ observe that on the set $\left\{x'_{\rho(i)} - x_i > \epsilon\right\} \cup \left\{x'_{\rho(j)} - x_j < -\epsilon\right\}$ at least one of these factors and so the whole contribution as $t \downarrow 0$ vanishes uniformly. On the other hand on the complement of this set we have $x'_{\rho(i)} \le x_i + \epsilon \le x_j + \epsilon \le x'_{\rho(j)} + 2\epsilon$. Since $\rho(j) < \rho(i)$ so that $x'_{\rho(j)} \le x'_{\rho(i)}$ we thus obtain that if x' is in the complement of $\left\{x'_{\rho(i)} - x_i > \epsilon\right\} \cup \left\{x'_{\rho(j)} - x_j < -\epsilon\right\}$ it also belongs to some 2ϵ neighbourhood of $\partial W^n(I^\circ)$ and hence outside the support of f. (13.55) then follows.

Now by multilinearity of the determinant the equation in $(0,\infty) \times \mathring{W}^n(I) \times \mathring{W}^n(I)$,

$$\partial_t s_t(x,x') = \sum_{i=1}^n L_{x_i}^{(k)} s_t(x,x'),$$

is satisfied since we have $\partial_t \mathcal{S}_t^{(k),j}(x,x') = L_x^{(k)}\mathcal{S}_t^{(k),j}(x,x')$ for all k. Here, $L_{x_i}^{(k)}$ is simply a copy of the differential operator $L^{(k)}$ acting in the x_i variable.

Moreover, for the Neumann/reflecting boundary conditions we need to check the following conditions $\partial_{x_i} s_t(x, x')|_{x_i=x_{i-1}} = 0$ for $i = 2, \cdots, n$.
This follows from,

$$\partial_{x_i} \mathcal{S}_t^{(i),i-j}(x_i, x_j')|_{x_i=x_{i-1}} = -e^{-c_{i-1,n}t} \mathcal{S}_t^{(i-1),i-1-j}(x_{i-1}, x_j').$$

This is true because of the following observations. For $j \leq -1$

$$\partial_z^{-j} p_t^{(i-1)}(x, z) = -e^{c_{i-1,n}t} \partial_z^{-j-1} \partial_x p_t^{(i)}(x, z).$$

For $j \geq 1$

$$\begin{aligned}
&\int_l^{x'} \frac{(x'-z)^{j-1}}{(j-1)!} p_t^{(i-1)}(x, z) dz \\
&= -e^{c_{i-1,n}t} \partial_x \int_l^{x'} \frac{(x'-z)^{j-1}}{(j-1)!} \int_l^z p_t^{(i)}(x, w) dw dz \\
&= -e^{c_{i-1,n}t} \partial_x \Bigg[\Bigg[-\frac{(x'-z)^j}{j!} \int_l^z p_t^{(k)}(x, w) dw \Bigg]_l^{x'} \\
&\quad - \int_l^{x'} -\frac{(x'-z)^j}{j!} p_t^{(i)}(x, z) dz \Bigg] \\
&= -e^{c_{i-1,n}t} \partial_x \int_l^{x'} \frac{(x'-z)^j}{j!} p_t^{(i)}(x, z) dz.
\end{aligned}$$

Hence $\mathcal{S}_t^{(i-1),j}(x, x') = -e^{c_{i-1,n}t} \partial_x \mathcal{S}_t^{(i),j+1}(x, x')$ and thus

$$\partial_{x_i} s_t(x, x')|_{x_i=x_{i-1}} = 0,$$

for $i = 2, \cdots, n$.

Define for f as in the first paragraph,

$$F(t, x) = \int_{W^n(I^\circ)} s_t(x, x') f(x') dx'.$$

Let $\mathbf{S}_x$ denote the law of $(X_1^{(1)}, \cdots, X_n^{(n)})$ started from $x = (x_1, \cdots, x_n) \in W^n$. Fixing T, ϵ and applying Ito's formula to the process $(F(T + \epsilon - t, x), t \leq T)$ we obtain that it is a local martingale and by virtue of boundedness indeed a true martingale. Hence,

$$F(T + \epsilon, x) = \mathbf{S}_x \left[F\left(\epsilon, \left(X_1^{(1)}(T), \cdots, X_n^{(n)}(T)\right)\right)\right].$$

Now letting $\epsilon \downarrow 0$ we obtain,

$$F(T, x) = \mathbf{S}_x \left[f \left(X_1^{(1)}(T), \cdots, X_n^{(n)}(T) \right) \right].$$

The result follows since the process spends zero Lebesgue time on the boundary so that in particular such f determine its distribution.

In the standard Brownian motion case with $p_t^{(k)}$ the heat kernel this recovers Proposition 8 from [66].

Now, we consider the interacting particle system at the other edge of the pattern with the ith particle getting reflected downwards from the $i-1$th, namely with $(x_1^1 \geq \cdots \geq x_1^n)$ this is given by the following system of $SDEs$ with reflection,

$$\begin{aligned} X_1^{(1)}(t) &= x_1^1 + \int_0^t \sqrt{2a(X_1^{(1)}(s))} d\gamma_1^1(s) + \int_0^t b^{(1)}(X_1^{(1)}(s)) ds, \\ &\vdots \\ X_1^{(m)}(t) &= x_1^m + \int_0^t \sqrt{2a(X_1^{(m)}(s))} d\gamma_1^m(s) + \int_0^t b^{(m)}(X_1^{(m)}(s)) ds - K_1^{m,+}(t), \\ &\vdots \\ X_1^{(n)}(t) &= x_1^n + \int_0^t \sqrt{2a(X_1^{(n)}(s))} d\gamma_1^n(s) + \int_0^t b^{(n)}(X_1^{(n)}(s)) ds - K_1^{n,+}(t), \end{aligned} \tag{13.56}$$

where γ_1^i are independent standard Brownian motions and $K_1^{i,+}$ are positive finite variation processes with the measure $dK_1^{i,+}$ supported on $\left\{ t : X_i^{(i)}(t) = X_{i-1}^{(i-1)}(t) \right\}$. Again see the figure below,

$$\overset{X_1^{(n)}}{\bullet} \longleftarrow \overset{X_1^{(n-1)}}{\bullet} \longleftarrow \overset{X_1^{(n-2)}}{\bullet} \cdots \overset{X_1^{(2)}}{\bullet} \longleftarrow \overset{X_1^{(1)}}{\bullet}.$$

Define,

$$\bar{\mathcal{S}}_t^{(k),j}(x, x') = \begin{cases} -\int_{x'}^{r} \frac{(x'-z)^{j-1}}{(j-1)!} p_t^{(k)}(x, z) dz & j \geq 1 \\ \partial_{x'}^{-j} p_t^{(k)}(x, x') & j \leq 0 \end{cases},$$

Then letting, with $x = (x_1, \cdots, x_n)$, $x' = (x_1', \cdots, x_n')$,

$$\bar{s}_t(x, x') = \det(\bar{\mathcal{S}}_t^{(i),i-j}(x_i, x_j'))_{i,j=1}^n, \tag{13.57}$$

we arrive at the following proposition.

Proposition 13.32 *Assume that the diffusion and drift coefficients of the generators $L^{(k)}$ are of the form $a(x) = a_0 + a_1 x + a_2 x^2$ and $b^{(k)}(x) = b_0 + (n-k)a_1 + (b_1 + 2(n-k)a_2)x$ and moreover assume that the boundaries of the state space are either natural or entrance for the $L^{(k)}$-diffusion. Then, the process $(X_1^{(1)}(t), \cdots, X_1^{(n)}(t))$ satisfying the SDEs (13.56), in which $X_1^{(k)}$ is an $L^{(k)}$-diffusion reflected off $X_1^{(k-1)}$, has transition densities $\bar{s}_t(x, x')$.*

Proof The key observation in this setting is the following relation between the transition kernels:

$$p_t^{(k)}(x, z) = e^{c_{k,n}t} \int_z^r \partial_x p_t^{(k+1)}(x, w) dw.$$

This is immediate from (13.54) since each diffusion process in this section is an honest Markov process.

Then, checking the parabolic equation with the correct spatial boundary conditions is as before. Now the $t = 0$ boundary condition, again follows from the fact that all contributions from off diagonal terms in the determinant have at least one term vanishing uniformly in this new domain $(x_1 \geq \cdots \geq x_n)$.

Via a simple integration, we obtain the following formulae for the distributions of the leftmost and rightmost particles in the Gelfand–Tsetlin pattern,

Corollary 13.4

$$\mathbb{P}_{x^{(0)}}(X_n^{(n)}(t) \leq z) = \det\left(\mathcal{S}_t^{(i),i-j+1}(x_i^{(0)}, z)\right)_{i,j=1}^n,$$

$$\mathbb{P}_{\bar{x}^{(0)}}(X_1^{(n)}(t) \geq z) = \det\left(-\bar{\mathcal{S}}_t^{(i),i-j+1}(\bar{x}_i^{(0)}, z)\right)_{i,j=1}^n,$$

where $x^{(0)} = (x_1^{(0)} \leq \cdots \leq x_n^{(0)})$ and $\bar{x}^{(0)} = (\bar{x}_1^{(0)} \geq \cdots \geq \bar{x}_n^{(0)})$.

For $p_t^{(k)}$ the heat kernel and $x^{(0)} = (0, \cdots, 0)$ this recovers a formula from [66]. In the $BESQ(d)$ case and $t = 1$ the above give expressions for the largest and smallest eigenvalues for the LUE ensemble. We obtain the analogous expressions in the Jacobi case as $t \to \infty$ since the JUE is the invariant measure of non-intersecting Jacobi processes.

13.5 Well-Posedness and Transition Densities for SDEs with Reflection

13.5.1 Well-Posedness of Reflecting SDEs

We will prove well-posedness (existence and uniqueness) for the systems of reflecting $SDEs$ (13.14), (13.20), (13.32), (13.52), and (13.56) considered in this work. It will be more convenient, although essentially equivalent for our purposes, to consider reflecting $SDEs$ for X in the time dependent domains (or between barriers) given by Y i.e. in the form of (13.28). More precisely we will consider $SDEs$ with reflection for a single particle X in the time dependent domain $[\mathsf{Y}^-, \mathsf{Y}^+]$ where Y^- is the lower time dependent boundary and Y^+ is the upper time dependent boundary. This covers all the cases of interest to us by taking $\mathsf{Y}^- = Y_{i-1}$ and $\mathsf{Y}^+ = Y_i$ with the possibility $\mathsf{Y}^- \equiv l$ and/or $\mathsf{Y}^+ \equiv r$.

We will first obtain weak existence, for coefficients $\sigma(x) = \sqrt{2a(x)}, b(x)$ continuous and of at most linear growth, the precise statement to found in Proposition 13.33 below. We begin by recalling the definition and some properties of the Skorokhod problem in a time dependent domain. We will use the following notation, $\mathbb{R}_+ = [0, \infty)$. Suppose we are given continuous functions $z, \mathsf{Y}^-, \mathsf{Y}^+ \in C(\mathbb{R}_+; \mathbb{R})$ such that $\forall T \geq 0$,

$$\inf_{t \leq T} \left(\mathsf{Y}^+(t) - \mathsf{Y}^-(t)\right) > 0,$$

a condition to be removed shortly by a stopping argument. We then say that the pair $(x, k) \in C(\mathbb{R}_+; \mathbb{R}) \times C(\mathbb{R}_+; \mathbb{R})$ is a solution to the Skorokhod problem for $\left(z, \mathsf{Y}^-, \mathsf{Y}^+\right)$ if for every $t \geq 0$ we have $x(t) = z(t) + k(t) \in [\mathsf{Y}^-(t), \mathsf{Y}^+(t)]$ and $k(t) = k^-(t) - k^+(t)$ where k^+ and k^- are non decreasing, in particular bounded variation functions, such that $\forall t \geq 0$,

$$\int_0^t \mathbf{1}\left(z(s) > \mathsf{Y}^-(s)\right) dk^-(s) = 0 \text{ and } \int_0^t \mathbf{1}\left(z(s) < \mathsf{Y}^+(s)\right) dk^+(s) = 0.$$

Observe that the constraining terms k^+ and k^- only increase on the boundaries of the time dependent domain, namely at Y^+ and Y^- respectively. Now, consider the *solution* map denoted by $\mathcal{S}$,

$$\mathcal{S} : C(\mathbb{R}_+; \mathbb{R}) \times C(\mathbb{R}_+; \mathbb{R}) \times C(\mathbb{R}_+; \mathbb{R}) \to C(\mathbb{R}_+; \mathbb{R}) \times C(\mathbb{R}_+; \mathbb{R})$$

given by,

$$\mathcal{S} : \left(z, \mathsf{Y}^-, \mathsf{Y}^+\right) \mapsto (x, k).$$

Then the key fact is that the map $\mathcal{S}$ is Lipschitz continuous in the supremum norm and there exists a unique solution to the Skorokhod problem, see for example Proposition 2.3 and Corollary 2.4 of [61] (also Theorem 2.6 of [14]). Below we will sometimes abuse notation and write $x = \mathcal{S}\left(z, \mathsf{Y}^-, \mathsf{Y}^+\right)$ just for the x-component of the solution (x, k).

Now suppose $\sigma : \mathbb{R} \to \mathbb{R}$ and $b : \mathbb{R} \to \mathbb{R}$ are Lipschitz continuous functions. Then by a classical argument based on Picard iteration, see for example Theorem 3.3 of [61], we obtain that there exists a unique strong solution to the $SDER$ (SDE

with reflection) for $\mathsf{Y}^-(0) \le X(0) \le \mathsf{Y}^+(0)$,

$$X(t) = X(0) + \int_0^t \sigma(X(s))\, d\beta(s) + \int_0^t b(X(s))\, ds + K^-(t) - K^+(t),$$

where β is a standard Brownian motion and $\left(K^+(t); t \ge 0\right)$ and $\left(K^-(t); t \ge 0\right)$ are non decreasing processes that increase only when $X(t) = \mathsf{Y}^+(t)$ and $X(t) = \mathsf{Y}^-(t)$ respectively so that for all $t \ge 0$ we have $X(t) \in [\mathsf{Y}^-(t), \mathsf{Y}^+(t)]$. Here, by strong solution we mean that on the filtered probability space $(\Omega, \mathcal{F}, \{\mathcal{F}_t\}, \mathbb{P})$ on which (X, K, β) is defined, the process (X, K) is adapted with respect to the filtration $\mathcal{F}_t^\beta$ generated by the Brownian motion β. Equivalently (X, K) where $K = K^+ - K^-$ solves the Skorokhod problem for $(z, \mathsf{Y}^-, \mathsf{Y}^+)$ where,

$$z(\cdot) \overset{def}{=} X(0) + \int_0^\cdot \sigma(X(s))\, d\beta(s) + \int_0^\cdot b(X(s))\, ds.$$

We write $\mathfrak{s}_L^R$ for the corresponding measurable solution map on path space, namely so that $X = \mathfrak{s}_L^R\left(\beta; \mathsf{Y}^-, \mathsf{Y}^+\right)$.

Now, suppose $\sigma : \mathbb{R} \to \mathbb{R}$ and $b : \mathbb{R} \to \mathbb{R}$ are merely continuous and of at most linear growth, namely:

$$|\sigma(x)|, |b(x)| \le C(1 + |x|),$$

for some constant C. We will abbreviate this assumption by (**CLG**). Then, we can still obtain weak existence using the following rather standard argument. Take $\sigma^{(n)} : \mathbb{R} \to \mathbb{R}$ and $b^{(n)} : \mathbb{R} \to \mathbb{R}$ to be Lipschitz, converging uniformly to σ and b and satisfying a uniform linear growth condition. More precisely:

$$\begin{aligned} &\sigma^{(n)} \overset{\text{unif}}{\longrightarrow} \sigma, \; b^{(n)} \overset{\text{unif}}{\longrightarrow} b, \\ &|\sigma^{(n)}(x)|, |b^{(n)}(x)| \le \tilde{C}(1 + |x|), \end{aligned} \tag{13.58}$$

for some constant $\tilde{C}$ that is independent of n. For example, we could take the mollification $\sigma^{(n)} = \phi_n * \sigma$, with $\phi_n(x) = n\phi(nx)$ where ϕ is a smooth bump function: $\phi \in C^\infty, \phi \ge 0, \int \phi = 1$ and $\text{supp}(\phi) \subset [-1, 1]$. Then, if $|\sigma(x)| \le C(1 + |x|)$ we easily get $|(\phi_n * \sigma)(x)| \le (2 + |x|)$ uniformly in n. Let $\left(X^{(n)}, K^{(n)}\right)$ be the corresponding strong solution to the $SDER$ above with coefficients $\sigma^{(n)}$ and $b^{(n)}$. Then the laws of,

$$X(0) + \int_0^\cdot \sigma^{(n)}\left(X^{(n)}(s)\right) d\beta(s) + \int_0^\cdot b^{(n)}\left(X^{(n)}(s)\right) ds,$$

are easily seen to be tight by applying Aldous' tightness criterion (see for example Chapter 16 of [43] or Chapter 3 of [29]) using the uniformity in n of the linear growth condition (13.58). Hence, from the Lipschitz continuity of $\mathcal{S}$ we obtain that the laws of $\left(X^{(n)}, K^{(n)}\right)$ are tight as well.

Thus, we can choose a subsequence $(n_i; i \geq 1)$ such that the laws of $\left(X^{(n_i)}, K^{(n_i)}\right)$ converge weakly to some (X, K). Using the Skorokhod representation theorem we can upgrade this to joint almost sure convergence on a new probability space $\left(\tilde{\Omega}, \tilde{\mathcal{F}}, \{\tilde{\mathcal{F}}_t\}, \tilde{\mathbb{P}}\right)$. More precisely, we can define processes $\left(\tilde{X}^{(i)}, \tilde{K}^{(i)}\right)_{i\geq 1}, \left(\tilde{X}, \tilde{K}\right)$ on $\left(\tilde{\Omega}, \tilde{\mathcal{F}}, \{\tilde{\mathcal{F}}_t\}, \tilde{\mathbb{P}}\right)$ so that:

$$\left(\tilde{X}^{(i)}, \tilde{K}^{(i)}\right) \stackrel{\mathrm{d}}{=} \left(X^{(n_i)}, K^{(n_i)}\right), \ \left(\tilde{X}, \tilde{K}\right) \stackrel{\mathrm{d}}{=} (X, K), \ \left(\tilde{X}^{(i)}, \tilde{K}^{(i)}\right) \xrightarrow{\text{a.s.}} (\tilde{X}, \tilde{K}).$$

Now, the stochastic processes:

$$M_n(t) = \tilde{X}^{(n)}(t) - \tilde{X}^{(n)}(0) - \int_0^t b^{(n)}\left(\tilde{X}^{(n)}(s)\right) ds - \left(\tilde{K}^{(n)}\right)^-(t) + \left(\tilde{K}^{(n)}\right)^+(t)$$

are martingales with quadratic variation:

$$\langle M_n, M_n\rangle(t) = \int_0^t \left(\sigma^{(n)}\left(\tilde{X}^{(n)}(s)\right)\right)^2 ds.$$

By the following convergences:

$$\sigma^{(n)} \xrightarrow{\text{unif}} \sigma, \ b^{(n)} \xrightarrow{\text{unif}} b, \ \left(\tilde{X}^{(i)}, \tilde{K}^{(i)}\right) \xrightarrow{\text{a.s.}} (\tilde{X}, \tilde{K})$$

we obtain that $M_n \xrightarrow{\text{a.s.}} M$ where,

$$M(t) = \tilde{X}(t) - \tilde{X}(0) - \int_0^t b\left(\tilde{X}(s)\right) ds - \tilde{K}^-(t) + \tilde{K}^+(t)$$

is a martingale with quadratic variation given by:

$$\langle M, M\rangle(t) = \int_0^t \sigma^2\left(\tilde{X}(s)\right) ds.$$

Then, by the martingale representation theorem there exists a standard Brownian motion $\tilde{\beta}$, that is defined on a possibly enlarged probability space, so that $M(t) = \int_0^t \sigma\left(\tilde{X}(s)\right) d\tilde{\beta}(s)$ and thus:

$$\tilde{X}(t) = \tilde{X}(0) + \int_0^t \sigma\left(\tilde{X}(s)\right) d\tilde{\beta}(s) + \int_0^t b\left(\tilde{X}(s)\right) ds + \tilde{K}^-(t) - \tilde{K}^+(t),$$

where again the non decreasing processes $\left(\tilde{K}^{+}(t); t \geq 0\right)$ and $\left(\tilde{K}^{-}(t); t \geq 0\right)$ increase only when $\tilde{X}(t) = \mathsf{Y}^{+}(t)$ and $\tilde{X}(t) = \mathsf{Y}^{-}(t)$ respectively so that $\tilde{X}(t) \in [\mathsf{Y}^{-}(t), \mathsf{Y}^{+}(t)]$ $\forall t \geq 0$. Hence, we have obtained the existence of a weak solution to the $SDER$ for σ and b continuous and of at most linear growth.

We now remove the condition that $\mathsf{Y}^{-}, \mathsf{Y}^{+}$ never collide by stopping the process at the first time $\tau = \inf\{t \geq 0 : \mathsf{Y}^{-}(t) = \mathsf{Y}^{+}(t)\}$ that they do. First we note that, there exists an extension to the Skorokhod problem and to $SDER$, allowing for reflecting barriers $\mathsf{Y}^{-}, \mathsf{Y}^{+}$ that come together, see [14, 61] for the detailed definition. Both results used in the previous argument, namely the Lipschitz continuity of the solution map, which we still denote by $\mathcal{S}$, and existence and uniqueness of strong solutions to $SDER$ extend to this setting, see e.g. Theorem 2.6, also Corollary 2.4 and Theorem 3.3 in [61]. The difference of the extended problem to the classical one described at the beginning, being that $k = k^{-} - k^{+}$ is allowed to have infinite variation. However, as proven in Proposition 2.3 and Corollary 2.4 in [14] (see also Remark 2.2 in [61]) the unique solution to the extended Skorokhod problem coincides with the one of the classical one in $[0, T]$ while $\inf_{t \leq T} \left(\mathsf{Y}^{+}(t) - \mathsf{Y}^{-}(t)\right) > 0$. Thus, by the previous considerations, for any $T < \tau$, we still have a weak solution to the $SDER$ above, with bounded variation local terms K; the final statement more precisely given as:

Proposition 13.33 *Assume $\mathsf{Y}^{-}, \mathsf{Y}^{+}$ are continuous functions such that $\mathsf{Y}^{-}(t) \leq \mathsf{Y}^{+}(t), \forall t \geq 0$ and let $\tau = \inf\{t \geq 0 : \mathsf{Y}^{-}(t) = \mathsf{Y}^{+}(t)\}$. Assume **(CLG)**, namely that $\sigma(\cdot), b(\cdot)$ are continuous functions satisfying an at most linear growth condition, for some positive constant C:*

$$|\sigma(x)|, |b(x)| \leq C(1 + |x|).$$

Then, there exists a filtered probability space $(\Omega, \mathcal{F}, \{\mathcal{F}_t\}, \mathbb{P})$ on which firstly an adapted Brownian motion β is defined (not necessarily generating the filtration). Moreover, for $\mathsf{Y}^{-}(0) \leq X(0) \leq \mathsf{Y}^{+}(0)$ the adapted process (X, K) satisfies:

$$\begin{aligned} X(t \wedge \tau) = X(0) &+ \int_0^{t \wedge \tau} \sigma(X(s))\, d\beta(s) \\ &+ \int_0^{t \wedge \tau} b(X(s))\, ds + K^{-}(t \wedge \tau) - K^{+}(t \wedge \tau), \end{aligned} \tag{13.59}$$

such that for all $t \geq 0$ we have $X(t \wedge \tau) \in [\mathsf{Y}^{-}(t \wedge \tau), \mathsf{Y}^{+}(t \wedge \tau)]$ and for any $T < \tau$ the non decreasing processes $\left(K^{+}(t); t \leq T\right)$ and $\left(K^{-}(t); t \leq T\right)$ increase only when $X(t) = \mathsf{Y}^{+}(t)$ and $X(t) = \mathsf{Y}^{-}(t)$ respectively.

We will now be concerned with pathwise uniqueness. Due to the intrinsic one-dimensionality of the problem we can fortunately apply a simple Yamada–Watanabe type argument. For the convenience of the reader we now recall assumption **(YW)**, defined in Sect. 13.2: Let I be an interval with endpoints $l < r$ and suppose ρ is a non-decreasing function from $(0, \infty)$ to itself such that $\int_{0+} \frac{dx}{\rho(x)} = \infty$. Consider, the following condition on functions $a : I \to \mathbb{R}_+$ and $b : I \to \mathbb{R}$, where we implicitly assume that a and b initially defined in I° can be extended continuously to the boundary points l and r (in case these are finite),

$$|\sqrt{a(x)} - \sqrt{a(y)}|^2 \le \rho(|x-y|),$$
$$|b(x) - b(y)| \le C|x-y|.$$

Moreover, we assume that $\sqrt{a(\cdot)}$ is of at most linear growth. Note that, for $b(\cdot)$ this is immediate by Lipschitz continuity.

Also, observe that since ρ is continuous at 0 with $\rho(0) = 0$ (the assumption on ρ implies this) we get that $\sqrt{a(\cdot)}$ is continuous. Thus, (**YW**) implies (**CLG**) and in particular the existence result above applies under (**YW**). We are now ready to state and prove our well-posedness result.

Proposition 13.34 *Under the* (**YW**) *assumption the SDER (13.59) with* $(\sigma, b) = (\sqrt{2a}, b)$ *has a pathwise unique solution.*

Proof Suppose that X and $\tilde{X}$ are two solutions of (13.59) with respect to the same noise. Then the argument given at Chapter IX Corollary 3.4 of [57] shows that $L^0(X_i - \tilde{X}_i) = 0$ where for a semimartingale Z, $L^a(Z)$ denotes its semimartingale local time at a (see for example Sect. 13.1 Chapter VI of [57]). Hence by Tanaka's formula we get,

$$\begin{aligned}
|X(t\wedge\tau) - \tilde{X}(t\wedge\tau)| &= \int_0^{t\wedge\tau} \mathrm{sgn}(X(s) - \tilde{X}(s))d(X(s) - \tilde{X}(s)) \\
&= \int_0^{t\wedge\tau} \mathrm{sgn}(X(s) - \tilde{X}(s)) \\
&\quad \times \left(\sqrt{2a(X(s))} - \sqrt{2a(\tilde{X}(s))}\right) d\beta(s) \\
&\quad + \int_0^{t\wedge\tau} \mathrm{sgn}(X(s) - \tilde{X}(s))(b(X(s)) - b(\tilde{X}(s)))ds \\
&\quad - \int_0^{t\wedge\tau} \mathrm{sgn}(X(s) - \tilde{X}(s))d(K^+(s) - \tilde{K}^+(s)) \\
&\quad + \int_0^{t\wedge\tau} \mathrm{sgn}(X(s) - \tilde{X}(s))d(K^-(s) - \tilde{K}^-(s)).
\end{aligned}$$

Note that $\mathsf{Y}^- \le X, \tilde{X} \le \mathsf{Y}^+$, dK^+ is supported on $\{t : X(t) = \mathsf{Y}^+(t)\}$ and $d\tilde{K}^+$ is supported on $\{t : \tilde{X}(t) = \mathsf{Y}^+(t)\}$. So if $\tilde{X} < X \le \mathsf{Y}^+$ then $dK^+ - d\tilde{K}^+ \ge 0$ and if $X < \tilde{X} \le \mathsf{Y}^+$ then $dK^+ - d\tilde{K}^+ \le 0$. Hence $\int_0^{t\wedge\tau} \mathrm{sgn}(X(s) - \tilde{X}(s))d(K^+(s) - \tilde{K}^+(s)) \ge 0$. With similar considerations

$\int_0^{t\wedge\tau} \operatorname{sgn}(X(s)-\tilde{X}(s))d(K^-(s)-\tilde{K}^-(s)) \leq 0$. Taking expectations we obtain,

$$\begin{aligned}
&\mathbb{E}[|X(t\wedge\tau)-\tilde{X}(t\wedge\tau)|] \\
&\leq \mathbb{E}\left[\int_0^{t\wedge\tau} \operatorname{sgn}(X(s)-\tilde{X}(s))(b(X(s))-b(\tilde{X}(s)))ds\right] \\
&\leq C\int_0^{t\wedge\tau} \mathbb{E}[|X(s)-\tilde{X}(s)|]ds.
\end{aligned}$$

The statement of the proposition then follows from Gronwall's lemma.

Under the pathwise uniqueness obtained in Proposition 13.34 above, if the evolution $\left(\mathsf{Y}^-(t\wedge\tau),\mathsf{Y}^+(t\wedge\tau);t\geq 0\right)$ is Markovian, then standard arguments (see for example Sect. 13.1 of Chapter IX of [57]) imply that $\left(\mathsf{Y}^-(t\wedge\tau),\mathsf{Y}^+(t\wedge\tau),X(t\wedge\tau);t\geq 0\right)$ is Markov as well. Moreover, under this (**YW**) condition we still have the solution map $X=\mathfrak{s}_L^R\left(\beta;\mathsf{Y}^-,\mathsf{Y}^+\right)$.

The reader should note that Proposition 13.34 covers in particular **all** the cases of Brownian motions, Ornstein–Uhlenbeck, $BESQ(d)$, $Lag(\alpha)$ and $Jac(\beta,\gamma)$ diffusions considered in the Applications and Examples section.

13.5.2 Transition Densities for SDER

The aim of this section is to prove under some conditions that $q_t^{n,n+1}$ and $q_t^{n,n}$ form the transition kernels for the two-level systems of $SDEs$ (13.14) and (13.20) in $W^{n,n+1}$ and $W^{n,n}$ respectively. For the sake of exposition we shall mainly restrict our attention to (13.14). In the sequel, τ will denote the stopping time $T^{n,n+1}$ (or $T^{n,n}$ respectively).

Throughout this section we assume (**R**) and (**BC**+) hold for the L-diffusion and (**YW**) holds for both the L and $\hat{L}$ diffusions. In particular, there exists a Markov semimartingale (X,Y) satisfying Eq. (13.14) (or respectively (13.20)).

To begin with we make a few simple but important observations. First, note that if the L-diffusion does not hit l (i.e. l is natural or entrance), then X_1 doesn't hit l either before being driven to l by Y_1 (in case l is exit for $\hat{L}$). Similarly, it is rather obvious, since the particles are ordered, that in case l is regular reflecting for the L-diffusion the time spent at l up to time τ by the $SDEs$ (13.14) is equal to the time spent by X_1 at l. This is in turn equal to the time spent at l by the excursions of X_1 between collisions with Y_1 (and before τ) during which the evolution of X_1 coincides with the unconstrained L-diffusion which spends zero Lebesgue time at l (e.g. see Chapter 2 paragraph 7 in [12]). Hence the system of reflecting $SDEs$ (13.14) spends zero Lebesgue time at either l or r up to time τ. Since in addition to this, the noise driving the $SDEs$ is uncorrelated and the diffusion coefficients do not vanish in I° we get that,

$$\int_0^\tau \mathbf{1}_{\partial W^{n,n+1}(I)}(X(t),Y(t))\,dt = 0 \quad \text{a.s.}\,. \tag{13.60}$$

We can now in fact relate the constraining finite variation terms K to the semimartingale local times of the gaps between particles (although this will not be essential in what follows). Using the observation (13.60) above and Exercise 1.16 (3°) of Chapter VI of [57], which states that for a positive semimartingale $Z = M + V \geq 0$ (where M is the martingale part) its local time at 0 is equal to $2\int_0^\cdot 1(Z_s = 0)\,dVs$, we get that for the $SDEs$ (13.14) the semimartingale local time of $Y_i - X_i$ at 0 up to time τ is,

$$2\int_0^{t\wedge\tau} \mathbf{1}(Y_i(s) = X_i(s))dK_i^+(s) = 2K_i^+(t\wedge\tau),$$

and similarly the semimartingale local time of $X_{i+1} - Y_i$ at 0 up to τ is,

$$2\int_0^{t\wedge\tau} \mathbf{1}(X_{i+1}(s) = Y_i(s))dK_{i+1}^-(s) = 2K_{i+1}^-(t\wedge\tau).$$

Now, we state a lemma corresponding to the *time 0* boundary condition.

Lemma 13.3 *For any $f : W^{n,n+1}(I^\circ) \to \mathbb{R}$ continuous with compact support we have,*

$$\lim_{t\to 0}\int_{W^{n,n+1}(I^\circ)} q_t^{n,n+1}((x,y),(x',y'))f(x',y')dx'dy' = f(x,y).$$

Proof This follows as in the proof of Lemma 1 of [66]. See also the beginning of the proof of Proposition 13.31.

We are now ready to prove the following result on the transition densities.

Proposition 13.35 *Assume* (**R**) *and* (**BC+**) *hold for the L-diffusion and* (**YW**) *holds for both the L and $\hat{L}$ diffusions. Moreover, assume that l and r are either natural or entrance for the L-diffusion. Then $q_t^{n,n+1}$ form the transition densities for the system of $SDEs$ (13.14).*

Proof Let $\mathbf{Q}_{x,y}^{n,n+1}$ denote the law of the process $(X_1, Y_1, \cdots, Y_n, X_{n+1})$ satisfying the system of $SDEs$ (13.14) and starting from (x,y). Define for f continuous with compact support,

$$F^{n,n+1}(t,(x,y)) = \int_{W^{n,n+1}(I^\circ)} q_t^{n,n+1}((x,y),(x',y'))f(x',y')dx'dy'.$$

Our goal is to prove that for fixed $T > 0$,

$$F^{n,n+1}(T,(x,y)) = \mathbf{Q}^{n,n+1}_{x,y}\big[f(X(T),Y(T))\mathbf{1}(T<\tau)\big]. \tag{13.61}$$

The result then follows since from observation (13.60) the only part of the distribution of $(X(T),Y(T))$ that charges the boundary corresponds to the event $\{T\geq\tau\}$.

In what follows we shall slightly abuse notation and use the same notation for both the scalar entries and the matrices that come into the definition of $q_t^{n,n+1}$. First, note the following with $x,y\in I^\circ$,

$$\partial_t A_t(x,x') = \mathcal{D}^x_m\mathcal{D}^x_s A_t(x,x')\ ,\quad \partial_t B_t(x,y') = \mathcal{D}^x_m\mathcal{D}^x_s B_t(x,y'),$$
$$\partial_t C_t(y,x') = \mathcal{D}^y_{\hat{m}}\mathcal{D}^y_{\hat{s}} C_t(y,x')\ ,\quad \partial_t D_t(y,y') = \mathcal{D}^y_{\hat{m}}\mathcal{D}^y_{\hat{s}} D_t(y,y').$$

To see the equation for $C_t(y,x')$ note that since $\mathcal{D}_{\hat{m}}=\mathcal{D}_s$ and $\mathcal{D}_{\hat{s}}=\mathcal{D}_m$ we have,

$$\begin{aligned}\partial_t C_t(y,x') &= -\mathcal{D}^y_s\partial_t p_t(y,x') = -\mathcal{D}^y_s\mathcal{D}^y_m\mathcal{D}^y_s p_t(y,x')\\ &= -\mathcal{D}^y_{\hat{m}}\mathcal{D}^y_{\hat{s}}\mathcal{D}^y_s p_t(y,x') = \mathcal{D}^y_{\hat{m}}\mathcal{D}^y_{\hat{s}} C_t(y,x').\end{aligned}$$

Hence, for fixed $(x',y')\in \mathring{W}^{n,n+1}(I^\circ)$ we have,

$$\partial_t q_t^{n,n+1}((x,y),(x',y')) = \left(\sum_{i=1}^{n+1}\mathcal{D}^{x_i}_m\mathcal{D}^{x_i}_s + \sum_{i=i}^{n}\mathcal{D}^{y_i}_{\hat{m}}\mathcal{D}^{y_i}_{\hat{s}}\right) q_t^{n,n+1}((x,y),(x',y')),$$
$$\text{in } (0,\infty)\times \mathring{W}^{n,n+1}(I^\circ).$$

Now, by definition of the entries A_t,B_t,C_t,D_t we have for $x,y\in I^\circ$,

$$\partial_x A_t(x,x')|_{x=y} = -\hat{m}(y)C_t(y,x'),$$
$$\partial_x B_t(x,y')|_{x=y} = -\hat{m}(y)D_t(y,y').$$

Hence for fixed $(x',y')\in W^{n,n+1}(I^\circ)$ by differentiating the determinant and since two rows are equal up to multiplication by a constant we obtain,

$$\partial_{x_i} q_t^{n,n+1}((x,y),(x',y'))|_{x_i=y_i}=0\ ,\ \partial_{x_i} q_t^{n,n+1}((x,y),(x',y'))|_{x_i=y_{i-1}}=0.$$

The Dirichlet boundary conditions for $y_i=y_{i+1}$ are immediate since again two rows of the determinant are equal. Furthermore, in case l or r are entrance boundaries for the L-diffusion the Dirichlet boundary conditions for $y_1=l$ and $y_n=r$ follow from the fact that (in the limit as $y\to l,r$),

$$D_t(y,y')|_{y=l,r}=0,\ C_t(y,x')|_{y=l,r}=\mathcal{D}^x_s A_t(x,x')|_{x=l,r}=0.$$

Fix $T, \epsilon > 0$. Applying Ito's formula we obtain that for each (x', y') the process,

$$\left(\mathfrak{Q}_t(x', y') : t \in [0, T]\right) = \left(q_{T+\epsilon-t}^{n,n+1}\left((X(t), Y(t)), (x', y')\right) : t \in [0, T]\right),$$

is a local martingale. Now consider a sequence of compact intervals J_k exhausting I as $k \to \infty$ and write τ_k for $\inf\{t : (X(t), Y(t)) \notin J_k\}$. Note that $\mathbf{1}(T < \tau \wedge \tau_k) \to \mathbf{1}(T < \tau)$ as $k \to \infty$ by our boundary assumptions, more precisely by making use of the observation that X does not hit l or r before Y does. Using the optional stopping theorem (since the stopped process $\left(\mathfrak{Q}_t^{\tau_k}(x', y') : t \in [0, T]\right)$ is bounded and hence a true martingale) and then the monotone convergence theorem we obtain,

$$q_{T+\epsilon}^{n,n+1}((x, y), (x', y')) = \mathbf{Q}_{x,y}^{n,n+1}\left[q_{\epsilon}^{n,n+1}((X(T), Y(T)), (x', y'))\mathbf{1}(T < \tau)\right].$$

Now multiplying by f continuous with compact support, integrating with respect to (x', y') and using Fubini's theorem to exchange expectation and integral we obtain,

$$F^{n,n+1}(T + \epsilon, (x, y)) = \mathbf{Q}_{x,y}^{n,n+1}\left[F^{n,n+1}(\epsilon, (X(T), Y(T))\mathbf{1}(T < \tau)\right].$$

By Lemma 13.3, we can let $\epsilon \downarrow 0$ to conclude,

$$F^{n,n+1}(T, (x, y)) = \mathbf{Q}_{x,y}^{n,n+1}\left[f(X(T), Y(T))\mathbf{1}(T < \tau)\right].$$

The proposition is proven.

Completely analogous arguments prove the following:

Proposition 13.36 *Assume* **(R)** *and* **(BC+)** *hold for the L-diffusion and* **(YW)** *holds for both the L and $\hat{L}$ diffusions. Moreover, assume that l is either natural or exit and r is either natural or entrance for the L-diffusion. Then $q_t^{n,n}$ form the transition densities for the system of SDEs (13.20).*

We note here that Propositions 13.35 and 13.36 apply in particular to the cases of Brownian motions with drifts, Ornstein–Uhlenbeck, $BESQ(d)$ for $d \geq 2$, $Lag(\alpha)$ for $\alpha \geq 2$ and $Jac(\beta, \gamma)$ for $\beta, \gamma \geq 1$ considered in the Applications and Examples section.

In the case l and/or r are regular reflecting boundary points we have the following proposition. This is where the non-degeneracy and regularity at the boundary in assumption **(BC+)** is used. This is technical but quite convenient since it allows for a rather streamlined rigorous argument. It presumably can be removed.

Proposition 13.37 *Assume* **(R)** *and* **(BC+)** *hold for the L-diffusion and* **(YW)** *holds for both the L and $\hat{L}$ diffusions. Moreover, assume that l and/or r are regular reflecting for the L-diffusion. Then $q_t^{n,n+1}$ form the transition densities for the system of SDEs (13.14).*

Proof The strategy is the same as in Proposition 13.35 above. We give the proof in the case that both l and r are regular reflecting for the L-diffusion (the other cases are analogous). First, recall that (**BC**+) in this case requires that $\lim_{x\to l,r} a(x) > 0$ and that the limits $\lim_{x\to l,r} b(x)$, $\lim_{x\to l,r} \big(a'(x) - b(x)\big)$ exist and are finite.

Now, note that by the non-degeneracy condition $\lim_{x\to l,r} a(x) > 0$ and since $\lim_{x\to l,r} b(x)$ is finite we thus obtain $\lim_{x\to l,r} s'(x) > 0$.

So for $x' \in I^\circ$ the relations,

$$\lim_{x\to l,r} \mathcal{D}_s^x A_t(x,x') = 0 \text{ and } \lim_{x\to l,r} \mathcal{D}_s^x B_t(x,x') = 0,$$

actually imply that for $x' \in I^\circ$,

$$\lim_{x\to l,r} \partial_x A_t(x,x') = 0 \text{ and } \lim_{x\to l,r} \partial_x B_t(x,x') = 0. \tag{13.62}$$

Moreover, by rearranging the backwards equations we have for fixed $y \in I^\circ$ that the functions,

$$\begin{aligned}
(t,x) \mapsto \partial_x^2 p_t(x,y) &= \frac{\partial_t p_t(x,y) - b(x)\partial_x p_t(x,y)}{a(x)},\\
(t,x) \mapsto \partial_x^2 \mathcal{D}_s^x p_t(x,y) &= \frac{\partial_t \mathcal{D}_s^x p_t(x,y) - \big(a'(x) - b(x)\big)\,\partial_x \mathcal{D}_s^x p_t(x,y)}{a(x)},\\
&= \frac{\partial_t \mathcal{D}_s^x p_t(x,y) - \big(a'(x) - b(x)\big)\, m(x)\partial_t p_t(x,y)}{a(x)},
\end{aligned}$$

and more generally for $n \ge 0$ and fixed $y \in I^\circ$,

$$(t,x) \mapsto \partial_t^n \partial_x^2 \mathcal{D}_s^x p_t(x,y) = \frac{\partial_t^{n+1} \mathcal{D}_s^x p_t(x,y) - \big(a'(x) - b(x)\big)\, m(x)\partial_t^{n+1} p_t(x,y)}{a(x)},$$

can be extended continuously to $(0,\infty) \times [l,r]$ (note the closed interval $[l,r]$). This is because every function on the right hand side can be extended by the assumptions of proposition and the fact that for $y \in I^\circ$, $\partial_t^n p_t(\cdot, y) \in Dom(L)$ (see Theorem 4.3 of [52] for example). Thus by Whitney's extension theorem, essentially a clever reflection argument in this case (see Section 3 of [37] for example), $q_t^{n,n+1}((x,y),(x',y'))$ can be extended as a $C^{1,2}$ function in $(t,(x,y))$ to the whole space. We can hence apply Ito's formula, and it is important to observe that the finite variation terms dK^l and dK^r at l and r respectively (corresponding to X_1 and X_{n+1}) vanish by the Neumann boundary conditions (13.62), from which we deduce as before that for fixed $T > 0$,

$$q_{T+\epsilon}^{n,n+1}((x,y),(x',y')) = \mathbf{Q}_{x,y}^{n,n+1}\big[q_\epsilon^{n,n+1}((X(T),Y(T)),(x',y'))\mathbf{1}(T<\tau)\big].$$

The conclusion then follows as in Proposition 13.35.

Completely analogous arguments give the following:

Proposition 13.38 *Assume* (**R**) *and* (**BC+**) *hold for the* L*-diffusion and* (**YW**) *holds for both the* L *and* $\hat{L}$ *diffusions. Moreover, assume that* l *is regular absorbing and/or* r *is regular reflecting for the* L*-diffusion. Then* $q_t^{n,n}$ *form the transition densities for the system of SDEs (13.20).*

These propositions cover in particular the cases of Brownian motions in the half line and in an interval considered in Sects. 13.3.2 and 13.3.3 respectively.

Acknowledgements Research of N.O'C. supported by ERC Advanced Grant 669306. Research of T.A. supported through the MASDOC DTC grant number EP/HO23364/1. We would like to thank an anonymous referee for many useful comments and suggestions which have led to many improvements in presentation.

Appendix

We collect here the proofs of some of the facts regarding conjugate diffusions that were stated and used in previous sections.

We first give the derivation of the table on the boundary behaviour of a diffusion and its conjugate. Keeping with the notation of Sect. 13.2 consider the following quantities with $x \in I^\circ$ arbitrary,

$$N(l) = \int_{(l^+,x]} (s(x) - s(y))M(dy) = \int_{(l^+,x]} (s(x) - s(y))m(y)dy,$$

$$\Sigma(l) = \int_{(l^+,x]} (M(x) - M(y))s(dy) = \int_{(l^+,x]} (M(x) - M(y))s'(y)dy.$$

We then have the following classification of the boundary behaviour at l (see e.g. [29]):

- l is an entrance boundary iff $N(l) < \infty$, $\Sigma(l) = \infty$.
- l is a exit boundary iff $N(l) = \infty$, $\Sigma(l) < \infty$.
- l is a natural boundary iff $N(l) = \infty$, $\Sigma(l) = \infty$.
- l is a regular boundary iff $N(l) < \infty$, $\Sigma(l) < \infty$.

From the relations $\hat{s}'(x) = m(x)$ and $\hat{m}(x) = s'(x)$ we obtain the following,

$$\hat{N}(l) = \int_{(l^+,x]} (\hat{s}(x) - \hat{s}(y))\hat{m}(y)dy = \Sigma(l),$$

$$\hat{\Sigma}(l) = \int_{(l^+,x]} (\hat{M}(x) - \hat{M}(y))\hat{s}'(y)dy = N(l).$$

These relations immediately give us the table on boundary behaviour, namely: If l is an entrance boundary for X, then it is exit for $\hat{X}$ and vice versa. If l is natural for X, then so it is for its conjugate. If l is regular for X, then so it is for its conjugate. In this instance as already stated in Sect. 13.2 we define the conjugate diffusion $\hat{X}$ to have boundary behaviour dual to that of X, namely if l is reflecting for X then it is absorbing for $\hat{X}$ and vice versa.

Proof (Proof of Lemma 2.1) There is a total number of 5^2 boundary behaviours (5 at l and 5 at r) for the L-diffusion (the boundary behaviour of $\hat{L}$ is completely determined from L as explained above) however since the boundary conditions for an entrance and regular reflecting ($\mathcal{D}_s v = 0$) and similarly for an exit and regular absorbing boundary ($\mathcal{D}_m \mathcal{D}_s v = 0$) are the same we can pair them to reduce to 3^2 cases $(\mathsf{b.c.}(l), \mathsf{b.c.}(r))$ abbreviated as follows:

$$(nat, nat), (ref, ref), (abs, abs), (nat, abs), (ref, abs),$$
$$(abs, ref), (abs, nat), (nat, ref), (ref, nat).$$

We now make some further reductions. Note that for $x, y \in I^\circ$,

$$\mathsf{P}_t \mathbf{1}_{[l,y]}(x) = \hat{\mathsf{P}}_t \mathbf{1}_{[x,r]}(y) \iff \mathsf{P}_t \mathbf{1}_{[y,r]}(x) = \hat{\mathsf{P}}_t \mathbf{1}_{[l,x]}(y).$$

After swapping $x \leftrightarrow y$ this is equivalent to,

$$\hat{\mathsf{P}}_t \mathbf{1}_{[l,y]}(x) = \mathsf{P}_t \mathbf{1}_{[x,r]}(y).$$

So we have a bijection that swaps boundary conditions with their duals $(\mathsf{b.c.}(l), \mathsf{b.c.}(r)) \leftrightarrow (\widehat{\mathsf{b.c.}(l)}, \widehat{\mathsf{b.c.}(r)})$. Moreover, if $\mathfrak{h} : (l, r) \to (l, r)$ is any homeomorphism such that $\mathfrak{h}(l) = r$, $\mathfrak{h}(r) = l$ and writing H_t for the semigroup associated with the $\mathfrak{h}(X)(t)$-diffusion and similarly $\hat{\mathsf{H}}_t$ for the semigroup associated with the $\mathfrak{h}(\hat{X})(t)$-diffusion we see that,

$$\mathsf{P}_t \mathbf{1}_{[l,y]}(x) = \hat{\mathsf{P}}_t \mathbf{1}_{[x,r]}(y) \ \forall x, y \in I^\circ \iff \mathsf{H}_t \mathbf{1}_{[l,y]}(x) = \hat{\mathsf{H}}_t \mathbf{1}_{[x,r]}(y) \ \forall x, y \in I^\circ.$$

And we furthermore observe that, the boundary behaviour of the $\mathfrak{h}(X)(t)$-diffusion at l is the boundary behaviour of the L-diffusion at r and its boundary behaviour at r is that of the L-diffusion at l and similarly for $\mathfrak{h}(\hat{X})(t)$. We thus obtain an equivalent problem where now $(\mathsf{b.c.}(l), \mathsf{b.c.}(r)) \leftrightarrow (\mathsf{b.c.}(r), \mathsf{b.c.}(l))$. Putting it all together, we reduce to the following 4 cases since all others can be obtained from the transformations above,

$$(nat, nat), (ref, nat), (ref, ref), (ref, abs).$$

The first case is easy since there are no boundary conditions to keep track of and is omitted. The second case is the one originally considered by Siegmund and studied extensively in the literature (see e.g. [21] for a proof). We give the proof for the last two cases.

First, assume l and r are regular reflecting for X and so absorbing for $\hat{X}$. Let $\mathcal{R}_\lambda$ and $\hat{\mathcal{R}}_\lambda$ be the resolvent operators associated with P_t and $\hat{\mathsf{P}}_t$ then with f being a continuous function with compact support in I° the function $u = \mathcal{R}_\lambda f$ solves Poisson's equation $\mathcal{D}_m\mathcal{D}_s u - \lambda u = -f$ with $\mathcal{D}_s u(l^+) = 0, \mathcal{D}_s u(r^-) = 0$. Apply $\mathcal{D}_m^{-1}$ defined by $\mathcal{D}_m^{-1} f(y) = \int_l^y m(z)f(z)dz$ for $y \in I^\circ$ to obtain $\mathcal{D}_s u - \lambda\mathcal{D}_m^{-1}u = -\mathcal{D}_m^{-1} f$ which can be written as,

$$\mathcal{D}_{\hat{m}}\mathcal{D}_{\hat{s}}\mathcal{D}_m^{-1}u - \lambda\mathcal{D}_m^{-1}u = -\mathcal{D}_m^{-1} f.$$

So $v = \mathcal{D}_m^{-1}u$ solves Poisson's equation with $g = \mathcal{D}_m^{-1} f$,

$$\mathcal{D}_{\hat{m}}\mathcal{D}_{\hat{s}}v - \lambda v = -g,$$

with the boundary conditions $\mathcal{D}_{\hat{m}}\mathcal{D}_{\hat{s}}v(l^+) = \mathcal{D}_s\mathcal{D}_m\mathcal{D}_m^{-1}u(l^+) = \mathcal{D}_s u(l^+) = 0$ and $\mathcal{D}_{\hat{m}}\mathcal{D}_{\hat{s}}v(r^-) = 0$. Now in the second case when l is reflecting and r absorbing we would like to check the reflecting boundary condition for $v = \mathcal{D}_m^{-1}u$ at r. Namely, that $(\mathcal{D}_{\hat{s}})v(r^-) = 0$ and note that this is equivalent to $(\mathcal{D}_m)v(r^-) = u(r^-) = 0$. This then follows from the fact that (since r is now absorbing for the L-diffusion) $(\mathcal{D}_m\mathcal{D}_s)u(r^-) = 0$ and that f is of compact support. The proof proceeds in the same way for both cases, by uniqueness of solutions to Poisson's equation (see e.g. Section 3.7 of [39]) this implies $v = \hat{\mathcal{R}}_\lambda g$ and thus we may rewrite the relationship as,

$$\mathcal{D}_m^{-1}\mathcal{R}_\lambda f = \hat{\mathcal{R}}_\lambda\mathcal{D}_m^{-1} f.$$

Let now f approximate δ_x with $x \in I^\circ$ to obtain with $r_\lambda(x, z)$ the resolvent density of $\mathcal{R}_\lambda$ with respect to the speed measure in $I^\circ \times I^\circ$,

$$\int_l^y r_\lambda(z, x)m(z)dz = m(x)\hat{\mathcal{R}}_\lambda\mathbf{1}_{[x,r]}(y).$$

Since $r_\lambda(z, x)m(z) = m(x)r_\lambda(x, z)$ we obtain,

$$\mathcal{R}_\lambda\mathbf{1}_{[l,y]}(x) = \hat{\mathcal{R}}_\lambda\mathbf{1}_{[x,r]}(y),$$

and the result follows by uniqueness of Laplace transforms.

It is certainly clear to the reader that the proof only works for x, y in the interior I°. In fact the lemma is not always true if we allow x, y to take the values l, r. To wit, first assume $x = l$ so that we would like,

$$\mathsf{P}_t \mathbf{1}_{[l,y]}(l) \stackrel{?}{=} \hat{\mathsf{P}}_t \mathbf{1}_{[l,r]}(y) = 1 \ \forall y.$$

This is true if and only if l is either absorbing, exit or natural for the L-diffusion (where in the case of a natural boundary we understand $\mathsf{P}_t \mathbf{1}_{[l,y]}(l)$ as $\lim_{x \to l} \mathsf{P}_t \mathbf{1}_{[l,y]}(x)$). Analogous considerations give the following: The statement of Lemma 13.1 remains true with $x = r$ if r is either a natural, reflecting or entrance boundary point for the L-diffusion. Enforcing the exact same boundary conditions gives that the statement remains true with y taking values on the boundary of I.

Remark 13.15 For the reader who is familiar with the close relationship between duality and intertwining first note that with the L-diffusion satisfying the boundary conditions in the paragraph above and denoting as in Sect. 13.2 by P_t the semigroup associated with an L-diffusion killed (not absorbed) at l our duality relation becomes,

$$P_t \mathbf{1}_{[x,r]}(y) = \hat{\mathsf{P}}_t \mathbf{1}_{[l,y]}(x).$$

It is then a simple exercise, see Proposition 5.1 of [16] for the general recipe of how to do this, that this is equivalent to the intertwining relation,

$$P_t \Lambda = \Lambda \hat{\mathsf{P}}_t,$$

where Λ is the unnormalized kernel given by $(\Lambda f)(x) = \int_l^x \hat{m}(z) f(z) dz$. This is exactly the intertwining relation obtained in (13.26) with $n_1 = n_2 = 1$.

Entrance Laws For $x \in I$ and $\mathfrak{h}_n$ a positive eigenfunction of P_t^n we would like to compute the following limit that defines our entrance law $\mu_t^x(\mathbf{y})$ (with respect to Lebesgue measure) and corresponds to starting the Markov process $P_t^{n,\mathfrak{h}_n}$ from $(x, \cdots, x)$,

$$\mu_t^x(\mathbf{y}) := \lim_{(x_1,\cdots,x_n) \to x\mathbf{1}} e^{-\lambda t} \frac{\mathfrak{h}_n(y_1, \cdots, y_n)}{\mathfrak{h}_n(x_1, \cdots, x_n)} \det\left(p_t(x_i, y_j)\right)_{i,j=1}^n .$$

Note that, since as proven in Sect. 13.3.10 all eigenfunctions built from the intertwining kernels are of the form $\det\left(h_i(x_j)\right)_{i,j=1}^n$ we will restrict to computing,

$$\mu_t^x(\mathbf{y}) := e^{-\lambda t} \det\left(h_i(y_j)\right)_{i,j=1}^n \lim_{(x_1,\cdots,x_n) \to x\mathbf{1}} \frac{\det\left(p_t(x_i, y_j)\right)_{i,j=1}^n}{\det\left(h_i(x_j)\right)_{i,j=1}^n}.$$

If we now assume that $p_t(\cdot, y) \in C^{n-1} \forall t > 0, y \in I^\circ$ and similarly $h_i(\cdot) \in C^{n-1}$ (in fact we only need to require this in a neighbourhood of x) we have,

$$\lim_{(x_1,\cdots,x_n)\to x\mathbf{1}} \frac{\det\left(p_t(x_i, y_j)\right)_{i,j=1}^n}{\det\left(h_i(x_j)\right)_{i,j=1}^n} = \lim_{(x_1,\cdots,x_n)\to x\mathbf{1}} \frac{\det\left(x_j^{i-1}\right)_{i,j=1}^n}{\det\left(h_i(x_j)\right)_{i,j=1}^n} \times \frac{\det\left(p_t(x_i, y_j)\right)_{i,j=1}^n}{\det\left(x_j^{i-1}\right)_{i,j=1}^n}$$

$$= \frac{1}{\det\left(\partial_x^{i-1} h_j(x)\right)_{i,j=1}^n} \times \det\left(\partial_x^{i-1} p_t(x, y_j)\right)_{i,j=1}^n .$$

For the fact that the Wronskian, $\det\left(\partial_x^{i-1} h_j(x)\right)_{i,j=1}^n > 0$ and in particular does not vanish see Sect. 13.3.10. Thus,

$$\mu_t^x(\mathbf{y}) = const_{x,t} \times \det\left(h_i(y_j)\right)_{i,j=1}^n \det\left(\partial_x^{i-1} p_t(x, y_j)\right)_{i,j=1}^n ,$$

is given by a biorthogonal ensemble as in (13.29). The following lemma, which is an adaptation of Lemma 3.2 of [47] to our general setting, gives some more explicit information.

Lemma 13.4 *Assume that for x' in a neighbourhood of x there is a convergent Taylor expansion $\forall t > 0, y \in I^\circ$,*

$$\frac{p_t(x', y)}{p_t(x, y)} = f(t, x') \sum_{i=0}^{\infty} \left(x' - x\right)^i \phi_i(t, y),$$

for some functions $f, \{\phi_i\}_{i\geq 0}$ that in particular satisfy $f(t, x)\phi_0(t, y) \equiv 1$. Then $\mu_t^x(\mathbf{y})$ is given by the biorthogonal ensemble,

$$const_{x,t} \times \det\left(h_i(y_j)\right)_{i,j=1}^n \det\left(\phi_{i-1}(t, y_j)\right)_{i,j=1}^n \prod_{i=1}^n p_t(x, y_i).$$

If moreover we assume that we have a factorization $\phi_i(t, y) = y^i g_i(t)$ then $\mu_t^x(\mathbf{y})$ is given by the polynomial ensemble,

$$const'_{x,t} \times \det\left(h_i(y_j)\right)_{i,j=1}^n \det\left(y_j^{i-1}\right)_{i,j=1}^n \prod_{i=1}^n p_t(x, y_i).$$

Proof By expanding the Karlin–McGregor determinant and plugging in the Taylor expansion above we obtain,

$$\frac{\det\left(p_t(x_i,y_j)\right)_{i,j=1}^n}{\prod_{i=1}^n p_t(x,y_i)} = \prod_{i=1}^n f(t,x_i) \sum_{k_1,\cdots,k_n\geq 0} \prod_{i=1}^n (x_i-x)^{k_i}$$
$$\times \sum_{\sigma\in\mathfrak{S}_n} sign(\sigma) \prod_{i=1}^n \phi_{k_i}(t,y_{\sigma(i)})$$
$$= \prod_{i=1}^n f(t,x_i) \sum_{k_1,\cdots,k_n\geq 0} \prod_{i=1}^n (x_i-x)^{k_i} \det\left(\phi_{k_i}(t,y_j)\right)_{i,j=1}^n .$$

First, note that we can restrict to $k_1,\cdots k_n$ distinct otherwise the determinant vanishes. Moreover, we can in fact restrict the sum over $k_1,\cdots,k_n \geq 0$ to $k_1,\cdots,k_n$ ordered by replacing $k_1,\cdots,k_n$ by $k_{\tau(1)},\cdots,k_{\tau(n)}$ and summing over $\tau\in\mathfrak{S}_n$ to arrive at the following expansion,

$$\frac{\det\left(p_t(x_i,y_j)\right)_{i,j=1}^n}{\prod_{i=1}^n p_t(x,y_i)} = \prod_{i=1}^n f(t,x_i)$$
$$\sum_{0\leq k_1<k_2<\cdots<k_n} \det\left(\left(x_j-x\right)^{k_i}\right)_{i,j=1}^n \det\left(\phi_{k_i}(t,y_j)\right)_{i,j=1}^n .$$

Now, write with $\mathbf{k}=(0\leq k_1<\cdots<k_n)$,

$$\chi_{\mathbf{k}}(z_1,\cdots,z_n) = \frac{\det\left(z_j^{k_i}\right)_{i,j=1}^n}{\det\left(z_j^{i-1}\right)_{i,j=1}^n},$$

for the Schur function and note that $\lim_{(z_1,\cdots,z_n)\to 0}\chi_{\mathbf{k}}(z_1,\cdots,z_n)=0$ unless $\mathbf{k}=(0,\cdots,n-1)$ in which case we have $\chi_{\mathbf{k}}\equiv 1$. We can now finally compute,

$$\lim_{(x_1,\cdots,x_n)\to x\mathbf{1}} \frac{\det\left(p_t(x_i,y_j)\right)_{i,j=1}^n}{\det\left(x_j^{i-1}\right)_{i,j=1}^n}$$
$$= \lim_{(x_1,\cdots,x_n)\to x\mathbf{1}} \frac{\det\left(p_t(x_i,y_j)\right)_{i,j=1}^n}{\det\left((x_j-x)^{i-1}\right)_{i,j=1}^n} = \prod_{i=1}^n p_t(x,y_i)$$
$$\times \lim_{(x_1,\cdots,x_n)\to x\mathbf{1}} \prod_{i=1}^n f(t,x_i)$$

$$\times \sum_{0 \le k_1 < k_2 < \cdots < k_n} \chi_{\mathbf{k}}(x_1 - x, \cdots, x_n - x) \det\left(\phi_{k_i}(t, y_j)\right)_{i,j=1}^n$$

$$= f^n(t, x) \times \prod_{i=1}^n p_t(x, y_i) \det\left(\phi_{i-1}(t, y_j)\right)_{i,j=1}^n .$$

The first statement of the lemma now follows with,

$$const_{x,t} = e^{-\lambda t} f^n(t, x) \frac{1}{\det\left(\partial_x^{i-1} h_j(x)\right)_{i,j=1}^n} .$$

The fact that when $\phi_i(t, y) = y^i g_i(t)$ we obtain a polynomial ensemble is then immediate.

References

1. M. Adler, E. Nordenstam, P. van Moerbeke, Consecutive minors for Dyson's Brownian motions. Stochastic Process. Appl. **124**(6), 2023–2051 (2014)
2. T. Assiotis, Hua-Pickrell diffusions and Feller processes on the boundary of the graph of spectra. Ann. Inst. Henri Poincaré Probab. Stat. (2019, accepted). https://arxiv.org/abs/1703.01813
3. T. Assiotis, Intertwinings for general β Laguerre and Jacobi processes. J. Theor. Probab. https://doi.org/10.1007/s10959-018-0842-0
4. T. Assiotis, Random growth and Karlin-McGregor polynomials. Electron. J. Probab. **23**, 106 (2018)
5. D. Bakry, I. Gentil, M. Ledoux, *Analysis and Geometry of Markov Diffusion Operators* (Springer, Berlin, 2014)
6. P. Biane, P. Bougerol, N. O'Connell, Littelmann paths and Brownian paths. Duke Math. J. **130**(1), 127–167 (2005)
7. F. Bornemann, P.J. Forrester, Singular values and evenness symmetry in random matrix theory. Forum Mathematicum (2015). https://www.degruyter.com/view/j/form.ahead-of-print/forum-2015-0055/forum-2015-0055.xml
8. F. Bornemann, M. LaCroix, The singular values of the GOE. Random Mat. Theory Appl. **4**(2), 1550009 (2015)
9. A. Borodin, P. Ferrari, Anisotropic growth of random surfaces in 2 + 1 dmensions. Commun. Math. Phys. **325**, 603–684 (2014)
10. A. Borodin, J. Kuan, Random surface growth with a wall and Plancherel measures for $O(\infty)$. Commun. Pure Appl. Math. **63**, 831–894 (2010)
11. A. Borodin, G. Olshanski, Markov processes on the path space of the Gelfand-Tsetlin graph and on its boundary. J. Funct. Anal. **263**, 248–303 (2012)
12. A.N. Borodin, P. Salminen, *Handbook of Brownian Motion-Facts and Formulae Second Edition*. Probability and Its Applications (Birkhauser, Basel, 2002)
13. M-F. Bru, Wishart processes. J. Theor. Probab. **4**(4), 725–751 (1991)
14. K. Burdzy, W. Wang, K. Ramanan, The Skorokhod problem in a time dependent interval. Stoch. Process. Appl. **119**(2), 428–452 (2009)
15. F. Calogero, Solution of the one-dimensional N-body problems with quadratic and/or inversely quadratic pair potentials. J. Math. Phys. **12**, 419–439 (1971)

16. P. Carmona, F. Petit, M. Yor, Beta-gamma random variables and intertwining relations between certain Markov processes. Rev. Mat. Iberoamericana **14**(2), 311–367 (1998)
17. M. Cerenzia, A path property of Dyson gaps, Plancherel measures for $Sp(\infty)$, and random surface growth. arXiv:1506.08742 (2015)
18. M. Cerenzia, J. Kuan, Hard-edge asymptotics of the Jacobi growth process (2016). https://arxiv.org/abs/1608.06384
19. L. Chen, D.W. Stroock, The fundamental solution to the Wright–Fisher equation. SIAM J. Math. Anal. **42**(2), 539–567 (2010)
20. B. Conrey, Notes on eigenvalue distributions for the classical compact groups, in *Recent Perspectives in Random Matrix Theory and Number Theory*, ed. by F. Mezzadri, N.C. Snaith. London Mathematical Society Lecture Note Series, vol. 332 (Cambridge University Press, Cambridge, 2005), pp. 111–146
21. T. Cox, U. Rösler, A duality relation for entrance and exit laws for Markov processes. Stochastic Processes Appl. **16**(2), 141–156 (1984)
22. C. Cuenca, Markov processes on the duals to infinite-dimensional classical lie groups (2016). http://arxiv.org/abs/1608.02281
23. M. Defosseux, Orbit measures, random matrix theory and interlaced determinantal processes. Annales de l'Institut Henri Poincare, Probabilites et Statistiques **46**(1), 209–249 (2010)
24. N. Demni, The Laguerre process and the generalized Hartman-Watson law. Bernoulli **13**(2), 556–580 (2007)
25. P. Diaconis, J.A. Fill, Strong stationary times via a new form of duality. Ann. Probab. **18**(4), 1483–1522 (1990)
26. T. Dieker, J. Warren, Determinantal transition kernels for some interacting particles on the line. Annales de l'Institut Henri Poincare, Probabilites et Statistiques **44**(6), 1162–1172 (2008)
27. T. Dieker, J. Warren, On the largest-eigenvalue process for generalized Wishart random matrices. ALEA **6**, 369–376 (2009)
28. Y. Doumerc, Matrices aleatoires, processus stochastiques et groupes de reflexions Ph.D. Thesis (2005). http://perso.math.univ-toulouse.fr/ledoux/files/2013/11/PhD-thesis.pdf
29. S. Ethier, T. Kurtz, *Markov Processes Characterization and Convergence*. Wiley Series in Probability and Statistics (Wiley, Hoboken, 1986)
30. P. Ferrari, R. Frings, On the partial connection between random matrices and interacting particle systems. J. Stat. Phys. **141**(4), 613–637 (2010)
31. P. Ferrari, R. Frings, Perturbed GUE minor process and Warren's process with drifts. J. Stat. Phys. **154**(1), 356–377 (2014)
32. J.A. Fill, V. Lyzinski, Strong stationary duality for diffusion processes. J. Theor. Probab. (2015). http://link.springer.com/article/10.1007/s10959-015-0612-1
33. P.J. Forrester, T. Nagao, Determinantal correlations for classical projection processes. J. Stat. Mech. Theory Exp. **2011**, 1–28 (2011)
34. P.J. Forrester, E.M. Rains, Inter-relationships between orthogonal, unitary and symplectic matrix ensembles, in *Random Matrix Models and Their Applications*, ed. by P.M. Bleher, A.R. Its, Mathematical Sciences Research Institute Publications, vol. 40 (Cambridge University Press, Cambridge, 2001)
35. A. Going-Jaeschke, M. Yor, A survey and some generalizations of Bessel processes. Bernoulli **9**(2), 313–349 (2003)
36. V. Gorin, Noncolliding Jacobi processes as limits of Markov chains on the Gelfand-Tsetlin graph. J. Math. Sci. **158**(6), 819–837 (2009)
37. M.R. Hestenes, Extension of the range of a differentiable function. Duke Math. J. **8**(1), 183–192 (1941)
38. J.R. Ipsen, H. Schomerus, Isotropic Brownian motions over complex fields as a solvable model for May-Wigner stability analysis (2016). https://arxiv.org/abs/1602.06364
39. K. Ito, H.P. McKean, *Diffusion Processes and Their Sample Paths*, Second Printing (Springer, Berlin, 1974)

40. K. Johansson, A multi-dimensional Markov chain and the Meixner ensemble. Ark. Mat. **46**(1), 79–95 (2010)
41. K. Johansson, Two time distribution in Brownian directed percolation. Commun. Math. Phys. **351**(2), 441–492 (2017)
42. L. Jones, N. O'Connell, Weyl chambers, symmetric spaces and number variance saturation. ALEA **2**, 91–118 (2006)
43. O. Kallenberg, *Foundations of Modern Probability*, Probability and its Applications, 2nd edn. (Springer, Berlin, 2002)
44. S. Karlin, *Total Positivity*, vol. 1 (Stanford University Press, Palo Alto, 1968)
45. S. Karlin, J. McGregor, Coincidence probabilities. Pac. J. Math. **9**(6), 1141–1164 (1959)
46. S. Karlin, H.M. Taylor, *A Second Course in Stochastic Processes* (Academic, Cambridge, 1981)
47. W. König, N. O'Connell, Eigenvalues of the laguerre process as non-colliding squared bessel processes. Electron. Commun. Probab. **6**, 107–114 (2001)
48. Y. Le Jan, O. Raimond, Flows, coalescence and noise. Ann. Probab. **32**(2), 1247–1315 (2004)
49. C. Lun, Eigenvalues of a Hermitian matrix of Brownian motions with drift. M.Sc. Thesis, University of Warwick (2012)
50. K. Matetski, J. Quastel, D. Remenik, The KPZ fixed point (2017). https://arxiv.org/abs/1701.00018
51. O. Mazet, Classification des semi-groupes de diffusion sur $\mathbb{R}$ associes a une famille de polynomes orthogonaux. Seminaire de probabilités de Strasbourg **31**, 40–53 (1997)
52. H.P. McKean, Elementary solutions for certain parabolic partial differential equations. Trans. Am. Math. Soc. **82**, 519–548 (1956)
53. E. Meckes, Concentration of measure and the compact classical matrix groups. Lecture Notes, IAS Program for Women and Mathematics (2014). https://www.math.ias.edu/files/wam/Haarnotes-revised
54. N. O'Connell, Directed polymers and the quantum Toda lattice. Ann. Probab. **40**(2), 437–458 (2012)
55. S. Pal, M. Shkolnikov, Intertwining diffusions and wave equations (2015). https://arxiv.org/abs/1306.0857
56. E.J. Pauwels, L.C.G. Rogers, Skew-product decompositions of Brownian motions, in *Geometry of Random Motion*, ed. by R. Durrett, M. Pinsky. Contemporary Mathematics, vol. 73 (American Mathematical Society, Providence, 1988), pp. 237–262
57. D. Revuz, M. Yor, *Continuous Martingales and Brownian Motion*. A Series of Comprehensive Studies in Mathematics, vol. 293, 3rd edn. (Springer, Berlin, 1999)
58. L.C.G. Rogers, J. Pitman, Markov functions. Ann. Probab. **9**(4), 573–582 (1981)
59. G. Schutz, Exact solution of the master equation for the asymmetric exclusion process. J. Stat. Phys. **88**(1), 427–445 (1997)
60. D. Siegmund, The equivalence of absorbing and reflecting barrier problems for stochastically Monotone Markov processes. Ann. Probab. **4**(6), 914–924 (1976)
61. L. Slominski, T. Wojciechowski, Stochastic differential equations with time-dependent reflecting barriers. Stochastics Int. J. Probab. Stochastic Process. **85**(1), 27–47 (2013)
62. D.W. Stroock, *Partial Differential Equations for Probabilists*. Cambridge Studies in Advanced Mathematics, vol. 112 (Cambridge University Press, Cambridge, 2008)
63. Y. Sun, Laguerre and Jacobi analogues of the Warren process (2016). https://arxiv.org/abs/1610.01635
64. B. Sutherland, Exact results for a quantum many-body problem in one dimension. Phys. Rev. A **4**, 2019–2021 (1971)
65. B. Tóth, Generalized ray–knight theory and limit theorems for self-interacting random walks on $\mathbb{Z}^1$. Ann. Probab. **24**(3), 1324–1367 (1996)
66. J. Warren, Dyson's Brownian motions,intertwining and interlacing. Electron. J. Probab. **12**, 573–590 (2007)

67. J. Warren, S. Watanabe, On the spectra of noises associated with Harris flows. Adv. Stud. Pure Math. **41**, 351–373 (2004)
68. J. Warren, P. Windridge, Some examples of dynamics for gelfand-tsetlin patterns. Electron. J. Probab. **14**, 1745–1769 (2009)
69. T. Weiss, P. Ferrari, H. Spohn, *Reflected Brownian Motions in the KPZ Universality Class*. Springer Briefs in Mathematical Physics, vol. 18, 1st edn. (Springer, Berlin, 2017)

Chapter 14
Brownian Sheet Indexed by $\mathbb{R}^N$: Local Time and Bubbles

Marguerite Zani and Wei Zhou

Abstract In this paper, we show a law of large numbers relating the bubbles of 1-dimensional, N-parameter Brownian Sheet on a bounded domain and the local time on that domain. This result generalizes the work of Mountford (Brownian sheet, local time and bubbles. In: *Séminaire de Probabilités, XXXVII*. Lecture Notes in Mathematics, vol 1832. Springer, Berlin, 2003, pp. 19–215).

Keywords Brownian Sheet · Bubbles · Local time

14.1 Introduction

The 1-dimensional, N-parameter Brownian sheet (or 1-dimensional Brownian sheet indexed by $\mathbb{R}_+^N$) is a Gaussian process with mean 0 and covariance function

$$\Sigma(\mathbf{s}, \mathbf{t}) = \prod_{l=1}^{N}(s_l \wedge t_l), \quad \mathbf{s}, \mathbf{t} \in \mathbb{R}_+^N .$$

We will use B.S. to refer to this process.

Definition 14.1 The white noise on $\mathbb{R}^N$—denoted by $\mathbb{W}_N$—is a zero mean Gaussian process of covariance function

$$\Sigma(A, B) = |A \cap B|$$

M. Zani (✉)
Institut Denis Poisson UMR 7013, Université d'Orléans, Université de Tours, CNRS, Orléans, France
e-mail: marguerite.zani@univ-orleans.fr

W. Zhou
Département de Mathématiques et Applications, École Normale Supérieure, Paris, France
e-mail: wei.zhou@ens.fr

C. Donati-Martin et al. (eds.), *Séminaire de Probabilités L*, Lecture Notes in Mathematics 2252, https://doi.org/10.1007/978-3-030-28535-7_14

where A, B are measurable sets on $\mathbb{R}^N$ and $|\cdot|$ is the Lebesgue measure.

We generalise the notation of an interval as follows

Definition 14.2 *Let* $\mathbf{u}, \mathbf{v}$ *be two vectors in* $\mathbb{R}^N$, *the* N*-th dimensional hypercube* $[\mathbf{u}, \mathbf{v}]$ *is defined by*

$$[\mathbf{u}, \mathbf{v}] := \prod_{i=1}^{N} [\min(u_i, v_i), \max(u_i, v_i)] .$$

By the Čentsov's representation, we can relate the B.S. to the white noise:

Theorem 14.1 (Čentsov) *Let* $\mathbb{W}_N$ *be the white noise on* $\mathbb{R}^N$. *Therefore, if* W *is the 1- dimensional brownian sheet indexed by* $\mathbb{R}_+^N$,

$$W(\mathbf{t}) = \mathbb{W}_N([0, \mathbf{t}]) , \quad \mathbf{t} \in \mathbb{R}_+^N .$$

In this paper we consider x bubbles for 1-dimensional and general N-parameter B.S. We generalize the work of Mountford [9] relating the number of bubbles to the local time of the brownian sheet. For previous works relating local time to Brownian excursions, see Revuz and Yor [11] or Rogers and Williams [12]. The asymptotics come from a first result similar to the master formula for the local time of a B.S. (see [11] and [13]), and from a convergence result on the most "frequent" bubbles. The strategy of Mountford relies on two different moves: firstly try to neglect the bubbles having asymptotically—when the size of the bubble tends to 0—a small contribution, and secondly compare the B.S. to a sum of Brownian Motions. As a matter of fact, the B.S. can be compared locally, i.e. considering an x-bubble for small x, to differences of independent Brownian Motions with different speeds. The time inhomogeneity in the result comes from this fact.

Studies of bubbles from differences of Brownian Motions has been precisely described in the work of Dalang and Walsh [2, 3]. For previous results on bubbles or level sets of N-dimensional B.S. see also Dalang and Mountford [1], Kendall [6] or Khoshnevisan [7].

This paper is organized as follows: in Sect. 14.2 we provide some definitions and give the main result for a general N-parameter, 1-dimensional B.S. Section 14.3 details the steps of the proof for $N = 3$. Section 14.4 presents the proof for the general $N > 3$ case. Finally the Appendix is devoted to technical results.

14.2 Main Result

We first give some basic definitions about bubbles and local time for the B.S. Let us call by "zero set" of W the set $W^{-1}(0) = \{\mathbf{t} \in \mathbb{R}_+^N , W(\mathbf{t}) = 0\}$. We know from [8] that the zero set of a N-parameter, d-dimensional B.S. is non trivial if and only if $d < 2N$ (trivial means $W^{-1}(0) = \partial\mathbb{R}_+^N$).

Definition 14.3 A bubble is a restriction of $\{W > 0\}$ to a single connected component.

We refer to an x-bubble ($x > 0$) a bubble which reaches its maximum in $[x, 2x]$.

In this setting (here $d = 1$ and $N \geq 1$), we have therefore the existence of bubbles. We define the local time at 0, see [8] Chapter 12 or [4] for the existence.

Definition 14.4 For any Borel set A of $\mathbb{R}_+^N$ we define the local time L of W at 0 by the following limit

$$L(A) := \lim_{\epsilon \to 0^+} L_\epsilon(A),$$

and L_ϵ is the approximate local time

$$L_\epsilon(A) = \frac{1}{2\epsilon} \int_A \mathbb{1}_{[-\epsilon,\epsilon]}(W(\mathbf{s}))d\mathbf{s}.$$

The main result of this paper is the following

Theorem 14.2 *Let N_x be the number of x-bubbles in $[0, 1]^N$ of a 1-dimensional Brownian sheet indexed by $\mathbb{R}_+^N$. Let L be its local time at 0 on $[0, 1]^N$. Then there exists a real positive constant k such that*

$$x^{2N-1} N_x \xrightarrow{P} k \int_{[0,1]^N} (\prod_{i=1}^N t_i)^{N-1} dL(\mathbf{t}), \text{ as } x \to 0^+, \tag{14.1}$$

where P stands for the convergence in probability.

To prove this theorem we follow the scheme of Mountford. We first start to show that we can neglect too small or too big bubbles, as well as bubbles close to the axes. Then we deal with average size bubbles comparing locally these bubbles to the ones obtained by summing independant standard Brownian Motions.

14.3 Proof in Dimension $N = 3$

In this paragraph we fix $N = 3$ and for simplicity we will denote by W the 3-parameter and 1 dimensional B.S. and by $\mathbb{W}$ the 3-dimensional white noise.

14.3.1 Number of "Rare" Bubbles

In this section, we look at the bubbles having small contribution in Theorem 14.2: we show that the expected number of bubbles with relatively small size or big size is small.

We start by looking at the bubbles of "small" size in the cube $[0, 1]^3$. More precisely we consider the x-bubbles which do not contain a cube of side length $x^2\epsilon$, for ϵ small enough.

Proposition 14.1 *For any $\alpha > 0$, there exists $\epsilon > 0$ such that the expected number of x-bubbles in $[0, 1]^3$ that do not contain a cube of side length $x^2\epsilon$ is less than α/x^5.*

The idea here is that for an x-bubble with side length of order smaller than ϵx^2 we can find a cube in the bubble with either:

- an edge which has a white noise contribution of order x
- a face which has a white noise contribution of order x
- a large white noise contribution for a certain cube inside the cube

Let us first show some technical results about bubbles of smaller dimensions. We define the following set

Definition 14.5 A 2-dimensional x-bubble $\mathscr{W}$ is in $A_n(x)$ if

$$\sup\{|\mathbb{W}_2(R)|\,,\ R \subset \mathscr{W}, R \text{ rectangle of side length } \leq 2^{-n}x^2\} > \frac{x}{18}\,,$$

where $\mathbb{W}_2$ is the 2-dimensional white noise.

Let us now define the number of bubbles that belong to $A_n(x)$ on the B.S :

Definition 14.6 For any $w \in [0, 1]$, let $Z(w, n)$ be the number of x-bubbles on the 2-parameter B.S. $W(r, s, w)$ which belong to $A_n(x)$. We denote by $X(u, n)$ and $Y(v, n)$ the analogous quantities when the first and second parameters are fixed in the B.S. W.

Lemma 14.1 *For $w \in [0, 1]$ we have*

$$E\left(\sum_{w\in x^2 2^{-n}\mathbb{Z}\cap[0,1]} Z(w, n)\right) \leq \frac{Ke^{-c2^n}}{x^5}\,. \tag{14.2}$$

We have analogous bounds for $E\left(\sum_{u\in x^2 2^{-n}\mathbb{Z}\cap[0,1]} X(u, n)\right)$ and $E\left(\sum_{v\in x^2 2^{-n}\mathbb{Z}\cap[0,1]} Y(v, n)\right)$.

Definition 14.7 For any $(v, w) \in [0, 1]^2$, we consider the Brownian motion obtained fixing two parameters in the B.S. W:

$$B^{X,v,w}(r) = W(r, v, w)\,, \quad r \in [0, 1]\,.$$

We denote by $X^{v,w}(x, n)$ the number of excursions of $B^{X,v,w}$ with maximum values in $[\frac{x}{2}, 2x]$ and such that

$$\sup \Big\{ |\mathbb{W}([r, r'] \times R)|, r < r' \in e; r' - r < 2^{-n}x^2; R \text{ rectangle in } [0, 1]^2 \\ \text{with bottom left vertex } (v, w) \text{ and of side length } \le 2^{-n}x^2 \Big\} > \frac{x}{18}. \tag{14.3}$$

Let $Y^{u,w}(x, n)$ and $Z^{u,v}(x, n)$ be the analogous quantities in directions Y and Z.

Lemma 14.2 *For $v, w \in [0, 1]$ fixed (respectively u, w and u, v)*

$$E(X^{v,w}(x, n)) \le \gamma e^{-c2^n},$$

for γ and c constants independent of $x \le 1$.

The proofs of Lemmas 14.1 and 14.2 are postponed to the Appendix.

Proof (Proof of Proposition 14.1) We set $\epsilon = 2^{-N}$ for some $N \in \mathbb{N}$. We consider an x-bubble $\mathscr{W}$ which does not contain a cube of side length $x^2 2^{-N}$. Let $\mathbf{t} = (t_1, t_2, t_3)$ be in $\mathscr{W}$ and such that $W(\mathbf{t}) = x$. We define:

- $v_1 = \inf\{t > 0, \exists R \text{ rectangle} \subset [t_1, t_1 + t] \times [t_2, t_2 + t] \times \{t_3\}; |\mathbb{W}(R)| > \frac{x}{9}$ or $\exists s \in [0, t], |W(\mathbf{t}) - W(\mathbf{t} + (s, 0, 0))| > \frac{x}{9}\}$
- $v_2 = \inf\{t > 0, \exists R \text{ rectangle} \subset [t_1, t_1 + t] \times \{t_2\} \times [t_3, t_3 + t]; |\mathbb{W}(R)| > \frac{x}{9}$ or $\exists s \in [0, t], |W(\mathbf{t}) - W(\mathbf{t} + (0, s, 0))| > \frac{x}{9}\}$
- $v_3 = \inf\{t > 0, \exists R \text{ rectangle} \subset \{t_1\} \times [t_2, t_2 + t] \times [t_3, t_3 + t]; |\mathbb{W}(R)| > \frac{x}{9}$ or $\exists s \in [0, t], |W(\mathbf{t}) - W(\mathbf{t} + (0, 0, s))| > \frac{x}{9}\}$.

And let $v = \min\{v_1, v_2, v_3\}$. We distinguish two cases:

(i) $v \le x^2 2^{-N}$
(ii) $v > x^2 2^{-N}$ and therefore there exists a $\mathbf{u} \in [t_1, t_1 + 2^{-N}x^2] \times [t_2, t_2 + 2^{-N}x^2] \times [t_3, t_3 + 2^{-N}x^2]$ such that $W(\mathbf{u}) = 0$.

Case (i) without loss of generality we can assume $v = v_1$. If v is determined by the Brownian motion $W(\mathbf{t} + (\cdot, 0, 0))$ we can use the argument of Mountford [9] for 2 dimensional bubbles to obtain a bound δ/x^5. Otherwise, we have for any $s \in [0, v]$

$$|W(\mathbf{t}) - W(\mathbf{t} + (s, 0, 0))| \le \frac{x}{9} \text{ and } \mathbb{W}([\mathbf{t}, \mathbf{t} + (v, v, 0)]) = \frac{x}{9}.$$

We split the interval into smaller ones and suppose

$$v \in (x^2 2^{-n-1}, x^2 2^{-n}] \text{ for } n \geq N\,.$$

Let s be a point of the grid $x^2 2^{n+1}\mathbb{N}^3$ contained in $[\mathbf{t}, \mathbf{t} + (v, v, v)]$. We notice that for any $\mathbf{t}' \in ([t_1, t_1 + v] \times \{t_2\} \times \{s_3\}) \cup (\{t_1\} \times [t_2, t_2 + v] \times \{s_3\})$,

$$|W(\mathbf{t}') - x| \leq 3\frac{x}{9} = \frac{x}{3}\,.$$

If there exists a rectangle R in $[(t_1, t_2, s_3), (t_1 + v, t_2 + v, s_3)]$ such that $|\mathbb{W}(R)| > x/18$, then there is a 2-dimensional bubble in the plane $\mathbb{R}^2 \times \{s_3\}$ containing $([t_1, t_1 + v] \times \{t_2\} \times \{s_3\}) \cup (\{t_1\} \times [t_2, t_2 + v] \times \{s_3\})$ in $A_n(x)$.

If conversely all rectangles in $[(t_1, t_2, s_3), (t_1+v, t_2+v, s_3)]$ containing (t_1, t_2, s_3) are of white noise $|\mathbb{W}(R)| \leq x/18$, then

$$|\mathbb{W}([t_1, t_1 + v] \times [t_2, t_2 + v] \times [t_3, s_3])| \geq \frac{x}{9} - \frac{x}{18}\,.$$

Notice that if there is a $w \in [t_3, s_3]$ such that $W(s_1, s_2, w) = 0$ then

$$|\mathbb{W}([t_1, s_1] \times [t_2, s_2] \times [t_3, w])| \geq x - 3\frac{x}{9} = \frac{2x}{3}\,,$$

and by taking the smallest w, we have an excursion with a contribution to $Z^{s_1,s_2}(x, n)$. In the other case, for any $w \in [t_3, s_3]$, $W(s_1, s_2, w) > 0$ then $[t_3, s_3]$ is within an excursion contributing to $Z^{s_1,s_2}(x, n)$.

In any case above, we can apply Lemma 14.1 or Lemma 14.2 to get the following bound for the expectation:

$$\frac{2^{2^n}}{x^4}\gamma e^{-c2^n} + \frac{Ke^{-c'2^n}}{x^5}\,.$$

Hence for $n \geq N$ and N large, we have the bound α/x^5.

Case (ii) as above, we assume $v = v_1$, and for $v_1 > x^2 2^{-N}$ we have for any $y \in [0, x^2 2^{-N}]$, $|W(\mathbf{t}) - W(\mathbf{t} + (y, 0, 0))| < x/9$ and for any R rectangle in $[t_1, t_1 + x^2 2^{-N}] \times [t_2, t_2 + x^2 2^{-N}] \times \{t_3\}$, $|\mathbb{W}(R)| < x/9$. Let s be the "smallest" point of the grid $x^2 2^{-N}\mathbb{Z}^3$ in $[\mathbf{t}, \mathbf{t} + x^2 2^{-N}(1, 1, 1)]$. Assume that for any y in $[0, x^2 2^{-N}]$, $W(s_1, s_2, t_3 + y) \in [11x/18, 25x/18]$ (if not, we have an excursion contributing to $Z^{s_1,s_2}(x, n)$).

Let u be a point in $[\mathbf{t}, \mathbf{t} + x^2 2^{-N}(1, 1, 1)]$ such that $W(\mathbf{u}) = 0$, therefore

$$\mathbb{W}([t_1, u_1] \times [t_2, u_2] \times [t_3, u_3]) \leq 0 - 6\frac{x}{9} = -\frac{2x}{3}\,.$$

It makes a contribution to $Z^{s_1,s_2}(x,n)$. In conclusion we have the bound $\frac{2^{2N}}{x^4}\gamma e^{-c2^N}$ and for N large, we get $\frac{\alpha}{x^5}$. This ends the proof of Proposition 14.1. □

This Proposition implies the two following Corollaries which deal with bubbles close to the axes and "big" bubbles repectively.

Corollary 14.1 *For any $\delta > 0$, there exists $\epsilon > 0$ such that the expected number of x-bubbles that reach x in $[0,\epsilon]\times[0,1]^2$ is less than δ/x^5.*

Proof Let $\epsilon > 0$ be such that the expected number of x-bubbles which do not contain a cube of side length ϵx^2 is less than $\frac{\delta}{x^5}$. We therefore bound following the dichotomy

$$\begin{aligned}\mathbb{E}[\text{number of } x\text{-bubbles}] &\le \frac{\delta}{x^5} + \mathbb{E}[\#\{x\text{-bubbles containing a cube of side length } \epsilon x^2\}]\\ &\le \frac{\delta}{x^5} + \frac{\mathbb{E}[|\{\mathbf{t}\in[0,1]^3, |W(\mathbf{t})| < 2x\}|]}{\epsilon^3 x^6}.\end{aligned}$$

To study the second term in the RHS above, we split the cube $[0,1]^3$ into smallest ones of side length ϵx. We denote by $K_{i,j,k}$ this cubes, where $(i,j,k)x\epsilon$ is the smallest vertex. Therefore we can bound

$$\begin{aligned}E[|\{\mathbf{t}\in[0,1]^3, |W(\mathbf{t})|<2x\}|] &= \sum_{1\le i,j,k\le 1/(x\epsilon)} E[|\{\mathbf{t}\in K_{i,j,k}; |W(\mathbf{t})|<2x\}|]\\ &= 2\sum_{1\le i,j,k\le 1/(x\epsilon)} \int_{K_{i,j,k}} \mathbb{P}(W(u,v,w)\in[0,2x])dudvdw\\ &\le 2\sum_{1\le i,j,k\le 1/(x\epsilon)} \epsilon^3x^3\mathbb{P}(W(ix\epsilon, jx\epsilon, kx\epsilon)\in[0,2x])\\ &= 2\sum_{1\le i,j,k\le 1/(x\epsilon)} \frac{\epsilon^3x^3}{\sqrt{\pi}}\int_0^{\sqrt{2}/\sqrt{ijkx\epsilon^3}} e^{-t^2}dt\\ &\le 2\frac{\epsilon^3x^3}{\sqrt{2\pi}}\sum_{1\le i,j,k\le 1/(x\epsilon)}\frac{2}{\sqrt{ijkx\epsilon}}\\ &= \frac{4x^{5/2}\epsilon^{5/2}}{\sqrt{2\pi}}\Big(\sum_{1\le x\le 1/(x\epsilon)}\frac{1}{\sqrt{i}}\Big)^3\\ &\le \frac{4x^{5/2}\epsilon^{5/2}}{\sqrt{2\pi}}\Big(\frac{2}{\sqrt{x\epsilon}}\Big)^3 = \kappa x\end{aligned}$$

for a constant κ. Therefore

$$\mathbb{E}[\text{number of } x\text{-bubbles}] \le \frac{\delta+\kappa}{x^5}.$$

Now, let us choose $\epsilon' > 0$, from the scaling property we know that $\frac{1}{\sqrt{\epsilon'}} W(\epsilon' u, v, w)$ is equal in law to W. Hence the expectation of bubbles attaining x in $[0, \epsilon'] \times [0, 1]^2$ is equal to the expectation of bubbles attaining $x/\sqrt{\epsilon'}$ in $[0, 1]^3$. From the inequality above, this last quantity is bounded by $K\epsilon'^{5/2}/x^5$. We choose ϵ' small and obtain the desired bound. □

Corollary 14.2 *For any $\delta > 0$, there exists $M > 0$ such that the expected number of x-bubbles of diameter bigger than Mx^2 is less that δ/x^5.*

For a fixed δ let us choose $\epsilon > 0$ such that both the expected number of x-bubbles that do not contain a cube of side length ϵx^2 and the expected number of x-bubbles which attain x in $[0, \epsilon] \times [0, 1]^2$ (or $[0, 1] \times [0, \epsilon] \times [0, 1]$ or $[0, 1]^2 \times [0, \epsilon]$) are bounded by δ/x^5. Consider now a x-bubble $\mathscr{W}$ which is not one of the mentioned above. Hence $\mathscr{W}$ contains a cube of side ϵx^2 and attains x in $[\epsilon, 1]^3$. Let us consider such a bubble with diameter bigger than Mx^2.

Let us define the set A of points $\boldsymbol{t}$ which satisfies the three following conditions:

(A_1) $0 < W(\mathbf{t}) < 2x$
(A_2) there is a cube containing $\boldsymbol{t}$ of side length ϵx^2 entirely positive
(A_3) all contours of type $\{\mathbf{t} + h\mathbf{s};\ h \in [0, \frac{Mx^2}{2}],\ \max(|s_i|) = 1\}$ are not entirely negative.

Note that the number of bubbles we want to estimate is less than $|A|/(\epsilon^3 x^6)$. Now we can write

$$E[|A|] = E[\int_{[0,1]^3} \mathbb{1}_A(\mathbf{s}) d\mathbf{s}] \le E(\int_{[\epsilon x^2,1]^3} \mathbb{1}_A(\mathbf{s}) d\mathbf{s} + 3\epsilon x^2).$$

Exchanging integral and expectation we have

$$E[|A|] \le \int_{[\epsilon x^2,1]^3} E(\mathbb{1}_A(\mathbf{s})) d\mathbf{s} + 3\epsilon x^2.$$

We can write conditionally to the value of $W(\mathbf{s})$:

$$E[\mathbb{1}_A(\mathbf{s})] = E[\mathbb{1}_{[0,2x]}(W(\mathbf{s}))\mathbb{1}_{A_2,A_3}(\mathbf{s})] = E[E[\mathbb{1}_{A_2,A_3}(\mathbf{s})|W(\mathbf{s}) \in [0, 2x]]].$$

In order to bound the RHS above, we consider two cases:

- The 2 and 3-dimensional contributions are large:

$$\sup_{R=[\mathbf{s},\mathbf{s}'],|\mathbf{s}-\mathbf{s}'|\le Mx^2/2,\ \mathbf{s} \text{ and } \mathbf{s}' \text{ have at least two different coordinates}} \{|W(R)|\} \ge \frac{x}{18}.$$

From previous arguments, we know that this probability is bounded by e^{-c/x^2}.

- The white noise contribution is small. We consider the event:

$$\mathscr{A}_1 = \{ \inf_{y \in [0, Mx^2/2]} W(\mathbf{s} + (y, 0, 0)) \in [-\frac{5x}{6}, 2x]\} .$$

We have

$$P(\mathscr{A}_1) = P(\mathscr{N}(0, \frac{1}{2}) \in [-\frac{5}{6\sqrt{M}}, \frac{2}{\sqrt{M}}]) = f(M) .$$

Performing identically in the two other directions, we get $P(\bar{A}_3) \geq (1 - f(M))^{12}$ since $\cap_i \mathscr{A}_i$ means we can find a box within $Mx^2/2$ of s whose surface is entirely negative. Summarizing, we get

$$E[\mathbb{1}_{A_2, A_3}(\mathbf{s}) | W(\mathbf{s}) \in [0, 2x]] \leq 1 - (1 - f(M))^6 + e^{-c/x^2} ,$$

and

$$E[|A|] \leq 3\epsilon x^2 + (1 - (1 - f(M))^6 + e^{-c/x^2}) \int_{[\epsilon,1]^3} P(W(\mathbf{s}) \in [0, 2x]) d\mathbf{s} .$$

From Khoshnevisan ([8], Proposition 3.4.1) we can bound $P(W(\mathbf{s}) \in [0, 2x])$ by Kx for some constant K independent of x. Hence for M large and small x we have $E[|A|] \leq \delta\epsilon^3 x$ and the Corollary 14.2 is proven. □

14.3.2 Estimates of Most Frequent Bubbles

We now consider a convergence result for most frequent bubbles, i.e. bubbles not too big or too small and not too close to the axes.

The main idea of this section is to approximate the B.S. locally by the sum of three independent Brownian Motions. It is important to remark that this is a local approximation.

We start with some general convergence result

Proposition 14.2 *If X^n is a process on the cube $[0, 1]^3$ such that*

$$X^n(u, v, w) = B_1^n(u) + B_2^n(v) + B_3^n(w) + V^n(u, v, w) + X^n(0, 0, 0)$$

where

- *B_1^n, B_2^n, B_3^n are independent and converge weakly to independent Brownian motions B_1, B_2 and B_3 respectively, in the space $(\mathscr{C}, \|\cdot\|_\infty)$.*
- *$X^n(0, 0, 0) = c_n \to c$ as $n \to +\infty$*
- *$\sup_{u,v,w} |V^n(u, v, w)| \to 0$ a.s.*

then the distribution of the number of 1*-bubbles of* X^n *on* $[0, m]^3$ *which have size at least* γ *and are contained in* $[0, m]^3$ *converge to the corresponding number for the process sum of the Brownian motions:* $X(u, v, w) = B_1(u) + B_2(v) + B_3(w) + c$.

Proof From Skorohod's theorem [5], we can consider that X^n and X are defined on some space $\tilde{\Omega}$ and that X^n tends to X uniformly on compact sets almost surely. Let A_n be the number of 1-bubbles contained in $[0, m]^3$ with size at least γ, and A the corresponding number for X. We use the following Lemma:

Lemma 14.3 *Let* $X(\omega)$ *be a sample path of* X *and* $G_1, \cdots, G_R$ *th* x*-bubbles of* $X(\omega)$ *of area at least* γx^6. *Suppose* $X_n(\omega) \to X(\omega)$ *in uniform norm. Then for all* i, $|G_i| \neq \gamma x^6$ *and for* n *large enough, we have:* $X_n(\omega)$ *has* R *bubbles* $G_1^n, \cdots, G_R^n$ *of area at least* γx^6 *such that:*

(i) For all i, $G_i \in (0, m)^3 \Leftrightarrow G_i^n \in (0, m)^3$
(ii) For all i, $G_i = \liminf_{n\to\infty} G_i^n = \limsup_{n\to\infty} G_i^n$
(iii) For all i, $|G_i^n| \to_{n\to\infty} |G_i|$. □

Proof (Proof of Lemma 14.3) It is easy to see that no bubble has area exactly γx^6. Let us show that the number of x-bubbles is fixed for n large enough. Let the (*a.s.* finite) x-bubbles of $X(\omega)$ be $G_1, \cdots, G_r, r \geq R$. Suppose that $m = 1$. We now use the following two lemmas which proofs are postponed to the Appendix.

Consider G a component of X defined above. If the maximum of G occurs at (t_1, t_2, t_3) then in a neighbourhood of t_1, B_1 must assume a local maximum at t_1. We define (s_l^1, s_u^1) as the largest interval where the process $s \to X(s, t_2, t_3)$ is greater than zero. Necessarily we have $B_1(s_l^1) = B_1(s_u^1) = -B_2(t_2) - B_3(t_3)$ and (s_l^1, s_u^1) is an excursion of B_1 above $-B_2(t_2) - B_3(t_3)$. Similarly, we define (s_l^2, s_u^2) and (s_l^3, s_u^3). We denote by $C = \{t_1\} \times \{t_2\} \times [s_l^3, s_u^3] \cup \{t_1\} \times [s_l^2, s_u^3] \times \{t_3\} \cup [s_l^1, s_u^1] \times \{t_2\} \times \{t_3\}$ and $K = [s_l^1, s_u^1] \times [s_l^2, s_u^2] \times [s_l^3, s_u^3]$ the cube generated by C. □

Lemma 14.4 *A.s. every positive bubble* G *is such that a.s. for all* $\beta > 0$ *there is a cube containing the cube* K *generated by* G *which is within* β *of* K *and on the surface of which* $X < 0$. □

Lemma 14.5 *For* X *restricted to a cube* K, *any disctinct two* x*-bubbles are a.s. non touching.* □

From Lemma 14.4 above, we know that G_i is contained in a cube inside each *a.s.* we can find a $\mathbf{t}_i \in G_i$ such that $X(\omega)(\mathbf{t}_i) > x$. Therefore for n large enough we have $X_n(\omega)(\mathbf{t}_i) > x$ and we can define G_i^n as the x-bubble containing $\mathbf{t}_i$.

Suppose now that there exists a distinct further x-bubble G_{r+1}^n. Taking a subsequence if required, we can assume that there exists $\mathbf{t}_{r+1}^n$ for each n such that $\mathbf{t}_{r+1}^n \notin \cup_{i=1}^r G_i^n$ and $X_n(\mathbf{t}_{r+1}^n) \geq x$. Let $\mathbf{t}_{r+1}$ be a limit point of $\mathbf{t}_{r+1}^n$, therefore $X(\omega)(\mathbf{t}_{r+1}) \geq x$ and $\mathbf{t}_{r+1}$ belongs to a certain G_i. Hence there exists a continuous path γ_i from $\mathbf{t}_i$ to $\mathbf{t}_{r+1}$ on which $X(\omega) > 0$. By uniform convergence, for n large enough, $X_n(\omega) > 0$ on γ_i and in a neighbourhood of $\mathbf{t}_{r+1}$. Hence for an infinity of n, $\mathbf{t}_{r+1}^n \in G_i^n$. This contradiction implies that there are only r disctinct bubbles for $X_n(\omega)$.

From the construction of G_i^n, we have (i).

To show (ii) let us first note that by uniform convergence, $G_i \subset \liminf G_i^n$. Let $y \in \limsup G_i^n$, and taking a subsequence, we can suppose $\forall n$, $y \in G_i^n$. Suppose that $y \notin G_i$. Then for any continuous path $\gamma : [0,1] \to \mathbb{R}^3$ there is a s_γ such that $\gamma(s_\gamma) \in \partial G_i$. From Lemma 14.5 any two $(X(\omega)(y)/2 \wedge x)$-bubbles are a.s. non touching and there exists $\alpha > 0$ such that $X(\omega)(s) < 0$ for any $s \in [s_\gamma + \alpha, s_\gamma + 2\alpha]$. By uniform convergence, for n large enough, $X_n(\omega) \circ \gamma < 0$ on this interval, which is contradictory. Hence $y \in G_i$ and $\limsup G_i^n \subset G_i$. (ii) is proved.

We have therefore $\limsup |G_i^n| \geq \liminf |G_i^n| \geq |G_i|$ by Fatou's lemma. Since $\liminf G_i^n = \limsup G_i^n = G_i$ we have $[0,1]^3 \setminus G_i = \liminf([0,1]^3 \setminus G_i^n)$ so that $1 - |G_i| \leq 1 - \limsup |G_i^n|$, hence $\limsup |G_i^n| = \liminf |G_i^n| = |G_i|$ and (iii) is proved. □

Hence from Lemma 14.3 above, since $X_n(\omega)$ converges uniformly to $X(\omega)$, we have $A^n(\omega) \to A(\omega)$, and A^n tends to A in law since their laws are entirely determined by the law of X^n and X. This ends the proof of Proposition 14.2. □

Let us denote by $g^\gamma(c, \mathbf{m})$ the number of bubbles described in Proposition 14.2 on $[0, m_1] \times [0, m_2] \times [0, m_3]$. We have the following results on g:

Lemma 14.6

(i) $g^\gamma(c, \mathbf{m})$ *is continuous a.s. in* $\mathbf{m}$ *for* c, γ *fixed.*
(ii) For c *such that* $(|c| - 2)^2 > m_1 + m_2 + m_3$,

$$E[g^\gamma(c, \mathbf{m})] \leq \frac{m_1 m_2 m_3}{\sqrt{2\pi}\gamma} \exp\left(\frac{-(|c|-2)^2}{2(m_1+m_2+m_3)}\right) \tag{14.4}$$

Lemma 14.7 *For* $\epsilon > 0$ *fixed, let* $\mathbf{t}^n \to \mathbf{t} \in [0,1]^3$, $x_n \to 0$, $(m_1^n, m_2^n, m_3^n) \to \mathbf{m}$, $c_n \to c$ *then for a Brownian sheet* W *the conditional number of* x_n*-bubbles contained in*

$$[\mathbf{t}^n, \mathbf{t}^n + x_n^2(\frac{m_1^n}{t_2^n t_3^n}, \frac{m_2^n}{t_1^n t_3^n}, \frac{m_3^n}{t_1^n t_2^n})]$$

of size at least $\gamma x_n^6/(t_1^n t_2^n t_3^n)^2$ *conditionally on* $W(\mathbf{t}^n) = c_n x_n$ *converges to* $g^\gamma(c, \mathbf{m})$.

From Lemma 14.6 we have the following

Corollary 14.3 *For any* $\mathbf{t} \in [\epsilon, 1]^3$, M, γ *fixed and* K *fixed big enough, we can bound the expected number of* x*-bubbles in* $[\mathbf{t}, \mathbf{t} + (\frac{M}{t_2t_3}, \frac{M}{t_1t_3}, \frac{M}{t_1t_2})]$ *of size at least* $\gamma x^6/(t_1t_2t_3)^2$ *conditionally on* $W(\mathbf{t}) = Kx$ *by*

$$\frac{cM^3}{\gamma} \exp\left(\frac{-(K-2)^2}{M}\right)$$

for c *a constant independent of* K, M.

14.3.3 Bubbles and Local Time

We fix $\delta > 0$ arbitrary small and choose ϵ to bound the expected number of x-bubbles reaching x in $[0, \epsilon] \times [0, 1]^2$ by δ/x^5. Let $M > 0$ be large enough such that the expected number of x-bubbles of diameter bigger than Mx^2 is less than δ/x^5, and fix N_0 large enough such that the number of bubbles that do not contain a cube of side length $2^{-N_0}x^2$ is less than δ/x^5. For the remaining of this part, if not mentionned, the bubbles are of diameter less than Mx^2 and volume greater than $2^{-3N_0}x^6$. We fix $\gamma, m > 0$ constants of Lemma 14.6 that will be chosen later on. We divide $[\epsilon, 1]^3$ into small cubes $C^i = [r_1^i, r_2^i] \times [s_1^i, s_2^i] \times [t_1^i, t_2^i]$ such that

$$\max(\frac{r_2^i}{r_1^i}, \frac{s_2^i}{s_1^i}, \frac{t_2^i}{t_1^i}) \leq \gamma + 1 .$$

We wish to show that the number of bubbles entirely contained in any C^i is of order $\int_{C^i} u^2v^2w^2 dL(u, v, w)$ in probability. We divide each cube C into smallest ones of size $u \times v \times w$ with

$$u := \inf\{y > \frac{mx^2}{s_1t_1}; \frac{r_2 - r_1}{y} \in \mathbb{Z}\}$$

$$v := \inf\{y > \frac{mx^2}{r_1t_1}; \frac{s_2 - s_1}{y} \in \mathbb{Z}\}$$

$$w := \inf\{y > \frac{mx^2}{r_1s_1}; \frac{t_2 - t_1}{y} \in \mathbb{Z}\}.$$

We also denote by u_i, v_j, w_k the intersection points of the grid and $\sigma_{i,j,k}$ the cube of the grid with (u_i, v_j, w_k) the smallest vertex. Let $\sigma'_{i,j,k}$ be a cube in $\sigma_{i,j,k}$with (u_i, v_j, w_k) the smallest vertex and of size $u\frac{s_1t_1}{v_jw_k} \times v\frac{r_1t_1}{u_iw_k} \times w\frac{r_1s_1}{u_iv_j}$. Let $N_{i,j,k}$ be the number of x-bubbles in $\sigma'_{i,j,k}$ of size at least $\gamma x^6/(u_iv_jw_k)^2$. Conditionally on $W(u_i, v_j, w_k) = cx$, we have the convergence described in Lemma 14.7. Let us define:

$$N^K_{i,j,k} = N_{i,j,k}\mathbb{1}_{[-Kx,Kx]}(W(u_i, v_j, w_k))$$

where K is a constant detailed further on. We now show that for a certain K most of the bubbles make a contribution to $N^K_{i,j,k}$.

Lemma 14.8 *For any γ, m of Lemma 14.6 there is a K depending on m, γ, ϵ such that*

$$E[\sum_{i,j,k}(N_{i,j,k} - N^K_{i,j,k}] \leq \frac{\delta}{x^5}|C|^2.$$

We consider the bubbles in C that do not contribute to $N_{i,j,k}$ and denote by N their number.

Lemma 14.9 *For $\epsilon > 0$, and M, N_0 chosen as before, we can choose γ small and m big enough so that we have*

$$E[N] \le \frac{\delta}{x^5}|C|.$$

We consider the bubbles which intersect the frontier of C and denote by Z the number of such bubbles. We can bound:

Lemma 14.10

$$E[Z] \le \frac{2\delta}{x^5}|C|\,.$$

We now estimate the sum $\sum_{i,j,k} N^K_{i,j,k}$ which contains most of the contributing bubbles of Theorem 14.2

Proposition 14.3 *For all C we have*

$$x^{10}E[(\sum_{i,j,k} N^N_{i,j,k} - g^\gamma(W(u_i, v_j, w_k)/x, \mathbf{m})\mathbb{1}_{|W(u_i,v_j,w_k)|\le Kx})^2] \underset{x\to 0}{\to} 0, \tag{14.5}$$

where $\mathbf{m} = [(0,0,0),(m,m,m)]$. Thus the convergence is in probability.

Proof (Proof of Proposition 14.3) Let us denote by

$$\Delta_{ijk} = N^N_{ijk} - g^\gamma(W(u_i, v_j, w_k)/x, \mathbf{m})\mathbb{1}_{|W(u_i,v_j,w_k)|\le Kx}\,.$$

We develop the RHS of (14.5):

$$x^{10}\sum_{i,j,k} E(\Delta^2_{i,j,k})$$
$$+ x^{10}\left[\sum_{i,j,k\neq k'} E(\Delta_{i,j,k}\Delta_{i,j,k'}) + \sum_{i,j\neq j',k} E(\Delta_{i,j,k}\Delta_{i,j',k})\right.$$
$$\left. + \sum_{i\neq i',j,k} E(\Delta_{i,j,k}\Delta_{i',j,k})\right]$$
$$+ x^{10}\left[\sum_{i,j\neq j',k\neq k'} E(\Delta_{i,j,k}\Delta_{i,j',k'}) + \sum_{i\neq i',j\neq j',k} E(\Delta_{i,j,k}\Delta_{i',j',k})\right.$$

$$+ \sum_{i\neq i',j,k\neq k'} E(\Delta_{i,j,k}\Delta_{i',j,k'})\Bigg]$$

$$+ x^{10} \sum_{i\neq i',j\neq j',k\neq k'} E(\Delta_{i,j,k}\Delta_{i',j',k'})\,. \tag{14.6}$$

Since N^K_{ijk} and $g^\gamma(W(u_i, v_j, w_k)/x, \mathbf{m})$ are bounded the first two lines of (14.6) above have bound $x^{10}(\frac{C}{x^6} + \frac{C'}{x^8}) = O(x^2)$. For the third line in (14.6), we use the total probability formula and get the bound

$$3C'' \sup_{i,j\neq j',k\neq k'} \big\{P(|W(u_i, v_j, w_k)| \leq Kx, |W(u_i, v_{j'}, w_{k'})| \leq Kx)\big\} = O(x)\,.$$

It remains to show that the fourth line of (14.6) tends to 0.

We first remark that for $i<i'$, $j<j'$, $k<k'$, conditionally on $W(u_{i'}, v_{j'}, w_{k'})$, the r.v. $N^K_{i',j',k'}$ is independant of $N^K_{i,j,k}$. By symmetry of the indexes, we only need to consider the case $i < i'$, $j > j'$, $k > k'$. We now consider $N^K_{i',j',k'}$ conditionned on

$$A = \big\{(W(r, s, t))_{,u_i\leq r\leq u_{i+1},v_j\leq s\leq v_{j+1},w_k\leq t\leq w_{k+1}}, W(u_{i'}, v_{j'}, w_{k'})\big\}\,. \tag{14.7}$$

For $r \in [u_{i'}, u_{i'+1}]$, $s \in [v_{j'}, v_{j'+1}]$ and $t \in [w_{k'}, w_{k'+1}]$ we decompose

$$W(r, s, t) = B_1(r)+B_2(s)+B_3(t)+D_1(r, s)+D_2(r, t)+D_3(s, t)+C(r, s, t)+W(u_{i'}, v_{j'}, w_{k'})\,,$$

where the Brownian motion B_1 is defined $B_1(r) = W(r, v_{j'}, w_{k'}) - W(u_{i'}, v_{j'}, w_{k'})$. We define similarly B_2 and B_3. Let D_1 be the 2-parameter B.S.

$$D_1(r, s) = W(r, s, w_{k'}) - W(r, v_{j'}, w_{k'}) - W(u_{i'}, s, w_{k'}) + W(u_{i'}, v_{j'}, w_{k'})\,,$$

and define similarly D_2, D_3. Finally, C is the white noise in the cube $[(u_{i'}, v_{j'}, w_{k'}), (r, s, t)]$. We see that B_1, D_1, D_2 and C are independant of A stated in (14.7). Let $\mathscr{A}$ be the intersection of the region where the white noise influences $N^K_{i,j,k}$ and $N^K_{i',j',k'}$ conditioned on $W(u_i, v_j, w_k)$ and $W(u_{i'}, v_{j'}, w_{k'})$. We wee that $\mathbb{W}(\mathscr{A})$ has variance of order $O(x^4)$. As x tends to 0, $|\mathbb{W}(\mathscr{A})|$ converges in probability to a Dirac at 0, and $B_2 + B_3 + D_3 - \mathbb{W}(\mathscr{A})$ converges in probability to $B + B' + D$, where B, B' are Brownian motions and D a 2-parameter B.S. (B, B', D, $\mathscr{A}$ are independent). From Proposition (14.2),

$$\sup\big\{E[\Delta_{i,j,k}\Delta_{i',j',k'}|W(u_i, v_j, w_k) \leq Kx, W(u_{i'}, v_{j'}, w_{k'}) \leq Kx]\big\} \to 0\,.$$

Bounding the whole expectation we have

$$\begin{aligned} E[\Delta_{i,j,k}\Delta_{i',j',k'}] \leq \sup\big\{E[\Delta_{i,j,k}\Delta_{i',j',k'}|W(u_i, v_j, w_k) \leq kx, W(u_{i'}, v_{j'}, w_{k'}) \leq Kx]\big\} \\ \times P(W(u_i, v_j, w_k) \leq Kx, W(u_{i'}, v_{j'}, w_{k'}) \leq Kx) \end{aligned}$$

Since $(W(u_i, v_j, w_k), W(u_{i'}, v_{j'}, w_{k'}))$is a Gaussian vector, when x tends to 0,

$$P(W(u_i, v_j, w_k) \leq Kx, W(u_{i'}, v_{j'}, w_{k'}) \leq Kx) = O(x^2) .$$

Hence we have shown that the fourth term in (14.6) tends to 0 □

14.3.4 *Proof of Theorem 14.2 when $N = 3$*

As defined previously, the local time at 0 of a 3-parameter B.S. is

$$L([x_1, x_2] \times [y_1, y_2] \times [z_1, z_2]) = \lim_{\epsilon \to 0} \frac{1}{2\epsilon} \int_{x_1}^{x_2} \int_{y_1}^{y_2} \int_{z_1}^{z_2} \mathbb{1}_{[-\epsilon,\epsilon]}(W(u, v, w))dudvdw .$$

We show the following convergence of Riemann sum type.

Proposition 14.4 *Let $f : \mathbb{R} \to \mathbb{R}$ be a bounded measurable function of compact support and K a cube in $]0, +\infty[^3$ with K_0 the bottom-left corner. We divide K into smallest cubes of size $c_1x^2 \times c_2x^2 \times c_3x^2$ and denote by $\mathbf{g}_{i,j,k}$ the point of coordinates $K_0 + (ic_1x^2, jc_2x^2, jc_3x^2)$. Hence*

$$x^5 \sum_{i,j,k} f(\frac{W(\mathbf{g}_{i,j,k})}{x}) \xrightarrow{P} \frac{1}{c_1c_2c_3} \int_{\mathbb{R}} f(t)dt L(K) \text{ as } x \to 0^+ . \quad (14.8)$$

Proof (Proof of Proposition 14.4) Let $[-A, A]$ be the support of f and let H be the set of bounded measurable functions having support $[-A, A]$ and satisfying convergence (14.8). By linearity, we see that H is a vector space. Actually, we only need to prove the result for functions of type $\mathbb{1}_{[a,b]}$ for $-A \leq a \leq b \leq A$. Without loss of generality we assume $0 \leq a \leq b$ and compute the expectation of the difference: denoting by $W_{i,j,k} = W(\mathbf{g}_{i,j,k})$ we have

$$E\left[\frac{1}{c_1c_2c_3}(b-a)L(K) - x^5 \sum_{i,j,k} \mathbb{1}_{[ax,bx]}(W_{i,j,k})\right]$$

$$= E\left[\frac{1}{c_1c_2c_3}bL(K) - \frac{x^5}{2} \sum_{i,j,k} \mathbb{1}_{[-bx,bx]}(W_{i,j,k})\right]$$

$$- E\left[\frac{1}{c_1c_2c_3}aL(K) - \frac{x^5}{2} \sum_{i,j,k} \mathbb{1}_{[-bx,bx]}(W_{i,j,k})\right] .$$

We only need to show that one of the above terms tends to zero while x tends to 0. For example, using the definition of local time with $\epsilon = bx$ we have

$$\lim_{x\to 0} E\left[\frac{1}{c_1c_2c_3}bL(K) - \frac{x^5}{2}\sum_{i,j,k}\mathbb{1}_{[-bx,bx]}(W_{i,j,k})\right]$$

$$= \lim_{x\to 0}\frac{1}{2c_1c_2c_3x}E\Big[\int_{x_1}^{x_2}\int_{y_1}^{y_2}\int_{z_1}^{z_2}\mathbb{1}_{[-bx,bx]}(W(u,v,w))dudvdw$$

$$- c_1x^2c_2x^2c_3x^2\sum_{i,j,k}\mathbb{1}_{[-bx,bx]}(W_{i,j,k})\Big]$$

$$= \lim_{x\to 0}\frac{1}{2c_1c_2c_3x}\sum_{i,j,k}\int_{K_{i,j,k}}P(W(u,v,w)\in[-bx,bx])$$

$$- P(W_{i,j,k}\in[-bx,bx])dudvdw.$$

Now we notice that $P(W(u,v,w)\in[-bx,bx]) - P(W_{i,j,k}\in[-bx,bx]) = O(x^3)$ and by variation of density, the difference is $O(x^2)$ and when x tends to 0 the difference tends to 0. Since $\mathbb{1}_{[a,b]}$ is positive, we have the convergence in L^1. The functions are bounded so we use the Lebesgue convergence theorem to conclude. □

Proof (Proof of Theorem 14.2) From Propositions 14.3 and 14.4, on a cube C we have

$$x^5\sum_{i,j,k}N_{i,j,k}^K \xrightarrow{P} \frac{(r_1s_1t_1)^2}{m^3}\int_{-K}^{K}g^\gamma(x,\mathbf{m})dxL(C) = k\int_C u^2v^2w^2dL(u,v,w)\,.$$

Notice that $\max(r_2/r_1, s_2/s_1, t_2/t_1) \le 1+\gamma$. Summing over all the cubes C_l we get

$$x^5\sum_l\sum_{i,j,k}(N_{i,j,k}^K)_l \xrightarrow{P} k\int_{[\epsilon,1]^3}u^2v^2w^2dL(u,v,w)\,.$$

Since we can choose ϵ arbitrary close to 0, we can replace $[\epsilon,1]^3$ by $[0,1]^3$ in the integral above. The remaining of the x-bubbles are in one of the following categories:

- x-bubbles of diameter greater than Mx^2, by Corollary 14.2 their expected number is less than δ/x^5;
- x-bubbles which do not contain a cube of size $2^{-3N}x^6$; by Proposition 14.1 their expected number is bounded by δ/x^5;
- x-bubbles which reach x in $[0,\epsilon]\times[0,1]^2\cup[0,1]\times[0,\epsilon]\times[0,1]\cup[0,1]^2\times[0,\epsilon]$; by Corollary 14.1, their expected number is bounded by δ/x^5;

- x-bubbles contained in a cube C_l but not in σ', by Lemma 14.9 their expected number is bounded by δ/x^5;
- x-bubbles which touch a frontier of a cube C_l, from Lemma 14.10 their expected number is bounded by δ/x^5.

By arbitrariness of δ, we have the wanted result. □

14.4 The General Case, $N \in \mathbb{N}$, $N > 3$

For the general case, we proceed as for $N = 3$. The analogue of Sect. 14.3.1 is proved by recurrence on N: we can show a result similar to Lemma 14.2 in dimension N and use it in dimension $N-1$ to prove Lemma 14.1. Then analogue of Proposition 14.1 is shown with bound α/x^{2N-1}. Concerning the analogue on the main contribution, we can show Lemma 14.7 for a general N replacing

$$[\mathbf{t}^n, \mathbf{t}^n + x_n^2(\frac{m_1^n}{t_2^n t_3^n}, \frac{m_2^n}{t_1^n t_3^n}, \frac{m_3^n}{t_1^n t_2^n})] \text{ by } [\mathbf{t}^n, \mathbf{t}^n + x_n^2(\frac{m_1^n}{t_2^n t_3^n \cdots t_N^n}, \frac{m_2^n}{t_1^n t_3^n \cdots t_N^n}, \cdots, \frac{m_N^n}{t_1^n t_2^n \cdots t_{N-1}^n})]$$

and the size is at least $\gamma x_n^{2N}/(t_1^n t_2^n \cdots t_N^n)^{N-1}$.

For the local time approximation, Proposition 14.4 is written as follows.

Proposition 14.5 *Let f be a measurable bounded function of compact support and K an hypercube in $]0, +\infty)^N$. We divide K into smallest hypercubes of size $\prod_{i=1}^N c_i x^2$, and we get*

$$x^{2N-1} \sum_{i_1,\cdots,i_N} f(W(i_1 c_1 x^2, \cdots, i_N c_N x^2)/x) \underset{x\to 0}{\longrightarrow} \frac{1}{c_1 \cdots c_N} \int_{\mathbb{R}} f(t) dt L(K). \tag{14.9}$$

Appendix: Proofs of Lemmas

Proof of Lemmas 14.1 and 14.2

We need an additional technical Lemma.

Lemma 14.11 *Fix an arbitrary $x > 0$. For $w \leq 1$, the probability that there exists an x-bubble on the sheet $(r, s) \to W(r, s, w)$ in $A_n(x)$ is bounded by e^{-k2^n} with k independent of w.*

Remark We remark that by rescaling properties, the probability for $w \leq 1$ that a rectangle has a large white noise contribution is smaller than this same probability for $w = 1$. This explains the uniformity of the constant k with respect to w.

Proof of Lemma 14.11 We fix $w = 1$ and consider a grid of steps $2^{-n}x^2$. Without loss of generality, we can extend the domain of (r, s) from $[0, 1]^2$ to $[0, x^2 2^{-n}(\lfloor 2^n/x^2 \rfloor + 1)]^2$. Let R_{ij} be the square $[i2^{-n}x^2, (i + 2)2^{-n}x^2] \times [j2^{-n}x^2, (j + 2)2^{-n}x^2]$. Remark that since R_{ij} is of side length $2^{-n+1}x^2$, any rectangle of side length less than $2^{-n}x^2$ is contained in a least one R_{ij}. We have the following bounds: for a rectangle R on the sheet $W(r, s, w)$,

$$
\begin{aligned}
&P(\sup_{R \text{ rectangle of side length } \leq 2^{-n}x^2} |\mathbb{W}(R)| \geq \frac{x}{18}) \\
&\leq \sum_{i,j} P(\sup_{R \subset R_{ij}\,,\ R \text{ rectangle of side length } \leq 2^{-n}x^2} |\mathbb{W}(R)| \geq \frac{x}{18}) \\
&\leq \sum_{i,j} P(\sup_{R \subset R_{i,j}} |\mathbb{W}(R)| \geq \frac{x}{18}).
\end{aligned}
$$

From Lemma 1.2 of [10]

$$
P(\sup_{R \subset R_{i,j}} |\mathbb{W}(R)| \geq \frac{x}{18}) \leq 16P(|\mathbb{W}(R_{i,j})| \geq \frac{x}{18}) = 16P(\mathcal{N}(0, 2^{-2n}x^4) \geq \frac{x}{18}),
$$

where $\mathcal{N}(m, \sigma^2)$ denotes a Gaussian random variable of mean m and variance σ^2. Let $\mathscr{W}$ be an x-bubble on the sheet $W(r, s, w)$. Hence we have

$$
P(\mathscr{W} \subset A_n(x)) \leq 16\frac{2^{2n}}{x^4} P(\mathcal{N}(0, 2^{-2n}x^4) \geq \frac{x}{18}),
$$

and we obtain the desired result by an elementary inequality on the error function.

□

We have a direct corollary of this result:

Corollary 14.4 *Fix an arbitrary $x > 0$. Almost surely, there exists a $N \geq 0$ such that for any rectangle of side length less than $2^{-N}x^2$ on $W(r, s, w)$, $|\mathbb{W}(R)| < \frac{x}{18}$.*

Proof (Proof of Corollary 14.4) Applying Borel-Cantelli Lemma, and since $\sum_n e^{-k2^n} < \infty$, we can conclude. □

Proof (Proof of Lemma 14.1) We consider the 2-parameter B.S. $W(r, s, w)$ and a grid of space $2^{-(n+1)}x^2$. Without loss of generality, we extend the domain of (r, s) from $[0, 1]^2$ to $[0, x^2 2^{-n}(\lfloor 2^n/x^2 \rfloor + 1)]^2$. We now consider a rectangle of side length less than $2^{-(n+1)}x^2$. It belongs at least to one of the two categories:

(1) It intersects the grid
(2) It is of side length less than $2^{-(n+1)}x^2$.

We consider a rectangle R in the bubble $\mathscr{W}$ of side length less than $2^{-n}x^2$ and such that $|\mathbb{W}(R)| > \frac{x}{18}$. We consider the case (1). We denote by $[r_1, r_2] \times \{r_0\}$ the intersection and $e^{s_0,w}$ the excursion of the Brownian motion $B^{s_0,w}(\cdot)$ containing $[r_1, r_2]$. From Lemma 3 of [9], we can bound the expected number of these excursions by $K\frac{2^n}{x^2}e^{-c\frac{2^n}{x^2}}$. If R is in case (2), $\mathscr{W}$ is in $A_{n+1}(x)$. We repeat the previous scheme and we note by Corollary 14.4 that this procedure ends almost surely and the bound for such bubbles $\mathscr{W}$ is

$$\sum_{k\geq n} K\,\frac{2^{2k}}{x^4}\,e^{-c\frac{2^{2k}}{x^2}}\,.$$

For small x the bound for $\mathbb{E}(Z(w,n))$ is $\frac{K}{x^3}e^{-c2^n}$. By rescaling properties, for $w\leq 1$ this result holds with the same constant. Actually, a bubble on a sheet of height less than 1 has a less important variation so the probability that it is in $A_n(x)$ is less than the same probability for a bubble of height 1. We sum on w and obtain the result.

□

Proof (Proof of Lemma 14.2) We define the stopping times $(T_i)_{i\geq 0}$ by

- $T_0 = 0$
- $T_{i+1} = \inf\{r > T_i\ ;\ \exists T_i < r' < r\ ,\ r - r' < 2^{-n}x^2,$

$$R \text{ rectangle of side length } \leq 2^{-n}x^2\,,\ |\mathbb{W}([r',r]\times R)| > \frac{x}{18}\}.$$

We can therefore bound $X^{v,w}(x,n)\ \leq\ \sup\{i, T_i\ <\ 1\}$. As previously done, we extend the domain of r from $[0,1]$ to $[0, x^2 2^{-n}(\lfloor 2^n/x^2\rfloor + 1)]$ and split to $\bigcup_i [i2^{-n}x^2, (i+1)2^{-n}x^2]\times[0,1]^2$. We note that any interval of length less than $2^{-n}x^2$ is included in an interval $I_i\ =\ [[i2^{-n}x^2, (i+2)2^{-n}x^2]$ for a certain i. Therefore we bound

$$\begin{aligned}
&P(T_1\leq u)\\
&\leq P\Big(\sup_{i2^{-n}x^2\leq u;\, r,r'\in I_i;\, 0<r-r'<2^{-n}x^2,\, R \text{ rectangle of side length } \leq 2^{-n}x^2} |\mathbb{W}([r',r]\times R)| > \frac{x}{18}\Big)\\
&\qquad\leq \sum_{i2^{-n}x^2\leq u} P\Big(\sup_{K\subset I_i\times[0,1]^2 \text{ cube of side length } \leq 2^{-n}x^2} \mathbb{W}(K) > \frac{x}{18}\Big).
\end{aligned}$$

Applying again the inequality of [10], we get

$$\begin{aligned}
&P\Big(\sup_{K\subset I_i\times[0,1]^2 \text{ cube of side length } \leq 2^{-n}x^2} \mathbb{W}(K) > \frac{x}{18}\Big)\\
&\qquad\leq 4^3 P(|\mathscr{N}(0, 2^{-3n}x^6))| > \frac{x}{18}) \leq k\exp(-\alpha\frac{2^{3n}}{x^6})\,.
\end{aligned}$$

Summing over i in the above inequality we get

$$P(T_1 \leq u) \leq k\frac{u^{2n}}{x^2}\exp(-\alpha\frac{2^{3n}}{x^6}).$$

Since the random variables $T_i - T_{i-1}$ are i.i.d. we compare them to a Poisson process and we can conclude. □

Proofs of Lemmas 14.4 and 14.5

Proof (Proof of Lemma 14.4) Let us assume that X reaches its maximum on (t_1, t_2, t_3) on G and define C and K as previously. Almost surely, for all $\beta > 0$, there exist $g_l^1 \in (s_l^1 - \beta, s_l^1)$ and $g_u^1 \in (s_u^1, s_u^1 + \beta)$ such that $B_1(g_l^1) + B_2(t_2) + B_3(t_3) < 0$ and $B_1(g_u^1) + B_2(t_2) + B_3(t_3) < 0$. Since t_i is the maximum of B_i on (s_l^i, s_u^i), $B_2(t_2) + B_3(t_3)$ is a maximum of $B_2 + B_3$ on $[s_l^2, s_u^2] \times [s_l^3, s_u^3]$. From the continuity of $B_2 + B_3$, for β small, $B_2(t_2) + B_3(t_3)$ is also a maximum on $[s_l^2 - \beta, s_u^2 + \beta] \times [s_l^3 - \beta, s_u^3 + \beta]$. We have therefore X stricl{}ty negative on $\{g_l^1\} \times [s_l^2 - \beta, s_u^2 + \beta] \times [s_l^3 - \beta, s_u^3 + \beta]$ and $\{g_u^1\} \times [s_l^2 - \beta, s_u^2 + \beta] \times [s_l^3 - \beta, s_u^3 + \beta]$.

Similarly we have g_l^2, g_u^2 for B_2 and g_l^3, g_u^3 for B_3 and X is stricly negative on

$[s_l^1 - \beta, s_u^1 + \beta] \times \{g_l^2\} \times [s_l^3 - \beta, s_u^3 + \beta]$ and $[s_l^1 - \beta, s_u^1 + \beta] \times \{g_u^2\} \times [s_l^3 - \beta, s_u^3 + \beta]$
$[s_l^1 - \beta, s_u^1 + \beta] \times [s_l^2 - \beta, s_u^2 + \beta] \times \{g_l^3\}$ and $[s_l^1 - \beta, s_u^1 + \beta] \times [s_l^2 - \beta, s_u^2 + \beta] \times \{g_u^3\}$.

So we can take the cube defined by the six faces

$$\begin{aligned}\{g_l^1\} \times [g_l^2, g_u^2] \times [g_l^3, g_u^3] \cup \{g_u^1\} \times [g_l^2, g_u^2] \times [g_l^3, g_u^3] \cup \\ [g_l^1, g_u^1] \times \{g_l^2\} \times [g_l^3, g_u^3] \cup [g_l^1, g_u^1] \times \{g_u^2\} \times [g_l^3, g_u^3] \cup \\ [g_l^1, g_u^1] \times [g_l^2, g_u^2] \times \{g_l^3\} \cup [g_l^1, g_u^1] \times [g_l^2, g_u^2] \times \{g_l^3\}.\end{aligned}$$

□

Proof (Proof of Lemma 14.5) The result of Lemma 14.5 is a consequence of Lemma 14.4. □

Proofs of Lemmas 14.6 and 14.7

Proof (Proof of Lemma 14.6) For part (i), since $|\{(u, v, w)\,;\ X(u, v, w) = 0\}| = 0$, $g^\gamma(c, \mathbf{m})$ is continuous almost surely.

Part (ii): the expectation of $g^\gamma(c, \mathbf{m})$ is less than

$$E|(u, v, w) \in [0, \mathbf{m}]; \ W(u, v, w) \in [0, 2]|/\gamma.$$

From a simple inequality on the error function, we get the result. □

Proof (Proof of Lemma 14.7) Let

$$X^n(u, v, w) = \frac{1}{x_n} W(t_1^n + x_n^2 \frac{u}{t_2^n t_3^n}, t_2^n + x_n^2 \frac{v}{t_1^n t_3^n}, t_3^n + x_n^2 \frac{w}{t_1^n t_2^n})$$

and by elementary formulas we have

$$\begin{aligned} X^n(u, v, w) = B_1^n(u) + B_2^n(v) + B_3^n(w) + \frac{1}{x_n}(E_1^n(x_n^2 \frac{v}{t_3^n}, x_n^2 \frac{w}{t_2^n}) \\ + E_2^n(x_n^2 \frac{u}{t_3^n}, x_n^2 \frac{w}{t_1^n}) + E_3^n(x_n^2 \frac{u}{t_2^n}, x_n^2 \frac{v}{t_1^n})) \\ + \frac{1}{x_n} W'(x_n^2 \frac{u}{t_2^n t_3^n}, x_n^2 \frac{v}{t_1^n t_3^n}, x_n^2 \frac{w}{t_1^n t_2^n}) + X^n(0, 0, 0). \end{aligned}$$

where B_i^n is the standard Brownian motion, E_i^n is a standard 2-parameter B.S., and W' is a 3-parameter B.S. They are mutually independent. We also note that on $[0, m_1^n] \times [0, m_2^n] \times [0, m_3^n]$, from the scaling properties,

$$\begin{aligned} &\frac{1}{x_n}(E_1^n(x_n^2 \frac{v}{t_3^n}, x_n^2 \frac{w}{t_2^n}) + E_2^n(x_n^2 \frac{u}{t_3^n}, x_n^2 \frac{w}{t_1^n}) + E_3^n(x_n^2 \frac{u}{t_2^n}, x_n^2 \frac{v}{t_1^n})) \\ &+ \frac{1}{x_n} W'(x_n^2 \frac{u}{t_2^n t_3^n}, x_n^2 \frac{v}{t_1^n t_3^n}, x_n^2 \frac{w}{t_1^n t_2^n}) \\ &= x_n(E_1^n(\frac{v}{t_3^n}, \frac{w}{t_2^n}) + E_2^n(\frac{u}{t_3^n}, \frac{w}{t_1^n}) + E_3^n(\frac{u}{t_2^n}, x_n^2 \frac{v}{t_1^n})) + x_n^2 W'(\frac{u}{t_2^n t_3^n}, \frac{v}{t_1^n t_3^n}, \frac{w}{t_1^n t_2^n}). \end{aligned} \tag{14.10}$$

Let us denote by $V^n(u, v, w)$ the RHS of (14.10) above. We have $\sup |V^n| \to 0$ almost surely, hence we can apply Proposition 14.2 □

Proofs of Lemmas 14.8, 14.9 and 14.10

Proof (Proof of Lemma 14.8) We compute the expectation $E[\sum_{i,j,k} N_{i,j,k} \mathbb{1}_{[2^p Kx, 2^{p+1} Kx)}(|W(u_i, v_j, w_k)|)]$ and sum over p. The inequality below from line 2 to 3 is a consequence of Corollary 14.3.

$$
\begin{aligned}
&E[\sum_{i,j,k} N_{i,j,k}\mathbb{1}_{[2^pKx,2^{p+1}Kx)}(|W(u_i,v_j,w_k)|)] \\
&\quad = \sum_{i,j,k} E[N_{i,j,k}\mathbb{1}_{[2^pKx,2^{p+1}Kx)}(|W(u_i,v_j,w_k)|)] \\
&\quad = \sum_{i,j,k} E\left[E\left[N_{i,j,k}/|W(u_i,v_j,w_k)| \in [2^pKx,2^{p+1}Kx)\right]\right. \\
&\qquad\qquad \left.\times \mathbb{1}_{[2^pKx,2^{p+1}Kx)}(|W(u_i,v_j,w_k)|)\right] \\
&\quad \le \sum_{i,j,k} \frac{cm^3}{\gamma}\exp\left(-\frac{(K2^p-2)^2}{m}\right) P(|W(u_i,v_j,w_k))| \in [2^pKx,2^{p+1}Kx)) \\
&\quad \le \sum_{i,j,k} \frac{cm^3 2^pKx}{\gamma\sqrt{2\pi u_i v_j w_k}}\exp\left(-\frac{(K2^p-2)^2}{m}\right) \\
&\quad \le \sum_{i,j,k} \frac{cm^3 2^pKx}{\gamma\sqrt{2\pi\epsilon^3}}\exp\left(-\frac{(K2^p-2)^2}{m}\right).
\end{aligned}
$$

We sum over i, j, k and from the construction of u, v, w we get

$$
\le \frac{c2^pK|C|^2}{\gamma x^5\sqrt{2\pi\epsilon^3}}\exp\left(-\frac{(K2^p-2)^2}{m}\right).
$$

If we sum over $p \ge 0$, for K big enough we have the final result since c is independent of K. □

Proof (Proof of Lemma 14.9) Let $D := C \setminus (\cup_{i,j,k}\sigma'_{i,j,k})$. Since the bubbles are of diameter less than Mx^2, a bubble not contained in $\sigma'_{i,j,k}$ is entirely contained in $\tilde{D} := \{\mathbf{x} \in [0,1]^3, d(\mathbf{x}, D) \le Mx^2\}$. We compute the Lebesgue measure of $\tilde{D}$

$$
\begin{aligned}
|\tilde{D}| &\le \sum_{i,j,k} 3(1-\frac{1}{(1+\gamma)^2})uvw + 2Mx^2uv + 2Mx^2vw + 2Mx^2uw \\
&\le 3(1-\frac{1}{(1+\gamma)^2})|C| + 6\frac{M}{m}|C|.
\end{aligned}
$$

Moreover, an easy bound on the density of $W(\mathbf{t})$ gives

$$
E|\{\mathbf{t} \in \tilde{D};\ W(\mathbf{t}) \in [0,2x]\}| \le \frac{x}{\epsilon^{3/2}}|\tilde{D}|.
$$

Finally we have

$$E(N) \leq (3(1 - \frac{1}{(1+\gamma)^2}) + 6\frac{M}{m})\frac{2^{3N_0}}{\epsilon^{3/2}x^5}|C| .$$

We then choose γ and m to have the result. □

Proof (Proof of Lemma 14.10) Proceeding as in the proof above, the bubbles intersecting the frontier of C are contained in $\tilde{\partial} C := \{\mathbf{t};\ d(\mathbf{t}, \partial C) \leq Mx^2\}$. We estimate the measure of $\tilde{\partial} C$:

$$|\tilde{\partial} C| \leq 4Mx^2[(r_2 - r_1)(s_2 - s_1) + (s_2 - s_1)(t_2 - t_1) + (r_2 - r_1)(t_2 - t_1)] \leq 12\frac{M}{m}|C| .$$

From the choice of m in Lemma 14.9, we get the result. □

Acknowledgements The authors want to thank the anonymous referee for valuable remarks that helped to improve this paper.

References

1. R.C. Dalang, T.S. Mountford, in *Level Sets, Bubbles and Excursions of a Brownian sheet.* Infinite dimensional stochastic analysis (Amsterdam, 1999), 117–128, Verh. Afd. Natuurkd. 1. Reeks. K. Ned. Akad. Wet., 52, R. Neth. Acad. Arts Sci., Amsterdam, 2000.
2. R.C. Dalang, J.B. Walsh, Geography of the level sets of the Brownian sheet. Prob. Theor. Rel. Fields **96**, 153–176 (1993)
3. R.C. Dalang, J.B. Walsh, The structure of the Brownian bubble. Prob. Theor. Rel. Fields **96**, 475–501 (1993)
4. W. Ehm, Sample function properties of multiparameter stable processes. Z. Wahrsch. Verw. Gebiete **56**, 195–228 (1981)
5. N. Ikeda, S. Watanabe, *Stochastic Differential Equations and Diffusion Processes.* North Holland Mathematical Library (Elsevier Science, Berlin, 1981)
6. W. Kendall, Contours of Brownian processes with several-dimensional time. Zeit. Wahr. Theorie **52**, 268–276 (1980)
7. D. Khoshnevisan, On the distribution of bubbles of the Brownian sheet. Ann. Probab. **23**, 786–805 (1995)
8. D. Khoshnevisan, *Multiparameter Processes. An introduction to Random Fields.* Springer Monographs in Mathematics (Springer, New York, 2002)
9. T.S. Mountford, Brownian sheet, local time and bubbles, in *Séminaire de Probabilités, XXXVII.* Lecture Notes in Mathematics, vol. 1832 (Springer, Berlin, 2003), 19–215
10. S. Orey, W.E. Pruitt, Sample functions of the N-parameter Wiener process. Ann. Probab. **1**, 138–163 (1973)
11. D. Revuz, M. Yor, *Continuous Martingales and Brownian Motion*, 3rd edn. (Springer, Berlin, 1999)
12. L.C.G. Rogers, D. Williams, in *Diffusions, Markov processes and martingales.* Ito Calculus, vol. 2 (Wiley, New York, 1987)
13. Y.-Y. Yen, M. Yor, in *Local Time and Excursion Theory for Brownian Motion.* Lecture Notes in Mathematics (Springer, Berlin, 2013), 2088

Chapter 15
Mod-ϕ Convergence, II: Estimates on the Speed of Convergence

Valentin Féray, Pierre-Loïc Méliot, and Ashkan Nikeghbali

Abstract In this paper, we give estimates for the speed of convergence towards a limiting stable law in the recently introduced setting of mod-ϕ convergence. Namely, we define a notion of *zone of control*, closely related to mod-ϕ convergence, and we prove estimates of Berry–Esseen type under this hypothesis. Applications include:

- the winding number of a planar Brownian motion;
- classical approximations of stable laws by compound Poisson laws;
- examples stemming from determinantal point processes (characteristic polynomials of random matrices and zeroes of random analytic functions);
- sums of variables with an underlying dependency graph (for which we recover a result of Rinott, obtained by Stein's method);
- the magnetization in the d-dimensional Ising model;
- and functionals of Markov chains.

15.1 Introduction

15.1.1 Mod-ϕ Convergence

Let $(X_n)_{n\in\mathbb{N}}$ be a sequence of real-valued random variables. In many situations, there exists a scale s_n and a limiting law ϕ which is infinitely divisible, such that $(X_n/s_n)_{n\in\mathbb{N}}$ converges in law towards ϕ. For instance, in the classical central limit theorem, if $X_n = \sum_{i=1}^{n} A_i$ is a sum of centered i.i.d. random variables with

V. Féray · A. Nikeghbali (✉)
Institut für Mathematik, Universität Zürich, Zürich, Switzerland
e-mail: Valentin.Feray@math.uzh.ch; ashkan.nikeghbali@math.uzh.ch

P.-L. Méliot
Université Paris-Sud - Faculté des Sciences d'Orsay, Institut de mathématiques d'Orsay, Orsay, France
e-mail: pierre-loic.meliot@math.u-psud.fr

C. Donati-Martin et al. (eds.), *Séminaire de Probabilités L*, Lecture Notes in Mathematics 2252, https://doi.org/10.1007/978-3-030-28535-7_15

$\mathbb{E}[(A_1)^2] < \infty$, then

$$s_n = \sqrt{n\,\mathbb{E}[(A_1)^2]}$$

and the limit is the standard Gaussian distribution $\mathcal{N}_{\mathbb{R}}(0, 1)$. In [27] and the subsequent papers [14, 21], the notion of mod-ϕ convergence was developed in order to get quantitative estimates on the convergence $\frac{X_n}{s_n} \rightharpoonup \phi$ (throughout the paper, $\rightharpoonup$ denotes convergence in distribution).

Definition 15.1 Let ϕ be an infinitely divisible probability measure, and $D \subset \mathbb{C}$ be a subset of the complex plane, which we assume to contain 0. We assume that the Laplace transform of ϕ is well defined over D, with Lévy exponent η:

$$\forall z \in D, \quad \int_{\mathbb{R}} \mathrm{e}^{zx}\, \phi(dx) = \mathrm{e}^{\eta(z)}.$$

We then say that $(X_n)_{n\in\mathbb{N}}$ converges mod-ϕ over D, with parameters $(t_n)_{n\in\mathbb{N}}$ and limiting function $\psi : D \to \mathbb{C}$, if $t_n \to +\infty$ and if, locally uniformly on D,

$$\lim_{n\to\infty} \mathbb{E}[\mathrm{e}^{zX_n}]\,\mathrm{e}^{-t_n\eta(z)} = \psi(z).$$

If $D = \mathrm{i}\mathbb{R}$, we shall just speak of mod-ϕ convergence; it is then convenient to use the notation

$$\theta_n(\xi) = \mathbb{E}[\mathrm{e}^{\mathrm{i}\xi X_n}]\,\mathrm{e}^{-t_n\eta(\mathrm{i}\xi)};$$
$$\theta(\xi) = \psi(\mathrm{i}\xi),$$

so that mod-ϕ convergence corresponds to $\lim_{n\to\infty}\theta_n(\xi) = \theta(\xi)$ (uniformly for ξ in compact subsets of $\mathbb{R}$). When nothing is specified, in this paper, we implicitly consider that $D = \mathrm{i}\mathbb{R}$. When $D = \mathbb{C}$ we shall speak of *complex* mod-ϕ convergence. In some situations, it is also appropriate to study mod-ϕ convergence on a band $\mathbb{R} \times \mathrm{i}[-b, b]$, or $[-c, c] \times \mathrm{i}\mathbb{R}$ (see [21, 44]).

Intuitively, a sequence of random variables $(X_n)_{n\in\mathbb{N}}$ converges mod-ϕ if it can be seen as a large renormalization of the infinitely divisible law ϕ, plus some residue which is asymptotically encoded in the Fourier or Laplace sense by the limiting function ψ. Then, ϕ will typically be:

1. in the case of lattice-valued distributions, a Poisson law or a compound Poisson law (cf. [5, 12, 21, 35]);
2. or, a *stable distribution*, for instance a Gaussian law.

In this paper, we shall only be interested in the second case. Background on stable distributions is given at the end of this introduction (Sect. 15.1.3). In particular we will see that, if ϕ is a stable distribution, then the mod-ϕ convergence of X_n

implies the convergence in distribution of a renormalized version Y_n of X_n to ϕ (Proposition 15.3).

We believe that mod-ϕ is a kind of universality class behind the central limit theorem (or its stable version) in the following sense. For many sequences $(X_n)_{n\in\mathbb{N}}$ of random variables that are proven to be asymptotically normal (or converging to a stable distribution), it is possible to prove mod-ϕ convergence; we refer to our monograph [21] or Sects. 15.3–15.5 below for such examples. These estimates on the Laplace transform/characteristic function can then be used to derive in an automatic way some *companion theorems*, refining the central limit theorem. In [21], we discuss in details the question of moderate/large deviation estimates and of finding the *normality zone*.

In the present paper, we shall be interested in the speed of convergence towards the Gaussian (or more generally the stable) distribution of the appropriate renormalization Y_n of X_n. To obtain sharp bounds on this speed of convergence, we do not work with mod-ϕ convergence, but we introduce the notion of *zone of control* for the renormalized characteristic function $\theta_n(\xi)$. In many examples, such a zone of control can be obtained by essentially the same arguments used to prove mod-ϕ convergence, and in most examples, mod-ϕ convergence actually holds.

15.1.2 Results and Outline of the Paper

We take as reference law a stable distribution ϕ of index $\alpha \in (0, 2]$. Let $(X_n)_{n\in\mathbb{N}}$ be a sequence of variables that admits a zone of control (this notion will be defined in Definition 15.5; this is closely related to the mod-ϕ convergence of $(X_n)_{n\in\mathbb{N}}$). As we will see in Proposition 15.6, this implies that some renormalization Y_n of X_n converges in distribution towards ϕ and we are interested in the speed of convergence for this convergence. More precisely, we are interested in upper bounds for the *Kolmogorov distance*

$$d_{\mathrm{Kol}}(Y_n, \phi) = \sup_{a\in\mathbb{R}} \left| \mathbb{P}[Y_n \leq a] - \int_{-\infty}^{a} \phi(dx) \right|.$$

The main theorem of Sect. 15.2 (Theorem 15.19) shows that this distance is $O(t_n^{-\gamma-1/\alpha})$, where γ is a parameter describing how large our zone of control is. We also obtain as intermediate result estimates for

$$\left| \mathbb{E}[f(Y_n)] - \int_{\mathbb{R}} f(x)\, \phi(dx) \right|,$$

where f lies in some specific set of tests functions (Proposition 15.15). A detailed discussion on the method of proof of these bounds can be found at the beginning of Sect. 15.2.

Section 15.3 gives some examples of application of the theoretical results of Sect. 15.2. The first one is a toy example, while the other ones are new to the best of our knowledge.

- We first consider sums of i.i.d. random variables with finite third moment. In this case, the classical Berry–Esseen estimate ensures that

$$d_{\mathrm{Kol}}(Y_n, \mathcal{N}_{\mathbb{R}}(0,1)) \leq \frac{3\,\mathbb{E}[|A_1|^3]}{\sigma^3\,\sqrt{n}},$$

  see [6] or [19, §XVI.5, Theorem 1]. Our general statement for variables with a zone of convergence gives essentially the same result, only the constant factor is not as good.
- We can extend the Berry–Esseen estimates to the case of independent but non identically distributed random variables. As an example, we look at the number of zeroes Z_r of a random analytic series that fall in a disc of radius r; it has the same law as a series of independent Bernoulli variables of parameters r^{2k}, $k \geq 1$. When the radius r of the disc goes to 1, one has a central limit theorem for Z_r, and the theory of zones of control yields an estimate $O((1-r)^{-1/2})$ on the Kolmogorov distance.
- We then look at the winding number φ_t of a planar Brownian motion starting at 1 (see Sect. 15.3.2 for a precise definition). This quantity has been proven to converge in the mod-Cauchy sense in [14], based on the computation of the characteristic function done by Spitzer [53]. The same kind of argument easily yields the existence of a zone of control and our general result applies: when t goes to infinity, after renormalization, φ_t converges in distribution towards a Cauchy law and the Kolmogorov distance in this convergence is $O((\log t)^{-1})$.
- In the third example, we consider *compound Poisson laws* (see [50, Chapter 1, §4]). These laws appear in the proof of the Lévy–Khintchine formula for infinitely divisible laws (*loc. cit.*, Chapter 2, §8, pp. 44–45), and we shall be interested in those that approximate the stable distributions $\phi_{c,\alpha,\beta}$. Again, establishing the existence of a zone of control is straight-forward and our general result shows that the speed of convergence is $O(n^{-1/\min(\alpha,1)})$ (Proposition 15.22), with an additional log factor if $\alpha = 1$ and $\beta \neq 0$ (thus exhibiting an interesting phase transition phenomenon).
- Ratios of Fourier transforms of probability measures appear naturally in the theory of self-decomposable laws and of the corresponding Ornstein–Uhlenbeck processes. Thus, any self-decomposable law ϕ is the limiting distribution of a Markov process $(U_t)_{t\geq 0}$, and when ϕ is a stable law, one has mod-ϕ convergence of an adequate renormalisation of U_t, *with a constant residue*. This leads to an estimate of the speed of convergence which depends on α, on the speed of $(U_t)_{t\geq 0}$ and on its starting point (Proposition 15.24).
- Finally, logarithms of characteristic polynomials of random matrices in a classical compact Lie group are mod-Gaussian convergent (see for instance [21, Section 7.5]), and one can compute a zone of control for this convergence,

which yields an estimate of the speed of convergence $O((\log n)^{-3/2})$. For unitary groups, one recovers [10, Proposition 5.2]. This example shows how one can force the index v of a zone of control of mod-Gaussian convergence to be equal to 3, see Remark 15.25.

The last two sections concentrate on the case where the reference law is Gaussian ($\alpha = 2$). In this case, we show that a sufficient condition for having a zone of control is to have *uniform bounds on cumulants* (see Definition 15.27 and Lemma 15.28). This is not surprising since such bounds are known to imply (with small additional hypotheses) mod-Gaussian convergence [21, Section 5.1]. Combined with our main result, this gives bounds for the Kolmogorov distance for variables having uniform bounds on cumulants—see Corollary 15.29. Note nevertheless that similar results have been given previously by Statulevičius [54] (see also Saulis and Statulevičius [52]). Our Corollary 15.29 coincides up to a constant factor to one of their result. Our contribution here therefore consists in giving a large variety of non-trivial examples where such bounds on cumulants hold:

- The first family of examples relies on a previous result by the authors [21, Chapter 9] (see Theorem 15.33 here), where bounds on cumulants for sums of variables with an underlying *dependency graph* are given. Let us comment a bit. Though introduced originally in the framework of the probabilistic method [1, Chapter 5], dependency graphs have been used to prove central limit theorems on various objects: random graphs [28], random permutations [9], probabilistic geometry [46], random character values of the symmetric group [21, Chapter 11]. In the context of Stein's method, we can also obtain bounds for the Kolmogorov distance in these central limit theorems [3, 47].

  The results of this paper give another approach to obtain bounds for this Kolmogorov distance for sums of bounded variables (see Sect. 15.4.2). The bounds obtained are, up to a constant, the same as in [47, Theorem 2.2]. Note that our approach is fundamentally different, since it relies on classical Fourier analysis, while Stein's method is based on a functional equation for the Gaussian distribution. We make these bounds explicit in the case of subgraph counts in Erdös–Rényi random graphs and discuss an extension to sum of unbounded variables.
- The next example is the finite volume magnetization in the Ising model on $\mathbb{Z}^d$. The Ising model is one of the most classical models of statistical mechanics, we refer to [23] and references therein for an introduction to this vast topic. The magnetization M_Δ (that is the sum of the spins in Δ) is known to have asymptotically normal fluctuations [45]. Based on a result of Duneau et al. [17], we prove that, if the magnetic field is non-zero or if the temperature is sufficiently large, M_Δ has uniform bounds on cumulants. This implies a bound on the Kolmogorov distance (Proposition 15.45):

$$d_{\mathrm{Kol}}\left(\frac{M_\Delta - \mathbb{E}[M_\Delta]}{\sqrt{\mathrm{Var}(M_\Delta)}},\ \mathcal{N}_{\mathbb{R}}(0,1)\right) \leq \frac{K}{\sqrt{|\Delta|}}.$$

It seems that this result improves on what is known so far. In [11], Bulinskii gave a general bound on the Kolmogorov distance for sums of *associated random variables*, which applied to M_Δ, yields a bound with an additional $(\log|\Delta|)^d$ factor comparing to ours. In a slightly different direction, Goldstein and Wiroonsri [24] have recently given a bound of order $O(|\Delta|^{1/(2d+2)})$ for the L^1-distance (the L^1-distance is another distance on distribution functions, which is a priori incomparable with the Kolmogorov distance; note also that their bound is only proved in the special case where $\Delta = \{-m, -m+1, \ldots, m\}^d$).

- The last example considers statistics of the form $S_n = \sum_{t=0}^{n} f_t(X_t)$, where $(X_t)_{t\geq 0}$ is an ergodic discrete time Markov chain on a finite space state. Again we can prove uniform bounds on cumulants and deduce from it bounds for the Kolmogorov distance (Theorem 15.51). The speed of convergence in the central limit theorem for Markov chains has already been studied by Bolthausen [8] (see also later contributions of Lezaud [40] and Mann [43]). These authors study more generally Markov chains on infinite space state, but focus on the case of a statistics f_t independent of the time. Except for these differences, the bounds obtained are of the same order; however our approach and proofs are again quite different.

It is interesting to note that the proofs of the bounds on cumulants in the last two examples are highly non trivial and share some common structure. Each of these statistics decomposes naturally as a sum. In each case, we give an upper bound for joint cumulants of the summands, which writes as a weighted enumeration of spanning trees. Summing terms to get a bound on the cumulant of the sum is then easy.

To formalize this idea, we introduce in Sect. 15.5 the notion of uniform weighted dependency graphs. Both proofs for the bounds on cumulants (for magnetization of the Ising model and functional of Markov chains) are presented in this framework. We hope that this will find further applications in the future.

15.1.3 Stable Distributions and Mod-Stable Convergence

Let us recall briefly the classification of stable distributions (see [50, Chapter 3]). Fix $c > 0$ (the *scale* parameter), $\alpha \in (0, 2]$ (the *stability* parameter), and $\beta \in [-1, 1]$ (the *skewness* parameter).

Definition 15.2 The stable distribution of parameters (c, α, β) is the infinitely divisible law $\phi = \phi_{c,\alpha,\beta}$ whose Fourier transform

$$\widehat{\phi}(\xi) = \int_{\mathbb{R}} \mathrm{e}^{\mathrm{i}x\xi}\, \phi(dx) = \mathrm{e}^{\eta(\mathrm{i}\xi)}$$

has for Lévy exponent $\eta(\mathrm{i}\xi) = \eta_{c,\alpha,\beta}(\mathrm{i}\xi) = -|c\xi|^\alpha\,(1 - \mathrm{i}\beta\, h(\alpha,\xi)\,\mathrm{sgn}(\xi))$, where

$$h(\alpha,\xi) = \begin{cases} \tan\left(\frac{\pi\alpha}{2}\right) & \text{if } \alpha \neq 1, \\ -\frac{2}{\pi}\log|\xi| & \text{if } \alpha = 1 \end{cases}$$

and $\mathrm{sgn}(\xi) = \pm 1$ is the sign of ξ.

The most usual stable distributions are (see Fig. 15.1):

- the standard Gaussian distribution $\frac{1}{\sqrt{2\pi}}\,\mathrm{e}^{-x^2/2}\,dx$ for $c = \frac{1}{\sqrt{2}}$, $\alpha = 2$ and $\beta = 0$;
- the standard Cauchy distribution $\frac{1}{\pi(1+x^2)}\,dx$ for $c = 1$, $\alpha = 1$ and $\beta = 0$;
- the standard Lévy distribution $\frac{1}{\sqrt{2\pi}}\,\frac{\mathrm{e}^{-1/2x}}{x^{3/2}}\,1_{x\geq 0}\,dx$ for $c = 1$, $\alpha = \frac{1}{2}$ and $\beta = 1$.

We recall that mod-ϕ convergence on an open subset D of $\mathbb{C}$ containing 0 can only occur when the characteristic function of ϕ is analytic around 0. Among stable distributions, only Gaussian laws (which correspond to $\alpha = 2$) satisfy this property. Mod-ϕ convergence on $D = \mathrm{i}\mathbb{R}$ can however be considered for any stable distribution ϕ.

Since $|\mathrm{e}^{\eta(\mathrm{i}\xi)}| = \mathrm{e}^{-|c\xi|^\alpha}$ is integrable, any stable law $\phi_{c,\alpha,\beta}$ has a density $m_{c,\alpha,\beta}(x)\,dx$ with respect to the Lebesgue measure. Moreover, the corresponding Lévy exponents have the following scaling property: for any $t > 0$,

$$t\,\eta_{c,\alpha,\beta}\left(\frac{\mathrm{i}\xi}{t^{1/\alpha}}\right) = \begin{cases} \eta_{c,\alpha,\beta}(\mathrm{i}\xi) & \text{if } \alpha \neq 1, \\ \eta_{c,\alpha,\beta}(\mathrm{i}\xi) - \left(\frac{2c\beta}{\pi}\log t\right)\mathrm{i}\xi & \text{if } \alpha = 1. \end{cases}$$

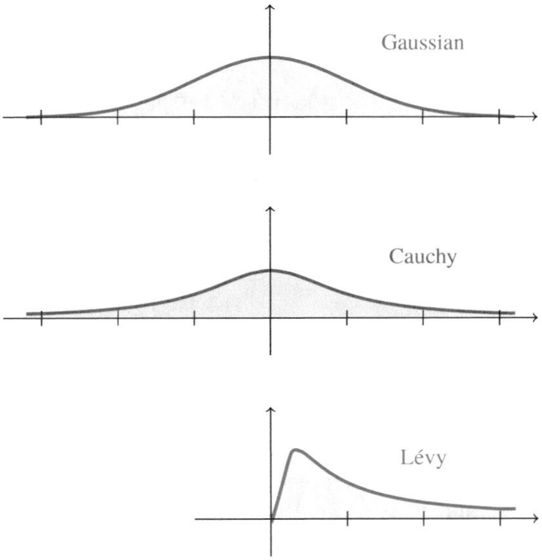

Fig. 15.1 Densities of the standard Gaussian, Cauchy and Lévy distribution

This will be used in the following proposition:

Proposition 15.3 *If $(X_n)_{n\in\mathbb{N}}$ converges in the mod-$\phi_{c,\alpha,\beta}$ sense, then*

$$Y_n = \begin{cases} \frac{X_n}{(t_n)^{1/\alpha}} & \text{if } \alpha \neq 1, \\ \frac{X_n}{t_n} - \frac{2c\beta}{\pi} \log t_n & \text{if } \alpha = 1 \end{cases}$$

converges in law towards $\phi_{c,\alpha,\beta}$.

Proof In both situations,

$$\mathbb{E}[\mathrm{e}^{\mathrm{i}\xi Y_n}] = \mathrm{e}^{\eta(\mathrm{i}\xi)}\, \theta_n\left(\frac{\xi}{(t_n)^{1/\alpha}}\right) = \mathrm{e}^{\eta(\mathrm{i}\xi)}\, \theta(0)\, (1+o(1)) = \mathrm{e}^{\eta(\mathrm{i}\xi)}\, (1+o(1))$$

thanks to the uniform convergence of θ_n towards θ, and to the scaling property of the Lévy exponent η. □

15.2 Speed of Convergence Estimates

The goal of this section is to introduce the notion of zone of control (Sect. 15.2.1) and to estimate the speed of convergence in the resulting central limit theorem. More precisely, we take as reference law a stable distribution $\phi_{c,\alpha,\beta}$ and a sequence $(X_n)_{n\in\mathbb{N}}$ that admits a zone of control (with respect to $\phi_{c,\alpha,\beta}$). As for mod-$\phi_{c,\alpha,\beta}$ convergent sequences, it is easy to prove that in this framework, an appropriate renormalization Y_n of X_n converges in distribution towards $\phi_{c,\alpha,\beta}$ (see Proposition 15.6 below).

If Y has distribution $\phi_{c,\alpha,\beta}$, we then want to estimate

$$d_{\mathrm{Kol}}(Y_n, Y) = \sup_{s\in\mathbb{R}} |\mathbb{P}[Y_n \leq s] - \mathbb{P}[Y \leq s]|. \tag{15.1}$$

To do this, we follow a strategy proposed by Tao (see [56, Section 2.2]) in the case of sums of i.i.d. random variables with finite third moment. The right-hand side of (15.1) can be rewritten as

$$\sup_{f\in\mathcal{F}} |\mathbb{E}[f(Y_n)] - \mathbb{E}[f(Y)]|,$$

where $\mathcal{F}$ is the class of measurable functions $y \mapsto 1_{y\leq s}$. Therefore, it is natural to approach the problem of speed of convergence by looking at general estimates on test functions. The basic idea is then to use the Parseval formula to compute the difference $\mathbb{E}[f(Y_n)] - \mathbb{E}[f(Y)]$, since we have estimates on the Fourier transforms of Y_n and Y. A difficulty comes from the fact that the functions $y \mapsto 1_{y\leq s}$ are not smooth, and in particular, their Fourier transforms are only defined in the sense of distributions. This caveat is dealt with by standard techniques of harmonic analysis (Sects. 15.2.2–15.2.4): namely, we shall work in a space of distributions instead of functions, and use an adequate smoothing kernel in order to be able to work with compactly supported Fourier transforms. Section 15.2.5 gathers all these tools to give an upper bound for (15.1). This is the main result of this section and can be found in Theorem 15.19.

Remark 15.4 An alternative way to get an upper bound for (15.1) from estimates on characteristic functions is to use the following inequality due to Berry (see [6] or [19, Lemma XVI.3.2]). Let X and Y be random variables with characteristic functions $f^*(\zeta)$ and $g^*(\zeta)$. Then, provided that Y has a density bounded by m, we have, for any $s \in \mathbb{R}$,

$$|\mathbb{P}[X \leq s] - \mathbb{P}[Y \leq s]| \leq \frac{1}{\pi} \int_{-T}^{T} \left| \frac{f^*(\zeta) - g^*(\zeta)}{\zeta} \right| d\zeta + \frac{24m}{\pi T}.$$

Using this inequality in our context should lead to similar estimates as the ones we obtain, possibly with different constants. The proof we use here however has the advantage of being more self-contained, and to provide estimates for test functions as intermediate results.

15.2.1 The Notion of Zone of Control

Definition 15.5 Let $(X_n)_{n\in\mathbb{N}}$ be a sequence of real random variables, $\phi_{c,\alpha,\beta}$ a reference stable law, and $(t_n)_{n\in\mathbb{N}}$ a sequence growing to infinity. Consider the following assertions:

(Z1) Fix $v > 0$, $w > 0$ and $\gamma \in \mathbb{R}$. There exists a zone $[-K(t_n)^\gamma, K(t_n)^\gamma]$ such that, for all ξ in this zone, if $\theta_n(\xi) = \mathbb{E}[\mathrm{e}^{\mathrm{i}\xi X_n}]\,\mathrm{e}^{-t_n\eta_{c,\alpha,\beta}(\mathrm{i}\xi)}$, then

$$|\theta_n(\xi) - 1| \leq K_1 |\xi|^v \exp(K_2 |\xi|^w)$$

for some positive constants K_1 and K_2 that are independent of n.

(Z2) One has

$$\alpha \leq w \quad ; \quad -\frac{1}{\alpha} \leq \gamma \leq \frac{1}{w-\alpha} \quad ; \quad 0 < K \leq \left(\frac{c^\alpha}{2K_2}\right)^{\frac{1}{w-\alpha}}.$$

Notice that (Z2) can always be forced by increasing w, and then decreasing K and γ in the bounds of Condition (Z1). If Conditions (Z1) and (Z2) are satisfied, then we say that we have a *zone of control* $[-K(t_n)^\gamma, K(t_n)^\gamma]$ with *index* (v, w).

Note that although the definition of zone of control depends on the reference law $\phi_{c,\alpha,\beta}$, the latter does not appear in the terminology (throughout the paper, it is considered fixed).

Proposition 15.6 *Let $(X_n)_{n\in\mathbb{N}}$ be a sequence of random variables, $\phi_{c,\alpha,\beta}$ a reference stable law, Y with distribution $\phi_{c,\alpha,\beta}$ and Y_n as in Proposition 15.3. Assume that $(X_n)_{n\in\mathbb{N}}$ has a zone of control $[-K(t_n)^\gamma, K(t_n)^\gamma]$ with index (v, w). If $\gamma > -\frac{1}{\alpha}$, then one has the convergence in law $Y_n \rightharpoonup Y$.*

Proof Condition (Z1) implies that, if Y_n is the renormalization of X_n and $Y \sim \phi_{c,\alpha,\beta}$, then for fixed ξ,

$$\left|\frac{\mathbb{E}[\mathrm{e}^{\mathrm{i}\xi Y_n}]}{\mathbb{E}[\mathrm{e}^{\mathrm{i}\xi Y}]} - 1\right| = \left|\theta_n\left(\frac{\xi}{(t_n)^{1/\alpha}}\right) - 1\right| \leq \frac{K_1|\xi|^v}{(t_n)^{v/\alpha}} \exp\left(\frac{K_2|\xi|^w}{(t_n)^{w/\alpha}}\right)$$

for t_n large enough, and the right-hand side goes to 0. This proves the convergence in law $Y_n \rightharpoonup Y$. □

The goal of the next few sections will be to get some speed of convergence estimates for this convergence in distribution.

Remark 15.7 In the definition of zone of control, we do not assume the mod-$\phi_{c,\alpha,\beta}$ convergence of the sequence $(X_n)_{n\in\mathbb{N}}$ with parameters $(t_n)_{n\in\mathbb{N}}$ and limit $\lim_{n\to\infty}\theta_n(\xi) = \theta(\xi)$. However, in almost all the examples considered, we shall indeed have (complex) mod-ϕ convergence (convergence of the residues θ_n), with the same parameters t_n as for the notion of zone of control. We shall then speak of *mod-ϕ convergence with a zone of convergence* $[-K(t_n)^\gamma, K(t_n)^\gamma]$ *and with index of control* (v, w). Mod-ϕ convergence implies other probabilistic results than estimates of Berry–Esseen type: central limit theorem with a large range of normality, moderate deviations (cf. [21]), local limit theorem [14], etc.

Remark 15.8 If one has mod-$\phi_{c,\alpha,\beta}$ convergence of $(X_n)_{n\in\mathbb{N}}$, then there is at least a zone of convergence $[-K, K]$ of index $(v, w) = (0, 0)$, with $\gamma = 0$; indeed, the residues $\theta_n(\xi)$ stay locally bounded under this hypothesis. Thus, Definition 15.5 is an extension of this statement. However, we allow in the definition the exponent γ to be negative (but not smaller than $-\frac{1}{\alpha}$). Indeed, in the computation of Berry–Esseen type bounds, we shall sometimes need to work with smaller zones than the one given by mod-ϕ convergence, see the hypotheses of Theorem 15.19, and Sects. 15.3.3 and 15.3.4 for examples.

Remark 15.9 In our definition of zone of control, we ask for a bound on $|\theta_n(\xi) - 1|$ that holds for any $n \in \mathbb{N}$. Of course, if the bound is only valid for $n \geq n_0$ large enough, then the corresponding bound on the Kolmogorov distance (Theorem 15.19) will only hold for $n \geq n_0$.

15.2.2 Spaces of Test Functions

Until the end of Sect. 15.2, all the spaces of functions considered will be spaces of complex valued functions on the real line. If $f \in \mathrm{L}^1(\mathbb{R})$, we denote its Fourier transform

$$\widehat{f}(\xi) = \int_{\mathbb{R}} \mathrm{e}^{\mathrm{i}x\xi}\, f(x)\, dx.$$

Recall that the Schwartz space $\mathscr{S}(\mathbb{R})$ is by definition the space of infinitely differentiable functions whose derivatives tend to 0 at infinity faster than any power of x. Restricted to $\mathscr{S}(\mathbb{R})$, the Fourier transform is an automorphism, and it satisfies the Parseval formula

$$\forall f, g \in \mathscr{S}(\mathbb{R}),\ \int_{\mathbb{R}} f(x)\,\overline{g(x)}\, dx = \frac{1}{2\pi} \int_{\mathbb{R}} \widehat{f}(\xi)\,\overline{\widehat{g}(\xi)}\, d\xi.$$

We refer to [38, Chapter VIII] and [48, Part II] for a proof of this formula, and for the theory of Fourier transforms. The Parseval formula allows to extend by duality and/or density the Fourier transform to other spaces of functions or distributions. In particular, if $f \in \mathrm{L}^2(\mathbb{R})$, then its Fourier transform $\widehat{f}$ is well defined in $\mathrm{L}^2(\mathbb{R})$, although in general the integral $\int_{\mathbb{R}} \mathrm{e}^{\mathrm{i}x\xi}\, f(x)\, dx$ does not converge; and we have again the Parseval formula

$$\forall f, g \in \mathrm{L}^2(\mathbb{R}),\ \int_{\mathbb{R}} f(x)\,\overline{g(x)}\, dx = \frac{1}{2\pi} \int_{\mathbb{R}} \widehat{f}(\xi)\,\overline{\widehat{g}(\xi)}\, d\xi,$$

which amounts to the fact that $f \mapsto \frac{1}{\sqrt{2\pi}} \widehat{f}$ is an isometry of $\mathrm{L}^2(\mathbb{R})$ (see [48, §7.9]).

We denote $\mathscr{M}^1(\mathbb{R})$ the set of probability measures on Borel subsets of $\mathbb{R}$. In the sequel, we will need to apply a variant of Parseval's formula, where $\overline{g(x)}\, dx$ is replaced by $\mu(dx)$, with μ in $\mathscr{M}^1(\mathbb{R})$. This is given in the following lemma (see [55, Lemma 2.3.3], or [41, p. 134]).

Lemma 15.10 *For any function $f \in \mathrm{L}^1(\mathbb{R})$ with $\widehat{f} \in \mathrm{L}^1(\mathbb{R})$, and any Borel probability measure $\mu \in \mathscr{M}^1(\mathbb{R})$, the pairing $\langle \mu \mid f \rangle = \int_{\mathbb{R}} f(x)\, \mu(dx)$ is well defined, and the Parseval formula holds:*

$$\int_{\mathbb{R}} f(x)\, \mu(dx) = \frac{1}{2\pi} \int_{\mathbb{R}} \widehat{f}(\xi)\, \widehat{\mu}(-\xi)\, d\xi,$$

where $\widehat{\mu}(\xi) = \int_{\mathbb{R}} \mathrm{e}^{\mathrm{i}\xi x}\, \mu(dx)$. The formula also holds for finite signed measures.

Let us now introduce two adequate spaces of test functions, for which we shall be able to prove speed of convergence estimates. We first consider functions $f \in \mathrm{L}^1(\mathbb{R})$ with compactly supported Fourier transforms:

Definition 15.11 We call *smooth test function of order* 0, or simply *smooth test function* an element $f \in L^1(\mathbb{R})$ whose Fourier transform is compactly supported. We denote $\mathscr{T}_0(\mathbb{R})$ the subspace of $L^1(\mathbb{R})$ that consists in smooth test functions; it is an ideal for the convolution product.

Example 15.1 If

$$\mathrm{sinc}(x) := \frac{\sin x}{x} = \frac{1}{2}\int_{-1}^{1} \mathrm{e}^{\mathrm{i}x\xi}\, d\xi,$$

then by Fourier inversion $\widehat{\mathrm{sinc}}(\xi) = \pi\, 1_{|\xi|\leq 1}$ is compactly supported on $[-1, 1]$. Therefore, $f(x) = (\mathrm{sinc}(x))^2$ is an element of $L^1(\mathbb{R})$ whose Fourier transform is compactly supported on $[-1, 1] + [-1, 1] = [-2, 2]$, and $f \in \mathscr{T}_0(\mathbb{R})$.

Let us comment a bit Definition 15.11. If f is in $\mathscr{T}_0(\mathbb{R})$, then its Fourier transform $\widehat{f}$ is bounded by $\|f\|_{L^1}$ and vanishes outside an interval $[-C, C]$, so $\widehat{f} \in L^1(\mathbb{R})$. Since f and $\widehat{f}$ are integrable, we can apply Lemma 15.10 with f. Moreover, f is then known to satisfy the Fourier inversion formula (see [48, §7.7]):

$$f(x) = \frac{1}{2\pi}\int_{\mathbb{R}} \widehat{f}(\xi)\, \mathrm{e}^{-\mathrm{i}\xi x}\, d\xi.$$

As the integral above is in fact on a compact interval $[-C, C]$, the standard convergence theorems ensure that f is infinitely differentiable in x, hence the term "smooth". Also, by applying the Riemann–Lebesgue lemma to the continuous compactly supported functions $\xi \mapsto (-\mathrm{i}\xi)^k\, \widehat{f}(\xi)$, one sees that $f(x)$ and all its derivatives $f^{(k)}(x)$ go to 0 as x goes to infinity. To conclude, $\mathscr{T}_0(\mathbb{R})$ is included in the space $\mathscr{C}_0^\infty(\mathbb{R})$ of smooth functions whose derivatives all vanish at infinity.

Actually, we will need to work with more general test functions, defined by using the theory of tempered distributions. We endow the Schwartz space $\mathscr{S}(\mathbb{R})$ of smooth rapidly decreasing functions with its usual topology of metrizable locally convex topological vector space, defined by the family of semi-norms

$$\|f\|_{k,l} = \sum_{a\leq k}\sum_{b\leq l} \sup_{x\in\mathbb{R}} |x^a\, (\partial^b f)(x)|.$$

We recall that a tempered distribution ψ is a continuous linear form $\psi : \mathscr{S}(\mathbb{R}) \to \mathbb{C}$. The value of a tempered distribution ψ on a smooth function f will be denoted $\psi(f)$ or $\langle \psi \mid f\rangle$. The space of all tempered distributions is classically denoted $\mathscr{S}'(\mathbb{R})$, and it is endowed with the $*$-weak topology. The spaces of integrable functions, of square integrable functions and of probability measures can all be embedded in the space $\mathscr{S}'(\mathbb{R})$ as follows: if f is a function in $L^1(\mathbb{R}) \cup L^2(\mathbb{R})$, or if μ is in $\mathscr{M}^1(\mathbb{R})$, then we associate to them the distributions

$$\langle f \mid g\rangle = \int_{\mathbb{R}} f(x)g(x)\, dx \quad ; \quad \langle \mu \mid g\rangle = \int_{\mathbb{R}} g(x)\, \mu(dx).$$

We then say that these distributions are represented by the function f and by the measure μ.

The Fourier transform of a tempered distribution ψ is defined by duality: it is the unique tempered distribution $\widehat{\psi}$ such that

$$\langle \widehat{\psi} \mid f \rangle = \langle \psi \mid \widehat{f} \rangle$$

for any $f \in \mathscr{S}(\mathbb{R})$. This definition agrees with the previous definitions of Fourier transforms for integrable functions, square integrable functions, or probability measures (all these elements can be paired with Schwartz functions). Similarly, if ψ is a tempered distribution, then one can also define by duality its derivative: thus, $\partial \psi$ is the unique tempered distribution such that

$$\langle \partial \psi \mid f \rangle = - \langle \psi \mid \partial f \rangle$$

for any $f \in \mathscr{S}(\mathbb{R})$. The definition agrees with the usual one when ψ comes from a derivable function, by the integration by parts formula. On the other hand, Fourier transform and derivation define linear endomorphisms of $\mathscr{S}'(\mathbb{R})$; also note that the Fourier transform is bijective.

Definition 15.12 A *smooth test function of order* 1, or *smooth test distribution* is a distribution $f \in \mathscr{S}'(\mathbb{R})$, such that ∂f is in $\mathscr{T}_0(\mathbb{R})$, that is to say that the distribution ∂f can be represented by an integrable function with compactly supported Fourier transform. We denote $\mathscr{T}_1(\mathbb{R})$ the space of smooth test distributions (Fig. 15.2).

We now discuss Parseval's formula for functions in $\mathscr{T}_1(\mathbb{R})$.

Proposition 15.13 *Any smooth test distribution $f \in \mathscr{T}_1(\mathbb{R})$ can be represented by a bounded function in $\mathscr{C}^\infty(\mathbb{R})$. Moreover, for any smooth test distribution in $\mathscr{T}_1(\mathbb{R})$:*

(TD1) If μ is a Borel probability measure, then the pairing $\langle \mu \mid f \rangle = \int_{\mathbb{R}} f(x)\, \mu(dx)$ is well defined.

(TD2) The tempered distribution $\widehat{f}$ can be paired with the Fourier transform $\widehat{\mu}$ of a probability measure with finite first moment, in a way that extends the pairing between $\mathscr{S}'(\mathbb{R})$ and $\mathscr{S}(\mathbb{R})$ when μ (and therefore $\widehat{\mu}$) is given by a Schwartz density.

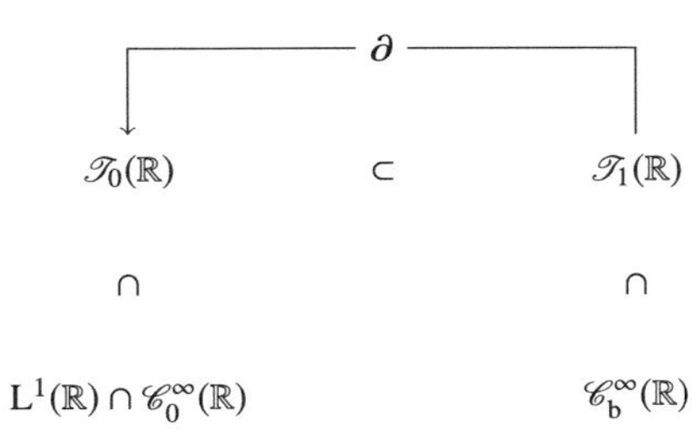

Fig. 15.2 The two spaces of test functions $\mathscr{T}_0(\mathbb{R})$ and $\mathscr{T}_1(\mathbb{R})$

(TD3) The Parseval formula holds: if $f \in \mathscr{T}_1(\mathbb{R})$ and μ has finite expectation, then

$$\langle \mu \mid f \rangle = \frac{1}{2\pi} \left\langle \widehat{f} \mid \overline{\widehat{\mu}} \right\rangle.$$

Proof We start by giving a better description of the tempered distributions f and $\widehat{f}$. Denote $\phi = \partial f$; by assumption, this tempered distribution can be represented by a function $\phi \in \mathscr{T}_0(\mathbb{R})$, which in particular is of class $\mathscr{C}^\infty$ and integrable. Set

$$\widetilde{f}(x) = \int_{y=0}^{x} \phi(y)\, dy.$$

This is a function of class $\mathscr{C}^\infty$, whose derivative is ϕ, and which is bounded since ϕ is integrable. Therefore, it is a tempered distribution, and for any $g \in \mathscr{S}(\mathbb{R})$,

$$\langle \partial f \mid g \rangle = \langle \phi \mid g \rangle = \left\langle \partial \widetilde{f} \mid g \right\rangle.$$

We conclude that $\partial(f - \widetilde{f})$ is the zero distribution. It is then a standard result that, given a tempered distribution ψ, one has $\partial \psi = 0$ if and only if ψ can be represented a constant. So,

$$f(x) = \int_{y=0}^{x} \phi(y)\, dy + f(0).$$

This shows in particular that f is a smooth bounded function.

A similar description can be provided for $\widehat{f}$. Recall that the principal value distribution, denoted $\mathrm{pv}(\frac{1}{x})$, is the tempered distribution defined for any $g \in \mathscr{S}(\mathbb{R})$ by

$$\left\langle \mathrm{pv}\left(\frac{1}{x}\right) \mid g \right\rangle = \lim_{\varepsilon \to 0} \left(\int_{|x| \geq \varepsilon} \frac{g(x)}{x}\, dx \right).$$

The existence of the limit is easily proved by making a Taylor expansion of g around 0. Denote

$$\mathscr{S}^{[1]} = \{g \in \mathscr{S}(\mathbb{R}) \mid g(x) = x\, h(x) \text{ with } h \in \mathscr{S}(\mathbb{R})\};$$
$$\mathscr{S}_{[1]} = \{g \in \mathscr{S}(\mathbb{R}) \mid g = \partial h \text{ with } h \in \mathscr{S}(\mathbb{R})\};$$

the Fourier transform establishes an homeomorphism between $\mathscr{S}^{[1]}$ and $\mathscr{S}_{[1]}$, and the restriction of $\mathrm{pv}(\frac{1}{x})$ to $\mathscr{S}^{[1]}$ is

$$\left\langle \mathrm{pv}\left(\frac{1}{x}\right) \mid g \right\rangle = \int_{\mathbb{R}} \frac{g(x)}{x}\, dx.$$

Let $\widehat{g}(\xi)$ be an element of $\mathscr{S}^{[1]}$, which we write as $\widehat{g}(\xi) = (-\mathrm{i}\xi)\,\widehat{h}(\xi)$ for some $h \in \mathscr{S}(\mathbb{R})$. This is equivalent to $g(x) = (\partial h)(x)$. Let us denote $g_-(x) = g(-x)$, $h_-(x) = h(-x)$, and $\mathrm{pv}(\frac{\mathrm{i}\widehat{\phi}(\xi)}{\xi})$ the tempered distribution defined by

$$\mathrm{pv}\left(\frac{\mathrm{i}\,\widehat{\phi}(\xi)}{\xi}\right) = \mathrm{i}\,\mathrm{pv}\left(\frac{1}{\xi}\right) \circ m_{\widehat{\phi}},$$

with $m_{\widehat{\phi}} : \mathscr{S}(\mathbb{R}) \to \mathscr{S}(\mathbb{R})$ equal to the multiplication by $\widehat{\phi}$. Then we can make the following computation:

$$\begin{aligned}\left\langle \widehat{f} \mid \widehat{g}\right\rangle &= \left\langle f \mid \widehat{\widehat{g}}\right\rangle = 2\pi\ \langle f \mid g_-\rangle = -2\pi\ \langle f \mid \partial h_-\rangle = 2\pi\ \langle \phi \mid h_-\rangle = \left\langle \phi \mid \widehat{\widehat{h}}\right\rangle \\ &= \left\langle \widehat{\phi} \mid \widehat{h}\right\rangle = \left\langle \widehat{\phi} \mid \frac{\mathrm{i}\widehat{g}(\xi)}{\xi}\right\rangle = \left\langle \mathrm{pv}\left(\frac{\mathrm{i}\,\widehat{\phi}(\xi)}{\xi}\right) \mid \widehat{g}\right\rangle.\end{aligned}$$

Thus, the tempered distributions $\widehat{f}$ and $\mathrm{pv}(\frac{\mathrm{i}\widehat{\phi}(\xi)}{\xi})$ agree on the codimension 1 subspace $\mathscr{S}^{[1]}$ of $\mathscr{S}(\mathbb{R})$. However, $\mathscr{S}^{[1]}$ is also the space of functions in $\mathscr{S}(\mathbb{R})$ that vanish at 0, so, if g_0 is any function in $\mathscr{S}(\mathbb{R})$ such that $g_0(0) = 1$, then for $g \in \mathscr{S}(\mathbb{R})$,

$$\begin{aligned}\left\langle \widehat{f} \mid g\right\rangle &= \left\langle \widehat{f} \mid g - g(0)g_0\right\rangle + g(0)\left\langle \widehat{f} \mid g_0\right\rangle \\ &= \left\langle \mathrm{pv}\left(\frac{\mathrm{i}\,\widehat{\phi}(\xi)}{\xi}\right) \mid g - g(0)g_0\right\rangle + \left\langle \widehat{f} \mid g_0\right\rangle\,\langle \delta_0 \mid g\rangle \\ &= \left\langle \mathrm{pv}\left(\frac{\mathrm{i}\,\widehat{\phi}(\xi)}{\xi}\right) \mid g\right\rangle + \left(\left\langle \widehat{f} - \mathrm{pv}\left(\frac{\mathrm{i}\,\widehat{\phi}(\xi)}{\xi}\right) \mid g_0\right\rangle\right)\,\langle \delta_0 \mid g\rangle \\ &= \left\langle \mathrm{pv}\left(\frac{\mathrm{i}\,\widehat{\phi}(\xi)}{\xi}\right) + L\,\delta_0 \mid g\right\rangle\end{aligned}$$

where L is some constant. Thus,

$$\widehat{f}(\xi) = \mathrm{pv}\left(\frac{\mathrm{i}\,\widehat{\phi}(\xi)}{\xi}\right) + L\,\delta_0$$

and a computation against test functions shows that

$$L = 2\pi f(0) - \mathrm{i}\left\langle \mathrm{pv}\left(\frac{1}{\xi}\right) \mid \widehat{\phi}\right\rangle.$$

The three parts of the proposition are now easily proven. For (TD1), since $f(x)$ is smooth and bounded, we can indeed consider the convergent integral $\int_{\mathbb{R}} f(x)\,\mu(dx)$. For (TD2), assuming that μ has a finite first moment, $\widehat{\mu}$ is a function of class $\mathscr{C}^1$

and with bounded derivative. The same holds for $\widehat{\phi\mu}$, and therefore, one can define

$$\begin{aligned}\int_{\mathbb{R}} \widehat{f}(\xi)\widehat{\mu}(-\xi)\,d\xi &= \mathrm{i}\left\langle \mathrm{pv}\left(\frac{1}{\xi}\right) \;\middle|\; \widehat{\phi}\,\overline{\widehat{\mu}}\right\rangle + L\\ &= \left(\lim_{\varepsilon\to\infty}\int_{|x|\geq\varepsilon}\frac{\mathrm{i}\,\widehat{\phi}(\xi)\,\widehat{\mu}(-\xi)}{\xi}\,d\xi\right) + L.\end{aligned}$$

Indeed, if $f \in \mathscr{C}^1(\mathbb{R})$, then $\lim_{\varepsilon\to\infty}\int_{1\geq|x|\geq\varepsilon}\frac{f(x)}{x}\,dx$ always exists, as can be seen by replacing f by its Taylor approximation at 0. Finally, let us prove the Parseval formula (TD3). The previous calculations show that

$$\begin{aligned}\frac{1}{2\pi}\int_{\mathbb{R}} \widehat{f}(\xi)\,\widehat{\mu}(-\xi)\,d\xi &= \frac{\mathrm{i}}{2\pi}\left\langle \mathrm{pv}\left(\frac{1}{\xi}\right) \;\middle|\; \widehat{\phi}\,\overline{\widehat{\mu}} - \widehat{\phi}\right\rangle + f(0)\\ &= \lim_{\varepsilon\to 0}\left(\frac{\mathrm{i}}{2\pi}\int_{|\xi|\geq\varepsilon}\widehat{\phi}(\xi)\left(\frac{\widehat{\mu}(-\xi)-1}{\xi}\right)d\xi\right) + f(0)\\ &= \frac{1}{2\pi}\int_{\mathbb{R}}\widehat{\phi}(\xi)\left(\frac{\widehat{\mu}(-\xi)-1}{-\mathrm{i}\xi}\right)d\xi + f(0).\end{aligned}$$

Indeed, the function $\xi \mapsto \frac{\widehat{\mu}(-\xi)-1}{-\mathrm{i}\xi}$ is continuous on $\mathbb{R}$ and bounded, with value $\frac{\widehat{\mu}'(0)}{\mathrm{i}} = \int_{\mathbb{R}} x\,\mu(dx)$ at $\xi = 0$; it can therefore be integrated against the function $\widehat{\phi}$ which is integrable (and even with compact support). On the other hand,

$$\begin{aligned}\int_{\mathbb{R}} f(x)\,\mu(dx) &= \int_{x\in\mathbb{R}}\int_{y=0}^{x}\phi(y)\,dy\,\mu(dx) + f(0)\\ &= \int_{(x,y)\in\mathbb{R}^2}(1_{x>y>0} - 1_{x<y<0})\,\phi(y)\,dy\,\mu(dx) + f(0)\\ &= \int_{y\in\mathbb{R}}\phi(y)\,F(y)\,dy + f(0), \quad \text{with } F(y) = \mu((y,\infty)) - 1_{y\leq 0}.\end{aligned}$$

One has $\int_{\mathbb{R}}|F(y)|\,dy = \int_{y=0}^{\infty}\mu((y,\infty)) + \int_{y=-\infty}^{0}\mu((-\infty,y)) = \int_{\mathbb{R}}|x|\,\mu(dx)$, which is finite. In the integral above, we can therefore consider $F(y)\,dy$ as a finite signed measure, and the Parseval formula applies by Lemma 15.10. One computes readily

$$\widehat{F}(\xi) = \frac{\widehat{\mu}(\xi)-1}{\mathrm{i}\xi},$$

which ends the proof. □

Remark 15.14 The Parseval formula of Proposition 15.13 extends readily to finite signed measures μ such that $\int_{\mathbb{R}}|x|\,|\mu|(dx) < +\infty$. Actually, it is sufficient to have a finite signed measure μ such that

$$\frac{\widehat{\mu}(\xi) - \widehat{\mu}(0)}{\xi^v}$$

is bounded in a vicinity of 0, for some $v > 0$. Then, $(\widehat{\mu}(\xi) - \widehat{\mu}(0))/\xi$ is integrable in a neighborhood of 0. This ensures that the distribution $f(x)$ (respectively, the distribution $\widehat{f}(\xi)$) can be evaluated against the measure $\mu(x)$ (respectively, against $\widehat{\mu}(\xi)$), and then, the proof of Parseval's formula is analogous to the previous arguments.

15.2.3 Estimates for Test Functions

We now give an estimate of $\mathbb{E}[f_n(Y_n)] - \mathbb{E}[f_n(Y)]$, where $(f_n)_{n\in\mathbb{N}}$ is a sequence of test functions in $\mathscr{T}_0(\mathbb{R})$ or $\mathscr{T}_1(\mathbb{R})$, and $(Y_n)_{n\in\mathbb{N}}$ is a sequence of random variables associated to a sequence $(X_n)_{n\in\mathbb{N}}$ which has a zone of control.

Proposition 15.15 *Let $(X_n)_{n\in\mathbb{N}}$ be a sequence of random variables, $\phi_{c,\alpha,\beta}$ a reference stable law, Y with law $\phi_{c,\alpha,\beta}$ and Y_n as in Proposition 15.3. We assume that:*

(1) $(X_n)_{n\in\mathbb{N}}$ has a zone of control $[-K(t_n)^\gamma, K(t_n)^\gamma]$ with index (v, w);
(2) $(f_n)_{n\in\mathbb{N}}$ is a sequence of smooth test functions in $\mathscr{T}_0(\mathbb{R})$, such that the support of $\widehat{f_n}$ is included into $[-K(t_n)^{\gamma+1/\alpha}, K(t_n)^{\gamma+1/\alpha}]$.

Then,

$$|\mathbb{E}[f_n(Y_n)] - \mathbb{E}[f_n(Y)]| \leq C_0(c, \alpha, v)\, K_1\, \frac{\|f_n\|_{\mathrm{L}^1}}{(t_n)^{v/\alpha}},$$

where $C_0(c, \alpha, v) = \dfrac{2^{\frac{v+1}{\alpha}}\, \Gamma((v+1)/\alpha)}{\pi \alpha\, c^{v+1}}$.
If instead of (2) we assume:

(2') $(f_n)_{n\in\mathbb{N}}$ is a sequence of smooth test distributions in $\mathscr{T}_1(\mathbb{R})$ such that if $\phi_n = \partial f_n$ is the derivative of the distribution f_n, then the support of $\widehat{\phi_n}$ is included into $[-K(t_n)^{\gamma+1/\alpha}, K(t_n)^{\gamma+1/\alpha}]$.

Then

$$|\mathbb{E}[f_n(Y_n)] - \mathbb{E}[f_n(Y)]| \leq C_1(c, \alpha, v)\, K_1\, \frac{\|\phi_n\|_{\mathrm{L}^1}}{(t_n)^{v/\alpha}},$$

where $C_1(c, \alpha, v) = \dfrac{2^{v/\alpha}\, \Gamma(v/\alpha)}{\pi \alpha\, c^v}$.

Proof Consider first a sequence $(f_n)_{n\in\mathbb{N}}$ of test functions in $\mathscr{T}_0(\mathbb{R})$, which satisfies (2). Using Parseval formula and the zone of control assumption, we have

$$\mathbb{E}[f_n(Y_n)] - \mathbb{E}[f_n(Y)] = \frac{1}{2\pi}\int_{-K(t_n)^{\gamma+\frac{1}{\alpha}}}^{K(t_n)^{\gamma+\frac{1}{\alpha}}} \widehat{f_n}(\xi)\, \mathrm{e}^{\eta(-\mathrm{i}\xi)} \left(\theta_n\left(-\xi/(t_n)^{\frac{1}{\alpha}}\right) - 1\right) d\xi;$$

$$|\mathbb{E}[f_n(Y_n)] - \mathbb{E}[f_n(Y)]| \leq \frac{K_1\|\widehat{f_n}\|_\infty}{2\pi\,(t_n)^{v/\alpha}} \int_{-K(t_n)^{\gamma+\frac{1}{\alpha}}}^{K(t_n)^{\gamma+\frac{1}{\alpha}}} |\xi|^v\, \mathrm{e}^{-|c\xi|^\alpha + K_2\left(\frac{|\xi|}{(t_n)^{1/\alpha}}\right)^w} d\xi.$$

For ξ in $[-K(t_n)^{\gamma+1/\alpha}, K(t_n)^{\gamma+1/\alpha}]$, since $(t_n)^{\gamma-1/(w-\alpha)} \leq 1$, the second term in the exponent can be bounded as follows:

$$K_2\left(\frac{|\xi|}{(t_n)^{1/\alpha}}\right)^w = K_2|\xi|^\alpha \left(\frac{|\xi|}{(t_n)^{\frac{1}{\alpha}+\frac{1}{w-\alpha}}}\right)^{w-\alpha} \leq K_2|\xi|^\alpha \left(K(t_n)^{\gamma-\frac{1}{w-\alpha}}\right)^{w-\alpha}$$
$$\leq \frac{|c\xi|^\alpha}{2}.$$

This is compensated by the term $-|c\xi|^\alpha$ and, therefore,

$$|\mathbb{E}[f_n(Y_n)] - \mathbb{E}[f_n(Y)]| \leq \frac{K_1\|\widehat{f_n}\|_\infty}{2\pi\,(t_n)^{v/\alpha}} \int_{\mathbb{R}} |\xi|^v \mathrm{e}^{-\frac{|c\xi|^\alpha}{2}}\, d\xi$$
$$\leq \frac{2^{\frac{v+1}{\alpha}}\, K_1}{\pi\alpha\, c^{v+1}(t_n)^{v/\alpha}}\, \Gamma\left(\frac{v+1}{\alpha}\right) \|f_n\|_{\mathrm{L}^1}.$$

This ends the proof of the first case. For test distributions $f_n \in \mathscr{T}_1(\mathbb{R})$ which satisfies the condition (2′), let us introduce the signed measure $\mu = \mathbb{P}_{Y_n} - \mathbb{P}_Y$. One has $\widehat{\mu}(0) = 0$, and by hypothesis,

$$\left|\frac{\widehat{\mu}(\xi)}{\xi}\right| \leq \frac{K_1\,|\xi|^{v-1}}{(t_n)^{v/\alpha}}\, \mathrm{e}^{-|c\xi|^\alpha + K_2\left(\frac{|\xi|}{(t_n)^{1/\alpha}}\right)^w}.$$

Remark 15.14 applies, and thus,

$$|\mathbb{E}[f_n(Y_n)]-\mathbb{E}[f_n(Y)]| = |\langle \mu \mid f_n\rangle| = \frac{1}{2\pi}\left|\langle \widehat{f_n} \mid \overline{\widehat{\mu}}\rangle\right| = \frac{1}{2\pi}\left|\int_{\mathbb{R}} \widehat{\phi_n}(\xi)\, \frac{\widehat{\mu}(-\xi)}{\xi}\, d\xi\right|.$$

From there, the computations are exactly the same as before, with an index $v - 1$ instead of v. □

15.2.4 Smoothing Techniques

We now explain how to relate the estimates on test functions or distributions to estimates on the Kolmogorov distance. The main tool with respect to this problem is the following:

Lemma 15.16 *There exists a function ρ (called kernel) on $\mathbb{R}$ with the following properties.*

1. *The kernel ρ is non-negative, with $\int_{\mathbb{R}} \rho(x)\,dx = 1$.*
2. *The support of $\widehat{\rho}$ is $[-1, 1]$ (hence, ρ is a test function in $\mathscr{T}_0(\mathbb{R})$).*
3. *The functions ρ and $\widehat{\rho}$ are even, and*

$$\rho(K) \leq \min\left(\frac{3}{8\pi}, \frac{96}{\pi\, K^4}\right).$$

Proof Set

$$\rho(x) = \frac{3}{8\pi}\left(\operatorname{sinc}\left(\frac{x}{4}\right)\right)^4.$$

It has its Fourier transform supported on $[-\frac{1}{4}, \frac{1}{4}] + [-\frac{1}{4}, \frac{1}{4}] + [-\frac{1}{4}, \frac{1}{4}] + [-\frac{1}{4}, \frac{1}{4}] = [-1, 1]$. On the other hand, an easy computation gives $\int_{\mathbb{R}} \rho(x)\,dx = 1$: use for example the Plancherel formula

$$\int_{\mathbb{R}} |f(x)|^2\,dx = \frac{1}{2\pi}\int_{\mathbb{R}} |\widehat{f}(\xi)|^2\,d\xi$$

with $f(x) = \operatorname{sinc}(x)^2$ and thus $\widehat{f}(\xi) = \frac{1}{2\pi}\,\widehat{\operatorname{sinc}} * \widehat{\operatorname{sinc}}(\xi) = \frac{\pi}{2}(2 - |\xi|)_+$. Finally, $\operatorname{sinc}(x) \leq \min(1, \frac{1}{|x|})$, which leads to the inequality stated for $\rho(K)$. □

In the following, for $\varepsilon > 0$, we set $\rho_\varepsilon(x) = \frac{1}{\varepsilon}\rho(\frac{x}{\varepsilon})$, which has its Fourier transform compactly supported on $[-\frac{1}{\varepsilon}, \frac{1}{\varepsilon}]$; see Fig. 15.3. We also denote $f_{a,\varepsilon}(x) = f_\varepsilon(x - a)$, where f_ε is the function $1_{(-\infty,0]} * \rho_\varepsilon$; cf. Fig. 15.4.

For all a, ε, $f_{a,\varepsilon}$ is an approximation of the Heaviside function $1_{(-\infty,a]}$, and one has the following properties:

Proposition 15.17 *The function $f_{a,\varepsilon}$ is a smooth test distribution in $\mathscr{T}_1(\mathbb{R})$ whose derivative $\partial f_{a,\varepsilon}$ has its Fourier transform compactly supported on $[-\frac{1}{\varepsilon}, \frac{1}{\varepsilon}]$, and satisfies $\|\partial f_{a,\varepsilon}\|_{L^1} = 1$. Moreover:*

1. *The function $f_{a,\varepsilon}$ has a non-positive derivative, and decreases from 1 to 0.*

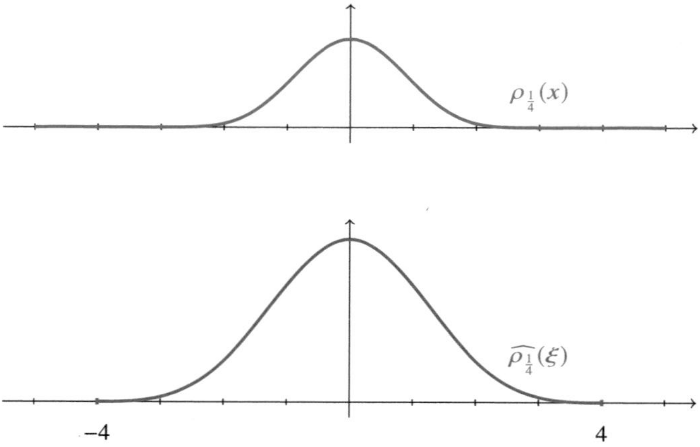

Fig. 15.3 The smoothing kernel $\rho_{\frac{1}{4}}$, and its Fourier transform which is supported on $[-4, 4]$

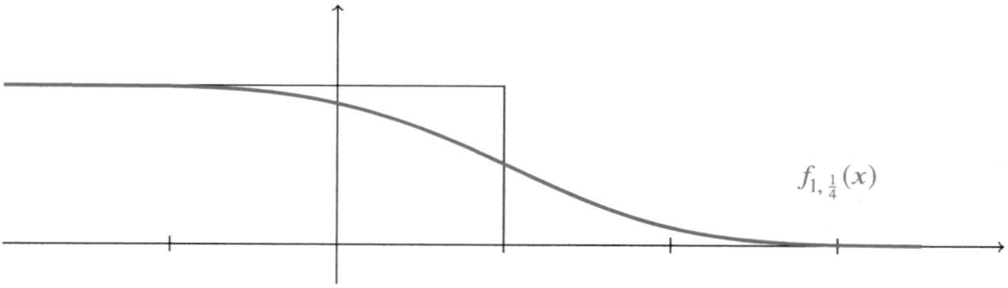

Fig. 15.4 The approximation $f_{1,\frac{1}{4}}$ of the Heaviside function $1_{(-\infty,1]}$

2. *One has* $f_1(x) = 1 - f_1(-x)$, *and for all* $K \geq 0$,

$$f_1(K) = \int_0^\infty \rho(K + y)\, dy \leq \frac{32}{\pi\, K^3};$$

$$\int_0^\infty f_1(u - K)\, du \leq K + \int_{w=0}^\infty \min\left(1, \frac{32}{\pi\, w^3}\right) dw = K + 3\sqrt[3]{\frac{3}{\pi}}.$$

Proof The derivative of $f_{a,\varepsilon}$ is

$$\begin{aligned}\partial(f_{a,\varepsilon})(x) &= \partial(1_{(-\infty,a]} * \rho_\varepsilon)(x) = \left(\partial(1_{(-\infty,a]}) * \rho_\varepsilon\right)(x) \\ &= (-\delta_a * \rho_\varepsilon)(x) = -\rho_\varepsilon(x - a),\end{aligned}$$

so it is indeed in $\mathscr{T}_0(\mathbb{R})$, and non-positive. Its Fourier transform is supported by $[-\frac{1}{\varepsilon}, \frac{1}{\varepsilon}]$, and $\|\partial f_{a,\varepsilon}\|_{\mathrm{L}^1} = \int_{\mathbb{R}} \rho_\varepsilon(x)\, dx = 1$. Then,

$$\lim_{x\to+\infty} f_{a,\varepsilon}(x) = \lim_{x\to+\infty} f_1\left(\frac{x-a}{\varepsilon}\right) = \lim_{y\to+\infty} f_1(y),$$

so $\lim_{x\to+\infty} f_{a,\varepsilon}(x) = 0$. Since by definition $f_1(y) = \int_y^\infty \rho(u)\, du$, the symmetry relation $f_1(x) = 1 - f_1(-x)$ follows from ρ even; it implies the other limit statement $\lim_{x\to-\infty} f_{a,\varepsilon}(x) = 1$. The inequalities in part ii) are immediate consequences of those of Lemma 15.16. □

Let us now state a result which converts estimates on smooth test distributions into estimates of Kolmogorov distances. It already appeared in [44, Lemma 16], and is inspired by [56, p. 87] and [19, Chapter XVI, §3, Lemma 1]:

Theorem 15.18 *Let X and Y be two random variables with cumulative distribution functions $F_X(a) = \mathbb{P}[X \leq a]$ and $F_Y(a) = \mathbb{P}[Y \leq a]$. Assume that for some $\varepsilon > 0$ and $B > 0$,*

$$\sup_{a\in\mathbb{R}} |\mathbb{E}[f_{a,\varepsilon}(X)] - \mathbb{E}[f_{a,\varepsilon}(Y)]| \leq B\varepsilon.$$

We also suppose that Y has a density w.r.t. Lebesgue measure that is bounded by m. Then, for every $\lambda > 0$,

$$\begin{aligned} d_{\mathrm{Kol}}(X,Y) &= \sup_{a\in\mathbb{R}} |F_X(a) - F_Y(a)| \\ &\leq (1+\lambda)\left(B + \frac{m}{\sqrt[3]{\pi}}\left(4\sqrt[3]{1+\frac{1}{\lambda}} + 3\sqrt[3]{3}\right)\right)\varepsilon. \end{aligned}$$

The choice of the parameter λ allows one to optimize constants according to the reference law of Y and to the value of B. A general bound is obtained by choosing $\lambda = \frac{1}{2}$; this gives after some simplifications

$$d_{\mathrm{Kol}}(X,Y) \leq \frac{3}{2}(B + 7m)\,\varepsilon,$$

which is easy to remember and manipulate.

Proof For the convenience of the reader, we reproduce here the proof given in [44]. Fix a positive constant K, and denote $\Delta = \sup_{a\in\mathbb{R}} |F_X(a) - F_Y(a)|$ the Kolmogorov distance between X and Y. One has

$$F_X(a) = \mathbb{E}[1_{X\leq a}] \leq \mathbb{E}[f_{a+K\varepsilon,\varepsilon}(X)] + \mathbb{E}[(1 - f_{a+K\varepsilon,\varepsilon}(X))\, 1_{X\leq a}]$$
$$\leq \mathbb{E}[f_{a+K\varepsilon,\varepsilon}(Y)] + \mathbb{E}[(1 - f_{a+K\varepsilon,\varepsilon}(X))\, 1_{X\leq a}] + B\varepsilon. \tag{15.2}$$

The second expectation writes as

$$\begin{aligned}\mathbb{E}[(1 - f_{a+K\varepsilon,\varepsilon}(X))\, 1_{X\leq a}] &= \int_{\mathbb{R}} (1 - f_{a+K\varepsilon,\varepsilon}(x))\, 1_{(-\infty,a]}(x)\, \mathbb{P}_X(dx) \\ &= -\int_{\mathbb{R}} ((1 - f_{a+K\varepsilon,\varepsilon}(x))\, 1_{(-\infty,a]}(x))'\, F_X(x)\, dx \\ &=: A_1 + A_2,\end{aligned}$$

where $A_1 = (1 - f_{a+K\varepsilon,\varepsilon}(a))\, F_X(a)$, $A_2 = \int_{\mathbb{R}} f'_{a+K\varepsilon,\varepsilon}(x)\, 1_{(-\infty,a]}(x)\, F_X(x)\, dx$. Indeed, in the space of tempered distributions, $((1 - f_{a+K\varepsilon,\varepsilon}(x))\, 1_{(-\infty,a]}(x))' = -(1 - f_{a+K\varepsilon,\varepsilon}(x))\, \delta_a(x) - f'_{a+K\varepsilon,\varepsilon}(x)\, 1_{(-\infty,a]}(x)$. We evaluate the two terms A_1 and A_2 as follows:

- Since $F_X(a) \leq F_Y(a) + \Delta$,

$$\begin{aligned}A_1 &\leq (1 - f_{a+K\varepsilon,\varepsilon}(a))\, F_Y(a) + (1 - f_{a+K\varepsilon,\varepsilon}(a))\, \Delta \\ &\leq \int_{\mathbb{R}} (1 - f_{a+K\varepsilon,\varepsilon}(x))\, \delta_a(x)\, F_Y(x)\, dx + (1 - f_1(-K))\, \Delta.\end{aligned}$$

- For A_2, since $F_X(x) \geq F_Y(x) - \Delta$ and the derivative of $f_{a+K\varepsilon,\varepsilon}$ is negative, an upper bound is obtained as follows:

$$\begin{aligned}A_2 &\leq \int_{\mathbb{R}} f'_{a+K\varepsilon,\varepsilon}(x)\, 1_{(-\infty,a]}(x)\, F_Y(x)\, dx - \Delta \int_{\mathbb{R}} f'_{a+K\varepsilon,\varepsilon}(x)\, 1_{(-\infty,a]}(x)\, dx \\ &= \int_{\mathbb{R}} f'_{a+K\varepsilon,\varepsilon}(x)\, 1_{(-\infty,a]}(x)\, F_Y(x)\, dx + (1 - f_{a+K\varepsilon,\varepsilon}(a))\, \Delta \\ &= \int_{\mathbb{R}} f'_{a+K\varepsilon,\varepsilon}(x)\, 1_{(-\infty,a]}(x)\, F_Y(x)\, dx + (1 - f_1(-K))\, \Delta.\end{aligned}$$

Therefore, by gathering the bounds on A_1 and A_2, we get

$$\mathbb{E}[(1 - f_{a+K\varepsilon,\varepsilon}(X))\, 1_{X\leq a}] \leq \mathbb{E}[(1 - f_{a+K\varepsilon,\varepsilon}(Y))\, 1_{Y\leq a}] + 2(1 - f_1(-K))\, \Delta. \tag{15.3}$$

On the other hand, if m is a bound on the density f_Y of Y, then

$$\begin{aligned}\mathbb{E}[f_{a+K\varepsilon,\varepsilon}(Y)\,1_{Y\geq a}] &= \int_a^\infty f_{a+K\varepsilon,\varepsilon}(y)\,f_Y(y)\,dy \\ &\leq m\int_a^\infty f_\varepsilon(y-a-K\varepsilon)\,dy = m\int_0^\infty f_\varepsilon(y-K\varepsilon)\,dy \\ &\leq m\varepsilon\int_0^\infty f_1(u-K)\,du \leq m\varepsilon\left(K+3\sqrt[3]{\frac{3}{\pi}}\right);\end{aligned}$$

and

$$\mathbb{E}[f_{a+K\varepsilon,\varepsilon}(Y)] \leq \mathbb{E}[f_{a+K\varepsilon,\varepsilon}(Y)\,1_{Y\leq a}] + m\left(K+3\sqrt[3]{\frac{3}{\pi}}\right)\varepsilon. \tag{15.4}$$

Putting together Eqs. (15.2), (15.3) and (15.4), we get

$$F_X(a) \leq F_Y(a) + \left(B + m\left(K+3\sqrt[3]{\frac{3}{\pi}}\right)\right)\varepsilon + \frac{64}{\pi\,K^3}\,\Delta.$$

Similarly, $F_X(a) \geq F_Y(a) - \left(B + m(K+3\sqrt[3]{\frac{3}{\pi}})\right)\varepsilon - \frac{64}{\pi\,K^3}\,\Delta$, so in the end

$$\Delta = \sup_{a\in\mathbb{R}} |F_X(a) - F_Y(a)| \leq \left(B + m\left(K+3\sqrt[3]{\frac{3}{\pi}}\right)\right)\varepsilon + \frac{64}{\pi\,K^3}\,\Delta.$$

As this is true for every K, setting $K = 4\sqrt[3]{\frac{1+\lambda}{\pi\lambda}}$ with $\lambda > 0$, one obtains

$$\Delta \leq (1+\lambda)\left(B + \frac{m}{\sqrt[3]{\pi}}\left(4\sqrt[3]{1+\frac{1}{\lambda}} + 3\sqrt[3]{3}\right)\right)\varepsilon.$$

In the next Sect. 15.2.5, we shall combine Theorem 15.18 and the estimates on smooth test distributions given by Proposition 15.15 to get a Berry–Esseen type bound on the Kolmogorov distances in the setting of mod-ϕ convergence.

15.2.5 Bounds on the Kolmogorov Distance

We are now ready to get an estimate for the Komogorov distance under a zone of control hypothesis.

Theorem 15.19 *Fix a reference stable distribution $\phi_{c,\alpha,\beta}$ and consider a sequence $(X_n)_{n\in\mathbb{N}}$ of random variables with a zone of control $[-K(t_n)^\gamma, K(t_n)^\gamma]$ of* index *(v, w). Assume in addition that $\gamma \leq \frac{v-1}{\alpha}$. As before, we denote Y a random variable with law $\phi_{c,\alpha,\beta}$, and Y_n the renormalization of X_n as in Proposition 15.3. Then, there exists a constant $C(\alpha, c, v, K, K_1)$ such that*

$$d_{\mathrm{Kol}}(Y_n, Y) \leq C(\alpha, c, v, K, K_1)\,\frac{1}{(t_n)^{1/\alpha+\gamma}}.$$

The constant $C(\alpha, c, v, K, K_1)$ is explicitly given by

$$\min_{\lambda>0}\left(\frac{1+\lambda}{\alpha\pi\, c}\left(\frac{2^{\frac{v}{\alpha}}\,\Gamma(\frac{v}{\alpha})\,K_1}{c^{v-1}} + \frac{\Gamma(\frac{1}{\alpha})}{\sqrt[3]{\pi}\,K}\left(4\sqrt[3]{1+\frac{1}{\lambda}} + 3\sqrt[3]{3}\right)\right)\right).$$

Note that the additional hypothesis $\gamma \leq \frac{v-1}{\alpha}$ can always be ensured by decreasing γ (but this makes the resulting bound weaker).

Proof We apply Proposition 15.15 with the smooth test distributions $f_n = f_{a,\varepsilon_n}$, with the value $\varepsilon_n := \frac{1}{K\,(t_n)^{1/\alpha+\gamma}}$; we know that $\|\partial f_n\|_{\mathrm{L}^1} = 1$ and that the Fourier transform $\widehat{f_n}$ is supported by the zone $[-K(t_n)^{1/\alpha+\gamma}, K(t_n)^{1/\alpha+\gamma}]$, so that

$$\left|\mathbb{E}[f_{a,\varepsilon_n}(Y_n)] - \mathbb{E}[f_{a,\varepsilon_n}(Y)]\right| \leq C_2(c, \alpha, v)\,\frac{K_1}{(t_n)^{v/\alpha}} \leq \frac{2^{\frac{v}{\alpha}}\,\Gamma(\frac{v}{\alpha})\,K_1\,K}{\alpha\pi\, c^v}\,\varepsilon_n.$$

This allows to apply Theorem 15.18 with a constant

$$B = \frac{2^{\frac{v}{\alpha}}\,\Gamma(\frac{v}{\alpha})\,K_1\,K}{\alpha\pi\, c^v},$$

and with $\varepsilon = \varepsilon_n = \frac{1}{K\,(t_n)^{1/\alpha+\gamma}}$. Indeed, note that the density of the law of Y is bounded by

$$m = \frac{1}{2\pi}\,\|\mathrm{e}^{\eta(\mathrm{i}\xi)}\|_{\mathrm{L}^1} = \frac{1}{\alpha\pi\, c}\,\Gamma\left(\frac{1}{\alpha}\right).$$

Remark 15.20 Suppose $\alpha = 2$, $c = \frac{1}{\sqrt{2}}$ (mod-Gaussian convergence), and $v = w = 3$. The maximal value allowed for the exponent γ in the size of the zone of control is then $\gamma = 1$, and later we shall encounter many examples of this situation. Then, we obtain

$$d_{\mathrm{Kol}}(Y_n, Y) \leq \frac{1+\lambda}{\sqrt{2\pi}}\left(2^{\frac{3}{2}}\,K_1 + \frac{1}{\sqrt[3]{\pi}\,K}\left(4\sqrt[3]{1+\frac{1}{\lambda}} + 3\sqrt[3]{3}\right)\right)\frac{1}{(t_n)^{3/2}}. \qquad (15.5)$$

In Sect. 15.4, we shall give conditions on cumulants of random variables that lead to mod-Gaussian convergence with a zone of control of size $O(t_n)$ and with index

(3, 3), so that (15.5) holds. We shall then choose K, K_1 and λ to make the constant in the right-hand side as small as possible.

Remark 15.21 In the general case, taking $\lambda = \frac{1}{2}$ in Theorem 15.19 leads to the inequality

$$d_{\mathrm{Kol}}(Y_n, Y) \leq C_3(\alpha, c, v, K_1, K)\,\frac{1}{(t_n)^{1/\alpha+\gamma}},$$

where $C_3(\alpha, c, v, K_1, K) = \frac{3}{2\pi\,\alpha\, c}\left(\frac{2^{\frac{v}{\alpha}}\,\Gamma(\frac{v}{\alpha})\,K_1}{c^{v-1}} + \frac{7\,\Gamma(\frac{1}{\alpha})}{K}\right)$.

15.3 Examples with an Explicit Fourier Transform

15.3.1 Sums of Independent Random Variables

As a direct application of Theorem 15.19, one recovers the classical Berry–Esseen estimates. Let $(A_n)_{n\in\mathbb{N}}$ be a sequence of centered i.i.d. random variables with a third moment. We denote $\mathbb{E}[(A_i)^2] = \sigma^2$ and $\mathbb{E}[|A_i|^3] = \rho$. Set $S_n = \sum_{i=1}^n A_i$, $X_n = S_n/(\sigma n^{1/3})$,

$$t_n = n^{1/3} \quad ; \quad K = \frac{\sigma^3}{\rho} \quad ; \quad v = w = 3 \quad ; \quad \gamma = 1.$$

Notice that $K \leq 1$ as a classical application of Hölder inequality. On the zone $\xi \in [-Kn^{1/3}, Kn^{1/3}]$, we have:

$$\begin{aligned}
|\theta_n(\xi) - 1| &= \left|\left(\mathbb{E}\left[\mathrm{e}^{\mathrm{i}\xi\frac{A_1}{\sigma n^{1/3}}}\right]\mathrm{e}^{\frac{\xi^2}{2n^{2/3}}}\right)^n - 1\right| \\
&\leq n\left|\mathbb{E}\left[\mathrm{e}^{\mathrm{i}\xi\frac{A_1}{\sigma n^{1/3}}}\right]\mathrm{e}^{\frac{\xi^2}{2n^{2/3}}} - 1\right|\left(\max\left(\left|\mathbb{E}\left[\mathrm{e}^{\mathrm{i}\xi\frac{A_1}{\sigma n^{1/3}}}\right]\mathrm{e}^{\frac{\xi^2}{2n^{2/3}}}\right|, 1\right)\right)^{n-1}.
\end{aligned}$$

For any t, $|\mathrm{e}^{\mathrm{i}t} - 1 - \mathrm{i}t + \frac{t^2}{2}| \leq \frac{|t^3|}{6}$, so

$$\begin{aligned}
&\left|\mathbb{E}\left[\mathrm{e}^{\mathrm{i}\xi\frac{A_1}{\sigma n^{1/3}}}\right]\mathrm{e}^{\frac{\xi^2}{2n^{2/3}}} - 1\right| \\
&\quad\leq \left|\mathbb{E}\left[\mathrm{e}^{\mathrm{i}\xi\frac{A_1}{\sigma n^{1/3}}}\right] - 1 + \frac{\xi^2}{2n^{2/3}}\right|\mathrm{e}^{\frac{\xi^2}{2n^{2/3}}} + \left|\mathrm{e}^{-\frac{\xi^2}{2n^{2/3}}} - 1 + \frac{\xi^2}{2n^{2/3}}\right|\mathrm{e}^{\frac{\xi^2}{2n^{2/3}}} \\
&\quad\leq \left(\frac{|\xi|^3}{6Kn} + \frac{\xi^4}{8n^{4/3}}\right)\mathrm{e}^{\frac{\xi^2}{2n^{2/3}}} \leq \frac{7\mathrm{e}^{1/2}}{24}\,\frac{|\xi|^3}{Kn}.
\end{aligned}$$

For the same reasons,

$$\left|\mathbb{E}\left[\mathrm{e}^{\mathrm{i}\xi\frac{A_1}{\sigma n^{1/3}}}\right]\mathrm{e}^{\frac{\xi^2}{2n^{2/3}}}\right| \le \frac{|\xi|^3}{6Kn}\,\mathrm{e}^{\frac{\xi^2}{2n^{2/3}}} + \left(1-\frac{\xi^2}{2n^{2/3}}\right)\mathrm{e}^{\frac{\xi^2}{2n^{2/3}}}$$
$$\le \frac{|\xi|^3}{6Kn}\,\mathrm{e}^{1/2}+1 \le \mathrm{e}^{\frac{\mathrm{e}^{1/2}}{6}\frac{|\xi|^3}{Kn}}$$

We conclude that

$$|\theta_n(\xi)-1| \le \frac{7\mathrm{e}^{1/2}}{24}\,\frac{|\xi|^3}{K}\,\mathrm{e}^{\frac{\mathrm{e}^{1/2}}{6}\frac{|\xi|^3}{K}}$$

on the zone of control $[-Kn^{1/3}, Kn^{1/3}]$. If we want Condition (Z2) to be satisfied, we need to change K and set

$$K = \frac{3}{2\mathrm{e}^{1/2}}\,\frac{\sigma^3}{\rho},$$

which is a little bit smaller than before. We then have a zone of control with constants $K_1 = \frac{7\mathrm{e}^{1/2}\rho}{24\sigma^3}$ and $K_2 = \frac{\mathrm{e}^{1/2}\rho}{6\sigma^3}$, and the inequality $K \le (\frac{c^\alpha}{2K_2})^{\frac{1}{w-\alpha}}$ is an equality. By Theorem 15.19,

$$d_{\mathrm{Kol}}(Y_n, \mathcal{N}_{\mathbb{R}}) \le \frac{1+\lambda}{\sqrt{2\pi}}\left(\frac{7}{24}\,2^{3/2}\mathrm{e}^{1/2} + \frac{2\mathrm{e}^{1/2}}{3\sqrt[3]{\pi}}\left(4\sqrt[3]{1+\frac{1}{\lambda}}+3\sqrt[3]{3}\right)\right)\frac{\rho}{\sigma^3\sqrt{n}}$$

with $Y_n = \frac{1}{\sigma\sqrt{n}}\sum_{i=1}^n A_i$. Taking $\lambda = 0.183$, we obtain a bound with a constant $C \le 4.815$, so we recover

$$d_{\mathrm{Kol}}(Y_n, \mathcal{N}_{\mathbb{R}}(0,1)) \le 4.815\,\frac{\rho}{\sigma^3\sqrt{n}},$$

which is almost as good as the statement in the introduction, where a constant $C = 3$ was given (the best constant known today is, as far as we know, $C = 0.4748$, see [34]). Of course, the advantage of our method is its large range of applications, as we shall see in the next sections.

Our notion of zone of control allows one to deal with sums of random variables that are independent but not identically distributed. As an example, consider for $r < 1$ a random series

$$Z_r = \sum_{k=1}^{\infty}\mathcal{B}(r^{2k}),$$

with Bernoulli variables of parameters r^{2k} that are independent. The random variable Z_r has the same law as the number of zeroes with module smaller than

r of a random analytic series $S(z) = \sum_{n=0}^{\infty} a_n\, z^n$, where the a_n's are independent standard complex Gaussian variables (see [21, Section 7.1]). If $h = \frac{4\pi r^2}{1-r^2}$ is the hyperbolic area of the disc of radius r and center 0, then we showed in *loc. cit.* that as h goes to infinity and r goes to 1, denoting $Z_r = Z^h$, the sequence

$$X_h = \frac{1}{h^{1/3}}\left(Z^h - \frac{h}{4\pi}\right)$$

is mod-Gaussian convergent with parameters $t_h = \frac{h^{1/3}}{8\pi}$ and limit $\theta(\xi) = \exp(\frac{(\mathrm{i}\xi)^3}{144\pi})$. Let us compute a zone of control for this mod-Gaussian convergence. We change a bit the parameters of the mod-Gaussian convergence and take

$$\widetilde{t}_h = \mathrm{Var}(X_h) = \frac{1}{h^{2/3}} \sum_{k=1}^{\infty} r^{2k}(1-r^{2k}) = \frac{h^{1/3}(h+4\pi)}{4\pi(2h+4\pi)}.$$

The precise reason for this small modification will be given in Remark 15.25. Then,

$$\theta_h(\xi) = \mathbb{E}[\mathrm{e}^{\mathrm{i}\xi X_h}]\,\mathrm{e}^{\frac{\widetilde{t}_h \xi^2}{2}} = \prod_{k=1}^{\infty}\left(1 + r^{2k}(\mathrm{e}^{\frac{\mathrm{i}\xi}{h^{1/3}}} - 1)\right)\,\mathrm{e}^{-\frac{r^{2k}\mathrm{i}\xi}{h^{1/3}} + \frac{r^{2k}(1-r^{2k})\xi^2}{2h^{2/3}}}.$$

Denote $\theta_{h,k}(\xi)$ the terms of the product on the right-hand side. For $|\xi| \leq \frac{h^{1/3}}{4}$, we are going to compute bounds on $|\theta_{h,k}(\xi)|$ and $|\theta_{h,k}(\xi) - 1|$. The holomorphic function

$$f_\alpha(z) = \log(1 + \alpha(\mathrm{e}^z - 1)) - \alpha z - \frac{\alpha(1-\alpha)\,z^2}{2}$$

has its two first derivatives at 0 that vanish, and its third complex derivative is

$$f_\alpha'''(z) = \alpha(1-\alpha)\,\mathrm{e}^z\,\frac{(1-\alpha(1+\mathrm{e}^z))}{(1+\alpha(\mathrm{e}^z-1))^3}.$$

If $|\xi| \leq \frac{h^{1/3}}{4}$, then $|\mathrm{e}^{\frac{\mathrm{i}\xi}{h^{1/3}}}| \leq \mathrm{e}^{1/4}$ and $|\mathrm{e}^{\frac{\mathrm{i}\xi}{h^{1/3}}} - 1| \leq \frac{1}{4}\,\mathrm{e}^{1/4} \leq \frac{1}{2}$, so

$$|\log \theta_{h,k}(\xi)| \leq \frac{|\xi|^3}{6h}\,r^{2k}(1-r^{2k})\,\mathrm{e}^{1/4}\,\frac{1+\frac{r^{2k}}{2}}{(1-\frac{1}{4}\mathrm{e}^{1/4}\,r^{2k})^3}$$

$$\leq \frac{|\xi|^3}{4h}\,\frac{\mathrm{e}^{1/4}\,r^{2k}}{(1-\frac{1}{4}\mathrm{e}^{1/4})^2} \leq \frac{|\xi|^3 r^{2k}}{h}.$$

Therefore, $|\theta_{h,k}(\xi)| \leq \exp(\frac{|\xi^3|r^{2k}}{h})$ and $|\theta_{h,k}(\xi) - 1| \leq \frac{|\xi^3|r^{2k}}{h} \exp(\frac{|\xi^3|r^{2k}}{h})$. We then obtain on the zone $|\xi| \leq \frac{h^{1/3}}{4}$

$$|\theta_h(\xi) - 1| \leq \sum_{k=1}^{\infty} |\theta_{h,k}(\xi) - 1| \prod_{j \neq k} |\theta_{h,j}(\xi)| \leq S \exp S$$

with $S = \sum_{k=1}^{\infty} \frac{|\xi^3|r^{2k}}{h} = \frac{|\xi^3|}{4\pi}$. The inequalities of Condition (Z2) forces us to look at a slightly smaller zone $\xi \in [-\pi \widetilde{t}_h, \pi \widetilde{t}_h]$; then, this zone of control has index $(3, 3)$ and constants $K_1 = K_2 = \frac{1}{4\pi}$. We can then apply Theorem 15.19, and we obtain for h large enough

$$d_{\mathrm{Kol}}\left(\frac{Z^h - \frac{h}{4\pi}}{\sqrt{\mathrm{Var}(Z^h)}}, \mathcal{N}_{\mathbb{R}}(0, 1)\right) \leq \frac{C}{\sqrt{h}}$$

with a constant $C \leq 166$.

15.3.2 Winding Number of a Planar Brownian Motion

In this section, we consider a standard planar Brownian motion $(Z_t)_{t \in \mathbb{R}_+}$ starting from $z = 1$. It is well known that, a.s., Z_t never touches the origin. One can thus write $Z_t = R_t\, e^{i\varphi_t}$, for continuous functions $t \mapsto R_t$ and $t \mapsto \varphi_t$, where $\varphi_0 = 0$, see Fig. 15.5.

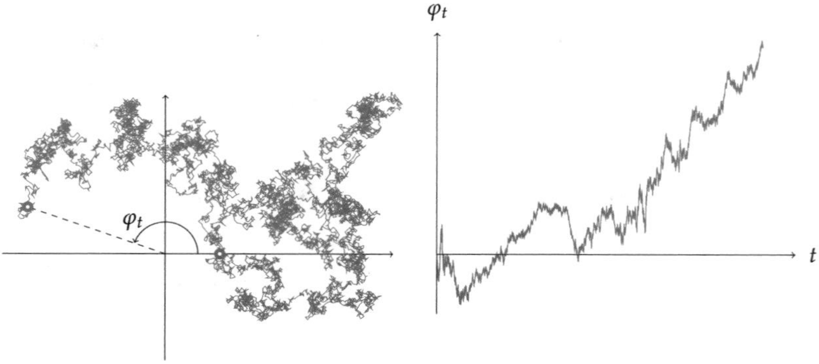

Fig. 15.5 Planar Brownian motion and its winding number; we will see that the latter is asymptotically mod-Cauchy

The Fourier transform of the *winding number* φ_t was computed by Spitzer in [53]:

$$\mathbb{E}[\mathrm{e}^{\mathrm{i}\xi\varphi_t}] = \sqrt{\frac{\pi}{8t}}\,\mathrm{e}^{-\frac{1}{4t}}\left(I_{\frac{|\xi|-1}{2}}\left(\frac{1}{4t}\right) + I_{\frac{|\xi|+1}{2}}\left(\frac{1}{4t}\right)\right),$$

where $I_\nu(z) = \sum_{k=0}^{\infty} \frac{1}{k!\,\Gamma(\nu+k+1)} \left(\frac{z}{2}\right)^{\nu+2k}$ is the modified Bessel function of the first kind. As a consequence, and as was noticed in [14, §3.2], $(\varphi_t)_{t\in\mathbb{R}_+}$ converges mod-Cauchy with parameters $\frac{\log 8t}{2}$ and limiting function $\theta(\xi) = \sqrt{\pi}\,\Gamma\left(\frac{|\xi|+1}{2}\right)^{-1}$. Indeed,

$$\mathbb{E}[\mathrm{e}^{\mathrm{i}\xi\varphi_t}]\exp\left(|\xi|\,\frac{\log 8t}{2}\right)$$
$$= \sqrt{\pi}\,\mathrm{e}^{-\frac{1}{4t}}\left(\sum_{k=0}^{\infty}\frac{1}{k!}\left(\frac{1}{8t}\right)^{2k}\left(\frac{1}{\Gamma(k+\frac{|\xi|+1}{2})} + \frac{1}{8t\,\Gamma(k+\frac{|\xi|+3}{2})}\right)\right),$$

and the limit of the power series as t goes to infinity is its constant term $\frac{1}{\Gamma(\frac{|\xi|+1}{2})}$.

Here, $|\theta(\xi) - 1|$ is of order $O(|\xi|)$ around 0, since the first derivative of Γ is not zero at $\frac{1}{2}$. Therefore, if the mod-convergence can be given a zone of control, then the index of this control will be $v = 1$, which forces for Berry–Esseen estimates $\gamma \leq 0$ since $\gamma \leq \min(\frac{v-1}{\alpha}, \frac{1}{w-\alpha})$. Conversely, for any ξ, notice that the function $x \mapsto \frac{1}{\Gamma(x+\frac{1}{2})}$ has derivative bounded on $\mathbb{R}_+$ by

$$-\frac{\Gamma'(\frac{1}{2})}{\left(\Gamma(\frac{1}{2})\right)^2} = 1.11_- < \frac{2}{\sqrt{\pi}},$$

and therefore that

$$|\theta_t(\xi) - 1| \leq 2\,\mathrm{e}^{-\frac{1}{4t}}\sum_{k=0}^{\infty}\frac{1}{k!}\left(\frac{1}{8t}\right)^{2k}\left(1+\frac{1}{8t}\right)\frac{|\xi|}{2}$$
$$\leq \mathrm{e}^{\left(\frac{1}{8t}\right)^2 - \frac{1}{4t}}\left(1+\frac{1}{8t}\right)|\xi| \leq |\xi|$$

for t large enough. So, in particular, one has mod-Cauchy convergence with index of control $(1, 1)$, zone of control $[-K, K]$ with K as large as wanted, and constants $K_1 = 1$ and $K_2 = 0$. It follows then from Theorem 15.19 that if C follows a standard Cauchy law, then

$$d_{\mathrm{Kol}}\left(\frac{2\varphi_t}{\log 8t}, C\right) \leq \frac{4}{\log 8t}$$

for t large enough. As far as we know, this estimate is new.

15.3.3 Approximation of a Stable Law by Compound Poisson Laws

Let $\phi_{c,\alpha,\beta}$ be a stable law; the Lévy–Khintchine formula for its exponent allows one to write

$$\mathrm{e}^{\eta_{c,\alpha,\beta}(\mathrm{i}\xi)} = \begin{cases} \mathrm{e}^{-|c\xi|^2} & \text{if } \alpha = 2, \\ \exp\left(\mathrm{i}m\xi + \int_{\mathbb{R}} (\mathrm{e}^{\mathrm{i}\xi x} - 1 - 1_{|x|<1}\mathrm{i}\xi x)\, \pi_{c,\alpha,\beta}(dx)\right) & \text{if } \alpha \in (0,2), \end{cases}$$

where $\pi_{c,\alpha,\beta}(dx)$ is the Lévy measure defined for $\alpha \in (0,2)$ by

$$\pi_{c,\alpha,\beta}(dx) = \frac{c_+ \, 1_{x>0}}{x^{1+\alpha}} + \frac{c_- \, 1_{x<0}}{|x|^{1+\alpha}},$$

with $m \in \mathbb{R}$ and $c_+, c_- \in \mathbb{R}_+$ related to (c, α, β) by $\beta = \frac{c_+ - c_-}{c_+ + c_-}$ and

$$m = \begin{cases} \frac{c_+ - c_-}{1-\alpha} & \text{if } \alpha \neq 1, \\ \left(\int_0^1 \frac{\sin t - t}{t^2}\, dt + \int_1^\infty \frac{\sin t}{t^2}\, dt\right)(c_+ - c_-) & \text{if } \alpha = 1; \end{cases}$$

$$c_+ + c_- = \begin{cases} \frac{\alpha\, c^\alpha}{\Gamma(1-\alpha)\, \sin\left(\frac{\pi(1-\alpha)}{2}\right)} & \text{if } \alpha \neq 1, \\ \frac{2c}{\pi} & \text{if } \alpha = 1. \end{cases}$$

The proof of the Lévy–Khintchine formula in the general case of an infinitely divisible law involves the following elementary fact (cf. [50, Chapter 2]): if μ is infinitely divisible and $\mu = (\rho_n)^{*n}$ for $n \geq 1$, then the compound Poisson law μ_n of intensity $n\,\rho_n$, which has Fourier transform

$$\widehat{\mu_n}(\xi) = \exp\left(\int_{\mathbb{R}} (\mathrm{e}^{\mathrm{i}x\xi} - 1)\, n\rho_n(dx)\right),$$

converges in law towards μ; thus, any infinitely divisible law is a limit of compound Poisson laws. In the case of stable laws, this approximation result can be precised in terms of Kolmogorov distances:

Proposition 15.22 *Let Y be a random variable with stable law $\phi_{c,\alpha,\beta}$, and Y_n be a random variable with the following compound Poisson distribution: if μ_n is the law of Y_n, then its Fourier transform is*

$$\widehat{\mu_n}(\xi) = \exp\left(\int_{\mathbb{R}} (\mathrm{e}^{\mathrm{i}x\xi} - 1)\, n\, \phi_{\frac{c}{n^{1/\alpha}},\alpha,\beta}(dx)\right).$$

The Kolmogorov distance between Y_n and Y is

$$d_{\mathrm{Kol}}(Y_n, Y) \leq \begin{cases} C(\alpha)\, n^{-1/\alpha} & \text{if } \alpha \in (1,2], \\ C(\alpha)\, n^{-1} & \text{if } \alpha \in (0,1) \text{ or } \alpha = 1, \beta = 0, \\ C'\,(\log n)^2\, n^{-1} & \text{if } \alpha = 1, \beta \neq 0, \end{cases}$$

with constants $C(\alpha)$ or C' that depend only on the exponent α.

We thus get a phase transition between the cases $\alpha > 1$ and $\alpha < 1$, with the case $\alpha = 1$ that exhibits distinct transition behaviors according to the value of β.

Proof Let us distinguish the following cases:

- Suppose first $\alpha \notin \{1, 2\}$. The definition of Y_n implies that

$$\mathbb{E}[\mathrm{e}^{\mathrm{i}\xi Y_n}] = \widehat{\mu_n}(\xi) = \exp\left(n\left(\mathrm{e}^{\frac{\eta_{c,\alpha,\beta}(\mathrm{i}\xi)}{n}} - 1\right)\right).$$

Set $X_n = n^{1/(2\alpha)} Y_n$, $t_n = \sqrt{n}$ and $\theta_n(\xi) = \mathbb{E}[\mathrm{e}^{\mathrm{i}\xi X_n}]\,\mathrm{e}^{-t_n\,\eta_{c,\alpha,\xi}(\mathrm{i}\xi)}$. We have

$$\theta_n(\xi) = \exp\left(n\left(\mathrm{e}^{\frac{\eta_{c,\alpha,\beta}(\mathrm{i}\xi)}{n^{1/2}}} - 1 - \frac{\eta_{c,\alpha,\beta}(\mathrm{i}\xi)}{n^{1/2}}\right)\right).$$

On the zone $[-K(t_n)^{1/\alpha}, K(t_n)^{1/\alpha}]$ with $K = \frac{|\cos(\frac{\pi\alpha}{2})|^{\frac{2}{\alpha}}}{c}$, we can use a Taylor formula with an integral form remainder:

$$n\left(\mathrm{e}^{\frac{\eta_{c,\alpha,\beta}(\mathrm{i}\xi)}{n^{1/2}}} - 1 - \frac{\eta_{c,\alpha,\beta}(\mathrm{i}\xi)}{n^{1/2}}\right) = (\eta_{c,\alpha,\beta}(\mathrm{i}\xi))^2 \left(\int_0^1 (1-u)\,\mathrm{e}^{\frac{u\eta_{c,\alpha,\beta}(\mathrm{i}\xi)}{n^{1/2}}}\,du\right)$$

$$\left|n\left(\mathrm{e}^{\frac{\eta_{c,\alpha,\beta}(\mathrm{i}\xi)}{n^{1/2}}} - 1 - \frac{\eta_{c,\alpha,\beta}(\mathrm{i}\xi)}{n^{1/2}}\right)\right| \leq \frac{1}{2}\,|\eta_{c,\alpha,\beta}(\mathrm{i}\xi)|^2 \leq \frac{1}{2}\left(\frac{c^\alpha}{\cos\left(\frac{\pi\alpha}{2}\right)}\right)^2 |\xi|^{2\alpha}.$$

We thus obtain a zone of control for $(X_n)_{n\in\mathbb{N}}$ with $v = w = 2\alpha$, $\gamma = \frac{1}{\alpha}$,

$$K_1 = K_2 = \frac{1}{2}\left(\frac{c^\alpha}{\cos\left(\frac{\pi\alpha}{2}\right)}\right)^2,$$

and one checks that

$$\left(\frac{c^\alpha}{2K_2}\right)^{\frac{1}{w-\alpha}} = \frac{|\cos(\frac{\pi\alpha}{2})|^{\frac{2}{\alpha}}}{c} = K.$$

Since we need $\gamma \leq \min\left(\frac{1}{w-\alpha}, \frac{v-1}{\alpha}\right)$ to obtain a bound on the Kolmogorov distance (see the hypotheses of Theorem 15.19), this leads to a reduction of γ when $\alpha < 1$:

$$\gamma + \frac{1}{\alpha} = \begin{cases} \frac{2}{\alpha} & \text{if } \alpha > 1, \\ 2 & \text{if } \alpha < 1. \end{cases}$$

With Theorem 15.19, we obtain the following upper bound for $d_{\mathrm{Kol}}(Y_n, Y)$:

$$\frac{1+\lambda}{\alpha\pi}\left(\frac{2}{(\cos(\frac{\pi\alpha}{2}))^2} + \frac{\Gamma(\frac{1}{\alpha})}{\sqrt[3]{\pi}\,\left|\cos(\frac{\pi\alpha}{2})\right|^{2/\alpha}}\left(4\sqrt[3]{1+\frac{1}{\lambda}} + 3\sqrt[3]{3}\right)\right)\frac{1}{n^{\frac{\gamma}{2}+\frac{1}{2\alpha}}},$$

Then, any choice of $\lambda > 0$ gives a constant $C(\alpha)$ that depends only on α.

- When $\alpha = 2$, the result follows from the usual Berry–Esseen estimates, since $\sqrt{n}\,Y_n$ has the law of a sum of n independent random variables with same law and finite variance and third moment.
- If $\alpha = 1$ and $\beta = 0$, then the same computations as above can be performed with a constant $K = \frac{1}{c}$, $v = w = 2$, $\gamma = 1$,

$$K_1 = K_2 = \frac{c^2}{2},$$

and this leads to

$$d_{\mathrm{Kol}}(Y_n, Y) \leq \frac{1+\lambda}{\pi}\left(2 + \frac{1}{\sqrt[3]{\pi}}\left(4\sqrt[3]{1+\frac{1}{\lambda}} + 3\sqrt[3]{3}\right)\right)\frac{1}{n},$$

and a constant $C = 3.04$ when $\lambda = 0.2$.

- Let us finally treat the case $\alpha = 1$, $\beta \neq 0$. Recall that we then have $\eta_{c,\alpha,\beta}(\mathrm{i}\xi) = -|c\xi|\,(1 + \frac{2\mathrm{i}\beta}{\pi}\,\mathrm{sgn}(\xi)\,\log|\xi|)$. We choose t_n such that $t_n \log t_n = \sqrt{n}$, and set

$$X_n = t_n\,Y_n + \frac{2c\beta}{\pi}\,\sqrt{n}.$$

We then have

$$\begin{aligned}\theta_n(\xi) &= \mathbb{E}[\mathrm{e}^{\mathrm{i}\xi X_n}]\,\mathrm{e}^{-t_n\,\eta_{c,\alpha,\beta}(\mathrm{i}\xi)} \\ &= \exp\left(\frac{2c\beta\mathrm{i}\xi}{\pi}\,t_n\log t_n + n\left(\mathrm{e}^{\frac{\eta_{c,\alpha,\beta}(\mathrm{i}t_n\xi)}{n}} - 1\right) - t_n\,\eta_{c,\alpha,\beta}(\mathrm{i}\xi)\right) \\ &= \exp\left(n\left(\mathrm{e}^{\frac{\eta_{c,\alpha,\beta}(\mathrm{i}t_n\xi)}{n}} - 1 - \frac{\eta_{c,\alpha,\beta}(\mathrm{i}t_n\xi)}{n}\right)\right),\end{aligned}$$

and the Taylor formula with integral remainder yields:

$$\begin{aligned}\left|n\left(\mathrm{e}^{\frac{\eta_{c,\alpha,\beta}(\mathrm{i}t_n\xi)}{n}}-1-\frac{\eta_{c,\alpha,\beta}(\mathrm{i}t_n\xi)}{n}\right)\right| &\leq \frac{1}{2n}\,|\eta_{c,\alpha,\beta}(\mathrm{i}t_n\xi)|^2 \\ &\leq \frac{c^2|\xi|^2}{2}\left(\frac{1+\frac{4}{\pi^2}\,(\log|t_n\xi|)^2}{(\log t_n)^2}\right).\end{aligned}$$

On the zone $[-\frac{t_n}{2c}, \frac{t_n}{2c}]$, we thus have

$$\begin{aligned}&\left|n\left(\mathrm{e}^{\frac{\eta_{c,\alpha,\beta}(\mathrm{i}t_n\xi)}{n}}-1-\frac{\eta_{c,\alpha,\beta}(\mathrm{i}t_n\xi)}{n}\right)\right| \\ &\quad\leq \frac{c^2|\xi|^2}{2}\left(\frac{1+\frac{4}{\pi^2}\,(2\log t_n-\log 2c)^2}{(\log t_n)^2}\right) \\ &\quad\leq c^2|\xi|^2 \quad \text{for } t_n \text{ large enough.}\end{aligned}$$

So, there is a zone of control with constants $K_1 = K_2 = c^2$, $v = w = 2$ and $\gamma = 1$, and $K = \frac{1}{2c}$. We thus get as before an estimate of $d_{\mathrm{Kol}}(Y_n, Y)$ of order $O((t_n)^{-2})$, and since $(t_n \log t_n)^2 = n$, $(t_n)^2$ is asymptotically equivalent to $\frac{4n}{\log^2 n}$. □

15.3.4 Convergence of Ornstein–Uhlenbeck Processes to Stable Laws

Another way to approximate a stable law $\phi_{c,\alpha,\beta}$ is by using the marginals of a random process of Ornstein–Uhlenbeck type. Consider more generally a self-decomposable law ϕ on $\mathbb{R}$, that is an infinitely divisible distribution such that for any $b \in (0, 1)$, there exists a probability measure p_b on $\mathbb{R}$ such that

$$\widehat{\phi}(\xi) = \widehat{\phi}(b\xi)\,\widehat{p_b}(\xi); \tag{15.6}$$

see [50, Chapter 3, Definition 15.1]. In Eq. (15.6), the laws p_b are the marginal laws of certain Markov processes. Fix a Lévy–Khintchine triplet $(l \in \mathbb{R},\ \nu^2 \in \mathbb{R}_+,\ \rho)$ with ρ probability measure on $\mathbb{R} \setminus \{0\}$ that integrates $\min(1, |x|^2)$, and consider the Lévy process $(Z_t)_{t\in\mathbb{R}_+}$ associated to this triplet:

$$\begin{aligned}\mathbb{E}[\mathrm{e}^{\mathrm{i}\xi Z_t}] &= \exp(t\psi(\mathrm{i}\xi)) \\ &= \exp\left(t\left(\mathrm{i}l\xi - \frac{\nu^2\xi^2}{2} + \int_{\mathbb{R}} (\mathrm{e}^{\mathrm{i}\xi x} - 1 - 1_{|x|<1}\mathrm{i}\xi x)\,\rho(dx)\right)\right).\end{aligned}$$

The Ornstein–Uhlenbeck process with triplet (l, ν^2, ρ), speed υ and starting point x is the solution $(U_t)_{t\geq 0}$ of the stochastic differential equation

$$U_t = \mathrm{e}^{-\upsilon t} x + \int_0^t \mathrm{e}^{-\upsilon(t-s)}\, dZ_s.$$

The Ornstein–Uhlenbeck process $(U_t)_{t\geq 0}$ can be shown to be a Markov process whose transition kernel $(P_t(x, dy))_{t\geq 0}$ satisfies:

$$\widehat{P_t(x,\cdot)}(\xi) = \int_{\mathbb{R}} \mathrm{e}^{\mathrm{i}\xi y}\, P_t(x, dy) = \exp\left(\mathrm{i}\xi \mathrm{e}^{-\upsilon t} x + \int_0^t \psi(\mathrm{i}\xi \mathrm{e}^{-\upsilon s})\, ds\right)$$

see [50, Lemma 17.1]. The connection with self-decomposable laws is provided by:

Theorem 15.23 (Sato and Yamazato [51]) *For any self-decomposable law ϕ and any fixed speed υ, there exists a unique Lévy–Khintchine triplet (l, ν^2, ρ) with $\int_{|x|\geq 1} \log|x|\, \rho(dx) < +\infty$, such that the associated Ornstein–Uhlenbeck process $(U_t)_{t\geq 0}$ with speed υ satisfies:*

$$\forall x \in \mathbb{R},\ \ P_t(x,\cdot) \rightharpoonup \phi.$$

If $\psi(\mathrm{i}\xi)$ is the exponent associated to (l, ν^2, ρ), then

$$\widehat{\phi}(\xi) = \exp\left(\int_{s=0}^{\infty} \psi(\mathrm{i}\xi \mathrm{e}^{-\upsilon s})\, ds\right).$$

We refer to [51] and [50, Theorem 17.5]. In the setting of Theorem 15.23, one has the relation

$$\widehat{\phi}(\xi) = \widehat{\phi}(\mathrm{e}^{-\upsilon t}\xi)\, \left(\widehat{P_t(x,\cdot)}(\xi)\, \widehat{\delta_{-\mathrm{e}^{-\upsilon t}x}}(\xi)\right),$$

so if $b \in (0, 1)$, setting $b = \mathrm{e}^{-\upsilon t}$, one recovers p_b as the law of $U_t - \mathrm{e}^{-\upsilon t} x$, where $(U_t)_{t\in\mathbb{R}_+}$ is the Ornstein–Uhlenbeck process that converges in distribution to ϕ and that has speed υ and starting point x.

Suppose that $\phi = \phi_{c,\alpha,\beta}$ is a stable law. Then, the previous computations can be reinterpreted in the framework of mod-ϕ convergence. We set

$$\theta(\xi) = \frac{\widehat{\delta_x}(\xi)}{\widehat{\phi}(\xi)} = \exp(\mathrm{i}\xi x - \eta(\mathrm{i}\xi)),$$

and

$$X_t = \begin{cases} \mathrm{e}^{\upsilon t} U_t & \text{if } \alpha \neq 1, \\ \mathrm{e}^{\upsilon t}\left(U_t - \frac{2c\beta \upsilon t}{\pi}\right) & \text{if } \alpha = 1. \end{cases}$$

Then,

$$\mathbb{E}[\mathrm{e}^{\mathrm{i}\xi X_t}] = \begin{cases} \widehat{\mu}(\mathrm{e}^{vt}\xi)\,\theta(\xi) & \text{if } \alpha \neq 1, \\ \widehat{\mu}(\mathrm{e}^{vt}\xi)\,\mathrm{e}^{-\mathrm{i}\xi\mathrm{e}^{vt}\frac{2c\beta vt}{\pi}}\,\theta(\xi) & \text{if } \alpha = 1, \end{cases}$$
$$= \mathrm{e}^{\mathrm{e}^{\alpha vt}\eta(\mathrm{i}\xi)}\,\theta(\xi),$$

so $(X_t)_{t\geq 0}$ converges mod-ϕ with parameters $\mathrm{e}^{\alpha vt}$, and with limit equal to the residue $\theta(\xi)$. Note that $\theta_t = \theta$ for any $t \geq 0$, so we are in a special situation where the residues are constant (time-independent). Assuming that $x \neq 0$, one has for any $\xi \in \mathbb{R}$

$$|\theta(\xi) - 1| \leq \begin{cases} K_1\,|\xi|\,\exp(K_2\,|\xi|^\alpha) & \text{if } \alpha \in (1, 2], \\ K_1\,|\xi|^\alpha\,\exp(K_2\,|\xi|^\alpha) & \text{if } \alpha \in (0, 1) \text{ or } \alpha = 1, \beta = 0, \\ K_1\,|\xi|\log|\xi|\,\exp(K_2\,|\xi|) & \text{if } \alpha = 1, \beta \neq 0. \end{cases}$$

For the two first cases, the condition $\gamma \leq \min(\frac{1}{w-\alpha}, \frac{v-1}{\alpha})$ in Theorem 15.19 imposes the following choices of γ when computing Berry–Esseen estimates: $\gamma = \frac{\alpha-1}{\alpha}$ when $\alpha \leq 1$, and $\gamma = 0$ when $\alpha \geq 1$. In these cases, one obtains:

$$d_{\mathrm{Kol}}(U_t, \phi_{c,\alpha,\beta}) = \begin{cases} O(\mathrm{e}^{-vt}) & \text{if } \alpha \in (1, 2], \\ O(\mathrm{e}^{-\alpha vt}) & \text{if } \alpha \in (0, 1) \text{ or } \alpha = 1, \beta = 0. \end{cases}$$

Because of the term $\log|\xi|$, the last case does not exactly fit the framework of zones of control, but it is easy to adapt the proofs and one gets an estimate $O(vt\,\mathrm{e}^{-vt})$. On the other hand, when $x = 0$, the only difference with the previous discussion is the case $\alpha \in (1, 2]$, where we obtain

$$|\theta(\xi) - 1| \leq K_1\,|\xi|^\alpha\,\exp(K_2\,|\xi|^\alpha)$$

and by Theorem 15.19, $d_{\mathrm{Kol}}(U_t, \phi_{c,\alpha,\beta}) = O(\mathrm{e}^{-\alpha vt})$, choosing $\gamma = \frac{\alpha-1}{\alpha}$. So, to summarise:

Proposition 15.24 *Let Y be a random variable with stable law $\phi_{c,\alpha,\beta}$, and $(U_t)_{t\geq 0}$ be the corresponding Ornstein–Uhlenbeck process with starting point x and speed v. We have:*

$$d_{\mathrm{Kol}}(U_t, Y) = \begin{cases} O(\mathrm{e}^{-vt}) & \text{if } \alpha \in (1, 2], x \neq 0, \\ O(\mathrm{e}^{-\alpha vt}) & \text{if } \alpha \in (0, 1) \text{ or } \alpha = 1, \beta = 0 \text{ or } \alpha \in (1, 2], x = 0, \\ O(vt\,\mathrm{e}^{-vt}) & \text{if } \alpha = 1, \beta \neq 0, \end{cases}$$

with constants in the $O(\cdot)$ depending only on x and α.

15.3.5 Logarithms of Characteristic Polynomials of Random Matrices

In [36, Sections 3 and 4] and [21, Section 7.5], the mod-Gaussian convergence of the random variables shown in Table 15.1 was proven.

Here, G is Barnes' function, which is the unique entire solution of the equations $G(1) = 1$ and $G(z+1) = G(z)\,\Gamma(z)$. Moreover, the mod-Gaussian convergence holds in fact on an half-plane $H = \{z \in \mathbb{C} \mid \mathrm{Re}(z) > -\alpha\}$. In the sequel, we denote X_n^{A}, X_n^{C} and X_n^{D} the mod-Gaussian convergent random variables, according to the type of the classical group (A for unitary groups, C for compact symplectic groups and D for even orthogonal groups). Before computing zones of control for these variables, let us make the following essential remark:

Remark 15.25 Let $(X_n)_{n\in\mathbb{N}}$ be a sequence of random variables that is mod-Gaussian convergent on a domain $D \subset \mathbb{C}$ which contains a neighborhood of 0 (this ensures that θ_n and all its derivatives converge towards those of θ). We denote $(t_n)_{n\in\mathbb{N}}$ the parameters of mod-Gaussian convergence of $(X_n)_{n\in\mathbb{N}}$. Then, without generality, one can assume $\theta_n'(0) = \theta_n''(0) = 0$ and $\theta'(0) = \theta''(0) = 0$. Indeed, set

$$\widetilde{X}_n = X_n + \mathrm{i}\theta_n'(0) \qquad ; \qquad \widetilde{t}_n = t_n - \theta_n''(0).$$

We then have

$$\widetilde{\theta}_n(\xi) := \mathbb{E}[\mathrm{e}^{\mathrm{i}\xi \widetilde{X}_n}]\,\mathrm{e}^{\widetilde{t}_n \frac{\xi^2}{2}} = \theta_n(\xi)\,\mathrm{e}^{-\theta_n'(0)\xi - \theta_n''(0)\frac{\xi^2}{2}}$$

and this new residue satisfies $\widetilde{\theta}_n'(0) = \widetilde{\theta}_n''(0) = 0$. For the construction of zones of control, it allows us to force $v = 3$, up to a translation of X_n and of the parameter t_n.

In the following, we only treat the case of unitary groups, the two other cases being totally similar (one could also look at the imaginary part of the log-characteristic polynomial). There is an exact formula for the Fourier transform of X_n^{A} [31, Formula (71)]:

$$\mathbb{E}[\mathrm{e}^{\mathrm{i}\xi X_n^{\mathrm{A}}}] = \prod_{k=1}^{n} \frac{\Gamma(k)\Gamma(k+\mathrm{i}\xi)}{\left(\Gamma(k+\frac{\mathrm{i}\xi}{2})\right)^2}.$$

Table 15.1 Mod-Gaussian convergence of the characteristic polynomials of Haar-distributed random matrices in compact Lie groups

Random matrix M_n	Random variable X_n	Parameters t_n	Residue $\theta(\xi)$
$\mathrm{Haar}(\mathrm{U}(n))$	$\mathrm{Re}(\log\det(I_n - M_n))$	$\frac{\log n}{2}$	$\frac{(G(1+\frac{\mathrm{i}\xi}{2}))^2}{G(1+\mathrm{i}\xi)}$
$\mathrm{Haar}(\mathrm{USp}(n))$	$\log\det(I_{2n} - M_n) - \frac{1}{2}\log\frac{\pi n}{2}$	$\log\frac{n}{2}$	$\frac{G(\frac{3}{2})}{G(\frac{3}{2}+\mathrm{i}\xi)}$
$\mathrm{Haar}(\mathrm{SO}(2n))$	$\log\det(I_{2n} - M_n) - \frac{1}{2}\log\frac{8\pi}{n}$	$\log\frac{n}{2}$	$\frac{G(\frac{1}{2})}{G(\frac{1}{2}+\mathrm{i}\xi)}$

We have $\mathbb{E}[X_n^{\mathrm{A}}] = 0$, and

$$\widetilde{t_n} = \mathbb{E}[(X_n^{\mathrm{A}})^2] = \frac{1}{2}\sum_{k=1}^{n} \frac{\Gamma''(k)}{\Gamma(k)} - \left(\frac{\Gamma'(k)}{\Gamma(k)}\right)^2 = \frac{1}{2}\sum_{k=1}^{n} \psi_1(k),$$

where $\psi_1(z)$ is the trigamma function $\frac{d^2}{dz^2}(\log \Gamma(z))$, and is given on integers by the remainder of the series $\zeta(2)$:

$$\psi_1(k) = \sum_{m=k}^{\infty} \frac{1}{m^2}.$$

Therefore, $\widetilde{t_n} = \frac{1}{2}\sum_{m=1}^{\infty} \frac{\min(n,m)}{m^2} = \frac{1}{2}(\log n + \gamma + 1 + O(n^{-1}))$. So, $(X_n^{\mathrm{A}})_{n\in\mathbb{N}}$ is mod-Gaussian convergent with parameters $(\widetilde{t_n})_{n\in\mathbb{N}}$ and limit

$$\widetilde{\theta}(\xi) = \frac{\left(G(1+\frac{\mathrm{i}\xi}{2})\right)^2}{G(1+\mathrm{i}\xi)}\, \mathrm{e}^{\frac{(\gamma+1)\xi^2}{4}},$$

which satisfies $\widetilde{\theta}'(0) = \widetilde{\theta}''(0) = 0$. With these conventions, we can write the residues $\widetilde{\theta}_n(\xi)$ as

$$\widetilde{\theta}_n(\xi) = \left(\prod_{k=1}^{n} \frac{\Gamma(k)\Gamma(k+\mathrm{i}\xi)}{\left(\Gamma(k+\frac{\mathrm{i}\xi}{2})\right)^2}\right) \mathrm{e}^{\frac{\widetilde{t_n}\xi^2}{2}} = \prod_{k=1}^{n}\left(\frac{\Gamma(k)\Gamma(k+\mathrm{i}\xi)}{\left(\Gamma(k+\frac{\mathrm{i}\xi}{2})\right)^2}\, \mathrm{e}^{\frac{\psi_1(k)\xi^2}{4}}\right).$$

Denote $\vartheta_k(\xi)$ the terms of the product on the right-hand side; we use a similar strategy as in Sect. 15.3.1 for computing a zone of control. The function $\varphi_k(\xi) = \log \vartheta_k(\xi)$ vanishes at 0, has its two first derivatives that also vanish at 0, and therefore writes as

$$\varphi_k(\xi) = \left(\int_0^1 \varphi_k'''(t\xi)\,(1-t)^2\,dt\right)\frac{\xi^3}{2}.$$

The third derivative of $\varphi_k(\xi)$ is given by

$$\varphi_k'''(\xi) = -\mathrm{i}\,\psi_2(k+\mathrm{i}\xi) + \frac{\mathrm{i}}{2}\,\psi_2\left(k+\frac{\mathrm{i}\xi}{2}\right),$$

with $\psi_2(z) = -2\sum_{j=0}^{\infty} \frac{1}{(j+z)^3}$. As a consequence, $\varphi_k'''(\xi)$ is *uniformly bounded* on $\mathbb{R}$ by

$$3\sum_{j=0}^{\infty} \frac{1}{(j+k)^3} \leq \frac{3\,\zeta(3)}{k^2}.$$

Therefore,

$$|\varphi_k(\xi)| \le \frac{\zeta(3)\,|\xi|^3}{2k^2};$$

$$|\vartheta_k(\xi)| \le \mathrm{e}^{\frac{\zeta(3)\,|\xi|^3}{2k^2}};$$

$$|\vartheta_k(\xi) - 1| \le \frac{\zeta(3)\,|\xi|^3}{2k^2}\,\mathrm{e}^{\frac{\zeta(3)\,|\xi|^3}{2k^2}}.$$

It follows that for any n and any $\xi \in \mathbb{R}$, $|\widetilde{\theta}_n(\xi) - 1| \le S \exp S$ with $S = \sum_{k=1}^{\infty} \frac{\zeta(3)\,|\xi|^3}{2k^2} = \frac{3\,\zeta(3)\,|\xi|^3}{\pi^2}$. Set $K_1 = K_2 = \frac{3\,\zeta(3)}{\pi^2}$, and $K = \frac{1}{4K_2} = \frac{\pi^2}{12\,\zeta(3)}$. We have a zone of control $[-Kt_n, Kt_n]$ of index $(3, 3)$, with constants K_1 and K_2. We conclude with Theorem 15.19:

Proposition 15.26 *Let M_n be a random unitary matrix taken according to the Haar measure. For n large enough,*

$$d_{\mathrm{Kol}}\left(\frac{\mathrm{Re}(\log\det(I_n - M_n))}{\sqrt{\mathrm{Var}(\mathrm{Re}(\log\det(I_n - M_n)))}},\ \mathcal{N}_{\mathbb{R}}(0,1)\right) \le \frac{C}{(\log n)^{3/2}}$$

with a constant $C \le 18$. Up to a change of the constant, the same result holds if one replaces $\mathrm{Re}(\log\det(I_n - M_n))$ *by* $\mathrm{Im}(\log\det(I_n - M_n))$*, or by*

$$\log\det(I_{2n} - P_n) - \mathbb{E}[\det(I_{2n} - P_n)],$$

with P_n Haar distributed in the unitary compact symplectic group $\mathrm{USp}(n)$ *or in the even special orthogonal group* $\mathrm{SO}(2n)$.

15.4 Cumulants and Dependency Graphs

15.4.1 Cumulants, Zone of Control and Kolmogorov Bound

In this section, we will see that appropriate bounds on the cumulants of a sequence of random variables $(S_n)_{n\in\mathbb{N}}$ imply the existence of a large zone of control for a renormalized version of S_n. In this whole section and in the next one, the reference stable law is the *standard Gaussian law.* We also assume that the random variables S_n are centered.

Let us first recall the definition of cumulants. If X is a real-valued random variable with exponential generating function

$$\mathbb{E}[\mathrm{e}^{zX}] = \sum_{r=0}^{\infty} \frac{\mathbb{E}[X^r]}{r!}\,z^r$$

convergent in a neighborhood of 0, then its *cumulants* $\kappa^{(1)}(X), \kappa^{(2)}(X), \dots$ are the coefficients of the series

$$\log \mathbb{E}[e^{zX}] = \sum_{r=1}^{\infty} \frac{\kappa^{(r)}(X)}{r!}\, z^r,$$

which is also well defined for z in a neighborhood of 0 (see for instance [39]). For example, $\kappa^{(1)}(X) = \mathbb{E}[X]$, $\kappa^{(2)}(X) = \mathbb{E}[X^2] - \mathbb{E}[X]^2 = \mathrm{Var}(X)$, and

$$\kappa^{(3)}(X) = \mathbb{E}[X^3] - 3\,\mathbb{E}[X^2]\,\mathbb{E}[X] + 2\,\mathbb{E}[X]^3.$$

We are interested in the case where cumulants can be bounded in an appropriate way.

Definition 15.27 Let $(S_n)_{n\in\mathbb{N}}$ be a sequence of (centered) real-valued random variables. We say that $(S_n)_{n\in\mathbb{N}}$ admits *uniform bounds on cumulants* with parameters (D_n, N_n, A) if, for any $r \geq 2$, we have

$$|\kappa^{(r)}(S_n)| \leq N_n\, r^{r-2}\, (2D_n)^{r-1}\, A^r.$$

In the following, it will be convenient to set $(\widetilde{\sigma}_n)^2 = \frac{\mathrm{Var}(S_n)}{N_n D_n}$. The inequality of Definition 15.27 with $r = 2$ gives $(\widetilde{\sigma}_n)^2 \leq 2A^2$.

Lemma 15.28 *Let $(S_n)_{n\in\mathbb{N}}$ be a sequence with uniform bounds on cumulants with parameters (D_n, N_n, A). Set*

$$X_n = \frac{S_n}{(N_n)^{1/3}\,(D_n)^{2/3}} \quad \text{and} \quad t_n = (\widetilde{\sigma}_n)^2 (\tfrac{N_n}{D_n})^{1/3} = \mathrm{Var}(X_n).$$

Then, we have for $(X_n)_{n\in\mathbb{N}}$ a zone of control $[-K\,t_n\,,\ K\,t_n]$ of index $(3,3)$, with the following constants:

$$K = \frac{1}{(8+4\mathrm{e})\,A^3}, \qquad K_1 = K_2 = (2+\mathrm{e})\,A^3.$$

Proof From the definition of cumulants, since X_n is centered and has variance t_n, we can write

$$\begin{aligned}\theta_n(\xi) &= \mathbb{E}[e^{i\xi X_n}]\, \exp\left(\frac{t_n\,\xi^2}{2}\right)\\ &= \exp\left(\sum_{r=3}^{\infty} \frac{\kappa^{(r)}(S_n)}{r!}\, \frac{(i\xi)^r}{(N_n (D_n)^2)^{r/3}}\right) = \exp(z),\end{aligned}$$

with

$$|z| \leq \frac{1}{2}\frac{N_n}{D_n}\sum_{r=3}^{\infty}\frac{r^{r-2}}{\mathrm{e}^r\, r!}\left(\left(\frac{D_n}{N_n}\right)^{1/3} 2\mathrm{e}A\,|\xi|\right)^r.$$

We set $y = (\frac{D_n}{N_n})^{1/3}\, 2\mathrm{e}A\,|\xi|$ and suppose that $y \leq 1$, that is to say that ξ is in the zone $[-L\,(\frac{N_n}{D_n})^{1/3}\,,\; L\,(\frac{N_n}{D_n})^{1/3}]$ with $L = \frac{1}{2\mathrm{e}A}$.

By Stirling's bound, the series $S(y) = \sum_{r=3}^{\infty}\frac{r^{r-2}}{r!\,\mathrm{e}^r}\, y^r$ is convergent for any $y \in [0, 1]$, and we have the inequality $S(y) \leq \frac{y^3}{2\mathrm{e}^3\,(1-y)}$, which implies

$$|z| \leq 2\,\frac{(A\,|\xi\,|)^3}{1-\left(\frac{D_n}{N_n}\right)^{1/3} 2\mathrm{e}A\,|\xi\,|}.$$

We now consider the zone of control $[-K t_n, K t_n]$ with $K = \frac{1}{(4\mathrm{e}+8)A^3}$. If ξ is in this zone, then we have indeed

$$|\xi| \leq \frac{(\widetilde{\sigma}_n)^2}{4\mathrm{e}A^3}\left(\frac{N_n}{D_n}\right)^{1/3} \leq \frac{1}{2\mathrm{e}A}\left(\frac{N_n}{D_n}\right)^{1/3} = L\left(\frac{N_n}{D_n}\right)^{1/3}$$

by using the remark just before the lemma. Then,

$$|z| \leq \frac{2A^3}{1-\frac{2\mathrm{e}A\,(\widetilde{\sigma}_n)^2}{(4\mathrm{e}+8)\,A^3}}\,|\xi\,|^3 \leq \frac{2A^3}{1-\frac{\mathrm{e}}{\mathrm{e}+2}}\,|\xi\,|^3 = (2+\mathrm{e})A^3\,|\xi\,|^3.$$

Thus, on the zone of control, $|\theta_n(\xi) - 1| = |\mathrm{e}^z - 1| \leq |z|\,\mathrm{e}^{|z|}$, whence a control of index $(3, 3)$ and with constants

$$K_1 = K_2 = (2+\mathrm{e})A^3.$$

We have chosen K so that $K = \frac{1}{4K_2}$, hence, the inequalities of Condition (Z2) are satisfied. □

Using the results of Sect. 15.2, we obtain:

Corollary 15.29 *Let S_n be a sequence with a uniform bounds on cumulants with parameters (D_n, N_n, A) and let $Y_n = \frac{S_n}{\sqrt{\mathrm{Var}(S_n)}}$. Then we have*

$$d_{\mathrm{Kol}}(Y_n\,,\; \mathcal{N}_{\mathbb{R}}(0,1)) \leq \frac{76.36\,A^3}{(\widetilde{\sigma}_n)^3}\sqrt{\frac{D_n}{N_n}}.$$

Proof We can apply Theorem 15.19, choosing $\gamma = 1$, and $\lambda = 0.193$. It yields a constant smaller than 77.911. It is however possible to get the better constant given above by redoing some of the computations in this specific setting. With $\rho > 4$, set $K = \frac{1}{(4\mathrm{e}+\rho)\,A^3}$, and $\varepsilon_n = \frac{1}{K\,(t_n)^{3/2}}$. On the zone $\xi \in [-\frac{1}{\varepsilon_n}, \frac{1}{\varepsilon_n}]$, we have a bound $|\theta_n(\xi) - 1| \leq M|\xi|^3 \exp(M|\xi|^3)$, this time with $M = (2 + \frac{8\mathrm{e}}{\rho})A^3$. Hence,

$$
\begin{aligned}
|\mathbb{E}[f_{a,\varepsilon_n}(Y_n)] - \mathbb{E}[f_{a,\varepsilon_n}(Y)]| &\leq \frac{M}{2\pi\,(t_n)^{3/2}} \int_{-\frac{1}{\varepsilon_n}}^{\frac{1}{\varepsilon_n}} |\xi|^2\, \mathrm{e}^{-\frac{\xi^2}{2} + M\frac{|\xi|^3}{(t_n)^{3/2}}}\, d\xi \\
&\leq \frac{M}{2\pi\,(t_n)^{3/2}} \int_{\frac{1}{\varepsilon_n}}^{\frac{1}{\varepsilon_n}} |\xi|^2\, \mathrm{e}^{-\xi^2\left(\frac{1}{2} - \frac{2}{\rho}\right)}\, d\xi \\
&\leq \frac{M}{\sqrt{2\pi}\,(t_n)^{3/2}\,(1-\frac{4}{\rho})^{3/2}} = \frac{1}{\sqrt{2\pi}}\,\frac{2}{\rho(1-\frac{4}{\rho})^{3/2}}\,\varepsilon_n.
\end{aligned}
$$

By Theorem 15.18, we get for any $\lambda > 0$:

$$
\begin{aligned}
&d_{\mathrm{Kol}}\left(\frac{S_n}{\sqrt{\mathrm{Var}(S_n)}}\,,\ \mathcal{N}_{\mathbb{R}}(0,1)\right) \\
&\leq \frac{(1+\lambda)}{\sqrt{2\pi}} \left(\frac{2}{\rho(1-\frac{4}{\rho})^{3/2}} + \frac{1}{\sqrt[3]{\pi}}\left(4\sqrt[3]{1+\frac{1}{\lambda}} + 3\sqrt[3]{3}\right)\right)\varepsilon_n.
\end{aligned}
$$

The best constant is then obtained with $\rho = 6.79$ and $\lambda = 0.185$. □

Remark 15.30 There is a trade-off in the bound of Corollary 15.29 between the a priori upper bound on $(\widetilde{\sigma}_n)^2$, and the constant C that one obtains such that the Kolmogorov distance is smaller than

$$
\frac{C\,A^3}{(\widetilde{\sigma}_n)^3}\sqrt{\frac{D_n}{N_n}}.
$$

This bound gets worse when $(\widetilde{\sigma}_n)^2$ is small, but on the other hand, the knowledge of a better a priori upper bound (that is precisely when $(\widetilde{\sigma}_n)^2$ is small) yields a better constant C. So, for instance, if one knows that $(\widetilde{\sigma}_n)^2 \leq A^2$ (instead of $2A^2$), then one gets a constant $C = 52.52$. A general bound that one can state and that takes into account this trade-off is:

$$
d_{\mathrm{Kol}}(Y_n\,,\ \mathcal{N}_{\mathbb{R}}(0,1)) \leq 27.55\left(\left(\frac{A}{\widetilde{\sigma}_n}\right)^3 + \frac{A}{\widetilde{\sigma}_n}\right)\sqrt{\frac{D_n}{N_n}}.
$$

We are indebted to Martina Dal Borgo for having pointed out this phenomenon. In the sequel, we shall freely use this small improvement of Corollary 15.29.

The above corollary ensures asymptotic normality with a bound on the speed of convergence when

$$\left(\frac{1}{(\widetilde{\sigma}_n)^3}\sqrt{\frac{D_n}{N_n}} \to 0\right) \iff \left((\widetilde{\sigma}_n)^2\left(\frac{N_n}{D_n}\right)^{1/3} \to +\infty\right).$$

Using a theorem of Janson, the asymptotic normality can be obtained under a less restrictive hypothesis, but without bound on the speed of convergence. Even if the main topic of the paper is to find bounds on the speed of convergence, we will recall the result here for the sake of completeness.

Proposition 15.31 *As above, let S_n be a sequence with a uniform bounds on cumulants with parameters (D_n, N_n, A) and assume that*

$$\lim_{n\to\infty} \frac{\mathrm{Var}(S_n)}{N_n\, D_n}\left(\frac{N_n}{D_n}\right)^{\varepsilon} = \lim_{n\to\infty} (\widetilde{\sigma}_n)^2\left(\frac{N_n}{D_n}\right)^{\varepsilon} = +\infty$$

for some parameter $\varepsilon \in (0, 1)$. Then,

$$\frac{S_n}{\sqrt{\mathrm{Var}(S_n)}} \rightharpoonup \mathcal{N}_{\mathbb{R}}(0, 1).$$

Proof Note that the bounds on cumulants can be rewritten as

$$|\kappa^{(r)}(Y_n)| \leq C_r \left(\frac{\mathrm{Var}(S_n)}{N_n\, D_n}\left(\frac{N_n}{D_n}\right)^{1-\frac{2}{r}}\right)^{-\frac{r}{2}}$$

for some constant C_r. Choosing r large enough so that $1 - \frac{2}{r} \geq \varepsilon$, we conclude that $\kappa^{(r)}(Y_n) \to 0$ for r large enough. This is a sufficient condition for the convergence in law towards a Gaussian distribution, see [28, Theorem 1] and [25]. □

Remark 15.32 Up to a change of parameters, it would be equivalent to consider bounds of the kind

$$|\kappa^{(r)}(S_n)| \leq (Cr)^r \alpha_n (\beta_n)^r, \quad \text{or} \quad \left|\kappa^{(r)}\left(\frac{S_n}{\sqrt{\mathrm{Var}(S_n)}}\right)\right| \leq \frac{r!}{\Delta_n^{r-2}},$$

as done by Saulis and Statulevičius in [52] or by the authors of this paper in [21]. In particular, it was proved in [21, Chapter 5], that under slight additional assumptions on the second and third cumulants, we have the following: the sequence $(X_n)_{n\in\mathbb{N}}$ defined in Lemma 15.28 converges in the mod-Gaussian sense with parameters $(t_n)_{n\in\mathbb{N}}$.

15.4.2 Dependency Graphs

In this paragraph, we will see that the uniform bounds on cumulants are satisfied for sums $S_n = \sum_{i=1}^{n} A_{i,n}$ of dependent random variables with specific dependency structure.

More precisely, if $(A_v)_{v \in V}$ is a family of real valued random variables, we call *dependency graph* for this family a graph $G = (V, E)$ with the following property: if V_1 and V_2 are two disjoint subsets of V with no edge $e \in E$ joining a vertex $v_1 \in V_1$ to a vertex $v_2 \in V_2$, then $(A_v)_{v \in V_1}$ and $(A_v)_{v \in V_2}$ are independent random vectors. For instance, let $(A_1, \ldots, A_7)$ be a family of random variables with dependency graph drawn on Fig. 15.6. Then the vectors $(A_1, A_2, A_3, A_4, A_5)$ and (A_6, A_7) corresponding to different connected components must be independent. Moreover, note that (A_1, A_2) and (A_4, A_5) must be independent as well: although they are in the same connected component of the graph G, they are not directly connected by an edge $e \in E$.

Theorem 15.33 (Féray–Méliot–Nikeghbali, See [21]) *Let $(A_v)_{v \in V}$ be a family of random variables, with $|A_v| \leq A$ a.s., for all $v \in V$. We suppose that $G = (V, E)$ is a dependency graph for the family and denote*

- $N = \frac{\sum_{v \in V} \mathbb{E}|A_v|}{A} \leq \operatorname{card} V$;
- *D the maximum degree of a vertex in G* plus one.

If $S = \sum_{v \in V} A_v$, then for all $r \geq 1$,

$$|\kappa^{(r)}(S)| \leq N\, r^{r-2}\, (2D)^{r-1}\, A^r.$$

Consider a sequence $(S_n)_{n \in \mathbb{N}}$, where each S_n writes as $\sum_{v \in V_n} A_{v,n}$, with the $A_{v,n}$ uniformly bounded by A (in a lot of examples, the $A_{v,n}$ are indicator variables, so that we can take $A = 1$). Set

$$N_n = \frac{\sum_{v \in V_n} \mathbb{E}|A_{v,n}|}{A}$$

and assume that, for each n, $(A_{v,n})_{v \in V_n}$ has a dependency graph of maximal degree $D_n - 1$. Then the sequence $(S_n)_{n \in \mathbb{N}}$ admits uniform bounds on cumulants with

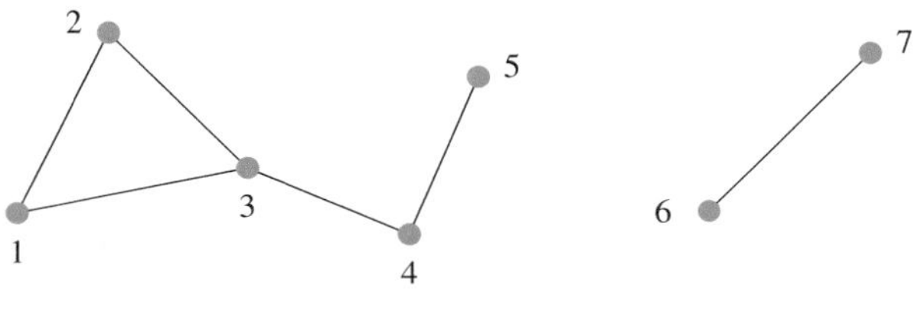

Fig. 15.6 A dependency graph for seven real-valued random variables

parameters (D_n, N_n, A) and the result of the previous section applies. Note that in this setting we have the bound $\widetilde{\sigma}_n \leq A$, so the bound of Corollary 15.29 holds with the better constant 52.52.

Remark 15.34 The parameter D is equal to the maximal number of neighbors of a vertex $v \in V$, plus 1. In the following, we shall simply call D the maximal degree, being understood that one always has to add 1. Another way to deal with this convention is to think of dependency graphs as having one loop at each vertex.

Example 15.2 The following example, though quite artificial, shows that one can construct families of random variables with arbitrary parameters N and D for their dependency graphs. Let $(U_k)_{k \in \mathbb{Z}/N\mathbb{Z}}$ be a family of independent Bernoulli random variables with $\mathbb{P}[U_k = 1] = 1 - \mathbb{P}[U_k = 0] = q$; and for $i \in \mathbb{Z}/N\mathbb{Z}$,

$$A_i = 2\left(\prod_{k=i+1}^{i+D} U_k\right) - 1.$$

Each A_i is a Bernoulli random variable with $\mathbb{P}[A_i = 1] = 1 - \mathbb{P}[A_i = -1] = q^D$, which we denote p (p is considered independent of N). We are interested in the fluctuations of $S = \sum_{i=1}^{N} A_i$. Note that the partial sums $\sum_{i=1}^{k \leq N} A_i$ correspond to random walks with correlated steps: as D increases, the consecutive steps of the random walk have a higher probability to be in the same direction, and therefore, the variance of the sum $S = \sum_{i=1}^{N} A_i$ grows. We refer to Fig. 15.7, where three such random walks are drawn, with parameters $p = \frac{1}{2}$, $N = 1000$ and $D \in \{5, 15, 30\}$.

If $d(i, j) \geq D$ in $\mathbb{Z}/N\mathbb{Z}$, then A_i and A_j do not involve any common variable U_k, and they are independent. It follows that if G is the graph with vertex set $\mathbb{Z}/N\mathbb{Z}$ and with an edge e between i and j if $d(i, j) \leq D$, then G is a dependency graph for the A_i's. This graph has N vertices, and maximal degree $2D - 1$. Moreover, one can compute exactly the expectation and the variance of

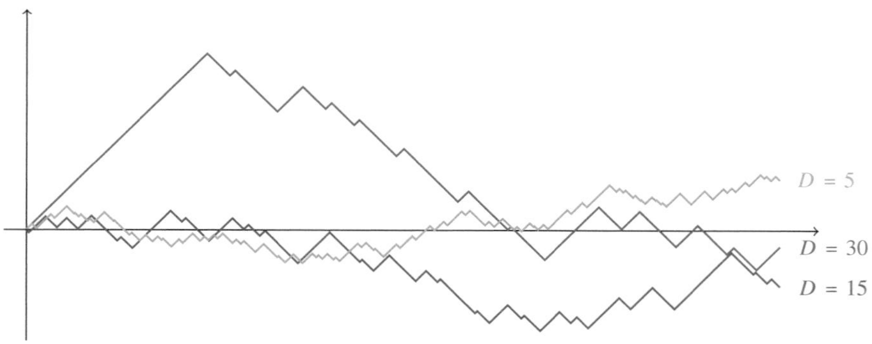

Fig. 15.7 Random walks with 1000 steps and correlation lengths $D = 5$, $D = 15$ and $D = 30$

$S = \sum_{i=1}^{N} A_i$:

$$\mathbb{E}[S] = N(2p-1);$$

$$\begin{aligned}\mathrm{Var}(S) &= 4\sum_{i,j=1}^{N} \mathrm{cov}(U_{i+1}\cdots U_{i+D}, U_{j+1}\cdots U_{j+N})\\ &= 4\sum_{i,j=1}^{N} \left(q^{D+\min(D,d(i,j))} - p^2\right) = 4Np \sum_{j=-(D-1)}^{D-1} (q^{|j|} - p)\\ &= 4Np\left(\frac{1+p^{\frac{1}{D}}-2p}{1-p^{\frac{1}{D}}} - (2D-1)p\right).\end{aligned}$$

If N and D go to infinity with $D = o(N)$, then $q = p^{\frac{1}{D}} = 1 + \frac{\log p}{D} + O(\frac{1}{D^2})$, and

$$\mathbb{E}[S] = N(2p-1);$$

$$\mathrm{Var}(S) = 8N(D+O(1))\,p\left(\frac{1-p}{-\log p} - p\right).$$

So, one can apply Corollary 15.29 to the sum S, and one obtains:

$$d_{\mathrm{Kol}}\left(\frac{S-\mathbb{E}[S]}{\sqrt{\mathrm{Var}(S)}},\ \mathcal{N}_{\mathbb{R}}(0,1)\right) \leq \frac{6}{\left(p\left(\frac{1-p}{-\log p} - p\right)\right)^{3/2}}\sqrt{\frac{D}{N}}.$$

Example 15.3 Fix $p \in (0,1)$, and consider a random *Erdös–Rényi graph* $G = G_n = G(n,p)$, which means that one keeps at random each edge of the complete graph K_n with probability p, independently from every other edge. Note that we only consider the case of fixed p here; for $p \to 0$, we would get rather weak bounds, see [21, Section 10.3.3] for a discussion on bounds on cumulants in this framework.

Let $H = (V_H, E_H)$ and $G = (V_G, E_G)$ be two graphs. The number of copies of H in G is the number of injections $i : V_H \to V_G$ such that, if $(h_1, h_2) \in E_H$, then $(i(h_1), i(h_2)) \in E_G$. In random graph theory, this is called the *subgraph count* statistics; we denote it by $I(H,G)$. We refer to Fig. 15.8 for an example, where $H = K_3$ is the triangle and G is a random Erdös–Rényi graph of parameters $n = 30$ and $p = \frac{1}{10}$.

One can always write $I(H,G)$ as a sum of dependent random variables. Identify V_H with $[\![1,k]\!]$ and V_G with $[\![1,n]\!]$, and denote $\mathfrak{A}(n,k)$ the set of arrangements $(a_1,\ldots,a_k)$ of size k in $[\![1,n]\!]$. Given such an arrangement, the *induced subgraph* $G[a_1,\ldots,a_k]$ is the graph with vertex set $[\![1,k]\!]$, and with an edge between i and j

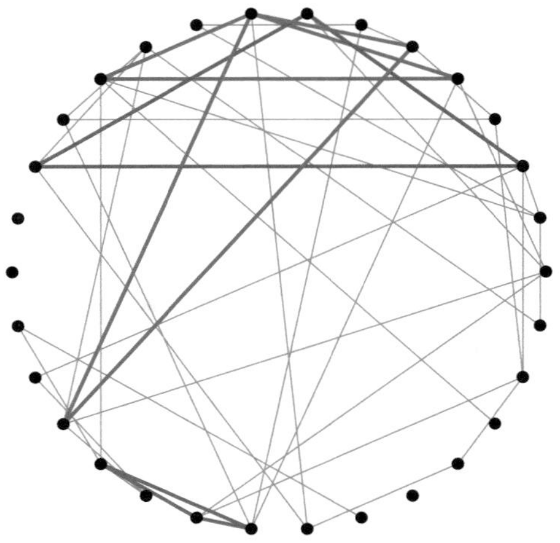

Fig. 15.8 Count of triangles in a random Erdös–Rényi graph of parameters $n = 30$ and $p = 0.1$. Here, there are $4 \times 3! = 24$ ways to embed a triangle in the graph

if $(a_i, a_j) \in E_G$. Then,

$$I(H, G) = \sum_{A \in \mathfrak{A}(n,k)} I_A(H, G),$$

where $I_A(H, G) = 1$ if $H \subset G[A]$, and 0 otherwise.

A dependency graph for the random variables $I_A(H, G_n)$ has vertex set $\mathfrak{A}(n, k)$ of cardinality $N_n = n^{\downarrow k} = n(n-1)(n-2)\cdots(n-k+1)$, and an edge between two arrangements $(a_1, \ldots, a_k)$ and $(b_1, \ldots, b_k)$ if they share at least two points (otherwise, the random variables $I_A(H, G_n)$ and $I_B(H, G_n)$ involve disjoint sets of edges and are therefore independent). As a consequence, the maximal degree of the graph is smaller than

$$D_n = \left(\binom{k}{2}^2 2(n-2)(n-3)\cdots(n-k+1)\right),$$

and of order n^{k-2}. Therefore, $\frac{D_n}{N_n} \leq \frac{2\binom{k}{2}^2}{n(n-1)} = O(\frac{1}{n^2})$, and on the other hand, if h is the number of edges of H, one can compute the asymptotics of the expectation and of the variance of $I(H, G_n)$:

$$\mathbb{E}[I(H, G_n)] = n^{\downarrow k} p^h;$$

$$\mathrm{Var}(I(H, G_n)) = 2h^2 p^{2h-1}(1-p)\, n^{2k-2} + O(n^{2k-3}),$$

see [21, Section 10] for the details of these computations. In particular,

$$\lim_{n\to\infty} \frac{\mathrm{Var}(S_n)}{N_n\, D_n} = p^{2h-1}(1-p)\left(\frac{h}{\binom{k}{2}}\right)^2 = \widetilde{\sigma}^2 > 0.$$

Thus, using Corollary 15.29, we get

$$d_{\mathrm{Kol}}\left(\frac{I(H,G_n)-\mathbb{E}[I(H,G_n)]}{\sqrt{\mathrm{Var}(I(H,G_n))}},\, \mathcal{N}_{\mathbb{R}}(0,1)\right) \leq \frac{4.65\,(k(k-1))^4}{p^{3h}\,(\frac{1}{p}-1)^{3/2}\,h^3}\,\frac{1}{n}$$

for n large enough. For instance, if $T_n = I(K_3, G_n)$ is the number of triangles in G_n, then

$$\mathbb{E}[T_n] = n^{\downarrow 3}\, p^3$$
$$\mathrm{Var}[T_n] = 18\, n^{\downarrow 4}\, p^5(1-p) + 6\, n^{\downarrow 3}\, p^3(1-p^3)$$

and

$$d_{\mathrm{Kol}}\left(\frac{T_n - n^{\downarrow 3} p^3}{\sqrt{18\, n^{\downarrow 4}\, p^5(1-p)+6\, n^{\downarrow 3}\, p^3(1-p^3)}},\, \mathcal{N}_{\mathbb{R}}(0,1)\right)$$
$$\leq \frac{234}{p^9(\frac{1}{p}-1)^{\frac{3}{2}}}\,\frac{1}{n}\,(1+O(\tfrac{1}{n})).$$

This result is not new, except maybe the explicit constant. We refer to [4] for an approach of speed of convergence for subgraph counts using Stein's method. More recently, Krokowski et al. [37] applied Malliavin calculus to the same problem. Our result corresponds to the case where p is constant of their Theorem 1. Similar bounds could be obtained by Stein's method, see [47].

To conclude our presentation of the convergence of sums of bounded random variables with sparse dependency graphs, let us analyse precisely the case of *uncorrelated* random variables.

Corollary 15.35 *Let $S_n = \sum_{i=1}^{N_n} A_{i,n}$ be a sum of centered and bounded random variables, that are uncorrelated and with $\mathbb{E}[(A_{i,n})^2] = 1$ for all i. We suppose that the random variables have a dependency graph of parameters $N_n \to +\infty$ and $s D_n$.*

1. *If $D_n = O((N_n)^{1/2-\varepsilon})$ for $\varepsilon > 0$, then $Y_n = \frac{S_n}{\sqrt{N_n}}$ converges in law towards the Gaussian distribution.*
2. *If $D_n = o((N_n)^{1/4})$, then the Kolmogorov distance between Y_n and $\mathcal{N}_{\mathbb{R}}(0,1)$ is a $O((D_n)^2/(N_n)^{1/2})$.*

Proof It is an immediate consequence of Corollary 15.29 and Proposition 15.31, since S_n admits uniform control on cumulants (see Theorem 15.33) and

$$\mathrm{Var}(S_n) = \sum_{i=1}^{N_n} \mathbb{E}[(A_{i,n})^2] = N_n.$$

□

15.4.3 Unbounded Random Variables and Truncation Methods

A possible generalization regards sums of *unbounded* random variables. In the following, we develop a truncation method that yields a criterion of asymptotic normality similar to Lyapunov's condition (see [7, Chapter 27]). A small modification of this method would similarly yield a Lindeberg type criterion. Let $S_n = \sum_{i=1}^{N_n} A_{i,n}$ be a sum of centered random variables, with

$$\left(\mathbb{E}[|A_{i,n}|^{2+\delta}]\right)^{\frac{1}{2+\delta}} \leq A$$

for some constant A independent of i and n, and some $\delta > 0$. We suppose as before that the family of random variables $(A_{i,n})_{i\in[\![1,N_n]\!]}$ has a (true) dependency graph G_n of parameters N_n and D_n. Note that in this case,

$$\begin{aligned}\mathrm{Var}(S_n) &= \sum_{i,j=1}^{N_n} \mathrm{cov}(A_{i,n}, A_{j,n}) \leq \sum_{i=1}^{N_n} \sum_{j\sim i} \|A_{i,n}\|_2 \, \|A_{j,n}\|_2 \\ &\leq \sum_{i=1}^{N_n} \sum_{j\sim i} \|A_{i,n}\|_{2+\delta} \, \|A_{j,n}\|_{2+\delta} \leq A^2 \, N_n \, D_n.\end{aligned}$$

We set

$$A_{i,n}^- = A_{i,n} \, 1_{|A_{i,n}|\leq L_n} \quad ; \quad A_{i,n}^+ = A_{i,n} \, 1_{|A_{i,n}|> L_n} \, ;$$

$$S_n^- = \sum_{i=1}^{N_n} A_{i,n}^- \quad ; \quad S_n^+ = \sum_{i=1}^{N_n} A_{i,n}^+ \, ;$$

where L_n is a truncation level, to be chosen later. Notice that G_n is still a dependency graph for the family of truncated random variables $(A_{i,n}^-)_{i\in[\![1,N_n]\!]}$. Therefore, we can apply the previously developed machinery (Theorem 15.33 and Corollary 15.29) to the scaled sum S_n^-/L_n. On the other hand, by Markov's inequality,

$$d_{\mathrm{Kol}}(S_n, S_n^-) = \sup_{s\in\mathbb{R}} |\mathbb{P}[S_n \geq s] - \mathbb{P}[S_n^- \geq s]| \leq \mathbb{P}[S_n^+ = 0]$$

$$\leq \sum_{i=1}^{N_n} \mathbb{P}[|A_{i,n}| \geq L_n] \leq N_n \left(\frac{A}{L_n}\right)^{2+\delta}.$$

Combining the two arguments leads to the following result (this replaces the previous assumption of boundedness $|A_{i,n}| \leq A$).

Theorem 15.36 *Let $(S_n = \sum_{i=1}^{N_n} A_{i,n})_{n\in\mathbb{N}}$ be a sum of centered random variables, with dependency graphs of parameters $N_n \to +\infty$ and D_n, and with*

$$\|A_{i,n}\|_{2+\delta} = (\mathbb{E}[|A_{i,n}|^{2+\delta}])^{1/(2+\delta)} \leq A$$

for all i, n and for some $\delta > 0$. We set $Y_n = S_n/\sqrt{\mathrm{Var}(S_n)}$. Recall that $(\widetilde{\sigma}_n)^2 = \frac{\mathrm{Var}(S_n)}{N_n D_n}$.

(U1) Set

$$V_n = (\widetilde{\sigma}_n)^2 \left(\frac{N_n}{D_n}\right)^{1/3} \frac{1}{(N_n)^{2/(2+\delta)}}$$

and suppose that $\lim_{n\to\infty} V_n = +\infty$ (which is only possible for $\delta > 4$). Then, for n large enough,

$$d_{\mathrm{Kol}}(Y_n, \mathcal{N}_{\mathbb{R}}(0,1)) \leq 78 \left(\frac{A^2}{V_n}\right)^{\frac{3(\delta+2)}{2(\delta+5)}} = o\left(\frac{1}{V_n}\right).$$

(U2) More generally, for $\varepsilon \in (\frac{2}{2+\delta}, 1)$, set

$$W_n = (\widetilde{\sigma}_n)^2 \left(\frac{N_n}{D_n}\right)^{\varepsilon} \frac{1}{(N_n)^{2/(2+\delta)}}$$

and suppose that $\lim_{n\to\infty} W_n = +\infty$. Then, $Y_n \rightharpoonup \mathcal{N}_{\mathbb{R}}(0,1)$.

Remark 15.37 It should be noticed that if $\delta \to +\infty$, then one essentially recovers the content of Corollary 15.29 (which can be applied because of Theorem 15.33). On the other hand, the inequality $\delta > 4$ amounts to the existence of bounded moments of order strictly higher than 6 for the random variables $A_{i,n}$. In practice, one can for instance ask for bounded moments of order 7 (i.e. $\delta = 5$), in which case the first condition (U1) reads

$$\lim_{n\to\infty} \frac{\mathrm{Var}(S_n)}{D_n\, N_n} \frac{(N_n)^{1/9}}{(D_n)^{1/3}} = +\infty.$$

Moreover, we will see in the proof of Theorem 15.36 that in this setting ($\delta = 5$), the constant 78 can be improved to 39, so that, for n large enough:

$$d_{\mathrm{Kol}}(Y_n,\ \mathcal{N}_{\mathbb{R}}(0,1)) \leq 39 \left(\frac{A^2}{V_n}\right)^{\frac{21}{20}}.$$

Proof We write as usual $Y_n = \frac{S_n}{\sqrt{\mathrm{Var}(S_n)}}$, and $Y_n^- = \frac{S_n^-}{\sqrt{\mathrm{Var}(S_n^-)}}$. In all cases, we have

$$\begin{aligned}
d_{\mathrm{Kol}}(Y_n,\ \mathcal{N}_{\mathbb{R}}(0,1)) &\leq d_{\mathrm{Kol}}(S_n, S_n^-) + d_{\mathrm{Kol}}\left(Y_n^-,\ \sqrt{\frac{\mathrm{Var}(S_n)}{\mathrm{Var}(S_n^-)}}\,\mathcal{N}_{\mathbb{R}}(0,1)\right)\\
&\leq N_n\left(\frac{A}{L_n}\right)^{2+\delta} + d_{\mathrm{Kol}}(Y_n^-, \mathcal{N}_{\mathbb{R}}(0,1)) + d_{\mathrm{Kol}}\left(\mathcal{N}_{\mathbb{R}}(0,1),\ \mathcal{N}_{\mathbb{R}}\left(0,\ \frac{\mathrm{Var}(S_n)}{\mathrm{Var}(S_n^-)}\right)\right)
\end{aligned}$$

by using the invariance of the Kolmogorov distance with respect to multiplication of random variables by a positive constant. In the sequel, we denote a, b and c the three terms on the second line of the inequality. The Kolmogorov distance between two Gaussian distributions is

$$\begin{aligned}
d_{\mathrm{Kol}}(\mathcal{N}_{\mathbb{R}}(0,1),\ \mathcal{N}_{\mathbb{R}}(0,\lambda^2)) &= \frac{1}{\sqrt{2\pi}} \sup_{s\in\mathbb{R}_+}\left(\int_s^{\lambda s} \mathrm{e}^{-\frac{u^2}{2}}\,du\right)\\
&\leq \frac{\lambda-1}{\sqrt{2\pi}} \sup_{s\in\mathbb{R}_+}\left(s\,\mathrm{e}^{-\frac{s^2}{2}}\right) = \sqrt{\frac{1}{2\pi\mathrm{e}}}\,|\lambda-1|
\end{aligned}$$

if $\lambda \geq 1$. One gets the same result if $\lambda \leq 1$, hence,

$$\begin{aligned}
d_{\mathrm{Kol}}\left(\mathcal{N}_{\mathbb{R}}(0,1),\ \mathcal{N}_{\mathbb{R}}\left(0,\ \frac{\mathrm{Var}(S_n)}{\mathrm{Var}(S_n^-)}\right)\right) &= d_{\mathrm{Kol}}\left(\mathcal{N}_{\mathbb{R}}(0,1),\ \mathcal{N}_{\mathbb{R}}\left(0,\ \frac{\mathrm{Var}(S_n^-)}{\mathrm{Var}(S_n)}\right)\right)\\
&\leq \sqrt{\frac{1}{2\pi\mathrm{e}}}\left|\sqrt{\frac{\mathrm{Var}(S_n^-)}{\mathrm{Var}(S_n)}} - 1\right|\\
&\leq \sqrt{\frac{1}{2\pi\mathrm{e}}}\,\frac{|\mathrm{Var}(S_n^-) - \mathrm{Var}(S_n)|}{\mathrm{Var}(S_n)}.
\end{aligned}$$

To evaluate the difference between the variances, notice that

$$\begin{aligned}
\mathrm{Var}(S_n^-) &= \mathrm{Var}\left(S_n - \sum_{i=1}^{N_n} A_{i,n}^+\right)\\
&= \mathrm{Var}(S_n) - 2\sum_{i,j=1}^{N_n} \mathrm{Cov}(A_{i,n}^+, A_{j,n}) + \sum_{i,j=1}^{N_n} \mathrm{Cov}(A_{i,n}^+, A_{j,n}^+)
\end{aligned}$$

If j is not connected to i in G_n, or equal to i, then $A_{i,n}$ and $A_{j,n}$ are independent, hence, $\mathrm{Cov}(A_{i,n}^+, A_{j,n}) = 0$. Otherwise, using Hölder and Bienaymé-Chebyshev inequalities,

$$\begin{aligned}|\mathrm{Cov}(A_{i,n}^+, A_{j,n})| &\leq \sqrt{\mathbb{E}[(A_{i,n})^2\, 1_{|A_{i,n}|\geq L_n}]\, \mathbb{E}[(A_{j,n})^2\, 1_{|A_{i,n}|\geq L_n}]} \\ &\leq \sqrt{\mathbb{E}[(A_{i,n})^{2+\delta}]^{\frac{2}{2+\delta}}\, \mathbb{P}[|A_{i,n}|\geq L_n]^{\frac{2\delta}{2+\delta}}\, \mathbb{E}[(A_{j,n})^{2+\delta}]^{\frac{2}{2+\delta}}} \\ &\leq A^2 \left(\frac{A}{L_n}\right)^{\delta}.\end{aligned}$$

Similarly,

$$\begin{aligned}|\mathrm{Cov}(A_{i,n}^+, A_{j,n}^+)| &\leq \mathbb{E}[|A_{i,n}^+ A_{j,n}^+|] + \mathbb{E}[|A_{i,n}^+|]\, \mathbb{E}[|A_{j,n}^+|] \\ &\leq A^2 \left(\left(\frac{A}{L_n}\right)^{\delta} + \left(\frac{A}{L_n}\right)^{2+2\delta}\right),\end{aligned}$$

hence, assuming that the level of truncation L_n is larger than A,

$$\frac{|\mathrm{Var}(S_n)^- - \mathrm{Var}(S_n)|}{\mathrm{Var}(S_n)} \leq \frac{N_n\, D_n}{\mathrm{Var}(S_n)}\, 3A^2 \left(\frac{A}{L_n}\right)^{\delta}.$$

Let us now place ourselves in the setting of Hypothesis (U1); we set

$$V_n = \frac{\mathrm{Var}(S_n)}{N_n\, D_n} \left(\frac{N_n}{D_n}\right)^{1/3} \frac{1}{(N_n)^{2/(2+\delta)}}.$$

Suppose that $L_n = K_n\, (N_n)^{\frac{1}{2+\delta}}$, with K_n going to infinity. We then have $a \leq \frac{A^{2+\delta}}{(K_n)^{2+\delta}}$, and on the other hand,

$$\begin{aligned}\left|\frac{\mathrm{Var}(S_n^-)}{\mathrm{Var}(S_n)} - 1\right| &\leq 3A^{2+\delta} \left(\frac{N_n\, D_n}{\mathrm{Var}(S_n)}\, \frac{1}{(N_n)^{\delta/(2+\delta)}}\, \frac{1}{(K_n)^{\delta}}\right) \\ &\leq 3A^{2+\delta} \left(\frac{1}{V_n\, (D_n)^{1/3}\, (N_n)^{2/3}\, (K_n)^{\delta}}\right) \to 0\end{aligned}$$

since by hypothesis, $\lim_{n\to\infty} \frac{1}{V_n} = 0$. So,

$$c \leq \frac{3A^{2+\delta}}{\sqrt{2\pi \mathrm{e}}}\, \frac{1}{V_n\, (D_n)^{1/3}\, (N_n)^{2/3}\, (K_n)^{\delta}} \leq \frac{3A^{2+\delta}}{\sqrt{2\pi \mathrm{e}}}\, \frac{1}{V_n\, (K_n)^{\delta}}.$$

Now, the sequence $(S_n^-/L_n)_{n\in\mathbb{N}}$ is a sequence of sums of centered random variables all bounded by 1, and to apply Corollary 15.29 to this sequence, we need

$$\lim_{n\to\infty} \frac{\mathrm{Var}(S_n^-)}{N_n\, D_n} \left(\frac{N_n}{D_n}\right)^{1/3} \frac{1}{(L_n)^2} = +\infty.$$

However, the previous computation shows that one can replace $\mathrm{Var}(S_n)$ by $\mathrm{Var}(S_n^-)$ in this expression without changing the asymptotic behavior, so

$$\lim_{n\to\infty} \frac{\mathrm{Var}(S_n^-)}{N_n\, D_n} \left(\frac{N_n}{D_n}\right)^{1/3} \frac{1}{(L_n)^2} = \lim_{n\to\infty} \frac{V_n}{(K_n)^2},$$

which is $+\infty$ if $(K_n)^2$ is not growing too fast to $+\infty$ (in comparison to the sequence V_n). Then, by Corollary 15.29,

$$b = d_{\mathrm{Kol}}(Y_n^-,\, \mathcal{N}_{\mathbb{R}}(0,1)) \leq 77\, \frac{(K_n)^3}{(V_n)^{3/2}}$$

for n large enough. Set $K_n = B(V_n)^{\frac{3}{2(5+\delta)}}$. Then,

$$\begin{aligned} d_{\mathrm{Kol}}(Y_n,\, \mathcal{N}_{\mathbb{R}}(0,1)) &\leq a+b+c \leq \left(\frac{A^{2+\delta}}{B^{2+\delta}} + 77\, B^3 + o(1)\right)\left(\frac{1}{V_n}\right)^{\frac{3(2+\delta)}{2(5+\delta)}} \\ &\leq \left(\left(\frac{231}{2+\delta}\right)^{\frac{2+\delta}{5+\delta}} + 77\left(\frac{2+\delta}{231}\right)^{\frac{3}{5+\delta}}\right)\left(\frac{A^2}{V_n}\right)^{\frac{3(2+\delta)}{2(5+\delta)}} \end{aligned}$$

for n large enough, and by choosing B in an optimal way. The term in parenthesis is maximal when $\delta = 229$, and is then equal to 78. This ends the proof of (U1), and one gets a better constant smaller than 39 when $\delta = 5$.

Under the Hypothesis (U2), we set

$$W_n = \frac{\mathrm{Var}(S_n)}{N_n\, D_n}\left(\frac{N_n}{D_n}\right)^{\varepsilon} \frac{1}{(N_n)^{2/(2+\delta)}}.$$

In order to prove the convergence in law $Y_n \rightharpoonup \mathcal{N}_{\mathbb{R}}(0,1)$, it suffices to have:

- $S_n - S_n^- = S_n^+$ that converges in probability to 0. This happens as soon as the level L_n is $K_n\,(N_n)^{1/(2+\delta)}$ with $K_n \to +\infty$.
- $\frac{|\mathrm{Var}(S_n)-\mathrm{Var}(S_n^-)|}{\mathrm{Var}(S_n)} \to 0$. With $L_n = K_n\,(N_n)^{1/(2+\delta)}$, the previous computations show that this quantity is a

$$O\left(\frac{1}{W_n\,(N_n)^{1-\varepsilon}\,(D_n)^{\varepsilon}\,(K_n)^{\delta}}\right),$$

which goes to 0.

- and by Theorem 15.31,

$$\frac{\mathrm{Var}(S_n)}{N_n\, D_n}\left(\frac{N_n}{D_n}\right)^{\varepsilon}\frac{1}{(L_n)^2} \to +\infty.$$

This follows from the Hypothesis (U2) if K_n is chosen to grow sufficiently slow.

Thus, the second part of Theorem 15.36 is proven. □

Example 15.4 Let $(X_i)_{i\in[\![1,N]\!]}$ be a centered Gaussian vector with $\mathbb{E}[(X_i)^2] = 1$ for any i, and with the covariance matrix $(\mathrm{Cov}(X_i, X_j))_{1\leq i,j\leq N}$ that is sparse in the following sense: for any i, the set of indices j such that $\mathrm{Cov}(X_i, X_j) \neq 0$ has cardinality smaller than D. We set $A_i = \exp(X_i)$; the random variables A_i follow the *log-normal distribution* of density

$$\frac{1}{\sqrt{2\pi}}\,\frac{1}{u^{1+\frac{\log u}{2}}}\,1_{u>0}\,du,$$

see Fig. 15.9. They have moments of all order: $\mathbb{E}[(A_i)^k] = \mathbb{E}[\mathrm{e}^{kX_i}] = \mathrm{e}^{\frac{k^2}{2}}$.

The variables A_i have the same dependency graph as the variables X_i. Moreover, if $\rho_{ij} = \mathrm{Cov}(X_i, X_j)$, then the covariance of two variables A_i and A_j is $\mathrm{e}(\mathrm{e}^{\rho_{ij}} - 1)$. Using moments of order $2+\delta$, we see that if

$$Y_N = \frac{\sum_{i=1}^{N}(A_i - \mathrm{e}^{\frac{1}{2}})}{\sqrt{\mathrm{e}\sum_{1\leq i,j\leq N}(\mathrm{e}^{\rho_{ij}} - 1)}}$$

$$V_{N,\delta} = \frac{\mathrm{e}\sum_{1\leq i,j\leq N}(\mathrm{e}^{\rho_{ij}} - 1)}{ND}\,\frac{N^{\frac{1}{3}-\frac{2}{2+\delta}}}{D^{\frac{1}{3}}} \to +\infty,$$

then $d_{\mathrm{Kol}}(Y_N,\, \mathcal{N}_{\mathbb{R}}(0,1)) \leq 78\left(\frac{\mathrm{e}^{\delta+2}}{V_{N,\delta}}\right)^{\frac{3(\delta+2)}{2(\delta+5)}}$ for N large enough.

To make this result more explicit, let us consider the following dependency structure for the Gaussian vector $X = (X_i)_{i\in[\![1,N]\!]}$:

$$\mathrm{Cov}(X) = \begin{pmatrix} 1 & * & \cdots & * \\ * & 1 & \ddots & \vdots \\ \vdots & \ddots & \ddots & * \\ * & \cdots & * & 1 \end{pmatrix},$$

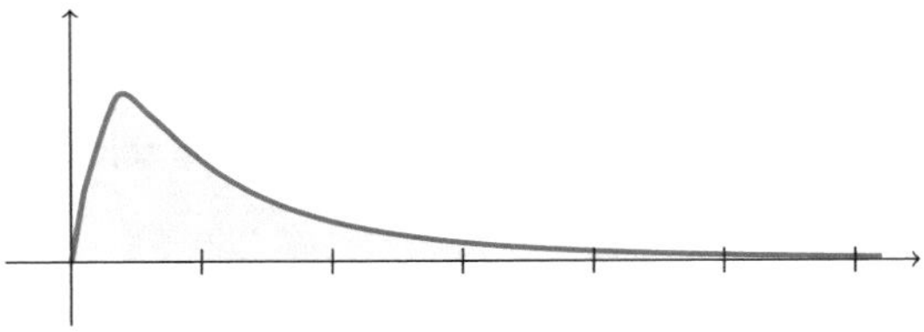

Fig. 15.9 The density of the log-normal distribution

where the non-diagonal terms $*$ are all smaller than $\frac{\rho}{D}$ in absolute value, and with less than D non-zero terms on each row or column. When $\rho \in [0, 1)$, the matrix is diagonally dominant, hence positive-definite, so there exists indeed a Gaussian vector X with these covariances. We then have

$$V_{N,\delta} \geq \mathrm{e}\,(1 - D(\mathrm{e}^{\frac{\rho}{D}} - 1))\, \frac{N^{\frac{1}{3} - \frac{2}{2+\delta}}}{D^{\frac{4}{3}}},$$

so if $1 \ll D \ll N^{\frac{1}{4}-\varepsilon}$, then one can apply Theorem 15.36 to get

$$d_{\mathrm{Kol}}(Y_N,\ \mathcal{N}_{\mathbb{R}}(0, 1)) \leq 78 \left(\frac{\mathrm{e}^{\frac{3}{2\varepsilon}-1}}{1-\rho}\right)^{\frac{3}{2}} \left(\frac{D}{N^{\frac{1}{4}-\varepsilon}}\right)^{\frac{2}{2\varepsilon+1}}$$

for N large enough. Moreover, as soon as $1 \ll D \ll N^{\frac{1}{2}-\varepsilon}$, $Y_N \rightharpoonup \mathcal{N}_{\mathbb{R}}(0, 1)$.

15.5 Ising Model and Markov Chains

In this section, we present examples of random variables that admit uniform bounds on cumulants, which do not come from dependency graphs. Their structure is nevertheless not so different since the variables that we consider write as sums of random variables that are *weakly dependent*. The technique to prove uniform bounds on cumulants relies then on the notion of *uniform weighted dependency graph*, which generalizes the notion of standard dependency graph (see Proposition 15.40).

15.5.1 Weighted Graphs and Spanning Trees

An *edge-weighted graph G*, or *weighted graph* for short, is a graph G in which each edge e is assigned a weight $w_G(e)$. Here we restrict ourselves to weights $w_G(e)$ with $w_G(e) \in \mathbb{R}_+$. Edges not in the graph can be thought of as edges of weight 0, all our definitions being consistent with this convention.

If B is a multiset of vertices of G, we can consider the graph $G[B]$ induced by G on B and defined as follows: the vertices of $G[B]$ correspond to elements of B (if B contains an element with multiplicity m, then m vertices correspond to this element), and there is an edge between two vertices if the corresponding vertices of G are equal or connected by an edge in G. This new graph has a natural weighted graph structure: put on each edge of $G[B]$ the weight of the corresponding edge in G (if the edge connects two copies of the same vertex of G, we put weight 1).

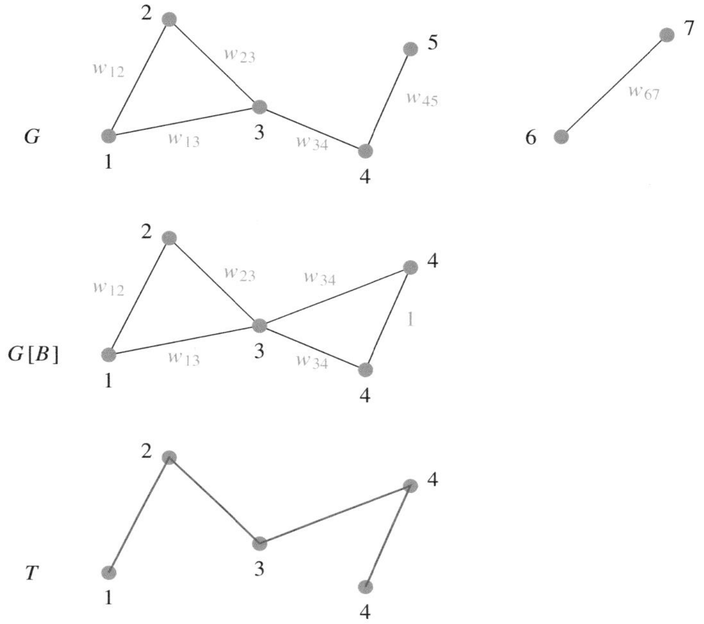

Fig. 15.10 A weighted dependency graph G for 7 random variables; the induced graph $G[B]$ with $B = \{1, 2, 3, 4, 4\}$; and a spanning tree T of $G[B]$, with $w(T) = w_{12}w_{23}w_{34}$

Definition 15.38 A spanning tree of a graph $G = (V, E)$ is a subset E' of E such that (V, E') is a tree. In other words, it is a subgraph of G that is a tree and covers all vertices.

The set of spanning trees of T is denoted $\mathrm{ST}(G)$. If G is a weighted graph, we say that the weight $w(T)$ of a spanning tree of G is defined as the *product* of the weights of the edges in T (Fig. 15.10).

15.5.2 Uniform Weighted Dependency Graphs

If $A_1, \ldots, A_r$ are real-valued random variables, there is a notion of joint cumulants that generalize the cumulants of Sect. 15.4:

$$\kappa(A_1, A_2, \ldots, A_r) = [z_1 z_2 \cdots z_r]\left(\log \mathbb{E}[\mathrm{e}^{z_1 A_1 + z_2 A_2 + \cdots + z_r A_r}]\right).$$

The joint cumulants are multilinear and symmetric functionals of $A_1, \ldots, A_r$. On the other hand,

$$\kappa^{(r)}(X) = \kappa(X, X, \ldots, X)$$

with r occurrences of X in the right-hand side. In particular, if $S = \sum_{v \in V} A_v$ is a sum of random variables, then

$$\kappa^{(r)}(S) = \sum_{v_1,\ldots,v_r \in V} \kappa(A_{v_1}, A_{v_2}, \ldots, A_{v_r}).$$

Definition 15.39 Let $(A_v)_{v \in V}$ be a family of random variables defined on the same probability space. A weighted graph $G = (V, E, w_G)$ is a C-uniform weighted dependency graph for $(A_v)_{v \in V}$ if, for any multiset $B = \{v_1, \ldots, v_r\}$ of elements of V, one has

$$\left|\kappa\big(A_v,\ v \in B\big)\right| \le C^{|B|} \sum_{T \in \mathrm{ST}(G[B])} w(T).$$

Proposition 15.40 *Let $(A_v)_{v \in V}$ be a finite family of random variables with a C-uniform weighted dependency graph G. Assume that G has $N = |V|$ vertices, and maximal weighted degree $D - 1$, that is:*

$$\forall v \in V, \quad \sum_{\substack{v' :\, v' \neq v \\ \{v, v'\} \in E_G}} w_G(\{v, v'\}) \le D - 1.$$

Then, for $r \ge 1$,

$$\left|\kappa^{(r)}\left(\sum_{v \in V} A_v\right)\right| \le N\, r^{r-2} D^{r-1}\, C^r.$$

Consider a sequence $(S_n)_{n \in \mathbb{N}}$, where each S_n writes as $\sum_{v \in V_n} A_{v,n}$. Set $N_n = |V_n|$ and assume that, for each n, $(A_v)_{v \in V_n}$ has C-uniform weighted dependency graph of maximal degree $D_n - 1$ (by assumption, C does not depend on n). Then the sequence (S_n) admits uniform bounds on cumulants with parameters $(\frac{D_n}{2}, N_n, C)$ and the results of Sect. 15.4, in particular Corollary 15.29, apply.

Proof By multilinearity and definition of a uniform weighted dependency graph, we have

$$\left|\kappa_r\left(\sum_{v \in V} A_v\right)\right| \le C^r \sum_{v_1,\ldots,v_r} \sum_{T \in \mathrm{ST}(G[v_1,\ldots,v_r])} w(T). \tag{15.7}$$

By possibly adding edges of weight 0, we may assume that $G[v_1, \ldots, v_r]$ is always the complete graph K_r so that $\mathrm{ST}(G[v_1, \ldots, v_r]) \simeq \mathrm{ST}(K_r)$ as sets. The weight of a tree T however depends on $v_1, \cdots, v_r$, namely

$$w(T) = \prod_{\{i,j\} \in T} w_G(\{v_i, v_j\}),$$

where $w_G(\{v_i, v_j\})$ is the weight of the edge $\{v_i, v_j\}$ in G (or 1 if $v_i = v_j$).

With this viewpoint, we can exchange the order of summation in (15.7). We claim that the contribution of a fixed tree $T \in \mathrm{ST}(K_r)$ can then be bounded as follows:

$$\sum_{v_1,\dots,v_r} w(T) = \sum_{v_1,\dots,v_r} \prod_{\{i,j\}\in T} w_G(\{v_i, v_j\}) \le N D^{r-1}. \tag{15.8}$$

Let us prove this claim by induction on r. The case $r = 1$ is trivial. Up to renaming the vertices of T, we may assume that r is a leaf of T so that T is obtained from a spanning tree $\widetilde{T}$ of K_{r-1} by adding an edge $\{i_0, r\}$ for some $i_0 < r$. Then

$$\sum_{v_1,\dots,v_r} \prod_{\{i,j\}\in T} w_G(\{v_i, v_j\})$$
$$= \sum_{v_1,\dots,v_{r-1}} \left(\prod_{\{i,j\}\in \widetilde{T}} w_G(\{v_i, v_j\}) \right) \left[\sum_{v_r \in V} w_G(\{v_{i_0}, v_r\}) \right].$$

The expression in square brackets is by definition smaller than D for all v_{i_0} (the sum for $v_r \neq v_{i_0}$ is smaller than $D - 1$ and the term for $v_r = v_{i_0}$ is 1). By induction hypothesis, the sum of the parenthesis is smaller than $N D^{r-2}$. This concludes the proof of (15.8) by induction. The lemma now follows immediately, since the number of spanning trees of K_r is well known to be r^{r-2}. □

Remark 15.41 A classical dependency graph with a uniform bound A on all variables A_v can be seen as a C-uniform weighted dependency graph for $C = 2A$ (all edges have weight 1); see [21, Section 9.4]. In this case, Proposition 15.40 reduces to Theorem 15.33. The proof of Proposition 15.40 given here is a simple adaptation of the second part of the proof of Theorem 15.33 (see [21, Chapter 9]) to the weighted context. The first, and probably the hardest part of the proof of Theorem 15.33 consisted in showing that a classical dependency graph is indeed a C-uniform weighted dependency graph.

Remark 15.42 In the case where the set V is a subset of $\mathbb{Z}^d$ and the weight function only depends on the distance, the notion of uniform weighted dependency graph coincides with the notion of *strong cluster properties*, proposed by Duneau, Iagolnitzer and Souillard in [16]. These authors also observed that this implies uniform bounds on cumulants when D is bounded, see [16, Eq. (10)].

Remark 15.43 A weaker notion of weighted dependency graph, where the bound on cumulant is not uniform on r, was recently introduced in [20]. This weaker notion only enables to prove central limit theorem, without normality zone or speed of convergence results. However, it seems to have a larger range of applications.

15.5.3 Magnetization of the Ising Model

We consider here the nearest-neighbour Ising model on a finite subset Λ of $\mathbb{Z}^d$ with a quadratic potential, i.e. for a *spin configuration* σ in $\{-1, +1\}^\Lambda$, its energy is given by

$$\mathcal{H}^{\Lambda}_{\beta,h}(w) = -\beta \sum_{\substack{i,j \in \Lambda \\ \{i,j\} \in E_{\mathbb{Z}^d}}} \sigma_i \sigma_j \; - h \sum_{i \in \Lambda} \sigma_i,$$

where $E_{\mathbb{Z}^d}$ is the set of edges of the lattice $\mathbb{Z}^d$ and h and β are two real parameters with $\beta > 0$. The probability $\mu_{\beta,h,\Lambda}[\sigma]$ of taking a configuration σ is then proportional to $\exp(-\mathcal{H}^{\Lambda}_{\beta,h}(\sigma))$.

We now want to make Λ grow to the whole lattice $\mathbb{Z}^d$. It is well known that for $h \neq 0$ or β smaller than a critical value $\beta_c(d)$ (thus, at high temperature), there is a unique limiting probability measure $\mu_{\beta,h}$ on spin configurations on $\mathbb{Z}^d$, see e.g. [23, Theorem 3.41]. In the following, we take parameters (β, h) in this domain of uniqueness and consider a random spin configuration σ, whose law is $\mu_{\beta,h}$.

In [17], Duneau et al. proved what they call the *strong cluster properties* for spins in the Ising model for $h \neq 0$ or sufficiently small β. Their result is actually more general (it holds for other models than the Ising model) but for simplicity, we stick to the Ising model here. Reformulated with the terminology of the present article, we have:

Theorem 15.44 (Duneau et al. [17]) *Fix the dimension $d \geq 1$ and $h \neq 0$, $\beta > 0$.*

1. *There exist $C = C(d, \beta, h)$ and $\varepsilon = \varepsilon(d, \beta, h) < 1$ such that under the probability measure $\mu_{\beta,h}$, the family $\{\sigma_i,\ i \in \mathbb{Z}^d\}$ has a C-uniform weighted dependency graph G, where the weight of the edge $\{i, j\}$ in G is $\varepsilon^{\|i-j\|_1}$.*
2. *The same holds for $h = 0$ and β is sufficiently small (i.e. $\beta < \beta_1(d)$, for some $\beta_1(d)$ depending on the dimension; this is sometimes referred to as the very high temperature regime).*

Note that the maximal weighted degree of this graph is a constant $C' < \infty$.

We now consider the magnetization in a finite box Δ defined as $M_\Delta = \sum_{i \in \Delta} \sigma_i$. We see M_Δ as a sequence of random variables (indexed by the countably many finite subsets of $\mathbb{Z}^d$). Restricting the uniform weighted dependency graph above to $\{\sigma_i,\ i \in \Delta\}$, each M_Δ is the sum of random variables with a C-uniform weighted dependency graph and maximal weighted degree at most C'. Therefore, using Proposition 15.40, we know that $M_{|\Delta|}$ admits uniform bounds on cumulants with parameters $(\frac{C'}{2}, |\Delta|, C)$. Moreover, since all spins are positively correlated by the FKG inequality (see [23, Section 3.6]), we have, using translation invariance

$$\mathrm{Var}(M_\Delta) \geq \sum_{i \in \Delta} \mathrm{Var}(\sigma_i) = \mathrm{Var}(\sigma_0)|\Delta|.$$

Note that $\mathrm{Var}(\sigma_0)$ is independent of Δ. With the notation of Sect. 15.4, this inequality ensures that $\widetilde{\sigma}_\Delta$ is bounded from below by a constant. Applying Corollary 15.29, we get:

Proposition 15.45 *Fix the dimension $d \geq 1$ and parameters $h \neq 0$, $\beta > 0$. The exists a constant $K = K(d, \beta, h)$ such that, for all subsets Δ of $\mathbb{Z}^d$, we have under $\mu_{\beta,h}$*

$$d_{\mathrm{Kol}}\left(\frac{M_\Delta - \mathbb{E}[M_\Delta]}{\sqrt{\mathrm{Var}(M_\Delta)}}\,,\ \mathcal{N}_{\mathbb{R}}(0,1)\right) \leq \frac{K}{\sqrt{|\Delta|}}.$$

The same holds for $h = 0$ and β sufficiently small (very high temperature).

Remark 15.46 In this remark, we discuss mod-Gaussian convergence in this setting. Consider a sequence Δ_n of subsets of $\mathbb{Z}^d$, tending to $\mathbb{Z}^d$ in the Van Hove sense (i.e. the sequence is increasing with union $\mathbb{Z}^d$, and the size of the boundary of Δ_n is asymptotic negligible, compared to the size of Δ_n itself). Then it is known from [18, Lemma 5.7.1] that

$$\lim_{n\to\infty} \tfrac{1}{|\Delta|}\,\mathrm{Var}(M_\Delta) = \sum_{k\in\mathbb{Z}^d} \mathrm{Cov}(\sigma_0, \sigma_k),$$

and the right-hand side has a finite value $\widetilde{\sigma}^2(\beta, h)$ for parameters (β, h) for which Theorem 15.44 applies. Similarly, we have

$$\lim_{n\to\infty} \tfrac{1}{|\Delta|}\,\kappa^{(3)}(M_\Delta) = \sum_{k,l\in\mathbb{Z}^d} \kappa(\sigma_0, \sigma_k, \sigma_l) < \infty.$$

We call $\rho(\beta, h)$ this quantity, and denote $L = \frac{\rho(\beta,h)}{\widetilde{\sigma}^3(\beta,h)}$. Let us then consider the rescaled variables

$$X_n = \frac{M_{\Delta_n} - \mathbb{E}[M_{\Delta_n}]}{(\mathrm{Var}(M_{\Delta_n}))^{1/3}}.$$

From [21, Section 5] (with $\alpha_n = \mathrm{Var}(M_\Delta)$ and $\beta_n = 1$), we know that X_n converges in the complex mod-Gaussian sense with parameters $t_n = (\mathrm{Var}(M_{\Delta_n}))^{1/3}$ and limiting function $\psi(z) = \exp(\frac{Lz^3}{6})$. This mod-Gaussian convergence takes place on the whole complex plane. Using the results of [21], this implies a normality zone for $(M_\Delta - \mathbb{E}[M_\Delta])/\sqrt{\mathrm{Var}(M_\Delta)}$ of size $o(|\Delta|^{1/6})$, see Proposition 4.4.1 in *loc. cit.*; and moderate deviation estimates at the edge of this normality zone, see Proposition 5.2.1.

Remark 15.47 For $h = 0$ and $\beta > \beta_c(d)$ (low temperature regime), there is no weighted dependency graph as above. Indeed, this would imply the analyticity of the partition function in terms of the magnetic field h, and the latter is known not to be analytic at $h = 0$ for $\beta > \beta_c(d)$; see [42, Chapter 6, §5] for details.

15.5.4 Functionals of Markov Chains

In this section, we consider a discrete time Markov chain $(M_t)_{t\geq 0}$ on a finite state space $\mathfrak{X}$, which is ergodic (irreducible and aperiodic) with invariant measure π. Its transition matrix is denoted P. To simplify the discussion, we shall also assume that the Markov chain is stationary, that is to say that the initial measure (i.e. the law of M_0) is π; most results have easy corollaries for any initial measure, using the fast mixing of such Markov chains.

Let us consider a sequence $(f_t)_{t\geq 0}$ of functions on $\mathfrak{X}$ that is uniformly bounded by a constant K. We set $Y_t = f_t(M_t)$. We will show that $\{Y_t\}_{t\in\mathbb{N}}$ admits a uniform weighted dependency graph. The proof roughly follows the one of [20, Section 10], where it was proved that it has a (non-uniform) weighted dependency graph, taking extra care of the dependence in the order r of the cumulant in the bounds. Instead of working directly with classical (joint) cumulants, we start by giving a bound for the so-called *Boolean cumulants*. Classical cumulants are then expressed in terms of Boolean cumulants thanks to a formula of Saulis and Statulevičius [52, Lemma 1.1]; see also a recent article of Arizmendi et al. [2] (we warn the reader that, in [52], Boolean cumulants are called centered moments).

Let $Z_1, \ldots, Z_r$ be random variables with finite moments defined on the same probability space. By definition, their Boolean (joint) cumulant is

$$B^{(r)}(Z_1, \ldots, Z_r)$$
$$= \sum_{l=0}^{r-1}(-1)^l \sum_{1\leq d_1<\cdots<d_l\leq r-1} \mathbb{E}[Z_1 \cdots Z_{d_1}]\,\mathbb{E}[Z_{d_1+1} \cdots Z_{d_2}] \cdots \mathbb{E}[Z_{d_l+1} \cdots Z_r].$$

While not at first sight, this definition is quite similar to the definition of classical (joint) cumulants, replacing the lattice of all set partitions by the lattice of interval set partitions; see [2, Section 2] for details. Note however that, unlike classical cumulants, Boolean cumulants are not symmetric functionals.

Proposition 15.48 *Let $r \geq 1$. With the above notation, there exists a constant θ_P depending on P with the following property. For any integers $t_1 \leq t_2 \leq \cdots \leq t_r$, we have*

$$\left|B^{(r)}(Y_{t_1}, \ldots, Y_{t_r})\right| \leq M^{\frac{r-1}{2}}\, K^r\, (\theta_P)^{t_r - t_1},$$

where $M = |\mathfrak{X}|$.

The proof of this bound relies on arguments due to Diaconis, Stroock and Fill, see [15, 22]. We also refer to [52, Section 4.1] for an alternate approach. Given an ergodic transition matrix P on $\mathfrak{X}$ with invariant measure π, we denote $\widetilde{P}$ the time reversal of P, which is the stochastic matrix defined by the equation

$$\widetilde{P}(x, y) = \frac{\pi(y)\, P(y, x)}{\pi(x)}.$$

This new transition matrix is again ergodic, with stationary measure π. The multiplicative reversiblization of P is the matrix $M(P) = P\widetilde{P}$. It is a stochastic matrix, which is ergodic with stationary measure π, and with all its eigenvalues that are real and belong to $[0, 1]$. Indeed, if D is the diagonal matrix $D = \mathrm{diag}(\pi)$, then $\widetilde{P} = D^{-1} P^t D$, and

$$\begin{aligned}\mathrm{Spec}(M(P)) &= \mathrm{Spec}\left(D^{1/2} P \widetilde{P} D^{-1/2}\right)\\ &= \mathrm{Spec}\left((D^{1/2} P D^{-1/2})(D^{-1/2} P^t D^{1/2})\right)\\ &= \mathrm{Spec}\left((D^{1/2} P D^{-1/2})(D^{1/2} P D^{-1/2})^t\right).\end{aligned}$$

Thus, $M(P)$ has the spectrum of a symmetric positive matrix, so it belongs to $\mathbb{R}_+$, and in fact to $[0, 1]$ since $M(P)$ is also stochastic. We denote

$$(\theta_P)^2 = \max\{|z| \mid z \text{ eigenvalue of } M(P),\ z \neq 1\}. \tag{15.9}$$

Notice that if P is reversible, then $\widetilde{P} = P$ and $M(P) = P^2$, so in this case

$$\theta_P = \max\{|z| \mid z \text{ eigenvalue of } P,\ z \neq 1\}.$$

In general, one can think of θ_P as the analogue of the second largest eigenvalue for non-reversible transition matrices. The following result estimates the rate of convergence of the Markov chain associated to P in terms of θ:

Theorem 15.49 (Fill [22]) *For any $x \in \mathfrak{X}$,*

$$\sum_{y \in \mathfrak{X}} |P^t(x, y) - \pi(y)| \leq \frac{(\theta_P)^t}{\sqrt{\pi(x)}};$$

$$\sum_{y \in \mathfrak{X}} \frac{|P^t(x, y) - \pi(y)|}{\sqrt{\pi(y)}} \leq \sqrt{M}\, \frac{(\theta_P)^t}{\sqrt{\pi(x)}}$$

where $M = |\mathfrak{X}|$.

Proof For completeness, we reproduce here the discussion of [22, Section 2], which relies on the following identity due to Mihail. If f is a function on $\mathfrak{X}$, we denote $\mathrm{Var}(f)$ its variance under the stationary probability measure π. We also introduce the Dirichlet form

$$\mathscr{E}(f, g) = \frac{1}{2} \sum_{x, y \in \mathfrak{X}} (f(x) - f(y))(g(x) - g(y))\, \pi(x)\, M(P)(x, y).$$

Then, for any function f,

$$\mathrm{Var}(f) = \mathrm{Var}(\widetilde{P} f) + \mathscr{E}(f, f).$$

Indeed, one can assume w.l.o.g. that $\pi(f) = \sum_{x\in\mathfrak{X}} \pi(x) f(x) = 0$. If $\langle f \mid g\rangle_\pi = \sum_{x\in\mathfrak{X}} \pi(x)\, f(x)\, g(x)$, then

$$\begin{aligned}\mathscr{E}(f, f) &= \langle f \mid (\mathrm{id} - M(P)) f\rangle_\pi = \langle f \mid f\rangle_\pi - \left\langle f \mid P\widetilde{P} f\right\rangle_\pi \\ &= \langle f \mid f\rangle_\pi - \left\langle \widetilde{P} f \mid \widetilde{P} f\right\rangle_\pi = \mathrm{Var}(f) - \mathrm{Var}(\widetilde{P} f)\end{aligned}$$

since $\widetilde{P}$ is the adjoint of P for the action on the left of functions and with respect to the scalar product $\langle \cdot \mid \cdot\rangle_\pi$. Consider now a Markov chain $(X_t)_{t\in\mathbb{N}}$ on $\mathfrak{X}$ with arbitrary initial distribution π_0, and denote $\pi_t = \pi_0 P^t$ the distribution at time t. We introduce the quantity

$$(\chi_t)^2 = \sum_{y\in\mathfrak{X}} \frac{(\pi_t(y) - \pi(y))^2}{\pi(y)}.$$

This is the variance of $f_t = \frac{\pi_t}{\pi}$ with respect to the probability measure π. By Mihail's identity,

$$(\chi_{t+1})^2 = \mathrm{Var}(f_{t+1}) = \mathrm{Var}(\widetilde{P} f_t) = \mathrm{Var}(f_t) - \mathscr{E}(f_t, f_t) = (\chi_t)^2 - \mathscr{E}(f_t, f_t).$$

By the minimax characterization of eigenvalues of symmetric positive matrices,

$$(\theta_P)^2 = 1 - \inf\left\{\frac{\mathscr{E}(f, f)}{\mathrm{Var}(f)},\ f \text{ non-constant}\right\}.$$

Therefore, $(\chi_{t+1})^2 \leq (\theta_P)^2\, (\chi_t)^2$, and $(\chi_t)^2 \leq (\theta_P)^{2t}\, (\chi_0)^2$ by induction on t. On the other hand, the Cauchy-Schwarz inequality yields

$$\sum_{y\in\mathfrak{X}} |\pi_t(y) - \pi(y)| \leq \sqrt{\sum_{y\in\mathfrak{X}} \pi(y)}\ \sqrt{\sum_{y\in\mathfrak{X}} \frac{(\pi_t(y) - \pi(y))^2}{\pi(y)}} = \chi_t.$$

If we choose $\pi_0 = \delta_x$, we finally obtain:

$$\sum_{y\in\mathfrak{X}} |P^t(x, y) - \pi(y)| \leq (\theta_P)^t\, \chi_0 = (\theta_P)^t \sqrt{\frac{1}{\pi(x)} - 1} \leq \frac{(\theta_P)^t}{\sqrt{\pi(x)}}.$$

Similarly,

$$\sum_{y\in\mathfrak{X}} \frac{|P^t(x,y)-\pi(y)|}{\sqrt{\pi(y)}} \leq \sqrt{M}\,\chi_t \leq \sqrt{M}\,(\theta_P)^t\,\chi_0 \leq \sqrt{M}\,\frac{(\theta_P)^t}{\sqrt{\pi(x)}}.$$

Proof (Proposition 15.48) If $f : \mathfrak{X} \to \mathbb{R}$, denote $D_f = \mathrm{diag}(f(x), x \in \mathfrak{X})$. Then, the Boolean cumulant has the following matrix expression:

$$B^{(r)}(Y_{t_1},\ldots,Y_{t_r}) = \pi\, D_{f_{t_1}}\,(P^{t_2-t_1} - \mathbf{1}\pi)\,\cdots\, D_{f_{t_{r-1}}}\,(P^{t_r-t_{r-1}} - \mathbf{1}\pi)\, D_{f_{t_r}}\,\mathbf{1},$$

where $\mathbf{1}$ is the column vector with all its entries equal to 1; see [20, Lemma 10.1]. If we expand this expression as a sum, and denote $Q_t = P^t - \mathbf{1}\pi$ and $\delta_i = t_{i+1} - t_i$, then

$$\begin{aligned}
&B^{(r)}(Y_{t_1},\ldots,Y_{t_r})\\
&\quad= \sum_{x_1,\ldots,x_r} \pi(x_1)\, f_{t_1}(x_1)\, Q_{t_2-t_1}(x_1,x_2)\, f_{t_2}(x_2)\,\cdots\, Q_{t_r-t_{r-1}}(x_{r-1},x_r)\, f_{t_r}(x_r)
\end{aligned}$$

and we obtain

$$\begin{aligned}
|B^{(r)}(Y_{t_1},\ldots,Y_{t_r})| &\leq K^r \sum_{x_1,\ldots,x_r} \pi(x_1)\,|Q_{\delta_1}(x_1,x_2)|\,\cdots\,|Q_{\delta_{r-1}}(x_{r-1},x_r)|\\
&\leq K^r\,(\theta_P)^{\delta_{r-1}} \sum_{x_1,\ldots,x_{r-1}} \pi(x_1)\,|Q_{\delta_1}(x_1,x_2)|\,\cdots\,\frac{|Q_{\delta_{r-2}}(x_{r-2},x_{r-1})|}{\sqrt{\pi(x_{r-1})}}\\
&\;\;\vdots\\
&\leq K^r\, M^{\frac{r-2}{2}}\,(\theta_P)^{t_r-t_1} \sum_{x_1}\sqrt{\pi(x_1)} \leq K^r M^{\frac{r-1}{2}}\,(\theta_P)^{t_r-t_1}.
\end{aligned}$$

Proposition 15.50 *The family of random variables* $\{Y_t\}_{t\in\mathbb{N}}$ *admits a* $K\sqrt{M}$*-uniform weighted dependency graph, where, for integers* $s < t$*, the weight between* Y_t *and* Y_s *is* $2(\theta_P)^{t-s}$*.*

Proof A lemma of Saulis and Statulevičius [52, Lemma 1.1] expresses usual cumulants in terms of Boolean cumulants:

$$\kappa^{(r)}(Y_{t_1},\ldots,Y_{t_r}) = \sum_{\pi\in\mathcal{P}([r])} (-1)^{|\pi|-1} N(\pi) \prod_{C\in\pi} B^{(|C|)}(Y_{t_j},\ j\in C). \tag{15.10}$$

Here, the sum runs over set-partitions π of $[r] := \{1,\ldots,r\}$; $|\pi|$ is the number of blocks of a set-partition π, the product runs over blocks C in π and $B^{(|C|)}(Y_{t_j},\ j\in C)$ is the Boolean cumulant of the subfamily (Y_{t_j}) indexed by integers j in the block C, with the times ordered in increasing order (recall that the Boolean are not symmetric functionals). Finally $N(\pi)$ is a combinatorial factor that can be computed as follows. For each block C of π, denote m_C and M_C its smallest and biggest elements; then call n_C the number of blocks $C' \neq C$ such that m_C is in the interval $[m_{C'}; M_{C'}]$. We finally define $N(\pi) = \prod_{C \in \pi; 1 \notin C} n_C$. In other terms, $N(\pi)$ counts the functions g mapping each block C of π (except the one containing 1) to a block $g(C) \neq C$ such that $m_C \in [m_{g(C)}; M_{g(C)}]$.

Let us make an observation. If π is a partition and k an integer such that each block of π either contains only numbers smaller than or equal to k or only numbers bigger than k (π is then said to be *disconnected*), then no function g as above exists (there is no possible image for the block C containing $k+1$) and $N(\pi) = 0$. On the other hand, for *connected* partitions π, we always have $N(\pi) > 0$, so that the sum in (15.10) is in effect a sum over connected partitions.

Equation (15.10) and Proposition 15.48 imply the bound

$$\left|\kappa^{(r)}(Y_{t_1}, \ldots, Y_{t_r})\right| \leq \left(K\sqrt{M}\right)^r \sum_{\pi \in \mathcal{P}([r])} N(\pi) \prod_{C \in \pi} (\theta_P)^{t_{M_C} - t_{m_C}}.$$

We would like to prove

$$\left|\kappa^{(r)}(Y_{t_1}, \ldots, Y_{t_r})\right| \leq 2^{r-1} \left(K\sqrt{M}\right)^r \sum_{T \in \mathrm{ST}(K_r)} w(T),$$

where $w(T) = \prod_{\{j,j'\} \in E_T,\, j<j'} (\theta_P)^{t_{j'} - t_j}$. Therefore it is sufficient for us to find an injective mapping η from pairs (π, g) as above to edge-bicolored spanning trees $\widetilde{T}$ such that

$$w\big(\eta(\pi, g)\big) = \prod_{C \in \pi} (\theta_P)^{t_{M_C} - t_{m_C}}; \tag{15.11}$$

here, by convention, the weight $w(\widetilde{T})$ of a colored tree is the weight $w(T)$ of its uncolored version. In the following, we describe such a mapping, concluding the proof of the proposition.

Let (π, g) be a pair of objects as above: π is a set-partition of $[r]$ and g is function mapping each block C of π (except the one containing 1) to a block $g(C) \neq C$ such that $m_C \in [m_{g(C)}; M_{g(C)}]$. For each block C of π, we consider the set

$$S(C) = C \cup \{m_{C'},\ C' \in g^{-1}(C)\}.$$

Let us call P_C the path with vertex-set $S(C)$, where the vertices are in increasing order along the path. We also color in blue (resp. in red) edges of this path whose extremity with smaller label is in C (resp. in $\{m_{C'},\ C' \in g^{-1}(C)\}$).

As an example, take $\pi = \{C_1, \cdots, C_6\}$ with $C_1 = \{1, 5, 10\}$, $C_2 = \{2, 11\}$, $C_3 = \{3, 9\}$, $C_4 = \{4, 6, 13\}$, $C_5 = \{7, 12\}$, $C_6 = \{8\}$. As function g, we take $g(C_2) = C_1$, $g(C_3) = C_1$, $g(C_4) = C_2$, $g(C_5) = C_1$ and $g(C_6) = C_4$. In this case,

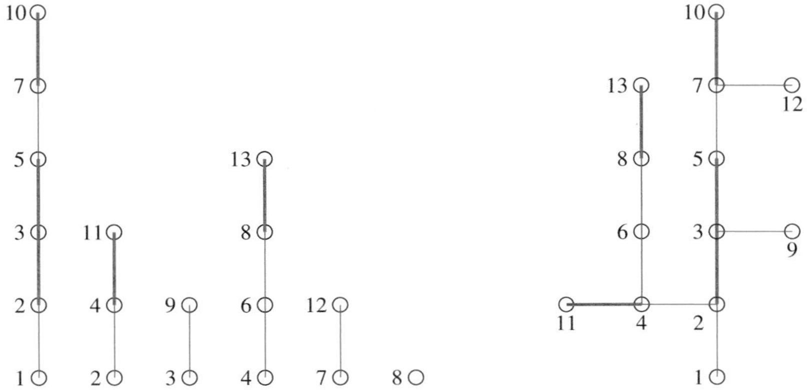

Fig. 15.11 Illustration of the construction η in the proof of Proposition 15.50: the path P_{C_i} and their gluing $\eta(\pi, g)$. For readers of a black-and-white printed version, red edges are thicker

we get $S(C_1) = \{1, 2, 3, 5, 7, 10\}$, $S(C_2) = \{2, 4, 11\}$, $S(C_4) = \{4, 6, 8, 13\}$ and $S(C_i) = C_i$ for $i \in \{3, 5, 6\}$. The associated bicolored paths are drawn in Fig. 15.11.

As in Fig. 15.11, we then take the union of the paths P_{C_i}, identifying vertices with the same label (the minimum $m_C \neq 1$ of a block C appears in the path $S(C)$ and in the path $S(g(C))$). Doing that, we get an edge-bicolored graph that we call $\eta(\pi, g)$. Let us first check that $\eta(\pi, g)$ is a tree. To this purpose, we order the blocks $C_1, \ldots, C_{|\pi|}$ of π in increasing order of their minima (as done in the example). Observe that this implies that $g(C_i) = C_j$ for some $j < i$. We will prove by induction that, for each $i \leq |\pi|$, $P_{C_1} \cup \cdots \cup P_{C_i}$ is a tree. The case $i = 1$ is trivial. For $i > 1$, the graph $P_{C_1} \cup \cdots \cup P_{C_i}$ is obtained by gluing the path P_{C_i} on the graph $P_{C_1} \cup \cdots \cup P_{C_{i-1}}$, identifying m_{C_i} which appears in both. Since $P_{C_1} \cup \cdots \cup P_{C_{i-1}}$ is a tree by induction hypothesis, the resulting graph $P_{C_1} \cup \cdots \cup P_{C_i}$ is a tree as well, concluding the induction. Thus $\eta(\pi, g) = P_{C_1} \cup \cdots \cup P_{C_{|\pi|}}$ is a tree.

The equality (15.11) is easy: since the edge set of $\eta(\pi, g)$ is the union of the edge sets of the P_{C_i}, we have

$$w\big(\eta(\pi, g)\big) = \prod_{i=1}^{|\pi|} w(P_{C_i}) = \prod_{i=1}^{|\pi|} (\theta_P)^{t_{M_{C_i}} - t_{m_{C_i}}}.$$

We finally need to prove that η is injective, i.e. that we can recover (π, g) from $\eta(\pi, g)$. We start by the following easy observation: in $\eta(\pi, g)$, vertices with a red incident edge going to a vertex with bigger label are exactly the vertices with a label which is the minimum $m_C \neq 1$ of a block C of π. By construction, such vertices have at most three incident edges, which are as follow:

(E1) as said above, a first one is red and goes to a vertex to bigger label;
(E2) a second one is either blue or red and goes to a vertex of lower label.

(E3) possibly, a last one is blue and goes to a vertex to bigger label (there is such an edge when m_C is not alone in its block);

Indeed, in the construction, edges (E1) and (E2) comes from $S_{g(C)}$ while edge (E3) comes from S_C. We split the vertex m_C into two, keeping edges (E1) and (E2) in the same component. Doing that for the $|\pi| - 1$ vertices $m_C \neq 1$, we inverse the gluing step of the construction of $\eta(\pi, g)$. It is then straightforward to recover (π, g). □

Theorem 15.51 *Let $(X_t)_{t\in\mathbb{N}}$ be an ergodic Markov chain on a finite state space $\mathfrak{X}$ of size M, and $\theta_P < 1$ the constant associated by (15.9) with the transition matrix P. We consider a sum $S_n = \sum_{t=1}^{n} f_t(X_t)$ with $\|f_t\|_\infty \leq K$. Then, for any $r \geq 1$,*

$$\left|\kappa^{(r)}(S_n)\right| \leq n\, r^{r-2} \left(2\,\frac{1+\theta_P}{1-\theta_P}\right)^{r-1} \left(K\sqrt{M}\right)^r. \tag{15.12}$$

As a consequence:

1. *When $\frac{\mathrm{Var}(S_n)}{n^{2/3}} \to +\infty$, we have*

$$d_{\mathrm{Kol}}\left(\frac{S_n - \mathbb{E}[S_n]}{\sqrt{\mathrm{Var}(S_n)}}, \mathcal{N}_{\mathbb{R}}(0,1)\right) \leq 76.36 \left(\frac{K\sqrt{M}}{\sqrt{\frac{\mathrm{Var}(S_n)}{n}}}\right)^3 \left(\frac{1+\theta_P}{1-\theta_P}\right)^2 \frac{1}{\sqrt{n}},$$

   and in particular, $\frac{S_n - \mathbb{E}[S_n]}{\sqrt{\mathrm{Var}(S_n)}}$ converges in law to a standard Gaussian.
2. *This convergence in law happens as soon as $\frac{\mathrm{Var}(S_n)}{n^{\varepsilon}} \to \infty$ for some $\varepsilon > 0$.*

Proof Combining Propositions 15.50 and 15.40, the sum S_n admits uniform bounds on cumulants with parameters

$$D_n = \left(1 + 2\sum_{t=1}^{\infty} (\theta_P)^t\right) = \frac{1+\theta_P}{1-\theta_P},$$

$N_n = n$ and $A = K\sqrt{M}$. Observe that D_n is here independent of n. We can apply Corollary 15.29 to get the first part of the theorem. The second follows from Theorem 15.31. □

Remark 15.52 A bound similar to Eq. (15.12) is given in [52, Theorem 4.19]. We believe however that the proof given there is not correct. Indeed, the proof associates with each partition π such that $N(\pi) \neq 0$ a sequence of number q_j; the authors then claim that "obviously $q_j \leq q_{j+1}$" (p. 93, after Eq. (4.53)). This is unfortunately not the case as can be seen on the example of partitions given p. 81 in *loc. cit.*: for this partition $q_3 = 3$, while $q_4 = 2$. As a consequence of this mistake, the authors forget many partitions π when expressing classical cumulants in terms of Boolean cumulants (since they encode partitions with only non-decreasing sequences q_i), which make the resulting bound on classical cumulants too sharp. We have not found a simpler way around this error than the use of uniform weighted dependency graphs presented here. Note nevertheless that our proof still uses several ingredients from [52]: the use of Boolean cumulants and the relation between Boolean and classical cumulants.

Remark 15.53 If the functions f_t are indicators $f_t(x) = 1_{x=s_t}$, then one can remove the size $(\sqrt{M})^3$ in the bound on the Kolmogorov distance. Indeed, in this case, we have

$$B^{(r)}(Y_{t_1}, \ldots, Y_{t_r}) = \pi(s_1)\, Q_{t_2-t_1}(s_1, s_2) Q_{t_3-t_2}(s_2, s_3) \cdots Q_{t_r-t_{r-1}}(s_{r-1}, s_r).$$

On the other hand, the individual terms of the matrix $Q_t(x, y)$ can be bounded by

$$|Q_t(x, y)| \leq \sqrt{\frac{\pi(y)}{\pi(x)}}\, (\theta_P)^t,$$

by adapting the proof of Theorem 15.49. Therefore,

$$\begin{aligned}\left|B^{(r)}(Y_{t_1}, \ldots, Y_{t_r})\right| &\leq (\theta_P)^{t_r-t_1}\, \pi(s_1) \sqrt{\frac{\pi(s_2)}{\pi(s_1)}} \cdots \sqrt{\frac{\pi(s_r)}{\pi(s_{r-1})}} \\ &\leq (\theta_P)^{t_r-t_1} \sqrt{\pi(s_1)\pi(s_r)} \leq (\theta_P)^{t_r-t_1}.\end{aligned}$$

Thus, in this case, one has the bound of Theorem 15.51 without the factor $(\sqrt{M})^3$.

15.5.5 The Case of Linear Functionals of the Empirical Measure

As a particular case of Theorem 15.51, one recovers the central limit theorem for linear functionals of empirical measures of Markov chains, that are random variables

$$Y_n = \frac{S_n - \mathbb{E}[S_n]}{\sqrt{n}} = \frac{1}{\sqrt{n}} \sum_{t=1}^{n} (f(X_t) - \pi(f))$$

with $f : \mathfrak{X} \to \mathbb{R}$ fixed function (independent of the time t). Thus, assuming for instance $\lim_{n\to\infty} \mathrm{Var}(Y_n) = \Sigma^2(f) > 0$, we have

$$d_{\mathrm{Kol}}\left(\frac{S_n - \mathbb{E}[S_n]}{\sqrt{\mathrm{Var}(S_n)}}, \mathcal{N}_{\mathbb{R}}(0,1)\right) \leq 77 \left(\frac{\|f\|_\infty \sqrt{M}}{\Sigma(f)}\right)^3 \left(\frac{1+\theta_P}{1-\theta_P}\right)^2 \frac{1}{\sqrt{n}} \tag{15.13}$$

for n large enough. We refer to [13, 26, 29, 33] and the references therein for the general background of this Markovian CLT, and to [8, 40, 43] for estimates of the Kolmogorov distance. It seems that we recover some results of [43] (see [49, §2.1.3]), but we were not able to find and read this paper. In this last paragraph, we discuss the problem of the variance $\mathrm{Var}(Y_n)$, giving sufficient conditions, which are simple to check on examples and ensure $\Sigma^2(f) > 0$. We also remark that, provided that $\Sigma^2(f) > 0$, we can also prove complex mod-Gaussian convergence, which implies a zone of normality result and moderate deviation estimates by [21].

Denote $g = f - \pi(f)$, which has expectation 0 under the stationary measure π. By eventually replacing f with g, we can assume that f is centered. The variance of Y_n tends to

$$\Sigma^2(f) = \mathbb{E}[(f(X_0))^2] + 2\sum_{t=1}^{\infty} \mathbb{E}[f(X_0)\, f(X_t)] < +\infty,$$

see [13, Lemma 3.3]. If $\Sigma^2(f) > 0$, then $\frac{\mathrm{Var}(S_n)}{n^{2/3}} = n^{1/3}\,\mathrm{Var}(Y_n) \to +\infty$ and Theorem 15.51 applies. Unfortunately, one can easily construct non-trivial examples with $\Sigma^2(f) = 0$. Thus, consider the Markov chain with three states and transition matrix

$$P = \begin{pmatrix} 0 & 1 & 0 \\ 1/2 & 0 & 1/2 \\ 1 & 0 & 0 \end{pmatrix};$$

it admits for invariant measure $\pi(1) = \pi(2) = \frac{2}{5}$ and $\pi(3) = \frac{1}{5}$. Set $f(1) = 1$, $f(2) = -1$ and $f(3) = 0$; one has $\pi(f) = 0$, and one computes

$$\mathbb{E}[f(X_0)f(X_k)] = \frac{1}{5}\left(\frac{2+\mathrm{i}}{(-1-\mathrm{i})^k} + \frac{2-\mathrm{i}}{(-1+\mathrm{i})^k}\right).$$

It follows that $\Sigma^2(f) = 0$, although f is non zero.

In the general case of an ergodic Markov chain, fix an element a of the state space $\mathfrak{X}$, and denote $\tau_a \geq 1$ the first hitting time of a by the Markov chain, which is almost surely finite and with expectation $1/\pi(a)$ when starting from a. Then, the asymptotic variance $\Sigma^2(f)$ can be rewritten as

$$\Sigma^2(f) = \pi(a)\; \mathbb{E}_a\left[\left(\sum_{k=1}^{\tau_a} g(X_k)\right)^2\right],$$

see [32, Chapter 4]. Therefore, a general condition in order to obtain the bound of Eq. (15.13) is:

Proposition 15.54 *We have $\Sigma^2(f) > 0$ if and only if there exists a cycle $(x_1, \ldots, x_n)$ in the graph of transitions of the Markov chain such that the sum $\sum_{i=1}^{n} g(x_i)$ of the values of $g = f - \pi(f)$ along this cycle is non-zero. In this case, the bound* (15.13) *holds.*

The proposition explains readily why the irreducible aperiodic Markov chain

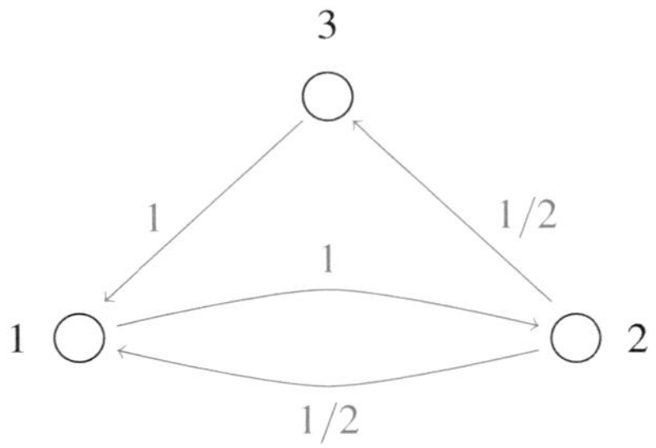

previously studied gives asymptotic variance 0 to the function $f(1) = 1$, $f(2) = -1$ and $f(3) = 0$: the minimal cycles of the chains are $(1, 2)$ and $(1, 2, 3)$, and the sum of the values of f along these cycles is always 0.

Another simple criterion to apply Theorem 15.51 to linear functionals of the empirical measure is:

Proposition 15.55 *Suppose that the ergodic Markov chain is reversible:*

$$\pi(x) P(x, y) = \pi(y) P(y, x)$$

for any x, y. Then, if f is a non-constant function, $\Sigma^2(f) > 0$ and the bound (15.13) *holds.*

Proof To say that P is reversible is equivalent to the fact that P is a symmetric operator of the Hilbert space $\mathrm{L}^2(\frac{1}{\pi})$. In particular, P has only real eigenvalues. Besides, the restriction of the operator $I + 2\sum_{k=1}^{\infty} P^k$ to the space of functions f with $\pi(f) = 0$ is well defined, and it is an auto-adjoint operator on this space with eigenvalues

$$\frac{1+\lambda_2}{1-\lambda_2}, \ldots, \frac{1+\lambda_M}{1-\lambda_M},$$

where $\lambda_2 \geq \lambda_3 \geq \cdots \geq \lambda_M$ are the real eigenvalues of P different from 1. The quantities above are all positive, and larger than $\frac{1-\theta_P}{1+\theta_P}$ (this value being obtained if $\lambda_M = -\theta_P$). Therefore,

$$\Sigma^2(f) = \left\langle f \,\middle|\, \left(I + 2\sum_{k=1}^{\infty} P^k\right) f \right\rangle_{\mathrm{L}^2(\pi)} \geq \frac{1-\theta_P}{1+\theta_P}\, \pi(f^2) > 0.$$

We then obtain

$$d_{\mathrm{Kol}}\left(\frac{S_n-\mathbb{E}[S_n]}{\sqrt{\mathrm{Var}(S_n)}},\,\mathcal{N}_{\mathbb{R}}(0,1)\right)\leq 77\left(\frac{\|g\|_\infty\sqrt{M}}{\sqrt{\pi(g^2)}}\right)^3\left(\frac{1+\theta_P}{1-\theta_P}\right)^{\frac{7}{2}}\frac{1}{\sqrt{n}}$$

for n large enough. □

Remark 15.56 In this remark, we discuss mod-Gaussian convergence for linear statistics of Markov chains. We use the above notation and assume that $\Sigma^2(f)>0$. Consider the third cumulant of S_n. One can easily prove that

$$\rho=\frac{1}{n}\lim_{n\to\infty}\kappa_3(S_n)=\sum_{j,k\in\mathbb{Z}}\kappa(f(X_0,X_j,X_k)),$$

where $(X_t)_{t\in\mathbb{Z}}$ is a bi-infinite stationary Markov chain with transition matrix P. (The sum on the right-hand side is finite, as consequence of Proposition 15.50). Let us call ρ this limit. We then consider the rescaled random variables

$$X_n=\left(\frac{S_n-\mathbb{E}[S_n]}{(\mathrm{Var}(S_n))^{1/3}}\right)_{n\in\mathbb{N}}.$$

As for the magnetization in the Ising model, from [21, Section 5] (with $\alpha_n=\mathrm{Var}(S_n)$ and $\beta_n=1$), we know that X_n converges in the mod-Gaussian sense with parameters $t_n=(\mathrm{Var}(S_n))^{1/3}$ and limit $\psi(z)=\exp(\frac{Lz^3}{6})$, with $L=\frac{\rho}{\Sigma^3(f)}$. Again, this mod-Gaussian convergence takes place on the whole complex plane and implies a normality zone for $S_n/\sqrt{\mathrm{Var}(S_n)}$ of size $o(n^{1/6})$, see Proposition 4.4.1 in *loc. cit.*; we also have moderate deviation estimates at the edge of this normality zone, see Proposition 5.2.1. This mod-Gaussian convergence could also have been proved by using an argument of the perturbation theory of linear operators, for which we refer to [30]. Indeed, the Laplace transform of X_n writes explicitly as

$$\mathbb{E}[\mathrm{e}^{zX_n}]=\pi\,(P_{z,f})^n\,1,$$

where 1 is the column vector with all its entries equal to 1, π is the stationary measure of the process, and $P_{z,f}$ is the infinitesimal modification of the transition matrix defined by

$$P_{z,f}(i,j)=P(i,j)\,\mathrm{e}^{\frac{z(f(j)-\pi(f))}{(\mathrm{Var}(S_n))^{1/3}}}.$$

For z in a zone of control of size $O(n^{1/3})$, one can give a series expansion of the eigenvalues and eigenvectors of $P_{z,f}$, which allows one to recover the mod-Gaussian convergence. The theory of cumulants and weighted dependency graphs allows one to bypass these difficult analytic arguments.

References

1. N. Alon, J. Spencer, *The Probabilistic Method.* Wiley-Interscience Series in Discrete Mathematics and Optimization, 3rd edn. (Wiley, New York, 2008)
2. O. Arizmendi, T. Hasebe, F. Lehner, C. Vargas, Relations between cumulants in noncommutative probability. Adv. Math. **282**, 56–92 (2015)
3. P. Baldi, Y. Rinott, On normal approximations of distributions in terms of dependency graphs. Ann. Probab. **17**(4), 1646–1650 (1989)
4. A.D. Barbour, M. Karoński, A. Ruciński, A central limit theorem for decomposable random variables with applications to random graphs. J. Combin. Theory B **47**(2), 125–145 (1989)
5. A.D. Barbour, E. Kowalski, A. Nikeghbali, Mod-discrete expansions. Probab. Theory Relat. Fields **158**(3), 859–893 (2009)
6. A.C. Berry, The accuracy of the Gaussian approximation to the sum of independent variates. Trans. Amer. Math. Soc. **49**(1), 122–136 (1941)
7. P. Billingsley, *Probability and Measure.* Wiley Series in Probability and Mathematical Statistics, 3rd edn. (Wiley, New York, 1995)
8. E. Bolthausen, The Berry–Esseen theorem for functionals of discrete Markov chains. Z. Wahr. Verw. Geb. **54**, 59–73 (1980)
9. M. Bóna, On three different notions of monotone subsequences, in *Permutation Patterns.* London Mathematical Society. Lecture Note Series, vol. 376 (Cambridge University Press, Cambridge, 2010), pp. 89–113
10. P. Bourgade, C. Hughes, A. Nikeghbali, M. Yor, The characteristic polynomial of a random unitary matrix: a probabilistic approach. Duke Math. J. **145**, 45–69 (2008)
11. A.V. Bulinskii, Rate of convergence in the central limit theorem for fields of associated random variables. Theory Probab. Appl. **40**(1), 136–144 (1996)
12. R. Chhaibi, F. Delbaen, P.-L. Méliot, A. Nikeghbali, Mod-ϕ convergence: approximation of discrete measures and harmonic analysis on the torus. arXiv:1511.03922 (2015)
13. R. Cogburn, The central limit theorem for Markov processes, in *Proceedings of the Sixth Annual Berkeley Symposium on Mathematical Statistics and Probability*, vol. 2 (University of California Press, Berkeley, 1972), pp. 485–512
14. F. Delbaen, E. Kowalski, A. Nikeghbali, Mod-ϕ convergence. Int. Math. Res. Not. **2015**(11), 3445–3485 (2015)
15. P. Diaconis, D. Stroock, Geometric bounds for eigenvalues of Markov chains. Ann. Appl. Probab. **1**(1), 36–61 (1991)
16. M. Duneau, D. Iagolnitzer, B. Souillard, Decrease properties of truncated correlation functions and analyticity properties for classical lattices and continuous systems. Commun. Math. Phys. **31**(3), 191–208 (1973)
17. M. Duneau, D. Iagolnitzer, B. Souillard, Strong cluster properties for classical systems with finite range interaction. Commun. Math. Phys. **35**, 307–320 (1974)
18. R. Ellis, *Entropy, Large Deviations, and Statistical Mechanics* (Springer, New York, 1985)
19. W. Feller, *An Introduction to Probability Theory and Its Applications.* Wiley Series in Probability and Mathematical Statistics, vol. II, 2nd edm. (Wiley, New York, 1971)
20. V. Féray, Weighted dependency graphs. arXiv preprint 1605.03836 (2016)
21. V. Féray, P.-L. Méliot, A. Nikeghbali, *Mod-ϕ convergence: Normality Zones and Precise Deviations.* Springer Briefs in Probability and Mathematical Statistics (Springer, Cham, 2016)
22. J.A. Fill, Eigenvalue bounds on convergence to stationarity for nonreversible Markov chains, with an application to the exclusion process. Ann. Appl. Probab. **1**(1), 62–87 (1991)
23. S. Friedli, Y. Velenik, *Statistical Mechanics of Lattice Systems: A Concrete Mathematical Introduction* (Cambridge University Press, Cambridge, 2017)
24. L. Goldstein, N. Wiroonsri, Stein's method for positively associated random variables with applications to Ising, percolation and voter models. Ann. Inst. Henri Poincaré **54**, 385–421 (2018). Preprint arXiv:1603.05322

25. G. Grimmett, Weak convergence using higher-order cumulants. J. Theor. Probab. **5**(4), 767–773 (1992)
26. O. Häggström, On the central limit theorem for geometrically ergodic Markov chains. Probab. Theory Relat. Fields **132**, 74–82 (2005)
27. J. Jacod, E. Kowalski, A. Nikeghbali, Mod-Gaussian convergence: new limit theorems in probability and number theory. Forum Math. **23**(4), 835–873 (2011)
28. S. Janson, Normal convergence by higher semi-invariants with applications to sums of dependent random variables and random graphs. Ann. Probab. **16**, 305–312 (1988)
29. G.L. Jones, On the Markov chain central limit theorem. Probab. Surv. **1**, 299–320 (2004)
30. T. Kato, *Perturbation Theory for Linear Operators*. Classics in Mathematics, 2nd edn. (Springer, Berlin, 1980)
31. J.P. Keating, N.C. Snaith, Random matrix theory and $\zeta(\frac{1}{2}+it)$. Commun. Math. Phys. **214**(1), 57–89 (2000)
32. J.G. Kemeny, J.L. Snell, *Finite Markov Chains*. Undergraduate Texts in Mathematics (Springer, Berlin, 1976)
33. C. Kipnis, S.R.S. Varadhan, Central limit theorem for additive functionals of reversible Markov processes and applications to simple exclusions. Commun. Math. Phys. **104**, 1–19 (1986)
34. V.Y. Korolev, I.G. Shevtsova, On the upper bound for the absolute constant in the Berry–Esseen inequality. Theory Probab. Appl. **54**(4), 638–658 (2010)
35. E. Kowalski, A. Nikeghbali, Mod-Poisson convergence in probability and number theory. Int. Math. Res. Not. **18**, 3549–3587 (2010)
36. E. Kowalski, A. Nikeghbali, Mod-Gaussian distribution and the value distribution of $\zeta(\frac{1}{2}+\mathrm{i}t)$ and related quantities. J. Lond. Math. Soc. **86**(2), 291–319 (2012)
37. K. Krokowski, A. Reichenbachs, C. Thäle, Discrete Malliavin–Stein method: Berry–Esseen bounds for random graphs, point processes and percolation. arXiv:1503.01029v1 [math.PR] (2015)
38. S. Lang, *Real and Functional Analysis*. Graduate Texts in Mathematics, 3rd edn., vol. 142 (Springer, New York, 1993)
39. V.P. Leonov, A.N. Shiryaev, On a method of calculation of semi-invariants. Theory Prob. Appl. **4**, 319–329 (1959)
40. P. Lezaud, Chernoff-type bound for finite Markov chains, PhD thesis, Université Paul Sabatier, Toulouse, 1996
41. P. Malliavin, *Integration and Probability*. Graduate Texts in Mathematics, vol. 157 (Springer, New York, 1995)
42. V.A. Malyshev, R.A. Minlos, *Gibbs Random Fields* (Springer, Dordrecht, 1991)
43. B. Mann, Berry–Esseen central limit theorems for Markov chains, PhD thesis, Harvard University, 1996
44. P.-L. Méliot, A. Nikeghbali, Mod-Gaussian convergence and its applications for models of statistical mechanics, in *In Memoriam Marc Yor – Séminaire de Probabilités XLVII*. Lecture Notes in Mathematics, vol. 2137 (Springer, Cham, 2015), pp. 369–425
45. C.M. Newman, Normal fluctuations and the FKG inequalities. Commun. Math. Phys. **74**(2), 119–128 (1980)
46. M. Penrose, J. Yukich, Normal approximation in geometric probability, in *Stein's Method and Applications*. Lecture Note Series, vol. 5 (Institute for Mathematical Sciences, National University of Singapore, Singapore, 2005), pp. 37–58
47. Y. Rinott, On normal approximation rates for certain sums of dependent random variables. J. Comput. Appl. Math. **55**(2), 135–143 (1994)
48. W. Rudin, *Functional Analysis*, 2nd edn. (McGraw-Hill, New York, 1991)
49. L. Saloff-Coste, Lectures on finite Markov chains, in *Lectures on Probability Theory and Statistics (Saint-Flour, 1996)*. Lecture Notes in Mathematics, vol. 1665 (Springer, Berlin, 1997), pp. 301–413
50. K.-I. Sato, *Lévy Processes and Infinitely Divisible Distributions*. Cambridge Studies in Advanced Mathematics, vol. 68 (Cambridge University Press, Cambridge, 1999)

51. K.-I. Sato, M. Yamazato, Stationary processes of ornstein-uhlenbeck type, in *Probability Theory and Mathematical Statistics, Fourth USSR-Japan Symposium*, ed. by K. Itô, J.V. Prokhorov. Lecture Notes in Mathematics, vol. 1021 (Springer, Berlin, 1983), pp. 541–551
52. L. Saulis, V.A. Statulevičius, *Limit Theorems for Large Deviations*. Mathematics and Its Applications (Soviet Series), vol. 73 (Kluwer Academic Publishers, Dordrecht, 1991)
53. F. Spitzer, Some theorems concerning 2-dimensional Brownian motion. Trans. Amer. Math. Soc. **87**, 187–197 (1958)
54. V.A. Statulevičius, On large deviations. Probab. Theory Relat. Fields **6**(2), 133–144 (1966)
55. D.W. Stroock, *Probability Theory: An Analytic View*, 2nd edn. (Cambridge University Press, Cambridge, 2011)
56. T. Tao, *Topics in Random Matrix Theory*. Graduate Studies in Mathematics, vol. 132 (American Mathematical Society, Providence, 2012)

Chapter 16
Random Flows Defined by Markov Loops

Yves Le Jan

Abstract We review some properties of the random networks defined by Markov loop ensembles and compute the distribution of random flows associated to them.

Keywords Free field · Markov loops · Networks · Flows

16.1 Introduction

We first present briefly the framework of our study, described in [2]. Consider a graph $\mathcal{G}$, i.e. a set of vertices X together with a set of non oriented edges E. We assume it is connected, and that there is no loop-edges nor multiple edges. The set of oriented edges, denoted E^o, is viewed as a subset of X^2. An oriented edge (x, y) is defined by the choice of an ordering in an edge $\{x, y\}$.

Given a graph $\mathcal{G} = (X, E)$, a set of non negative conductances $C_{x,y} = C_{y,x}$ indexed by the set of edges E and a non negative killing measure κ on the set of vertices X, we can associate to them an energy (or Dirichlet form) $\mathcal{E}$, which we will assume to be positive definite, a transience assumption. For any function f on X, we have:

$$\mathcal{E}(f, f) = \frac{1}{2} \sum_{x,y} C_{x,y}(f(x) - f(y))^2 + \sum_x \kappa_x f(x)^2.$$

There is a duality measure λ defined by $\lambda_x = \sum_y C_{x,y} + \kappa_x$. Let $G^{x,y}$ denote the symmetric Green's function associated with $\mathcal{E}$. Its inverse equals $M_\lambda - C$ with M_λ denoting the diagonal matrix defined by λ.

Y. Le Jan (✉)
NYU Shanghai, Shanghai, China

Département de Mathématique, Université Paris-Sud, Orsay, France
e-mail: yl57@nyu.edu; yves.lejan@math.u-psud.fr

C. Donati-Martin et al. (eds.), *Séminaire de Probabilités L*, Lecture Notes in Mathematics 2252, https://doi.org/10.1007/978-3-030-28535-7_16

The associated symmetric continuous time Markov process can be obtained from the Markov chain defined by the transition matrix $P^x_y = \frac{C_{x,y}}{\lambda_y}$ by adding independent exponential holding times of mean 1 before each jump. If P is submarkovian, the chain is absorbed at a cemetery point Δ. If X is finite, the transition matrix is necessarily submarkovian.

We denote by μ the loop measure associated with this symmetric Markov process. (see [2]).

The Poissonian loop ensemble $\mathcal{L}_\alpha$ is the Poisson process of loops of intensity $\alpha\mu$, constructed in such a way that the set of loops $\mathcal{L}_\alpha$ increases with α. We denote by $\hat{\mathcal{L}}_\alpha$ the occupation field associated with $\mathcal{L}_\alpha$ i.e. the total time spent in x by the loops of $\mathcal{L}_\alpha$, normalized by λ_x.

The complex (respectively real) free field is the complex (real) Gaussian field on X whose covariance function is $2G$ (G). We will denote it by φ (respectively $\varphi^{\mathbb{R}}$).

It has been shown in [6] (see also [2]) that the fields $\hat{\mathcal{L}}_1$ and $\frac{1}{2}\varphi^2$ have the same distribution. The same property holds for $\hat{\mathcal{L}}_{\frac{1}{2}}$ and $\frac{1}{2}(\varphi^{\mathbb{R}})^2$.

Given any oriented edge (x, y) of the graph, we denote by $N^{(\alpha)}_{x,y}$ the total number of jumps made from x to y by the loops of $\mathcal{L}_\alpha$ and for the non oriented edges, we set $N^{(\alpha)}_{\{x,y\}} = N^{(\alpha}_{x,y} + N^{(\alpha}_{y,x}$. Recall (see sections 2–3 and 2–4 in [2]) that $E(\hat{\mathcal{L}}_\alpha) = \alpha G^{x,x}$ and $E(N^{(\alpha)}_{x,y}) = \alpha C_{x,y} G^{x,y}$.

Recall finally the relation between φ and the pair of fields $(\hat{\mathcal{L}}_1, (N^{(1)})$ given in Remark 12 of [2], Chapter 6, and the relation between $\varphi^{\mathbb{R}}$ and the pair of fields $(\hat{\mathcal{L}}_1, (N^{(\frac{1}{2})}_{\{\}})$ given in Remark 11:

For any complex matrix $s_{x,y}$ with finitely many non zero entries, all of modulus less or equal to 1, and any finitely supported positive measure χ,

$$E(\prod_{x,y} s_{x,y}^{N^{(1)}_{x,y}} \prod_x e^{-\sum_x \chi_x \hat{\mathcal{L}}^x_1}) = E(e^{\frac{1}{2}\sum_{x,y} C_{x,y}(s_{x,y}-1)\varphi_x\bar{\varphi}_y - \frac{1}{2}\sum \chi_x \phi_x \bar{\phi}_x}). \tag{16.1}$$

For any real symmetric matrix $s_{x,y}$ with finitely many non zero entries, all in $[0, 1)$, and any finitely supported positive measure χ,

$$E(\prod_{x,y} s_{x,y}^{N^{(\frac{1}{2})}_{x,y}} \prod_x e^{-\sum_x \chi_x \hat{\mathcal{L}}^x_{\frac{1}{2}}}) = E(e^{\frac{1}{2}\sum_{x,y} C_{x,y}(s_{x,y}-1)\varphi^{\mathbb{R}}_x \varphi^{\mathbb{R}}_y - \frac{1}{2}\sum \chi_x (\varphi^{\mathbb{R}}_x)^2}). \tag{16.2}$$

Remarks

– A consequence of (16.1) is that for any set (x_i, y_i) of distinct oriented edges, and any set z_l of distinct vertices,

$$E(\prod_i N^{(1)}_{(x_i,y_i)} \prod_l (N^{(1)}_{z_l} + 1)) = E(\prod_i \frac{1}{2} C_{(x_i,y_i)} \varphi_{x_i} \bar{\varphi}_{y_i} \prod_l \frac{1}{2} \lambda_{z_l} \varphi_{z_l} \bar{\varphi}_{z_l}). \tag{16.3}$$

- In particular, if X is assumed to be finite, if $[D_{N^{(1)}}]_{(x,y)} = 0$ for $x \neq y$ and $[D_{N^{(1)}}]_{(x,x)} = 1 + N_x^{(1)}$, for all $\chi \geqslant \lambda$, then (cf. [3], with a minor correction here), denoting by $\mathrm{Per}(G)$ the permanent of G:

$$\begin{aligned} E(\det(M_\chi D_{N^{(1)}} - N^{(1)})) &= 2^{-|X|} E(\det(M_\varphi (M_\chi - C) M_{\bar{\varphi}})) \\ &= \det(M_\chi - C) E(\prod_{x_1,x_2} \varphi_{x_1} \bar{\varphi}_{x_2}) \\ &= \det(M_\chi - C)\mathrm{Per}(G). \end{aligned}$$

- A consequence of (16.2) is that for any set $\{x_i, y_i\}$ of distinct edges, and any set z_l of distinct vertices,

$$E(\prod_i N^{(\frac{1}{2})}_{\{x_i,y_i\}} \prod_l (N^{(\frac{1}{2})}_{z_l} + 1)) = E(\prod_i C_{(x_i,y_i)} \varphi^{\mathbb{R}}_{x_i} \varphi^{\mathbb{R}}_{y_i} \prod_l \frac{1}{2} \lambda_{z_l} (\varphi^{\mathbb{R}}_{z_l})^2). \tag{16.4}$$

- In particular, if X is assumed to be finite, if $[D_{N^{(\frac{1}{2})}}]_{(x,y)} = 0$ for $x \neq y$ and $[D_{N^{(\frac{1}{2})}}]_{(x,x)} = 1 + N_x^{(\frac{1}{2})}$, for all $\chi \geqslant \lambda$, then

$$E(\det(2M_\chi D_{N^{(\frac{1}{2})}} - N^{(\frac{1}{2})}_{\{\}})) = E(\det(M_{\varphi^{\mathbb{R}}}(M_\chi - C)M_{\varphi^{\mathbb{R}}})) = \det(M_\chi - C)\mathrm{Per}(G).$$

  Similar expressions can be given for the minors.
- Using (for example) Eqs. (16.3) and (16.4), we can study correlation decays. For example, if (x, y) and (u, v) are non adjacent edges,

$$Cov(N^{(1)}_{x,y}, N^{(1)}_{u,v}) = E(:\varphi_x \bar{\varphi}_y :: \varphi_u \bar{\varphi}_v :) = G^{x,v} G^{y,u}$$

- Note that a natural coupling of the free field with the loop ensemble of intensity $\frac{1}{2}\mu$ has been given by Lupu [7], using the vertex occupation field and the partition of X defined by the zeros of the edge occupation field.
- Most results are proved for finite graphs with non zero killing measure. Their extension to infinite transient graphs can be done by considering the restriction of the energy to functions vanishing outside a finite set D, i.e. the Markov chain killed at the exit of D and letting D increase to X, so that the Green function of the restriction, denoted G_D, increases to G.

16.2 Networks

We define a network to be a $\mathbb{N}$-valued function defined on oriented edges of the graph. It is given by a matrix k with $\mathbb{N}$-valued coefficients which vanish on the diagonal and on entries (x, y) such that $\{x, y\}$ is not an edge of the graph. We say that k is Eulerian if

$$\sum_y k_{x,y} = \sum_y k_{y,x}.$$

For any Eulerian network k, we define k_x to be $\sum_y k_{x,y} = \sum_y k_{y,x}$. It is obvious that the field $N^{(\alpha)}$ defines a random network which verifies the Eulerian property.

Note that $\sum_y N^{(\alpha)}_{\{x,y\}}$ is always even. We call even networks the sets of numbers attached to non oriented edges such that $k_x = \frac{1}{2}\sum_y k_{\{x,y\}}$ is an integer.

The cases $\alpha = 1$ and $\alpha = \frac{1}{2}$ are of special interest. We need to recall the results of [3] and [4] which can be deduced from Eqs. (16.1) and (16.2) (with a minor correction here in (ii)):

Theorem 16.1

(i) For any Eulerian network k,

$$P(N^{(1)} = k) = \det(I - P)\frac{\prod_x k_x!}{\prod_{x,y} k_{x,y}!}\prod_{x,y} P_{x,y}^{k_{x,y}}.$$

(ii) For any Eulerian network k, and any nonnegative function ρ on X

$$\begin{aligned} &P(N^{(1)} = k\ ,\ \hat{\mathscr{L}}_1 \in (\rho, \rho + d\rho)) \\ &= \det(I - P)\prod_{x,y}\frac{(\sqrt{\rho_x}C_{x,y}\sqrt{\rho_y})^{k_{x,y}}}{k_{x,y}!}\prod_x \lambda_x e^{-\lambda_x\rho_x}d\rho_x \\ &= \frac{1}{det(G)}\prod_{x,y}\frac{(\sqrt{\rho_x}C_{x,y}\sqrt{\rho_y})^{k_{x,y}}}{k_{x,y}!}\prod_x e^{-\lambda_x\rho_x}d\rho_x. \end{aligned}$$

(iii) For any even network k,

$$P(N^{(\frac{1}{2})}_{\{\}} = k) = \sqrt{\det(I - P)}\frac{\prod_x 2k_x!}{\prod_x 2^{k_x}k_x!\prod_{\{x,y\}} k_{\{x,y\}}!}\prod_{x,y} P_{x,y}^{k_{x,y}}.$$

(iv) For any even network k, and any nonnegative function ρ on X

$$P(N_{\{\}}^{(\frac{1}{2})}=k\ ,\ \hat{\mathscr{L}}_{\frac{1}{2}}\in(\rho,\rho+d\rho))$$
$$=\sqrt{\det(I-P)}\prod_{\{x,y\}}\frac{(\sqrt{\rho_x}C_{x,y}\sqrt{\rho_y})^{k_{\{x,y\}}}}{k_{\{x,y\}}!}\prod_x\frac{\sqrt{\lambda_x}}{\sqrt{2\pi\rho_x}}e^{-\lambda_x\rho_x}d\rho_x$$
$$=\frac{1}{\sqrt{\det(G)}}\prod_{\{x,y\}}\frac{(\sqrt{\rho_x}C_{x,y}\sqrt{\rho_y})^{k_{\{x,y\}}}}{k_{\{x,y\}}!}\prod_x\frac{1}{\sqrt{2\pi\rho_x}}e^{-\lambda_x\rho_x}d\rho_x.$$

Let us give the proof of (iv) which was not detailed in [4].

Denote by $\mathfrak{V}$ the set of even networks. Note that on one hand, for any symmetric matrix S

$$E(\prod_{\{x,y\}}S_{x,y}^{N_{\{x,y\}}^{(\frac{1}{2})}}e^{-\sum_x\chi_x\hat{\mathscr{L}}_{\frac{1}{2}}^x})=\sum_{k\in\mathfrak{V}}E(e^{-\sum_x\chi_x\hat{\mathscr{L}}_{\frac{1}{2}}^x}1_{N_{\{\}}^{(\frac{1}{2})}=k})\prod_{\{x,y\}}S_{x,y}^{k_{\{x,y\}}}.$$

On the other hand, from the previous lemma:

$$E(\prod_{\{x,y\}}S_{x,y}^{N_{\{x,y\}}^{(\frac{1}{2})}-\sum_x\chi_x\hat{\mathscr{L}}_{\frac{1}{2}}^x})=E(e^{\sum_{x\neq y}(\frac{1}{2}C_{x,y}(S_{x,y}-1)\varphi_x^{\mathbb{R}}\varphi_y^{\mathbb{R}})}e^{-\frac{1}{2}(\sum_x(\lambda_x+\chi_x)(\varphi_x^{\mathbb{R}}))^2})$$
$$=\frac{1}{\sqrt{\det(G)}}\int_{\mathbb{R}^{|X|}}e^{-\frac{1}{2}(\sum_x(\lambda_x+\chi_x){u_x}^2-\sum_{x\neq y}C_{x,y}S_{x,y}u_xu_y)}$$
$$\times\prod_x\frac{du_x}{\sqrt{2\pi}}.$$

Expanding the exponential of the double sum, and noting that only monomials with even degree in each u_x contribute to the integral, we get that this expression equals

$$\frac{1}{\sqrt{\det(G)}}\int_{\mathbb{R}^{|X|}}e^{-\frac{1}{2}\sum_x(\lambda_x+\chi_x)u_x^2}\sum_{n\in\mathfrak{V}}\prod_{\{x,y\}}\frac{1}{n_{\{x,y\}}!}C_{x,y}(S_{x,y}u_xu_y)^{n_{\{x,y\}}}\prod_x\frac{du_x}{\sqrt{2\pi}}.$$

It follows that for any functional F on $\mathbb{R}^{|X|}$, $E(\prod_{x\neq y}S_{x,y}^{N_{x,y}^{(1)}}F(\hat{\mathscr{L}}_1))$ equals

$$\frac{1}{\det(G)}\int_{\mathbb{R}^{|X|}}e^{-\frac{1}{2}\sum_x(\lambda_x)u_x^2}\sum_{n\in\mathfrak{V}}\prod_{\{x,y\}}\frac{1}{n_{\{x,y\}}!}(C_{x,y}S_{x,y}u_xu_y)^{n_{\{x,y\}}}F(\frac{1}{2}u^2)\prod_x\frac{du_x}{\sqrt{2\pi}}.$$

We conclude the proof of the proposition by performing a change of variable $\rho = \frac{1}{2}r^2$, letting F be an infinitesimal indicator function and by identifying the coefficients of $\prod_{\{x,y\}} S_{x,y}^{n_{\{x,y\}}}$. Then, (iii) follows by integration.

Remarks

- It follows from (iv) that the symmetrized $N^{(\frac{1}{2})}$ field conditioned by the vertex occupation field is, as it was observed by Werner in [8], a random current model.
- These results can be extended to infinite transient graphs as follows: For (i) given any finitely supported Eulerian network k and any bounded function F on Eulerian networks:

$$E(F(N^{(1)} + k))$$

$$= E(1_{\{N^{(1)} \geqslant k\}} F(N^{(1)}) \prod_x \frac{(N_x^{(1)} - k_x)!}{N_x^{(1)}!} \prod_{x,y} \frac{N_{x,y}^{(1)}!}{(N_{x,y}^{(1)} - k_{x,y})!} \prod_{x,y} P_{x,y}^{-k_{x,y}}.$$

- This quasi invariance property can be used to prove the closability and express the generator of some Dirichlet forms defined naturally on the space $\mathfrak{E}$ of Eulerian networks: If ν is a bounded measure and G a bounded function on $\mathfrak{E}$, the energy of G can be defined as $\int E((G(N^{(1)} + k) - G(N^{(1)}))^2)\nu(dk)$. They define stationary processes on $\mathfrak{E}$, with invariant distribution given by the distribution of $N^{(1)}$.

  Similar quasi invariance properties and stationary processes can be derived from (ii), (iii) and (iv).

Markov Property

From theorem 16.1 follows a Markov property which generalizes the reflection positivity property proved in chapter 9 of [2]:

Theorem 16.2 *Let X be the disjoint union of X_i, $i = 1, 2$ and $\mathcal{G}_i$ be the restriction of $\mathcal{G}$ to X_i.*

(i) Given the values of $N_{x,y}^{(1)}$ and $N_{y,x}^{(1)}$ for $x \in X_1$ and $y \in X_2$, the restrictions of the fields $(N^{(1)}, \widehat{\mathcal{L}_1})$ to $\mathcal{G}_1$ and $\mathcal{G}_2$ are independent.

(ii) Given the values of $N_{\{x,y\}}^{(\frac{1}{2})}$ for $x \in X_1$ and $y \in X_2$, the restrictions of the fields $(N_{\{\}}^{(\frac{1}{2})}, \widehat{\mathcal{L}_{\frac{1}{2}}})$ to $\mathcal{G}_1$ and $\mathcal{G}_2$ are independent.

In both cases, we can check on the expressions given in 16.1 (ii) and (iv) that after fixing the values of the conditioning, the joint density function factorizes. See [8] and also [1] in the context of non backtracking loops.

In the case $\alpha = 1$, the Markov property of these fields is preserved if we modify P by a factor of the form $\prod_x e^{-\beta \Phi_x}$ and a normalization constant, with Φ_x a non-negative function of $\widehat{\mathcal{L}_1^x}$ and of $\{N_{x,y}^{(1)}, N_{y,x}^{(1)}, y \neq x\}$

In the case $\alpha = \frac{1}{2}$, the Markov property of these fields is preserved if we modify P by a factor of the form $\prod_x e^{-\beta \Phi_x}$ and a normalization constant, with $\beta > 0$ and Φ_x a non-negative function of $\widehat{\mathcal{L}_{\frac{1}{2}}^x}$ and of $\{N_{\{x,y\}}^{(\frac{1}{2})},\ y \neq x\}$

Remark An example of interest (considering the remark at the end of the next section), in the case $\alpha = 1$, is $\Phi_x = R_x := (N_x^{(1)})^2 - \sum_y (N_{x,y}^{(1)})^2$.

16.3 Random Flows

We say that a Eulerian network j is a flow if it defines an orientation on edges on which it does not vanish, i.e., more precisely, iff $j_{x,y}\, j_{y,x} = 0$ for all edges $\{x, y\}$. It is easy to check that the measure $j_x = \sum_y j_{x,y}$ is preserved by the Markovian matrix q defined as follows: $q_y^x = \frac{j_{x,y}}{j_x}$ if $j_x > 0$, $q_y^x = \delta_y^x$ if $j_x = 0$.

We can define the stochasticity of the flow at x to be $S_x = j_x^2 - \sum_y j_{x,y}^2$. If it vanishes everywhere, the Markovian transition matrix is a permutation of X.

We can define the flow $J(k)$ associated to a Eulerian network k by

$$J(k)_{x,y} = 1_{\{k_{x,y} - k_{y,x} > 0\}}[k_{x,y} - k_{y,x}].$$

We now show that a simple expression of the joint distribution of the flow $J(N^{(1)})$ and the vertex occupation field $\hat{\mathcal{L}_1}$ can be derived from Theorem 16.1.

Let $\mathfrak{C}$ be the set of $\mathbb{N}$-valued sets of conductances on $\mathcal{G}$. For any flow h on $\mathcal{G}$:

$$\{J(N^{(1)}) = h\} = \bigcup_{c \in \mathfrak{C}} \{ \bigcap_{\{x,y\} \in E} \{N_{x,y}^{(1)} = c_{\{x,y\}} + h_{x,y},\, N_{y,x}^{(1)} = c_{\{x,y\}} + h_{y,x}\}\}.$$

From 16.1 (ii), it follows that:

$$P(J(N^{(1)}) = h,\ \hat{\mathcal{L}_1} \in (\rho, \rho + d\rho))$$

$$= \sum_{k \in \mathfrak{K}} \det(I - P) \prod_{x,y} \frac{(\sqrt{\rho_x} C_{x,y} \sqrt{\rho_y})^{k_{x,y} + h_{x,y}}}{(k_{x,y} + h_{x,y})!} \prod_x \lambda_x e^{-\lambda_x \rho_x} d\rho_x .$$

Recall the definition of the modified Bessel function:

$$I_\nu(x) = \sum_{m=0}^{\infty} \frac{1}{m! \Gamma(\nu + m + 1)} \left(\frac{x}{2}\right)^{2m+\nu} .$$

From this follows:

Theorem 16.3 *For any flow h, setting $h_{\{x,y\}} = \max(h_{x,y}, h_{y,x})$*

$$P(J(N^{(1)}) = h,\ \hat{\mathscr{L}}_1 \in (\rho, \rho + d\rho)) \\ = \det(I - P) \prod_{\{x,y\}} I_{h_{\{x,y\}}}(2\sqrt{\rho_x} C_{x,y} \sqrt{\rho_y}) \prod_x \lambda_x e^{-\lambda_x \rho_x} d\rho_x \,.$$

Remark Recall that we defined $R_x := (N_x^{(1)})^2 - \sum_y (N_{x,y}^{(1)})^2$. Note that $R = 0$ implies that the flow defined by $N^{(1)}$ has zero stochasticity (but the converse is not true). Hence, as β increases to infinity, the probability modified by $\prod_x e^{-\beta R_x}$ concentrates on the set of flows of null stochasticity (as it has positive P-probability). These results, among others, were presented by the author in [5].

References

1. F. Camia, M. Lis, Non-backtracking loop soups and statistical mechanics on spin networks. Ann. Inst. Henri Poincaré Probab. Stat. **18**, 403–433 (2017)
2. Y. Le Jan, Markov paths, loops and fields, in *École d'Été de Probabilités de Saint-Flour XXXVIII – 2008*. Lecture Notes in Mathematics, vol. 2026 (Springer, Berlin, 2011)
3. Y. Le Jan, Markov loops, Free field and Eulerian networks. J. Math. Soc. Jpn. **67**(4), 1671–1680 (2015)
4. Y. Le Jan, Markov loops, coverings and fields. Ann. Fac. Sci. Toulouse Math. (6) **26**(2), 401–416 (2017)
5. Y. Le Jan, On Markovian random networks. arXiv 1802.01032
6. Y. Le Jan, Markov loops, determinants and Gaussian fields. arXiv:math/0612112
7. T. Lupu, From loop clusters and random interlacement to the free field. Ann. Probab. **44**(3), 2117–2146 (2016)
8. W. Werner, On the spatial Markov property of soups of unoriented and oriented loops, in *Séminaire de de Probabilités*. XLVIII, Lecture Notes in Mathematics, vol. 2168 (Springer, Berlin, 2016). pp. 481–503

Chapter 17
Brownian Winding Fields

Yves Le Jan

Abstract The purpose of the present note is to review and improve the convergence of the renormalized winding fields introduced in Camia et al. (Nucl Phys B 902:483–507, 2016) and van de Brug et al. (Electron J Probab 23(81):17, 2018).

Keywords Brownian loops · Windings

In the seminal work of Symanzik [10], Poisson ensembles of Brownian loops were implicitly used. Since the work of Lawler and Werner [4] on "loop soups", these ensembles have also been the object of many investigations.

Windings of two dimensional random paths have been widely studied. Let us mention the seminal work of Spitzer [9] for Brownian paths, and Schramm [8] for SLE. The purpose of the present note is to review and improve the convergence of the renormalized winding fields introduced in [1] and [11], using a martingale convergence argument.The result is somewhat reminiscent of Gaussian multiplicative chaos ([3]). In the context of Brownian loop ensembles, a different type of renormalization was used to define the occupation field and its powers (see chapter 10 in [5]). The method here is related to zeta renormalization used in [6, 7] to study the homology of Brownian loops defined on manifolds.

We consider a bounded open subset of the plane, denoted D. We denote by D_R the disc of radius R centered at 0. For any point x in D, let j_x be a uniformizing map mapping D onto D_1 and x to 0 and for $\delta < 1$, by $B(x, \delta)$ the pullback of D_δ in D.

Y. Le Jan (✉)
NYU Shanghai, Shanghai, China

Département de Mathématique, Université Paris-Sud, Orsay, France
e-mail: yl57@nyu.edu; yves.lejan@math.u-psud.fr

C. Donati-Martin et al. (eds.), *Séminaire de Probabilités L*, Lecture Notes in Mathematics 2252, https://doi.org/10.1007/978-3-030-28535-7_17

The σ-finite measure μ on the set of Brownian loops and the Poisson process of Brownian loops are defined in the same way as Lawler and Werner "loop soup" (cf [4]). More precisely, denoting by dA the area measure, $\mu = \int_{x\in X}\int_0^\infty \frac{1}{t}\mathbb{M}_t^{x,x}dt\ dA(x)$ where $\mathbb{M}_t^{x,y}$ denotes the distribution of the Brownian bridge in D between x and y, multiplied by the heat kernel density $p_t(x,y)$.

For any positive α, the Poisson process of loops of intensity $\alpha\mu$ is denoted $\mathcal{L}_\alpha$. If U is an open subset of D, we denote by $\mathcal{L}_\alpha^U$ the set of loops in $\mathcal{L}_\alpha$ contained in U.

Almost surely, for a given x, the loops of $\mathcal{L}_\alpha$ do not visit x. We denote by $n_x(l)$ the winding number around 0 of the pullback of a loop l in $\mathcal{L}_\alpha$. As the Brownian loops, as Brownian paths, have vanishing Lebesgue measure, $n_x(l)$ is defined almost everywhere in x, almost surely.

Let β denote any $[0,2\pi)$-valued function defined on D. Let h be any bounded function with compact support in D. For any $\delta < 1$, define

$$W_x^{\beta_x,\delta,\alpha} = \prod_{l\in\mathcal{L}_\alpha\setminus\mathcal{L}_\alpha^{B(x,\delta)}} e^{i\beta_x n_x(l)}$$

The winding field $W^{\beta,\alpha}(h)$ is defined as follows:

Theorem 17.1 *For δ_n decreasing to zero, $\int_D h(x)\delta_n^{-\alpha\, a(\beta_x)}W_x^{\beta_x,\delta_n,\alpha}dA(x)$ is a martingale, with $a(\beta_x) = \frac{\beta_x(2\pi-\beta_x)}{4\pi^2} \leqslant \frac{1}{4}$. For $\alpha < 4$, it converges a.s. and in L^p for all $p \geqslant 1$ towards a limit denoted $W^{\beta,\alpha}(h)$.*

Remarks In contrast with Gaussian multiplicative chaos, moments of all order are defined for any $\alpha < 4$. The question of determining the behaviour of these martingales for $\alpha \geqslant 4$ seems open. As mentioned in [1], one may also investigate the possibility of finding a characterization of the distribution of the winding field, in terms of conformal field theory.

Proof For $0 < R \leqslant \infty$, let $\mathbb{M}_t^{R,x,y}$ denote the distribution of the Brownian bridge in D_R multiplied by the heat kernel density, μ^R the associated loop measure and $\mathcal{L}_\alpha^R$ the corresponding loop ensemble. Up to time change (under which winding indices are invariant), $\mathcal{L}_\alpha^1$ is the image of $\mathcal{L}_\alpha$ under any uniformizing map.

Lemma 17.1 $\int_{\mathbb{C}} dA(z)\mathbb{P}_1^{\infty,z,z}(n_0 = k) = \dfrac{1}{2\pi^2k^2}$ □

This result was established in [2], with reference to [13]. Let us outline briefly its proof, for the convenience of the reader:

In polar coordinates, a well known consequence of the skew-product decomposition of the Brownian bridge measure is that

$$\int e^{iu\, n_0(l)}\mathbb{M}_1^{\infty,z,z}(dl) = \int e^{iu\int_l d\theta}\mathbb{M}_1^{\infty,z,z}(dl) = E(e^{-\frac{u^2}{2}\int_0^1 \rho_s^2 ds})q_1(z,z)$$

in which ρ_s denotes a Bessel(0) bridge from $|z|$ to $|z|$ and q_t the Bessel(0) transition kernels semigroup. It follows from Feynman-Kac formula and Bessel differential equation that this expression equals $e^{-|z|^2} I_{|u|}(|z|)$

As the Dirac measure at $2\pi n$ is the Fourier transform of $e^{-i2\pi nu}$, we get that for any $r > 0$

$$\mathbb{M}_1^{\infty,r,r}(n_0 = n) = 2e^{-r^2} \int_0^\infty I_{|u|}(r)\cos(2\pi nu)du$$

From this, as observed by Yor in [13], using the expression of the modified Bessel function $I_{|u|}$ as a contour integral, we obtain that:

$$\mathbb{M}_1^{\infty,r,r}(n_0 = n)$$
$$= e^{-r^2} \int_0^\infty e^{-r^2\cosh(t)} \left[\frac{2n-1}{r^2 + (2n-1)^2\pi^2} - \frac{2n+1}{r^2 + (2n+1)^2\pi^2} \right] dt$$

Hence, integrating with respect to $2\pi r dr$,

$$\int_{\mathbb{C}} dA(z)\mathbb{M}_1^{\infty,z,z}(n_0 = n)$$
$$= \pi \int_0^\infty \frac{dt}{1+\cosh(t)} \left[\frac{2n-1}{r^2 + (2n-1)^2\pi^2} - \frac{2n+1}{r^2 + (2n+1)^2\pi^2} \right] dt.$$

As observed in [2], the final result follows from a residue calculation yielding telescopic series.

Lemma 17.2 $\mu^R(l \not\subseteq D_\delta, n_0(l) = k) = \frac{1}{2\pi^2 k^2}\log(\frac{R}{\delta})$.

To prove this lemma, we use the zeta regularisation method, which, in this context, allows to introduce a $T(l)^s$ factor under μ^R, and let s decrease to zero. ($T(l)$ denoting the loop time length).

$\mu^R(l \not\subseteq D_\delta,\ n_0(l) = k)$ is the limit as $s \downarrow 0$ of $\int T(l)^s \mathbb{1}_{l \not\subseteq D_\delta}\mathbb{1}_{n_0(l)=k}\mu^R(dl)$

$$= \int_0^\infty \int_{D_R} \mathbb{M}_t^{R,z,z}(n_0 = k)dA(z)t^{s-1}dt - \int_0^\infty \int_{D_\delta} \mathbb{M}_t^{\delta,z,z}(n_0 = k)dA(z)t^{s-1}dt$$
$$= \int_0^\infty \int_{D_R} \mathbb{M}_t^{R,z,z}(n_0 = k)dA(z)t^{s-1}dt - \int_0^\infty \int_{D_R} \mathbb{M}_{t(R/\delta)^2}^{R,z,z}(n_0 = k)dA(z)t^{s-1}dt$$
$$= \frac{1-(\delta/R)^{2s}}{s} \int_0^\infty \int_{D_R} \mathbb{M}_t^{R,z,z}(n_0 = k)dA(z)st^{s-1}dt$$

From Lemma 17.1, for η arbitrarily small, we can choose $\epsilon > 0$ such that for $u < \epsilon$, $|\int_{D_{R/u}} \mathbb{M}_1^{R/u,z,z}(n_0 = k)dA(z) - \dfrac{1}{2\pi^2 k^2}| < \eta$.

Then $\frac{1-(\delta/R)^{2s}}{s}\int_0^{\epsilon}\int_{D_R}\mathbb{M}_t^{R,z,z}(n_0 = k)dA(z)st^{s-1}dt = \frac{1-(\delta/R)^{2s}}{s}\int_0^{\epsilon}\int_{D_{R/t}}$ $\mathbb{M}_t^{R/t,z,z}(n_0 = k)dA(z)st^{s-1}dt$ is arbitrarily close from $\frac{1}{2\pi^2k^2}\log(\frac{R}{\delta})$ for ϵ and s small enough.

To prove that $\frac{1-(\delta/R)^{2s}}{s}\int_{\epsilon}^{\infty}\int_{D_R}\mathbb{M}_t^{R,z,z}(n_0 = k)dA(z)st^{s-1}dt$ converges to zero with s, note that $\int_{D_R}\mathbb{M}_t^R(z,z)(n_0 = k)dA(z) \leqslant \int_{D_R}P_t^R(z,z)dA(z)$, denoting by $P_t^R(x,y)$ the heat kernel on the disc of radius R. It follows from Weyl asymptotics that this trace can be bounded by $Ce^{-\lambda_0 t}t$, λ_0 denoting the ground state eigenvalue on D_R and C a positive constant. The result follows as the resulting gamma density converges to zero on $[\epsilon,\infty)$ and this concludes the proof of the second lemma. □

Lemma 17.3 $E(W_x^{\beta_x,\delta,\alpha}) = \delta^{\alpha\, a(\beta_x)}$.

This result follows by bounded convergence from Lemma 17.2 and from the Fourier series identity $\sum_1^{\infty}\frac{1}{\pi^2k^2}(1-\cos(k\beta)) = \frac{\beta(2\pi-\beta)}{4\pi^2}$ as

$$E(W_x^{\beta_x,\delta,\alpha}) = \lim_{N\to\infty} E(\prod_{k=-N}^{N} e^{ik\beta_x|\{l\in\mathcal{L}_\alpha\, l\not\subseteq B(x,\delta), n_x(l)=k\}|})$$

$$= \lim_{N\to\infty} E(\prod_{k=-N}^{N} e^{ik\beta_x|\{l\in\mathcal{L}_\alpha^1\, l\not\subseteq D_\delta, n_0(l)=k\}|})$$

$$= \lim_{N\to\infty} \exp(\alpha\log(\delta)\sum_1^N\frac{1}{\pi^2k^2}(1-\cos(k\beta_x)))$$

To complete the proof of the theorem, remark first that it follows from the independence property of a Poisson point process that for δ_n decreasing to 0, and for any x, $\frac{W_x^{\beta_x,\delta_n,\alpha}}{E(W_x^{\beta_x,\delta_n,\alpha})} = \delta_n^{-\alpha\, a(\beta_x)}W_x^{\beta,\delta_n,\alpha}$ is a martingale with independent multiplicative increments. We denote it by $Z_{n,x}^{\beta_x,\alpha}$. Hence, the martingale property of the integral $\int_D h(x)Z_{n,x}dA(x)$ is obvious. To show the convergence, we need a uniform bound on its L_{2p} norm, for any integer $p \geqslant 1$.

Given $2p$ distinct points x_l in a compact $K \subset D$ supporting h, for $\delta_{l,n} < \delta_{l,0} = \sup(\{\epsilon,\ B(x_l,\epsilon)\cap B(x_k,\epsilon) = \emptyset$ for any $k \neq l\})$ decreasing to zero, all $B(x_l,\delta_{l,0})$ are disjoint and the product $\prod_{l\leqslant 2p}\delta_{l,n}^{-\alpha\, a(\beta_{x_l})}W_{x_l}^{\beta_{x_l},\delta_{l,n},\alpha}$ is a martingale. Its expectation is bounded by $\prod_{l\leqslant 2p}\delta_{l,0}^{-\alpha\, a(\beta_{x_l})}$.

For some multiplicative constant, $c > 0$ depending on the compact support K of h, $\delta_{l,0} \leqslant c\min\{d(x_l,x_{l'}),\ l'\neq l\}$ for all $l \leqslant 2p$. It follows in particular that

$$E(|\int_D h(x)Z_{n,x}^{\beta_x,\alpha}dA(x)|^2) = E(\int_{D^2} h(x)Z_{n,x}^{\beta_x,\alpha}h(y)Z_{n,y}^{-\beta_y,\alpha}dA(x)dA(y) \leqslant c^2\|h\|_\infty^2 I$$

with $I = \int\int_{D^2} d(x_1, x_2)^{-\alpha/2} dA(x_1)dA(x_2)$, which proves the L_2 and a.s. convergence.

More generally, for any integer $p > 1$, the $2p$-th moment $E(|\int_D h(x)Z_{n,x}dA(x)|^{2p})$ is bounded by $(c\|h\|_\infty)^{2p} \int\int_{D^{2p}} \prod_{l\leqslant 2p} \min_{l'\neq l} d(x_{l'}, x_l)^{-\alpha/4} dA(x_1)\ldots dA(x_{2p})$. To see this expression is finite for $\alpha < 4$, we will consider only the case $p = 2$ as the general proof is similar. The term with highest singularity comes from the case where, up to a permutation, the smallest distances are $d(x_1, x_2)$ and $d(x_3, x_4)$. Then the integral on that sector of D^4 can be bounded by $(I)^2$. In the other cases, i.e. when, up to a permutation, the smallest distances are $d(x_1, x_2)$, $d(x_3, x_1)$ and $d(x_4, x_1)$, or $d(x_1, x_2)$, $d(x_3, x_1)$ and $d(x_4, x_2)$, the integral on the corresponding sector can be bounded by $C^2\ I$, with $C = \sup_{x\in K} \int_D d(x, y)^{-\alpha/4} dA(y)$. □

Remarks

(1) It can be shown that the martingales $Z_{n,x}^{\beta_x,\alpha}$ do not converge, consequently, $W^{\beta,\alpha}(h)$ is a generalized field. The class of test functions h can actually be extended to integrals of delta functions along a smooth curve segment if $\alpha < 4$.
(2) It follows from theorem 7 in chapter 9 of [5] (see also the Markov property in [12]) that the discrete analogue of $W^{\beta,\alpha}$ verifies reflection positivity for $\alpha = 1$, 2, or 3 in case D is invariant under some reflection. This property should extend to the Brownian case.

Acknowledgements I thank Federico Camia and Marci Lis for interesting discussions and the referee for helpful remarks.

References

1. F. Camia, A. Gandolfi, M. Kleban, Conformal correlation functions in the Brownian loop soup, Nucl. Phys. **B 902**, 483–507 (2016)
2. C. Garban, J. T. Ferreras, The expected area of a filled planar Brownian loop is Pi/5. Commun. Math. Phys. **264**, 797–810 (2006)
3. J.-P. Kahane, Sur le chaos multiplicatif. Ann. Sci. Math. Québec **9**, 105–150 (1985)
4. G. Lawler, W. Werner, The Brownian loop soup. PTRF **128**, 565–588 (2004)
5. Y. Le Jan, Markov paths, loops and fields, in *École d'Été de Probabilités de Saint-Flour XXXVIII – 2008*. Lecture Notes in Mathematics, vol. 2026 (Springer, Berlin, 2011)
6. Y. Le Jan, Brownian loops topology. Potential Anal. (2019). https://doi.org/10.1007/s11118-019-09765-z
7. Y. Le Jan. Homology of Brownian loops. arXiv: 1610.09784
8. O. Schramm. Scaling Limits of Loop-Erased Random Walks and Uniform Spanning Trees. Israel J. Math. **118**, 221–288 (2000)
9. F. Spitzer, Some theorems concerning 2-dimensional Brownian motion. Trans. Amer. Math. Soc. **87**, 187–197 (1958)
10. K. Symanzik, Euclidean quantum field theory, in *Scuola intenazionale di Fisica "Enrico Fermi". XLV Corso* (Academic, Cambridge, 1969), pp. 152–223
11. T. van de Brug, F. Camia, M. Lis, Spin systems from loop soups. Electron. J. Probab. **23**(81), 77 (2018)

12. W. Werner. On the spatial Markov property of soups of unoriented and oriented loops. in *Séminaire de de Probabilités XLVIII*. Lecture Notes in Mathematics, vol. 2168 (Springer, Berlin, 2016), pp. 481–503
13. M. Yor, Loi de l'indice du lacet Brownien, et distribution de Hartman-Watson, Z. Wahrsch. Verw. Gebiete **53**, 71–95 (1980)

Chapter 18
Recurrence and Transience of Continuous-Time Open Quantum Walks

Ivan Bardet, Hugo Bringuier, Yan Pautrat, and Clément Pellegrini

Abstract This paper is devoted to the study of continuous-time processes known as continuous-time open quantum walks (CTOQW). A CTOQW represents the evolution of a quantum particle constrained to move on a discrete graph, but which also has internal degrees of freedom modeled by a state (in the quantum mechanical sense). CTOQW contain as a special case continuous-time Markov chains on graphs. Recurrence and transience of a vertex are an important notion in the study of Markov chains, and it is known that all vertices must be of the same nature if the Markov chain is irreducible. In the present paper we address the corresponding result in the context of irreducible CTOQW. Because of the "quantum" internal degrees of freedom, CTOQW exhibit non standard behavior, and the classification of recurrence and transience properties obeys a "trichotomy" rather than the classical dichotomy. Essential tools in this paper are the so-called "quantum trajectories" which are jump stochastic differential equations which can be associated with CTOQW.

18.1 Introduction

Open quantum walks (OQW) have been developed originally in [1, 2]. They are natural quantum extensions of classical Markov chains and, in particular, any classical discrete-time Markov chain on a finite or countable set can be obtained as a particular case of OQW. Roughly speaking, OQW are random walks on a

I. Bardet
Institut des Hautes Études Scientifiques, Université Paris-Saclay, Bures-sur-Yvette, France

H. Bringuier · C. Pellegrini
Institut de Mathématiques de Toulouse, UMR5219, UPS IMT, Toulouse Cedex, France
e-mail: hugo.bringuier@math.univ-toulouse.fr; clement.pellegrini@math.univ-toulouse.fr

Y. Pautrat (✉)
Laboratoire de Mathématiques d'Orsay, Univ. Paris-Sud, CNRS, Université Paris-Saclay, Orsay, France
e-mail: yan.pautrat@math.u-psud.fr

C. Donati-Martin et al. (eds.), *Séminaire de Probabilités L*, Lecture Notes in Mathematics 2252, https://doi.org/10.1007/978-3-030-28535-7_18

graph where, at each step, the walker jumps to the next position following a law which depends on an internal degree of freedom, the latter describing a quantum-mechanical state. From a physical point of view, OQW are simple models offering different possibilities of applications (see [28, 29]). From a mathematical point of view, their properties can been studied in analogy with those of classical Markov chain. In particular, usual notions such as irreducibility, period, ergodicity, have been investigated in [3, 8–10, 20]. For example, the notions of transience and recurrence have been studied in [5], proper definitions of these notions have been developed in this context and the analogues of transient or recurrent points have been characterized. An interesting feature is that the internal degrees of freedom introduce a source of memory which gives rise to a specific non-Markovian behavior. Recall that, in the classical context (see [22]), an exact dichotomy exists for irreducible Markov chains: a point is either recurrent or transient, and the nature of a point can be characterized in terms of first return time, or in terms of number of visits. In contrast, irreducible open quantum walks exhibit three possibilities regarding the behavior of return time and number of visits. In this article, we study the recurrence and transience, as well as their characterizations, for continuous-time versions of OQW.

In the same way that open quantum walks are quantum extensions of discrete-time Markov chains, there exist natural quantum extensions of continuous-time Markov processes. One can point to two different types of continuous-time evolutions with a structure akin to open quantum walks. The first (see [6]) is a natural extension of classical Brownian motion and is called open quantum Brownian motion; it is obtained by considering OQW in the limit where both time and space are properly rescaled to continuous variables. The other type of such evolution (see [25]) is an analogue of continuous-time Markov chains on a graph, is obtained by rescaling time only, and is called continuous-time open quantum walks (CTOQW). In this article we shall concentrate on the latter.

Roughly speaking CTOQW represents a continuous-time evolution on a graph where a "walker" jumps from node to node at random times. The intensity of jumps depends on the internal degrees of freedom; the latter are modified by the jump, but also evolve continuously between jumps. In both cases the form of the intensity, as well as the evolution of the internal degrees of freedom at jump times and between them, can be justified from a quantum mechanical model.

As is well-known, in order to study a continuous-time Markov chain, it is sufficient to study the value of the process at the jump times. Indeed, the time before a jump depends exclusively on the location of the walker, and the destination of the jump is independent of that time. As a consequence, the process restricted to the sequence of jump times is a discrete-time Markov chain, and all the properties of that discrete-time Markov chain such as irreducibility, period, transience, recurrence, are transferred to the continuous-time process. This is not the case for OQW. In particular, a CTOQW restricted to its jump times is not a (discrete-time) open quantum walk. Therefore, the present study of recurrence and transience cannot be directly derived from the results in [5]. Nevertheless, we can still adopt a similar approach and, for instance, we study irreducibility of CTOQW in connection to that of quantum dynamical systems as in [11]. Note that general notions of recurrence and transience are developed in [17] for general quantum Markov semigroups with unbounded generators. The work elaborated in [17] is based on potential theory and we explicit the connection between the notions of recurrence and transience of CTOQW and those in [17]. Finally, as in the discrete case, we obtain a trichotomy, in the sense that irreducible CTOQW can be classified into three different classes, depending on the properties of the associated return time and number of visits.

The paper is structured as follows: in Sect. 18.2, we recall the definition of continuous-time open quantum walks and in particular introduce useful classical processes attached to CTOQW; Sect. 18.3 is devoted to the notion of irreducibility for CTOQW; in Sect. 18.4, we address the question of recurrence and transience and give the classification of CTOQW mentioned above.

18.2 Continuous Time Open Quantum Walks and Their Associated Classical Processes

This section is devoted to the introduction of continuous-time open quantum walks (CTOQW). In Sect. 18.2.1, we introduce CTOQW as a special instance of quantum Markov semigroups (QMS) with generators preserving a certain block structure. Section 18.2.2 is devoted to the exposition of the Dyson expansion associated with a QMS, which will be a relevant tool in all remaining sections. It also allows us to introduce the relevant probability space. Finally, in Sect. 18.2.3 we associate to this stochastic process a Markov process called *quantum trajectory* which has an additional physical interpretation, and which will be useful in its analysis.

18.2.1 Definition of Continuous-Time Open Quantum Walks

Let V denotes a set of vertices, which may be finite or countably infinite. CTOQW are quantum analogues of continuous-time Markov semigroups acting on the set $\mathrm{L}^{\infty}(V)$ of bounded functions on V. They are associated with stochastic processes evolving in the composite system

$$\mathcal{H} = \bigoplus_{i \in V} \mathfrak{h}_i \,, \tag{18.1}$$

where the $\mathfrak{h}_i$ are separable Hilbert spaces. This decomposition has the following interpretation: the label i in V represents the position of a particle and, when the particle is located at the vertex $i \in V$, its internal state is encoded in the space $\mathfrak{h}_i$ (see below). Thus, in some sense, the space $\mathfrak{h}_i$ describes the internal degrees of freedom of the particle when it is sitting at site $i \in V$. When $\mathfrak{h}_i$ does not depend on i, that is if $\mathfrak{h}_i = \mathfrak{h}$, for all $i \in V$, one has the identification $\mathcal{H} \simeq \mathfrak{h} \otimes \ell^2(V)$ and then it is natural to write $\mathfrak{h}_i = \mathfrak{h} \otimes |i\rangle$ (we use here Dirac's notation where the *ket* $|i\rangle$ represents the i-th vector in the canonical basis of $\ell^\infty(V)$, the *bra* $\langle i|$ represents the associated linear form, and $|i\rangle\langle j|$ represents the linear map $\varphi \mapsto \langle j|\varphi\rangle\, |i\rangle$). We will adopt the notation $\mathfrak{h}_i \otimes |i\rangle$ to denote $\mathfrak{h}_i$ in the general case (i.e. when $\mathfrak{h}_i$ depends on i) to emphasize the position of the particle, using the identification $\mathfrak{h}_i \otimes \mathbb{C} \simeq \mathfrak{h}_i$. We thus write:

$$\mathcal{H} = \bigoplus_{i \in V} \mathfrak{h}_i \otimes |i\rangle\ . \tag{18.2}$$

Last, we denote by $\mathcal{I}_1(\mathcal{K})$ the two-sided ideal of trace-class operators on a given Hilbert space $\mathcal{K}$ and by $\mathcal{S}_{\mathcal{K}}$ the space of density matrices on $\mathcal{K}$, defined by:

$$\mathcal{S}_{\mathcal{K}} = \{\rho \in \mathcal{I}_1(\mathcal{K}) \mid \rho^* = \rho, \rho \geq 0, \mathrm{Tr}(\rho) = 1\}.$$

A *faithful* density matrix is an invertible element of $\mathcal{S}_{\mathcal{K}}$, which is therefore a trace-class and positive-definite operator. Following quantum mechanical fashion, we will use the word "state" interchangeably with "density matrix".

We recall that a quantum Markov semigroup (QMS) on $\mathcal{I}_1(\mathcal{K})$ is a semigroup $\mathcal{T} := (\mathcal{T}_t)_{t\geq 0}$ of completely positive maps on $\mathcal{I}_1(\mathcal{K})$ that preserve the trace. The QMS is said to be uniformly continuous if $\lim_{t\to 0}\|\mathcal{T}_t - \mathrm{Id}\| = 0$ for the operator norm on $\mathcal{B}(\mathcal{K})$. It is then known (see [21]) that the semigroup $(\mathcal{T}_t)_{t\geq 0}$ has a generator $\mathcal{L} = \lim_{t\to\infty}(\mathcal{T}_t - \mathrm{Id})/t$ which is a bounded operator on $\mathcal{I}_1(\mathcal{K})$, called the Lindbladian, and Lindblad's theorem characterizes the structure of such generators. One consequently has $\mathcal{T}_t = \mathrm{e}^{t\mathcal{L}}$ for all $t \geq 0$, where the exponential is understood as the limit of the norm-convergent series.

Continuous-time open quantum walks are particular instances of uniformly continuous QMS on $\mathcal{I}_1(\mathcal{H})$, for which the Lindbladian has a specific form. To make this more precise, we define the following set of block-diagonal density matrices of $\mathcal{H}$:

$$\mathcal{D} = \Big\{\mu \in \mathcal{S}(\mathcal{H})\, ;\ \mu = \sum_{i\in V} \rho(i) \otimes |i\rangle\langle i|\Big\}\ .$$

In particular, for $\mu \in \mathcal{D}$ with the above definition, one has $\rho(i) \in \mathcal{I}_1(\mathfrak{h}_i)$, $\rho(i) \geq 0$ and $\sum_{i\in V} \mathrm{Tr}\big(\rho(i)\big) = 1$. In the sequel, we use the usual notations $[X, Y] = XY - YX$ and $\{X, Y\} = XY + YX$, which stand respectively for the commutator and anticommutator of two operators $X, Y \in \mathcal{B}(\mathcal{H})$.

Definition 18.1 Let $\mathcal{H}$ be a Hilbert space that admits a decomposition of the form (18.1). A *continuous-time open quantum walk* is a uniformly continuous quantum Markov semigroup on $\mathcal{I}_1(\mathcal{H})$ such that its Lindbladian $\mathcal{L}$ can be written:

$$\begin{aligned}\mathcal{L} : \mathcal{I}_1(\mathcal{H}) &\to \mathcal{I}_1(\mathcal{H})\\ \mu\ &\mapsto -\mathrm{i}[H, \mu] + \sum_{i,j\in V} \mathbb{1}_{i\neq j}\big(S_i^j \mu S_i^{j*} - \frac{1}{2}\{S_i^{j*} S_i^j, \mu\}\big)\ ,\end{aligned} \tag{18.3}$$

where H and $(S_i^j)_{i,j}$ are bounded operators on $\mathcal{H}$ that take the following form:

- $H = \sum_{i \in V} H_i \otimes |i\rangle\langle i|$, with H_i bounded self-adjoint operators on $\mathfrak{h}_i$, i in V;
- for every $i \neq j \in V$, S_i^j is a bounded operator on $\mathcal{H}$ with support included in $\mathfrak{h}_i$ and range included in $\mathfrak{h}_j$, and such that the sum $\sum_{i,j \in V} S_i^{j*} S_i^j$ converges in the strong sense. Consistently with our notation, we can write $S_i^j = R_i^j \otimes |j\rangle\langle i|$ for bounded operators $R_i^j \in \mathcal{B}(\mathfrak{h}_i, \mathfrak{h}_j)$.

We will say that the open quantum walk is *semifinite* if $\dim \mathfrak{h}_i < \infty$ for all $i \in V$.

From now on we will use the convention that $S_i^i = 0$, $R_i^i = 0$ for any $i \in V$. As one can immediately check, the Lindbladian $\mathcal{L}$ of a CTOQW preserves the set $\mathcal{D}$. More precisely, for $\mu = \sum_{i \in V} \rho(i) \otimes |i\rangle\langle i| \in \mathcal{D}$, we have $\mathcal{T}_t(\mu) =: \sum_{i \in V} \rho_t(i) \otimes |i\rangle\langle i|$ for all $t \geq 0$, with

$$\frac{\mathrm{d}}{\mathrm{d}t}\rho_t(i) = -\mathrm{i}[H_i, \rho_t(i)] + \sum_{j \in V} \left(R_j^i \rho_t(j) R_j^{i*} - \frac{1}{2}\{R_i^{j*} R_i^j, \rho_t(i)\} \right) .$$

18.2.2 *Dyson Expansion and Associated Probability Space*

In this article, our main focus is on a stochastic process $(X_t)_{t \geq 0}$ that informally represents the position of a particle or walker constrained to move on V. In order to rigorously define this process and its associated probability space, we need to introduce the *Dyson expansion* associated with a CTOQW. In particular, this allows to define a probability space on the possible trajectories of the walker. We will recall the result for general QMS as we will use it in the next section. The application to CTOQW is described shortly afterwards.

Let $(\mathcal{T}_t)_{t \geq 0}$ be a uniformly continuous QMS with Lindbladian $\mathcal{L}$ on $\mathcal{I}_1(\mathcal{K})$ for some separable Hilbert space $\mathcal{K}$. By virtue of Lindblad's Theorem [21], there exists a bounded self-adjoint operator $H \in \mathcal{B}(\mathcal{K})$ and bounded operators L_i on $\mathcal{K}$ $(i \in I)$ such that for all $\mu \in \mathcal{I}_1(\mathcal{K})$,

$$\mathcal{L}(\mu) = -\mathrm{i}[H, \mu] + \sum_{i \in I} \left(L_i \mu L_i^* - \frac{1}{2}\{L_i L_i^*, \mu\} \right),$$

where I is a finite or countable set and where the series is strongly convergent. The first step is to give an alternative form for the Lindbladian. First introduce

$$G := -\mathrm{i}H - \frac{1}{2}\sum_{i \in I} L_i^* L_i ,$$

so that for any $\mu \in \mathcal{D}$,

$$\mathcal{L}(\mu) = G\mu + \mu G^* + \sum_{i\in I} L_i \,\mu\, L_i^* \,. \tag{18.4}$$

Remark that $G + G^* + \sum_{i\in I} L_i^* L_i = 0$ (the form described in (18.4) is actually the general form of the generator of a QMS given by Lindblad [21]). The operator $-(G+G^*)$ is positive semidefinite and $t \mapsto \mathrm{e}^{tG}$ defines a one-parameter semigroup of contractions on $\mathcal{K}$ by a trivial application of the Lumer–Phillips theorem (see e.g. Corollary 3.17 in [14]). We are now ready to give the Dyson expansion of the QMS.

Proposition 18.1 *Let $(\mathcal{T}_t)_{t\geq 0}$ be a QMS with Lindbladian $\mathcal{L}$ as given above. For any initial density matrix $\mu \in \mathcal{S}_{\mathcal{K}}$, one has*

$$\mathcal{T}_t(\mu) = \sum_{n=0}^{\infty} \sum_{i_1,\dots,i_n\in I} \int_{0<t_1<\dots<t_n<t} \zeta_t(\xi)\, \mu\, \zeta_t(\xi)^* \,\mathrm{d}t_1 \cdots \mathrm{d}t_n \,, \tag{18.5}$$

where $\zeta_t(\xi) = \mathrm{e}^{(t-t_n)\,G} L_{i_n} \cdots L_{i_1} \mathrm{e}^{t_1\,G}$ *for* $\xi = (i_1, \dots, i_n; t_1, \dots, t_n)$.

We now turn to applying this to CTOQW. Due to the block decomposition of H and of the S^i_j, one can write $G = \sum_{i\in V} G_i \otimes |i\rangle\langle i|$, where (recall that $R^i_i = 0$)

$$G_i = -\mathrm{i}H_i - \frac{1}{2}\sum_j R_i^{j*} R_i^j \,, \tag{18.6}$$

so that $G_i + G_i^* = -\sum_j R_i^{j*} R_i^j$. From Proposition 18.1 we then get the following expression for the Lindbladian: for all $\mu = \sum_{i\in V} \rho(i) \otimes |i\rangle\langle i|$ in $\mathcal{D}$,

$$\mathcal{L}(\mu) = \sum_{i\in V} \Big(G_i \rho(i) + \rho(i) G_i^* + \sum_j R^i_j\, \rho(j)\, R^{i*}_j\Big) \otimes |i\rangle\langle i| \,. \tag{18.7}$$

Corollary 18.1 *Let $(\mathcal{T}_t)_{t\geq 0}$ be a CTOQW with Lindbladian $\mathcal{L}$ given by (18.7). For any initial density matrix $\mu \in \mathcal{D}$, one has*

$$\mathcal{T}_t(\mu) = \sum_{n=0}^{\infty} \sum_{i_0,\dots,i_n\in V} \int_{0<t_1<\dots<t_n<t} T_t(\xi)\, \rho(i_0) T_t(\xi)^* \mathrm{d}t_1 \cdots \mathrm{d}t_n \otimes |i_n\rangle\langle i_n| \,, \tag{18.8}$$

where, for $\xi = (i_0, \dots, i_n; t_1, \dots, t_n)$ with $i_0, \dots, i_n \in V^{n+1}$ and $0 < t_1 < \dots < t_n$,

$$T_t(\xi) := \mathrm{e}^{(t-t_k)G_{i_k}} R^{i_k}_{i_{k-1}} \mathrm{e}^{(t_k-t_{k-1})G_{i_{k-1}}} \cdots \mathrm{e}^{(t_2-t_1)G_{i_1}} R^{i_1}_{i_0} \mathrm{e}^{t_1 G_{i_0}}. \tag{18.9}$$

if k is the largest element such that $t_k \leq t$.

Note the small discrepancy between ξ in (18.5) and ξ in (18.8): ξ contains an additional index i_0, which is due to the decomposition of μ.

Remark 18.1 Equation (18.5) is also called an unravelling of the QMS. It was first introduced in [12, 30], with the heuristic interpretation of giving an expression for $\mathcal{T}_t(\mu)$ as the average, over all possible trajectories $\xi = (i_0, \ldots, i_n; t_1, \ldots, t_n)$, of the evolution of μ "when it follows the trajectory ξ". We will discuss connections with an operational interpretation of $T_t(\xi)\rho(i_0)T_t(\xi)^*$ in Sect. 18.2.4.

The decomposition described in (18.9) will allow us to give a rigorous definition of the probability space associated with the evolution of the particle on V. The goal is to introduce the probability measure $\mathbb{P}_\mu$ that models the law of the position of the particle, when the initial density matrix is $\mu \in \mathcal{D}$. The following is inspired by [4, 7, 19].

First define the set of all possible trajectories up to time $t \in [0, \infty]$ as $\Xi_t := \bigsqcup_{n\in\mathbb{N}} \Xi_t^{(n)}$, where $\Xi_t^{(n)}$ is the set of trajectories on V up to time t comprising n jumps:

$$\Xi_t^{(n)} := \{\xi = (i_0, \ldots, i_n; t_1, \ldots, t_n) \in V^{n+1} \times \mathbb{R}^n,\ 0 < t_1 < \cdots < t_n < t\}\ .$$

For $t \in \mathbb{R}_+$, the set $\Xi_t^{(n)}$ is equipped with the σ-algebra $\Sigma_t^{(t)}$ and with the measure $\nu_t^{(n)}$, which is induced by the map

$$\begin{aligned} I_n : \big(V^{n+1} \times [0,t)^n, \mathcal{P}(V^{n+1}) \times \mathcal{B}([0,t)^n), \delta^{n+1} \times \tfrac{1}{n!}\lambda_n\big) &\to \big(\Xi_t^{(n)}, \Sigma_t^{(n)}, \nu_t^{(n)}\big)\ , \\ (i_0, \ldots, i_n; s_1, \ldots, s_n) &\mapsto (i_0, \ldots, i_n; s_{\min}, \ldots, s_{\max}) \end{aligned}$$

where δ is the counting measure on V, $\mathcal{B}([0,t)^n)$ is the Borel σ-algebra on $[0,t)^n$ and λ_n is the Lebesgue measure on $\mathcal{B}([0,t)^n)$ for all $n \geq 0$. These measures are σ-finite and this allows us to apply Carathéodory's extension Theorem. We first define the σ-algebra $\Sigma_t := \sigma(\Sigma_t^{(t)}, n \in \mathbb{N})$ and the measure ν_t on Ξ_t such that $\nu_t = \nu_t^{(n)}$ on $\Xi_t^{(n)}$. For a given $\mu = \sum_{i\in V} \rho(i) \otimes |i\rangle\langle i|$ in $\mathcal{D}$, one can then define the probability measure $\mathbb{P}_\mu^t$ on (Ξ_t, Σ_t) such that, for all $E \in \Sigma_t$,

$$\begin{aligned} \mathbb{P}_\mu^t(E) &:= \int_E \mathrm{Tr}\big(T_t(\xi)\, \mu\, T_t(\xi)^*\big)\, \mathrm{d}\nu_t(\xi) \\ &= \sum_{n=0}^{\infty} \sum_{i_0,\ldots,i_n\in V} \int_{0<t_1<\cdots<t_n<t} \mathbb{1}_{\xi\in E}\, \mathrm{Tr}\big(T_t(\xi)\rho(i_0)T_t(\xi)^*\big)\, \mathrm{d}t_1 \cdots \mathrm{d}t_n\ , \end{aligned}$$

where $\xi = (i_0, \ldots, i_n; t_1, \ldots, t_n)$ and where $T_t(\xi)$ is defined by Eq. (18.9). The measure $\mathbb{P}_\mu^t$ is a probability measure as one can check that $\mathbb{P}_\mu^t(\Xi_t) = \mathrm{Tr}\big(\mathrm{e}^{t\mathcal{L}}(\mu)\big) = 1$. The family of probability measures $\big(\mathbb{P}_\mu^t\big)_{t\geq 0}$ is consistent, as (18.9) and (18.2.2) show that if $E \in \Sigma_t$,

$$\mathbb{P}_\mu^{t+s}(E) = \sum_{n=0}^{\infty} \sum_{i_0,\dots,i_n \in V} \int_{0<t_1<\dots<t_n<t} \mathbb{1}_{\xi \in E} \operatorname{Tr}\big(e^{s\mathcal{L}}\big(T_t(\xi)\, \rho(i_0)\, T_t(\xi)^*\big)\big)\, dt_1 \cdots dt_n$$
$$= \mathbb{P}_\mu^t(E)$$

for all $t, s \geq 0$. Hence, Kolmogorov's consistency Theorem allows us to extend $(\mathbb{P}_\mu^t)_{t\geq 0}$ to a probability measure $\mathbb{P}_\mu$ on $(\Xi_\infty, \Sigma_\infty)$ where $\Sigma_\infty = \sigma(\Sigma_t, t \in \mathbb{R}_+)$.

In most of our discussions below we will specialize to the case where μ is of the form $\mu = \rho \otimes |i\rangle\langle i|\ Q_{<2\alpha}ii$. In such a case, we denote by $\mathbb{P}_{i,\rho}$ the probability $\mathbb{P}_\mu$.

18.2.3 Quantum Trajectories Associated with CTOQW

Quantum trajectories are another convenient way to describe the distribution of the process $(X_t, \rho_t)_{t\geq 0}$ associated with the CTOQW. Actually, the combination of quantum trajectories and of the Dyson expansion will be essential tools for the main result of this article. Formally speaking, quantum trajectories model the evolution of the state when a continuous measurement of the position of the particle is performed. The state at time t can be described by a pair (X_t, ρ_t) with $X_t \in V$ the position of the particle at time t (as recorded by the measuring device) and $\rho_t \in \mathcal{S}_\mathcal{H}$ the density matrix describing the internal degrees of freedom, given by the wave function collapse postulate and thus constrained to have support on $\mathfrak{h}_i$ alone. The stochastic process $(X_t, \rho_t)_{t\geq 0}$ is then a Markov process, and this will allow us to use the standard machinery for such processes. However, their rigorous description is less straightforward than the one for discrete-time OQW. It makes use of stochastic differential equations driven by jump processes. We refer to [25] for the justification of the below description and for the link between discrete and continuous-time models. Remark that we denote by the same symbol the stochastic process $(X_t)_{t\geq 0}$ appearing in this and the previous section. This will be justified in Remark 18.2.4 below.

In order to present the stochastic differential equation satisfied by the pair $(X_t, \rho_t)_{t\geq 0}$ we need a usual filtered probability space $\big(\Omega, \mathcal{F}, (\mathcal{F}_t)_{t\geq 0}, \mathbb{P}\big)$, where we consider independent Poisson point processes $N^{i,j}, i, j \in V, i \neq j$ on $\mathbb{R}^2$ (again we take $N^{i,i} = 0$ by convention). These Poisson point processes will govern the jump from site i to site j on the graph V.

Definition 18.2 Let $(\mathcal{T}_t)_{t\geq 0}$ be a CTOQW with Lindbladian $\mathcal{L}$ of the form (18.3) and let $\mu = \sum_{i\in V} \rho(i) \otimes |i\rangle\langle i|$ be an initial density matrix in $\mathcal{D}$. The quantum trajectory describing the indirect measurement of the position of the CTOQW is the Markov process $(\mu_t)_{t\geq 0}$ taking values in the set $\mathcal{D}$ such that

$$\mu_0 = \rho_0 \otimes |X_0\rangle\langle X_0|\,,$$

where X_0 and ρ_0 are random with distribution

$$\mathbb{P}\Big((X_0,\rho_0)=\big(i,\frac{\rho(i)}{\mathrm{Tr}(\rho(i))}\big)\Big)=\mathrm{Tr}\big(\rho(i)\big) \text{ for all } i\in V$$

and such that $\mu_t =: \rho_t \otimes |X_t\rangle\langle X_t|$ satisfies for all $t \geq 0$ the following stochastic differential equation:

$$\begin{aligned}\mu_t &= \mu_0 + \int_0^t \mathcal{M}(\mu_{s-})\,\mathrm{d}s \\ &\quad + \sum_{i,j}\int_0^t\int_{\mathbb{R}}\left(\frac{S_i^j\,\mu_{s-}\,S_i^{j*}}{\mathrm{Tr}(S_i^j\,\mu_{s-}S_i^{j*})}-\mu_{s-}\right)\mathbb{1}_{0<y<\mathrm{Tr}(S_i^j\,\mu_{s-}S_i^{j*})}N^{i,j}(\mathrm{d}y,\mathrm{d}s)\end{aligned} \tag{18.10}$$

where

$$\mathcal{M}(\mu)=\mathcal{L}(\mu)-\sum_{i,j}\big(S_i^j\,\mu\,S_i^{j*}-\mu\,\mathrm{Tr}(S_i^j\,\mu\,S_i^{j*})\big)$$

so that for $\mu=\sum_i \rho(i)\otimes|i\rangle\langle i|\in\mathcal{D}$,

$$\mathcal{M}(\mu)=\sum_i\big(G_i\rho(i)+\rho(i)G_i^*-\rho(i)\,\mathrm{Tr}\big(G_i\rho(i)+\rho(i)G_i^*\big)\big)\otimes|i\rangle\langle i|\,.$$

Remark 18.2 An interesting fact has been pointed out in [25]: continuous-time classical Markov chains can be realized within this setup by considering $\mathfrak{h}_i=\mathbb{C}$ for all $i\in V$.

Let us briefly describe the evolution of the solution $(\mu_t)_{t\geq 0}$ of (18.10), and in particular explain why μ_t is of the form $\rho_t\otimes|X_t\rangle\langle X_t|$. Assume that $X_0=i_0$ for some $i_0\in V$ and consider ρ_0 a state on $\mathfrak{h}_{i_0}$. Remark that for any state ρ on $\mathfrak{h}_{i_0}$, $\mathcal{M}(\rho\otimes|i_0\rangle\langle i_0|)$ is of the form $\rho'\otimes|i_0\rangle\langle i_0|$. We then consider the solution, for all $t\geq 0$, of

$$\eta_t=\rho_0+\int_0^t\big(G_{i_0}\eta_s+\eta_sG_{i_0}^*-\eta_s\mathrm{Tr}(G_{i_0}\eta_s+\eta_sG_{i_0}^*)\big)\,\mathrm{d}s\,.$$

We stress the fact that the solution of this equation takes its values in the set on states of $\mathfrak{h}_{i_0}$ (this nontrivial fact is well-known in the theory of quantum trajectories, see [24] for further details). Now let us define the first jump time. To this end we introduce for $j\neq i_0$

$$T_1^j=\inf\Big\{t\geq 0;\; N^{i_0,j}\big(\{u,y\,|\,0\leq u\leq t,\;0\leq y\leq \mathrm{Tr}(R_{i_0}^j\eta_uR_{i_0}^{j\,*})\}\big)\geq 1\Big\}\,.$$

The random variables T_1^j are nonatomic, and mutually independent. Therefore, if we let $T_1 = \inf_{j \neq i_0}\{T_1^j\}$ then there exists a unique $j \in V$ such that $T_1^j = T_1$. In addition,

$$\begin{aligned}
\mathbb{P}\big(T_1 \leq \varepsilon\big) &\leq \sum_{j \neq i_0} \mathbb{P}\big(T_1^j \leq \varepsilon\big) \\
&= \sum_{j \neq i_0} \big(1 - \mathrm{e}^{-\int_0^\varepsilon \mathrm{Tr}(R_{i_0}^j \eta_u R_{i_0}^{j*})\,\mathrm{d}u}\big) \\
&\leq \sum_{j \neq i_0} \int_0^\varepsilon \mathrm{Tr}(R_{i_0}^j \eta_u R_{i_0}^{j*})\,\mathrm{d}u \\
&\leq \varepsilon \sum_{j \neq i_0} \| R_{i_0}^{j*} R_{i_0}^j \|
\end{aligned} \tag{18.11}$$

where the sums are over all j in V with $j \neq i_0$. Now remark that our assumption that $\sum_{i,j} S_i^{j*} S_i^j$ converges strongly implies that the sum $\sum_{j \neq i} \| R_i^{j*} R_i^j \|$ is finite for all i in V, so that Eq. (18.11) implies $\mathbb{P}(T_1 > 0) = 1$. On $[0, T_1]$ we then define the solution $(X_t, \rho_t)_{t \geq 0}$ as

$$(X_t, \rho_t) = (i_0, \eta_t) \text{ for } t \in [0, T_1) \quad \text{and}$$

$$(X_{T_1}, \rho_{T_1}) = \big(j, \frac{R_i^j \eta_{T_1-} R_i^{j*}}{\mathrm{Tr}(R_i^j \eta_{T_1-} R_i^{j*})}\big) \text{ if } T_1 = T_1^j.$$

We then solve

$$\eta_t = \rho_{T_1} + \int_0^t \big(G_j \eta_s + \eta_s G_j^* - \eta_s \mathrm{Tr}(G_j \eta_s + \eta_s G_j^*)\big)\,\mathrm{d}s\ ,$$

and define a new jumping time T_2 as above. By this procedure we define an increasing sequence $(T_n)_n$ of jumping times. We show that $T := \lim_n T_n = +\infty$ almost surely: we introduce

$$N_t = \sum_{i,j} \big(\int_0^{t \wedge T} \int_{\mathbb{R}} \mathbb{1}_{0 < y < \mathrm{Tr}(S_i^j \mu_{s-} S_i^{j*})} N^{i,j}(\mathrm{d}y, \mathrm{d}s)\big)$$

(the sum is over all i, j with $i \neq j$) which counts the number of jumps before t. In particular $N_{T_p} = p$ for all $p \in \mathbb{N}$. Now from the properties of the Poisson processes we have for all $p \in \mathbb{N}$ and all $m \in \mathbb{N}$,

$$\mathbb{E}(N_{T_p \wedge m}) \leq \mathbb{E}\big(N_m\big) = \sum_{i,j} \mathbb{E}\big(\int_0^{m \wedge T} \mathrm{Tr}(S_i^j \mu_{s-} S_i^{j*})\,\mathrm{d}s\big) \leq m \sum_{i,j} \| S_i^{j*} S_i^j \|\ .$$

Denoting $C = \sum_{i,j} \|S_i^{j*} S_i^j\|$ (which is finite) the inequality $p\,\mathbb{P}(T_p \leq m) \leq \mathbb{E}(N_{T_p \wedge m})$ implies

$$\mathbb{P}(T_p \leq m) \leq \frac{m}{p} C\,.$$

This implies that $\mathbb{P}(\lim_p T_p \leq m) = 0$ for all $m \in \mathbb{N}$ so that $\lim_p T_p = +\infty$ almost surely. Therefore, the above considerations define (X_t, ρ_t) for all $t \in \mathbb{R}_+$.

18.2.4 Connection Between Dyson Expansion and Quantum Trajectories

The connection between the process $(X_t, \rho_t)_{t\geq 0}$ defined in this section and the Dyson expansion has been deeply studied in the literature. We do not give all the details of this construction and instead refer to [4, 7] for a complete and rigorous justification. The main point is that the process $(X_t, \rho_t)_{t\geq 0}$ defined in Sect. 18.2.3 can be constructed explicitly on the space $(\Xi^\infty, \Sigma^\infty, \mathbb{P})$, as we now detail.

Recall the interpretation of $\xi = (i_0, \dots, i_n; t_1, \dots, t_n)$ as the trajectory of a particle, initially at i_0 and jumping to i_k at time t_k. First, on $(\Xi_\infty, \Sigma_\infty, \mathbb{P})$ define the random variable $\tilde{N}_t^{i,j}$ by

$$\tilde{N}_t^{i,j}(\xi) = \mathrm{card}\big\{k = 0, \dots, n-1 \mid t_{k+1} \leq t \text{ and } (i_k, i_{k+1}) = (i, j)\big\}$$

for $\xi = (i_0, \dots, i_n; t_1, \dots, t_n)$ as above. Now, let

$$\begin{aligned} \tilde{X}_t(\xi) &= \begin{cases} i_k & \text{if } t_k \leq t < t_{k+1} \\ i_n & \text{if } t_n \leq t. \end{cases} \\ \tilde{\rho}_t(\xi) &= \frac{T_t(\xi)\rho(i_0)\,T_t(\xi)^*}{\mathrm{Tr}(T_t(\xi)\rho(i_0)T_t(\xi)^*)} \end{aligned} \tag{18.12}$$

(recall that $T_t(\xi)$ is defined in (18.9)) and

$$\tilde{\mu}_t = \tilde{\rho}_t \otimes |\tilde{X}_t\rangle\langle\tilde{X}_t|\,.$$

Differentiating (18.12), one can show that the process $(\tilde{\mu}_t)_{t\geq 0}$ satisfies

$$\mathrm{d}\tilde{\mu}_t = \mathcal{M}(\tilde{\mu}_{t-})\,\mathrm{d}t + \sum_{i,j}\Big(\frac{S_i^j\,\tilde{\mu}_{s-}S_i^{j*}}{\mathrm{Tr}(S_i^j\,\tilde{\mu}_{t-}S_i^{j*})} - \tilde{\mu}_{t-}\Big)\,\mathrm{d}\tilde{N}^{i,j}(t)\,. \tag{18.13}$$

It is proved in [4] that the processes

$$(\tilde{N}_t^{i,j})_{t\geq 0} \quad \text{and} \quad \Big(\int_0^t \int_{\mathbb{R}} \mathbb{1}_{0<y<\mathrm{Tr}(S_i^j \mu_{s-} S_i^{j*})} N^{i,j}(\mathrm{d}y, \mathrm{d}s)\Big)_{t\geq 0}$$

(for $(\mu_t)_{t\geq 0}$ and $N^{i,j}$ defined in the previous section) have the same distribution. Therefore, $(\tilde{\mu}_t)_{t\geq 0}$ and $(\mu_t)_{t\geq 0}$ have the same distribution. For this reason, we will denote the random variables $\tilde{\eta}_t$, $\tilde{X}_t$, $\tilde{\rho}_t$ by η_t, X_t, ρ_t, i.e. we identify the random variables obtained by the construction in Sect. 18.2.3 and those defined by (18.12). In addition, from expression (18.8) for $\mathcal{T}_t$ and (18.12) for ρ_t, X_t we recover immediately that $\mu_t = \rho_t \otimes |X_t\rangle\langle X_t|$ satisfies

$$\mathbb{E}_{\mu_0}(\mu_t) = \mathcal{T}_t(\mu_0)$$

where $\mathbb{E}_{\mu_0}$ is the expectation with respect to the probability $\mathbb{P}_{\mu_0}$ defined in Sect. 18.2.2. This identity shows that the quantum Markov semigroup $(\mathcal{T}_t)_{t\geq 0}$ plays for the process $(X_t, \rho_t)_{t\geq 0}$ the same role as the Markov semigroup in the classical case. Because a notion of irreducibility is naturally associated with such a semigroup (see [11] for the original definition and [16] for general considerations on the irreducibility of Lindbladians), this will allow us to associate a notion of irreducibility to a continuous-time open quantum walk.

Now note that expressions (18.12) give an interpretation of X_t and ρ_t in terms of quantum measurement. Indeed, one can see the operator $T_t(\xi)$ for $\xi = (i_0, \ldots, i_n; t_1, \ldots, t_n)$ (or, rather, the map $\rho \mapsto T_t(\xi)\rho T_t(\xi)^*$) as describing the effect of the trajectory where jumps (up to time t) occur at times $t_1,\ldots,t_n$ and $i_0,\ldots,i_n$ is the sequence of updated positions: as long as the particle sits at $i_k \in V$, the evolution of its internal degrees of freedom is given by the semigroup of contraction $(\mathrm{e}^{t\, G_{i_k}})_{t\geq 0}$ and, as the particle jumps to i_{k+1}, it undergoes an instantaneous transformation governed by $R_{i_k}^{i_{k+1}}$ (this $T_t(\xi)$ is then the analogue for continuous-time OQW of the operator L_π of [9]). Therefore, the expression for $\rho_t(\xi)$ in (18.12) encodes the effect of the reduction postulate, or postulate of the collapse of the wave function, on the state of a particle initially at i_0 and with internal state ρ_0. This rigorous connection of the unravelling (18.9) to (indirect) measurement was first described in [4] (see also [23, 24], as well as [13] for a connection to two-time measurement statistics).

To summarize this section and the preceding one, we have defined a Markov process $(\mu_t)_t$ as $\mu_t = \rho_t \otimes |X_t\rangle\langle X_t|$, where $X_t \in V$ and $\rho_t \in \mathcal{S}_{\mathfrak{h}_{X_t}}$, of which the law can be computed in two ways: either by the Dyson expansion of the CTOQW as in (18.2.2) or by use of the stochastic differential equation (18.10).

18.3 Irreducibility of Quantum Markov Semigroups

In this section, we state the equivalence between different notions of irreducibility for general quantum Markov semigroup. Our main motivation is the fact that we could not find a complete proof in the case of an infinite-dimensional Hilbert space, as is required e.g. for CTOQW with infinite V. We then discuss irreducibility for CTOQW.

Theorem 18.1 *Let $\mathcal{T} := (\mathcal{T}_t)_{t\geq 0}$ be a quantum Markov semigroup with Lindbladian*

$$\mathcal{L}(\mu) = G\mu + \mu G^* + \sum_{i\in I} L_i \, \mu \, L_i^* \, . \tag{18.14}$$

The following assertions are equivalent:

1. *$\mathcal{T}$ is* positivity improving*: for all $A \in \mathcal{I}_1(\mathcal{K})$ with $A \geq 0$ and $A \neq 0$, there exists $t > 0$ such that $\mathrm{e}^{t\mathcal{L}}(A) > 0$.*
2. *For any $\varphi \in \mathcal{K}\backslash\{0\}$, the set $\mathbb{C}[\mathcal{L}]\,\varphi$ is dense in $\mathcal{K}$ where $\mathbb{C}[\mathcal{L}]$ is the set of polynomials in e^{tG} for $t > 0$ and in L_i for $i \in I$.*
3. *For any $\varphi \in \mathcal{K}\backslash\{0\}$, the set $\mathbb{C}[G, L]\,\varphi$ is dense in $\mathcal{K}$ where $\mathbb{C}[G, L]$ is the set of polynomials in G and in L_i for $i \in I$.*
4. *$\mathcal{T}$ is irreducible, i.e. there exists $t > 0$ such that $\mathcal{T}_t$ admits no non-trivial projection $P \in \mathcal{B}(\mathcal{K})$ with $\mathcal{T}_t\big(P\mathcal{I}_1(\mathcal{K})P\big) \subset P\mathcal{I}_1(\mathcal{K})P$.*

From now on, any quantum Markov semigroup which satisfies any one of the equivalent statements of Theorem 18.1 is simply called *irreducible*.

Remark 18.3 Positivity improving maps are also called primitive. We therefore call *primitivity* the property of being positivity improving. Remark also that one can replace "there exists $t > 0$" by "for all $t > 0$" in assertions 1. and 4. above to get another equivalent formulation of irreducibility and primitivity. This follows from the observation that assertion 3. does not depend on t.

Proof We first prove the equivalence of 1. and 2. Note that 1. holds if and only if for every $\varphi_0 \neq 0$, there exists $t_0 > 0$ such that $\langle\varphi, \mathrm{e}^{t\mathcal{L}}(|\varphi_0\rangle\langle\varphi_0|)\varphi\rangle > 0$ for all $\varphi \neq 0$. Now remark that from Eq. (18.8),

$$\langle\varphi, \mathrm{e}^{t\mathcal{L}}(|\varphi_0\rangle\langle\varphi_0|)\varphi\rangle = \sum_{n=0}^{\infty} \sum_{i_0,\ldots,i_n\in I} \int_{0<t_1<\cdots<t_n<t} |\langle\varphi, \zeta_t(\xi)\varphi_0\rangle|^2 \, \mathrm{d}t_1 \cdots \mathrm{d}t_n \tag{18.15}$$

where $\xi = (i_1, \ldots, i_n; t_1, \ldots, t_n)$. Assume 1. and fix $\varphi_0 \neq 0$. If for some $t \geq 0$, the left-hand side of (18.15) is positive for any $\varphi \neq 0$, then for any such $\varphi \neq 0$ there exists ξ with $\langle\varphi, \zeta_t(\xi)\varphi_0\rangle \neq 0$. Since $\zeta_t(\xi)\varphi_0 \in \mathbb{C}[\mathcal{L}]\varphi_0$ and the latter is a vector space, this implies that $\mathbb{C}[\mathcal{L}]\varphi_0$ is dense in $\mathcal{K}$. Now assume

2. and fix $\varphi_0 \neq 0$. Since $\mathbb{C}[\mathcal{L}]\varphi_0$ is dense in $\mathcal{K}$, for any $\varphi \neq 0$ there exists an element $\psi = \mathrm{e}^{s_n G} L_{i_n} \cdots L_{i_1} \mathrm{e}^{s_1 G}\varphi_0$ such that $\langle \varphi, \psi \rangle \neq 0$. However, for $t \geq s_1 + \ldots + s_n$, ψ is of the form $\zeta_t(\xi)\varphi_0$ for some $\xi = (i_1, \ldots, i_n; t_1, \ldots, t_n)$. By continuity of ζ in $t_1, \ldots, t_n$, the right-hand side of (18.15) is positive and this proves 1.

To prove the equivalence of 2. and 3., we use the fact that $G = \lim_{t\to 0}(\mathrm{e}^{tG} - \mathrm{Id})/t$, which implies that for any $\varphi \in \mathcal{K}\backslash\{0\}$,

$$\mathbb{C}[G, L]\varphi \subset \overline{\mathbb{C}[\mathcal{L}]\varphi} \subset \overline{\mathbb{C}[\mathcal{L}]\varphi} .$$

Since $\mathrm{e}^{tG} = \lim_{n\to\infty} \sum_{k=0}^{n} t^k G^k/k!$, for any $\varphi \in \mathcal{K}\backslash\{0\}$ we also have

$$\mathbb{C}[\mathcal{L}]\varphi \subset \overline{\mathbb{C}[G, L]\varphi} \subset \overline{\mathbb{C}[G, L]\varphi} .$$

Therefore, for any $\varphi \in \mathcal{K}\backslash\{0\}$,

$$\mathbb{C}[\mathcal{L}]\varphi \text{ is dense in } \mathcal{K} \Leftrightarrow \mathbb{C}[G, L]\varphi \text{ is dense in } \mathcal{K} . \tag{18.16}$$

That 1. implies 4. is obvious. It remains to prove that 4. implies 2. To this end, suppose that $\mathcal{T}$ is irreducible. Let $\varphi \in \mathcal{K}\backslash\{0\}$ and denote by P the orthogonal projection on $\overline{\mathbb{C}[\mathcal{L}]\varphi}$. The goal is to prove that $P = \mathrm{Id}$. For all $\psi \in \mathcal{K}\backslash\{0\}$,

$$\begin{aligned}
\mathrm{e}^{t\mathcal{L}}(P|\psi\rangle\langle\psi|P) &= \sum_{n=0}^{\infty} \sum_{i_0,\ldots,i_n \in I} \int_{0<t_1<\cdots<t_n<t} \zeta_t(\xi) P|\psi\rangle\langle\psi|P\zeta_t(\xi)^* \mathrm{d}t_1 \cdots \mathrm{d}t_n \\
&= \sum_{n=0}^{\infty} \sum_{i_0,\ldots,i_n \in I} \int_{0<t_1<\cdots<t_n<t} |\zeta_t(\xi)P\psi\rangle\langle\zeta_t(\xi)P\psi|\, \mathrm{d}t_1 \cdots \mathrm{d}t_n ,
\end{aligned}$$

Since $\zeta_t(\xi) \in \mathbb{C}[\mathcal{L}]$ and $P\psi \in \overline{\mathbb{C}[\mathcal{L}]\varphi}$, we have $\zeta_t(\xi)P\psi \in \overline{\mathbb{C}[\mathcal{L}]\varphi}$ and thus

$$\mathcal{T}_t(P|\psi\rangle\langle\psi|P) = P\,\mathcal{T}_t(P|\psi\rangle\langle\psi|P)P .$$

Since $\mathcal{T}_t$ is irreducible by assumption, P must be trivial. As it is non-zero, $P = \mathrm{Id}$. Since P is the orthogonal projection on $\overline{\mathbb{C}[\mathcal{L}]\varphi}$, this shows that $\mathbb{C}[\mathcal{L}]\varphi$ is dense in $\mathcal{K}$. □

Remark 18.4 An immediate corollary of Theorem 18.1 is that a quantum Markov semigroup $\mathcal{T} = (\mathcal{T}_t)_t$ is irreducible if and only if its adjoint $\mathcal{T}^* = (\mathcal{T}_t^*)_t$ is irreducible.

We now introduce the notion of irreducibility of a CTOQW, focusing on the trajectorial formulation. Let $\mathcal{T} := (\mathcal{T}_t)_{t\geq 0}$ be a CTOQW on a set V. For i, j in V

and $n \in \mathbb{N}$, we denote by $\mathcal{P}^n(i, j)$ the set of continuous-time trajectories going from i to j in n jumps:

$$\mathcal{P}^n(i, j) = \{\xi = (i_0, \dots, i_n; t_1, \dots, t_n) \in \Xi_\infty^{(n)} \mid i_0 = i,\ i_n = j\}$$

and we set $\mathcal{P}(i, j) = \cup_{n\in\mathbb{N}}\mathcal{P}^n(i, j)$. For any $\xi = (i, \dots, j; t_1, \dots, t_n)$ in $\mathcal{P}(i, j)$, we recall that the operator $T_t(\xi)$ from $\mathfrak{h}_i$ to $\mathfrak{h}_j$ is defined by

$$T_t(\xi) = \mathrm{e}^{(t-t_n)G_{i_n}} R^j_{i_{n-1}} \mathrm{e}^{(t_n-t_{n-1})G_{i_{n-1}}} \dots \mathrm{e}^{(t_2-t_1)G_{i_1}} R^{i_1}_i \mathrm{e}^{t_1 G_i} .$$

The following proposition is a direct application of Theorem 18.1, and will constitute our definition of irreducibility for continuous-time open quantum walks. The criterion here is equivalent to any other formulation proposed in Theorem 18.1.

Proposition 18.2 *A CTOQW with Lindbladian (18.3) is* irreducible *if and only if, for every i and j in V, and for any φ in $\mathfrak{h}_i\setminus\{0\}$, the set*

$$\left\{T_t(\xi)\,\varphi,\ t \geq 0,\ \xi \in \mathcal{P}(i, j)\right\}$$

is total in $\mathfrak{h}_j$.

Remark 18.5 From Theorem 18.1, an equivalent condition of irreducibility is that for every i and j in V and for any φ in $\mathfrak{h}_i\setminus\{0\}$, the set of all $G^{k_n}_{i_n} R^j_{i_{n-1}} G^{k_{n-1}}_{i_{n-1}} \cdots G^{k_1}_{i_1} R^{i_1}_i G^{k_0}_i\,\varphi$ for any $i_0, i_1, \dots, i_n$ with $i_0 = i$ and $i_n = j$, and any $k_0, \dots, k_n$ in $\mathbb{N}_0$, is total in $\mathfrak{h}_j$. This immediately implies that a CTOQW is irreducible if, for every i and j in V and φ in $\mathfrak{h}_i\setminus\{0\}$, the set

$$\left\{R^{i_n}_{i_{n-1}} \dots R^{i_1}_{i_0}\varphi,\ i_0, i_1, \dots, i_n \in V,\ i_0 = i, i_n = j,\ n \in \mathbb{N}_0\right\} \tag{18.17}$$

is total in $\mathfrak{h}_j$. This is equivalent to saying that the completely positive map induced by the off-diagonal terms of $\mathcal{L}$ (i.e. the map $\mu \mapsto \sum_{i,j}(R^i_j \otimes |i\rangle\langle j|)\mu(R^i_j \otimes |i\rangle\langle j|)^*)$ is irreducible as a (discrete-time) completely positive map (see [11, 15]). This of course is true for continuous-time Markov chains, which are irreducible if the discrete-time map induced by the off-diagonal terms is irreducible. In the case of CTOQW, however, this is not true, as the next example shows.

Example 18.1 Consider the OQW with $V = \{1, 2\}$ and $\mathfrak{h}_1 = \mathfrak{h}_2 = \mathbb{C}^2$, and Lindbladian defined by (18.7) with:

$$G_1 = G_2 = \frac{1}{2}\begin{pmatrix} -1 & 2 \\ -2 & -1 \end{pmatrix},\quad R^2_1 = R^1_2 = \begin{pmatrix} 0 & 1 \\ 1 & 0 \end{pmatrix}.$$

One can easily check that $\left\{T_t(\xi)\,\varphi,\ t \geq 0,\ \xi \in \mathcal{P}(i, j)\right\} = \mathfrak{h}_j$ for all $i, j \in \{1, 2\}$ and $\varphi \in \mathfrak{h}_i \setminus \{0\}$, so that the CTOQW is irreducible, but the criterion in (18.17) in terms of R^2_1 and R^1_2 is not satisfied.

18.4 Transience and Recurrence of Irreducible CTOQW

In the classical theory of Markov chains on a finite or countable graph, an irreducible Markov chain can be either transient or recurrent. Transience and recurrence issues are central to the study of Markov chains and help describe the Markov chain's overall structure. In the case of CTOQW, transience and recurrence notions are made more complicated by the fact that the process $(X_t)_{t\geq 0}$ alone is not a Markov chain.

In the present section, we define the notion of recurrence and transience of a vertex in our setup and prove a dichotomy similar to the classical case, based on the average occupation time at a vertex. However, compared to the classical case, the relationship between the occupation time and the first passage time at the vertex is less straightforward. Recall that the first passage time at a given vertex $i \in V$ is defined as

$$\tau_i = \inf\{t \geq T_1 | X_t = i\}$$

where T_1 is defined in Sect. 18.2.3. Similarly the occupation time is given by

$$n_i = \int_0^{\infty} \mathbb{1}_{X_t=i}\, dt \ .$$

In the discrete-time and irreducible case (Theorem 3.1. of [5]), the authors prove that there exists a trichotomy rather than the classical dichotomy. We state a similar result for continuous-time semifinite open quantum walks (we recall that an OQW is semifinite if $\dim \mathfrak{h}_i < \infty$ for all $i \in V$).

Theorem 18.2 *Consider a semifinite irreducible continuous-time open quantum walk. Then we are in one (and only one) of the following situations:*

1. *For any i, j in V and ρ in $\mathcal{S}_{\mathfrak{h}_i}$, one has $\mathbb{E}_{i,\rho}(n_j) = \infty$ and $\mathbb{P}_{i,\rho}(\tau_j < \infty) = 1$.*
2. *For any i, j in V and ρ in $\mathcal{S}_{\mathfrak{h}_i}$, one has $\mathbb{E}_{i,\rho}(n_j) < \infty$ and $\mathbb{P}_{i,\rho}(\tau_i < \infty) < 1$.*
3. *For any i, j in V and ρ in $\mathcal{S}_{\mathfrak{h}_i}$, one has $\mathbb{E}_{i,\rho}(n_j) < \infty$, but there exist i in V and ρ, ρ' in $\mathcal{S}_{\mathfrak{h}_i}$ (ρ necessarily non-faithful) such that $\mathbb{P}_{i,\rho}(\tau_i < \infty) = 1$ and $\mathbb{P}_{i,\rho'}(\tau_i < \infty) < 1$.*

Note that in the sequel we only focus on the semifinite case. Recall that when $\mathfrak{h}_i$ is one-dimensional for all $i \in V$, we recover classical continuous-time Markov chains. In this case, the Markov chain falls in one of the first two categories of this theorem; that is, the third category is a specifically quantum situation.

The rest of this section is dedicated to the proof of Theorem 18.2. More precisely, in Sect. 18.4.1 we prove the dichotomy between infinite and finite average occupation time. This allows us to define transience and recurrence of CTOQW. We also give examples of CTOQW that fall in each of the three classes of Theorem 18.2. In Sect. 18.4.2 we state technical results that give closed expressions for the occupation time and the first passage time. Finally, the proof of Theorem 18.2 is given in Sect. 18.4.3.

18.4.1 Definition of Recurrence and Transience

We begin by proving that for an irreducible CTOQW, the average occupation time $\mathbb{E}_{i,\rho}(n_j)$ of site j starting from site i is either finite for all i, j or infinite for all i, j.

Proposition 18.3 *Consider a semifinite irreducible continuous-time open quantum walk. Suppose furthermore that there exist $i_0, j_0 \in V$ and $\rho_0 \in \mathcal{S}_{\mathfrak{h}_{i_0}}$ such that $\mathbb{E}_{i_0,\rho_0}(n_{j_0}) = \infty$. Then, for all $i, j \in V$ and $\rho \in \mathcal{S}_{\mathfrak{h}_i}$ one has $\mathbb{E}_{i,\rho}(n_j) = \infty$.*

Proof Fix $i, j \in V$ and $\rho \in \mathcal{S}_{\mathfrak{h}_i}$. Then one has

$$\mathbb{E}_{i,\rho}(n_j) = \int_0^\infty \mathbb{P}_{i,\rho}(X_t = j)\,\mathrm{d}t = \int_0^\infty \mathrm{Tr}\big(\mathrm{e}^{t\mathcal{L}}(\rho \otimes |i\rangle\langle i|)(\mathrm{Id} \otimes |j\rangle\langle j|)\big)\,\mathrm{d}t\ .$$

By hypothesis, $(\mathcal{T}_t)_{t\geq 0}$ is irreducible and thus positivity improving by Theorem 18.1; by Remark 18.4 the same is true of $(\mathcal{T}_t^*)_{t\geq 0}$. Therefore, since for any $i \in V$, $\mathfrak{h}_i$ is finite-dimensional, for any $s > 0$ there exist scalars $\alpha, \beta > 0$ such that

$$\mathrm{e}^{s\mathcal{L}}(\rho \otimes |i\rangle\langle i|) \geq \alpha\, \rho_0 \otimes |i_0\rangle\langle i_0| \ \text{ and } \ \mathrm{e}^{s\mathcal{L}^*}(\mathrm{Id} \otimes |j\rangle\langle j|) \geq \beta\, \mathrm{Id} \otimes |j_0\rangle\langle j_0|\ .$$

We then have, fixing $s > 0$,

$$\begin{aligned}\mathbb{E}_{i,\rho}(n_j) &\geq \int_{2s}^\infty \mathrm{Tr}\big(\mathrm{e}^{(t-2s)\mathcal{L}}\big(\mathrm{e}^{s\mathcal{L}}(\rho \otimes |i\rangle\langle i|)\big)\,\mathrm{e}^{s\mathcal{L}^*}(\mathrm{Id} \otimes |j\rangle\langle j|)\big)\,\mathrm{d}t \\ &\geq \alpha\beta \int_0^\infty \mathrm{Tr}\big(\mathrm{e}^{u\mathcal{L}}(\rho_0 \otimes |i_0\rangle\langle i_0|)(\mathrm{Id} \otimes |j_0\rangle\langle j_0|)\big)\,\mathrm{d}u \\ &\geq \alpha\beta\, \mathbb{E}_{i_0,\rho_0}(n_{j_0})\ .\end{aligned}$$

This concludes the proof. □

This proposition leads to a natural definition of recurrent and transient vertices of V:

Definition 18.3 For any continuous-time open quantum walk, we say that a vertex i in V is:

- recurrent if for any $\rho \in \mathcal{S}_{\mathfrak{h}_i}$, $\mathbb{E}_{i,\rho}(n_i) = \infty$;
- transient if there exists $\rho \in \mathcal{S}_{\mathfrak{h}_i}$ such that $\mathbb{E}_{i,\rho}(n_i) < \infty$.

Thus, by Proposition 18.3, for an irreducible CTOQW, either all vertices are recurrent, in which case we say that the CTOQW is recurrent; or all vertices are transient, in which case we say that it is transient. Furthermore, in the transient case, $\mathbb{E}_{i,\rho}(n_i) < \infty$ for all ρ in $\mathcal{S}_{\mathfrak{h}_i}$.

As already mentioned in the introduction, a general notion of recurrence and transience of quantum dynamical semigroups has been defined by Fagnola and Rebolledo in [17] (see also [18]). It is natural to wonder if this general notion reduces to ours in the case of CTOQW. When applied to the semigroup $(\mathcal{T}_t)_{t\geq 0}$, the definition of recurrence in [17] (denoted FR-recurrence in [5]) is that for any positive semidefinite operator A of $\mathcal{B}(\mathcal{H})$, the set

$$D(\mathfrak{U}(A)) = \Big\{\varphi = \sum_{i\in V} \varphi_i \otimes |i\rangle \text{ s.t. } \int_0^\infty \langle \varphi, \mathcal{T}_s^*(A)\,\varphi\rangle \,\mathrm{d}s < \infty\Big\} .$$

equals $\{0\}$. As we have

$$\mathbb{E}_{i,\rho}(n_i) = \int_0^\infty \mathrm{Tr}\big(\rho\,\mathcal{T}_s^*(\mathrm{Id}_{\mathfrak{h}_i} \otimes |i\rangle\langle i|)\big)\,\mathrm{d}s ,$$

we see that our definition of recurrence for CTOQW is equivalent to the fact that for any $i \in V$, $D(\mathfrak{U}(\mathrm{Id}_{\mathfrak{h}_i} \otimes |i\rangle\langle i|)) = \{0\}$. Consequently, it is clear that if the CTOQW is FR-recurrent, then it is recurrent in our sense. Conversely, if the CTOQW is recurrent in our sense, then for any definite-positive A and any $\varphi = \sum_{i\in V} \varphi_i \otimes |i\rangle$, there exists i such that $\varphi_i \neq 0$, and if the CTOQW is semifinite, then $A \geq \lambda_i \mathrm{Id}_{\mathfrak{h}_i} \otimes |i\rangle\langle i|$ for some $\lambda_i > 0$. We then have

$$\int_0^\infty \langle \varphi, \mathcal{T}_s^*(A)\,\varphi\rangle \,\mathrm{d}s \geq \lambda_i\, \mathbb{E}_{i,|\varphi_i\rangle\langle\varphi_i|}(n_i) = +\infty .$$

By Theorem 2 of [17], the quantum dynamical semigroup $(\mathcal{T}_t)_{t\geq 0}$ is not transient, and by Proposition 5 of the same reference, it must be recurrent if $(\mathcal{T}_t)_{t\geq 0}$ is irreducible. Therefore, for irreducible semifinite CTOQW our notion of recurrence and FR-recurrence are equivalent. We refer to [5] for a more complete discussion of the different notions of recurrence that appear in the literature for OQW. Note that the notion of FR-recurrence is more general since it encompasses the case of unbounded generators (the approach of [17] derives from potential theory); here we are essentially interested in semifinite CTOQW in order to have a clear trichotomy, so that our S_i^j are automatically bounded.

We conclude this section by illustrating Theorem 18.2 with simple examples. The n-th example below corresponds to the n-th situation in Theorem 18.2.

Example 18.2

1. For $V = \{0, 1\}$ and $\mathfrak{h}_0 = \mathfrak{h}_1 = \mathbb{C}$, consider the CTOQW characterized by the following operators:

$$G_0 = G_1 = -\frac{1}{2}, \qquad R_0^1 = R_1^0 = 1 .$$

Then the process $(X_t)_{t\geq 0}$ is a classical continuous Markov chain on $\{0, 1\}$, where the walker jumps from one site to the other after an exponential time of parameter 1.

2. For $V = \mathbb{Z}$ and $\mathfrak{h}_i = \mathbb{C}$ for all $i \in \mathbb{Z}$, consider the CTOQW described by the transition operators:

$$G_i = -\frac{1}{2}, \quad R_i^{i+1} = \frac{\sqrt{3}}{2}, \quad R_i^{i-1} = \frac{1}{2} \text{ for all } i \in \mathbb{Z}.$$

The process $(X_t)_{t\geq 0}$ is a classical continuous Markov chain on $\mathbb{Z}$ where after an exponential time of parameter 1, the walker jumps to the right with probability $\frac{3}{4}$ or to the left with probability $\frac{1}{4}$.

3. Consider the CTOQW defined by $V = \mathbb{N}$ with $\mathfrak{h}_1 = \mathbb{C}^2$ and $\mathfrak{h}_0 = \mathfrak{h}_i = \mathbb{C}$ for $i \geq 2$, and

$$G_0 = -\frac{1}{2}, \quad G_1 = -\frac{1}{2} I_2,$$

$$R_0^1 = \frac{1}{\sqrt{5}} \begin{pmatrix} 2 \\ 1 \end{pmatrix}, \quad R_1^0 = \begin{pmatrix} 0 & 1 \end{pmatrix}, \quad R_1^2 = \begin{pmatrix} 1 & 0 \end{pmatrix}, \quad R_2^1 = \frac{1}{2\sqrt{2}} \begin{pmatrix} 1 \\ 1 \end{pmatrix},$$

$$G_i = -\frac{1}{2}, \quad R_i^{i+1} = \frac{\sqrt{3}}{2}, \quad R_{i+1}^i = \frac{1}{2} \text{ for } i \geq 2.$$

This is an example of positivity improving CTOQW where, for $\rho = \begin{pmatrix} 0 & 0 \\ 0 & 1 \end{pmatrix}$, one has $\mathbb{P}_{1,\rho}(\tau_1 < \infty) = 1$ but $\mathbb{P}_{i,\rho'}(\tau_i < \infty) < 1$ for any $\rho' \neq \begin{pmatrix} 0 & 0 \\ 0 & 1 \end{pmatrix}$. This example therefore exhibits “specifically quantum” behavior. It is inspired from [5].

18.4.2 Technical Results

Proposition 18.4 below is essential, as it expresses the probability of reaching a site in finite time as the trace of the initial state, evolved by a certain operator.

Proposition 18.4 *For any continuous-time open quantum walk, there exists a completely positive linear operator $\mathfrak{P}_{i,j}$ from $\mathcal{I}(\mathfrak{h}_i)$ to $\mathcal{I}(\mathfrak{h}_j)$ such that for every $i, j \in V$ and $\rho \in \mathcal{S}_{\mathfrak{h}_i}$,*

$$\mathbb{P}_{i,\rho}(\tau_j < \infty) = \mathrm{Tr}\big(\mathfrak{P}_{i,j}(\rho)\big).$$

Furthermore, the map $\mathfrak{P}_{i,j}$ can be expressed by:

$$\mathfrak{P}_{i,j}(\rho) = \sum_{n=0}^{\infty} \sum_{\substack{i_1,\ldots,i_{n-1} \in V\setminus\{j\} \\ i_0=i, i_n=j}} \int_{0<t_1<\ldots<t_n<\infty} R(\xi)\, \rho\, R(\xi)^* \mathrm{d}t_1 \ldots \mathrm{d}t_n,$$

where $\xi=(i_0,\dots,i_n;t_1,\dots,t_n)$ *and* $R(\xi)=R_{i_{n-1}}^{i_n}\,\mathrm{e}^{(t_n-t_{n-1})G_{i_{n-1}}}\,R_{i_{n-2}}^{i_{n-1}}\dots R_{i_0}^{i_1}\,\mathrm{e}^{t_1G_{i_0}}$.

Note that we do not require the $\mathfrak{h}_i$ to be finite-dimensional here.

Proof We have the trivial identity:

$$\mathbb{P}_{i,\rho}(\tau_j<t)=\sum_{n=0}^{\infty}\sum_{\substack{i_1,\dots,i_{n-1}\in V\setminus\{j\}\\ i_0=i,i_n=j}}\times\int_{0<t_1<\dots<t_n<t}\mathrm{Tr}\big(\mathrm{e}^{(t-t_n)\mathcal{L}}\big(R(\xi)\,\rho\,R(\xi)^*\otimes|j\rangle\langle j|\big)\big)\,\mathrm{d}t_1\dots\mathrm{d}t_n. \tag{18.18}$$

Then, since $\mathrm{e}^{(t-t_n)\mathcal{L}}$ is trace preserving,

$$\mathbb{P}_{i,\rho}(\tau_j<t)=\sum_{n=0}^{\infty}\sum_{\substack{i_1,\dots,i_{n-1}\in V\setminus\{j\}\\ i_0=i,i_n=j}}\int_{0<t_1<\dots<t_n<t}\mathrm{Tr}\big(R(\xi)\,\rho\,R(\xi)^*\big)\,\mathrm{d}t_1\dots\mathrm{d}t_n\ ,$$

and since both sides of the identity are nondecreasing in t, taking the limit $t\to+\infty$ yields

$$\mathbb{P}_{i,\rho}(\tau_j<\infty)=\sum_{n=0}^{\infty}\sum_{\substack{i_1,\dots,i_{n-1}\in V\setminus\{j\}\\ i_0=i,i_n=j}}\int_{0<t_1<\dots<t_n<\infty}\mathrm{Tr}\big(R(\xi)\,\rho\,R(\xi)^*\big)\,\mathrm{d}t_1\dots\mathrm{d}t_n\ .$$

It remains to show that $\mathfrak{P}_{i,j}$ is well defined. Let us denote by $(V_n)_{n\in\mathbb{N}}$ an increasing sequence of subsets of V such that $|V_n|=\min(n,|V|)$ and $\bigcup_{n\in\mathbb{N}}V_n=V$. For any $X\in\mathcal{I}(\mathfrak{h}_i)\setminus\{0\}$ write the canonical decomposition $X=X_1-X_2+\mathrm{i}X_3-\mathrm{i}X_4$ of X as a linear combination of four nonnegative operators. We get

$$\begin{aligned}&\mathrm{Tr}\Big|\sum_{n=0}^{N}\sum_{\substack{i_1,\dots,i_{n-1}\in V_N\setminus\{j\}\\ i_0=i,i_n=j}}\int_{0<t_1<\dots<t_n<N}R(\xi)\,X\,R(\xi)^*\,\mathrm{d}t_1\dots\mathrm{d}t_n\Big|\\
&\le\sum_{m=1}^{4}\mathrm{Tr}\Big|\sum_{n=0}^{N}\sum_{\substack{i_1,\dots,i_{n-1}\in V_N\setminus\{j\}\\ i_0=i,i_n=j}}\int_{0<t_1<\dots<t_n<N}R(\xi)\,X_m\,R(\xi)^*\,\mathrm{d}t_1\dots\mathrm{d}t_n\Big|\\
&\le\sum_{m=1}^{4}\mathrm{Tr}\,X_m\times\sum_{n=0}^{N}\sum_{\substack{i_1,\dots,i_{n-1}\in V_N\setminus\{j\}\\ i_0=i,i_n=j}}\int_{0<t_1<\dots<t_n<N}\mathrm{Tr}\Big(R(\xi)\,\frac{X_m}{\mathrm{Tr}(X_m)}\,R(\xi)^*\Big)\\
&\quad\times\mathrm{d}t_1\dots\mathrm{d}t_n\end{aligned}$$

$$\leq \sum_{m=1}^{4} \operatorname{Tr} X_m \times \mathbb{P}_{i,\frac{X_m}{\operatorname{Tr}(X_m)}}(\tau_j < N)$$
$$\leq \sum_{m=1}^{4} \operatorname{Tr} X_m$$
$$\leq 2\operatorname{Tr}|X|$$

(alternatively apply Theorem 5.17 in [31] to $X_1 - X_2$ and $X_3 - X_4$). Then

$$\sup_N \operatorname{Tr}\Big|\sum_{n=0}^{N} \sum_{\substack{i_1,\ldots,i_{n-1}\in V_N\setminus\{j\}\\ i_0=i, i_n=j}} \int_{0<t_1<\ldots<t_n<N} R(\xi)\, X\, R(\xi)^*\, \mathrm{d}t_1 \ldots \mathrm{d}t_n\Big| < \infty\,.$$

Consequently, by the Banach–Steinhaus Theorem, the operator on $\mathcal{I}(\mathfrak{h}_i)$ to $\mathcal{I}(\mathfrak{h}_j)$ defined by

$$\mathfrak{P}_{i,j}(X) = \sum_{n=0}^{\infty} \sum_{\substack{i_1,\ldots,i_{n-1}\in V\setminus\{j\}\\ i_0=i, i_n=j}} \int_{0<t_1<\cdots<t_n<\infty} R(\xi)\, X\, R(\xi)^*\, \mathrm{d}t_1 \cdots \mathrm{d}t_n$$

is everywhere defined and bounded. □

As a corollary, using the definition of the operator $\mathfrak{P}_{i,j}$ for $i, j \in V$, we obtain a useful expression for $\mathbb{E}_{i,\rho}(n_j)$:

Corollary 18.2 *For every $i, j \in V$ and $\rho \in \mathcal{S}_{\mathfrak{h}_i}$, we have*

$$\mathbb{E}_{i,\rho}(n_j) = \sum_{k=0}^{\infty} \operatorname{Tr}\big(\mathfrak{P}_{j,j}^k \circ \mathfrak{P}_{i,j}(\rho)\big)\,. \tag{18.19}$$

Proof Let $i, j \in V$ and $\rho \in \mathcal{S}_{\mathfrak{h}_i}$. Then

$$\mathbb{E}_{i,\rho}(n_j) = \int_0^\infty \mathbb{P}_{i,\rho}(X_t = j)\, \mathrm{d}t = \int_0^\infty \operatorname{Tr}\big(\mathrm{e}^{t\mathcal{L}}(\rho \otimes |i\rangle\langle i|)(\mathrm{Id} \otimes |j\rangle\langle j|)\big)\, \mathrm{d}t$$
$$= \operatorname{Tr}\Big(\sum_{n=0}^{\infty}\sum_{k=0}^{n} \sum_{\substack{m_1,\ldots,m_k\in\mathbb{N}\\ m_1<\cdots<m_k=n}} \sum_{\substack{i_1,\ldots,i_{m_1-1},i_{m_1+1},\ldots,i_{m_k-1}\neq j\\ i_0=i,\, i_{m_1}=i_{m_2}=\cdots=i_{m_k}=j}}$$
$$\times \int_{0<t_1<\cdots<t_n<t} \Upsilon\rho\Upsilon^*\, \mathrm{d}t_1 \cdots \mathrm{d}t_n\, \mathrm{d}t\Big),$$

where $\Upsilon = R_{i_{n-1}}^{i_n} \mathrm{e}^{(t_n - t_{n-1})G_{i_{n-1}}} R_{i_{n-2}}^{i_{n-1}} \cdots R_{i_{m_1}}^{i_{m_1+1}} \mathrm{e}^{(t_{m_1+1}-t_{m_1})G_{i_{m_1}}} R_{i_{m_1-1}}^{i_{m_1}} \cdots R_{i_0}^{i_1} \mathrm{e}^{t_1 G_{i_0}}$. The above expression corresponds to a decomposition of any path from i to j as a concatenation of a path from i to j, and k paths from j to j which do not go through j except at their start- and endpoints. This yields Eq. (18.19). □

The next corollary allows us to link the quantity $\mathbb{P}_{i,\rho}(\tau_j < \infty)$ to the adjoint of the operator $\mathfrak{P}_{i,j}$. In particular, as we shall see, it is a first step towards linking the properties of $\mathbb{P}_{i,\rho}(\tau_j < \infty)$ and $\mathbb{E}_{i,\rho}(n_j)$.

Corollary 18.3 *Let i and j be in V. One has*

$$\mathbb{P}_{i,\rho}(\tau_j < \infty) = 1 \Leftrightarrow \mathfrak{P}^*_{i,j}(\mathrm{Id}) = \begin{pmatrix} \mathrm{Id} & 0 \\ 0 & * \end{pmatrix}$$

in the decomposition $\mathfrak{h}_i = \mathrm{Ran}\,\rho \oplus (\mathrm{Ran}\,\rho)^\perp$.

In particular, if there exists a faithful ρ in $\mathcal{S}_{\mathfrak{h}_i}$ such that $\mathbb{P}_{i,\rho}(\tau_j < \infty) = 1$, then one has $\mathbb{P}_{i,\rho'}(\tau_j < \infty) = 1$ for any ρ' in $\mathcal{S}_{\mathfrak{h}_i}$.

Proof By Proposition 18.4, one has $\mathbb{P}_{i,\rho}(\tau_j < \infty) = \mathrm{Tr}\big(\rho\, \mathfrak{P}^*_{i,j}(\mathrm{Id})\big)$. Therefore, if $\mathbb{P}_{i,\rho}(\tau_j < \infty) = 1$, then $\mathfrak{P}^*_{i,j}(\mathrm{Id})$ has the following form in the decomposition $\mathfrak{h}_i = \mathrm{Ran}\,\rho \oplus (\mathrm{Ran}\,\rho)^\perp$:

$$\mathfrak{P}^*_{i,j}(\mathrm{Id}) = \begin{pmatrix} \mathrm{Id} & A \\ A & B \end{pmatrix}.$$

Besides, the fact that $\mathrm{Id} \geq \mathfrak{P}^*_{i,j}(\mathrm{Id})$ forces A to be null. In particular, if ρ is faithful, then $\mathfrak{P}^*_{i,j}(\mathrm{Id}) = \mathrm{Id}$ and therefore $\mathbb{P}_{i,\rho'}(\tau_j < \infty) = 1$ for any ρ' in $\mathcal{S}_{\mathfrak{h}_i}$. □

18.4.3 Proof of Theorem 18.2

Let i and j be in V. As we can see in Corollary 18.3, if we suppose that $\mathbb{P}_{i,\rho}(\tau_j < \infty) = 1$ for a faithful density matrix ρ, we necessarily have $\mathfrak{P}^*_{i,j}(\mathrm{Id}) = \mathrm{Id}$. This will be used in the following proposition, which in turn explains the statement regarding non-faithfulness in the third category of Theorem 18.2.

Proposition 18.5 *Let i be in V. If there exists a faithful ρ in $\mathcal{S}_{\mathfrak{h}_i}$ such that $\mathbb{P}_{i,\rho}(\tau_i < \infty) = 1$, then one has $\mathbb{E}_{i,\rho'}(n_i) = \infty$ for any ρ' in $\mathcal{S}_{\mathfrak{h}_i}$.*

Proof We set $\tau_1^i = \tau_i$ and, for all $n > 1$, we define $\tau_i^{(n)}$ as the time at which $(X_t)_{t \geq 0}$ reaches i for the n-th time:

$$\tau_i^{(n)} = \inf\{t > \tau_{n-1}^i \,|\, X_t = i \text{ and } X_{t-} \neq i\}.$$

From Corollary 18.3, one has $\mathbb{P}_{i,\rho'}(\tau_i < \infty) = 1$ for all ρ' in $\mathcal{S}_{\mathfrak{h}_i}$. This implies that for all $n > 0$, $\tau_i^{(n)}$ is $\mathbb{P}_{i,\rho'}$-almost finite for any $\rho' \in \mathcal{S}_{\mathfrak{h}_i}$. For $n \geq 0$, let T_n^i be the occupation time in i between $\tau_i^{(n)}$ and $\tau_i^{(n+1)}$:

$$T_n^i = \inf\{u > 0 \,|\, X_{\tau_i^{(n)}+u} \neq i\}$$

with the convention that $\tau_i^{(0)} = 0$. Since we have

$$\mathbb{E}_{i,\rho'}(n_i) \geq \mathbb{E}_{i,\rho'}\Big(\sum_{n\geq 1} T_n^i\Big) \geq \sum_{n\geq 1} \inf_{\hat{\rho}\in\mathcal{S}_{\mathfrak{h}_i}} \mathbb{E}_{i,\hat{\rho}}(T_n^i)\ ,$$

it will be enough to obtain a lower bound for $\mathbb{E}_{i,\hat{\rho}}(T_n^i)$ which is uniform in n and in $\hat{\rho}$. To this end, we use the quantum trajectories defined in (18.10). We first compute $\mathbb{P}_{i,\hat{\rho}}(T_n^i > t)$ for all $t \geq 0$. To treat the case of $n = 1$ we consider the solution of

$$\eta_t^{\hat{\rho}} = \hat{\rho} + \int_0^t \big(G_i\,\eta_s^{\hat{\rho}} + \eta_s^{\hat{\rho}}\,G_i^* - \eta_s^{\hat{\rho}}\,\mathrm{Tr}(G_i\,\eta_s^{\hat{\rho}} + \eta_s^{\hat{\rho}}\,G_i^*)\big)\,\mathrm{d}s\ . \tag{18.20}$$

Using the independence of the Poisson processes $N^{i,j}$ involved in (18.10) we get

$$\begin{aligned}
\mathbb{P}_{i,\hat{\rho}}(T_1^i > t) &= \mathbb{P}_{i,\hat{\rho}}\big(\text{no jump has occurred before time t}\big)\\
&= \mathbb{P}_{i,\hat{\rho}}\Big(N^{i,j}\big(\{u, y \,|\, 0 \leq u \leq t,\ 0 \leq y \leq \mathrm{Tr}(R_i^j\,\eta_u^{\hat{\rho}}\,R_i^{j*})\}\big) = 0\ \forall j \neq i\Big)\\
&= \prod_{j\neq i} \mathbb{P}_{i,\hat{\rho}}\Big(N^{i,j}\big(\{u, y \,|\, 0 \leq u \leq t,\ 0 \leq y \leq \mathrm{Tr}(R_i^j\,\eta_u^{\hat{\rho}}\,R_i^{j*})\}\big) = 0\Big)\\
&= \prod_{j\neq i} \exp\Big(-\int_0^t \mathrm{Tr}(R_i^j\,\eta_s^{\hat{\rho}}\,R_i^{j*})\,\mathrm{d}s\Big)\\
&= \exp\Big(\int_0^t \mathrm{Tr}\big((G_i + G_i^*)\,\eta_s^{\hat{\rho}}\big)\,\mathrm{d}s\Big)
\end{aligned} \tag{18.21}$$

where we used relation (18.6). Similarly, using the strong Markov property,

$$\begin{aligned}
\mathbb{P}_{i,\hat{\rho}}(T_n^i > t) &= \mathbb{E}_{i,\hat{\rho}}(\mathbb{1}_{T_n^i > t})\\
&= \mathbb{E}_{i,\hat{\rho}}\big(\mathbb{E}_{i,\rho_i^{(n)}}(\mathbb{1}_{T_1^i > t})\big) \quad \text{where } \rho_i^{(n)} := \rho_{\tau_i^{(n)}}\\
&= \mathbb{E}_{i,\hat{\rho}}\Big(\exp\Big(\int_0^t \mathrm{Tr}\big((G_i + G_i^*)\,\eta_s^{\rho_i^{(n)}}\big)\,\mathrm{d}s\Big)\Big)\\
&\geq \mathrm{e}^{-t\|G_i + G_i^*\|_\infty}\ .
\end{aligned}$$

Now, using the fact that $\mathbb{E}_{i,\hat{\rho}}(T_n^i) = \int_0^\infty \mathbb{P}_{i,\hat{\rho}}(T_n^i > t)\,\mathrm{d}t$, this gives us the expected lower bound:

$$\mathbb{E}_{i,\hat{\rho}}(T_n^i) \geq \frac{1}{\|G_i + G_i^*\|_\infty} .$$

This concludes the proof. □

The next proposition is connected to the first point of Theorem 18.2.

Proposition 18.6 *Consider a semifinite irreducible continuous-time open quantum walk. If there exist i, j in V and $\rho \in \mathcal{S}_{\mathfrak{h}_i}$ such that $\mathbb{E}_{i,\rho}(n_j) = \infty$, then one has $\mathbb{P}_{j,\rho'}(\tau_j < \infty) = 1$ for any ρ' in $\mathcal{S}_{\mathfrak{h}_j}$.*

Proof By Proposition 18.2, there is no nontrivial invariant subspace of $\mathfrak{h}_j$ left invariant by $R(\xi)$ for all $\xi \in \mathcal{P}(j, j)$. Since any such ξ is a concatenation of paths from j to j that remain in $V \setminus \{j\}$ except for their start- and endpoints, there is also no nontrivial projection P_j of $\mathfrak{h}_j$ such that $\mathfrak{P}_{j,j}(P_j \mathcal{I}_1(\mathfrak{h}_j) P_j) \subset P_j \mathcal{I}_1(\mathfrak{h}_j) P_j$ (where $\mathfrak{P}_{j,j}$ is the operator of Proposition 18.4). The latter is therefore a completely positive irreducible map acting on the set of trace-class operators on $\mathfrak{h}_j$. By the Russo–Dye Theorem (see [26]), one has $\|\mathfrak{P}_{j,j}\| = \|\mathfrak{P}_{j,j}^*(\mathrm{Id})\| \leq 1$, so that the spectral radius λ of $\mathfrak{P}_{j,j}$ satisfies $\lambda \leq 1$. By the Perron–Frobenius Theorem of Evans and Hoegh-Krøhn (see [15] or alternatively Theorem 3.1 in [27]), there exists a faithful density matrix ρ' on $\mathfrak{h}_j$ such that $\mathfrak{P}_{j,j}(\rho') = \lambda \rho'$. If $\lambda < 1$, then by Corollary 18.2 one has $\mathbb{E}_{j,\rho'}(n_j) < \infty$, but then Proposition 18.3 contradicts our running assumption that $\mathbb{E}_{i,\rho}(n_j) = \infty$. Therefore $\lambda = 1$ and ρ' is a faithful density matrix such that $\mathbb{P}_{j,\rho'}(\tau_j < \infty) = \mathrm{Tr}\big(\mathfrak{P}_{j,j}(\rho')\big) = \mathrm{Tr}(\rho') = 1$. We then conclude by Corollary 18.3. □

Proposition 18.7 *Consider a semifinite irreducible continuous-time open quantum walk; if there exists $i \in V$ such that for all $\rho' \in \mathcal{S}_{\mathfrak{h}_i}$ one has $\mathbb{P}_{i,\rho'}(\tau_i < \infty) = 1$, then $\mathbb{P}_{i,\rho}(\tau_j < \infty) = 1$ for any $j \in V$ and $\rho \in \mathcal{S}_{\mathfrak{h}_i}$.*

Proof Fix i and j in V. Observe first that, by irreducibility, for any ρ in $\mathcal{S}_{\mathfrak{h}_i}$, there exists

$$\xi = (i = i_0, i_1, \ldots, i_{n-1}, i_n = j; t_1, \ldots, t_n)$$

such that $\mathrm{Tr}\big(R(\xi)\rho R(\xi)^*\big) > 0$. We denote by $t(\xi)$ the element t_n of ξ. Using the continuity of $\mathrm{Tr}\big(R(\xi)\rho R(\xi)^*\big)$ in ρ and the compactness of $\mathcal{S}_{\mathfrak{h}_i}$, we obtain a finite family $\xi_1, \ldots, \xi_p$, of paths, again going from i to j, such that

$$\inf_{\rho \in \mathcal{S}_{\mathfrak{h}_i}} \max_{k=1,\ldots,p} \mathrm{Tr}\big(R(\xi_k)\rho R(\xi_k)^*\big) > 0 .$$

Let $\delta > \max_{k=1,\dots,p} t(\xi_k)$. By continuity of each $\mathrm{Tr}\big(R(\xi_i)\rho R(\xi_i)^*\big)$ in the underlying jump times $t_1, \dots, t_n$ and using expression (18.18), we have

$$\alpha := \inf_{\rho\in\mathcal{S}_{\mathfrak{h}_i}} \mathbb{P}_{i,\rho}(\tau_j \leq \delta) > 0\,.$$

Now, if $\mathbb{P}_{i,\rho}(\tau_i < \infty) = 1$ for all ρ in $\mathcal{S}_{\mathfrak{h}_i}$, then the discussion in Sect. 18.2.3 implies that almost-surely one can find an increasing sequence $(\tau_{i,n})_n$ of times with $\tau_{i,n} \to \infty$ and $x_{\tau_{i,n}} = i$. Choose a subsequence $(\tau_{i,\varphi(n)})_n$ such that $\tau_{i,\varphi(n)} - \tau_{i,\varphi(n-1)} > \delta$ for all n. Since never reaching j means in particular not reaching j between $\tau_{i,\varphi(n)}$ and $\tau_{i,\varphi(n+1)}$ for $n = 1, \dots, k$, the Markov property of $(X_t, \rho_t)_{t\geq 0}$ and the lower bound $\tau_{i,\varphi(n)} - \tau_{i,\varphi(n-1)} > \delta$ imply that for all $\rho \in \mathcal{S}_{\mathfrak{h}_i}$,

$$\begin{aligned}\mathbb{P}_{i,\rho}(\tau_j = \infty) &\leq \mathbb{P}_{i,\rho}\big(\forall n \in \{0, \dots, k\},\ \forall t \in [\tau_{i,\varphi(n)}, \tau_{i,\varphi(n+1)}]\,,\ X_t \neq j\big)\\ &\leq \big(\sup_{\rho\in\mathfrak{h}_i} \mathbb{P}_{i,\rho}(\tau_j > \delta)\big)^k\\ &\leq (1-\alpha)^k\,.\end{aligned}$$

Since the above is true for all k, we have $\mathbb{P}_{i,\rho}(\tau_j < \infty) = 1$. □

Now we combine all the results of Sect. 18.4.2 to prove Theorem 18.2.

Proof (Proof of Theorem 18.2) Proposition 18.3 shows that either $\mathbb{E}_{i,\rho}(n_j) = \infty$ for all i, j and ρ, or $\mathbb{E}_{i,\rho}(n_j) < \infty$ for all i, j and ρ. Proposition 18.6 combined with Proposition 18.7 shows that in the former case, $\mathbb{P}_{i,\rho}(\tau_j < \infty) = 1$ for all i, j and ρ as well. Proposition 18.5 shows that, in the latter case, $\mathbb{P}_{i,\rho}(\tau_j < \infty) = 1$ may only occur for non-faithful ρ, and this concludes the proof. □

Acknowledgements All four authors are supported by ANR grant StoQ (ANR-14-CE25-0003-01). The research of Y.P. is also supported by ANR grant NONSTOPS (ANR-17-CE40-0006-01, ANR17-CE40-0006-02, ANR-17-CE40-0006-03).

References

1. S. Attal, F. Petruccione, C. Sabot, I. Sinayskiy, Open quantum random walks. J. Stat. Phys. **147**(4), 832–852 (2012)
2. S. Attal, F. Petruccione, I. Sinayskiy, Open quantum walks on graphs. Phys. Lett. A **376**(18), 1545–1548 (2012)
3. S. Attal, N. Guillotin-Plantard, C. Sabot, Central limit theorems for open quantum random walks and quantum measurement records. Ann. Henri Poincaré **16**(1), 15–43 (2015)
4. A. Barchielli, V.P. Belavkin, Measurements continuous in time and a posteriori states in quantum mechanics. J. Phys. A **24**(7), 1495–1514 (1991)
5. I. Bardet, D. Bernard, Y. Pautrat, Passage times, exit times and Dirichlet problems for open quantum walks. J. Stat. Phys. **167**(2), 173–204 (2017)
6. M. Bauer, D. Bernard, A. Tilloy, The open quantum Brownian motions. J. Stat. Mech. Theory Exp. **2014**(9), p09001, 48 (2014)

7. L. Bouten, M. Guţă, H. Maassen, Stochastic Schrödinger equations. J. Phys. A **37**(9), 3189–3209 (2004)
8. R. Carbone, Y. Pautrat, Homogeneous open quantum random walks on a lattice. J. Stat. Phys. **160**(5), 1125–1153 (2015)
9. R. Carbone, Y. Pautrat, Open quantum random walks: reducibility, period, ergodic properties. Ann. Henri Poincaré **17**(1), 99–135 (2016)
10. S.L. Carvalho, L.F. Guidi, C.F. Lardizabal, Site recurrence of open and unitary quantum walks on the line. Quantum Inf. Process. **16**(1), 32 (2017). Art. 17
11. E.B. Davies, Quantum stochastic processes. II. Commun. Math. Phys. **19**, 83–105 (1970)
12. E.B. Davies, *Quantum Theory of Open Systems* (Academic/Harcourt Brace Jovanovich Publishers, London/New York, 1976)
13. J. Dereziński, W. De Roeck, C. Maes, Fluctuations of quantum currents and unravelings of master equations. J. Stat. Phys. **131**(2), 341–356 (2008)
14. K.-J. Engel, R. Nagel, *One-parameter Semigroups for Linear Evolution Equations*. Graduate Texts in Mathematics, vol. 194 (Springer, New York, 2000). With contributions by S. Brendle, M. Campiti, T. Hahn, G. Metafune, G. Nickel, D. Pallara, C. Perazzoli, A. Rhandi, S. Romanelli and R. Schnaubelt
15. D.E. Evans, R. Høegh-Krohn, Spectral properties of positive maps on C^*-algebras. J. Lond. Math. Soc. (2) **17**(2), 345–355 (1978)
16. F. Fagnola, R. Rebolledo, Subharmonic projections for a quantum Markov semigroup. J. Math. Phys. **43**(2), 1074–1082 (2002)
17. F. Fagnola, R. Rebolledo, Transience and recurrence of quantum Markov semigroups. Probab. Theory Relat. Fields **126**(2), 289–306 (2003)
18. F. Fagnola, R. Rebolledo, Notes on the qualitative behaviour of quantum Markov semigroups, in *Open Quantum Systems. III*. Lecture Notes in Mathematics, vol. 1882 (Springer, Berlin, 2006), pp. 161–205
19. V. Jakšić, C.-A. Pillet, M. Westrich, Entropic fluctuations of quantum dynamical semigroups. J. Stat. Phys. **154**(1–2), 153–187 (2014)
20. C.F. Lardizabal, R.R. Souza, Open quantum random walks: ergodicity, hitting times, gambler's ruin and potential theory. J. Stat. Phys. **164**(5), 1122–1156 (2016)
21. G. Lindblad, On the generators of quantum dynamical semigroups. Commun. Math. Phys. **48**(2), 119–130 (1976)
22. J.R. Norris, *Markov Chains*. Cambridge Series in Statistical and Probabilistic Mathematics, vol. 2 (Cambridge University Press, Cambridge, 1998). Reprint of 1997 original
23. C. Pellegrini, Existence, uniqueness and approximation of a stochastic Schrödinger equation: the diffusive case. Ann. Probab. **36**(6), 2332–2353 (2008)
24. C. Pellegrini, Poisson and diffusion approximation of stochastic master equations with control. Ann. Henri Poincaré **10**(5), 995–1025 (2009)
25. C. Pellegrini, Continuous time open quantum random walks and non-Markovian Lindblad master equations. J. Stat. Phys. **154**(3), 838–865 (2014)
26. B. Russo, H.A. Dye, A note on unitary operators in C^*-algebras. Duke Math. J. **33**, 413–416 (1966)
27. R. Schrader, Perron-Frobenius theory for positive maps on trace ideals (2000). Preprint. arXiv: math-ph/0007020
28. I. Sinayskiy, F. Petruccione, Efficiency of open quantum walk implementation of dissipative quantum computing algorithms. Quantum Inf. Process. **11**(5), 1301–1309 (2012)
29. I. Sinayskiy, F. Petruccione, Quantum optical implementation of open quantum walks. Int. J. Quantum Inf. **12**(2), 1461010, 8 (2014)
30. M.D. Srinivas, E.B. Davies, Photon counting probabilities in quantum optics. Opt. Acta **28**(7), 981–996 (1981)
31. M.M. Wolf, Quantum channels & operations: Guided tour (2012). http://www-m5.ma.tum.de/foswiki/pub/M5/Allgemeines/MichaelWolf/QChannelLecture.pdf. Lecture notes based on a course given at the Niels-Bohr Institute

Chapter 19
Explicit Speed of Convergence of the Stochastic Billiard in a Convex Set

Ninon Fétique

Abstract In this paper, we are interested in the speed of convergence of the stochastic billiard evolving in a convex set K. This process can be described as follows: a particle moves at unit speed inside the set K until it hits the boundary, and is randomly reflected, independently of its position and previous velocity. We focus on convex sets in $\mathbb{R}^2$ with a curvature bounded from above and below. We give an explicit coupling for both the continuous-time process and the embedded Markov chain of hitting points on the boundary, which leads to an explicit speed of convergence to equilibrium.

19.1 Introduction

In this paper, our goal is to give explicit bounds on the speed of convergence of a process, called "stochastic billiard", towards its invariant measure, under some assumptions that we will detail further. This process can be informally described as follows: a particle moves at unit speed inside a domain until it hits the boundary. At this time, the particle is reflected inside the domain according to a random distribution on the unit sphere, independently on its position and previous velocity.

The stochastic billiard is a generalisation of shake-and-bake algorithm (see [1]), in which the reflection law is the cosine law. In that case, it has been proved that the Markov chain of hitting points on the boundary has a uniform stationary distribution. In [1], the shake-and-bake algorithm is introduced for generating uniform points on the boundary of bounded polyhedra. More generally, stochastic billiards can be used for sampling from a bounded set or the boundary of such a set, through the Markov Chain Monte Carlo algorithms. In that sense, it is therefore important to have an idea of the speed of convergence of the process towards its invariant distribution.

N. Fétique (✉)
Institut Denis Poisson, Université de Tours, CNRS, Tours, France

Institut Denis Poisson, Université d'Orléans, CNRS, Orléans, France
e-mail: ninon.fetique@lmpt.univ-tours.fr

C. Donati-Martin et al. (eds.), *Séminaire de Probabilités L*, Lecture Notes in Mathematics 2252, https://doi.org/10.1007/978-3-030-28535-7_19

Stochastic billiards have been studied a lot, under different assumptions on the domain in which it lives and on the reflection law. Let us mention some of these works. In [5], Evans considers the stochastic billiard with uniform reflection law in a bounded d-dimensional region with C^1 boundary, and also in polygonal regions in the plane. He proves first the exponentially fast total variation convergence of the Markov chain, and moreover the uniform total variation Césaro convergence for the continuous-time process. In [3], the authors only consider the stochastic billiard Markov chain, in a bounded convex set with curvature bounded from above and with a cosine distribution for the reflection law. They give a bound for the speed of convergence of this chain towards its invariant measure, that is the uniform distribution on the boundary of the set, in order to get a bound for the number of steps of the Markov chain required to sample approximatively the uniform distribution. Finally, let us mention the work of Comets, Popov, Schütz and Vachkovskaia [2], in which some ideas have been picked and used in the present paper. They study the convergence of the stochastic billiard and its associated Markov chain in a bounded domain in $\mathbb{R}^d$ with a boundary locally Lipschitz and almost everywhere C^1. They consider the case of a reflection law which is absolutely continuous with respect to the Haar measure on the unit sphere of $\mathbb{R}^d$, and supported on the whole half-sphere that points into the domain. They show the exponential ergodicity of the Markov chain and the continuous-time process and also their Gaussian fluctuations. The particular case of the cosine reflection law is discussed. Even if they do not give speeds of convergence, their proofs could lead to explicit speeds if we write them in particular cases (as for the stochastic billiard in a disc of $\mathbb{R}^2$ for instance). However, as we will mention in Sect. 19.2.3, the speed of convergence obtained in particular cases will not be relevant, since their proof is adapted to their very general framework, and not for more particular and simple domains.

The goal of this paper is to give explicit bounds on the speed of convergence of the stochastic billiard and its embedded Markov chain towards their invariant measures. For that purpose, we are going to give an explicit coupling of which we can estimate the coupling time.

In a first part, we study the particular case of the billiard in a disc. In that case, everything is quite simple since all the quantities can be explicitly expressed.

Then, in a second part, we extend the results for the case of the stochastic billiard in a compact convex set of $\mathbb{R}^2$ with curvature bounded from above and below. In that case, we can no more do explicit computations on the quantities describing the process, since we do not know exactly the geometry of the convex set. However, thanks to the assumptions on the curvature, we are able to estimate the needed quantities.

In both cases, the disc and the convex set, we suppose that the reflection law has a density function which is bounded from below by a strictly positive constant on a part of the sphere. The speed of convergence will obviously depend on it. However, for the convergence of the stochastic billiard process in a convex set, we will need to suppose that the reflection law is supported on the whole half sphere that points inside the domain.

At the end of this paper, we briefly discuss the extension of the results to higher dimensions.

19.1.1 Notations

We introduce some notations used in the paper:

- for $A \subset \mathbb{R}$, $\mathbf{1}_A$ denotes the indicator function of the set A, that is $\mathbf{1}_A(x)$ is equal to 1 if $x \in A$ and 0 otherwise;
- for $x \in \mathbb{R}$, $\lfloor x \rfloor$ denotes the floor of the real x;
- for $x, y \in \mathbb{R}^2$, we note by $\|x\|$ the euclidean norm of x and we write $\langle x, y \rangle$ for the scalar product of x and y;
- for $A \subset \mathbb{R}^2$, ∂A denotes the boundary of the set A;
- $\mathcal{B}_r$ denotes the closed ball of $\mathbb{R}^2$ centred at the origin with radius r, i.e. $\mathcal{B}_r = \{x \in \mathbb{R}^2 : \|x\| \leq r\}$, and $\mathbb{S}^1$ denotes the unit sphere of $\mathbb{R}^2$, i.e. $\mathbb{S}^1 = \{x \in \mathbb{R}^2 : \|x\| = 1\}$;
- for $\mathcal{I} \subset \mathbb{R}$, $|\mathcal{I}|$ denotes the Lebesgue measure of the set $\mathcal{I}$;
- for $K \subset \mathbb{R}^2$ a compact convex set, we consider the 1-dimensional Hausdorff measure in $\mathbb{R}^2$ restricted to ∂K. Therefore, if $A \subset \partial K$, $|A|$ denotes this Hausdorff measure of A;
- for $A \subset \mathbb{R}^2$, if $x \in \partial A$, we write n_x the unitary normal vector of ∂A at x pointing to the interior of A and we define $\mathbb{S}_x$ the set of vectors that point to the interior of A: $\mathbb{S}_x = \{v \in \mathbb{S}^1 : \langle v, n_x \rangle \geq 0\}$;
- if two random variables X and Y are equal in law we write $X \overset{\mathcal{L}}{=} Y$, and we write $X \sim \mu$ to say that the random variable X has μ for law, or simply $\mathcal{L}(X)$ to nominate the law of X;
- we denote by $\mathcal{G}(p)$ the geometric law with parameter p.

19.2 Coupling for the Stochastic Billiard

19.2.1 Generalities on Coupling

In order to describe the way we will prove the exponential convergences and obtain bounds on the speeds of convergence, we first need to introduce some notions.

Let ν and $\widetilde{\nu}$ be two probability measures on a measurable space E. We say that a probability measure on $E \times E$ is a coupling of ν and $\widetilde{\nu}$ if its marginals are ν and $\widetilde{\nu}$. Denoting by $\Gamma(\nu, \widetilde{\nu})$ the set of all the couplings of ν and $\widetilde{\nu}$, we say that two random variables Y and $\widetilde{Y}$ satisfy $(Y, \widetilde{Y}) \in \Gamma(\nu, \widetilde{\nu})$ if ν and $\widetilde{\nu}$ are the respective laws of Y and $\widetilde{Y}$. The total variation distance between these two probability measures is then defined by

$$\|\nu - \widetilde{\nu}\|_{TV} = \inf_{(Y,\widetilde{Y}) \in \Gamma(\nu,\widetilde{\nu})} \mathbb{P}(Y \neq \widetilde{Y}).$$

For other equivalent definitions of the total variation distance and its properties, see for instance [6].

Let $(Y_t)_{t\geq 0}$ and $(\tilde{Y}_t)_{t\geq 0}$ be two Markov processes. A coupling $((Y_t, \tilde{Y}_t))_{t\geq 0}$ is called a coalescent coupling if there exists an almost surely finite random time T, such that $Y_{T+s} = \tilde{Y}_{T+s}$ for all $s \geq 0$. In that case, $T_c = \inf\left\{t \geq 0 : Y_t = \tilde{Y}_t\right\}$ is called the coupling time of Y and $\tilde{Y}$, and from the definition of the total variation distance, it immediately follows that

$$\|\mathcal{L}(Y_t) - \mathcal{L}(\tilde{Y}_t)\|_{TV} \leq \mathbb{P}(T_c > t).$$

Therefore, let T^* be a random variable stochastically bigger than T_c, $T_c \leq_{st} T^*$, which means that $\mathbb{P}(T_c \leq t) \geq \mathbb{P}(T^* \leq t)$ for all $t \geq 0$. If T^* has a finite exponential moment, Markov's inequality gives then, for any λ such that the Laplace transform of T^* is well defined:

$$\|\mathcal{L}(Y_t) - \mathcal{L}(\tilde{Y}_t)\|_{TV} \leq \mathbb{P}\left(T^* > t\right) \leq \mathrm{e}^{-\lambda t}\mathbb{E}\left[\mathrm{e}^{\lambda T^*}\right].$$

Thus, if we manage to stochastically bound the coupling time of two stochastic billiards by a random time whose Laplace transform can be estimated, we get an exponential bound for the speed of convergence of the stochastic billiard towards its invariant measure.

Let us now speak about maximal coupling, a result that we will use a lot in the proofs of our main results. Let us consider μ and ν two probability distributions on $\mathbb{R}^d$ with respective density functions f and g with respect to the Lebesgue measure. Let us suppose that there exists a constant $c > 0$ and an interval I such that for all $x \in I$, $f(x) \geq c$ and $g(x) \geq c$. Then, there exists a coupling (X, Y) (called a maximal coupling) of μ and ν such that $\mathbb{P}(X = Y) \geq c|I|$. For more details, see for instance Section 4 of Chapter 1 of [7].

We end this part with a definition that we will use throughout this paper.

Definition 19.1 Let $K \subset \mathbb{R}^2$ be a compact convex set.

We say that a pair of random variables (X, T) living in $\partial K \times \mathbb{R}^+$ is α-continuous on the set $A \times B \subset \partial K \times \mathbb{R}^+$ if for any measurable $A_1 \subset A$, $B_1 \subset B$:

$$\mathbb{P}(X \in A_1, T \in B_1) \geq \alpha|A_1||B_1|.$$

We can also adapt this definition for a single random variable.

19.2.2 Description of the Process

Let us now give a precise description of the stochastic billiard $((X_t, V_t))_{t\geq 0}$ in a set K.

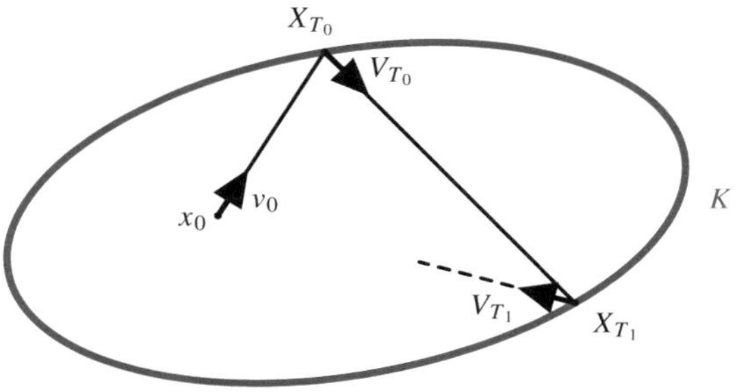

Fig. 19.1 A trajectory of the stochastic billiard in a set K, starting in the interior of K

We assume that $K \subset \mathbb{R}^2$ is a compact convex set with a boundary at least C^1.

Let $e = (1, 0)$ be the first coordinate vector of the canonical basis of $\mathbb{R}^2$. We consider a law γ on the half-sphere $\mathbb{S}_e = \{v \in \mathbb{S}^1 : e \cdot v \geq 0\}$. Let moreover $(U_x, x \in \partial K)$ be a family of rotations of $\mathbb{S}^1$ such that $U_x e = n_x$, where we recall that n_x is the normal vector of ∂K at x pointing to the interior of K.

Let $(\eta_n)_{n\geq 0}$ be a sequence of i.i.d. random variables on $\mathbb{S}_e$ with law γ.

Given $(x_0, v_0) \in K \times \mathbb{S}^1$, we consider the process $((X_t, V_t))_{t\geq 0}$ living in $K \times \mathbb{S}^1$ constructed as follows (see Fig. 19.1):

- If $x_0 \in K \setminus \partial K$, let $(X_0, V_0) = (x_0, v_0)$ and let $T_0 = \inf\{t > 0 : x_0 + tv_0 \notin K\}$. For $t \in [0, T_0)$ let then $X_t = x_0 + tv_0$ and $V_t = v_0$. Else, i.e. if $x_0 \in \partial K$, let $T_0 = 0$.
- Let $X_{T_0} = x_0 + T_0 v_0$, and $V_{T_0} = U_{X_0}\eta_0$.
- Let $\tau_1 = \inf\{t > 0 : X_{T_0} + tV_{T_0} \notin K\}$ and define $T_1 = \tau_1 + T_0$. We put $X_t = X_{T_0} + tV_{T_0}$, $V_t = V_{T_0}$ for $t \in [T_0, T_1)$, and $X_{T_1} = X_{T_0} + \tau_1 V_{T_0}$.
  Then, let $V_{T_1} = U_{X_{T_1}}\eta_1$.
- Let $\tau_2 = \inf\{t > 0 : X_{T_1} + tV_{T_1} \notin K\}$ and define $T_2 = T_1 + \tau_2$. We put $X_t = X_{T_1} + tV_{T_1}$, $V_t = V_{T_1}$ for $t \in [T_1, T_2)$, and $X_{T_2} = X_{T_1} + \tau_2 V_{T_1}$.
  Then, let $V_{T_2} = U_{X_{T_2}}\eta_2$.
- And we start again …

As mentioned in the introduction $(X_{T_n})_{n\geq 0}$ is a Markov chain living in ∂K and the process $((X_t, V_t))_{t\geq 0}$ is a Markov process living in $K \times \mathbb{S}^1$.

For $x \in \partial K$, it is equivalent to consider the new speed in $\mathbb{S}_x$ or to consider the angle in $\left[-\frac{\pi}{2}, \frac{\pi}{2}\right]$ between this vector speed and the normal vector n_x. For $n \geq 1$, we thus denote by Θ_n the random variable in $\left[-\frac{\pi}{2}, \frac{\pi}{2}\right]$ such that $r_{X_{T_n},\Theta_n}(n_{X_{T_n}}) \overset{\mathcal{L}}{=} V_{T_n}$, where for $x \in \partial K$ and $\theta \in \mathbb{R}$, $r_{x,\theta}$ denotes the rotation with center x and angle θ.

We make the following assumption on γ (see Fig. 19.2):

Assumption $(\mathcal{H})$

The law γ has a density function ρ with respect to the Haar measure on $\mathbb{S}_e$, which satisfies: there exist $\mathcal{J}$ an open subset of $\mathbb{S}_e$, containing e and symmetric with respect to e, and $\rho_{\min} > 0$ such that:

$$\rho(u) \geq \rho_{\min}, \quad \text{for all } u \in \mathcal{J}.$$

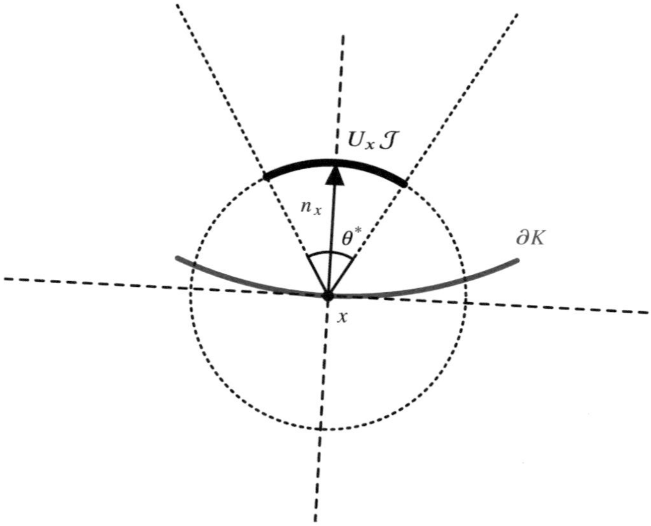

Fig. 19.2 Illustration of Assumptions $(\mathcal{H})$ and $(\mathcal{H}')$

This assumption is equivalent to the following one on the variables $(\Theta_n)_{n\geq 0}$:

Assumption $(\mathcal{H}')$

The variables Θ_n, $n \geq 0$, have a density function f with respect to the Lebesgue measure on $\left[-\frac{\pi}{2}, \frac{\pi}{2}\right]$ satisfying: there exist $f_{\min} > 0$ and $\theta^* \in (0, \pi)$ such that:

$$f(\theta) \geq f_{\min}, \quad \text{for all } \theta \in \left[-\frac{\theta^*}{2}, \frac{\theta^*}{2}\right].$$

In fact, since these two assumptions are equivalent, we have

$$\rho_{\min} = f_{\min} \quad \text{and} \quad |\mathcal{J}| = \theta^*.$$

In the sequel, we will use both descriptions of the speed vector depending on which is the most suitable.

19.2.3 A Coupling for the Stochastic Billiard

Let us now informally describe the idea of the couplings used to explicit the speeds of convergence of our processes to equilibrium. They will be explained explicitly in Sects. 19.3 and 19.4.

To get a bound on the speed of convergence of the Markov chain recording the location of hitting points on the boundary of the stochastic billiard, the strategy is the following. We consider two stochastic billiard Markov chains with different initial conditions. We estimate the number of steps that they have to do before they have a strictly positive probability to arrive on the same place at a same step. In particular, it is sufficient to know the number of steps needed before the position of each chain charges the half of the boundary of the set on which they evolve. Then, their coupling time is stochastically smaller than a geometric time whose Laplace transform is known.

The case of the continuous-time process is a bit more complicated. To couple two stochastic billiards, it is not sufficient to make them cross in the interior of the set where they live. Indeed, if they cross with a different speed, then they will not be equal after. So the strategy is to make them arrive at the same place on the boundary of the set at the same time, and then they can always keep the same velocity and stay equal. We will do this in two steps. First, we will make the two processes hit the boundary at the same time, but not necessarily at the same point. This will take some random time, that we will be able to quantify. And secondly, with some strictly positive probability, after two bounces, the two processes will have hit the boundary at the same point at the same time. We repeat the scheme until the first success. This leads us to a stochastic upper bound for the coupling time of two stochastic billiards.

Obviously, the way that we couple our processes is only one way to do that, and there are many as we want. Let us for instance describe the coupling constructed in [2]. Consider two stochastic billiard processes evolving in the set K with different initial conditions. Their first step is to make the processes hit the boundary in the neighbourhood of a good $x_1 \in \partial K$. This can be done after n_0 bounces, where n_0 is the minimum number of bounces needed to connect any two points of the boundary of K. Once the two processes have succeeded, they are in the neighbourhood of x_1, but at different times. Then, the strategy used by the authors of [2] is to make the two processes do round trips between the neighbourhood of x_1 and the neighbourhood of another good $y_1 \in \partial K$. Thereby, if the point y_1 is well chosen, the time difference between the two processes decreases gradually, while the positions of the processes stay the same after one round trip. However, the number of round trips needed to compensate for the possibly big difference of times could be very high. This particular coupling is therefore well adapted for sets whose boundary can be quite "chaotic", but not for convex sets with smooth boundary as we consider in this paper.

19.3 Stochastic Billiard in the Disc

In this section, we consider the particular case where K is a ball: $K = \mathcal{B}_r$, for some fixed $r > 0$.

In that case, for each $n \geq 0$, the couple $(X_{T_n}, V_{T_n}) \in \partial\mathcal{B}_r \times \mathbb{S}^1$ can be represented by a couple $(\Phi_n, \Theta_n) \in [0, 2\pi) \times \left[-\frac{\pi}{2}, \frac{\pi}{2}\right]$ as follows (see Fig. 19.3):

- to a position x on $\partial\mathcal{B}_r$ corresponds a unique angle $\phi \in [0, 2\pi)$. The variable Φ_n nominates this unique angle associated to X_{T_n}, i.e. (r, Φ_n) are the polar coordinates of X_{T_n}.

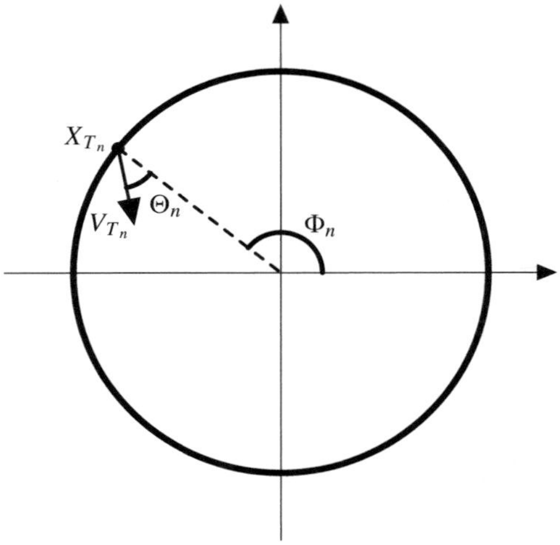

Fig. 19.3 Definition of the variables Φ_n and Θ_n in bijection with the variables X_{T_n} and V_{T_n}

- at each speed V_{T_n} we associate the variable Θ_n introduced in Sect. 19.2.2, satisfying Assumption $(\mathcal{H}')$.

Remark that for all $n \geq 0$, the random variable Θ_n is independent of Φ_k for all $k \in \{0, \cdots, n\}$. We also recall that the variables Θ_n, $n \geq 0$, are all independent.

In the sequel, we do not care about the congruence modulo 2π : it is implicit that when we write Φ, we consider its representative in $[0, 2\pi)$.

Let us state the following proposition that links the different random variables together.

Proposition 19.1 *For all $n \geq 1$ we have:*

$$\tau_n = 2r\cos(\Theta_{n-1}) \quad \textit{and} \quad \Phi_n = \pi + 2\Theta_{n-1} + \Phi_{n-1}. \tag{19.1}$$

Proof The relationships are immediate with geometric considerations. □

19.3.1 The Embedded Markov Chain

In this section, the goal is to obtain a control of the speed of convergence of the stochastic billiard Markov chain on the circle. For this purpose, we study the distribution of the position of the Markov chain at each step.

Let $\Phi_0 = \phi_0 \in [0, 2\pi)$.

Proposition 19.2 *Let $(\Phi_n)_{n\geq 0}$ be the stochastic billiard Markov chain evolving on $\partial\mathcal{B}_r$, satisfying assumption $(\mathcal{H}')$. Let us denote by f_{Φ_n} the density function of Φ_n, for $n \geq 1$.*

We have

$$f_{\Phi_1}(u) \geq \frac{f_{\min}}{2}, \quad \forall u \in \mathcal{I}_1 = \left[\pi - \theta^* + \phi_0, \pi + \theta^* + \phi_0\right].$$

Moreover, for all $n \geq 2$, *for all* $\eta_2, \cdots, \eta_n$ *such that* $\eta_2 \in (0, 2\theta^*)$, *and for* $k \in \{2, \cdots, n-1\}$, $\eta_{k+1} \in \left(0, \min\left\{2\theta^*; k\theta^* - \sum_{\ell=2}^{k} \eta_\ell\right\}\right)$, *we have*

$$f_{\Phi_n}(u) \geq \left(\frac{f_{\min}}{2}\right)^n \prod_{k=2}^{n} \eta_k,$$

$$\forall u \in \mathcal{I}_n = \left[n(\pi - \theta^*) + \phi_0 + \sum_{k=2}^{n} \eta_k, n(\pi + \theta^*) + \phi_0 - \sum_{k=2}^{n} \eta_k\right].$$

Proof Since the Markov chain is rotationally symmetric, we do the computations with $\phi_0 = 0$.

- Case $n = 1$:

  We have, thanks to (19.1), $\Phi_1 = \pi + 2\Theta_0 + \phi_0 = \pi + 2\Theta_0$. Thus, for any measurable bounded function g, we get:

$$\begin{aligned}\mathbb{E}\left[g(\Phi_1)\right] = \mathbb{E}\left[g\left(\pi + 2\Theta_0\right)\right] &= \int_{-\frac{\pi}{2}}^{\frac{\pi}{2}} g\left(\pi + 2x\right) f(x)\mathrm{d}x \\ &\geq f_{\min} \int_{-\frac{\theta^*}{2}}^{\frac{\theta^*}{2}} g\left(\pi + 2x\right)\mathrm{d}x = \frac{f_{\min}}{2} \int_{\pi-\theta^*}^{\pi+\theta^*} g(u)\mathrm{d}u.\end{aligned}$$

  We deduce:

$$f_{\Phi_1}(u) \geq \frac{f_{\min}}{2}, \quad \forall u \in \left[\pi - \theta^*, \pi + \theta^*\right].$$

- Induction: let us suppose that for some $n \geq 1$, $f_{\Phi_n}(u) \geq c_n$ for all $u \in [a_n, b_n]$. Then, using the relationship (19.1) and the independence between Θ_n and Φ_n we have, for any measurable bounded function g:

$$\begin{aligned}\mathbb{E}\left[g(\Phi_{n+1})\right] &= \mathbb{E}\left[g(\pi + 2\Theta_n + \Phi_n)\right] \\ &\geq f_{\min} c_n \int_{-\frac{\theta^*}{2}}^{\frac{\theta^*}{2}} \int_{a_n}^{b_n} g(\pi + 2\theta + x)\mathrm{d}x\mathrm{d}\theta.\end{aligned}$$

  Using the substitution $u = \pi + 2\theta + x$ in the integral with respect to x and Fubini's theorem, we have:

$$\mathbb{E}\left[g(\Phi_{n+1})\right] \geq f_{\min} c_n \int_{\pi-\theta^*+a_n}^{\pi+\theta^*+b_n} \left(\int_{-\frac{\theta^*}{2}}^{\frac{\theta^*}{2}} \mathbf{1}_{\frac{1}{2}(u-\pi-b_n)\leq\theta\leq\frac{1}{2}(u-\pi-a_n)}\mathrm{d}\theta\right) g(u)\mathrm{d}u,$$

and we deduce the following lower bound of the density function $f_{\Phi_{n+1}}$ of Φ_{n+1}:

$$f_{\Phi_{n+1}}(u) \geq f_{\min} c_n \left\| \left[-\frac{\theta^*}{2}, \frac{\theta^*}{2} \right] \cap \left[\frac{1}{2}(u - \pi - b_n), \frac{1}{2}(u - \pi - a_n) \right] \right\|,$$

for all $u \in [\pi - \theta^* + a_n, \pi + \theta^* + b_n]$.

When u is equal to one extremal point of this interval, this lower bound is equal to 0. However, let $\eta_{n+1} \in \left(0, \min\left\{2\theta^*; \frac{1}{2}(b_n - a_n)\right\}\right)$. Then the intersection $\left[-\frac{\theta^*}{2}, \frac{\theta^*}{2}\right] \cap \left[\frac{1}{2}(u - \pi - b_n), \frac{1}{2}(u - \pi - a_n)\right]$ is non-empty, and we have, for
$u \in [\pi - \theta^* + a_n + \eta_{n+1}, \pi + \theta^* + b_n - \eta_{n+1}]$:

$$f_{\Phi_{n+1}}(u) \geq f_{\min} c_n \min\left\{\theta^*; \frac{\eta_{n+1}}{2}\right\} = f_{\min} c_n \frac{\eta_{n+1}}{2}.$$

The result follows immediately. □

By choosing a constant sequence for the η_k, $k \geq 2$ in the Proposition 19.2, we immediately deduce:

Corollary 19.1 *For all $n \geq 2$, for all $\varepsilon \in (0, \theta^*)$, we have*

$$f_{\Phi_n}(u) \geq \left(\frac{f_{\min}}{2}\right)^n \varepsilon^{n-1},$$

$\forall u \in \mathcal{J}_n = [n(\pi - \theta^*) + \phi_0 + (n-1)\varepsilon, n(\pi + \theta^*) + \phi_0 - (n-1)\varepsilon]$.

Let $(\mathcal{J}_n)_{n\geq 2}$ defined as in Corollary 19.1. We put $\mathcal{J}_1 = \mathcal{I}_1$ with $\mathcal{I}_1$ defined in Proposition 19.2.

Theorem 19.1 *Let $(\Phi_n)_{n\geq 0}$ be the stochastic billiard Markov chain on the circle $\partial\mathcal{B}_r$, satisfying Assumption ($\mathcal{H}'$).*

There exists a unique invariant probability measure ν on $[0, 2\pi)$ for the Markov chain $(\Phi_n)_{n\geq 0}$, and we have:

1. *if $\theta^* > \frac{\pi}{2}$, for all $n \geq 0$,*

$$\|\mathbb{P}(\Phi_n \in \cdot) - \nu\|_{TV} \leq \left(1 - f_{\min}(2\theta^* - \pi)\right)^{n-1},$$

2. *if $\theta^* \leq \frac{\pi}{2}$, for all $n \geq 0$ and all $\varepsilon \in (0, \theta^*)$,*

$$\|\mathbb{P}(\Phi_n \in \cdot) - \nu\|_{TV} \leq (1 - \alpha)^{\frac{n}{n_0} - 1},$$

where

$$n_0 = \left\lfloor \frac{\pi - 2\varepsilon}{2(\theta^* - \varepsilon)} \right\rfloor + 1 \quad \textit{and} \quad \alpha = \left(\frac{\varepsilon}{2}\right)^{n_0 - 1} {f_{\min}}^{n_0} \left(2n_0\theta^* - 2(n_0 - 1)\varepsilon - \pi\right).$$

Proof The existence of the invariant measure is immediate thanks to the compactness of $\partial\mathcal{B}_r$ (see [4]). The following proof leads to its uniqueness and the speed of convergence.

Let $((\Phi_n, \Theta_n))_{n\geq 0}$ and $((\tilde{\Phi}_n, \tilde{\Theta}_n))_{n\geq 0}$ be two versions of the process described above, with initial positions ϕ_0 and $\tilde{\phi}_0$ on $\partial\mathcal{B}_r$.

In order to couple Φ_n and $\tilde{\Phi}_n$ at some time n, it is sufficient to show that the intervals $\mathcal{J}_n$ and $\tilde{\mathcal{J}}_n$ corresponding to Corollary 19.1 have a non empty intersection. Since these intervals are included in $[0, 2\pi)$, a sufficient condition to have $\mathcal{J}_n \cap \tilde{\mathcal{J}}_n \neq \emptyset$ is that the length of these two intervals is strictly bigger than π.

Let $\varepsilon \in (0, \theta^*)$. We have

$$|\mathcal{J}_1| = |\tilde{\mathcal{J}}_1| = 2\theta^*,$$

and for $n \geq 2$,

$$|\mathcal{J}_n| = |\tilde{\mathcal{J}}_n| = 2n\theta^* - 2(n-1)\varepsilon.$$

Therefore the length of $\mathcal{J}_n$ is a strictly increasing function of n (which is intuitively clear).

- Case 1: $\theta^* > \frac{\pi}{2}$. In that case we have $|\mathcal{J}_1| = |\tilde{\mathcal{J}}_1| > \pi$. Therefore we can construct a coupling $\left(\Phi_1, \tilde{\Phi}_1\right)$ (see the reminder on maximal coupling in Sect. 19.2.1) such that we have, using Proposition 19.2:

$$\begin{aligned}\mathbb{P}\left(\Phi_1 = \tilde{\Phi}_1\right) &\geq \frac{f_{\min}}{2}\left|\mathcal{J}_1 \cap \tilde{\mathcal{J}}_1\right| \\ &\geq \frac{f_{\min}}{2} 2(2\theta^* - \pi) \\ &= f_{\min}(2\theta^* - \pi).\end{aligned}$$

- Case 2: $\theta^* \leq \frac{\pi}{2}$. Here we need more jumps before having a positive probability to couple Φ_n and $\tilde{\Phi}_n$. Let us thus define

$$n_0 = \min\{n \geq 2 : 2n\theta^* - 2(n-1)\varepsilon > \pi\} = \left\lfloor \frac{\pi - 2\varepsilon}{2(\theta^* - \varepsilon)} \right\rfloor + 1.$$

  Using the lower bound of the density function of Φ_{n_0} obtained in Corollary 19.1, we deduce that we can construct a coupling $\left(\Phi_{n_0}, \tilde{\Phi}_{n_0}\right)$ such that:

$$\begin{aligned}\mathbb{P}\left(\Phi_{n_0} = \tilde{\Phi}_{n_0}\right) &\geq \left(\frac{f_{\min}}{2}\right)^{n_0} \varepsilon^{n_0 - 1}\left|\mathcal{J}_{n_0} \cap \tilde{\mathcal{J}}_{n_0}\right| \\ &\geq \left(\frac{f_{\min}}{2}\right)^{n_0} \varepsilon^{n_0-1} 2\left(2n_0\theta^* - 2(n_0 - 1)\varepsilon - \pi\right) \\ &= \left(\frac{\varepsilon}{2}\right)^{n_0-1} (f_{\min})^{n_0}\left(2n_0\theta^* - 2(n_0-1)\varepsilon - \pi\right).\end{aligned}$$

To treat both cases together, let us define

$$m_0 = \mathbf{1}_{\theta^*>\frac{\pi}{2}} + \left(\left\lfloor \frac{\pi - 2\varepsilon}{2(\theta^* - \varepsilon)} \right\rfloor + 1\right) \mathbf{1}_{\theta^* \leq \frac{\pi}{2}}.$$

and

$$\alpha = f_{\min}(2\theta^* - \pi)\mathbf{1}_{\theta^*>\frac{\pi}{2}} + \left(\frac{\varepsilon}{2}\right)^{m_0 - 1} (f_{\min})^{m_0} \left(2m_0\theta^* - 2(m_0 - 1)\varepsilon - \pi\right) \mathbf{1}_{\theta^* \leq \frac{\pi}{2}}.$$

Since our processes $(\Phi_n)_{n\geq 0}$ and $(\tilde{\Phi}_n)_{n\geq 0}$ are Markovian processes, once they are equal, we can let them equal afterwards. And then we get:

$$\begin{aligned}
\|\mathbb{P}(\Phi_n \in \cdot) - \nu\|_{TV} &\leq \mathbb{P}\left(\Phi_n \neq \tilde{\Phi}_n\right) \\
&\leq \mathbb{P}\left(\Phi_{\lfloor \frac{n}{m_0} \rfloor m_0} \neq \tilde{\Phi}_{\lfloor \frac{n}{m_0} \rfloor m_0}\right) \\
&\leq (1 - \alpha)^{\lfloor \frac{n}{m_0} \rfloor} \\
&\leq (1 - \alpha)^{\frac{n}{m_0} - 1}.
\end{aligned}$$

19.3.2 The Continuous-Time Process

We assume here that the constant θ^* introduced in Assumption $(\mathcal{H}')$ satisfies

$$\theta^* \in \left(\frac{2\pi}{3}, \pi\right).$$

This condition on θ^* is essential in the proof of Theorem 19.2 to couple our processes with "two jumps". However, if $\theta^* \in \left(0, \frac{2\pi}{3}\right]$ we can adapt our method (see Remark 19.2).

Notation We define a sequence $(S_n)_{n\geq 0}$ of random times by

$$S_0 = 0 \quad \text{and for } n \geq 1, \; S_n = T_n - T_0.$$

By this way, we avoid the presence of the time T_0 in the computations, and the law of the random time S_n is the law of the instant of the nth bounce, when starting on the boundary of $\mathcal{B}_r$. If $T_0 = 0$, that is if the process starts on the boundary of $\mathcal{B}_r$, then $(S_n)_{n\geq 0} = (T_n)_{n\geq 0}$.

We observe that thanks to the rotational symmetry of the process, for all $m, n, p \geq 0$ we have the following equality in law: $S_{n+m} - S_m \overset{\mathcal{L}}{=} S_n \overset{\mathcal{L}}{=} T_{n+p} - T_p$.

Proposition 19.3 *Let $((X_t, V_t))_{t\geq 0}$ be the stochastic billiard process in the ball $\mathcal{B}_r$ satisfying Assumption $(\mathcal{H}')$ with $\theta^* \in \left(\frac{2\pi}{3}, \pi\right)$.*

We denote by f_{S_2} the density function of S_2. Let $\eta \in \left(0, 2r\left(1-\cos\left(\frac{\theta^*}{2}\right)\right)\right)$. We have

$$f_{S_2}(x) \geq \delta \quad \text{for all } x \in [4r\cos\left(\frac{\theta^*}{2}\right) + \eta, 4r - \eta],$$

where

$$\delta = \frac{2f_{\min}^2}{r\sin\left(\frac{\theta^*}{2}\right)} \min\left\{\frac{\theta^*}{2} - \arccos\left(\cos\left(\frac{\theta^*}{2}\right) + \frac{\eta}{2r}\right); \arccos\left(1 - \frac{\eta}{2r}\right)\right\}. \tag{19.2}$$

Proof If the density function f is supported on $\left[-\frac{\theta^*}{2}, \frac{\theta^*}{2}\right]$, it is immediate to observe that $4r\cos\left(\frac{\theta^*}{2}\right) \leq S_2 \leq 4r$. But let us be more precise.

Let $g : \mathbb{R} \to \mathbb{R}$ be a bounded measurable function. Thanks to (19.1) we have $S_2 = 2r\,(\cos(\Theta_0) + \cos(\Theta_1))$ with Θ_0, Θ_1 two independent random variables with density function f. Therefore, using Assumption $(\mathcal{H}')$ we have:

$$\begin{aligned}
\mathbb{E}\left[g(S_2)\right] &= \mathbb{E}\left[g\left(2r\left(\cos(\Theta_0) + \cos(\Theta_1)\right)\right)\right] \\
&\geq f_{\min}^2 \int_{-\frac{\theta^*}{2}}^{\frac{\theta^*}{2}} \int_{-\frac{\theta^*}{2}}^{\frac{\theta^*}{2}} g\left(2r\left(\cos(u) + \cos(v)\right)\right) du dv \\
&= 4f_{\min}^2 \int_0^{\frac{\theta^*}{2}} \int_0^{\frac{\theta^*}{2}} g\left(2r\left(\cos(u) + \cos(v)\right)\right) du dv.
\end{aligned}$$

The substitution $x = 2r\,(\cos(u) + \cos(v))$ in the integral with respect to u gives then:

$$\mathbb{E}\left[g(S_2)\right] \geq 4f_{\min}^2 \int_0^{\frac{\theta^*}{2}} \int_{2r\left(\cos\left(\frac{\theta^*}{2}\right)+\cos(v)\right)}^{2r(1+\cos(v))} g(x) \frac{1}{2r\sin\left(\arccos\left(\frac{x}{2r} - \cos(v)\right)\right)} dx dv.$$

Fubini's theorem leads to

$$\begin{aligned}
&\mathbb{E}\left[g(S_2)\right] \\
&\geq \frac{2f_{\min}^2}{r} \int_{4r\cos\left(\frac{\theta^*}{2}\right)}^{4r} \\
&\left(\int_0^{\frac{\theta^*}{2}} \frac{1}{\sqrt{1-\left(\frac{x}{2r} - \cos(v)\right)^2}} \mathbf{1}_{\frac{x}{2r}-1<\cos(v)<\frac{x}{2r}-\cos\left(\frac{\theta^*}{2}\right)} dv\right) g(x) dx.
\end{aligned}$$

We then deduce a lower-bound for the density function of S_2:

$$
\begin{aligned}
&f_{S_2}(x)\\
&\geq \frac{2f_{\min}^2}{r}\int_0^{\frac{\theta^*}{2}} \frac{1}{\sqrt{1-\left(\frac{x}{2r}-\cos(v)\right)^2}}\mathbf{1}_{\frac{x}{2r}-1<\cos(v)<\frac{x}{2r}-\cos\left(\frac{\theta^*}{2}\right)}dv\mathbf{1}_{x\in\left(4r\cos\left(\frac{\theta^*}{2}\right),4r\right)}.
\end{aligned}
$$

Let $x \in \left(4r\cos\left(\frac{\theta^*}{2}\right), 4r\right)$. Cutting the interval $\left(4r\cos\left(\frac{\theta^*}{2}\right), 4r\right)$ at point $2r\left(1+\cos\left(\frac{\theta^*}{2}\right)\right)$, we get, defining the intervals $A_1 = \left(4r\cos\left(\frac{\theta^*}{2}\right), 2r\left(1+\cos\left(\frac{\theta^*}{2}\right)\right)\right]$ and $A_2 = \left[2r\left(1+\cos\left(\frac{\theta^*}{2}\right)\right), 4r\right)$:

$$
\begin{aligned}
f_{S_2}(x) \geq\ & \frac{2f_{\min}^2}{r}\int_0^{\frac{\theta^*}{2}} \frac{1}{\sqrt{1-\left(\frac{x}{2r}-\cos(v)\right)^2}}\mathbf{1}_{\frac{x}{2r}-1<\cos(v)<\frac{x}{2r}-\cos\left(\frac{\theta^*}{2}\right)}dv\mathbf{1}_{x\in A_1}\\
&+\frac{2f_{\min}^2}{r}\int_0^{\frac{\theta^*}{2}} \frac{1}{\sqrt{1-\left(\frac{x}{2r}-\cos(v)\right)^2}}\mathbf{1}_{\frac{x}{2r}-1<\cos(v)<\frac{x}{2r}-\cos\left(\frac{\theta^*}{2}\right)}dv\mathbf{1}_{x\in A_2}\\
=\ & \frac{2f_{\min}^2}{r}\int_{\arccos\left(\frac{x}{2r}-\cos\left(\frac{\theta^*}{2}\right)\right)}^{\frac{\theta^*}{2}} \frac{1}{\sqrt{1-\left(\frac{x}{2r}-\cos(v)\right)^2}}dv\mathbf{1}_{x\in A_1}\\
&+\frac{2f_{\min}^2}{r}\int_0^{\arccos\left(\frac{x}{2r}-1\right)} \frac{1}{\sqrt{1-\left(\frac{x}{2r}-\cos(v)\right)^2}}dv\mathbf{1}_{x\in A_2}.
\end{aligned}
$$

Then, for $v \in \left(\arccos\left(\frac{x}{2r}-\cos\left(\frac{\theta^*}{2}\right)\right), \frac{\theta^*}{2}\right)$ we have $\cos(v) \leq \frac{x}{2r}-\cos\left(\frac{\theta^*}{2}\right)$, and for

$v \in \left(0, \arccos\left(\frac{x}{2r}-1\right)\right)$ we have $\cos(v) \leq 1$. We thus have:

$$
\begin{aligned}
f_{S_2}(x) \geq\ & \frac{2f_{\min}^2}{r\sin\left(\frac{\theta^*}{2}\right)}\left(\frac{\theta^*}{2}-\arccos\left(\frac{x}{2r}-\cos\left(\frac{\theta^*}{2}\right)\right)\right)\mathbf{1}_{x\in A_1}\\
&+\frac{2f_{\min}^2}{r}\frac{\arccos\left(\frac{x}{2r}-1\right)}{\sqrt{\frac{x}{r}\left(1-\frac{x}{4r}\right)}}\mathbf{1}_{x\in A_2}.
\end{aligned}
$$

We can observe that the lower bound of f_{S_2} is strictly positive for $x \in \left(4r\cos\left(\frac{\theta^*}{2}\right), 4r\right)$, but is equal to 0 when x is one of the extremal points of this interval. Let us therefore introduce $\eta \in \left(0, 2r\left(1-\cos\left(\frac{\theta^*}{2}\right)\right)\right)$. We have:

- for $x \in [4r\cos\left(\frac{\theta^*}{2}\right) + \eta, 2r\left(1+\cos\left(\frac{\theta^*}{2}\right)\right)]$ we have

$$\frac{2f_{\min}^2}{r\sin\left(\frac{\theta^*}{2}\right)}\left(\frac{\theta^*}{2} - \arccos\left(\frac{x}{2r} - \cos\left(\frac{\theta^*}{2}\right)\right)\right)$$

$$\geq \frac{2f_{\min}^2}{r\sin\left(\frac{\theta^*}{2}\right)}\left(\frac{\theta^*}{2} - \arccos\left(\frac{4r\cos\left(\frac{\theta^*}{2}\right)+\eta}{2r} - \cos\left(\frac{\theta^*}{2}\right)\right)\right)$$

$$= \frac{2f_{\min}^2}{r\sin\left(\frac{\theta^*}{2}\right)}\left(\frac{\theta^*}{2} - \arccos\left(\cos\left(\frac{\theta^*}{2}\right) + \frac{\eta}{2r}\right)\right);$$

- for $x \in \left[2r\left(1+\cos\left(\frac{\theta^*}{2}\right)\right), 4r - \eta\right]$ we have

$$\frac{2f_{\min}^2}{r}\frac{\arccos\left(\frac{x}{2r} - 1\right)}{\sqrt{\frac{x}{r}\left(1-\frac{x}{4r}\right)}} \geq \frac{2f_{\min}^2}{r}\frac{\arccos\left(\frac{4r-\eta}{2r} - 1\right)}{\sqrt{\frac{2r\left(1+\cos\left(\frac{\theta^*}{2}\right)\right)}{r}\left(1 - \frac{2r\left(1+\cos\left(\frac{\theta^*}{2}\right)\right)}{4r}\right)}}$$

$$= \frac{2f_{\min}^2}{r}\frac{\arccos\left(1-\frac{\eta}{2r}\right)}{\sqrt{\left(1+\cos\left(\frac{\theta^*}{2}\right)\right)\left(1-\cos\left(\frac{\theta^*}{2}\right)\right)}}$$

$$= \frac{2f_{\min}^2}{r\sin\left(\frac{\theta^*}{2}\right)}\arccos\left(1-\frac{\eta}{2r}\right).$$

The result follows then immediately. □

Notation For $x \in \partial\mathcal{B}_r$, we denote by φ_x the unique angle in $[0, 2\pi)$ describing the position of x on $\partial\mathcal{B}_r$. Moreover, we write Φ_n^x for the position of the Markov chain after n steps, and that started at position x on $\partial\mathcal{B}_r$.

Let us remark that thanks to the rotational symmetry of the process in the disc, we do not have to take care of the starting position on ∂K when we look at the inter-jump times.

Proposition 19.4 *Let $((X_t, V_t))_{t\geq 0}$ be the stochastic billiard process in $\mathcal{B}_r$ satisfying Assumption $(\mathcal{H}')$ with $\theta^* \in \left(\frac{2\pi}{3}, \pi\right)$.*

For all $\varepsilon \in \left(0, \frac{\theta^*}{4}\right)$, the pair $\left(\Phi_2^x, S_2\right)$ is $\frac{f_{\min}^2}{2r\sin\left(\frac{\theta^*}{4}\right)}$-continuous on $(\varphi_x - \theta^* + 4\varepsilon, \varphi_x + \theta^* - 4\varepsilon) \times \left(4r\cos\left(\frac{\theta^*}{4}\right), 4r\cos\left(\frac{\theta^*}{4} - \varepsilon\right)\right)$ for all $x \in \partial\mathcal{B}_r$.

Proof By symmetry of the process, it is sufficient to prove the lemma for $x \in \partial\mathcal{B}_r$ such that $\varphi_x = 0$, what we do.

Let $\varepsilon \in \left(0, \frac{\theta^*}{4}\right)$, $A \subset (-\theta^* + 4\varepsilon, \theta^* - 4\varepsilon)$ and $(r_1, r_2) \subset \left(2r\cos\left(\frac{\theta^*}{4}\right), 2r\cos\left(\frac{\theta^*}{4} - \varepsilon\right)\right)$.

Let us recall that $\Phi_2^0 = 2\Theta_0 + 2\Theta_1$ and $S_2 = 2r(\cos(\Theta_0) + \cos(\Theta_1))$, where Θ_0, Θ_1 are independent variables with density function f. We thus have:

$$\begin{aligned}
&\mathbb{P}\left(\Phi_2^0 \in A, S_2 \in (r_1, r_2)\right)\\
&= \mathbb{P}\left(2\Theta_0 + 2\Theta_1 \in A, 2r(\cos(\Theta_0) + \cos(\Theta_1)) \in (r_1, r_2)\right)\\
&= \int_{-\frac{\pi}{2}}^{\frac{\pi}{2}} \int_{-\frac{\pi}{2}}^{\frac{\pi}{2}} \mathbf{1}_{2u+2v \in A} \mathbf{1}_{\cos(u)+\cos(v) \in \left(\frac{r_1}{2r}, \frac{r_2}{2r}\right)} f(u) f(v) \mathrm{d}u \mathrm{d}v\\
&\geq f_{\min}^2 \int_{-\frac{\theta^*}{2}}^{\frac{\theta^*}{2}} \int_{-\frac{\theta^*}{2}}^{\frac{\theta^*}{2}} \mathbf{1}_{\frac{u+v}{2} \in \frac{A}{4}} \mathbf{1}_{\cos\left(\frac{u+v}{2}\right)\cos\left(\frac{u-v}{2}\right) \in \left(\frac{r_1}{4r}, \frac{r_2}{4r}\right)} \mathrm{d}u \mathrm{d}v.
\end{aligned}$$

Let us consider

$$g : (u, v) \in \left[-\frac{\theta^*}{2}, \frac{\theta^*}{2}\right]^2 \longmapsto \left(\frac{u+v}{2}, \frac{u-v}{2}\right).$$

We have

$$\left[-\frac{\theta^*}{4}, \frac{\theta^*}{4}\right]^2 \subset g\left(\left[-\frac{\theta^*}{2}, \frac{\theta^*}{2}\right]^2\right),$$

and

$$\left|\det \mathrm{Jac}_g\right| = \frac{1}{2}.$$

With this substitution, and using Fubini's theorem, we get:

$$\begin{aligned}
&\mathbb{P}\left(\Phi_2^0 \in A, S_2 \in (r_1, r_2)\right)\\
&\geq f_{\min}^2 \int_{-\frac{\theta^*}{4}}^{\frac{\theta^*}{4}} \int_{-\frac{\theta^*}{4}}^{\frac{\theta^*}{4}} \mathbf{1}_{x \in \frac{A}{4}} \mathbf{1}_{\cos(x)\cos(y) \in \left(\frac{r_1}{4r}, \frac{r_2}{4r}\right)} 2 \mathrm{d}x \mathrm{d}y\\
&= 4 f_{\min}^2 \int_{-\frac{\theta^*}{4}}^{\frac{\theta^*}{4}} \int_0^{\frac{\theta^*}{4}} \mathbf{1}_{\cos(x)\cos(y) \in \left(\frac{r_1}{4r}, \frac{r_2}{4r}\right)} \mathrm{d}y \mathbf{1}_{x \in \frac{A}{4}} \mathrm{d}x.
\end{aligned}$$

We now do the substitution $z = \cos(x)\cos(y)$ in the integral with respect to $\mathrm{d}y$:

$$\begin{aligned}
&\mathbb{P}\left(\Phi_2^0 \in A, S_2 \in (r_1, r_2)\right)\\
&\geq 4 f_{\min}^2 \int_{-\frac{\theta^*}{4}}^{\frac{\theta^*}{4}} \int_{\cos\left(\frac{\theta^*}{4}\right)\cos(x)}^{\cos(x)} \mathbf{1}_{z\in\left(\frac{r_1}{4r}, \frac{r_2}{4r}\right)} \frac{1}{\sqrt{\cos^2(x) - z^2}} \mathrm{d}z \mathbf{1}_{x\in\frac{A}{4}} \mathrm{d}x\\
&\geq 4 f_{\min}^2 \int_{-\frac{\theta^*}{4}}^{\frac{\theta^*}{4}} \int_{\cos\left(\frac{\theta^*}{4}\right)\cos(x)}^{\cos(x)} \mathbf{1}_{z\in\left(\frac{r_1}{4r}, \frac{r_2}{4r}\right)} \frac{1}{\sin\left(\frac{\theta^*}{4}\right)} \mathrm{d}z \mathbf{1}_{x\in\frac{A}{4}} \mathrm{d}x\\
&\geq \frac{4 f_{\min}^2}{\sin\left(\frac{\theta^*}{4}\right)} \int_{-\frac{\theta^*}{4}+\varepsilon}^{\frac{\theta^*}{4}-\varepsilon} \int_{\cos\left(\frac{\theta^*}{4}\right)}^{\cos\left(\frac{\theta^*}{4}-\varepsilon\right)} \mathbf{1}_{z\in\left(\frac{r_1}{4r}, \frac{r_2}{4r}\right)} \mathrm{d}z \mathbf{1}_{x\in\frac{A}{4}} \mathrm{d}x\\
&= \frac{f_{\min}^2}{4r \sin\left(\frac{\theta^*}{4}\right)} (r_2 - r_1) \left|A\right|,
\end{aligned}$$

where we have used for the last equality the fact that $A \subset [-\theta^* + 4\varepsilon, \theta^* - 4\varepsilon)$ and $(r_1, r_2) \subset \left(4r\cos\left(\frac{\theta^*}{4}\right), 4r\cos\left(\frac{\theta^*}{4} - \varepsilon\right)\right)$.

This ends the proof. □

Let us fix $\eta \in (0, r\left(1 - 2\cos\left(\frac{\theta^*}{2}\right)\right))$ and $\varepsilon \in \left(0, \frac{2\theta^* - \pi}{8}\right)$ (the condition $\theta^* > \frac{2\pi}{3}$ ensures that we can take such η and ε).

Let us define

$$h = 4r\left(1 - \cos\left(\frac{\theta^*}{2}\right)\right) - 2\eta - 2r = 2r\left(1 - 2\cos\left(\frac{\theta^*}{2}\right)\right) - 2\eta > 0 \tag{19.3}$$

and

$$\begin{aligned}
\alpha &= \frac{f_{\min}^2}{2r\sin\left(\frac{\theta^*}{4}\right)}(4\theta^* - 2\pi - 16\varepsilon)4r\left(\cos\left(\frac{\theta^*}{4} - \varepsilon\right) - \cos\left(\frac{\theta^*}{4}\right)\right)\\
&= \frac{2 f_{\min}^2}{\sin\left(\frac{\theta^*}{4}\right)}(4\theta^* - 2\pi - 16\varepsilon)\left(\cos\left(\frac{\theta^*}{4} - \varepsilon\right) - \cos\left(\frac{\theta^*}{4}\right)\right).
\end{aligned} \tag{19.4}$$

Theorem 19.2 *Let $((X_t, V_t))_{t\geq 0}$ be the stochastic billiard process in $\mathcal{B}_r$ satisfying Assumption $(\mathcal{H}')$ with $\theta^* \in \left(\frac{2\pi}{3}, \pi\right)$.*

There exists a unique invariant probability measure χ on $\mathcal{B}_r \times \mathbb{S}^1$ for the process $((X_t, V_t))_{t\geq 0}$.

Moreover let $\eta \in (0, r\left(1 - 2\cos\left(\frac{\theta^*}{2}\right)\right))$ and $\varepsilon \in \left(0, \frac{2\theta^* - \pi}{8}\right)$. For all $t \geq 0$ and all $\lambda < \lambda_M$ we have

$$\|\mathbb{P}(X_t \in \cdot, V_t \in \cdot) - \chi\|_{TV} \leq C_\lambda e^{-\lambda t},$$

where

$$\lambda_M = \min\left\{\frac{1}{4r}\log\left(\frac{1}{1-\delta h}\right)\right.$$
$$\left.; \frac{1}{4r}\log\left(\frac{-(1-\delta h)+\sqrt{(1-\delta h)^2+4\delta h(1-\alpha)}}{2\delta h(1-\alpha)}\right)\right\}. \quad (19.5)$$

and

$$C_\lambda = \frac{\alpha\delta h e^{10\lambda r}}{1-e^{4\lambda r}(1-\delta h)-e^{8\lambda r}\delta h(1-\alpha)},$$

with δ*,* h *and* α *respectively given by* (19.2)–(19.4).

Remark 19.1 The following proof of this theorem is largely inspired by the proof of Theorem 2.2 in [2].

Proof The existence of the invariant probability measure comes from the compactness of the space $\mathcal{B}_r \times \mathbb{S}^1$. The following proof shows its uniqueness and gives the speed of convergence of the stochastic billiard to equilibrium.

Let $((X_t, V_t))_{t\geq 0}$ and $((\tilde{X}_t, \tilde{V}_t))_{t\geq 0}$ be two versions of the stochastic billiard with $(X_0, V_0) = (x_0, v_0) \in \mathcal{B}_r \times \mathbb{S}^1$ and $(\tilde{X}_0, \tilde{V}_0) = (\tilde{x}_0, \tilde{v}_0) \in \mathcal{B}_r \times \mathbb{S}^1$. We are going to construct these two processes until they become equal.

We recall the definition of T_0 and $\tilde{T}_0$:

$$T_0 = \inf\{t \geq 0 : x_0 + tv_0 \notin K\}, \quad \text{and} \quad \tilde{T}_0 = \inf\{t \geq 0 : \tilde{x}_0 + t\tilde{v}_0 \notin K\}.$$

We are going to couple (X_t, V_t) and $(\tilde{X}_t, \tilde{V}_t)$ in two steps: we first couple the times, so that the two processes hit $\partial\mathcal{B}_r$ at a same time, and then we couple both position and time.

Step 1 Proposition 19.3 ensures that S_2 and $\tilde{S}_2$ are both δ-continuous on $[4r\cos\left(\frac{\theta^*}{2}\right)+\eta, 4r-\eta]$. Therefore, the variables T_2 and $\tilde{T}_2$ are δ-continuous on $[T_0+4r\cos\left(\frac{\theta^*}{2}\right)+\eta, T_0+4r-\eta] \cap [\tilde{T}_0+4r\cos\left(\frac{\theta^*}{2}\right)+\eta, \tilde{T}_0+4r-\eta]$, with $\left|[T_0+4r\cos\left(\frac{\theta^*}{2}\right)+\eta, T_0+4r-\eta] \cap [\tilde{T}_0+4r\cos\left(\frac{\theta^*}{2}\right)+\eta, \tilde{T}_0+4r-\eta]\right|$ $\geq h$ since $|T_0-\tilde{T}_0| \leq 2r$. Note that the condition $\theta^* > \frac{2\pi}{3}$ has been introduced to ensure that this intersection is non-empty.

Thus, there exists a coupling of T_2 and $\tilde{T}_2$ such that

$$\mathbb{P}(E_1) \geq \delta h,$$

where

$$E_1 = \left\{T_2 = \tilde{T}_2\right\}.$$

On the event E_1 we define $T_c^1 = T_2$.

On the event E_1^c, we can suppose, by symmetry that $T_2 \leq \tilde{T}_2$. In order to try again to couple the hitting times, we need to begin at times whose difference is smaller than $2r$. Let us thus define

$$m_1 = \min\left\{n > 2 : T_n > \tilde{T}_2\right\} \text{ and } \tilde{m}_1 = 2.$$

We then have, by construction of m_1 and $\tilde{m}_1$, $\left|T_{m_1} - \tilde{T}_{\tilde{m}_1}\right| \leq 2r$. Therefore, as previously, there exists a coupling of T_{m_1+2} and $\tilde{T}_{\tilde{m}_1+2}$ such that

$$\mathbb{P}\left(E_2 | E_1^c\right) \geq \delta h,$$

where

$$E_2 = \left\{T_{m_1+2} = \tilde{T}_{\tilde{m}_1+2}\right\}.$$

On the event $E_1^c \cap E_2$ we define $T_c^1 = T_{m_1+2}$.

We then repeat the same procedure. We thus construct two sequences of stopping times $(m_k)_{k\geq 1}$, $(\tilde{m}_k)_{k\geq 1}$ and a sequence of events $(E_k)_{k\geq 1}$, with

$$E_k = \left\{T_{m_{k-1}+2} = \tilde{T}_{\tilde{m}_{k-1}+2}\right\},$$

satisfying

$$\mathbb{P}\left(E_k | E_1^c \cap \cdots \cap E_{k-1}^c\right) \geq \delta h.$$

On the event $E_1^c \cap \cdots \cap E_{k-1}^c \cap E_k$ we define $T_c^1 = T_{m_k+2}$. By construction, T_c^1 is the coupling time of the hitting times of the boundary.

Since the inter-jump times are directly linked to the speeds of the process, and thus to its positions, we can construct both stochastic billiards (X_t, V_t) and $(\tilde{X}_t, \tilde{V}_t)$ until time T_c^1, and along with T_c^1.

We observe that by this construction of T_c^1, we have

$$T_c^1 \leq_{st} T_0 + \sum_{l=1}^{G^1} S^{1,l} \tag{19.6}$$

with $G^1 \sim \mathcal{G}(\delta h)$ and $S^{1,l}$, $l \geq 1$, independent random times with distribution f_{S_2}, and independent of G^1.

Step 2 Let us now suppose that T_c^1, $((X_t, V_t))_{0 \leq t \leq T_1^c}$ and $((\tilde{X}_t, \tilde{V}_t))_{0 \leq t \leq T_1^c}$ are constructed as described above.

We define $y = X_{T_c^1}$ and $\tilde{y} = \tilde{X}_{T_c^1}$, which are by construction of T_c^1 on $\partial \mathcal{B}_r$. We also define $N_c^1 = \min\left\{n > 0 : X_{T_n} = y\right\}$, which is deterministic conditionally to T_c^1.

Proposition 19.4 ensures that the couples $\left(X_{T_{N_c^1+2}}, T_{N_c^1+2} - T_{N_c^1}\right)$ and $\left(\tilde{X}_{T_{N_c^1+2}}, \tilde{T}_{N_c^1+2} - \tilde{T}_{N_c^1}\right)$ are both $\frac{f_{\min}^2}{2r\sin\left(\frac{\theta^*}{4}\right)}$-continuous on the set $A \times B$ where $A = (\varphi_y - \theta^* + 4\varepsilon, \varphi_y + \theta^* - 4\varepsilon) \cap (\varphi_{\tilde{y}} - \theta^* + 4\varepsilon, \varphi_{\tilde{y}} + \theta^* - 4\varepsilon)$ and $B = \left(4r\cos\left(\frac{\theta^*}{4}\right), 4r\cos\left(\frac{\theta^*}{4} - \varepsilon\right)\right)$, with $|A| \geq 4\theta^* - 2\pi - 16\varepsilon$ (let us mention that these intervals are seen in $[0, 2\pi]/(2\pi\mathbb{Z})$ since they are intervals of angles). Note that the condition $\theta^* > \frac{2\pi}{3}$ implies in particular that the previous intersection in non-empty.

Therefore we can construct a coupling such that

$$\mathbb{P}\left(F | E_1^c \cap \cdots \cap E_{N_c^1-1}^c \cap E_{N_c^1}, T_c^1\right) \geq \alpha,$$

where

$$F = \left\{X_{T_{N_c^1+2}} = \tilde{X}_{T_{N_c^1+2}} \text{ and } T_{N_c^1+2} = \tilde{T}_{N_c^1+2}\right\}.$$

On the event F we define $T_c = T_{N_c^1+2}$, and we construct $((X_t, V_t))_{T_c^1 \leq t \leq T_c}$ and $((\tilde{X}_t, \tilde{V}_t))_{T_c^1 \leq t \leq T_c}$ along with the coupling of $(X_{T_{N_c^1+2}}, T_{N_c^1+2})$ and $(\tilde{X}_{T_{N_c^1+2}}, \tilde{T}_{N_c^1+2})$.

If F does not occur, we can not directly try to couple both position and time since the two processes have not necessarily hit $\partial \mathcal{B}_r$ at the same time. We thus have to couple first the hitting times, as we have done in step 1.

Let us suppose that on $\left(E_1^c \cap \cdots \cap E_{N_c^1-1}^c \cap E_{N_c^1}\right) \cap F^c$, we have $T_{N_c^1+2} \leq \tilde{T}_{N_c^1+2}$ (the other case can be treated in the same way thanks to the symmetry of the problem). Let us define

$$\ell = \min\left\{n > N_c^1 + 2 : T_n > \tilde{T}_{N_c^1+2}\right\} \text{ and } \tilde{\ell} = N_c^1 + 2$$

We clearly have $\left|T_\ell - \tilde{T}_{\tilde{\ell}}\right| \leq 2r$. Therefore, we can start again: we try to couple the times at which the two processes hit the boundary, and then to couple the positions and times together.

Finally, the probability that we succeed to couple the positions and times in "one step" (Step 1 and Step 2) is:

$$\begin{aligned}
&\mathbb{P}\left(\left(\bigcup_{k\geq 1}\left(E_1^c\cap\cdots\cap E_{k-1}^c\cap E_k\right)\right)\cap F\right)\\
&=\mathbb{P}\left(F\,\middle|\,\bigcup_{k\geq 1}\left(E_1^c\cap\cdots\cap E_{k-1}^c\cap E_k\right)\right)\mathbb{P}\left(\bigcup_{k\geq 1}\left(E_1^c\cap\cdots\cap E_{k-1}^c\cap E_k\right)\right)\\
&=\mathbb{P}\left(F\,\middle|\,\bigcup_{k\geq 1}\left(E_1^c\cap\cdots\cap E_{k-1}^c\cap E_k\right)\right)\\
&\geq\alpha.
\end{aligned}$$

Therefore, the coupling time $\hat{T}$ of the couples position-time satisfies:

$$\hat{T}\leq_{st} T_0+\sum_{k=1}^{G}\left(\left(\sum_{l=1}^{G^k}S^{k,l}\right)+S^k\right)$$

where $G\sim\mathcal{G}(\alpha)$, $G^1, G^2,\cdots\sim\mathcal{G}(\delta h)$ are independent geometric variables, and $\left(S^{k,l}\right)_{k,l\geq 1}$, $\left(S^k\right)_{k\geq 1}$ are independent random variables, independent from the geometric variables, with distribution f_{S_2}.

Let $\lambda\in(0,\lambda_M)$, with λ_M defined in Eq. (19.5). Since all the random variables $S^{k,l}$ and S^k, $k,l\geq 1$, are almost surely smaller than two times the diameter of the ball $\mathcal{B}_r$, and since T_0 is almost surely smaller than this diameter, we have:

$$\begin{aligned}
\mathbb{P}\left(\hat{T}>t\right)&\leq \mathrm{e}^{-\lambda t}\mathbb{E}\left[\mathrm{e}^{\lambda\hat{T}}\right]\\
&\leq \mathrm{e}^{\lambda(T_0-t)}\mathbb{E}\left[\exp\left(\lambda\sum_{k=1}^{G}\left(\left(\sum_{l=1}^{G^k}S^{k,l}\right)+S^k\right)\right)\right]\\
&\leq \mathrm{e}^{\lambda(2r-t)}\mathbb{E}\left[\prod_{k=1}^{G}\left(\left(\prod_{l=1}^{G^k}\exp\left(\lambda 4r\right)\right)\exp\left(\lambda 4r\right)\right)\right]\\
&= \mathrm{e}^{\lambda(2r-t)}\mathbb{E}\left[\prod_{k=1}^{G}\mathbb{E}\left[\mathrm{e}^{4\lambda r(G^k+1)}\right]\right].
\end{aligned}$$

Now, using the expression of generating function of a geometric random variable we get:

$$\begin{aligned}
\mathbb{P}\left(\hat{T}>t\right)&\leq \mathrm{e}^{\lambda(2r-t)}\mathbb{E}\left[\prod_{k=1}^{G}\left(\sum_{l=1}^{\infty}\mathrm{e}^{4\lambda r(l+1)}\delta h(1-\delta h)^{l-1}\right)\right]\\
&= \mathrm{e}^{\lambda(2r-t)}\mathbb{E}\left[\left(\frac{\mathrm{e}^{8\lambda r}\delta h}{1-\mathrm{e}^{4\lambda r}(1-\delta h)}\right)^{G}\right]
\end{aligned}$$

$$= e^{\lambda(2r-t)} \frac{\alpha e^{8\lambda r}\delta h}{1 - e^{4\lambda r}(1-\delta h)} \frac{1}{1 - \frac{e^{8\lambda r}\delta h(1-\alpha)}{1-e^{4\lambda r}(1-\delta h)}}$$

$$= e^{-\lambda t} \frac{\alpha e^{10\lambda r}\delta h}{1 - e^{4\lambda r}(1-\delta h) - e^{8\lambda r}\delta h(1-\alpha)}.$$

These calculations are valid for $\lambda > 0$ such that the generating functions are well defined, that is for $\lambda > 0$ satisfying

$$e^{4\lambda r}(1-\delta h) < 1 \text{ and } \frac{e^{8\lambda r}\delta h(1-\alpha)}{1-e^{4\lambda r}(1-\delta h)} < 1.$$

The first condition is equivalent to $\lambda < \frac{1}{4r}\log\left(\frac{1}{1-\delta h}\right)$.

The second condition is equivalent to $\delta h(1-\alpha)s^2 + (1-\delta h)s - 1 < 0$ with $s = e^{4\lambda r}$. It gives $s_1 < s < s_2$ with $s_1 = \frac{-(1-\delta h)-\sqrt{\Delta}}{2\delta h(1-\alpha)} < 0$ and $s_2 = \frac{-(1-\delta h)+\sqrt{\Delta}}{2\delta h(1-\alpha)} > 1$ where $\Delta = (1-\delta h)^2 + 4\delta h(1-\alpha) > 0$. And finally we get $\lambda < \frac{1}{4r}\log(s_2)$.

Therefore, the estimation for $\mathbb{P}\left(\hat{T} > t\right)$ is indeed valid for all $\lambda \in (0, \lambda_M)$. The conclusion of the theorem follows immediately.

Remark 19.2 If $\theta^* \in \left(0, \frac{2\pi}{3}\right]$, Step 1 of the proof of Theorem 19.2 fails: the intervals on which the random variables S_2 and $\tilde{S}_2$ are continuous can have an empty intersection. Similarly, in Step 2, the intersection of the intervals on which the couples $\left(X_{T_{N_c^1+2}}, T_{N_c^1+2} - T_{N_c^1}\right)$ and $\left(\tilde{X}_{T_{N_c^1+2}}, \tilde{T}_{N_c^1+2} - \tilde{T}_{N_c^1}\right)$ are continuous can be empty depending on the value of θ^*.

However, instead of trying to couple the times or both positions and times in two jumps, we just need more jumps to do that. Therefore, the method and the results are similar in the case $\theta^* \leq \frac{2\pi}{3}$, the only difference is that the computations and notations will be much more awful.

19.4 Stochastic Billiard in a Convex Set with Bounded Curvature

We make the following assumption on the set K in which the stochastic billiard evolves:

Assumption $(\mathcal{K})$

K is a compact convex set with curvature bounded from above by $C < \infty$ and bounded from below by $c > 0$.

This means that for each $x \in \partial K$, there is a ball B_1 with radius $\frac{1}{C}$ included in K and a ball B_2 containing K, so that the tangent planes of K, B_1 and B_2 at x coincide

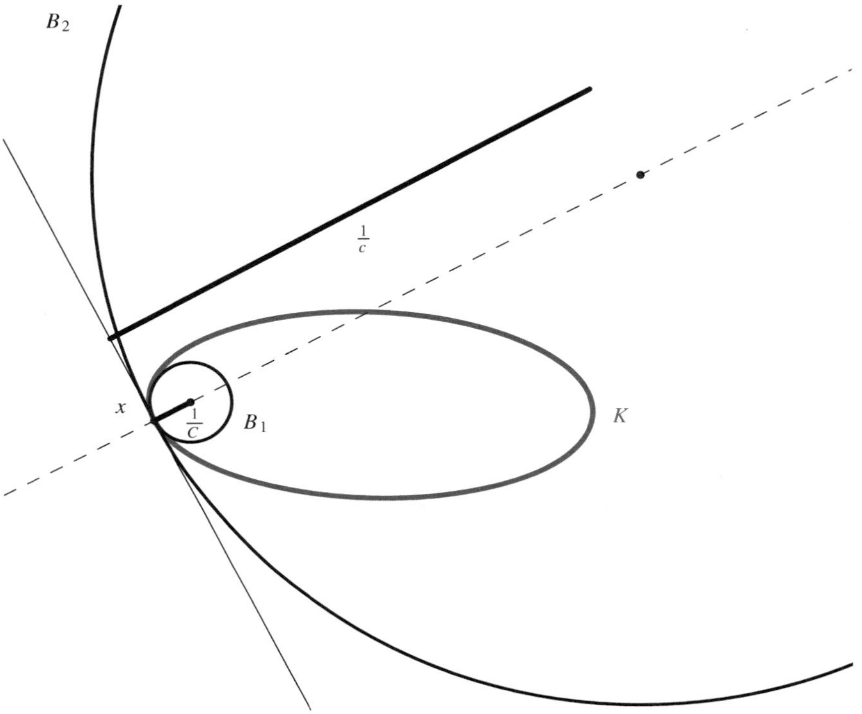

Fig. 19.4 Illustration of Assumption ($\mathcal{K}$)

(see Fig. 19.4). In fact, for $x \in \partial K$, the ball B_1 is the ball with radius $\frac{1}{C}$ and with center the unique point at distance $\frac{1}{C}$ from x in the direction of n_x. And B_2 is the one with the center at distance $\frac{1}{c}$ from x in the direction of n_x.

In this section, we consider the stochastic billiard in such a convex K.

Let us observe that the case of the disc is a particular case. Moreover, Assumption ($\mathcal{K}$) excludes in particular the case of the polygons: because of the upper bound C on the curvature, the boundary of K can not have "corners", and because of the lower bound c, the boundary can not have straight lines.

In the following, D will denote the diameter of K, that is

$$D = \max\{\|x - y\|: x, y \in \partial K\}.$$

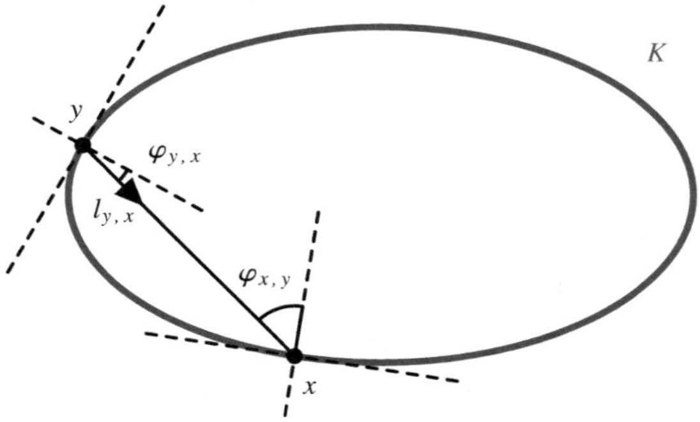

Fig. 19.5 Definition of the quantities $\varphi_{x,y}$ and $l_{y,x}$ for $x, y \in \partial K$

19.4.1 The Embedded Markov Chain

Notation We define $l_{x,y} = \frac{y-x}{\|x-y\|} = -l_{y,x}$ and we denote by $\varphi_{x,y}$ the angle between $l_{x,y}$ and the normal n_x to ∂K at the point x (see Fig. 19.5).

The following property, proved by Comets and al. in [2], gives the dynamics of the Markov chain $(X_{T_n})_{n\geq 0}$ defined in Sect. 19.2.2.

Proposition 19.5 *The transition kernel of the chain* $(X_{T_n})_{n\geq 0}$ *is given by:*

$$\mathbb{P}\left(X_{T_{n+1}} \in A \,\middle|\, X_{T_n} = x\right) = \int_A Q(x, y)\mathrm{d}y$$

where

$$Q(x, y) = \frac{\rho(U_x^{-1} l_{x,y}) \cos(\varphi_{y,x})}{\|x - y\|}.$$

This proposition is one of the main ingredients to obtain the exponentially-fast convergence of the stochastic billiard Markov chain towards its invariant probability measure.

Theorem 19.3 *Let* $K \subset \mathbb{R}^2$ *satisfying Assumption* $(\mathcal{K})$ *with diameter D. Let* $(X_{T_n})_{n\geq 0}$ *be the stochastic billiard Markov chain on* ∂K *verifying Assumption* $(\mathcal{H})$.

There exists a unique invariant measure ν *on* ∂K *for* $(X_{T_n})_{n\geq 0}$.

Moreover, recalling that $\theta^* = |\mathcal{J}|$ *in Assumption* $(\mathcal{H})$, *we have:*

1. *if* $\theta^* > \frac{C|\partial K|}{8}$, *for all* $n \geq 0$,

$$\|\mathbb{P}\left(X_{T_n} \in \cdot\right) - \nu\|_{TV} \leq \left(1 - q_{\min}\left(\frac{8\theta^*}{C} - |\partial K|\right)\right)^{n-1};$$

2. *if* $\theta^* \le \frac{C|\partial K|}{8}$, *for all* $n \ge 0$ *and all* $\varepsilon \in \left(0, \frac{2\theta^*}{C}\right)$,

$$\|\mathbb{P}\left(X_{T_n} \in \cdot\right) - \nu\|_{TV} \le (1-\alpha)^{\frac{n}{n_0}-1}$$

where

$$n_0 = \left\lfloor \frac{\frac{|\partial K|}{2} - 2\varepsilon}{\frac{4\theta^*}{C} - 2\varepsilon} \right\rfloor + 1 \text{ and}$$

$$\alpha = (\frac{4\theta^*}{C})^{n_0-1} {q_{\min}}^{n_0} \left(4\left(\frac{2n_0\theta^*}{C} - (n_0-1)\varepsilon\right) - |\partial K|\right)$$

with

$$q_{\min} = \frac{c\rho_{\min}\cos\left(\frac{\theta^*}{2}\right)}{CD}.$$

Proof Once more, the existence of the invariant measure is immediate since the state space ∂K of the Markov chain is compact. The following shows its uniqueness and gives the speed of convergence of $(X_{T_n})_{n\ge 0}$ towards ν.

Let $(X_{T_n})_{n\ge 0}$ and $(\tilde{X}_{T_n})_{n\ge 0}$ be two versions of the Markov chain with initial conditions x_0 and $\tilde{x}_0$ on ∂K. In order to have a strictly positive probability to couple X_{T_n} and $\tilde{X}_{T_n}$ at time n, it is sufficient that their density functions are bounded from below on an interval of length strictly bigger than $\frac{|\partial K|}{2}$. Let us therefore study the length of set on which the density function $f_{X_{T_n}}$ of X_{T_n} is bounded from below by a strictly positive constant.

Let $x \in \partial K$. For $v \in \mathbb{S}_x$, we denote by $h_x(v)$ the unique point on ∂K seen from x in the direction of v. We firstly get a lower bound on $|h_x(U_x\mathcal{J})|$, the length of the subset of ∂K seen from x with a strictly positive density.

It is easy to observe, with a drawing for instance, the following facts:

- $|h_x(U_x\mathcal{J})|$ increases when $\|x - h_x(n_x)\|$ increases,
- $|h_x(U_x\mathcal{J})|$ decreases when the curvature at $h_x(n_x)$ increases,
- $|h_x(U_x\mathcal{J})|$ decreases when $|\varphi_{h_x(n_x),x}|$ increases.

Therefore, $|h_x(U_x\mathcal{J})|$ is minimal when $\|x - h_x(n_x)\|$ is minimal, when the curvature at $h_x(n_x)$ is maximal, and then equal to C, and finally when $\varphi_{h_x(n_x),x} = 0$. Moreover, the minimal value of $\|x - h_x(n_x)\|$ is $\frac{2}{C}$ since C is the upper bound for the curvature of ∂K. The configuration that makes the quantity $|h_x(U_x\mathcal{J})|$ minimal is thus the case where x and $h_x(n_x)$ define a diameter on a circle of diameter $\frac{2}{C}$ (see Fig. 19.6). We immediately deduce a lower bound for $|h_x(U_x\mathcal{J})|$:

$$|h_x(U_x\mathcal{J})| \ge 2\theta^* \times \frac{2}{C} = \frac{4\theta^*}{C}.$$

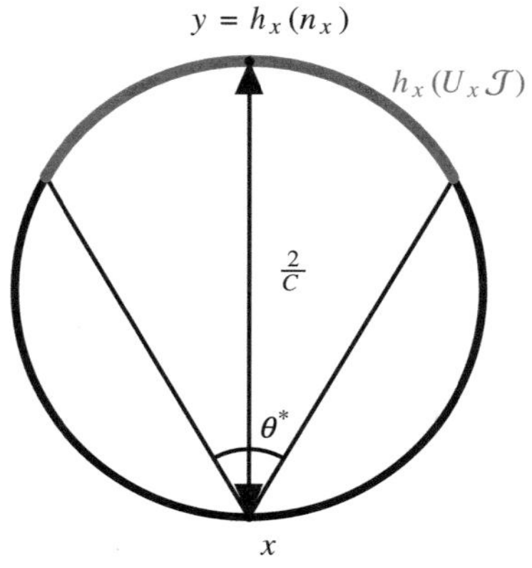

Fig. 19.6 Worst scenario for the length of $h_x(U_x\mathcal{J})$

This means that the density function $f_{X_{T_1}}$ of X_{T_1} is strictly positive on a subset of ∂K of length at least $\frac{4\theta^*}{C}$.

Let now $\varepsilon \in \left(0, \frac{2\theta^*}{C}\right)$. As it has been done in Sect. 19.3 for the disc, we can deduce that for all $n \geq 2$, the density function $f_{X_{T_n}}$ is strictly positive on a set of length at least $2n\theta^* \frac{2}{C} - 2(n-1)\varepsilon = \frac{4n\theta^*}{C} - 2(n-1)\varepsilon$.

Let us define, for $x \in \partial K$ and $n \geq 1$, $\mathcal{J}_x^n$ the set of points of ∂K that can be reached from x in n bounces by picking for each bounce a velocity in $\mathcal{J}$.

We now separate the cases where we can couple in one jump, and where we need more jumps.

- Case 1: $\theta^* > \frac{C|\partial K|}{8}$. In that case we have, for all $x \in \partial K$, $|\mathcal{J}_x^1| \geq \frac{4\theta^*}{C} > \frac{|\partial K|}{2}$, and we can thus construct a coupling $(X_{T_1}, \tilde{X}_{T_1})$ such that:

$$\mathbb{P}\left(X_{T_1} = \tilde{X}_{T_1}\right) \geq q_{\min} \left|\mathcal{J}_{x_0}^1 \cap \tilde{\mathcal{J}}_{\tilde{x}_0}^1\right| \geq q_{\min} \times 2\left(\frac{4\theta^*}{C} - \frac{|\partial K|}{2}\right)$$
$$= q_{\min}\left(\frac{8\theta^*}{C} - |\partial K|\right),$$

where $q_{\min}$ is a uniform lower bound of $Q(a, b)$ with $a \in \partial K$ and $b \in h_a(U_a\mathcal{J})$, i.e.

$$q_{\min} \leq \min_{a \in \partial K, b \in h_a(U_a\mathcal{J})} Q(a, b).$$

Let us thus give an explicit expression for $q_{\min}$. Let $a \in \partial K$ and $b \in h_a(U_a\mathcal{J})$. We have

$$Q(a, b) \geq \frac{\rho_{\min} \cos\left(\varphi_{b,a}\right)}{D}.$$

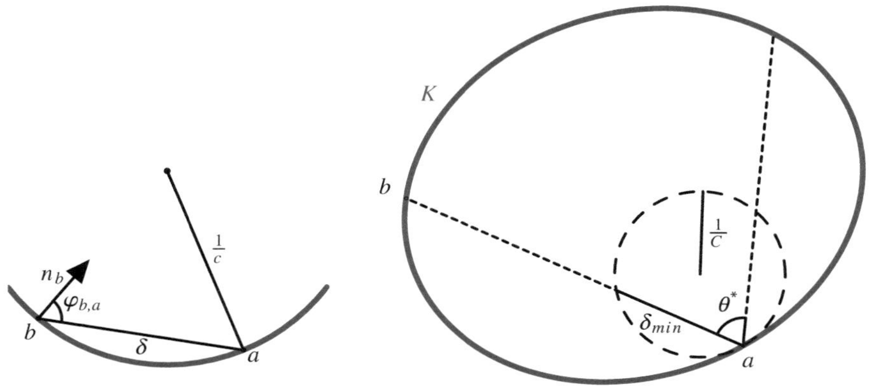

Fig. 19.7 Illustration for the calculation of a lower bound for $\cos(\varphi_{b,a})$ with $a \in \partial K$ and $b \in h_a(U_a\mathcal{J})$

We could have $\cos(\varphi_{b,a}) = 0$ if a and b were on a straight part of ∂K, which is not possible since the curvature of K is bounded from below by c. Thus, the quantity $\cos(\varphi_{b,a})$ is minimal when a and b are on a part of a disc with curvature c. In that case, $\cos(\varphi_{b,a}) = \frac{\delta c}{2}$, where δ is the distance between a and b (see the first picture of Fig. 19.7). Since $b \in h_a(U_a\mathcal{J})$, we have $\delta \geq \delta_{\min} := \frac{2\cos\left(\frac{\theta^*}{2}\right)}{C}$ (see the second picture of Fig. 19.7). Finally we get

$$Q(a,b) \geq \frac{c\rho_{\min}\cos\left(\frac{\theta^*}{2}\right)}{CD} =: q_{\min}.$$

- Case 2: $\theta^* \leq \frac{C|\partial K|}{8}$. In that case, we need more than one jump to couple the two Markov chains. Therefore, defining

$$n_0 = \min\left\{n \geq 2 : \frac{4n\theta^*}{C} - 2(n-1)\varepsilon > \frac{\partial K}{2}\right\} = \left\lfloor \frac{\frac{|\partial K|}{2} - 2\varepsilon}{\frac{4\theta^*}{C} - 2\varepsilon} \right\rfloor + 1,$$

we get that the intersection $\mathcal{J}_{x_0}^{n_0} \cap \tilde{\mathcal{J}}_{\tilde{x}_0}^{n_0}$ is non-empty, and then we can construct $X_{T_{n_0}}$ and $\tilde{X}_{T_{n_0}}$ such that the probability $\mathbb{P}\left(X_{T_{n_0}} = \tilde{X}_{T_{n_0}}\right)$ is strictly positive. It remains to estimate a lower bound of this probability.

First, we have

$$\left|\mathcal{J}_{x_0}^{n_0} \cap \tilde{\mathcal{J}}_{\tilde{x}_0}^{n_0}\right| \geq 2\left(\frac{4n_0\theta^*}{C} - 2(n_0-1)\varepsilon - \frac{|\partial K|}{2}\right)$$
$$= 4\left(\frac{2n_0\theta^*}{C} - (n_0-1)\varepsilon\right) - |\partial K|.$$

Moreover, let $x \in \{x_0, \tilde{x}_0\}$ and $y \in \mathcal{J}_{x_0}^{n_0} \cap \tilde{\mathcal{J}}_{\tilde{x}_0}^{n_0}$. We have:

$$Q^{n_0}(x,y) \geq \int_{h_x(U_x\mathcal{J})} \int_{h_{z_1}(U_{z_1}\mathcal{J})} \cdots \int_{h_{z_{n-2}}(U_{z_{n-2}}\mathcal{J})}$$
$$Q(x,z_1)Q(z_1,z_2)\cdots Q(z_{n_0-1},y)\mathrm{d}z_1\mathrm{d}z_2\cdots\mathrm{d}z_{n_0-1}$$
$$\geq (\frac{4\theta^*}{C})^{n_0-1}{q_{\min}}^{n_0}.$$

We thus deduce:

$$\mathbb{P}\left(X_{T_{n_0}} = \tilde{X}_{T_{n_0}}\right) \geq (\frac{4\theta^*}{C})^{n_0-1}{q_{\min}}^{n_0}\left|\mathcal{J}_{x_0}^{n_0} \cap \tilde{\mathcal{J}}_{\tilde{x}_0}^{n_0}\right|$$
$$\geq (\frac{4\theta^*}{C})^{n_0-1}{q_{\min}}^{n_0}\left(4\left(\frac{2n_0\theta^*}{C} - (n_0-1)\varepsilon\right) - |\partial K|\right).$$

We can now conclude, including the two cases: let us define

$$m_0 = \mathbf{1}_{\theta^* > \frac{C|\partial K|}{8}} + \left(\left\lfloor \frac{\frac{|\partial K|}{2} - 2\varepsilon}{\frac{4\theta^*}{C} - 2\varepsilon} \right\rfloor + 1\right)\mathbf{1}_{\theta^* \leq \frac{C|\partial K|}{8}}$$

and

$$\alpha = q_{\min}\left(\frac{8\theta^*}{C} - |\partial K|\right)\mathbf{1}_{\theta^* > \frac{C|\partial K|}{8}}|$$
$$+ (\frac{4\theta^*}{C})^{m_0-1}{q_{\min}}^{m_0}\left(4\left(\frac{2m_0\theta^*}{C} - (m_0-1)\varepsilon\right) - |\partial K|\right)\mathbf{1}_{\theta^* \leq \frac{C|\partial K|}{8}}.$$

We have proved that we can construct a coupling $\left(X_{T_{m_0}}, \tilde{X}_{T_{m_0}}\right)$ such that $\mathbb{P}\left(X_{T_{m_0}} = \tilde{X}_{T_{m_0}}\right) \geq \alpha$, and then we get

$$\|\mathbb{P}\left(X_{T_n} \in \cdot\right) - \nu\|_{TV} \leq (1-\alpha)^{\frac{n}{m_0}-1}.$$

19.4.2 The Continuous-Time Process

In this section, we will work in the case $|\mathcal{J}|= \theta^* = \pi$.

Notations We still use the following notation, already introduced in the case of the disc:

$$S_0 = 0 \quad \text{and} \quad \text{for } n \geq 1, \quad S_n = T_n - T_0.$$

Moreover, for $x \in \partial K$, we write S_n^x an $X_{S_n}^x$ respectively for the nth hitting time of ∂K and the position of the Markov chain after n steps, and that started at position $x \in \partial K$.

Thereby, we have the following equalities: $\mathcal{L}(X_{S_n}^x) = \mathcal{L}(X_{T_n}|X_{T_0} = x)$ and $\mathcal{L}(S_n^x) = \mathcal{L}(T_n - T_0|X_{T_0} = x)$.

Proposition 19.6 *Let $K \subset \mathbb{R}^2$ satisfying Assumption $(\mathcal{K})$. Let $((X_t, V_t))_{t\geq 0}$ the stochastic billiard process evolving in K and verifying Assumption $(\mathcal{H})$ with $|\mathcal{J}|= \pi$.*

For all $x \in \partial K$, the random time S_1^x, which is the first hitting time of ∂K when starting at point x, is $c\rho_{\min}$-continuous on $\left[0, \frac{2}{C}\right]$.

Proof Let $x \in \partial K$. Let us recall that the curvature of K is bounded from above by C, which means that for each $x \in \partial K$, there is a ball B_1 with radius $\frac{1}{C}$ included in K so that the tangent planes of K and B_1 at x coincide. Therefore, starting from x, the maximal time to go on another point of ∂K is bigger than $\frac{2}{C}$ (the diameter of the ball B_1).

That is why we are going to prove the continuity of S_1^x on the interval $\left[0, \frac{2}{C}\right]$. Let thus $0 \leq r \leq R \leq \frac{2}{C}$.

Let Θ be a random variable living in $\left[-\frac{\pi}{2}, \frac{\pi}{2}\right]$ such that the velocity vector $(\cos(\Theta), \sin(\Theta))$ follows the law γ.

The time S_1^x being completely determined by the velocity V_{T_0} and thus by its angle with respect to n_x, it is clear that there exist $-\frac{\pi}{2} \leq \theta_1 \leq \theta_2 \leq \theta_3 \leq \theta_4 \leq \frac{\pi}{2}$ such that we have:

$$\mathbb{P}\left(S_1^x \in [r, R]\right) = \mathbb{P}\left(\Theta \in [\theta_1, \theta_2] \cup [\theta_3, \theta_4]\right).$$

Then, thanks to Assumption $(\mathcal{H})$ on the law γ, and since we assume here that $|\mathcal{J}|= \pi$, the density function of Θ is bounded from below by $\rho_{\min}$ on $\left[-\frac{\pi}{2}, \frac{\pi}{2}\right]$. It gives:

$$\mathbb{P}\left(S_1^x \in [r, R]\right) \geq \rho_{\min}\left(\theta_2 - \theta_1 + \theta_4 - \theta_3\right).$$

Moreover, since the curvature is bounded from below by c, there exists a ball B_2 with radius $\frac{1}{c}$ containing K so that the tangent planes of K and B_2 at x coincide. And it is easy to see that the differences $\theta_2 - \theta_1$ and $\theta_4 - \theta_2$ are larger than the difference $\alpha_2 - \alpha_1$ where α_1 and α_2 are the angles corresponding to the distances r and R starting from x and to arrive on the ball B_2.

The time of hitting the boundary of B_2 is equal to $d \in \left[0, \frac{2}{C}\right]$ if the angle between n_x and the velocity is equal to $\arccos\left(\frac{cd}{2}\right)$. We thus deduce:

$$\begin{aligned}\mathbb{P}\left(S_1^x \in [r, R]\right) &\geq 2\rho_{\min}\left(\arccos\left(\frac{cr}{2}\right) - \arccos\left(\frac{cR}{2}\right)\right)\\ &\geq 2\rho_{\min}\left|\frac{cr}{2} - \frac{cR}{2}\right|\\ &= \rho_{\min} c\,(R-r)\,,\end{aligned}$$

where we have used the mean value theorem for the second inequality.

Let us introduce some constants that will appear in the following results.
Let $\beta > 0$ and $\delta > 0$ such that $\frac{|\partial K|}{3} - \max\{2\delta; \beta + \delta\} > 0$.
Let $\varepsilon \in \left(0, \min\{\beta; \frac{2}{C}\}\right)$ such that $h > 0$ where

$$h = \frac{\delta}{D}\left(\frac{\beta c}{2}\right)^2 - \varepsilon M, \tag{19.7}$$

with

$$M = 2\left(\frac{1}{\frac{1}{C} - \varepsilon} + \frac{1}{\beta - \varepsilon} + C\right). \tag{19.8}$$

Let us remark that M is non decreasing with ε, so that it is possible to take ε small enough to have $h > 0$.

Proposition 19.7 *Let $K \subset \mathbb{R}^2$ satisfying Assumption ($\mathcal{K}$) with diameter D. Let $((X_t, V_t))_{t\geq 0}$ the stochastic billiard process evolving in K and verifying Assumption ($\mathcal{H}$) with $|\mathcal{J}| = \pi$.*

Let $x, \tilde{x} \in \partial K$ with $x \neq \tilde{x}$.

There exist $R_1 > 0$, $R_2 > 0$ and $J^* \subset \partial K$, with $|J^*| < h\varepsilon$, such that the couples $(X^x_{S_2}, S^x_2)$ and $(\tilde{X}^{\tilde{x}}_{\tilde{S}_2}, \tilde{S}^{\tilde{x}}_2)$ are both η-continuous on $J^* \times (R_1, R_2)$, with

$$\eta = \frac{1}{2}\left(\frac{c\rho_{\min}}{2D}\right)^2\left(\frac{1}{C} - \varepsilon\right)(\beta - \varepsilon)\,.$$

Moreover we have $R_2 - R_1 \geq 2\,(h\varepsilon - |J^*|)$.

Remark 19.3 The following proof is largely inspired by the proof of Lemma 5.1 in [2].

Proof Let $x, \tilde{x} \in \partial K$, $x \neq \tilde{x}$. Let us denote by $\Delta_{x\tilde{x}}$ the bisector of the segment defined by the two points x and $\tilde{x}$. The intersection $\Delta_{x\tilde{x}} \cap \partial K$ contains two points, let us thus define $\bar{y}$ the one which achieves the larger distance towards x and $\tilde{x}$ (we consider this point of intersection since we need in the sequel to have a lower bound on $\|x - \bar{y}\|$ and $\|\tilde{x} - \bar{y}\|$).

Let $t \in I \mapsto g(t)$ be a parametrization of ∂K with $g(0) = \bar{y}$, such that $\|g'(t)\|= 1$ for all $t \in I$. Consequently, the length of an arc satisfies length$(g_{|[s,t]}) = ||g(t) - g(s)|| = |t - s|$. We can thus write $I = [0, |\partial K|]$, and $g(0) = g(|\partial K|)$. Note that the parametrization g is C^2 thanks to Assumption $(\mathcal{K})$.

In the sequel, for $z \in \partial K$, we write s_z (or t_z) for the unique $s \in I$ such that $g(s) = z$. And for $A \subset \partial K$, we define $I_A = \{t \in I : g(t) \in A\}$.

Let us define, for $s, t \in I$ and $w \in \{x, \tilde{x}\}$:

$$\varphi_w(s, t) = \|w - g(s)\|+\|g(s) - g(t)\|.$$

Lemma 19.1 *There exists an interval $I^*_{\beta,\delta} \subset I$, satisfying $|I^*_{\beta,\delta}|< h\varepsilon$, such that for $w \in \{x, \tilde{x}\}$:*

$$|\partial_s \varphi_w(s, t)|\geq h, \quad \textit{for } s \in B^{\varepsilon}_{\bar{y}} \textit{ and } t \in I^*_{\beta,\delta},$$

where $B^{\varepsilon}_{\bar{y}} = \{s \in I;\ |s - s_{\bar{y}}|\leq \varepsilon\}$.

We admit this lemma for the moment and prove it after the end of the current proof.

Let us suppose for instance that $\partial_s \varphi_w(s, t)$ is positive for $s \in B^{\varepsilon}_{\bar{y}}$ and $t \in I^*_{\beta,\delta}$, for $w = x$ and $w = \tilde{x}$. If one or both of $\partial_s \varphi_x(s, t)$ and $\partial_s \varphi_{\tilde{x}}(s, t)$ are negative, we just need to consider $|\varphi_x|$ or $|\varphi_{\tilde{x}}|$, and everything works similarly.

We thus have, by the lemma:

$$\partial_s \varphi_w(s, t) \geq h, \quad \text{for } s \in B^{\varepsilon}_{\bar{y}} \text{ and } t \in I^*_{\beta,\delta}.$$

Let us now define:

$$r_1 = \sup_{t\in I^*_{\beta,\delta}} \inf_{s\in B^{\varepsilon}_{\bar{y}}} \varphi_x(s, t) \quad \text{and} \quad r_2 = \inf_{t\in I^*_{\beta,\delta}} \sup_{s\in B^{\varepsilon}_{\bar{y}}} \varphi_x(s, t)$$

and

$$\tilde{r_1} = \sup_{t\in I^*_{\beta,\delta}} \inf_{s\in B^{\varepsilon}_{\bar{y}}} \varphi_{\tilde{x}}(s, t) \quad \text{and} \quad \tilde{r_2} = \inf_{t\in I^*_{\beta,\delta}} \sup_{s\in B^{\varepsilon}_{\bar{y}}} \varphi_{\tilde{x}}(s, t).$$

Since $s \mapsto \varphi_x(s, t)$ and $s \mapsto \varphi_{\tilde{x}}(s, t)$ are strictly increasing on $B^{\varepsilon}_{\bar{y}}$ for all $t \in I^*_{\beta,\delta}$, we deduce that, considering $B^{\varepsilon}_{\bar{y}}$ as the interval (s_1, s_2),

$$r_1 = \sup_{t\in I^*_{\beta,\delta}} \varphi_x(s_1, t), \quad r_2 = \inf_{t\in I^*_{\beta,\delta}} \varphi_x(s_2, t), \quad \tilde{r_1} = \sup_{t\in I^*_{\beta,\delta}} \varphi_{\tilde{x}}(s_1, t), \quad \tilde{r_2} = \inf_{t\in I^*_{\beta,\delta}} \varphi_{\tilde{x}}(s_2, t).$$

Lemma 19.2 *We have $r_1 < r_2$ and $\tilde{r}_1 < \tilde{r}_2$.*

Moreover, there exist R_1, R_2 with $0 \leq R_1 < R_2$ satisfying $R_2 - R_1 \geq 2(h\varepsilon - |I_{\beta,\delta}^*|)$, such that $(r_1, r_2) \cap (\tilde{r}_1, \tilde{r}_2) = (R_1, R_2)$.

We admit this result to continue the proof, and will give a demonstration later.

We can now prove that the pairs $\left(X_{S_2}^x, S_2^x \right)$ and $\left(\tilde{X}_{\tilde{S}_2}^{\tilde{x}}, \tilde{S}_2^{\tilde{x}} \right)$ are both η-continuous on $I_{\beta,\delta}^* \times (R_1, R_2)$ with some $\eta > 0$ that we are going to define after the computations.

We first prove that $\left(X_{S_2}^x, S_2^x \right)$ is η-continuous on $I_{\beta,\delta}^* \times (r_1, r_2)$. By the same way we can prove that $\left(\tilde{X}_{\tilde{S}_2}^{\tilde{x}}, \tilde{S}_2^{\tilde{x}} \right)$ is η-continuous on $I_{\beta,\delta}^* \times (\tilde{r}_1, \tilde{r}_2)$. These two facts imply immediately the continuity with (R_1, R_2) since the interval (R_1, R_2) is included in (r_1, r_2) and $(\tilde{r}_1, \tilde{r}_2)$.

Let $(u_1, u_2) \subset (r_1, r_2)$ and $A \subset I_{\beta,\delta}^*$. We have:

$$\mathbb{P}\left(X_{S_2}^x \in A,\, S_2^x \in (u_1, u_2) \right)$$

$$\geq \int_{I_A} \int_{B_{\bar{y}}^\varepsilon} Q(x, g(s)) Q(g(s), g(t)) \mathbf{1}_{\varphi_x(s,t) \in (u_1, u_2)} \mathrm{d}s \mathrm{d}t.$$

Let $s \in B_{\bar{y}}^\varepsilon$ and $t \in I_{\beta,\delta}^*$. We now give a lower bound of $Q(x, g(s))$ and $Q(g(s), g(t))$.

Proposition 19.5 gives:

$$Q(x, g(s)) = \frac{\rho(U_x^{-1} l_{x,g(s)}) \cos\left(\varphi_{g(s),x} \right)}{\|x - g(s)\|}$$

$$\geq \frac{c\rho_{\min}}{2D} \left(\frac{1}{C} - \varepsilon \right),$$

where we have used the same method as in he proof of Theorem 19.3 (with Fig. 19.7) to get that $\cos\left(\varphi_{g(s),x} \right) \geq \frac{\|x - g(s)\| c}{2}$, and then the fact that $\|x - g(s)\| \geq \frac{1}{C} - \varepsilon$. Let us prove this latter. With the notations of Fig. 19.8, by Pythagore's theorem we have, for $\bar{y} \in \{\bar{y}_1, \bar{y}_2\}$, $\|x - \bar{y}\|^2 = \left(\frac{\|x - \tilde{x}\|}{2} \right)^2 + \|u - \bar{y}\|^2$. Moreover, since the curvature of K is bounded by C, it follows that $\|\bar{y}_1 - \bar{y}_2\| \geq \frac{2}{C}$, and then $\max\{\|u - \bar{y}_1\|;\, \|u - \bar{y}_2\|\} \geq \frac{1}{C}$. We deduce: $\max\{\|x - \bar{y}_1\|;\, \|x - \bar{y}_2\|\} \geq \frac{1}{C}$. Therefore, by the definition of $\bar{y}$, we have $\|x - \bar{y}\| \geq \frac{1}{C}$. Thus, the reverse triangle inequality gives, for $s \in B_{\bar{y}}^\varepsilon$, $\|x - g(s)\| \geq \frac{1}{C} - \varepsilon$.

By the same way, since $\|g(t) - g(s)\| \geq \beta - \varepsilon$, we have:

$$Q(g(s), g(t)) \geq \frac{c\rho_{\min}}{2D} (\beta - \varepsilon).$$

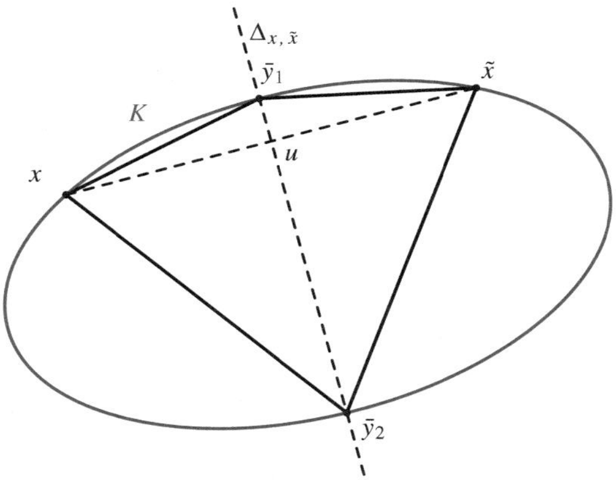

Fig. 19.8 Upper bound for the distance $\|w - \bar{y}\|$, $w \in \{x, \tilde{x}\}$

Therefore we get:

$$\mathbb{P}\left(X^x_{S_2} \in A, S^x_2 \in (u_1, u_2)\right) \geq a \int_{I_A} \int_{B^\varepsilon_{\bar{y}}} \mathbf{1}_{\varphi_x(s,t)\in(u_1,u_2)} \mathrm{d}s\mathrm{d}t,$$

with

$$a = \left(\frac{c\rho_{\min}}{2D}\right)^2 \left(\frac{1}{C} - \varepsilon\right)(\beta - \varepsilon). \tag{19.9}$$

Let us define, for $t \in I^*_{\beta,\delta}$:

$$M_{x,t}(u_1, u_2) := \left\{s \in B^\varepsilon_{\bar{y}} : \varphi_x(s, t) \in (u_1, u_2)\right\}.$$

Using the fact that $s \mapsto \varphi_x(s, t)$ is strictly increasing on $B^\varepsilon_{\bar{y}}$ for $t \in I^*_{\beta,\delta}$ we get ($\varphi_w^{-1}(s, t)$ stands for the inverse function of $s \mapsto \varphi_x(s, t)$):

$$\begin{aligned}\left|M_{x,t}(u_1, u_2)\right| &= \left|\left\{s \in B^\varepsilon_{\bar{y}} : s \in \left(\varphi_x^{-1}(u_1, t), \varphi_x^{-1}(u_2, t)\right)\right\}\right| \\ &= \left|(s_1, s_2) \cap \left(\varphi_x^{-1}(u_1, t), \varphi_x^{-1}(u_2, t)\right)\right|.\end{aligned}$$

By definition of r_1 and r_2, and since $(u_1, u_2) \subset (r_1, r_2)$ we have:

$$\varphi_x(s_1, t) \leq r_1 \leq u_1 \quad \text{and} \quad \varphi_x(s_2, t) \geq r_2 \geq u_2,$$

and since $s \mapsto \varphi_x(s, t)$ is strictly increasing:

$$s_1 \leq \varphi_x^{-1}(u_1, t) \quad \text{and} \quad s_2 \geq \varphi_x^{-1}(u_2, t).$$

Therefore we deduce:

$$\begin{aligned}\left|M_{x,t}(u_1,u_2)\right| &= \left|\left(\varphi_x^{-1}(u_1,t),\varphi_x^{-1}(u_2,t)\right)\right| \\ &= \left|\varphi_x^{-1}\left((u_1,u_2),t\right)\right| \\ &\geq \frac{1}{2}(u_2-u_1).\end{aligned}$$

For the last inequality we have used the following property. Let $\psi : \mathbb{R} \mapsto \mathbb{R}$ a function. If for all $x \in [a_1, a_2]$ we have $c_1 < \psi'(x) < c_2$ with $0 < c_1 < c_2 < \infty$, then for any interval $I \subset [\psi(a_1), \psi(a_2)]$, we have $c_2^{-1}|I| \leq |\psi^{-1}(I)| \leq c_1^{-1}|I|$. In our case, the Cauchy–Schwarz inequality gives $\partial_s\varphi_x(s,t) \leq 2$ (see Eq. (19.10) for the expression of $\partial_s\varphi_x(s,t)$).

Finally we get, with a given by (19.9):

$$\begin{aligned}\mathbb{P}\left(X^x_{S_2} \in A, S^x_2 \in (u_1,u_2)\right) &\geq a\int_A \frac{1}{2}(u_2-u_1)\mathrm{d}z \\ &= \frac{a}{2}(u_2-u_1)|A|,\end{aligned}$$

which proves that $\left(X^x_{S_2}, S^x_2\right)$ is $\frac{a}{2}$-continuous on $I^*_{\beta,\delta} \times (\tilde{r}_1, \tilde{r}_2)$.

Thanks to the remarks before, the proof is completed with $\eta = \frac{a}{2}$ and $J = I^*_{\beta,\delta}$.

Let us now give the proofs of Lemmas 19.1 and 19.2 that we have admitted so far.

Proof (Proof of Lemma 19.1) We use the notations introduced at the end of the proof of Proposition 19.7.

We have, for $s, t \in I$:

$$\partial_s\varphi_w(s,t) = \left\langle \frac{g(s)-w}{\|g(s)-w\|} + \frac{g(s)-g(t)}{\|g(s)-g(t)\|}, g'(s)\right\rangle. \tag{19.10}$$

By the definition of g, we note that $g'(s)$ is a director vector of the tangent line of ∂K at point $g(s)$.

It is easy to verify that for $w \in \{x, \tilde{x}\}$, there exists a unique $t \in I \setminus \{s_{\bar{y}}\}$ such that

$$\partial_s\varphi_w(s_{\bar{y}},t) = 0. \tag{19.11}$$

For $w = x$ (resp. $w = \tilde{x}$), we denote by t_{z_x} (resp. $t_{z_{\tilde{x}}}$) this unique element of I. With our notations we thus have $g(t_{z_x}) = z_x$ and $g(t_{z_{\tilde{x}}}) = z_{\tilde{x}}$.

Let $w \in \{x, \tilde{x}\}$. We have:

$$\begin{aligned}
&\partial_t \partial_s \varphi_w(s,t) \\
&= \partial_t \left(\left\langle \frac{g(s)-g(t)}{\|g(s)-g(t)\|}, g'(s) \right\rangle \right) \\
&= \frac{1}{\|g(t)-g(s)\|} \left(-\langle g'(t), g'(s)\rangle + \left\langle \frac{g(t)-g(s)}{\|g(t)-g(s)\|}, g'(t) \right\rangle \right. \\
&\left. \left\langle \frac{g(t)-g(s)}{\|g(t)-g(s)\|}, g'(s) \right\rangle \right).
\end{aligned}$$

Let us look at the term in parenthesis. Let us denote by $[u, v]$ the oriented angle between the vectors $u, v \in \mathbb{R}^2$. We have:

$$\begin{aligned}
&-\langle g'(t), g'(s)\rangle + \left\langle \frac{g(t)-g(s)}{\|g(t)-g(s)\|}, g'(t) \right\rangle \left\langle \frac{g(t)-g(s)}{\|g(t)-g(s)\|}, g'(s) \right\rangle \\
&= -\cos\left([g'(t), g'(s)]\right) + \cos\left([g(t)-g(s), g'(t)]\right) \cos\left([g(t)-g(s), g'(s)]\right) \\
&= -\cos\left([g'(t), g'(s)]\right) + \frac{1}{2}\cos\left([g(t)-g(s), g'(t)] - [g(t)-g(s), g'(s)]\right) \\
&\quad + \frac{1}{2}\cos\left([g(t)-g(s), g'(t)] + [g(t)-g(s), g'(s)]\right) \\
&= -\cos\left([g'(t), g'(s)]\right) + \frac{1}{2}\cos\left([g'(s), g'(t)]\right) \\
&\quad + \frac{1}{2}\cos\left([g(t)-g(s), g'(t)] + [g(t)-g(s), g'(s)]\right) \\
&= -\frac{1}{2}\cos\left([g'(t), g'(s)]\right) + \frac{1}{2}\cos\left([g(t)-g(s), g'(t)] + [g(t)-g(s), g'(s)]\right) \\
&= -\sin\left(\frac{1}{2}\left([g(t)-g(s), g'(t)] + [g(t)-g(s), g'(s)] + [g'(t), g'(s)]\right)\right) \\
&\quad \times \sin\left(\frac{1}{2}\left([g(t)-g(s), g'(t)] + [g(t)-g(s), g'(s)] - [g'(t), g'(s)]\right)\right) \\
&= -\sin\left([g(t)-g(s), g'(s)]\right) \sin\left([g(t)-g(s), g'(t)]\right).
\end{aligned}$$

Therefore we get

$$\partial_t \partial_s \varphi_w(s,t) = -\frac{1}{\|g(t)-g(s)\|} \sin\left([g(t)-g(s), g'(s)]\right) \sin\left([g(t)-g(s), g'(t)]\right),$$

and then

$$|\partial_t \partial_s \varphi_w(s,t)| = \frac{1}{\|g(t)-g(s)\|} \left|\sin\left([g(t)-g(s), g'(s)]\right) \sin\left([g(t)-g(s), g'(t)]\right)\right|$$
$$= \frac{1}{\|g(t)-g(s)\|} \left|\cos\left(\varphi_{g(s),g(t)}\right) \cos\left(\varphi_{g(t),g(s)}\right)\right|.$$

Let $t \in I$ such that $|t - s_{\bar{y}}| \geq \beta$ (we recall that β is introduced at the beginning of the section). Using once more Fig. 19.7, we get, as we have done in the proof of Theorem 19.3:

$$\begin{aligned} |\partial_t \partial_s \varphi_w(s,t)| &\geq \frac{1}{\|g(t)-g(s)\|} \left(\frac{\beta c}{2}\right)^2 \\ &\geq \frac{1}{D}\left(\frac{\beta c}{2}\right)^2. \end{aligned} \tag{19.12}$$

Using Eqs. (19.11) and (19.12), the mean value theorem gives: for $t \in I$ such that $|t - s_{\bar{y}}| \geq \beta$ and $|t - t_{z_w}| \geq \delta$ (δ is introduced at the beginning of the section),

$$\left|\partial_s \varphi_w(s_{\bar{y}}, t)\right| = \left|\partial_s \varphi_w(s_{\bar{y}}, t) - \partial_s \varphi_w(s_{\bar{y}}, t_{z_w})\right| \geq \frac{1}{D}\left(\frac{\beta c}{2}\right)^2 |t - t_{z_w}| \geq \frac{\delta}{D}\left(\frac{\beta c}{2}\right)^2. \tag{19.13}$$

We want now such an inequality for $s \in I$ near from $s_{\bar{y}}$. We thus compute:

$$\begin{aligned} &\partial_s^2 \varphi_w(s,t) \\ &= \frac{1}{\|w-g(s)\|} + \frac{1}{\|g(s)-g(t)\|} + \left\langle \frac{g(s)-w}{\|g(s)-w\|} + \frac{g(s)-g(t)}{\|g(s)-g(t)\|}, g''(s) \right\rangle \\ &\quad - \frac{1}{\|w-g(s)\|} \left\langle \frac{w-g(s)}{\|w-g(s)\|}, g'(s) \right\rangle^2 \\ &\quad - \frac{1}{\|g(s)-g(t)\|} \left\langle \frac{g(s)-g(t)}{\|g(s)-g(t)\|}, g'(s) \right\rangle^2. \end{aligned}$$

We immediately deduce, using the Cauchy–Schwarz inequality, and the fact that $\|g'(s)\| = 1$ for all $s \in I$:

$$\begin{aligned} &|\partial_s^2 \varphi_w(s,t)| \\ &\leq \frac{1}{\|w-g(s)\|} + \frac{1}{\|g(s)-g(t)\|} + 2\|g''(s)\| + \frac{1}{\|w-g(s)\|} + \frac{1}{\|g(s)-g(t)\|} \\ &\leq 2\left(\frac{1}{\|w-g(s)\|} + \frac{1}{\|g(s)-g(t)\|} + C\right), \end{aligned}$$

where we recall that C is the upper bound on the curvature of K.

Let now $t \in I$ such that $|t - s_{\bar{y}}| \geq \beta$ and $|t - t_{z_w}| \geq \delta$, and let $s \in I$ such that $|s - s_{\bar{y}}| \leq \varepsilon$. With such s and t we have $|t - s| \geq \beta - \varepsilon$. Moreover, we have already seen in proof of Proposition 19.7 (with Fig. 19.8) that $\|w - g(s)\| \geq \frac{1}{C} - \varepsilon$ for $s \in B_{\bar{y}}^{\varepsilon}$. Therefore, for such s and t:

$$|\partial_s^2 \varphi_w(s,t)| \leq 2\left(\frac{1}{\frac{1}{C} - \varepsilon} + \frac{1}{\beta - \varepsilon} + C\right) = M > 0. \tag{19.14}$$

Using once again the mean value theorem with Eqs. (19.13) and (19.14), we deduce that for all $t \in I$ such that $|t - s_{\bar{y}}| \geq \beta$ and $|t - t_{z_w}| \geq \delta$, and for all $s \in I$ such that $|s - s_{\bar{y}}| \leq \varepsilon$:

$$|\partial_s \varphi_w(s,t)| \geq \frac{\delta}{D}\left(\frac{\beta c}{2}\right)^2 - \varepsilon M = h > 0.$$

Let us now take $I_{\beta,\delta}^* \subset I \setminus \{s_{\bar{y}}, t_{z_x}, t_{z_{\bar{x}}}\}$ an interval of length strictly smaller than $h\varepsilon$ (this condition on the length of $I_{\beta,\delta}^*$ is not necessary for the lemma, but for the continuation of the proof of the proposition), and such that for all $t \in I_{\beta,\delta}^*$, $|t - t_{z_x}| \geq \delta$, $|t - t_{z_{\bar{x}}}| \geq \delta$ and $|t - s_{\bar{y}}| \geq \beta$. In order to ensure that $I_{\beta,\delta}^*$ is not empty, we take β and δ such that $\frac{|\partial K|}{3} - \max\{2\delta; \beta + \delta\} > 0$. Indeed, it is necessary that one of the intervals "$(t_{z_x}, t_{z_{\bar{x}}})$", "$(t_{z_x}, s_{\bar{y}})$" and "$(s_{\bar{y}}, t_{z_{\bar{x}}})$" at which we removes a length β or δ on the good extremity, is not empty. And since the larger of these intervals has a length at least $\frac{\partial K}{3}$, we obtain the good condition on β and δ.

We thus just proved that $|\partial_s \varphi_w(s,t)| \geq h$ for $s \in B_{\bar{y}}^{\varepsilon}$ and $t \in I_{\beta,\delta}^*$, which is the result of the lemma. □

Proof (Proof of Lemma 19.2) Let us first prove that $r_1 < r_2$. We do it only for r_1 and r_2 since it is the same for $\tilde{r_1}$ and $\tilde{r_2}$. We have:

$$\begin{aligned} r_2 - r_1 &= \inf_{t \in I_{\beta,\delta}^*} \varphi_x(s_2,t) - \sup_{t \in I_{\beta,\delta}^*} \varphi_x(s_1,t) \\ &= \inf_{t \in I_{\beta,\delta}^*} \varphi_x(s_2,t) - \inf_{t \in I_{\beta,\delta}^*} \varphi_x(s_1,t) - \left(\sup_{t \in I_{\beta,\delta}^*} \varphi_x(s_1,t) - \inf_{t \in I_{\beta,\delta}^*} \varphi_x(s_1,t)\right) \\ &\geq h(s_2 - s_1) - \left(\sup_{t \in I_{\beta,\delta}^*} |\partial_t \varphi_x(s_1,t)|\right) \left|I_{\beta,\delta}^*\right| \\ &\geq 2h\varepsilon - \left|I_{\beta,\delta}^*\right|, \end{aligned}$$

and this quantity is strictly positive since $|I_{\beta,\delta}^*| < h\varepsilon$ by construction.

For the first inequality, we have used the mean value theorem twice, and for the last inequality, we have used the fact that $\sup_{t\in I^*_{\beta,\delta}} |\partial_t \varphi_x(s_1,t)| = \sup_{t\in I^*_{\beta,\delta}} \left|\left\langle \frac{g(t)-g(s_1)}{\|g(t)-g(s_1)\|}, g'(t)\right\rangle\right| \leq 1$ thanks to the Cauchy–Schwarz inequality.

Let us now prove that the intersection $(r_1, r_2) \cap (\tilde{r}_1, \tilde{r}_2)$ is not empty.

Let $t \in I^*_{\beta,\delta}$, we have:

$$
\begin{aligned}
r_2 - \varphi_x(s_{\bar{y}}, t) &= \inf_{t\in I^*_{\beta,\delta}} \varphi_x(s_2,t) - \varphi_x(s_{\bar{y}}, t) \\
&= \inf_{t\in I^*_{\beta,\delta}} \varphi_x(s_2,t) - \inf_{t\in I^*_{\beta,\delta}} \varphi_x(s_{\bar{y}}, t) - \left(\varphi_x(s_{\bar{y}}, t) - \inf_{t\in I^*_{\beta,\delta}} \varphi_x(s_{\bar{y}}, t)\right) \\
&\geq h(s_2 - s_{\bar{y}}) - |I^*_{\beta,\delta}| \\
&= h\varepsilon - |I^*_{\beta,\delta}| \\
&> 0,
\end{aligned}
$$

once again thanks to the mean value theorem. Similarly we have

$$
\begin{aligned}
\varphi_x(s_{\bar{y}}, t) - r_1 &= \varphi_x(s_{\bar{y}}, t) - \sup_{t\in I^*_{\beta,\delta}} \varphi_x(s_1,t) \\
&= \varphi_x(s_{\bar{y}}, t) - \sup_{t\in I^*_{\beta,\delta}} \varphi_x(s_{\bar{y}}, t) - \left(\sup_{t\in I^*_{\beta,\delta}} \varphi_x(s_1,t) - \sup_{t\in I^*_{\beta,\delta}} \varphi_x(s_{\bar{y}}, t)\right) \\
&\geq -|I^*_{\beta,\delta}| + h(s_{\bar{y}} - s_1) \\
&= h\varepsilon - |I^*_{\beta,\delta}| \\
&> 0.
\end{aligned}
$$

Moreover, since $\bar{y} \in \Delta_{x,\tilde{x}}$, we have $\varphi_x(s_{\bar{y}}, t) = \varphi_{\tilde{x}}(s_{\bar{y}}, t)$, and we thus can prove the same inequalities with $\tilde{r}_1$ and $\tilde{r}_2$ instead of r_1 and r_2.

Finally we thus get that the interval $(R_1, R_2) = (r_1, r_2) \cap (\tilde{r}_1, \tilde{r}_2)$ is well defined and

$$
R_2 - R_1 \geq 2\left(h\varepsilon - |I^*_{\beta,\delta}|\right).
$$

Remark 19.4 The fact that $|\mathcal{J}| = \pi$ is here to ensure that the process can go from x and $\tilde{x}$ to $\bar{y}$ in the proof of Proposition 19.7. If $|\mathcal{J}| < \pi$, since x and $\tilde{x}$ are unspecified and $\bar{y}$ can therefore be everywhere on ∂K, nothing ensures that this path is available.

We can now state the following theorem on the speed of convergence of the stochastic billiard in the convex set K.

Theorem 19.4 *Let $K \subset \mathbb{R}^2$ satisfying Assumption $(\mathcal{K})$ with diameter D. Let $((X_t, V_t))_{t\geq 0}$ the stochastic billiard process evolving in K and verifying Assumption $(\mathcal{H})$ with $|\mathcal{J}| = \pi$.*

There exists a unique invariant probability measure χ on $K \times \mathbb{S}$ for the process $((X_t, V_t))_{t\geq 0}$.

Moreover, let us define n_0 and p by (19.15) *and* (19.16) *with $\zeta \in \left(0, \frac{2}{C}\right)$. Let us consider η, $I^*_{\beta,\delta}$, R_1, R_2 as in Proposition 19.7 and Lemma 19.1, and let us define κ by* (19.17).

For all $t \geq 0$ and all $\lambda < \lambda_M$:

$$\|\mathbb{P}(X_t \in \cdot, V_t \in \cdot) - \chi\|_{TV} \leq C_\lambda e^{-\lambda t},$$

where

$$\lambda_M = \min\left\{\frac{1}{2D}\log\left(\frac{1}{1-p}\right); \frac{1}{2D}\log\left(\frac{-(1-p)+\sqrt{(1-p)^2+4p(1-\kappa)}}{2p(1-\kappa)}\right)\right\}$$

and

$$C_\lambda = \frac{p\kappa e^{5\lambda D}}{1 - e^{2\lambda D}(1-p) - e^{4\lambda D}p(1-\kappa)}.$$

Proof As previously, the existence of an invariant probability measure for the stochastic billiard process comes from the compactness of $K \times \mathbb{S}^1$. The following proof ensures its uniqueness and gives an explicit speed of convergence.

Let $((X_t, V_t))_{t\geq 0}$ and $((\tilde{X}_t, \tilde{V}_t))_{t\geq 0}$ be two versions of the stochastic billiard with $(X_0, V_0) = (x_0, v_0) \in K \times \mathbb{S}^1$ and $(\tilde{X}_0, \tilde{V}_0) = (\tilde{x}_0, \tilde{v}_0) \in K \times \mathbb{S}^1$.

We define (or recall the definition for T_0 and $\tilde{T}_0$):

$$T_0 = \inf\{t \geq 0, x_0 + tv_0 \notin K\}, \qquad w = x_0 + T_0 v_0 \in \partial K,$$

and

$$\tilde{T}_0 = \inf\{t \geq 0, \tilde{x}_0 + t\tilde{v}_0 \notin K\}, \qquad \tilde{w} = \tilde{x}_0 + \tilde{T}_0 \tilde{v}_0 \in \partial K.$$

Step 1 From Proposition 19.6, we deduce that for all $x \in \partial K$ and all $\zeta \in \left(0, \frac{1}{C}\right)$, S^x_n is $(c\rho_{\min})^n \zeta^{n-1}$-continuous on the interval $\Gamma_n = \left[(n-1)\zeta, \frac{nC}{2} - (n-1)\zeta\right]$.

Let thus $\zeta \in \left(0, \frac{1}{C}\right)$ and let us define

$$n_0 = \min\{n \geq 1 : |\Gamma_n| > D\} = \left\lfloor \frac{D - 2\zeta}{2\left(\frac{1}{C} - 1\right)} \right\rfloor + 1. \tag{19.15}$$

The variables T_{n_0} and $\tilde{T}_{n_0}$ are both $(c\rho_{\min})^{n_0}\zeta^{n_0-1}$-continuous on $[T_0+(n_0-1)\zeta, T_0+\frac{nC}{2}-(n_0-1)\zeta] \cap \left[\tilde{T}_0+(n_0-1)\zeta, \tilde{T}_0+\frac{nC}{2}-(n_0-1)\zeta\right]$. Since $\left|T_0-\tilde{T}_0\right| \leq D$, this intersection is non-empty and its length is larger that $\frac{2n_0}{C}-2(n_0-1)\zeta-D$.

Let us define

$$p=(c\rho_{\min})^{n_0}\zeta^{n_0-1}\left(\frac{2n_0}{C}-2(n_0-1)\zeta-D\right). \tag{19.16}$$

Using the fact that the for all $x\in\partial K$, $S^x_{n_0}\leq n_0 D$ almost surely, we deduce that we can construct a coupling such that the coupling-time T^1_c of T_{n_0} and $\tilde{T}_{n_0}$ satisfies:

$$T^1_c \leq_{st} T_0+n_0 D G^1$$

with $G^1\sim \mathcal{G}(p)$.

Step 2 Let us now suppose that T^1_c, $((X_t, V_t))_{0\leq t\leq T^c_1}$ and $((\tilde{X}_t, \tilde{V}_t))_{0\leq t\leq T^c_1}$ are constructed as described above.

We define $y=X_{T^1_c}$ and $\tilde{y}=\tilde{X}_{T^1_c}$, which are by construction of T^1_c on ∂K. We also define $N^1_c=\min\left\{n>0: X_{T_n}=y\right\}$, which is deterministic conditionally to T^1_c.

By the Proposition 19.7, we can construct a coupling of $(X^y_{S_2}, S^y_2)$ and $(\tilde{X}^{\tilde{y}}_{\tilde{S}_2}, \tilde{S}^y_2)$ such that

$$\mathbb{P}\left(X^y_{S_2}=\tilde{X}^{\tilde{y}}_{\tilde{S}_2} \text{ and } S^y_2=\tilde{S}^y_2\right)\geq \eta|I^*_{\beta,\delta}|(R_2-R_1).$$

Therefore we can construct $((X_t, V_t))$ and $((\tilde{X}_t, \tilde{X}_t))$ until time $T_{N^1_c+2}$ such that

$$\mathbb{P}\left(X_{T_{N^1_c+2}}=\tilde{X}_{T_{N^1_c+2}} \text{ and } T_{N^1_c+2}=\tilde{T}_{N^1_c+2}\right)\geq \eta|I^*_{\beta,\delta}|(R_2-R_1).$$

Defining

$$\kappa=\eta|I^*_{\beta,\delta}|(R_2-R_1), \tag{19.17}$$

we get that the entire coupling-time of the two processes satisfies:

$$\hat{T}\leq_{st} T_0+\sum_{l=1}^{G}\left(n_0 D G^l+n_0 D\right)=T_0+\sum_{l=1}^{G}\left(n_0 D(G^l+1)\right)$$

where $G\sim\mathcal{G}(\kappa)$ and the $(G^l)_{l\geq 1}$ are independent $\mathcal{G}(p)$ distributed, and independent of G.

Finally, we get

$$\mathbb{P}\left(\hat{T} > t\right) \leq e^{-\lambda t} \frac{p\kappa e^{5\lambda D}}{1 - e^{2\lambda D}(1-p) - e^{4\lambda D} p(1-\kappa)},$$

for all $\lambda \in (0, \lambda_M)$.

19.5 Discussion

All the results presented in this paper are in dimension 2. However, the ideas developed here can be adapted to higher dimensions. Let us briefly explain it.

19.5.1 Stochastic Billiard in a Ball of $\mathbb{R}^d$

Let us first look at the stochastic billiard (X, V) in a ball $\mathcal{B} \subset \mathbb{R}^d$ with $d \geq 2$.

As we have done in Sect. 19.3, we can represent the Markov chain $(X_{T_n}, V_{T_n})_{n \geq 0}$ by another Markov chain. Indeed, for $n \geq 1$, the position $X_{T_n} \in \partial\mathcal{B}$ can be uniquely represented by its hyperspherical coordinates: a $(d-1)$-tuple $(\Phi_n^1, \cdots, \Phi_n^{d-1})$ with $\Phi_n^1, \cdots, \Phi_n^{d-2} \in [0, \pi)$ and $\Phi_n^{d-1} \in [0, 2\pi)$.

Similarly, for $n \geq 1$, the vector speed $V_{T_n} \in \left\{v \in \mathbb{S}^{d-1} : v \cdot n_{X_{T_n}} \geq 0\right\}$ can be represented by its hyperspherical coordinates.

Thereby, we can give relations between the different random variables as in Proposition 19.1, and in theory, we can do explicit computations to get lower bounds on the needed density function. Then the same coupling method in two steps can be applied. Nevertheless, it could be difficult to manage the computations in practice when the dimension increases.

19.5.2 Stochastic Billiard in a Convex Set $K \subset \mathbb{R}^d$

To get bounds on the speed of convergence of the stochastic billiard (X, V) in a convex set $K \subset \mathbb{R}^d$, $d \geq 2$, satisfying Assumption $(\mathcal{K})$, we can apply exactly the same method as in Sect. 19.4. The main difficulty could be the proof of the equivalent of Proposition 19.7. But it can easily be adapted, and we refer to the proof of Lemma 5.1 in [2], where the authors lead the proof in dimension $d \geq 3$.

Acknowledgements I thank Hélène Guérin and Florent Malrieu for their help in this work. I also thank the anonymous referee for her/his careful proofreading of this paper and her/his relevant and useful comments.

This work was supported by the Agence Nationale de la Recherche project PIECE 12-JS01-0006-01.

References

1. C.G.E. Boender, R.J. Caron, J.F. McDonald, A.H.G. Rinnooy Kan, H.E. Romeijn, R.L. Smith, J. Telgen, A.C.F. Vorst, Shake-and-bake algorithms for generating uniform points on the boundary of bounded polyhedra. Oper. Res. **39**(6), 945–954 (1991)
2. F. Comets, S. Popov, G.M. Schütz, M. Vachkovskaia, Billiards in a general domain with random reflections. Arch. Ration. Mech. Anal. **191**(3), 497–537 (2009)
3. A.B. Dieker, S.S. Vempala, Stochastic billiards for sampling from the boundary of a convex set. Math. Oper. Res. **40**(4), 888–901 (2015)
4. S.N. Ethier, T.G. Kurtz, *Markov Processes – Characterization and Convergence*. Wiley Series in Probability and Mathematical Statistics: Probability and Mathematical Statistics (Wiley, New York, 1986)
5. S.N. Evans, Stochastic billiards on general tables. Ann. Appl. Probab. **11**(2), 419–437 (2001)
6. T. Lindvall, *Lectures on the Coupling Method*. Dover Books on Mathematics Series (Dover Publications, Mineola, 2002)
7. H. Thorisson, *Coupling, Stationarity, and Regeneration*. Dover Books on Mathematics Series (Springer, New York, 2000)

LECTURE NOTES IN MATHEMATICS

Springer

Editorial Policy

1. Lecture Notes aim to report new developments in all areas of mathematics and their applications – quickly, informally and at a high level. Mathematical texts analysing new developments in modelling and numerical simulation are welcome.

   Manuscripts should be reasonably self-contained and rounded off. Thus they may, and often will, present not only results of the author but also related work by other people. They may be based on specialised lecture courses. Furthermore, the manuscripts should provide sufficient motivation, examples and applications. This clearly distinguishes Lecture Notes from journal articles or technical reports which normally are very concise. Articles intended for a journal but too long to be accepted by most journals, usually do not have this "lecture notes" character. For similar reasons it is unusual for doctoral theses to be accepted for the Lecture Notes series, though habilitation theses may be appropriate.

2. Besides monographs, multi-author manuscripts resulting from SUMMER SCHOOLS or similar INTENSIVE COURSES are welcome, provided their objective was held to present an active mathematical topic to an audience at the beginning or intermediate graduate level (a list of participants should be provided).

   The resulting manuscript should not be just a collection of course notes, but should require advance planning and coordination among the main lecturers. The subject matter should dictate the structure of the book. This structure should be motivated and explained in a scientific introduction, and the notation, references, index and formulation of results should be, if possible, unified by the editors. Each contribution should have an abstract and an introduction referring to the other contributions. In other words, more preparatory work must go into a multi-authored volume than simply assembling a disparate collection of papers, communicated at the event.

3. Manuscripts should be submitted either online at www.editorialmanager.com/lnm to Springer's mathematics editorial in Heidelberg, or electronically to one of the series editors. Authors should be aware that incomplete or insufficiently close-to-final manuscripts almost always result in longer refereeing times and nevertheless unclear referees' recommendations, making further refereeing of a final draft necessary. The strict minimum amount of material that will be considered should include a detailed outline describing the planned contents of each chapter, a bibliography and several sample chapters. Parallel submission of a manuscript to another publisher while under consideration for LNM is not acceptable and can lead to rejection.

4. In general, **monographs** will be sent out to at least 2 external referees for evaluation.

   A final decision to publish can be made only on the basis of the complete manuscript, however a refereeing process leading to a preliminary decision can be based on a pre-final or incomplete manuscript.

   Volume Editors of **multi-author works** are expected to arrange for the refereeing, to the usual scientific standards, of the individual contributions. If the resulting reports can be forwarded to the LNM Editorial Board, this is very helpful. If no reports are forwarded or if other questions remain unclear in respect of homogeneity etc, the series editors may wish to consult external referees for an overall evaluation of the volume.

5. Manuscripts should in general be submitted in English. Final manuscripts should contain at least 100 pages of mathematical text and should always include

   - a table of contents;
   - an informative introduction, with adequate motivation and perhaps some historical remarks: it should be accessible to a reader not intimately familiar with the topic treated;
   - a subject index: as a rule this is genuinely helpful for the reader.
   - For evaluation purposes, manuscripts should be submitted as pdf files.

6. Careful preparation of the manuscripts will help keep production time short besides ensuring satisfactory appearance of the finished book in print and online. After acceptance of the manuscript authors will be asked to prepare the final LaTeX source files (see LaTeX templates online: https://www.springer.com/gb/authors-editors/book-authors-editors/manuscriptpreparation/5636) plus the corresponding pdf- or zipped ps-file. The LaTeX source files are essential for producing the full-text online version of the book, see http://link.springer.com/bookseries/304 for the existing online volumes of LNM). The technical production of a Lecture Notes volume takes approximately 12 weeks. Additional instructions, if necessary, are available on request from lnm@springer.com.

7. Authors receive a total of 30 free copies of their volume and free access to their book on SpringerLink, but no royalties. They are entitled to a discount of 33.3 % on the price of Springer books purchased for their personal use, if ordering directly from Springer.

8. Commitment to publish is made by a *Publishing Agreement*; contributing authors of multiauthor books are requested to sign a *Consent to Publish form*. Springer-Verlag registers the copyright for each volume. Authors are free to reuse material contained in their LNM volumes in later publications: a brief written (or e-mail) request for formal permission is sufficient.

Addresses:

Professor Jean-Michel Morel, CMLA, École Normale Supérieure de Cachan, France
E-mail: moreljeanmichel@gmail.com

Professor Bernard Teissier, Equipe Géométrie et Dynamique,
Institut de Mathématiques de Jussieu – Paris Rive Gauche, Paris, France
E-mail: bernard.teissier@imj-prg.fr

Springer: Ute McCrory, Mathematics, Heidelberg, Germany,
E-mail: lnm@springer.com

Printed in the United States
By Bookmasters